Understanding the Universe

George Greenstein is the Sidney Dillon Professor of Astronomy, Emeritus at Amherst College, Massachusetts. He is an accomplished writer, having written one textbook, three books on science for the general public and numerous magazine articles. One of his books won both the American Institute of Physics/US Steel science-writing award and the Phi Beta Kappa Award in Science. Professor Greenstein is a recognized leader in the American Astronomical Society's effort to reform astronomy education in the United States. Some time ago he co-organized a series of workshops for department chairs of the most prestigious universities in the country, which led to a set of proposed goals for reform of introductory astronomy courses nationwide. Professor Greenstein's field of research interest is theoretical astrophysics.

'... an inquiry approach that explores the nature of scientific research sets this book apart from other textbooks. The readings and exercises are scaffolded to allow the students to build their own understanding of the big ideas in astronomy. The separation into different math levels makes it appropriate for a wide range of classes. I really liked the problem sets that required students to describe the logic behind their solutions.'

Mary Kay Hemenway,
University of Texas at Austin

'... a compelling and powerful introduction to astronomy, laying bare the fundamentals of scientific arguments and the scientific process.'

Steven Furlanetto,
University of California-Los Angeles

'... delivers on its promises. It is indeed inquiry based, and overtly uses astronomy as a means to explore the nature of science ... [this text] does not merely tell students about the Universe; it helps them *understand* the Universe.'

Bruce Partridge,
Haverford College

'George Greenstein has done an excellent job of clearly explaining the most important aspects of astronomy. His book brings the reader along on a journey of discovery and treats "what we know" and "how we know" as equally important. Exhorting students to actively participate rather than passively memorize reinforces a message that almost all instructors send. I encourage my colleagues teaching introductory astronomy to consider this book carefully.'

Pauline Barmby,
Western University

'This unique text provides a superb framework for introducing students to the approaches scientists take to solving problems. By posing a variety of "mysteries" faced by both ancient and contemporary astronomers, gathering and presenting data, searching for patterns, asking questions, posing and testing hypotheses both qualitatively and quantitatively, Greenstein introduces his readers to the tools of the detective-scientist.'

Stephen Strom,
National Optical Astronomy Observatory

Understanding

the Universe

An Inquiry Approach to Astronomy and the Nature of Scientific Research

GEORGE GREENSTEIN

Amherst College, Massachusetts

CAMBRIDGE
UNIVERSITY PRESS

CAMBRIDGE UNIVERSITY PRESS
Cambridge, New York, Melbourne, Madrid, Cape Town,
Singapore, São Paulo, Delhi, Mexico City

Cambridge University Press
The Edinburgh Building, Cambridge CB2 8RU, UK

Published in the United States of America by Cambridge University Press, New York

www.cambridge.org
Information on this title: www.cambridge.org/9780521145329

First published 2013

Printed in the United States by Edwards Brothers

A catalog record for this publication is available from the British Library

Library of Congress Cataloging-in-Publication Data

Greenstein, George, 1940–
Understanding the Universe: An Inquiry Approach to Astronomy and the Nature of Scientific
Research / George Greenstein.
        p.   cm.
ISBN 978-0-521-19259-0 (Hardback) – ISBN 978-0-521-14532-9 (Paperback)   1. Astronomy.
2. Research–Methodology.   I. Title.
QB61.G744 2012
520–dc23   2011042772

ISBN 978-0-521-19259-0 Hardback
ISBN 978-0-521-14532-9 Paperback

Additional resources for this publication at www.cambridge.org/greenstein

To all my students

It was you who taught me how to teach

CONTENTS

Color plate section between pages 368 and 369.

To the instructor

The philosophy behind this book

When I was in college studying science, I found the experience fundamentally unsatisfying. I was continually oppressed by the feeling that my only role was to "shut up and learn." I felt there was nothing I could say to my instructors that they would find interesting. Nor did I feel that there was anything I could tell my fellow-students that they would find interesting. As I sat in the science lecture hall, I was utterly silent. That's not a good state to be in when you are 19 years old.

Doubly galling was the fact that at the same time my roommate was taking a history course. One day he came back to our dorm room filled with excitement over a class discussion. (The question was whether President Truman was right to have dropped the atom bomb on Hiroshima.) Another friend at the time was taking a literature course, and he mentioned to me that, during a class discussion, he had made a point the instructor himself had found striking.

Meanwhile, I was busy with Ampère's law. We never had any fascinating class discussions about this law. No one, teacher or student, ever asked me what I thought about it.

We professors have a tendency to think that independent, creative thinking cannot be done by non-science students, and that only advanced science majors have learned enough of the material to think critically about it. I believe this attitude is false. This book is designed to move beyond a "shut up and learn" format, and to challenge students to think for themselves – even at the beginning level. It asks students to use their native intelligence to actually confront subtle scientific issues.

Unique features of this book

As the title suggests, this book emphasizes *student-active learning*. Rather than emphasizing the facts of astronomy, it emphasizes how we know them, and it regularly involves the student in the chain of arguments that lead to them. Although the book's mathematical level is appropriate for non-scientists, it asks a good deal of the reader, and it wrestles with conflicting theories, incomplete evidence and hypothesis testing. We hope that, ten years from now, our students will remember what we taught them about the Universe – but it is also important that they remember the habits of mind that have allowed us to discover these facts, and that they followed with comprehension and interest the development of our understanding.

The book covers a smaller number of topics than most texts, strictly confining attention to those most essential to the field. Recognizing that this may be the only science course the student ever takes, it devotes greater than usual attention to *how we know what we know*. Recognizing that few students are taking this course in order to prepare for another, it makes no attempt to cover every astronomical subject. Rather, it spends as much time as needed to develop a full understanding of each topic.

Most students find it hard to believe that scientists think intuitively. Rather they feel that science involves the manipulation of abstract, meaningless symbols. Far too often students turn off their native intelligence and abandon their intuitive understanding when approaching such a strange, unfamiliar topic as science. As much as possible, this book is written in such a manner as to resist this tendency. Thus *mathematics* is often used in order to make an intuitive point. The first use of Newton's law of gravity, for example, is to calculate the gravitational attraction of two people, in order to illustrate vividly why we are not aware in daily life of the mutual attraction between every pair of bodies.

The mathematics is never beyond the level of simple arithmetic. *A two-track system of mathematics* is used, in which the *logic of the calculation* is first analyzed, and the *detailed calculation* always comes second and is placed in a sidebar. Problems at the end of each chapter employ the same system: if the instructor wishes, students can be asked to perform only the first step.

Throughout the book the treatment is informed by the rubric, supported by the field of Science Education Research, that "you can only learn what you already almost know." The treatment of gravitation in the Solar System begins by reminding students about what they already know of the everyday experience of throwing things, then analyzes this in terms of Newton's laws, and only then moves on to the subject of orbits.

Inquiry teaching

This book is written in an "inquiry" mode. You may not be used to this form of instruction. There is no hiding the fact that it can be an unnerving way to teach. But it is only unnerving at first. And it is also a delightful way to teach. It can be fun for the students: I find that the energy level in the classroom goes up dramatically when I introduce one of discussion topics found in this text. And it can be fun for the instructor as well.

My advice would be that, if you find this method of instruction appealing, start slowly. The first time you try it, continue with your traditional method of teaching, and add in just a little bit of this new method. As time passes and you get used to it, gradually add more and more to the mix. This book is here to help you as you do.

To the student

Throughout this book, we will be doing two things at once.

- We will grapple with the phenomena of the astronomical Universe, seeking to understand the cosmos in which we live.
- We will step back and watch ourselves as we do this, and we will explore the mental procedures scientists go through in their work.

As its title suggests, the book emphasizes *student-active learning*. Rather than emphasizing the facts of astronomy, it emphasizes how we know them, and it regularly involves you in the chain of arguments that lead to them. What will you remember of your astronomy course ten years from now? Certainly few of the detailed facts you will encounter. But if this book does its job right, you will remember the habits of mind that have allowed scientists to discover these facts – and you will remember that you followed with comprehension and interest the development of our understanding.

The mathematics we will use is never beyond the level of simple arithmetic. *A two-track system* is used, in which the *logic of the calculation* is first analyzed, and the *detailed calculation* always comes second and is placed in a sidebar. Problems at the end of each chapter employ the same system. Take a look at Appendix II for mathematical help.

You may find it hard to believe that scientists think intuitively. Rather you may feel that science involves the manipulation of abstract, meaningless symbols. Nothing could be farther from the truth. Far too often it is easy to turn off our native intelligence and abandon our intuitive understanding when approaching such a strange, unfamiliar topic as science. As much as possible, this book is written in such a manner as to resist this tendency. For this reason, mathematics is often used – not to get a definite result, but to make an intuitive point.

"You must decide"

To give you some practice in thinking creatively about science, there is a series of questions in which you will be asked to make a firm choice concerning an issue for which there is no clear "right answer" – and to defend your choice in a well-reasoned essay. For example, one essay asks what balance NASA should strike between supporting ground-based and orbiting telescopes. Another asks you to identify the research program that has the best chance of identifying dark matter.

"Detectives on the case"

This book pays careful attention one of the most important aspects of science: the creation of new theories. How do scientists go about devising their theories? I like to think of the method as being much like that of a detective working to solve a crime. This topic is returned to in a variety of contexts, deepening and extending your understanding with each repetition. Here is a list.

DETECTIVES ON THE CASE		
Title	Location	
The reasons for the seasons	Chapter 1	The sky
The paradox of weightlessness	Chapter 3	Newton's laws: gravity and orbits
What causes tides?	Chapter 7	The inner Solar System
Why is Io so hot?	Chapter 8	The outer Solar System
Craters on the moons of Jupiter	Chapter 8	The outer Solar System
What are Saturn's rings?	Chapter 8	The outer Solar System
What are the comets?	Chapter 9	Smaller bodies in the Solar System
Limb darkening	Chapter 11	Our Sun
Parallax	Chapter 12	A census of stars
How can we understand the orbits of the planets?	Chapter 13	The formation of stars and planets
What powers the shining of the stars?	Chapter 14	Stellar structure
What are planetary nebulae?	Chapter 15	Stellar evolution and death
What are the pulsars?	Chapter 15	Stellar evolution and death
High- and low-velocity stars, and stellar populations	Chapter 16	The Milky Way Galaxy
What are the spiral nebulae?	Chapter 17	Galaxies
What powers radio galaxies and quasars?	Chapter 17	Galaxies

"The nature of science"

One of the most important elements of this book is the effort to understand science in general. It seeks to acquaint you with science as a way of thinking, a way of looking at the world, that is unique in the history of thought. What has made science such a powerful agent of change in modern society?

This "chapter" will not be found at any particular place. Rather it is scattered throughout the book. Two reasons guided this choice.

- Were this discussion confined to a particular chapter, there is some danger that you might read it but then forget it. By returning to the subject again and again, we reinforce its importance.
- Each element of the nature of science is introduced in the context of a specific astronomical topic. This gives the discussion a significance an abstract presentation would have lacked.

Nevertheless, "The nature of science" is a coherent whole, and it can be read as such. For those wishing to do so, its sections are as follows.

THE NATURE OF SCIENCE		
Title	Location	
Hypothesis testing in science: why does the Sun rise and set?	Chapter 1	The sky
The importance of skepticism: testing our theory of the Moon's phases	Chapter 1	The sky
The importance of skepticism and a test of astrology	Chapter 2	The origins of astronomy
The design of experiments	Chapter 2	The origins of astronomy
Lessons from history	Chapter 2	The origins of astronomy
Certainty and uncertainty in science	Chapter 3	Newton's laws: gravity and orbits
Big science	Chapter 5	The astronomers' tools: telescopes and space probes
The role of luck in scientific discovery	Chapter 8	The outer Solar System
Science is abstract	Chapter 9	Smaller bodies in the Solar System
Science and public policy	Chapter 9	Smaller bodies in the Solar System
Certainty and uncertainty in science	Chapter 9	Smaller bodies in the Solar System
The design of observations	Chapter 10	Planets beyond the Solar System
The importance of accuracy	Chapter 10	Planets beyond the Solar System
Indirect evidence	Chapter 10	Planets beyond the Solar System
The understanding that science brings	Chapter 11	Our Sun
Scientists change their minds	Chapter 11	Our Sun
Representative samples and observational selection	Chapter 12	A census of stars
Theory and observation	Chapter 13	The formation of stars and planets
The nature of scientific theories	Chapter 16	The Milky Way Galaxy
Scientists need lots of data	Chapter 16	The Milky Way Galaxy
The process of discovery in science	Chapter 17	Galaxies
How much weight should we give evidence?	Chapter 17	Galaxies
Uncertainty in science	Chapter 18	Cosmology
The design of observations	Chapter 18	Cosmology

Three BIG FACTS about the Universe

Throughout all the hundreds of pages of this book, you may find it difficult to "see the forest for the trees": to separate the fundamentally important issues from all the details. To guide you in your thinking, here is my "short list" of the truly essential facts about astronomy. Keep them in mind as you read the book.

The Universe is very big

It is probably impossible to appreciate the immensity of the astronomical Universe. If we represent the entire Earth by a dot a mere one 25th of an inch across, the Sun would be 40 feet away, and the nearest star a full 1840 miles distant. Our Milky Way Galaxy would be an astonishing 46 million miles in diameter. Beyond this lies the void of intergalactic space and untold billions of other galaxies. We have never found an end to these oceanic immensities. Indeed, the Universe might be infinite in extent.

The Universe is very old

It is also probably impossible to appreciate the immensity of the age of the cosmos. Our Earth is more than four billion years old: that is thousands of times longer than the span of time our human race has been in existence. If we shrink the lifetime of a person to a single minute, the Big Bang (about 13 billion years ago) occurred nearly four centuries ago.

We are not the center of the Universe

Nothing about the Earth is unique. Our home planet lies in the outskirts of our Galaxy. We revolve about the Sun, which orbits about the Galaxy, which itself moves through space. Immense numbers of other planets revolve around their home stars.

Three BIG FACTS about the nature of science

And here is my "short list" of the truly essential facts about the nature of science.

The Universe is knowable

It is actually possible to find out something about the cosmos.

We do this by making observations and formulating theories to explain them

These observations require ever-more sensitive telescopes and ever-more sophisticated techniques. The theories often involve concepts unfamiliar to us in daily life.

These theories are tested

Once we have formulated a theory, we do not simply believe in it. Rather, we test it, and the tests are repeated over and over again. The more tests the better: the more different kinds of tests the better. Only those theories that withstand this process are accepted. There is a great deal of evidence in their favor. Nevertheless, we are always learning new things.

Before we start

Take a few minutes to write yourself a letter in which you discuss (1) why you have decided to study astronomy and (2) what you hope to get out of this study.

Keep your letter in a safe place. At the conclusion of the course you will be asked to take it out and read it, and to answer a few questions about it.

ACKNOWLEDGMENTS

This book is the product of my entire career. Throughout this career my understanding of astronomy, and of the means we use to understand it, has changed radically. These changes are due to all the scientists and educators I have interacted with over the years. Each one of them contributed – sometimes overtly, sometimes invisibly – to my development. I cannot hope to name them all, but it was they who helped me become the person who would write this book.

Part I

Introducing steps to astronomy

Astronomy was the first science. Indeed, it is older than science. Thousands of years before the scientific revolution, thousands of years before telescopes and modern chemistry, geology and physics, people gazed at the sky and realized there was a lot going on up there to think about.

We begin our study of astronomy by considering what you can see with your naked eye. The daily passage of the Sun across the sky, the phases of the Moon, eclipses and the migration of the Sun across the constellations – all these regularities cry aloud for explanation, and they hint of a great cosmic structure. Early ideas of this structure – we now call it the Solar System – were formulated by ancient peoples, and they persisted for millennia.

Eventually these ideas were overthrown in the scientific revolution. We will trace briefly the course of this revolution, but in doing so our concern is not really historical. Our actual concern is to illuminate the nature of science through a study of its origins. Science is a way of thinking, a way of looking at the world, that was unique in the history of thought. Nothing more vividly illustrates the remarkable nature of science than a study of how it differs from what came before.

With the work of Isaac Newton the scientific revolution reached its climax. In his magisterial *Mathematical Principles of Natural Philosophy* this extraordinary genius set forth principles that govern the workings of the cosmos. We will devote an entire chapter to Newton's laws of motion and of gravitation, the single most important force operating in the astronomical universe.

Astronomy faces a difficulty not shared by other sciences: we cannot get our hands on what we study. The geologist can pick up a rock and examine it: the biologist can dissect an animal. But among all the objects in the Universe, only four – the Moon, Venus, Mars and a moon of Saturn – have actually been landed upon by spacecraft. The rest of the cosmos we are forced to study from afar.

Luckily, the Universe is continually broadcasting information to us, coded into light. It is by studying this light that we gain information about the cosmos. Indeed, until the advent of the space program, this was the *only* means we had of gathering information about the cosmos.

Telescopes are the very symbol of astronomy, the most important instruments at our disposal. For centuries they functioned as what might be called "giant eyes," operating as they did in the visual region of the electromagnetic spectrum. More recently new instruments such as radio telescopes and X-ray telescopes have been invented, capable of "looking" at the sky in entirely new wavelengths. Some sit on the ground: others orbit in space. Each has changed the way we do astronomy.

In recent decades the space program has made it possible to study the Moon and planets by actually visiting them. Astronauts have walked on the Moon, and robotic space probes have visited every planet of the Solar System. This entirely new way of studying the Solar System has brought back a wealth of information.

1

The sky

Astronomy belongs to everyone. The Universe is here for all of us to see. Its study is not just the province of astronomers, with their expensive telescopes and strange, unfamiliar mathematics. In this chapter, we are concerned with astronomy that you can do with your naked eye.

Some of the most universal aspects of our lives are influenced by astronomical phenomena. Imagine, for instance, a world in which day did not turn into night, or one in which there were no seasons! As we think about these, we will quickly realize that they are more subtle than perhaps we had thought. Indeed, even so simple a thing as the daily path of the Sun across the sky was historically explained in several different ways.

So too with eclipses and the phases of the Moon, the measurement of time and the drifting of the Sun along the zodiac – we begin our voyage through the Universe with these, some of the most fundamental aspects of our everyday environment.

Rising and setting: the rotation of the Earth

Perhaps the most basic of all astronomical observations is the simple fact that day turns into night and then day again in a never-ending cycle. This perpetual alteration, caused by the passage of the Sun across the sky, is so familiar that we hardly ever stop to pay attention to it. But in fact there is more to it than many people think.

Let us begin our study of astronomy with this, perhaps the simplest of all astronomical observations: the study of the Sun's path across the sky. To perform this study you will need no advanced scientific equipment. Simply step outside just before dawn, face east, and watch what happens. What you see depends on where you live: we will concentrate on the view of the sky from the mid northern hemisphere.

Many people believe that the Sun moves straight up as it rises. Does it? You can answer this question by mentally marking the location on the horizon at which it rises – just to the right of that house across the street, perhaps, or directly over that distant tree. An hour or so later, when the Sun has risen higher, step outside again and note its new position. Does it lie directly above the point at which it rose? You will find that it does not. In fact the Sun has moved along a slanting path, upwards and to the right as sketched in Figure 1.1.

Many people also believe that at noon the Sun is directly overhead. Here too, it is worthwhile to actually make the observation. You will find that it is not: the Sun at noon lies somewhere between the overhead point and the southern horizon. At this

east

Figure 1.1 **The rising Sun moves up and to the right (in the northern hemisphere).**

south

Figure 1.2 **At noon the Sun moves horizontally to the right (in the northern hemisphere).**

west

Figure 1.3 **The setting Sun moves down and to the right (in the northern hemisphere).**

time it is moving from left to right, parallel to the horizon as sketched in Figure 1.2.

Finally, watch the Sun set in the west. Perhaps you will not be surprised to observe that the Sun slants downwards and to the right as it sets, as sketched in Figure 1.3.

Why does the Sun rise and set?

How can we understand this intriguing behavior? We all know, of course, that the motion of the Sun is an illusion. In reality the Sun is holding still: it is *we* who move, carried by the rotation of the Earth. It is as if we were on some gigantic conveyer belt that steadily carries us up and over a hill. In fact this is a good analogy: the "hill" is the curving bulk of the Earth itself, and the "conveyer belt" is the Earth's rotation. Because the Sun rises in the east, we know that the rotation of the Earth carries us from west to east.

This simple analogy, however, does not help us understand the complexities of the Sun's path across the sky. The problem is that the analogy fails to capture the way in which we move as we are carried along by the rotation of the Earth. This motion is illustrated in Figure 1.4.

As we can see, people located on different parts of the Earth move in different ways. A person standing at the north pole spins about like a figure skater. A person standing at the equator executes a sort of "tumbling" motion. And finally, a person situated between pole and equator traces out a curious cone-shaped path over the course of one day.

The complexities of the apparent path of the Sun in the sky arise from these complex motions of the people who are observing it. Perhaps the simplest motion to think about is that of a person at the north pole: a simple spinning on one's axis. You can duplicate this motion for yourself by standing up on your toes and spinning about. If you look at a lamp as you do this, you will see it appear to move horizontally (Figure 1.5), mimicking the motion of the Sun as seen from the pole.

To mimic the motion of an observer at the equator, imagine what you would see if you were to gaze at a distant street light as you walked up and over a hill, if it were possible to remain not vertical but *perpendicular to the hill* (Figure 1.6). You would see the street light move straight up.

Figure 1.4 **The rotation of the Earth moves you in different ways depending on your location.**

Figure 1.5 **Simulating the motion of an observer at the pole.**

Figure 1.6 **Simulating the motion of an observer at the equator.**

And finally, the motion of a person at some location intermediate between equator and pole is intermediate between these two situations, and yields the observed slanting path of the Sun.

THE NATURE OF SCIENCE

HYPOTHESIS TESTING IN SCIENCE: WHY DOES THE SUN RISE AND SET?

It is important to emphasize that what we have described in the above section is a *theory*, one designed to explain the observations we have made of the Sun's path across the sky. Let us call it "the Round Earth theory":

- *the Round Earth theory*: the Earth is round, and it is spinning on its axis. The Sun holds still.

But a different idea has also been held:

- *the Flat Earth theory*: the Earth is flat, and it is holding still. It is the Sun that moves.

Of course we all know perfectly well that the Round Earth theory is true and the Flat Earth theory is false. We have all seen images of the spherical globe of the Earth sent back to us from space. But let us try to ignore this knowledge for a moment, and try to put ourselves in the shoes of people who lived centuries ago, before the dawn of the space age. They might have regarded the Flat Earth theory as being perfectly good. After all, in our daily lives the ground beneath our feet certainly *looks* flat. What could they have done when faced with two different explanations for the same thing?

Scientists are continually encountering such situations. We discover something interesting, and then we think up a theory to account for it. The problem is that often we are able to think up *lots* of theories! What then? *How can we choose between competing hypotheses?*

The answer is that we ask each hypothesis to make a *prediction* concerning something that has not yet been observed. The different theories will make different

east

Figure 1.7 **The rising Sun moves straight up as seen from the equator, according to the Round Earth theory.**

predictions. We then go out and *conduct new observations*, testing to see which prediction turns out to be true.

So to choose between our two theories of the Sun's path, let us ask each of them to predict *the Sun's path across the sky as seen from different locations on the Earth*.

Predictions of the Round Earth theory. Imagine that you were to get in a plane, travel thousands of miles and then track the motion of the Sun. What would you observe according to the Round Earth theory? As you can see from our discussion in the previous section, you would find that the slant angle at which the Sun rises and sets depends on your location. The farther south you journey, the more nearly vertical would this angle become. Indeed, if you were to travel all the way to the equator you would find the Sun rising straight up (Figure 1.7) and setting straight down.

Similarly, if you were to journey to the north pole you would find that as 24 hours pass the Sun would *never* rise or set. Rather it would move in a circle from left to right, parallel to the horizon (Figure 1.8).

Predictions of the Flat Earth theory. The Flat Earth theory makes a different prediction. If the Earth were flat and the Sun moved across the sky, the Sun's path would be the same no matter where you were. So this theory predicts that if you were to get in a plane, travel thousands of miles and then track the motion of the Sun, you would find everything to be the same.

Testing the predictions. Experience confirms that the predictions of the Round Earth theory are borne out, and the predictions of the Flat Earth theory are not. The path of the Sun across the sky does indeed depend on where you are.

In this way we accept one hypothesis and reject the other. And in later chapters, we will find that the process is always the same: scientists are forever testing their theories by forcing them to make predictions, and they choose the best theory by finding the one that makes the most successful predictions.

A SCIENTIFIC THEORY MUST MAKE PREDICTIONS

It is essential to hypothesis testing that *a theory has to make a definite prediction*. If you have an idea that does not make a prediction, your idea is not a theory.

As an example, consider yet a third idea about the Sun's path.

Figure 1.8 **The Sun neither rises nor sets as seen from the north pole, according to the Round Earth theory.**

- *The Free Will Idea*: the Sun moves across the sky because it wants to.

We would not say that this is a legitimate theory: it is not a hypothesis but a vague notion. It makes no predictions about anything. Even if we granted that the Sun could "want" something, there is no way we could test this idea.

It is also essential to hypothesis testing that the theory must be able to make a prediction about something *that has not yet been observed*. In our present case, we must ask each theory to make its prediction before we had actually gotten in a plane and traveled to a new location. Only in such a circumstance would we have conducted a fair test of the competing hypotheses.

north

This is easy for us nowadays, but in ancient times it was an exceedingly difficult proposition. In the past, cultures that did not engage in long-distant voyages of exploration had no way to decide between the Round Earth and Flat Earth hypotheses. Similarly, in modern science we often encounter situations in which we are not yet able to conduct observations that would allow us to choose between competing theories. We therefore live in a state of perpetual uncertainty, in which we are forced to suspend judgment, and tentatively hold in our minds a set of mutually exclusive explanations.

Hypothesis testing is a universal aspect of science. We will encounter it again and again throughout this book.

"Rotation" of the sky: circumpolar stars

It is not only the Sun that rises and sets. Every object in the sky – the Moon, the planets, the stars – appears to move. You can prove this to yourself by going outside at night and glancing at the sky to the East – and then returning a few hours later and looking again. You will find that everything that used to be close to the eastern horizon has now risen higher and to the right. A little thought shows why this should be so: we view the entire sky from the rotating Earth, and everything in it undergoes the same kinds of motion as does the Sun.

But this does not mean that everything rises and sets. Indeed, there is a group of stars – so-called "circumpolar stars" – that never rise or set. They are always above the horizon.

Figure 1.5 illustrated how a person at the north pole moves, carried along by the rotation of the Earth. In Figure 1.9 we illustrate how the sky appears to move as seen by this person. As you can see, this person sees every star moving in a circle. The center of this circle, the so-called *celestial pole*, lies directly above the observer. It is the location on the sky toward which the rotation axis of the Earth points. We can think of the sky as a gigantic dome peppered with stars: as seen from us on Earth, this dome appears to rotate about an axis passing through the celestial pole. The North Star lies almost exactly at this pole. Notice that in Figure 1.9 stars appear to move parallel to the horizon. So none of them rise or set. We see that from the Earth's poles *every* star is circumpolar.

Figure 1.10 does the same for a person located on the Earth's equator. A person on the equator is tilted by 90° relative to one at the pole, so that as seen from the equator the dome representing the sky appears to be tilted by 90°. Thus from the equator the celestial pole lies on the horizon to the north. As illustrated, from the equator the stars appear to move perpendicularly to the horizon, so that they rise straight up. Furthermore, all of them rise and set: from the Earth's equator *no* star is circumpolar.

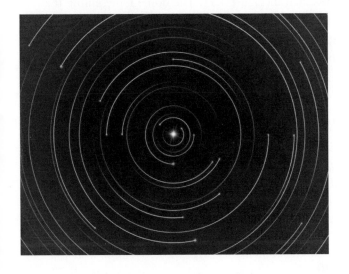

Figure 1.9 **"Rotation" of the sky as seen from the north pole. We are looking straight up.**

Figure 1.10 **"Rotation" of the sky as seen from the equator.**

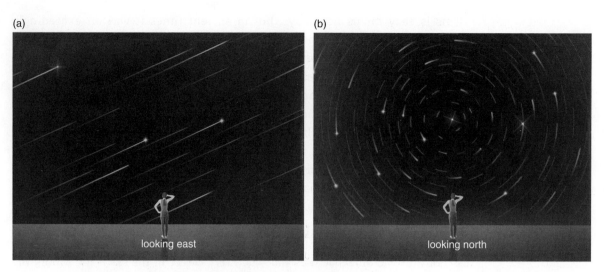

(a)   looking east

(b)   looking north

Figure 1.11 **"Rotation" of the sky as seen from intermediate latitudes.**

The "rotation" of the sky as seen from somewhere lying in between these two extremes is intermediate. It is illustrated in Figure 1.11. The dome representing the sky appears to be tilted, with the celestial pole lying above the northern horizon. If we look to the east we see stars rising, but if we look to the north we see stars appearing to rotate about the celestial pole. Those stars whose paths never pass below the northern horizon are circumpolar.

How rapidly does the Earth rotate?

Although we do not feel it, we are all moving, carried along by the rotation of the Earth. Let us calculate how rapidly we are moving.

The calculation is easiest if we concentrate on a person situated on the equator. As you can see from Figure 1.4, this person is moving in a circle whose radius is the radius of the Earth.

The logic of the calculation

Velocity is distance divided by time. The distance the person moves is the circumference of the Earth, and the time required to move this distance is one day. So the steps in the calculation are as follows.

Step 1. Find the circumference of the Earth.
Step 2. Divide this by one day to find the velocity.

As you can see from the detailed calculation, a person at the equator is moving at a bit more than a thousand miles per hour.

Now let us transfer our attention from the equator to the north pole. As we can see from Figure 1.4, this person is not moving in a circle at all – she is simply spinning about on an axis. So the velocity at the north pole is not a thousand miles per hour, but zero.

This is enough to tell us that the velocity at which we are moving must depend on our location. It is greatest at the equator, and zero at the pole. In Problem 8 at the end of this chapter we study this further.

Detailed calculation

We will begin with a rough calculation, and then do it more carefully.

Rough calculation

Step 1. Find the circumference of the Earth.

The Earth's circumference is about 24 000 miles.

Step 2. Divide this by one day.

There are 24 hours in a day, so the velocity is roughly

$$V = 24\,000 \text{ miles}/24 \text{ hours} = 1000 \text{ miles/hour}.$$

Careful calculation

Our above value for the circumference of the Earth was not exact: we should do the calculation more carefully. Furthermore, throughout this book we will be using the MKS system (the meter/kilogram/second system), so we need to work in these units.

Step 1. Find the circumference of the Earth.

The Earth's radius R is 6378 kilometers, or 6.378×10^6 meters. Its circumference is then

$$C = 2\pi R = (2)\,\pi\,(6.378 \times 10^6 \text{ meters}) = 4.01 \times 10^7 \text{ meters}.$$

Step 2. Divide this by one day.

We need to express the length of the day in seconds: we get one day = (24 hours/day)(60 minutes/hour)(60 seconds/minute) = 8.64×10^4 seconds.

So the velocity is

$$V = 4.01 \times 10^7 \text{ meters}/8.64 \times 10^4 \text{ seconds} = 464 \text{ meters/second}.$$

How good was our rough calculation?

We can find out by converting the result of our careful calculation into miles per hour: we find

464 meters/second = 1038 miles/hour.

Our rough calculation had been pretty good!

NOW YOU DO IT
Measure the diameter of a basketball. Now spin it once per second. How rapidly is a point on its equator travelling? Can you run this fast?
(A) Explain the logic of the calculation you will do.
(B) Do the detailed calculation.

The Sun is 6.96×10^8 meters in radius. And although it is not obvious to the naked eye, the Sun actually rotates on its axis: it does so once every 25.4 days. How rapidly is a point on its equator moving? (A) Explain the logic of the calculation you will do. (B) Do the detailed calculation.

The phases and motion of the Moon

Our simple observations of the Sun have led to some interesting results. Now transfer attention to the Moon. As we have already emphasized, the Moon, like the Sun, rises and sets, and it follows a path across the sky that depends on your location on the Earth. But there is more to the Moon than this.

Not long ago I decided to study the Moon, and I observed it carefully over the course of a month. Here is what I saw.

First observation. Early one morning I stepped outside just before dawn. A brilliant glow marked the point at which the Sun was about to rise. Close to the Sun was a thin crescent Moon (Figure 1.12)

Second observation. This was performed about a week later. At about noon I noticed that the Moon was rising in the east. It was in its half phase (Figure 1.13).

Third observation. The was performed about a week after the second. As the Sun was setting I noticed that the Moon was rising in the east. It was full (Figure 1.14).

The fourth observation was performed about a week after the third. I noticed that the Moon was rising in the east at about midnight. It was in its half phase (Figure 1.15).

The entire cycle repeats endlessly, taking just 29½ days to be completed – the *lunar month.* Indeed, our word "month" comes from the word "moon."

These observations tell us a number of things about the Moon.

(1) *The bright portion of the Moon is always oriented toward the Sun*, as you can see from Figures 1.12–1.15.
(2) *The Moon rises at different times of day.* Indeed, the time of moonrise undergoes a regular progression as the lunar month proceeds. Initially (Figure 1.12) the Moon rises more or less when the Sun rises. But as the month rolls by, the Moon rises later and later each day.

Figure 1.12 **The crescent Moon rises at dawn.**

looking east

Figure 1.13 **The half Moon rises around noon.**

Figure 1.14 **The full Moon rises around sunset.**

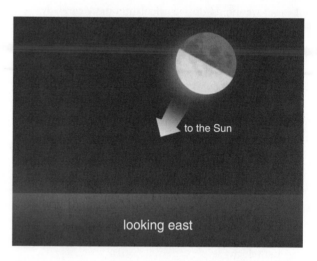

Figure 1.15 **The next half Moon rises around midnight.**

(3) *The phase of the Moon is connected to when it rises.* As you can see from the above observations, the phase of the Moon undergoes a regular progression, synchronized with the progression in its time of rising.

Let us see if we can understand these three points.

The part of the Moon that we can see is the part that is illuminated by the Sun. If you could go to the Moon and stand on the bright part, you would be in daytime. Similarly, if you could stand on its dark part, you would be in nighttime. Clearly, at the various phases of the Moon we are seeing differing portions of its daylight and night-time parts. When the Moon is in its crescent phase, we are seeing mostly its night side and only a little bit of its day side. Similarly, when the Moon is in its half phase, we are seeing half of its dark side and half of its day side.

Let us make a model of the Moon illuminated by the Sun. The Moon, of course, is spherical, so our model must be too. Let's say that our model is an orange. In your room, turn out all the lights but one, and study how this orange is illuminated by that one remaining light. As illustrated in Figure 1.16, *half of any sphere is illuminated by a distant light, and half is not.* This is true for your orange, and it is true of the Moon as well.

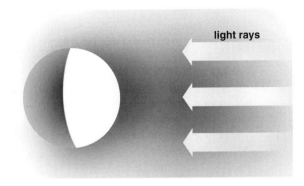

Figure 1.16 **Half of a sphere is illuminated by a distant light, and half is not.**

NOW YOU DO IT
Consider an egg illuminated by a distant light bulb. Imagine that you turn the egg this way and that. Is half of the egg always illuminated? Or are there some configurations in which more than half is illuminated?

Suppose now you walk about your room, looking at the orange. From some locations, you will see only its illuminated half, from other locations you will see only its dark half, and from yet other locations something in between. These are the "phases of the orange," as shown in Figure 1.17. Notice in Figure 1.16 that the illuminated portion of your orange is always oriented toward your light bulb. Thus we understand point (1) above.

Now let us pass on to the time of moonrise.

In Figure 1.18 the rotation of the Earth is carrying the observer toward the Sun, which is just about to appear over the observer's horizon. This figure therefore illustrates the moment of sunrise. As we have seen, sometimes the Moon rises at this moment. In this configuration, it must lie in the same direction as the Sun, as indicated in Figure 1.18.

On other days, the Moon must lie in some other direction. In Figure 1.19 the Moon is about to *appear* over the horizon just as the Sun is setting. This figure illustrates a configuration in which the Moon *rises* as the Sun sets – a configuration reached just half a lunar month after the configuration of Figure 1.18.

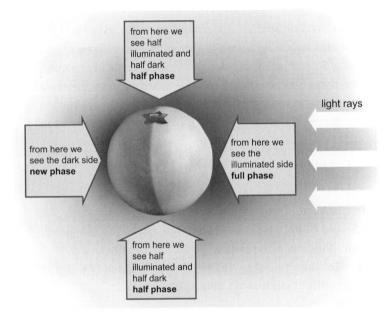

Figure 1.17 **Phases of an orange as illuminated by a distant light bulb.**

Notice that in Figure 1.19 the Moon is in a different direction than in Figure 1.18. It must have moved. And of course it has: it is orbiting about us! *The steady progression in the time of moonrise arises because of the Moon's orbit about the Earth* (Figure 1.20).

Let us now add to this the understanding we have reached of the phases of the Moon. Figure 1.21 diagrams how the Moon's phases are related to its position in its orbit about the Earth.

Figure 1.18 **If the Moon rises when the Sun rises, it must lie in the *same* direction as the Sun.**

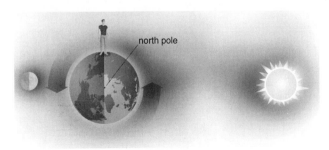

Figure 1.19 **If the Moon rises when the Sun sets, it must lie in the *opposite* direction to the Sun.**

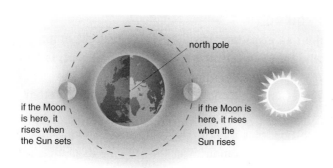

Figure 1.20 **The Moon orbits the Earth.**

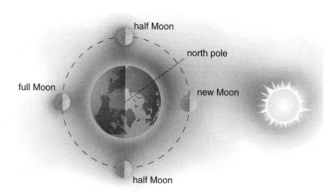

Figure 1.21 **The Moon's phases are related to its position in its orbit.**

Only one more step is needed to complete our understanding of the phases and the motion of the Moon. Let us return to our discussion of the "phases of an orange" and modify our exercise somewhat. In our initial exercise you walked about the orange, examining it from various directions. Now sit down, to make yourself into a model of the *Earth*, and have a friend carry *the orange* about you in a circle, mimicking the orbit of the Moon. As you look at the orange, you will see it go through all the phases we observe in the Moon (Figure 1.22).

NOW YOU DO IT
Suppose that you live in Los Angeles and you observe the Moon to be a thin crescent. You telephone a friend who lives in Santiago, Chile. She comments that she can see the Moon. Does she also see it to be a thin crescent?

Imagine a telephone pole illuminated by the setting Sun. Since the Sun is close to the horizon, the configuration looks like this.

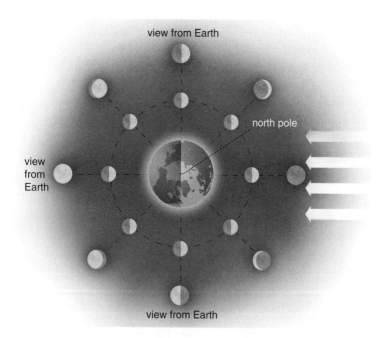

Figure 1.22 **Phases of the Moon.**

To the Sun →

In some sense this telephone pole has "phases" – after all, part of it is illuminated by the Sun and part is not. Imagine that you were to walk around the pole. From where would you see a "new pole?" From where would you see a "full pole?" From where a "half pole?"

THE NATURE OF SCIENCE

THE IMPORTANCE OF SKEPTICISM: TESTING OUR THEORY OF THE MOON'S PHASES

⇒ **Looking forward**
We will study the scientific revolution in Chapter 2.

Skepticism is the most cherished scientific virtue. Every scientist is skeptical, and every scientific theory is provisional and subject to doubt. The scientific revolution was a revolution of skepticism. As we will see in the next chapter, in Copernicus' time the Ptolemaic model of the Solar System had dominated thinking for nearly 15 centuries – contrast this with Einstein's relativity theory, which has dominated our thinking for only one century! Nevertheless, Copernicus felt free to discard it for a model he thought better. Similarly, Kepler felt free to discard the Greek idea of uniform circular motion, and Galileo to discard the Greek idea of the perfection of the heavens.

The scientist who thinks she has made a discovery is immediately skeptical. She tests it in every way she can. She thinks of every possible error she might have made. She thinks of every possible alternative interpretation. And once she announces her new discovery, the skepticism only increases. Other scientists attack it. They too search for errors she might have made. The whole process can be very bruising to the ego.

The more important the discovery, the more severe the skepticism. Thus Einstein's theory of relativity is constantly being tested. There is a whole program of searching for errors Einstein might have made. Highly accurate experiments are done, searching for the tiniest deviations from his predictions. Some widely respected scientists have spent many years of their careers in this effort to de-throne relativity. These people are not regarded as renegades, and their work is not regarded as disrespectful of Einstein. Just the opposite: the very intensity of their effort is testimony to relativity's importance.

What is the reason for this great emphasis on skepticism? Why are scientists seemingly so intent on destroying their very own theories? We can illustrate the reason by means of an analogy.

If you need to climb a tall ladder, it is wise to shake it just before starting up. Rather than trusting to the ladder right away, a prudent person will rock it to and fro. The point is not to knock the ladder down: the point is to make sure that nothing *else* can knock it down. You are testing your ladder to make sure that it is strong. You need to be confident that your ladder can be trusted.

In just this way, we want our scientific knowledge to be strong. We want to be confident of it. And the only way to do this is to attack it: to be skeptical, and to see if it withstands our attacks. The theories that we accept today are those that have survived this continual testing and re-testing. They have been severely shaken and proved themselves strong. Over and over again we have probed them, searching for weaknesses; and over and over again these theories have been proved to be worthy of our trust. We have found many reasons to be confident of them.

Let us therefore be careful, and test our theory of the Moon's phases. As we saw in our discussion of hypothesis testing, every scientific theory must make *a definite prediction about something that has not yet been observed*. Can we think of a prediction of our theory that could be tested in this way?

We can! According to our theory, *every* spherical body illuminated by the Sun should show phases. And just as the Moon is a sphere, so too is the Earth. So our

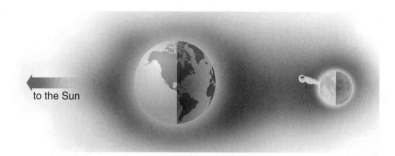

Figure 1.23 **Phases of the Earth and Moon.**

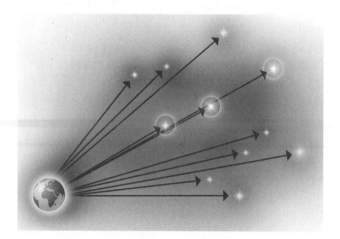

Figure 1.24 **Constellations are not real physical structures. The indicated stars lie close to one another as seen from our vantage point. But because they are at different distances from us, they are actually far away from each another.**

theory's prediction is that *the Earth has phases too.* Imagine, then, that the Moon happens to be full, and that an astronaut journeys to the Moon and then looks back at the Earth. Figure 1.23 illustrates the configuration. As you can see, our theory predicts that the astronaut would see the Earth to be in its new phase.

NOW YOU DO IT
Suppose the Moon is in its half phase. What is the phase of the Earth according to that astronaut?
    Suppose the astronaut moves to a point half way between the Moon's equator and its north pole. Does the phase of the Earth as seen by the astronaut change?

As we commented in our discussion of hypothesis testing, sometimes it is very hard to test a theory's predictions. Not until the space program did it become possible to test our theory of the Moon's phases in this way!

The constellations

One of the great pleasures of naked-eye astronomy is identifying constellations. Orion the hunter, Ursa Major and Minor, the bears: these are lifetime companions. But it is important to emphasize that constellations are not real physical structures. Stars are scattered more or less randomly across the sky. At least in our vicinity, there is no overall pattern to their arrangement. Nevertheless, purely by accident, some of them seem to lie in recognizable patterns that we refer to as constellations. As an analogy, imagine that you have a handful of pebbles and you toss them on the floor. Bending down to look at their arrangement, you might find that here and there, purely by chance, a few happen to lie in some sort of pattern.

Figure 1.24 illustrates how we know that the constellations are not actual structures. If we measure the distances to the various stars in a constellation, we find that they are all different. The stars therefore do not actually lie close to one another: they only appear to as seen from our perspective. (There are, however, a few exceptions to this general rule. One well-known instance is the Pleiades. Measurements of their distances reveal that they do in fact lie close to one another. The Pleiades is a cluster of stars gravitationally bound to one another.)

The motion of the Sun against the stars

Visibility of the constellations

People who know the constellations are aware that different constellations are visible in different seasons. Libra is a summer constellation, while Taurus is visible in winter. This is reflected in the constellation maps you will find in Appendix VII. There is a different map for every season of the year. A map of the United States is just as valid in June as it is in February – but a map of the sky is not.

Why should this should be so? What happens to Libra to render it invisible in winter, and to Taurus to render it invisible in summer? We can answer this question by fixing attention on some particular constellation, and following it as the months pass. When does it rise and set?

To be specific, let us concentrate on Libra. In May, this constellation rises about 7 in the evening, and it sets at 5 in the morning. So in May, Libra is visible throughout the night: it rises around sunset and it sets around sunrise. By midsummer, on the other hand, it is already high in the sky at sunset, and it sets around midnight. So it is only visible for the first half of the night. And by November, Libra sets just as night begins: it is not visible at all.

....

NOW YOU DO IT
Consult Appendix VII.
   In what months is Pisces visible?
   Identify three constellations that are visible in September.

....

Clearly Libra has not magically ceased to exist during the winter. Rather, its absence arises from the fact that it can be seen only at night, and that it rises at night only during certain times of year.

Like the Moon, at any given moment half the Earth is illuminated by sunlight and the remainder is not. This means that at any given moment the constellations are visible from only half the Earth – the nighttime half. But there is another consideration: even for those people for whom it is night, only half the constellations are visible. The rest lie beneath the horizon and are blocked from view by the Earth's solid bulk (Figure 1.25). To solve the mystery of the "disappearance" of Libra, we should ask *when Libra lies in the visible half of the sky as seen from the nighttime half of the Earth*.

Imagine two astronomers who live on opposite sides of the Earth. Let us name them Alice and Bob: Alice lives in North America, while Bob lives in Australia. Because they live on opposite sides of the Earth, it is day for Alice when it is night for Bob and vice versa. So at any given moment, only one of these two astronomers can see the constellations.

In the configuration of Figure 1.26 that person is Alice. It is she who is experiencing nighttime. Furthermore, Libra lies above her horizon. Thus, in the configuration illustrated in this figure, Alice can see Libra. Furthermore, if we let 12 hours pass, the rotation of the Earth will have moved Bob into Alice's former position, and Alice into Bob's: then it will be Bob who can see the constellation. We conclude that in the configuration of Figure 1.26, *Libra is visible to everybody* (although not at the same moment).

But now consider the different configuration illustrated in Figure 1.27. Here it is Bob who is experiencing darkness and can see the constellations. Unfortunately, at Bob's location Libra lies below the horizon and is invisible. So

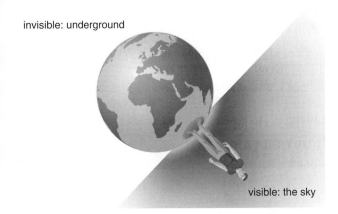

invisible: underground

visible: the sky

Figure 1.25 **Only half the sky is visible. The other half is obscured by the bulk of the Earth.**

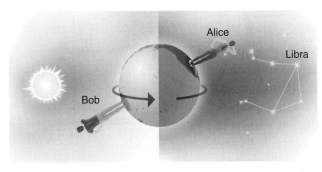

Alice

Libra

Bob

Figure 1.26 **Libra is visible to Alice but not to Bob. Twelve hours later, Bob will be able to see it and Alice won't.**

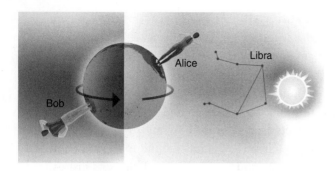

Figure 1.27 **Libra is invisible to both Alice and Bob.**

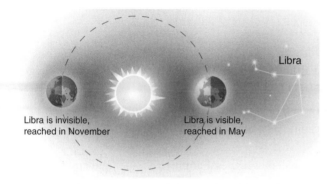

Figure 1.28 **Libra's visibility and the orbit of the Earth about the Sun.**

neither Alice nor Bob can see Libra. Furthermore, if they wait 12 hours, the Earth's rotation will have exchanged Alice's and Bob's positions – but it still remains true that neither person can see the constellation. We conclude that *Libra is visible to nobody* in the configuration of Figure 1.27.

NOW YOU DO IT
The direction to the constellation of Capricornus is more or less perpendicular to the direction to Libra. Repeat the above analysis, focusing now on Capricornus. In what configurations is it visible? In what configurations is it invisible?

The slow progression from a configuration in which Libra can be seen to one in which it cannot must correspond to a progression from Figure 1.26 to Figure 1.27. But what would make the configuration change in this way? Looking at these two diagrams, we might get the impression that the Sun has moved from the one to the other. But in fact it is not the Sun that moved – it is the Earth! During the time interval separating Figure 1.26 from Figure 1.27, the Earth moved to the other side of the Sun.

The regular cycle, in which each constellation passes from visibility to invisibility and back again, is caused by the orbit of the Earth about the Sun. In Figure 1.28 we expand our view somewhat and illustrate this motion. The configuration of Figure 1.26, in which Libra is visible, is reached when the Earth lies on the side of its orbit toward that constellation. That of Figure 1.27 is reached six months later, when we are on the other side of our orbit.

We now understand why each constellation requires just one year to pass through its cycle of visibility to invisibility and back again. This is how long it takes the Earth to orbit the Sun.

NOW YOU DO IT
Draw a figure, similar to Figure 1.28, illustrating the visibility of Capricornus.

The motion of the Sun through the constellations: the ecliptic

Let us return to the configuration as sketched in Figure 1.27, in which Libra cannot be seen because it lies in the same direction as the Sun. Suppose that, by some remarkable feat of magic, we could reduce the brightness of the Sun until the stars could be seen in broad daylight. Then we would be able to see Libra after all! Indeed, our line of sight to the Sun would continue on toward that constellation. So *in this configuration, the Sun is in the constellation of Libra.*

Of course we cannot reduce the brightness of the Sun – but we *can* look west just after it has set and the stars are beginning to come out. As sketched in Figure 1.29,

Figure 1.29 **Finding the position of the Sun against the constellations.**

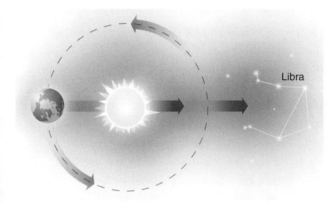

Figure 1.30 **Our line of sight to the Sun swings about as we orbit. Thus the Sun appears to drift across the constellations.**

we can in this way estimate the location of the Sun against the distant backdrop of the constellations.

In Figure 1.30 we expand our view, and study the Earth in its orbit about the Sun. In the configuration illustrated, the Sun lies in the constellation of Libra. But it does not remain there. As the Earth orbits the Sun, our line of sight to it steadily swings about. So the Sun appears to drift steadily across the constellations. Since the Earth requires a year to orbit once about the Sun, the Sun takes a year to complete this motion against the stars. We emphasize that this motion is entirely illusory: in reality it is not the Sun but the Earth that is moving. The apparent passage of the Sun against the constellations is an effect of perspective, caused by the steadily shifting vantage point from which we view it.

Let us chart the Sun's apparent motion. We can do so by carrying out the above procedure night after night, determining the Sun's location against the constellations throughout an entire year. Each night, the Earth has moved somewhat in its orbit; therefore, each night the Sun is in a slightly different location among the stars.

Once we have done this, we need a means of representing our result. We need a map of the constellations that indicates them all, without regard to whether they are visible or not on any given night. Such a map is shown in Figure 1.31. On this map, the constellations that lie close to the Sun are invisible, while those that lie far from it are visible.

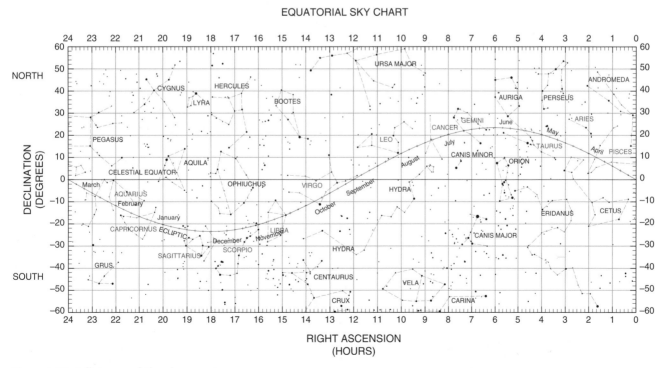

Figure 1.31 **A flat map of the sky.**

Figure 1.32 **A celestial globe.**

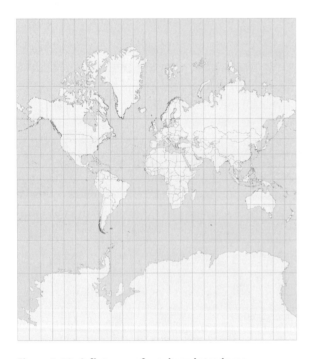

Figure 1.33 **A flat map of a sphere introduces distortions. This map of the Earth represents equatorial regions correctly. But Greenland is too large, and the poles are stretched out into lines.**

On this figure the Sun's location at various months is indicated. We see that the Sun follows a quite definite path through the constellations, indicated by the curving dotted line. *The Sun's path against the constellations is known in astronomy as the ecliptic. In astrology it is known as the zodiac.* In Chapter 2 we will discuss the astrological significance of the zodiac.

The celestial sphere

Perhaps when you were a child you looked up at the sky and felt that you were looking at the inside of a dome. This is indeed just what the night sky looks like. This imaginary dome is known as *the celestial sphere.* And as the spherical Earth can be represented by a globe, the celestial sphere can be represented by a *celestial globe* (Figure 1.32).

Celestial globes are useful for the same reason that Earthly globes are: they avoid the distortions caused by representing a three-dimensional sphere on a two-dimensional surface. We are all familiar with these distortions. Figure 1.33 shows a map of the world. It represents equatorial regions correctly – but Greenland and Antarctica are too large, and the poles, mere mathematical points, are stretched out into lines. Similarly, the constellations are distorted on the flat map shown in Figure 1.31.

A celestial globe has no distortions. It shows not continents and seas, but stars and constellations. One imagines the Earth as being located at its very center. The constellations are drawn upon the celestial globe so that, were we to climb inside it and view them from its center, they would appear as they do in reality as we see them each night. This means that the constellations are drawn backward upon a celestial globe as we look at it in Figure 1.32.

Just as the Earth has poles, so the celestial globe has *celestial poles.* The north celestial pole is directly above the Earth's north pole. If you were to go to the Earth's north pole, the north celestial pole would be straight overhead. The pole star, Polaris, is located almost exactly at this point. Similarly, the south celestial pole lies directly overhead as seen from the Earth's south pole. Unfortunately there is no bright star located there, so residents of the southern hemisphere have no south pole star with which to reckon directions.

Just as the Earth's equator is an imaginary circle lying midway between the poles, so the *celestial equator* lies midway between the celestial poles. This imaginary circle drawn on the celestial sphere lies directly overhead as seen by people living along the Earth's equator.

The *ecliptic*, the Sun's apparent path across the sky, is also an imaginary circle drawn on the celestial sphere. It does not coincide with the equator. However, these two circles do intersect at two points. Thus, as the months roll by and the Sun appears to move along the ecliptic, twice a year it crosses the celestial equator. As we will see in the next section, in these configurations we have an equinox, in which days and nights are 12 hours long. For this reason, these intersection points between the celestial equator and the ecliptic are known as *the vernal and autumnal equinoxes*.

As you can see from Figure 1.31, on June 21 the Sun is as far north of the celestial equator as it ever gets. We will see in the next section that when the Sun reaches this point in its yearly passage we are at *the summer solstice*, in which the days are longest in the northern hemisphere. Similarly, on December 21 the Sun is as far south as it ever gets: this is *the winter solstice*, in which the days are shortest in the northern hemisphere.

All these features are also represented, without distortions, on a celestial sphere.

Detectives on the case

The reasons for the seasons

A recurring theme throughout this book is that scientists create theories to explain their observations. I like to think of scientists as being like detectives: they come upon something strange, and they create a theory to account for it. And just like detectives, they use clues to solve their mysteries. These clues are provided by the Universe itself. Scientific observations are always revealing strange and unexpected things about the cosmos – and these can guide us as we devise our theories.

Over and over again throughout this book, we will watch this in action. In doing so, our goal is to *understand the process a scientist goes through in creating a theory*. As our first example we will study the process whereby we come to an understanding of the seasons.

As you read this section, be aware that we will be doing two things at once.

- We will grapple with the seasons, seeking to understand what causes them.
- We will step back and watch ourselves as we do this, and we will try to understand the mental procedure we go through in our effort to create a theory.

So let us turn our attention to the regular progression of the seasons. Because this progression follows a cycle of just one year, which is also the time required for the Earth to orbit the Sun, we know the seasons must be connected in some way with our orbit. What might the connection be? Let us try out a few theories.

First hypothesis. One possible theory is that the Earth is farther from the Sun in winter than in summer. And it is indeed the case that the Earth's orbit about the Sun is not exactly circular (see Chapter 2, where we discuss Kepler's laws). But for two reasons we know that this cannot be the explanation for the seasons.

(1) Seasons in the northern hemisphere are opposite to those in the southern hemisphere. When it is summer in the north, it is winter in the south. But when we are farthest from the Sun, the *entire* Earth receives less heat from it, so that this theory predicts the northern hemisphere to experience winter when the southern hemisphere does.

(2) The Earth is indeed in an elliptical orbit, in which its distance from the Sun varies. But take a look at Table 1.1 listing this distance. The Earth is *not* farthest

Table 1.1. **Our distance from the Sun.**

| Greatest distance | July 14 | 94 514 000 miles | Difference = 3 102 000 miles |
| Least distance | January 5 | 91 412 000 miles | |

from the Sun in December. It is farthest in July! Indeed, we are more than three million miles farther from the Sun in July than in January. How can the northern hemisphere experience summer in spite of this extra distance in July? The only possible conclusion is that three million miles, while it looks like a great distance to us, is in reality not big enough to have any appreciable effect on the seasons.

Let us then try a *second hypothesis*. Perhaps in June the *northern hemisphere* is closer to the Sun than the southern hemisphere. After all, everybody knows that the Earth's axis of rotation is tilted. Figure 1.34 illustrates this possibility.

This hypothesis nicely accounts for the fact that the seasons in the northern and southern hemispheres are opposite to one another: when the northern hemisphere is closer to the Sun, the southern hemisphere is farther. Unfortunately, however, this hypothesis too will not work. This is because the difference in distance is too small. This difference is indicated by the thick white line in Figure 1.34. As you can see, it is roughly the radius of the Earth. But the Earth's radius is a mere four thousand miles – incomparably less than three million miles, which itself is too small to have an effect!

We have now tried and rejected two theories. The third theory we will try is one that we now know to be correct. *The true reasons for the seasons* do indeed have to do with the fact that we receive more warmth from the Sun in summer than in winter. But this is not because we are closer to it. Rather it arises for two reasons.

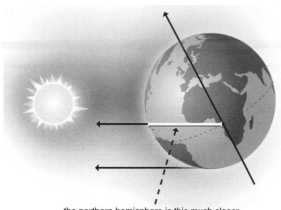

the northern hemisphere is this much closer
to the Sun than the southern hemisphere

Figure 1.34 **An incorrect theory of the seasons. Our theory is that, in the northern hemisphere's summer, that hemisphere is closer to the Sun than the southern. The difference in distance is roughly the Earth's radius. This figure is drawn for the northern hemisphere's summer.**

(A) *In summer the days are long, while in winter they are short.* It is immediately obvious why this is important for understanding the warming effects of the Sun: the longer the Sun is above the horizon, the more time is available for it to heat us.

(B) Every day, noon marks the highest point the Sun ever reaches in its motion across the sky. As we saw in our study of the Sun's path across the sky, if you were to step outside at noon and glance at the Sun, you would find that it is not directly overhead. It turns out that this high point is different from summer to winter: *in summer the Sun is high in the sky at noon, while in winter it is low in the sky at noon.* Furthermore, as illustrated in Figure 1.35, the higher the Sun is in the sky, the greater are its heating effects. In this figure we see a patch of ground, and an imaginary tube that contains the rays of sunlight destined to reach it. All the rays contained in this tube strike this patch and serve to warm it: all the rays that lie outside the tube miss it and do not. We can clearly see in Figure 1.35 that the lower the sun is in the sky, the larger the patch of ground the rays reach: the rays are spread over a larger region, and the ground is heated less.

if the Sun were high in the sky

if the Sun were low in the sky

Figure 1.35 **Heating effect of the Sun's rays is greater if the Sun is high in the sky than if it is low in the sky.**

For both these reasons, we receive more warmth from the Sun in summer than in winter.

How scientists create their theories

Now let us stop thinking about the seasons, and start thinking about the process we have just gone through. How did we come up with our theory of the seasons?

We did so by trial and error. We tried this and we tried that. And as we did so, we constantly asked each idea to make definite predictions, and we compared these predictions with observation. Only by doing so were we able to realize that our first two ideas were wrong.

There's nothing so terrible about being wrong – scientists invent wrong theories every day of the week. The important thing is to learn from one's mistakes. Indeed, mistakes can be helpful, for by thinking about the errors we made, we realize how to correct them.

These comments return us to two points that we have already emphasized: *the importance of skepticism* and the fact that *science is testable*. It is not enough to propose a theory. It is not even enough that the theory makes sense. Because we are skeptical, we insist that every proposed theory makes a specific prediction that can be tested.

What leads us, finally, to our correct theory? Scientists never invent their theories out of nothing. The Universe is continually giving us clues. In this case the clue is the fact that the cycle of seasons takes place over a year, and the year is the amount of time required for the Earth to orbit about the Sun. It is as if the Universe were giving us advice, and whispering in our ear "think about the motion of the Earth about the Sun."

Theory creation is a fundamental aspect of science. We will return to it again and again throughout this book.

Why are there seasons?

We have reached a preliminary understanding of the reasons for the seasons. Let us now flesh out this understanding. To understand the seasons in full, we will need to know the origin of the seasonal variation in (A) the length of the day and (B) the angle of the Sun's rays. It turns out that both are caused by the same thing: *the changing orientation of the incoming sunlight relative to the rotation axis of the Earth.*

Let us first concentrate on (A): the length of the day. To make things definite, let us consider some particular city – Chicago, say – and ask how long it is in daylight. Imagine that we were looking down on the Earth from space, and that we watched Chicago as the Earth's rotation carried it from the sunlit to the nighttime portions of the Earth. It takes the Earth 24 hours to turn about once. For what fraction of this time is Chicago in daylight?

In Figure 1.36 we illustrate an orientation in which the Earth's axis is perpendicular to the Sun's rays. As we can see from this figure, half of Chicago's path is in daylight and half in darkness. Thus, in this configuration, Chicago's days and nights are both 12 hours long.

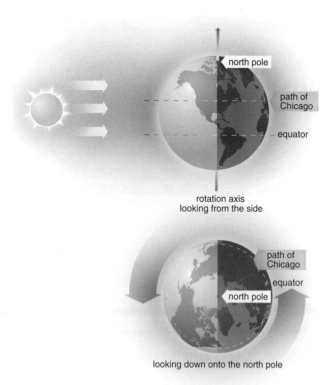

Figure 1.36 **On the equinoxes the Earth's rotation axis is perpendicular to the Sun's rays. Half of Chicago's path is in daylight, so Chicago receives sunlight for 12 hours.**

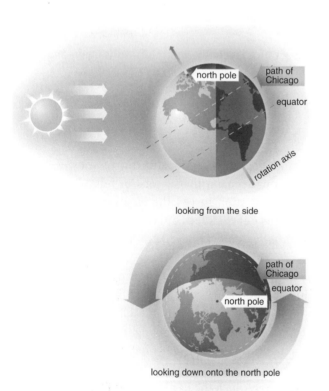

Figure 1.37 **In the northern summer configuration the Earth's rotation axis points toward the Sun. Since more than half of Chicago's path is in daylight, Chicago receives sunlight for more than 12 hours.**

NOW YOU DO IT
Draw a figure, analogous to Figure 1.36, illustrating the path of Miami. Are days and nights in Miami 12 hours long in this configuration?

This figure, therefore, must illustrate the configuration at the spring and fall equinoxes, which occur in March and September. Notice that in this figure the Sun lies directly above the Earth's equator. This means that it lies on the celestial equator. Recalling from the previous section that (by definition) the Sun always lies on the ecliptic, we realize that the equinoxes occur when the Sun, on its journey along the ecliptic, crosses the celestial equator.

Figure 1.37 illustrates a different configuration: one in which the Earth's rotation axis is tilted toward the Sun. More than half of Chicago's path is in daylight, so that day in Chicago is more than 12 hours long.

NOW YOU DO IT
Draw a figure, analogous to Figure 1.37, illustrating the path of Miami. Are days in Miami more than 12 hours long in this configuration?

This is the northern summer configuration. Conversely, Figure 1.38 illustrates the northern winter configuration, in which less than half of Chicago's path is in daylight and the day is less than 12 hours long.

NOW YOU DO IT
Draw a figure, analogous to Figure 1.38, illustrating the path of Miami. Are days in Miami less than 12 hours long in this configuration?

Let us now pass on to (B): the position of the Sun in the sky, and the angle between the Sun's rays and the vertical. On a spherical Earth, "up" is the direction pointing away from the Earth's center: we are interested in the angle made to this direction by the incoming Sun's rays. As illustrated in Figures 1.39 and 1.40, this angle is less in summer than winter. Therefore, the Sun is higher in the sky at noon in summer than in winter.

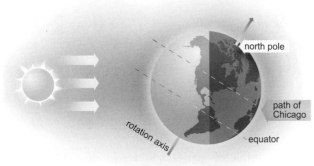

looking from the side

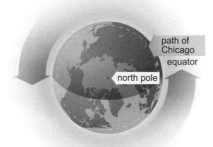

looking down onto the north pole

Figure 1.38 **In the northern winter configuration the Earth's rotation axis points away from the Sun. Since more than half of Chicago's path is in darkness, Chicago receives sunlight for less than 12 hours.**

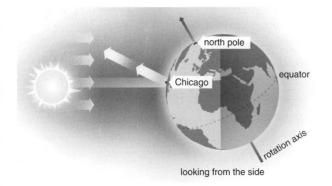

looking from the side

Figure 1.39 **The angle of the Sun's rays is fairly high at noon in midsummer in Chicago.**

NOW YOU DO IT
Draw figures analogous to Figures 1.39 and 1.40, illustrating the angle between the vertical and the Sun's rays as seen from a city lying in the southern hemisphere. Imagine that this city is just as far south of the equator as Chicago is north of it.

We now understand how the seasons arise from the changing orientation of the incoming sunlight relative to the rotation axis of the Earth. But why does this orientation change? It changes because, as the Earth orbits the Sun, *our rotation axis keeps pointing in the same direction* (Figure 1.41). Therefore, as the Earth orbits, sometimes the rotation axis points toward the Sun and sometimes away from it.

It is interesting to ask why the Earth's axis of rotation always points in the same direction. What keeps it from swinging about, so that it keeps tilting toward the Sun (Figure 1.42)? In this case, Chicago would experience summer forever.

NOW YOU DO IT
Suppose the configuration illustrated in Figure 1.42 were actually possible. Describe the passage of day and night, and the passage of the seasons, that would result.

One way to answer this question is to ask yourself whether you have ever seen a spinning object whose axis keeps pointing in the same direction. Probably you have: *the gyroscope.* Indeed, it can be shown from the principles of physics that the axis of every spinning body keeps pointing in the same direction, unless some force actively intervenes to twist it about. The seasons arise because the whole planet Earth is a single, gigantic gyroscope.

NOW YOU DO IT
The asteroids lie in a belt lying between the orbits of Mars and Jupiter. As opposed to the planets, whose orbits are nearly perfectly circular, some are in very elliptical orbits. Imagine an asteroid whose axis of rotation happens to be tilted in exactly the same way as the Earth's – but whose orbit is very elliptical. Describe the passage of day and night, and passage of the seasons, on such an asteroid.

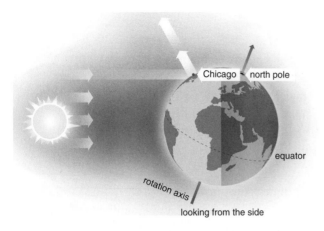

Figure 1.40 **The angle of the Sun's rays is nowhere near vertical at noon in midwinter in Chicago.**

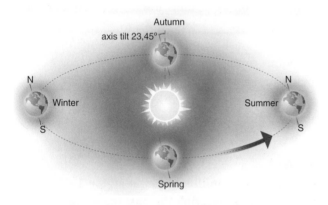

Figure 1.41 **The origin of the seasons. As the Earth orbits the Sun, its rotation axis always points in the same direction. So in some portions of the orbit this axis points toward the Sun, and in other portions away from it.**

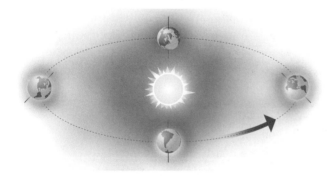

Figure 1.42 **An impossible configuration in which the Earth's rotation axis remains tilting toward the Sun as it orbits.**

Eclipses

An eclipse of the Sun occurs when the Moon passes between it and us. Because the Sun's light has been blocked, night briefly falls. Similarly, an eclipse of the Moon occurs when the Earth passes between it and the Sun. An astronaut standing on the Moon would see the Earth eclipsing the Sun: here on the Earth, we see the Moon darken as the light to it is cut off.

We can also think of eclipses in terms of shadows. A solar eclipse occurs when the Earth enters the shadow cast by the Moon: a lunar eclipse when the Moon enters the shadow cast by the Earth.

Returning to Figure 1.21, illustrating how the Moon's phases are related to its position in its orbit, we see that the configuration of a solar eclipse is one in which the Moon is new. Similarly, the configuration of a lunar eclipse is one in which the Moon is full. Why, then, do we not have eclipses each month? Because the orbit of the Moon does not happen to lie in the same plane as that of the Earth. As illustrated in Figure 1.43, for half of its orbit the Moon lies above the plane containing the Earth and the Sun, and for the other half below this plane. In the configuration illustrated, the new Moon lies slightly above the Sun, and so fails to eclipse it. Similarly, the full Moon passes slightly below the Earth's shadow, failing to be eclipsed by it.

But the Moon's orbit changes as time passes (because the Moon is influenced by gravity from both the Earth and Sun). In particular, the plane of the Moon's orbit steadily shifts. Eventually it reaches the configuration illustrated in Figure 1.44. In this configuration, we do get a solar eclipse when the Moon is new, and a lunar eclipse when it is full.

NOW YOU DO IT
Suppose the plane containing the orbit of the Moon were not tilted relative to the the plane containing the orbit of the Earth. Would there be eclipses? How often?

Eclipses of the Sun

In a solar eclipse, the Moon passes between the Earth and the Sun.

As it appears in the sky, the Sun is almost exactly the same size as the Moon. In reality, of course, the Sun is

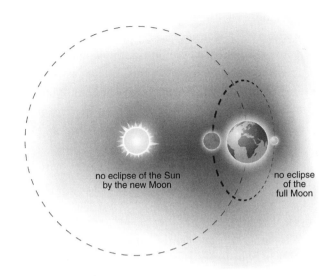

Figure 1.43 **No eclipses occur when the Moon's orbit is tilted as shown.**

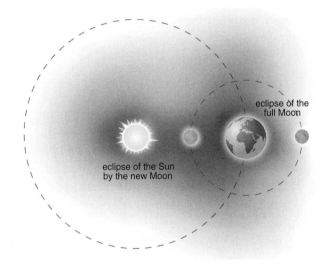

Figure 1.44 **Eclipses occur when the Moon's orbit is tilted as shown.**

much larger than the Moon – but it is also much farther away, just the right distance to make their apparent sizes equal. It is a wonderful coincidence, for it means that in a solar eclipse the Moon just covers the main body of the Sun, while failing to cover its magnificent corona (Chapter 11). Indeed, it is known that the Moon is moving away from the Earth, so that in the distant future there will be no total solar eclipses.

NOW YOU DO IT

(1) Draw a diagram, illustrating what we would see were the Moon to pass between us and the Sun in the far distant future.

(2) In the distant past, did eclipses of the Sun allow its corona to be visible?

(3) The Moon is currently 384 000 kilometers from the Earth, and it is receding from us by 0.038 meters each year. If it continues receding at this rate, how long will it take to double its distance from us?

The process begins with a partial eclipse, as the Moon slowly begins to drift across the Sun's face (Figure 1.45). The darkness gradually deepens. As the Sun changes from a broad disk to a slender crescent, the shadows it casts become ever more sharp. In the final stages a remarkable phenomenon appears in the dappled shade cast by a tree: every tiny gap among the leaves functions as a pinhole camera, and every patch of sunlight upon the ground becomes a tiny image of the crescent Sun.

Toward the climax the darkness gathers rapidly. Evening birds can be heard to sing, and bats have been observed to commence their nightly feeding. In the final seconds brilliant points of light sometimes appear around the Moon's periphery. This phenomenon, known as Baily's beads, arises as sunlight streams through valleys along the Moon's edge. If you are standing atop a mountain you might be lucky enough to see the shadow of the Moon racing toward you across the landscape: its speed exceeds one thousand miles per hour.

In totality the Sun is replaced by an utterly black disk – many people feel that it looks like a hole punched into the sky. Surrounding the disk in a luminescent pearly halo is the corona, the Sun's outer atmosphere. Arching high above its surface you might glimpse prominences, great plumes of superheated gas ejected from the Sun: they look like nothing so much as enormous flames frozen into stillness.

The spectacle lasts at most a matter of minutes. Then the Moon drifts yet farther, Baily's beads briefly reappear, and the Sun returns as a crescent facing the other

Figure 1.45 **An eclipse of the Sun.** 👁 **(Also see color plate section.)**

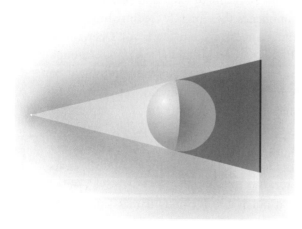

Figure 1.46 **The shadow cast by a point source of light has sharp edges.**

direction. The final partial phase brings the drama to a close.

A total eclipse of the Sun is the single most magnificent phenomenon naked-eye astronomy has to offer. A partial solar eclipse, or an eclipse of the Moon, pales in comparison to it. So impressive are total eclipses that it is worth traveling great distances to see one. If you do, be sure to study the local map with care, for totality is only visible from a very narrow band. If you are even a few miles off, you will miss it.

To observe an eclipse it is essential to protect your eyes. Dark glasses are *not* sufficient. The most convenient protection is provided by glasses specifically designed for the purpose: they are quite inexpensive and can be purchased from any science store, or from a variety of supply houses of astronomical equipment.

Eclipses of the Moon

An eclipse of the Moon occurs when the Moon enters the shadow cast by the Earth. An astronaut standing on the Moon would see an eclipse of the Sun, caused by the Earth drifting across its face. We on the Earth see the Moon darken, as the sunlight falling upon it is cut off.

The Earth's shadow is somewhat complex, owing to the fact that the Sun is not a point of light, but rather a broad disk. Figure 1.46 illustrates the shadow cast by a sphere illuminated by a point source of light. It has sharp edges. If you are in the illuminated region of this figure, you can see the light: if you are in the dark region, your view of it is blocked. In contrast, Figure 1.47(a) shows a different situation, in which the source of light is a broad disk. This shadow consists of a pitch-black *umbra* surrounded by a less dark *penumbra*. If you are located in the umbra, your view of the light source is entirely blocked: from the illuminated region, it is entirely unblocked. But if you are located in the penumbra, your view is partially blocked, and you see a crescent Sun. Figure 1.47(b) applies this to the actual situation of the Earth and Sun, and it illustrates the appearance of the Sun from these three vantage points.

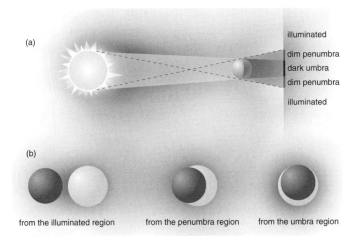

(a)

illuminated
dim penumbra
dark umbra
dim penumbra
illuminated

(b)

from the illuminated region   from the penumbra region   from the umbra region

Figure 1.47 **The shadow cast by a broad source of light. In (a) we show that it consists of a central dark area (the umbra) surrounded by a region of partial illumination (the penumbra). In (b) we illustrate the appearance of the Sun and Earth as seen by an astronaut standing on the Moon.**

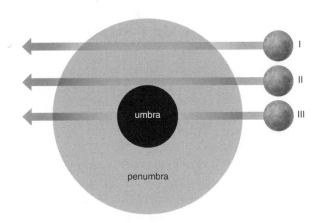

Figure 1.48 **Possible lunar eclipses.**

Figure 1.49 **A partial eclipse of the Moon.**

NOW YOU DO IT

You can perform an experiment illustrating these principles. Turn off all the lights but one at night, and leave a single light bulb shining. Remove the lampshade, so it is the bulb itself that casts the illumination. Notice that the light bulb is a broad disk, and that shadows are blurred.

But now take a sheet of cardboard and punch into it a tiny pinhole. Hold this cardboard up to the light bulb. Now what little illumination there is is provided by a tiny point of light (the light that makes it through the pinhole). Notice that now the shadows may be very faint – but they have sharp edges, and are no longer blurred.

Figure 1.48 illustrates the umbra and penumbra as seen from the Earth: in this figure we are looking down along our shadow toward the Moon. Superimposed on this shadow we have drawn three possible paths that the Moon might take through our shadow as it travels about the Earth in its orbit.

If the Moon follows path I, we have a *penumbral eclipse*, in which the Moon enters only the penumbra. An astronaut standing on the Moon would see the Sun partially blocked by the Earth, so that sunlight on the Moon is partially reduced in brightness. We on the Earth see that the Moon has grown dimmer. In such an eclipse, the dimming of the Moon can be so slight that it is hardly noticeable.

If the Moon follows path II it first enters the penumbra for the penumbral phase of the eclipse, and it then grazes the umbra. But only part of the Moon actually enters the umbra. This a *partial eclipse*, as illustrated in Figure 1.49. In viewing the curving edge of the Earth's shadow in such an eclipse, we are directly witnessing the spherical shape of the Earth.

Finally, if the Moon follows path III, it first enters the penumbra and then the umbra. While in the umbra, our astronaut on the Moon would see the Sun entirely blocked: we on the Earth see that the Moon has grown far dimmer in the *total phase* of the eclipse. Even during totality, however, the Moon can still be faintly seen. This arises from sunlight refracted around the Earth. The astronaut would see the Earth's edge glowing a deep red along a thin curving line stretching along the planet's perimeter.

While an eclipse of the Sun can only be viewed from a narrow strip, an eclipse of the Moon can be viewed from anywhere, so long as the Moon is above the horizon.

Time zones

By international agreement, a number of time zones have been established across the Earth. Within each zone, clocks are synchronized with each other – that is, they all read the same time. But each zone differs by one hour from those alongside it. Thus, if clocks read noon in one zone

- clocks in the zone to the east of it read 1 p.m.,
- clocks in the zone to the west of it read 11 a.m.

Let us explore why this curious arrangement has been reached.

It is obviously important for us to synchronize our clocks. As an example of what can happen if we don't, suppose that Alice makes a date to meet Bob at noon. If her wristwatch read noon when his read 1 p.m., she would show up for the meeting to find him long gone. Only if both their watches read the same will they will they arrive at the meeting place together. This seems to argue that people should synchronize their clocks to read the same time.

But we must not synchronize our clocks over the entire Earth! To see why, imagine what would happen if we did. Suppose, for instance that Alice were to get on the telephone and talk to a friend, Claire say, who lives on the opposite side of the world. Suppose further that it were the middle of the day for Alice, and her clock read noon. In this case Claire's clock would also read noon – but it would be the middle of the night for her! Clearly, it makes no sense for Alice's and Claire's clocks to be synchronized.

What is the difference between Alice and Bob on the one hand, and Alice and Claire on the other? The difference is that Alice and Bob are close together, but Alice and Claire are not. *It makes sense for people close to one another to synchronize their clocks. But people far apart need to take into account the essentially spherical shape of the Earth in setting their clocks.*

This is why the time zones have been established. Figure 1.50 is a map of the world showing the internationally agreed upon time zones. There are 24 of them, each differing by one hour in clock settings. Because there are 360 degrees of longitude about the Earth, each zone is $360/24 = 15$ degrees of longitude in width.

Daylight savings time further complicates the situation. In summer the Sun can rise very early – as early as 4 a.m. in the northern United States. In an effort to allow more daylight during the normal working day, many nations introduce daylight savings time, in which clocks are set ahead by one hour during these seasons.

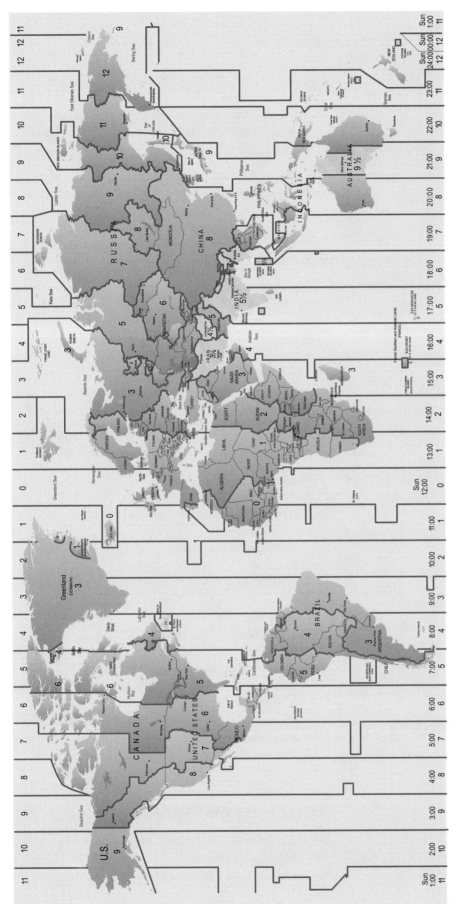

Figure 1.50 **Time zones.**

Summary

Rising and setting of the Sun
- The Sun moves in a slanting path across the sky.
- This path depends on your latitude on the Earth.
- The Sun's apparent motion arises because the Earth is a rotating sphere.
- Our velocity due to this motion is quite large.

Circumpolar stars
- Stars also move across the sky.
- Stars close to the celestial poles move in circles about it, and do not rise or set.

Phases of the Moon
- The Moon's phases are correlated with its position relative to the Sun.
- Half the Moon is illuminated by the Sun.
- The Moon's phases arise because we see various parts of its illuminated portion at various times.
- Which portion we see depends upon where the Moon is in its orbit about the Earth.

Constellations
- Constellations are not real structures: they are patterns we see in the random distribution of stars in the sky.

Motion of the Sun across the sky
- Each constellation is only visible at certain times of year.
- This arises because the Sun appears to migrate against the backdrop of the stars.
- This migration arises because of the Earth's motion in its orbit about the Sun.
- The path of this migration is known as the ecliptic or zodiac.

The celestial sphere
- The celestial sphere is a globe representing the sky as seen from the Earth.
- The celestial poles are those points on the sphere above our poles.
- The celestial equator is the line above the Earth's equator.

The reasons for the seasons
- Seasons arise for two reasons.
  - The varying length of the day.
  - The varying angle at which the Sun's rays strike the ground.

- These variations arise because the Earth's axis of rotation keeps pointing in the same direction as it orbits the Sun.

Eclipses of the Sun
- In an eclipse of the Sun, the Moon's shadow falls upon the Earth.
- Such an eclipse is very spectacular.
- It can only be seen from specific locations on the Earth's surface.

Eclipses of the Moon
- In an eclipse of the Moon, the Earth's shadow falls on the Moon.
- It can be seen from anywhere from which the Moon is visible.
- The Earth's shadow consists of a central dark umbra, and a surrounding partially illuminated penumbra.

Time zones
- Time zones are established so that we may synchronize our clocks around an essentially spherical Earth.

The nature of science
- Hypothesis testing in science.
  - If an idea does not make predictions, it is not a scientific theory.
  - Two theories have been held about why the Sun rises and sets.
  - We force each theory to make a prediction. We then decide between these theories by comparing their predictions with observation.
- The importance of skepticism.
  - Skepticism is the most cherished scientific virtue: we are continually testing our theories.
  - We test our theories in order to be sure we can depend on them.
  - We do this by forcing our theories to make predictions about things that have not yet been observed.
  - Then we test to see if the predictions are borne out.

DETECTIVES ON THE CASE

We considered three hypotheses to account for the seasons. We concluded that the first two made predictions in conflict with observations, and so could not be valid.

Problems

(1) In all of Figures 1.1–1.3 the Sun moves from left to right. Explain why it would move in this direction as seen from anywhere in the northern hemisphere.

(2) Figure 1.9 illustrated the "rotation" of the sky as seen by an observer at the north pole. Draw a similar figure illustrating the sky's "rotation" as seen by an observer at the south pole.

(3) Suppose the Earth suddenly started spinning in the opposite direction. In what way would the apparent motion of the sky be altered? In your analysis, consider observers at various locations about the Earth.

(4) Do people living in the southern hemisphere have circumpolar stars? Explain your reasoning.

(5) Suppose Alice lives in Miami and Bob lives in a city exactly as far south of the equator as Miami is north of it. Which of them has more circumpolar stars? Explain your reasoning.

(6) Suppose Alice lives Chicago, and Bob lives in a city exactly to the west of Chicago. Which of them has more circumpolar stars? Explain your reasoning.

(7) The radius of Mars is 3397 kilometers, and it rotates on its axis once every 24 hours 37 minutes. How rapidly is a point on its equator moving? (A) Explain the logic of the calculation you will do. (B) Do the detailed calculation.

(8) How rapidly does a person move who is situated in the northern hemisphere, somewhere between the Earth's equator and its pole? Use Figure 1.51 to explain why the answer decreases from a maximum at the equator to zero at the pole.

(9) Consider a person who lives exactly as far south of the equator as Miami is north of it. Does this person travel exactly as rapidly as people living in Miami?

(10) Hold up a book to a distant light bulb. Turn the book this way and that. Is half the book always illuminated? Or are there ways you can hold it such that more than half is illuminated?

(11) Go back to the illuminated book of the previous problem. From where would you only see its dark side? From where would you see only its illuminated side? Is there such a thing as a "half book"?

(12) As we discussed, just as the Moon has phases so too does the Earth. Imagine an astronaut standing on the Moon looking at the Earth. Describe the phases of the Earth that the astronaut would see. What would this astronaut see the Earth's phase to be
- when we see the Moon as a crescent,
- when we see a new Moon.

(13) Referring to the constellation maps in Appendix VII, in which months is Leo visible?

(14) Referring to the constellation maps in Appendix VII, identify three constellations that are visible in March.

(15) Figure 1.26 indicates the direction to Libra, which is visible in summer. Given that Taurus is visible in winter, indicate in Figure 1.26 the direction to Taurus.

(16) Recall that in May, Libra rises around sunset and sets around sunrise. Given that in February it is Leo that rises at around sunset and sets around sunrise, indicate in Figure 1.26 the direction to Leo.

(17) Referring to Figure 1.31: in what constellation does the Sun lie in January? In June? In September? Where does the Sun lie on the fourth of July? On Christmas Day?

(18) Suppose the Sun is in Aquarius. What month is it?

(19) Figures 1.36–1.38 illustrate the path of Chicago at various configurations: from these configurations we can deduce the length of Chicago's day. Draw figures, analogous to these, appropriate to a city in the southern hemisphere. Suppose this city is just as far south of the equator as Chicago is north of it. Comment on the lengths of this city's days at various seasons, as compared to those of Chicago.

(20) Figure 1.39 illustrates that the Sun is fairly high in the sky from Chicago at noon in midsummer. Use this diagram to explain the following.
- Why is it that, for people to the south of Chicago, the Sun is higher in the sky than from Chicago?
- Why is it that, for people to the north of Chicago, the Sun is lower in the sky than from Chicago?

Figure 1.51 **Motion of a point lying in the northern hemisphere.**

- Indicate on this diagram where you would have to go for the Sun to be directly overhead. (This is the Tropic of Cancer.)
- Indicate on this diagram where you would have to go for the Sun to be on the horizon. (This is the southern Arctic circle.)

(21) Give a discussion similar to that of the previous problem, but referring to midwinter (Figure 1.40).

(22) The axis of the planet Uranus is tilted at 90° to its orbit, so that sometimes the axis points *directly toward* the Sun. As illustrated in Figure 1.52, in this configuration a location on Uranus' northern hemisphere is in perpetual daylight and never experiences night.

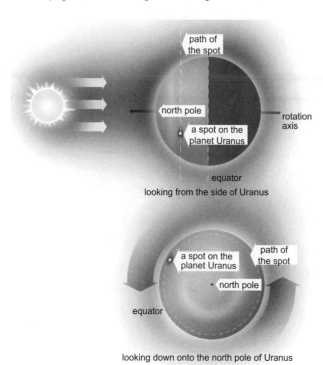

Figure 1.52 **Rotation of Uranus.**

As we discussed, however, a planet is a gigantic gyroscope whose axis keeps pointing in the same direction as it orbits. So at other points in Uranus' orbit, its axis no longer points toward the Sun. Discuss the passage of day and night, and the passage of the seasons, on Uranus. (For your information, Uranus requires 84 years to orbit the Sun.)

(23) Imagine a hypothetical planet whose axis of rotation was perpendicular to the plane of its orbit (Figure 1.53). Describe the passage of day and night, and passage of the seasons, on such a planet.

(24) Imagine that the Moon were to become suddenly *much* bigger than it actually is. What would an eclipse of the Sun be like? Suppose alternatively that the Moon were to become suddenly *much* smaller than it actually is. What would an eclipse of the Sun be like?

(25) Use the principles illustrated in Figures 1.46 and 1.47 to explain why it is that
- if you went to a planet farther from the Sun than Earth, shadows would have sharper edges;
- if you went to a planet closer to the Sun than Earth, shadows would have more blurred edges.

(26) Suppose you lived on a planet that had a moon and was very far from the Sun. Would there be such a thing as a penumbral eclipse?

Figure 1.53 **A hypothetical planet.**

You must decide

(1) Herodotus, who lived in the fifth century BC, has written that an eclipse of the Sun occurred during a battle between the Lydians and the Medians: the combatants were so shaken by the event that they put down their weapons and declared peace.

Clearly the combatants, although they disagreed sufficiently to go to war, agreed on the significance of the eclipse. Let us call their attitude toward eclipses a "divinity view."
- The *Divinity View of Eclipses*: an eclipse is a signal that a supernatural being is watching what you are doing and is displeased.

Write an essay in which you discuss the following.
- Is this a scientific theory?
- Does it make predictions?
- How would you test these predictions?

(2) Consider the following theory of the seasons.
- *The Punishment Theory of Seasons*: God makes it cold to punish humanity for its evil ways, and He makes it warm to show His benevolence.

Write an essay in which you discuss the same three questions as in (1).

2

The origins of astronomy

As we emphasized in the previous chapter, the Universe is here for everyone to see. Furthermore, regularities in the Universe give hints of a vast cosmic structure in which we all live. In this chapter we will follow the first stages in the historical development of people's attempts to understand this structure.

Certain primitive societies built monuments whose purpose appears to have been at least partially astronomical. Others built theories. Astrology is one of these theories, now discredited. The ancient Greeks built other theories, and they were among the first to develop mathematical methods of studying the structure of what we now term the Solar System. Their ideas were so powerful as to last for millennia, until overthrown in the scientific revolution. Copernicus, Brahe, Kepler and Galileo are the giants who created our modern idea of science. The final genius, Isaac Newton, was so important as to merit an entire chapter in his own right.

Throughout our historical survey we will make many detours. Indeed, our primary interest in this chapter is not really historical at all. Rather our goal is to garner insights into the nature of astronomy, and indeed into the nature of science in general, as they can be gleaned from a study of its origins.

Ancient astronomy

Astrology

In the previous chapter we considered observations of the sky that you can make with your naked eye. Many ancient peoples made such observations. Indeed, they were often exceedingly careful observers of the sky. The Babylonians and Egyptians, for instance, kept detailed records of celestial events, and they searched these records for regularities.

To some degree these studies were conducted for purposes of prediction. As we have seen in the previous chapter, the month of the year can be determined by observing when the constellations rise. And many important events occur at definite times of the year. The annual flooding of the Nile, for instance, always occurs toward the end of June. But in June, Aquila rises at sunset. So by observing Aquila, trained observers of the sky were able to predict this all-important event.

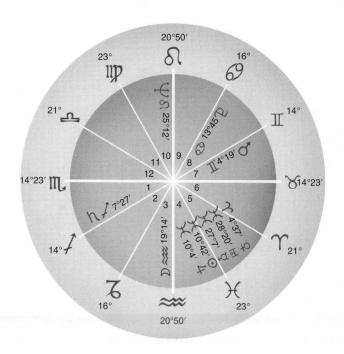

Figure 2.1 **A horoscope.**

But these astronomical studies were also conducted for other reasons. The Babylonians associated their gods with various astronomical objects: Shamash was connected with the Sun, and Ishtar with Venus. Astronomical observations therefore were bound up with religious practices. Because these gods were thought to concern themselves with Earthly conditions, the study of the sky became all-important.

Indeed, in early times there was no clear distinction between astronomy and astrology. Many thinkers whom we now regard as astronomers were interested in astrology. Ptolemy, the Greek astronomer whose *Almagest* was one of the most important scientific works of all time, also wrote a treatise on astrology. Kepler, one of the fathers of the scientific revolution, earned his living for a time as an astrologer.

The central tenet of astrology is that our personalities are strongly influenced by the positions of the Sun, Moon and planets at the time of our birth. An astrologer wishing to predict someone's personality will need to know her birth time and place, and will then record the astronomical positions on a chart known as a horoscope.

A representative horoscope is shown in Figure 2.1. The outer perimeter of this figure shows the ecliptic, or "zodiac" as it is known in astrology. As we saw in Chapter 1, this is the path followed by the Sun across the constellations. We will shortly see that not just the Sun, but also the Moon and planets move along the ecliptic. So all the astrologically significant bodies stay on the zodiac. We might imagine a time-lapse movie, showing their motions as the weeks roll by. The horoscope is a "freeze frame," in which this motion is captured at the instant of a person's birth.

Astrology divides the zodiac into 12 *signs*, such as Libra, Virgo, and so on. Because the zodiac is a circle, and a circle has 360 degrees, each sign is 360/12 = 30 degrees wide. As the Earth rotates on its axis and the constellations rise in the east and set in the west, the zodiac with its signs rotates overhead. Astrology also divides the sky into 12 *houses*, each 30 degrees wide, but these do *not* rotate overhead. The first house is that segment of the sky that lies within 30 degrees of the eastern horizon: this contains the objects that are about to rise. Objects in the second house, lying just below the first, will rise next. *Cusps* are the boundaries between houses.

The horoscope of Figure 2.1 contains a good deal of information. In it the outer circle is the zodiac with its signs, and the "spokes of the wheel" are the cusps that delineate the houses. Within the houses are symbols indicating where the Sun, Moon and planets were at the instant of this person's birth: beside each is a number indicating its angular distance from the beginning of the sign in which it resides. Based on such a horoscope, the astrologer will then make his predictions. (The astrological columns we read in newspapers are far cruder, and are based only on the position of the Sun.)

Astrologers refer to someone born between June 21 and July 22 as a Cancer, between certain other dates as a Libra, and so on. By this they mean that the Sun is in these signs between these dates. It is important to realize that this is not the case. As we saw in the previous chapter, the position of the Sun against the constellations can be easily determined by looking for the constellations just after sunset. If you look, you will not find the Sun where astrology claims it to be. This is because the signs have "slipped" relative to the actual constellations in the several thousand years since astrology was formalized. This slippage is due to the slow precession of the rotation axis of the Earth, similar to the precession of a spinning top.

As we mentioned, in its early stages there was no clear distinction between astronomy and astrology. But by now we have much evidence for the falsity of the claims of astrology. We now turn to this topic.

THE NATURE OF SCIENCE

THE IMPORTANCE OF SKEPTICISM AND A TEST OF ASTROLOGY

As we emphasized in the previous chapter, skepticism is the most cherished scientific virtue. Every scientist is skeptical, and every scientific theory is provisional and subject to numerous tests. In order to understand how scientists test a theory, we describe in detail a study that was made to assess the validity of astrology.[1]

A large number of volunteers submitted the times and places of their birth. These were given to a team of widely respected astrologers. The astrologers prepared horoscopes for each volunteer. Then, based on the horoscopes, they prepared interpretations of each volunteer's personality.

Two experiments were done with these interpretations.

- *Experiment 1.* Each volunteer received in the mail three different interpretations. One was the interpretation of their personality done by an astrologer: the other two were interpretations of *other people's* personalities. They were asked to choose which interpretation fit their personality best.
- *Experiment 2.* Each astrologer received in the mail results of a standard and widely respected psychological test of three different people. One was of a volunteer whose horoscope they had interpreted: the other two were results for *other people's* personality tests. They were asked to choose which psychological test matched the horoscope.

If astrology were false, would the correct choices never be made? Not at all! If astrology were false, the choices would be made randomly, which means that the correct choices would be made ⅓rd of the time. If, on the other hand, astrology were true, the correct choices would be made more often than that.

---

[1] Shawn Carlson, A double-blind test of astrology, *Nature*, **318**, 419–425 (1985).

IS ASTROLOGY CORRECT?

The experiment's results were that the correct choices were made ⅓rd of the time, within errors introduced by random behavior. This result was very strong evidence against the claims of astrology.

This is but one of a large number of studies that have been done testing the validity of astrology. None of them has unearthed any reason to accept its claims. Nevertheless there is a widespread acceptance of astrology by the general public. This is deeply distressing to scientists, who readily drop a theory that fails so many tests. As we will emphasize below, to a scientist, data are paramount. What data tell you must always be accepted. No matter how strongly you believe something to be true, if the observations tell you that it is not, then you must give up your cherished notion. Scientists cannot understand why astrology continues to enjoy widespread public support despite its failure in so many careful studies. Scientists regard people who believe in astrology as being like those who keep on investing in a stock that continues to lose money, and that has never performed well.

THE NATURE OF SCIENCE

THE DESIGN OF EXPERIMENTS

Let us return to the test of astrology that we have just described. A careful look at the test will reveal two important elements of every scientific study.

(1) UNCONSCIOUS BIAS

One element of this study was that it was "double blind." By this we mean that neither the experimenter nor the astrologers knew the identities of the volunteers. Elaborate procedures were devised to ensure that the people whose horoscopes were prepared and interpreted were anonymous. Other elaborate procedures were followed to ensure that the astrologers had no contact with the experimenter. But why? What was the reason for all these complicated procedures? The answer is that their function was *to guard against the danger of bias in the experiment.*

The person conducting the study was a scientist, and might well have been convinced that astrology was false. For this reason, he might have sought to influence the study's results. He might have subtly passed misleading hints to the astrologers, seeking to throw off their interpretations. The astrologers, on the other hand, believed astrology to be true, and they might have altered their interpretations in the other direction, based on their personal knowledge of the volunteers.

This kind of "cheating" is relatively easy to guard against. Far more difficult to deal with is *unconscious* bias. Very likely both the experimenter and the astrologers were trying scrupulously to be honest – but they both had their preconceptions, and entirely unconsciously they might have been behaving in ways that influenced the results of the experiment. We all know how difficult it is to avoid such bias. The "double blind" procedures were the only way to guard against it.

Unconscious bias is a perpetual problem in every scientific study. The danger of such bias is not just limited to this one study. Consider, for example, the program of testing Einstein's theory of relativity. Because Einstein is so widely revered, there is a danger that the people conducting the tests are unconsciously biased toward finding relativity correct. Alternatively, some experimenters could be biased against relativity. The scientist who proves Einstein wrong will not be reviled: he will be instantly famous. She very well might get a Nobel prize. This points to an important element of science: the delight scientists feel in overthrowing their most cherished theories. Scientific revolutions are not times of destruction: on the contrary, they are times of hope and progress. For all these reasons, the scientist conducting a test of relativity might harbor an unconscious wish to find it false.

Because unconscious bias is so insidious, and because it is sometimes so difficult even to know which way a scientist is biased, all scientific studies are designed as carefully as possible to be immune to it.

(2) SCIENTISTS NEED LARGE AMOUNTS OF DATA

A second important element of the study was that it involved large numbers of volunteers. But why? Wouldn't the test have been valid using only a few volunteers? The answer is that it would not. Let us investigate why.

It is possible to toss a fair coin and get nothing but heads. Similarly, just by sheer luck, the correct choices might have been made even if astrology were false. How could the experiment guard against this? By using many volunteers. It's not so unusual to toss a coin four times and get only heads. But it is very unusual to toss it 40 times and get only heads – and it is *very* unusual to get only heads in 400 tosses. This is why scientists always need large amounts of data.

Let us investigate this more carefully. We will focus on the first of the above two experiments, in which each volunteer was sent three horoscopes and was asked to identify the correct one. If astrology is false, each volunteer is going to make his or her choice randomly. But this does not mean that the choices will always be wrong! It is entirely possible that each person will make the correct choice, and so provide evidence that astrology is valid even though it is not. In this case, the experiment will have given misleading results.

Of course we do not want this to happen. If astrology really is false, we want our experiment to provide evidence for this fact. Let us investigate whether this would have been possible had there been only a small number of volunteers.

We will begin with the simplest case: an experiment consisting of only one volunteer. Let us call this volunteer Alice. In the experiment, Alice receives in the mail three horoscopes, say of persons "A," "B" and "C." In reality horoscope "C" is hers – but she does not know this fact, and she is asked to choose which is hers. Table 2.1 lists the various choices she might make.

Since Alice chooses randomly, she is just as likely to choose any of the entries in Table 2.1. As we can see, there is one correct choice out of three entries, so that the chances are one out of three that she will choose the correct horoscope.

Table 2.1. **An experiment with one volunteer.**

Alice chooses horoscope
A
B
C*

* This is the correct horoscope.

Table 2.2. **An experiment with two volunteers.**

Alice chooses horoscope	Bob chooses horoscope
A	a
A	b
A	c*
B	a
B	b
B	c*
C*	a
C*	b
C*	c*

* These are the correct horoscopes.

This means that the chances are one out of three that the experiment gives misleading results.

Now pass on to an experiment with more data – with two volunteers rather than one. Now both Alice and Bob receive horoscopes: Alice gets horoscopes "A," "B" and "C," and Bob gets horoscopes "a," "b" and "c." Now, when each person is asked to choose, the possibilities are more complicated. Alice, for instance, could choose "A" and Bob "a" – but Alice might choose "A" while Bob chooses "b" and so forth. The full list of possibilities is given in Table 2.2.

In this table there is still only one entry in which both volunteers chose correctly – but now there are nine entries, so that the chances are one out of nine that both will choose the correct horoscope.

Let us collect our results:

- an experiment consisting of a single volunteer has a probability of 1/3 of yielding evidence that astrology is correct even though it is not;
- an experiment consisting of two volunteers has a probability of 1/9 of yielding evidence that astrology is correct even though it is not.

Notice that the chances are fairly great that a one-volunteer experiment will yield misleading results – but a two-volunteer experiment has a far smaller chance of doing so. Similarly, in a three-volunteer experiment the chances are very small of yielding misleading results. We have now demonstrated the point we wish to emphasize: *in order for an experiment to guard against misleading results, it must involve great amounts of data.*

· ·

NOW YOU DO IT

Let us now illustrate this general principle with regard to tossing a coin. Suppose you have a coin that in fact is fair – i.e. it is equally likely to come up heads or tails. How likely is it that an experiment will reveal *misleading data, namely that the coin always turns up heads*?

Begin with an experiment consisting of a very small amount of data, namely a single toss. When you toss the coin there are two possibilities.

Heads*
Tails

* This experiment yields heads.

As you can see, the chances are ½ that a one-toss experiment yields misleading results. Your task in this problem is to study a *two-toss experiment*. What are the chances that the experiment yields misleading data?

· ·

So far in our discussion we have been asking what would happen were astrology to be false. But our conclusion would have been the same had we assumed astrology to be true. In this case each volunteer would have chosen the correct horoscope. But had there been a small number of volunteers, the experimenters would have known that the chances of doing so randomly were fairly large, and that the correct choices might have been arrived at by chance. So the

Figure 2.2 **Stonehenge.**

experimenters *would have not trusted their experiment's conclusions*. Here, too, the only way they could be sure is to collect great amounts of data.

This is an important element of all scientific work. Scientists need to repeat their observations over and over again, in order to guard against this kind of error introduced by random behavior. This is the primary reason why scientists are continually crying out for more and more data.

Archaeoastronomy

Archaeoastronomy is the combination of *archaeo*logy and *astronomy*. Monuments and other artifacts left by prehistoric societies give us evidence of their astronomical interests.

Perhaps the best known archaeoastronomical site is Stonehenge, in England (Figure 2.2). Stonehenge must have been extraordinarily difficult to build: some of its stones weigh 50 tons and they were brought there from quarries hundreds of miles away. It was built in a long effort between about 2800 BC and 1100 BC: that is roughly the time span from the fall of the Roman empire to now. It is difficult for us to imagine such a concerted effort extending over so great a span of time.

Many alignments can be identified within the monument. Most of these have no known astronomical reference, and there is strong disagreement about how the monument as a whole is to be interpreted. But there is no question that at least some of these alignments refer to specific astronomical configurations (Figure 2.3). On the longest day of the year, the Sun rises directly over the so-called "heelstone" as seen from the monument's center. Similarly, from the center the lines of sight to two markers along the outer ring point to the least southernmost moonrise and the least northernmost moonset.

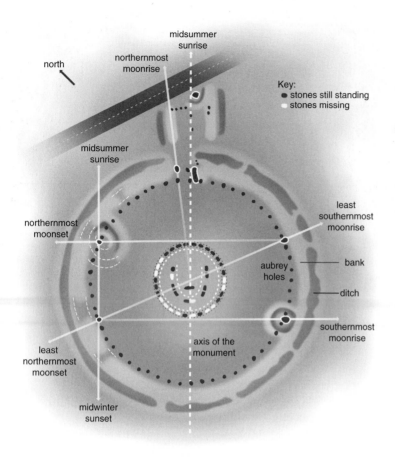

Figure 2.3 **Alignments at Stonehenge that point to astronomically important phenomena.**

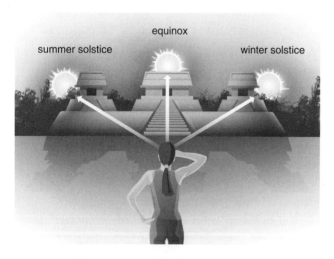

Figure 2.4 **Sight lines at Uaxactun mark the positions of sunrise at the summer solstice, equinox and winter solstice.**

Similar monuments exist throughout the world. In what is now Guatemala the Mayas built at Uaxactun three temples, oriented such that sight lines to them from a fourth pointed toward the sunrise on the summer and winter solstices, and the equinox (Figure 2.4).

The builders of most of these structures had no written language, and we know very little about them. But their monuments tell us two irrefutable things about them all: that they observed the sky with care, and that what they saw in it was very important to them.

Ancient Greek astronomy

With the culture of ancient Greece, enormous advances in astronomy were achieved. We will consider here the work of three Greek astronomers: two who invented ways to survey the heavens, and one who built a model of the cosmos that set the tone of astronomical thought for nearly 1500 years.

Aristarchus and the distances to the Sun and Moon

We begin with an achievement of Aristarchus of Samos (*c.* 310 BC–230 BC). Aristarchus was concerned with the phases of the Moon. As we discussed in Chapter 1, the Moon passes through its cycle of phases in 29½ days. But does it pass through this cycle uniformly? Aristarchus thought that it did not.

In Figure 2.5 we indicate on a line the times of the various phases of the Moon *according to Aristarchus*. He claimed that these times did not progress at a regular rate. Specifically, he wrote that the interval of time from half through *new* to half phase was shorter, by about one day, than the following interval from half through *full* to half.

We now know that this is entirely untrue. In reality these intervals of time are essentially equal. But let us continue with Aristarchus' argument. What would account for this curious irregularity?

We know from Chapter 1 that, at half Moon, our line of sight to the Moon is perpendicular to the Sun's rays (Figure 2.6). In Figure 2.7 we indicate the two points in the Moon's orbit at which this configuration is reached: these mark the two half Moons. How could the time intervals

Figure 2.5 **Timing of the Moon's phases is irregular, according to Aristarchus. While this is in fact not the case, Aristarchus' method was an important step in the triangulation of the heavens.**

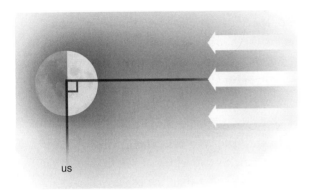

Figure 2.6 **At half phase the sight line to the Moon is perpendicular to the Sun's rays.**

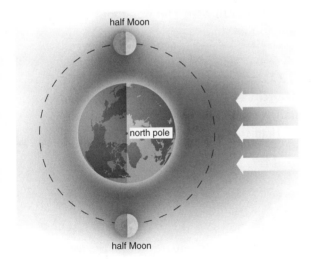

Figure 2.7 **Half moons occur at the two indicated points if the Sun is very far away.**

between them be unequal? One thought is that the Moon might speed up and slow down in its orbit. But this possibility Aristarchus would not accept. Indeed, it was so unacceptable to him that he did not even mention it in his writings.

Rather, he turned to another possibility. There is an assumption buried in Figure 2.7, one so obviously true that we did not even bother to think about it: that the incoming rays of sunlight are parallel. They are parallel if the Sun is very far away. But if the Sun is close, they are not (Figure 2.8).

Figure 2.7 is valid only if the Sun is exceedingly distant. If it is not, we need another picture. Aristarchus thought in terms of Figure 2.9. This figure naturally explains the curious irregularity with which he was concerned. It arises because the Moon travels *a shorter distance* from the first half phase to the second, and a longer distance from the second back to the first.

Aristarchus realized that this allowed him to do some celestial surveying. He drew triangles (Figure 2.10), and he used these triangles to compare the distance of the Moon (R) to that of the Sun (D). In this figure "A" is the path the Moon travels from its first to second half phases, and "B" is the path from the second back to the first. Clearly, B is longer than A. Aristarchus' claim was that the time required to travel B was one day longer than that to travel A:

$$\text{Time (B)} = \text{Time (A)} + 1 \text{ day}.$$

But he knew the month was 29½ days long – and the month was the *sum* of the two times:

$$\text{Time (B)} + \text{Time (A)} = 29\tfrac{1}{2} \text{ days}$$

Can you think of two numbers, differing by 1, that add to 29½? The answer is

$$\text{Time (A)} = 14\tfrac{1}{4} \text{ days},$$

$$\text{Time (B)} = 15\tfrac{1}{4} \text{ days}.$$

NOW YOU DO IT
Suppose Aristarchus had found that the interval of time from half through new to half phase was one week shorter than that from half through full to half. Find Time (A) and Time (B).

Aristarchus then needed to find the angle a subtended by the arc A. He reasoned that it must be in the same proportion to a full circle (360 degrees) that Time (A) was to the time for the full cycle (29½ days):

$$a/360 = [14\tfrac{1}{4}]/[29\tfrac{1}{2}]$$

from which he found that a was 174 degrees.

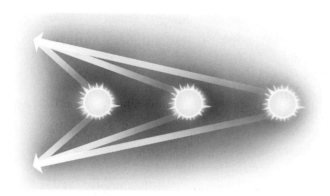

Figure 2.8 **The more distant the Sun, the more nearly parallel are its rays.**

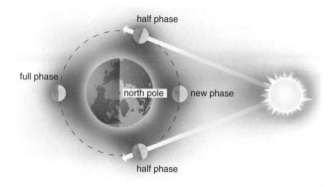

Figure 2.9 **If the Sun is close, half Moons occur at the two indicated points.**

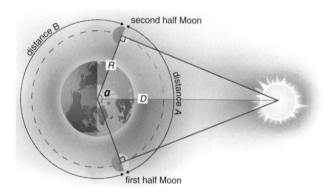

Figure 2.10 **Celestial surveying of the distances to the Moon (R) and Sun (D) according to Aristarchus.**

NOW YOU DO IT
Suppose Aristarchus had found that the interval of time from half through new to half phase was one week shorter than that from half through full to half. Find the angle "a."

Finally, Aristarchus considered the triangle highlighted in Figure 2.10. It was a right triangle, of which the base was R (the distance of the Moon) and the hypotenuse was D (the distance of the Sun). He also knew the angle formed by sight lines to the Sun and Moon: it was $a/2$, or 87 degrees. Using these facts, he could then calculate that, in such a triangle, the ratio D/R was roughly 20.

There are two ways to do this.

(1) Take a piece of paper and on it draw a right triangle with one angle of 87 degrees. Then, using a ruler, measure the triangle's two legs: you will find that one is about 20 times greater than the other.

(2) If you are familiar with trigonometry, you will recall that the cosine of ($a/2$) is R/D. Since the cosine of 87 degrees is 0.05, we find $D/R = 1/0.05$, or 20.

Aristarchus had found that the Sun was about 20 times farther away than the Moon.

NOW YOU DO IT
Suppose Aristarchus had found that the interval of time from half through new to half phase was one week shorter than that from half through full to half. Find the ratio D/R.

Comments

We have already mentioned that Aristarchus' starting point – the irregularity in the progression in the Moon's phases – is in fact erroneous. It is not the case that the time interval "A" is a day shorter than "B." While this does not diminish the magnitude of his achievement, it does mean that his particular conclusion is wrong. The Sun is a good deal more than 20 times farther from us than the Moon. The correct factor, in fact, is about 390 (which means that, although there is in fact some irregularity in the Moon's phases, it is too small to be noticeable).

Aristarchus' rejection of the hypothesis that the Moon sped up and slowed down in its orbit was in keeping with Greek philosophy. To the Greeks the heavens were unchanging, and it was for this reason that they insisted that celestial bodies moved with steady, unchanging speeds – we will see this with a vengeance when we consider the work of Ptolemy below. Similarly, to the Greeks the heavens were perfect – and to them the circle, endlessly closing in on itself, forever maintaining the same distance from its center, was the most perfect geometrical figure. Thus Aristarchus assumed the Moon's orbit to be perfectly circular, whereas in fact it is not.

Figure 2.11 **The Sun is not directly overhead at noon.**

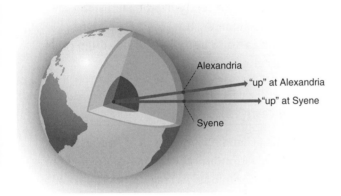

Figure 2.12 **"Up" is different at different places if the Earth is round.**

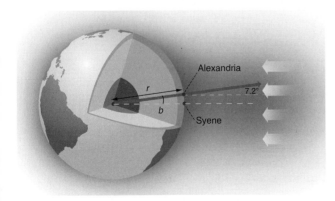

Figure 2.13 **Eratosthenes' determination of the radius of the Earth.**

Eratosthenes and the size of the Earth

The second celestial surveyor we will study is Eratosthenes (276 BC–196 BC), who invented a way to measure the size of the Earth. While Aristarchus was concerned with the phases of the Moon, Eratosthenes considered *the position of the Sun in the sky at noon on the longest day of the year.*

Recall from Chapter 1 that even at noon on the longest day of the year the Sun is not directly overhead. Where is it, then? Where Eratosthenes lived – Alexandria, Egypt, which was part of Greek civilization in those days – it was 7.2 degrees away from directly overhead (Figure 2.11). But from travelers Eratosthenes had heard that in the city of Syene, far to the south, the Sun *was* directly overhead. How could this be?

It is a common misconception that ancient peoples thought the world was flat. The ancient Greeks knew it was round. Eratosthenes realized that in this case "overhead" was *a different direction* at Syene than at Alexandria. Figure 2.12 illustrates the situation. "Up" is away from the Earth's center – and on a round Earth, this differs from place to place.

To this picture Eratosthenes added the Sun's rays (he took them to be parallel because, unlike Aristarchus, he thought in terms of a very distant Sun). The result was Figure 2.13. In this figure, "b" is the angle subtended at the center of the Earth by Alexandria and Syene. Because the Sun's rays were parallel, Eratosthenes knew from geometry that b was also 7.2 degrees.

He therefore knew that the distance from Alexandria to Syene, divided by the distance all the way around the world, must equal 7.2 degrees divided by 360 degrees:

$$\frac{\text{distance from Alexandria to Syene}}{\text{circumference of the Earth}} = \frac{7.2}{360}.$$

From this proportion, Eratosthenes was able to find the radius of the Earth. Let us see how.

The logic of the calculation

In this proportion there is only one unknown: the Earth's circumference. So we can solve the proportion to find it. We know that the circumference of a circle is $2\pi r$, where r is the Earth's radius. So the steps in the calculation are as follows.

Detailed calculation

Step 1. Solve the proportion for the Earth's circumference.
Let us call the circumference C and the distance from Alexandria to Syene d. Then the proportion is

$$d/C = 7.2°/360°.$$

Solving we get

$$C = d \,(360°/7.2°).$$

Step 2. Set the circumference equal to $2\pi r$.
We get $2\pi r = d\,(360°/7.2°)$.

Step 3. Divide by 2π to find a formula for the Earth's radius r.
We get $r = (d/2\pi)\,(360°/7.2°)$.

Step 4. Plug in the distance from Alexandria to Syene and find r.
In his writings, Eratosthenes states that the distance from Alexandria to Syene was about 5000 *stadia*. The *stadium* was a unit of distance in ancient Greece. Unfortunately, we do not know exactly how long it was. Our best guess for the length of a stadium is ⅙th of a kilometer, or 0.17 kilometer. In this case we calculate the radius of the Earth to be

$$r = \frac{(5000)(0.17 \text{ kilometer})}{(2\pi)} \frac{360°}{7.2°} = 6760 \text{ kilometers}.$$

Step 1. Solve the proportion for the Earth's circumference.
Step 2. Set the circumference equal to $2\pi r$.
Step 3. Divide by 2π to find a formula for the Earth's radius r.
Step 4. Plug in the distance from Alexandria to Syene and find r.

As you can see from the detailed calculation, this depends on the length of the Greek *stadium*, which we do not know exactly. Taking our best guess for it, we find that the Earth's radius is about 6760 kilometers. This is close to the correct answer of 6378 kilometers.

NOW YOU DO IT
Suppose our historians are wrong, and the stadium were actually half a kilometer in length. What would Eratosthenes have found for the radius of the Earth?

Suppose Eratosthenes had measured the Sun to be 15 degrees from directly overhead at noon on the first day of summer. How big would he have calculated the Earth to be? (Take the stadium to be ⅙th of kilometer.)

Ptolemy's model of the Solar System

One of the greatest astronomers of antiquity, Claudius Ptolemy, lived in Alexandria in the middle of the second century AD. His astronomical work has come down to us through his mighty compendium known in Arabic as "Al Magisti," the "greatest": we know it today as the *Almagest*.

The *Almagest* is concerned with what we now call the Solar System: the Sun, Moon and planets. How are they arranged? Ptolemy's answer to this question was so comprehensive that it set the tone for astronomy for nearly 1500 years, up to the scientific revolution.

The motion of the planets

Before we describe Ptolemy's model, let us discuss the facts it was designed to account for.

To the naked eye – and the ancients had no telescopes – planets look like stars. There are only two differences.

- Planets change their brightness as the seasons roll by. They grow brighter and dimmer as the months pass.
- Planets continually change their positions relative to the stars. They move across the sky in complex ways. Indeed, the very word "planet" is Greek for "wanderer."

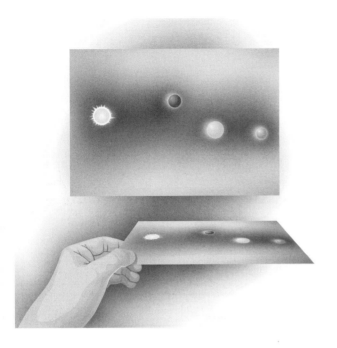

Figure 2.14 **The Sun, Moon and planets must lie in a plane.**

But the planets do not move randomly. In fact, there are many striking facts about their motion. We will focus on five.

- *Fact 1*. The most striking regularity of all is that *planetary motion takes place along a line*. The planets do not wander across the entire sky. Rather they are roughly confined to the ecliptic. The same is true of the Moon. This is quite a remarkable fact. We have already seen that the ecliptic is the Sun's path against the stars – but why should it also be roughly the paths of the Moon and planets?

This tells us that the Sun, Moon and planets *all lie in a plane* – a plane also containing the Earth. To see why this must be so, do the following experiment. Take a sheet of paper, and on it draw dots representing the Sun, Moon and planets. Then (Figure 2.14) orient this paper so that your eye lies in the plane of the paper. You will see that the dots you have drawn all lie in a line.

In Chapter 1 we defined the ecliptic as tracing out the Sun's path across the sky. We now realize that it has a greater significance. *The ecliptic is the plane of the Solar System, seen edge-on.*

There are other striking facts about the motions of the planets.

- *Fact 2*. Most of the time the planets *move from west to east* against the backdrop of the stars. Strikingly, this is the same direction as that of the Sun and Moon.
- *Fact 3*. From time to time a planet will *reverse its motion and briefly move from east to west*. This is known as "retrograde motion."
- *Fact 4*. For Mars, Jupiter and Saturn this retrograde motion *always occurs when the planet is brightest*.
- *Fact 5*. Among the planets, *Mercury and Venus never stray far from the Sun*. Mercury's greatest angular distance from the Sun is 28 degrees, Venus' is 47 degrees. The other planets, however, can be found at any angular separation from the Sun.

Clearly, there is much to be explained here. Ptolemy constructed a model designed to do just this.

Within Ptolemy's model the Earth was at the center of the Universe while the Moon, Sun, planets and fixed stars revolved about it. In keeping with the Greek attitude that the heavens were perfect and unchanging, his model was composed of circles rotating with constant velocity. Each planet was assumed to be attached to a rotating circle, the *epicycle*, whose center was carried along on a second rotating circle, the *deferent* (Figure 2.15). Depending on how rapidly the epicycle rotated relative to the deferent, a variety of motions could be produced. Next Ptolemy added the *eccentric*: the Earth was not placed at the center of the deferent, but off to one side (Figure 2.16). And finally, Ptolemy added the *equant*: the deferent rotated uniformly not with respect to the Earth, nor with respect

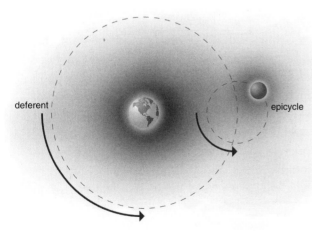

Figure 2.15 **Ptolemy's deferent and epicycle.**

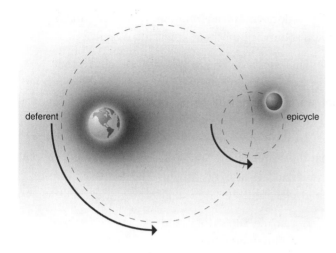

Figure 2.16 **Ptolemy's eccentric.**

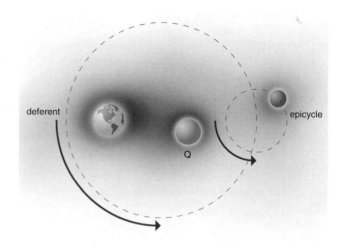

Figure 2.17 **Ptolemy's equant.**

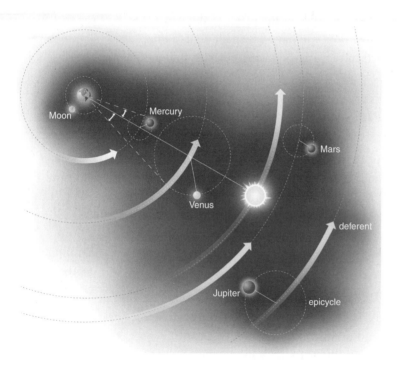

Figure 2.18 **A simplified version of Ptolemy's model of the Solar System.**

to its center, but with respect to yet another off-center point "Q" (Figure 2.17).

A simplified version of Ptolemy's full model of the Solar System is shown in Figure 2.18. Motionless at its center lay the Earth. Closest to us lay the Moon, then Mercury and Venus, and then the Sun. Beyond the Sun were Mars and the rest of the planets. Finally, encircling them all lay the sphere of the fixed stars, rotating once a day.

Let us study Ptolemy's model, and ask how it accounts for the four regularities we have noted in the motions of the planets.

All his circles were taken to lie in a plane, the plane of the Solar System: thus it accounts for Fact 1, that the Moon and planets never leave the ecliptic. All rotated in the same sense, accounting for Fact 2, the general west-to-east motion of the Sun, Moon and planets.

How did Ptolemy's model account for Fact 3, the occasional retrograde motion? All of his epicycles and deferents rotated in the same sense – counterclockwise in Figure 2.18. But look closely at Jupiter. At its location in that figure, the motion of its epicycle is *opposing* that of its deferent. While Jupiter's deferent is carrying it counterclockwise, its epicycle is carrying it backward in a clockwise sense. Ptolemy arranged for the speed of the epicycle to be faster than that of the deferent, producing Jupiter's retrograde motion. Notice also that, at this point, Jupiter is closest to the Earth, thus explaining why a planet is brightest when its motion is retrograde (Fact 4).

Ptolemy located the centers of the epicycles of Mercury and Venus on the line joining the Earth to the Sun, and he had their deferents rotate at the same rate as the

Sun – once a year. As illustrated in Figure 2.18, this ensured that Mercury and Venus would always remain close to the Sun as seen from the Earth, which is Fact 5.

Ptolemy's model of the Solar System was extraordinarily influential. It dominated astronomical thought for nearly 1500 years – until the time of Copernicus, and the scientific revolution.

The scientific revolution

Nicolaus Copernicus

The scientific revolution began with the work of the Polish astronomer Nicolaus Copernicus (1473–1543): Figure 2.19. His theory that the Earth moved about the Sun, rather than being the center of the Universe, was one of the most important scientific advances of all time.

The implications of Copernicus' proposal are breathtaking. Take a look out the window. All those buildings, the trees and sky and the very ground itself – is it really possible that they are moving? We certainly do not *feel* the motion! Furthermore, although Copernicus did not know it at the time, the Earth is moving with great speed: 30 kilometers per second, or 67 000 miles per hour.

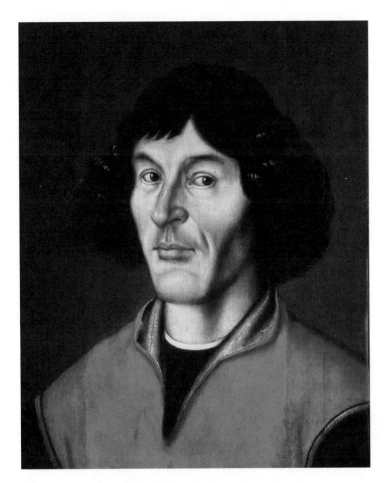

Figure 2.19 **Portrait of Nicolaus Copernicus (1473–1543) (oil on canvas), Pomeranian School (sixteenth century).**

WHAT DO YOU THINK?
As we are emphasizing, Copernicus' theory that the Earth moves is entirely contradicted by everyday experience. Nevertheless, nowadays almost everyone accepts it as being true. Presumably you do too.

Write a brief essay in which you discuss the following.
- How do you reconcile this theory with your daily experience?
- What other things do you yourself accept as being true – even though they are contradictory to what you experience?

Copernicus' proposal removed the Earth from a privileged position at the center of the cosmos. This opened up an extraordinary range of possibilities. If our world was not unique, then perhaps there were other worlds out there. Perhaps these worlds possessed life. Not only that, but by implication his proposal was removing humanity from a central position within creation. It is not too much to say that the seeds of the Darwinian revolution can be found in Copernicus' work.

There were numerous objections to his thesis that the Earth moved. It was argued that the oceans, air and clouds would be ripped away from the Earth by its great speed. To this Copernicus responded that the oceans, air and clouds shared in the Earth's motion. It was also argued that the Earth would be torn apart by the speed of its rotation. To this Copernicus responded that the same objection applied to the rotation of the sphere of the fixed stars required by Ptolemy's model – indeed, with greater force, for Ptolemy's sphere of fixed stars was far larger than the size of the Earth's orbit. But it is important to note that by these arguments Copernicus was not *proving* the Earth moved. He was merely showing there were no objections to the hypothesis. Indeed, actual evidence of our motion was not found until 1729.

Remarkably, Copernicus had not been impelled by a desire to revolutionize our conception of ourselves and our place in the Universe. His motivation was far more abstract. He was dissatisfied with two technical details about Ptolemy's model of the cosmos.

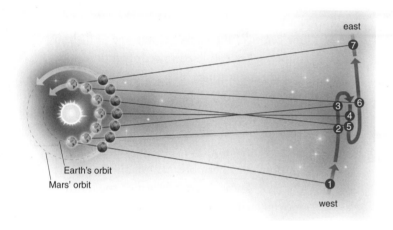

• Ptolemy's use of the equant struck Copernicus as repugnant. He believed that celestial circles should rotate uniformly in actual fact, not merely as seen from some off-center location.
• Ptolemy had found it necessary to use several *different* sets of circles to account for different aspects of the motion of *the same* planet. Clearly, therefore, Ptolemy had not invented a single, unified model of the Solar System.

In many ways Copernicus' approach was quite conservative. He mimicked Ptolemy in adopting the deferent, epicycle and eccentric; and in insisting on the principle of uniform circular motion. But his model differed considerably from Ptolemy's in how it accounted for the observed motion of the planets.

Consider for example Fact 3, retrograde motion. Recall that Ptolemy had accounted for this by carefully adjusting the speed of an epicycle so as to compensate for that of a deferent. But within the Copernican system the explanation is simpler (Figure 2.20). For Copernicus, retrograde motion occurs because we are observing the heavens from a moving Earth. We move faster than an outer planet, so that from time to time we catch up and pass it. As this happens, the outer planet briefly appears to move backward. Notice too that this model nicely accounts for Fact 4, that planets are brightest while undergoing this motion: it only occurs when we are closest to them.

Figure 2.20 Copernicus' explanation for retrograde motion.

Copernicus' explanation for Fact 5, that Mercury and Venus never stray far from the Sun, is likewise simpler than Ptolemy's. Recall from Figure 2.18 that Ptolemy accounted for this by forcing these planets' deferents to lie on the line joining the Earth to the Sun. Copernicus' explanation is that we are farther from the Sun than they, so they necessarily appear close to it (see Figure 2.21).

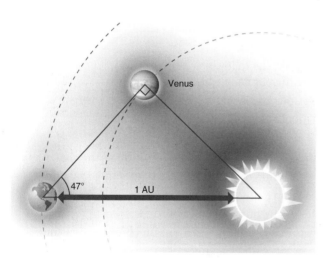

Figure 2.21 Mercury and Venus in the Copernican system never stray far from the Sun. In the configuration illustrated such a planet is at its maximum angular deviation from the Sun.

THE NATURE OF SCIENCE

HYPOTHESIS TESTING IN SCIENCE

We are now faced with an interesting question. We have two theories, Ptolemy's and Copernicus', both of which explain perfectly well the observed motions of the planets. But the theories are completely different from one another! What do we do in such a situation? If both explain the facts, how do we decide between them?

It might be thought that Copernicus' model was *simpler* than Ptolemy's and therefore should be adopted. In fact this is not the case. Copernicus was forced to use roughly as many circles as Ptolemy.

It might also be thought that, because the Earth is in reality not the center of the Universe, Copernicus' model must be *more accurate* than Ptolemy's. Remarkably, this too is not the case! Predictions of the locations of planets based on Copernicus' theory are no more accurate than those based on Ptolemy's. Perhaps this is not surprising: after all, we now know that planets move on ellipses, not on epicycles, deferents and eccentrics.

In the preceding section we stated that Copernicus offered *simpler explanations* than Ptolemy for the observed facts of planetary motion. Is this a good reason to prefer the Copernican theory? Many scientists believe that it is. Simplicity and elegance have a place in science, just as in any other field of human endeavor.

But while simplicity and elegance are important, they do not constitute absolute proof. As we have emphasized, skepticism is the hallmark of all scientific endeavor. No scientist would accept one theory merely because it was simpler than another. To be certain that a theory is correct, scientists insist on finding something more definite. What could this be? Can we find a clear difference between Ptolemy's and Copernicus' models of the Solar System – some situation in which they make *different predictions*? If so, we could test to see which prediction turns out to be correct.

It turns out that we can. The difference has to do with the *phases* of the inner planets, Mercury and Venus. Although so far we have only discussed the phases of the Moon (Chapter 1), a little thought shows that every planet has phases – after all a planet, just like the Moon, has a daytime hemisphere and a nighttime hemisphere. These are illustrated for Ptolemy's model in Figure 2.22, and for Copernicus' model in Figure 2.23. If you study these, you will find that in the Ptolemaic model we see at most only a small part of the daylight hemisphere. So *in the Ptolemaic model, Mercury and Venus are in either the new or crescent phase*. But in the Copernican model we can see either the dark or sunlit hemisphere, depending on where it is in its orbit. So *in the Copernican model, they can reach full phase*.

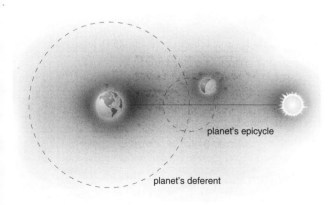

planet's epicycle

planet's deferent

Figure 2.22 **Mercury's phases in Ptolemy's system are either crescent or new.**

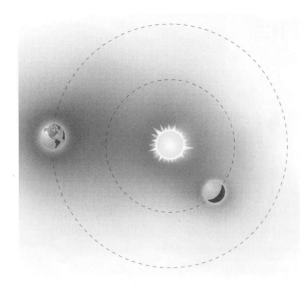

Figure 2.23 **Mercury can have any phase in the Copernican system.**

NOW YOU DO IT
On Figure 2.22, draw little dots at various points along the planet's orbit, and indicate the phases of the planet at these locations. Now do the same on Figure 2.23 for the Copernican system.

Suppose you lived on Mercury, and you were observing the Earth. Discuss the phases of the Earth that you would see, in both the Ptolemaic and the Copernican systems.

This is the difference we are looking for. To decide which model of the Solar System is correct, all we need do is observe the phases of Mercury or Venus. Of course, this is easier said than done: to see the phases we need a telescope, which at the time of Copernicus had not yet been invented. It was Galileo (see below) who first turned a telescope on the heavens, and it was he who first observed the phases of the planets and so provided decisive confirmation of Copernicus' theory of the Solar System.

A second test. As we emphasized, scientists continually need more and more data. One test is never enough: the more tests we can find, the better. Can we think of a *second* difference between the Ptolemaic and Copernican systems that would help us decide between them? We can. The difference is that *in the Ptolemaic system the Earth is motionless, but in the Copernican system it is moving.* If we could find a way actually to detect this motion we could decide between the two systems.

As with the first test, this is hard to do. We need some sort of "speedometer" with which to measure the velocity of the Earth. In Chapter 4 we will describe such a speedometer: the Doppler effect. Using it, we have been able to observe directly the motion of the Earth. Here, too, the Copernican system is found to be correct, and the Ptolemaic false.

This is the first example we have found of the *confirmation of a scientific theory*. We will encounter many more throughout the book.

Science is testable

This kind of testing of hypotheses is a vital aspect of science. Scientists are constantly faced with competing theories, and they are constantly searching for ways to decide which one to accept. The method is always the same: they find some instance in which the two theories make different predictions, and then they conduct an experiment to see which prediction is borne out.

It important to realize that, in order for this process to take place, *a scientific theory must be capable of being tested*. It must make predictions that we can check. If a theory does not make any such predictions, it isn't a scientific theory.

There is nothing unusual about this. You yourself do it all the time. Suppose you are indoors and you have a "theory" that it is raining. You can test your theory by looking out the window. Suppose your hypothesis is that Jane is a very good poker player. You can find out if you are right by engaging her in a few games.

Scientists are lucky in having this means of testing their ideas. Other fields of study are not so fortunate. Philosophers, for instance, have no means of finding out if they are right or wrong. No experiment can decide whether it is ethical to eat meat. No observation can tell us whether we have free will. For this reason, philosophical problems are much harder to solve than scientific ones. Philosophers have been discussing these questions for millennia, and they will probably continue to do so into the indefinite future.

WHAT DO YOU THINK?

Consider the theory that "Singer A" is a very fine musician, while "Singer B" is not. Write a brief essay in which you discuss whether this theory is capable of being tested.

In Chapter 1, in our discussion of the phases of the Moon, we advanced the theory that *half of the Earth is always in daylight, and half is not*. Write a brief essay in which you discuss whether this theory is capable of being tested. In your discussion, consider two different scientists: one who lived at the time of Copernicus, and one who lives now.

We will return to these issues again and again throughout this book.

Tycho Brahe

Tycho Brahe the son of a Danish nobleman, had a remarkable youth (1546–1601; see Figure 2.24). As a youngster he was abducted by a childless uncle, who eventually managed to persuade Brahe's parents to allow him to raise the boy. Brahe seems to have been a fiery and difficult individual: in his youth he fought a duel with a fellow-student over who was the better mathematician and, as a consequence, wore a silver nose throughout his life. As an adult he was imperious and irascible.

Brahe constructed on the island of Hven the extraordinary observatory of *Uraniborg* (Figure 2.25). Only slightly smaller than a football field, it was a nobleman's estate devoted to astronomy. It had its own library, paper mill and printing press, servants' quarters – and even a prison. Uraniborg on the island of Hven became a scientific center, and was regularly visited by Europe's premier astronomers. The operation was supported by rents and taxes from the island's farms, over which Brahe ruled like a lord.

The instruments Brahe constructed at Uraniborg (Figure 2.26) were utterly unprecedented for his day. Brahe's great innovation was to conduct highly accurate observations over long periods of time – some 20 years, in fact. The data

Figure 2.24 **Tycho Brahe.**

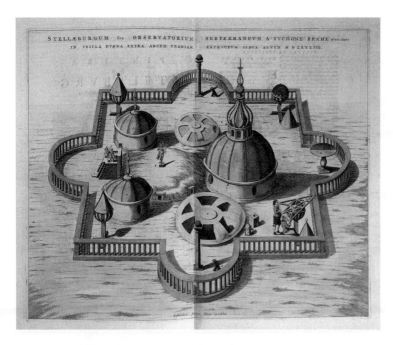

Figure 2.25 **Tycho Brahe's observatory at Uraniborg.**

Figure 2.26 **Brahe observing at his mural quadrant.**

he obtained on planetary motions turned out to be absolutely vital to the history of science: only observations of such high accuracy were capable of disproving the antiquated notions of deferents, epicycles and eccentrics; and leading Kepler to his realization that planets moved on ellipses.

Johannes Kepler

Johannes Kepler is one of the most fascinating figures of the scientific revolution (1571–1630; see Figure 2.27). An adept mathematician, he earned his living for a while as an astrologer; a firm believer in the concept that celestial motions have physical causes, he also wrote a treatise on music and its realization in the motion of the planets.

Kepler was employed for a time by Brahe, who wished him to produce a new model of the Solar System based on his observations. Brahe, however, refused to allow Kepler to see most of his data. Only after Brahe's death did the young mathematician gain access to his treasure trove of observations. Kepler's study of these data led him to his famous three laws of planetary motion, the first two published in 1610, the third in 1618.

Kepler's first law states that *the orbits of the planets are ellipses, with the Sun at one focus*. Furthermore, although Kepler did not know it, his first law applies not just to planets, but to every orbiting body.

What is an ellipse? It is a generalization of the concept of a circle. To draw a circle we can think of tying one end of a string to a nail, the other to a pencil, and swinging the pencil about while making sure to pull the string tight (Figure 2.28). The path traced out is a circle. To draw an ellipse we modify this by tying the string to *two* nails, called the foci of the ellipse (Figure 2.29).

Just as an ellipse is a generalization of a circle, so the major and minor axes of an ellipse are generalizations of the circle's diameter (Figure 2.30). And just as the radius is half the diameter, so the *semimajor and semiminor axes* are half the major and minor axes.

Depending on where we place the two nails – the two foci – we get various ellipses. The *eccentricity* of an ellipse is a measure of how far apart we place them. The larger the eccentricity the farther apart they are (Figure 2.31). If we place the two foci on top of one another we get a circle, which is an ellipse of zero eccentricity (Figure 2.32).

We can think of an ellipse as a "squashed circle": the more squashed it is, the greater its eccentricity.

Kepler's first law states that every planet of the Solar System follows an elliptical path as it orbits the Sun, and that the Sun lies at one of the ellipse's two foci (Figure 2.33). Remarkably, nothing whatsoever occupies the other focus.

Notice one consequence of this law: as it orbits, a planet moves closer to and farther from the Sun. We might think that this is the reason for our seasons. As we discussed in Chapter 1, this is not the case: in fact the Earth reaches its closest point in midwinter, and its most distant point in midsummer. However, the asteroids also move in elliptical orbits, and in them the seasons could indeed arise from this cause.

Kepler's second law states that as a planet moves, *a line drawn from it to the Sun sweeps out equal areas in equal times.* This law also applies to every orbiting body. In Figure 2.34, we indicate the areas swept out in two 30-day intervals. Kepler's second law states that these areas are equal.

As illustrated in the figure, in order that the various triangles have equal area, their bases must be unequal. But these bases are the distances traveled by the planet in these intervals of time. So this law is telling us how the body speeds up and slows down as it orbits the Sun. It implies that the body moves rapidly when close to the Sun, and slowly when far from it.

Figure 2.27 **Johannes Kepler.**

NOW YOU DO IT
Recall that a circle is just a special kind of ellipse: one of zero eccentricity. Draw your own version of Figure 2.34 for a circular orbit, and use it to explain why *a body in a circular orbit moves with constant velocity*, neither speeding up nor slowing down.

As we will see in Chapter 9, comets move in very elliptical orbits about the Sun. Furthermore, their semimajor axes are quite large – far larger than that of the Earth in its orbit. Use Kepler's second law to explain why a comet spends most of its time being very cold, and only a short amount of time being significantly warmed by sunlight.

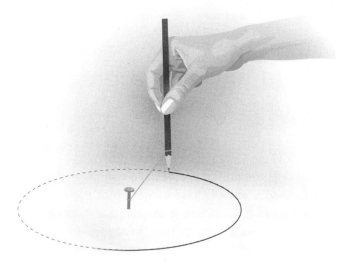

Figure 2.28 **Drawing a circle.**

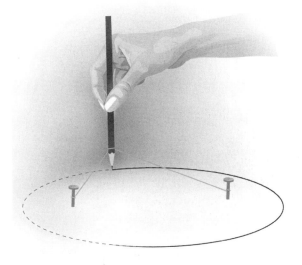

Figure 2.29 **Drawing an ellipse.**

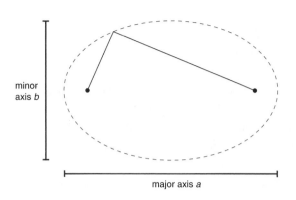

Figure 2.30 **Axes of an ellipse.**

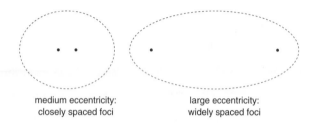

medium eccentricity:
closely spaced foci

large eccentricity:
widely spaced foci

Figure 2.31 **The eccentricity of an ellipse is a measure of the distance between its foci.**

Kepler's third law is concerned with the length of a planet's "year" – the time required to complete one orbit about the Sun. The farther a planet is from the Sun, the longer it takes to orbit. For example, the length of Pluto's "year" is 248 Earth years, while Mercury's is only 88 days.

Kepler's third law quantifies this. If we call the planet's *period* "P" the length of time required to complete an orbit – the length of its "year" – then this law states that *the period squared is proportional to the semimajor axis "a" cubed*:

$$P^2 \text{ is proportional to } a^3$$

or alternatively

$$P^2/a^3 = K,$$

where the constant K is the same for all planets.

What is the value of this constant K? It depends on the units we use to measure P and a. A convenient unit for a is the *astronomical unit*, or AU. The AU is *the average distance between the Earth and the Sun*. If the Earth's orbit were perfectly circular, this would be the radius of our orbit. Since in fact our orbit is slightly elliptical, the correct definition is that *the AU is the semimajor axis of our orbit*. Measurements reveal that

1 astronomical unit $= 1.496 \times 10^{11}$ meters (about 93 million miles).

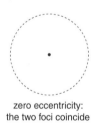

zero eccentricity:
the two foci coincide

Figure 2.32 **A circle is an ellipse of zero eccentricity.**

Suppose we decide to measure a planet's period in Earth years, and its semimajor axis in astronomical units. To find the value of K, we simply need to choose some planet and plug its P and a into the above equation. We could use any planet, but the easiest to use is the Earth: for it $P = 1$ year and $a = 1$ astronomical unit, hence

$$K = (1)^2/(1)^3 = 1,$$

so that Kepler's third law states

$$P^2 = a^3$$

if P is in years and a in astronomical units.

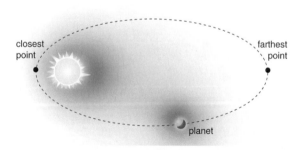

closest
point

farthest
point

planet

Figure 2.33 **An elliptical orbit about the Sun.**

....

NOW YOU DO IT
Appendix III lists the orbital periods and semimajor axes for the planets. Choose one of these planets (not the Earth – we've already done it!) and use it to calculate K. (Don't worry if you don't get *exactly* 1: this is because our measurements of P and a are not perfectly accurate!)
....

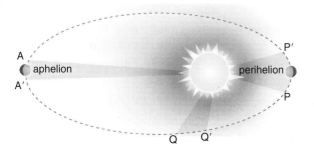

Figure 2.34 **Kepler's second law states that the area of the triangle swept out in, say, 30 days does not change as the body orbits. Therefore, the farther a body is from the Sun, the slower it moves.**

We have seen that Kepler's first two laws apply to every orbiting body. As explored in Problem (11) at the end of this chapter, his third law also applies to *every body orbiting the Sun* – i.e. comets and asteroids. But if the body is a moon orbiting a planet, it is still the case that $P^2/a^3 = K$, but the constant K is no longer equal to 1 (Problem (12)). In the following chapter we will understand why all this is so.

Galileo Galilei

In 1608 the Italian astronomer Galileo Galilei (1564–1642; see Figure 2.35) learned of an optical instrument that had recently been invented in Holland, and that made distant objects appear closer. He built his own (Figure 2.36), and thus began one of the most important sagas in the history of science.

Galileo was not the inventor of the telescope. Nor was he the first to train it on the heavens. But Galileo was both an extraordinary scientist and a brilliant writer. He was the first to use the telescope to conduct systematic, long-term observations of the sky – and, equally importantly, he was the first to publicize widely what he saw.

Galileo's astronomical discoveries

To appreciate the significance of Galileo's discoveries using the telescope, consider for a moment what the nighttime sky looks like to the naked eye.

Without a telescope, the Moon is essentially the only interesting object in the sky: it has phases, and its surface is mottled in interesting ways. The Sun, on the other hand, is an entirely smooth disk, and the stars and planets are merely featureless points of light.

In contrast, what Galileo saw through his telescope was rich and varied beyond all expectation. On the Moon he saw mountains, broad plains and craters. On the Sun he saw spots. He saw that these spots grew larger and smaller, subtly changing their shape from one day to the next; and he saw that they moved across the Sun's visible disk, suggesting that the Sun rotated on its axis. Turning his telescope to Venus, he saw that it had phases, like the Moon – and he also saw Venus reaching full phase, so conclusively demonstrating the truth of Copernicus' theory. He discovered four moons of Jupiter – these were the first new moons ever to be discovered. Furthermore, he

Figure 2.35 **Galileo Galilei. Painting by Justus Sustermans.**

Figure 2.36 **Galileo's telescope alongside a modern equivalent.** 👁 **(Also see color plate section.)**

realized that as they orbited he was seeing before his very eyes a new, miniature replica of the Solar System. And he saw that Saturn was not a perfect sphere, but had strange protuberances (they later turned out to be Saturn's rings: his telescope had not been sufficiently powerful to make them out).

The cosmos was turning out to be far more interesting a place than anybody had guessed. In 1610 Galileo published *The Starry Messenger*, in which he described these discoveries. This book was unique for its time. Previous works describing astronomical discoveries had been large, technical treatises, meant to be read only by other astronomers. Copernicus' and Kepler's works were full of calculations, mathematical formulas and tables of data, and they were written in Latin, the language of scholarship. But Galileo's book was explicitly meant to be read by every educated person. It was short – a mere 24 pages – and it was written in Italian. Most important of all it was engagingly written, and it was a pleasure to read. *The Starry Messenger* was an instant sensation throughout Europe.

Galileo and the Inquisition

Galileo's discoveries convinced him that the Earth really did move, and he advocated this with all his skill. By this time Copernicus' theory had been around for 70 years without causing much of a stir. But Galileo's insistent and brilliant arguments raised the issue to a head. In 1616 the Catholic Church issued a statement, declaring the doctrine that the Earth moves to be "false and absurd," and Galileo was ordered not to defend it.

But for years Cardinal Barberini had been a personal friend and admirer of Galileo's, and in 1623 he was named Pope Urban VIII. On six separate occasions the Pope received Galileo, and eventually gave him permission to write a book about the Copernican theory. The Pope carefully warned him to treat the theory only hypothetically, but Galileo seems to have felt that he could defy the order with impunity. In 1632 he published the *Dialog Concerning the Two Chief World Systems*.

In this book, three fictional characters debate the Ptolemaic and Copernican models of the Solar System. In its preface, Galileo states that his purpose was not to defend the Copernican theory but merely to examine it, but readers quickly saw that nothing could have been further from the truth: the book is a brilliant and persuasive argument that the Earth moves. The very name Galileo gives to the *Dialog*'s defender of the Ptolemaic system, *Simplicio*, is a dead giveaway.

Even though he was infirm and in ill health, Galileo was called before the Roman Inquisition. There he was tried and found guilty, and forced to recant his arguments and formally renounce the doctrine that the Earth moved. His sentence was commuted to house arrest for the remainder of his life.

The *Dialog* joined works by Copernicus and Kepler on the index of forbidden books, from which it was only removed in 1835. Since that time, the Church has steadily moved to renounce the position it had taken with respect to Galileo: in 1979 Pope John Paul II stated that Galileo had been unjustly treated by the

Church, and a statement by the Pope in 1992 has been interpreted by many as a full rehabilitation of Galileo.

. . . .

WHAT DO YOU THINK?

Certain people have argued that Galileo did a grave disservice to science by writing the *Dialog Concerning the Two Chief World Systems*. They argue that, by doing so, he forced the Church to react to the Copernican hypothesis. Conversely, had Galileo not promulgated Copernicus' theory so strongly, the Church would not have acted, and no opposition would have built up to the idea that the Earth moved.

    Write a brief essay in which you discuss this argument. Do you agree with it? Disagree? Be sure to give reasons for your position!

. . . .

THE NATURE OF SCIENCE

LESSONS FROM HISTORY

We have now completed most of our survey of the history of the development of astronomy. We will finish this survey in the next chapter. But we are already able to look back and gather some lessons from history. Science is a way of thinking, a way of looking at the world, that is unique in the history of thought. Scanning over our story of its development, we can see a number of science's most important characteristics.

RESPECT FOR DATA

In science, data are paramount. What data tell you must always be accepted. No matter how strongly you believe something to be true, if the observations tell you that it is not, then you must give up your cherished notion.

    Historically, this is a relatively recent idea. The ancient Greeks did not believe in it. To them, observational data were not very important at all. We saw this twice in our discussion of Greek astronomy.

- In our treatment of Aristarchus, we commented that his claimed irregularity in the progression of the Moon's phases simply does not exist. As a matter of fact, we do not have the slightest indication of where Aristarchus got this curious "fact." In his writings he cited no observations, gave no data. Indeed, he appeared to regard his calculation as a mathematical exercise, rather than an attempt to find out something true about the Universe.
- Ptolemy's model of the Solar System was not very accurate. Often a planet would be found as much as 5 degrees of arc away from its position as predicted by his theory – and 5 degrees is a big angle in astronomical terms: ten times the diameter of the Moon. To us this would be an intolerable discrepancy, but it appears not to have bothered Ptolemy at all. Nor did it bother his successors for nearly 1500 years.

Even Copernicus shared in this view. His model of the Solar System was no more accurate than Ptolemy's. Historically, it was Tycho Brahe who first insisted on systematic, highly accurate observations. And it was Kepler who first exemplified our modern attitude of respect for the results of such observations. For Kepler the slightest discrepancy between theory and observation was reason enough to toss out the theory. At one point, he abandoned a theory after 900 pages of calculation when the result disagreed with data by a mere 8 minutes of arc! Eight arc minutes was an unheard-of accuracy at the time, far better than anyone had ever achieved – but it was not good enough for Kepler. For him – and for modern science – no amount of philosophical argument, no amount of theoretical justification, can rescue a theory that conflicts with observation to even the slightest degree.

SCIENCE IS AN ATTEMPT TO UNDERSTAND REALITY

We have already mentioned that Ptolemy's model of the Solar System used several *different* sets of circles to account for different aspects of the motion of *the same* planet. Clearly, therefore, Ptolemy had not invented a single, unified model of the Solar System. He could not have believed that planets actually rode along his deferents and epicycles – for how would a planet "know which deferent and epicycle to ride"? Rather, Ptolemy's aim had been to invent a set of geometrical constructions that crudely matched the observed motion of the planets, without claiming any physical reality for these constructions.

In contrast, we do insist on the physical reality of our theories. We believe that Mars really does move along an elliptical orbit. We believe that the Earth is actually not the center of the Universe.

SCIENTISTS ARE SKEPTICAL, AND ARE CONSTANTLY TESTING THEIR THEORIES

If the test turns out negative, we abandon the theory. If it turns out positive, the theory has been confirmed and we have greater reason to accept it. Of course, for this to happen these theories must be capable of being tested! Something that cannot be checked against observation is not a scientific theory.

SCIENCE AND RELIGION ARE SEPARATE

Centuries of history have taught us that religion must not dictate what to think concerning matters of fact. The experience of Galileo is an example of what can happen if religion oversteps this boundary. Galileo felt that there was strong evidence that the Earth was not the center of the Universe, and that it orbited the Sun. But this ran counter to the doctrine of the Catholic Church, and the Church forced him to recant these ideas – ideas that we now know to be true. It is no exaggeration to call this a shameful episode in the history of thought.

As we have noted, the Church has long since accepted the idea that the Earth moves. But our concern here is not with the details of this particular episode. It is with the question of religion and science in general.

Science is concerned with the quest for facts about the Universe. If this is our aim, we must not let ourselves be swayed by anything other than evidence. To repeat: in science, data are paramount. What the data tell us we must accept. Conversely, in science, religious belief is *not* paramount, and what belief tells us we are not bound to accept. No matter how strongly faith declares some theory to be true, if the observations say otherwise then we cannot accept that theory. In science, no amount of religious conviction can rescue a theory that conflicts with observations.

⇐ **Looking backward**
We likened a scientist to a detective in "Detectives on the case" in Chapter 1.

In the previous chapter we argued that a scientist working to understand the Universe is like a detective working to solve a mystery. What would we think if the detective has found that clue after clue points in some particular direction – but the government, or the suspect's best friend, is trying to persuade her otherwise?

Throughout this book we will encounter scientific findings that have strong implications for religion.

- It is not merely that the Earth is no longer thought to be the center of the Solar System – we are no longer thought to be the center of *anything*. The Earth does not lie at the center of our Galaxy, and our Galaxy does not lie at the center of the Universe.
- Our planet is not unique. There are many other planets, some orbiting our Sun, some orbiting others.
- Our Sun is not unique. It is an average star.
- The Earth originated as the result of purely natural causes.
- Although we do not know how life arose, we have no reason to think that it arose as the result of supernatural causes.
- Our Sun will eventually go out, and all life on Earth will come to an end.

We all must find our own spiritual responses to these important truths. But as scientists we must not allow them to influence our search for more truths.

Many scientists feel that there is no conflict between science and religion. This is because they regard science and religion as dealing with completely different things. Science is concerned with matters of fact: religion, in contrast, is concerned with matters of spirituality and faith. The same view is held by the major world religions, which find no conflict between their tenets and the findings of science.

Furthermore, many scientists are comfortable with religion: a study conducted in 1997 found that 40% of American scientists believed in a personal God. These scientists regard their scientific work as being utterly separated from their religious belief.

Two striking sayings might throw light on how many scientists view religion.

- "The more science teaches us, the more we realize that God has been seriously underestimated."
- "Science tells us the facts. Religion tells us what these facts mean for our lives."

Summary

Astrology

- Ancient peoples studied the sky closely and noted its regularities.
- Astrology was their attempt to construct a theory of what they saw.
- Astrology had both practical and religious significance.
- Horoscopes are tools astrologers used to predict a person's personality.
- Initially there was no clear distinction between astrology and astronomy.
- Tests of astrology have shown that its predictions are no more accurate than random guesses.

Archaeoastronomy

- Stonehenge contains some alignments that have clear astronomical significance.
- This is not the only example of ancient astronomical monuments.

Ancient Greek astronomy

- Aristarchus devised a method for comparing the distance to the Moon with that to the Sun.
- While correct in principle, his method relied on an erroneous observation.
- Eratosthenes devised a means of determining the size of the Earth.
- Ptolemy constructed a model of the Solar System with the Earth at the center, and the planets moving about it.
- His model assumed circular motion, and contained epicycles, equants and deferents.

Copernicus

- Copernicus devised a model in which the Earth moved about the Sun.
- There were no good arguments against this hypothesis.
- On the other hand, not for many years was it confirmed by observation.
- His model also assumed circular motion, and contained epicycles, equants and deferents.

Brahe

- Brahe was the first scientist to perform high-accuracy measurements.
- His observations of the motions of the planets provided the basis for Kepler's model of the Solar System.

Kepler

- Kepler's model of the Solar System rejected the idea of circular motion.
- Kepler discovered three laws of planetary motion:
  - the first law: planets move in elliptical orbits,
  - the second law: a line from a planet to the Sun sweeps out equal areas in equal times,
  - the third law: the period of an orbit squared is proportional to its semimajor axis cubed.

Galileo

- Galileo used a telescope to make numerous discoveries:
  - mountains and craters on the Moon,
  - phases of Venus,
  - moons of Jupiter.
- His writings brought his discoveries, and his defense of the Copernican system, to a wide audience.
- He was brought before the Inquisition, which forced him to recant his views.

The nature of science

- Designing tests.
  - In designing experimental tests, great care must be taken to eliminate unconscious bias on the part of the experimenter.
  - Scientists need large amounts of data in order to be sure of their conclusions.
- Deciding between competing theories.
  - We decide between the Ptolemaic and Copernican theories of the Solar System by asking each to predict the phases of Venus: the predictions of the Copernican theory are borne out and those of the Ptolemaic model are not.
  - It is important to have multiple tests of a theory: the Copernican theory has been additionally confirmed by directly observing the motion of the Earth.
- Lessons from history.
  - In science, data are paramount: a theory that disagrees with observation, no matter how slightly, must be abandoned.
  - Science is an attempt to understand reality.
  - Science is testable.
  - Religion must not pronounce upon scientific issues.

Problems

(1) In our discussion of the design of experiments, we studied why scientists need large amounts of data. Let us now study this with regard to tossing a coin. Suppose you have a coin that in fact is fair – i.e. it is equally likely to come up heads or tails. How likely is it that an experiment will reveal *misleading data, namely that the coin always turns up heads*?

Begin with an experiment consisting of a very small amount of data, namely a single toss. When you toss the coin there are two possibilities.

| Heads* |
| Tails |

* This experiment yields heads.

As you can see, the chances are ½ that a one-toss experiment yields misleading results. You have already studied a two-toss experiment. Now your task is to study a *three-toss experiment*. What are the chances that these experiments yield misleading data?

(2) In our discussion of the design of experiments, we studied why scientists need large amounts of data. Let us now study this with regard to tossing a die. Suppose you have a die that in fact is fair – i.e. it is equally likely to come up 1, 2, 3, 4, 5 or 6. How likely is it that an experiment will reveal *misleading data, namely that the die always turns up "6"*?

Begin with an experiment consisting of a very small amount of data, namely a single toss. When you toss the die, there are six possibilities.

| 1 |
| 2 |
| 3 |
| 4 |
| 5 |
| 6* |

* This experiment yields a "6".

As you can see, the chances are 1/6 that a one-toss experiment yields misleading results. Your task in this problem is to study a *two-toss experiment*. What are the chances that it yields misleading data?

(3) Suppose Aristarchus had found that the interval of time from half through new to half phase was shorter by *two* days than the interval from half through full to half. (i) Find Time (A) and Time (B). (ii) Find the angle "*a*" in the triangle of Figure 2.12. (iii) Find the ratio D/R.

(4) Suppose Eratosthenes had measured the Sun to be almost exactly overhead at noon on the first day of summer in Alexandria. Would he have concluded that the radius of the Earth was very large or very small? Explain your answer.

(5) Suppose Eratosthenes had measured the Sun to be 20 degrees from directly overhead at noon on the first day of summer. How big would he have calculated the Earth to be?

(6) Suppose our historians are wrong, and the stadia were actually three kilometers in length. What would Eratosthenes have found for the radius of the Earth?

(7) Suppose that somehow Eratosthenes already knew the radius of the Earth – but did not know the distance from Alexandria to Syene. (1) Describe *the logic of the calculation* he would have used to find this distance. (2) Taking the radius of the Earth to be 6378 kilometers, carry through *the detailed calculation* to find this distance.

(8) We will repeatedly emphasize throughout this book that every scientific theory must be tested, by comparing its predictions with observations. Let us apply this to Eratosthenes' theory that the Earth is round, with a radius of 6378 kilometers. We will require this theory to *predict the angle between the Sun's rays and the vertical at a point 2000 kilometers north of Alexandria*. Then we will go there and measure it. (A) Describe the logic of the calculation you will use to predict this angle. (B) Perform the detailed calculation. [For your information, measurements do indeed confirm Eratosthenes' theory.]

(9) (A) Consider the motion of Jupiter in the Ptolemaic system. Draw your own version of Figure 2.20, and on this figure, draw little dots at various points along the planet's orbit, and indicate the phases of the planet at these locations as seen from the Earth. (B) Now do the same adopting the Copernican system. Could observations of the phases of Jupiter help decide between the Ptolemaic and the Copernican theories?

(10) Figure 2.36 illustrates that Kepler's second law implies that a body speeds up and slows down in its elliptical orbit about the Sun. Recall that the greater the eccentricity of an ellipse, the more "squashed" it is. Draw your own versions of Figure 2.36 for ellipses of greater and greater eccentricity, and use them to

explain why *a body in a moderately eccentric orbit varies its speed by a little, while a body in a very eccentric orbit varies its speed by a lot.*

(11) Recall that Kepler's third law states that P^2/a^3 also equals 1 for any body that orbits the Sun – comets and asteroids as well as planets. Here are the periods and semimajor axes of a few comets and asteroids. Choose one of them and verify this. (Don't worry if you don't get *exactly* 1: this is because our measurements of P and a are not perfectly accurate!)

Object	Period (years)	Semimajor axis (AU)
Halley's comet	76	17.9
Comet Encke	3.3	2.2
Asteroid Ceres	4.6	2.77
Asteroid Hektor	11.6	5.15

(12) Recall that Kepler's third law states that P^2/a^3 no longer equals 1 if we consider moons orbiting planets. Here are the periods and semimajor axes of a few of Jupiter's moons. Choose *two* of them and verify that (i) P^2/a^3 no longer equals 1, and (ii) you get the same value for P^2/a^3 no matter what moon you choose. (Don't worry if you don't get *exactly* the answers you expect: this is because our measurements of P and a are not perfectly accurate!)

Moon	Period (years)	Semimajor axis (AU)
Io	0.0049	0.0028
Ganymede	0.02	0.0071
Leda	0.66	0.074
Sinope	2.1	0.16

. . . .

WHAT DO YOU THINK?

(1) Science and religion

There are three attitudes a person might take toward science and religion.

(1) Science denigrates a person's religious beliefs.
(2) Science enhances a person's religious beliefs.
(3) Science has nothing to do with a person's religious beliefs.

Which of these best exemplifies your own personal attitude? Explain why.

(2) Science is testable

We have discussed philosophy as an example of a field that cannot be tested. Give other examples of fields whose theories cannot be proven right or wrong.

. . . .

3

Newton's laws: gravity and orbits

We have not yet finished our survey of the origins of astronomy. With the work of Isaac Newton (1642–1727) we reach the culmination of the scientific revolution. Because the laws he set forth are so central to an understanding of astronomy, we will devote an entire chapter to them.

Gravitation is the single most important factor that determines the structure of the astronomical Universe. There is not a single aspect of astronomy for which gravity fails to play a crucial role. Newton codified the idea that gravitation is a force, and he gave a formula describing its strength. According to his formula, every body in the Universe exerts a force of gravitational attraction on every other, the strength depending on their masses and the distance between them. He also created laws describing what happens to a body when a force acts upon it, his famous three laws of motion.

The theory Newton created more than three centuries ago is still the backbone of much of our understanding of the cosmos. It describes the orbits of the planets about the Sun and of stars in the Galaxy, the force compressing a star to thermonuclear temperatures and the force that governed the slow process of amalgamation whereby the Earth was formed. And it guides our engineers today as they plot the paths our space probes will take on their journeys to the distant reaches of the Solar System.

Isaac Newton

Newton (Figure 3.1), the son of a farmer, was born in Woolsthorpe, in Lincolnshire, England, in 1642, the year of Galileo's death. A student at Cambridge University, he was sent back home for two years when the university closed as a consequence of an outbreak of the bubonic plague. While at home, he developed the main outlines of his theories of gravitation and motion.

But Newton did not pursue this work when he returned to Cambridge as a professor in 1669. It was, in fact, something of an accident that he was induced to do so. By this time it was known that an inverse square law of force from the Sun could explain circular orbits. But would such a force give rise to the elliptical orbits Kepler had discovered? Several of the greatest minds of the day had attempted to

Figure 3.1 **Isaac Newton.**
◉ (Also see color plate section.)

answer this question, but to no avail. In 1684 the astronomer Edmund Halley (of comet fame) consulted Newton about this problem, which he and others had found too difficult to solve. Newton mentioned that he had already solved it, but had neglected to tell anybody.

Indeed, Newton seems to have stuffed the pages containing these calculations into a drawer and then lost them. He was, however, able to re-do the work. Halley prevailed upon Newton to publish this and his other researches. The result was the *Philosophiae Naturalis Principia Mathematica* (*The Mathematical Principles of Natural Philosophy*), one of the most important books in the history of science. Halley personally paid to have it published.

In this chapter we will present Newton's laws of gravitation and motion, and then go on to show how these laws can be used to understand orbits. We will close with a discussion of the specialized orbits that are being used to send space probes to the farthest reaches of the Solar System.

Universal gravitation

To appreciate how all-important gravity is, it is helpful to imagine what would happen were it magically to disappear. If, by turning some imaginary switch, we could suddenly make the force of gravitation cease, what would happen?

Gravitation is what attracts us to the Earth. If this attraction were to cease, all of us would become weightless. The tiniest push would send us floating upwards into the air. This sounds like a pleasant experience – but how far upwards would we float? We often fail to realize just how nearby is the deadly vacuum of interplanetary space. On the summit of Mount Everest, the air is so thin that most climbers require oxygen – and this summit lies less than six miles above sea level. Were gravity to vanish, all of us would have suffocated by the time we had floated upwards a mere ten miles or so.

Were gravitation to vanish, it is not just we who would drift off into interplanetary space. So would everything not actually fastened to the Earth's surface. In particular, the air we breathe would fly away, so that the vacuum of space would penetrate down to the Earth's surface. Indeed, since our atmosphere is a gas, and a gas exerts a pressure, the atmosphere would not merely float away gently: it would explode off into space.

Gravitation is what keeps the Earth in its orbit about the Sun. Were gravitation to cease, our planet would fly away from its orbit and off into space. But the Sun is what keeps us warm: were the Earth to fly off in this fashion, its temperature would quickly drop to that of interstellar space, which is close to absolute zero, or 459 degrees Fahrenheit below zero (273 degrees Celsius below zero)!

The very existence of the Sun, and every other star, is crucially dependent on gravity. As we will see in our study of stellar structure, stars are balls of superheated gas. Because they are so hot they exert an enormous pressure: were gravitation to magically cease, every star in the Universe would violently explode. And finally galaxies, giant rotating disks shot through with spiral arms, are also held together by gravity. If gravity were to cease, galaxies would fly apart.

It is a lucky thing that this little nightmare will never come to pass. But it is enough to make vivid how much the Universe as we know it depends upon gravitation.

Newton's law of gravitation

Newton's law of universal gravitation is as follows.

> Universal gravitation: every object attracts every other object with a force proportional to the product of their masses and inversely proportional to the square of the distance between them.

Let us examine each element of this law in turn.

- Gravitation is *universal* because it influences everything. There is not a single object in existence that is not influenced by gravity. This is in contrast to other forces of nature, which only affect certain objects. The magnetic force, for example, influences magnets but not wood.
- Gravitation is *an attraction*, and never a repulsion. This too is in contrast to other forces of nature, which can be either attractive or repulsive. Magnets, for instance, can either attract or repel one another.

The mathematical statement of Newton's law of universal gravitation is

$$F = GMm/R^2.$$

Here F is the gravitational force, M is the mass of one object, m is the mass of the other and R is the distance between them. The constant G is known as *Newton's constant of gravitation*.

The numerical value of G depends on the system of units we use. In this book we are using the MKS system (the "meter–kilogram–second" system). In the MKS system the unit with which distance is measured is the meter, the unit of mass is the kilogram and the unit of force is the newton. The newton is somewhat smaller than the unit of force with which we are all more familiar, the pound: one newton is 0.22 pounds. In this system of units

$$G = 6.67 \times 10^{-11} \text{ newton meter}^2/\text{kilogram}^2.$$

Intuitive mathematics

Let us examine Newton's law of gravitation, to see what it says about gravitation intuitively.

(1) Notice first that the law tells us to multiply by G. But since the numerical value of G is very small, the resulting force will be small. This tells us that *the gravitational force is relatively weak.*

(2) Notice also that the law tells us to multiply together the masses of the two bodies. If these masses are small, the resulting force will be small, and if they are

large the result will be large. This tells us that *the gravitational force between low-mass bodies is small; that between high-mass bodies is large.* Since in daily life the objects we encounter are of relatively low mass – people, cars, etc. – this means that, in daily life, gravitation is not very important. But in astronomy the masses are large – planets, stars, etc. Thus we see that gravitation is very important in astronomy.

(3) Notice finally that the law tells us to divide by the distance between the two bodies. If this distance is large, the result will be small. Similarly, if the distance is small, the result will be large. This tells us that *the gravitational force is larger for bodies near one another than for bodies far apart from each other.*

Let us now use Newton's law of gravitation to calculate the force of attraction between various objects. We do so for two reasons: first, to get some practice using the law, and second to draw various lessons about gravitation.

We will begin with an everyday example: the force between two people standing fairly close to one another. We will take them to be at arm's length.

The *logic of the calculation* is as follows.

Step 1. Choose the masses of the two people and their distance apart.
Step 2. Plug these into Newton's formula to find the force.

Detailed calculation

Step 1. Choose the masses of the two people and their distance apart.
The mass of a typical person is roughly 70 kilograms. We will take both people to have this mass. Let us imagine that they are one meter apart – this is about arm's length.

Step 2. Plug these into Newton's formula to find the force.
We plug these values, and that for G, into Newton's formula

$F = (6.67 \times 10^{-11}$ newton meter^2/kilogram^2) (70 kilograms) (70 kilograms) / (1 meter)^2.

Using a calculator we find

$F = 3.3 \times 10^{-7}$ newtons.

Step 3. Convert the force in MKS units into pounds.
The newton is somewhat smaller than the unit of force with which we are all more familiar, the pound: one newton is 0.22 pounds. To convert from newtons to pounds we multiply the number of newtons by 0.22:

$F = (3.3 \times 10^{-7}$ newtons)(0.22 pounds/newton)
   $= 7.2 \times 10^{-8}$ pounds, or one 14 millionth of a pound.

Notice that this conforms to our *two first intuitive results*: the gravitational force between our two relatively low-mass bodies is indeed very small. Notice also that it conforms to our third intuitive result: had we considered them to be closer together, we would have divided by a smaller number and obtained a larger value for the force.

Recall that in this book we are using MKS units. So the answer we get from our calculation will be in these units as well. This will be a perfectly valid answer – but it is hard to understand intuitively what it means. Accordingly, the last step in our calculation should be to convert the answer in step 2 to more intuitively comprehensible units.

Step 3. Convert the force in MKS units into pounds.

As you can see from the detailed calculation in the insert, the answer we get turns out to be one 14 millionth of a pound. This is an exceedingly small force – and that is the lesson we draw from this first example. We have learned that *the gravitational attraction between two people is too weak to notice.* That is why you do not actually *feel* the force of attraction whenever you approach a friend. More generally, the example shows that *the gravitational attraction between any two objects in our daily experience is too weak to notice.*

NOW YOU DO IT

Find the force of gravity (in pounds) between a 70-kilogram person and a building (roughly 100 000 kilograms) one meter away.

Under what circumstances does gravity become strong? According to (2) of our intuitive look at Newton's law, the force of gravity is strong if we consider a large mass. Let us then turn our attention to the force of attraction exerted by an exceedingly large mass: the entire Earth itself. Let us calculate the force of attraction between *the Earth* and a person (Figure 3.2).

The logic of the calculation

We proceed in exactly the same way as before, but we use the mass of the Earth, not of a person, for M.

As you can see from the detailed calculation below, we get a large force. But our calculation has taught us far more than this. Our result of 150 pounds is the weight of that 70-kilogram person! Similarly, your weight is the force of gravity that the Earth exerts upon you. The second lesson we draw from this example is that *what we experience in our daily lives as weight is in reality the force of gravity attracting us toward the Earth.*

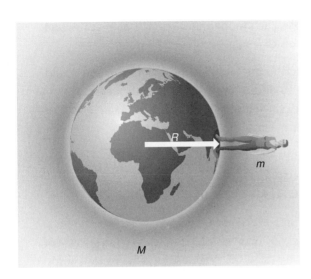

Figure 3.2 **The force of gravity between the Earth and a person gives rise to the person's weight.**

....

NOW YOU DO IT
Find the force of gravity in pounds between the Earth and a 50-kilogram person.

....

As our next example let us consider *the weight of that 70-kilogram person on the Moon.*

The logic of the calculation

We proceed in exactly the same way as before, but we use the mass of the Moon, not of the Earth, for M.

As you can see from the detailed calculation overleaf, the person weighs less on the Moon than on the Earth. Notice, however, that the person's *mass* is just the same – namely, 70 kilograms. This is the lesson we draw from this third example: *mass is not the same thing as weight.* The *mass* of an object measures how much matter it contains, but the *weight* of that object measures the force of gravitational attraction acting upon it. Indeed, if we were to transport that 70-kilogram person far off in interstellar space, where there are no nearby bodies to exert their forces of attraction, that person's weight would be zero! But his mass would still be 70 kilograms.

....

NOW YOU DO IT
Find the weight in pounds of a 70-kilogram person on Mars. (Mars' radius is 3397 kilometers; its mass is 0.11 that of the Earth.)

Detailed calculation

The mass M of the Earth is 6×10^{24} kilograms. In calculating the distance between the person and the Earth, we take the person's distance to the center of the Earth, which is the Earth's radius (6378 kilometers, or about 6.4×10^6 meters). We obtain

$F = (6.67 \times 10^{-11}$ newton meter^2/kilogram^2) $(6 \times 10^{24}$ kilograms) (70 kilograms)/$(6.4 \times 10^6$ meters)^2

$F = 684$ newtons,

which is 150 pounds.

Notice that this *conforms to our intuitive expectation (2)*: 150 pounds is, as expected, a significant force.

Detailed calculation

The calculation is exactly analogous to what we did for the Earth: the only difference is that we use the Moon's mass (7.35×10^{22} kilograms) and radius (1738 kilometers, or 1.738×10^6 meters) instead of the Earth's. We obtain $F = 114$ newtons, or a mere 25 pounds.

Notice that this *conforms to our intuitive expectation (2)*: because the Moon's mass is less than the Earth's, gravity there is weaker. Everything weighs less on the Moon than here on the Earth.

Suppose you stepped on the bathroom scale, and it told you that you weighed 120 pounds.

(A) Describe the logic of the calculation you would use to determine from this your mass.

(B) Carry through the detailed calculation and find your mass.

Newton's laws of motion

Once a force is exerted on a body, it moves: the body is pushed around by the force. Newton's laws of motion describe how it moves: they describe the response of a body to a force. There are three such laws; we will consider each in turn.

The first law: inertia

Newton's first law of motion, the law of inertia, states that *every body continues what it had been doing unless compelled to change its state of motion by a force.* Thus, if the body was initially at rest, it remains at rest. If it was initially moving, it continues moving in the same direction with the same speed.

In some ways, the principle of inertia makes intuitive sense. If a marble lies at rest upon the floor, it does not suddenly start rolling of its own accord. Rather, a force is required to set it moving. But in other ways, the principle of inertia does not make intuitive sense. Suppose we set that marble on the floor to rolling, and then step back and watch what happens. Once we have ceased pushing upon the marble, the law of inertia decrees that it should continue rolling forever. But as we all know, it does not. The marble slows and soon comes to rest.

Is the marble violating Newton's first law of motion? Of course not. Actually there *is* a force acting on the marble: friction against the floor.

Only if we can reduce friction (and air resistance) to zero is it possible to see the law of inertia in action. One common instance occurs when you are driving in winter and hit a patch of ice on the road. The ice drastically reduces the friction between the tires and the road. In response, the car now obeys the law of inertia: it continues along at the same speed in a straight line, no matter how frantically you hit the brakes or turn the steering wheel.

You may have seen in physics labs a demonstration of another means of eliminating friction: the air track (Figure 3.3). Air from a compressor is blown into the track and out through a network of tiny holes. The cart is

Figure 3.3 **An air track reduces friction. The carts upon it therefore continue moving with constant speed for great periods of time.**

slightly raised up off the track, and it floats on a cushion of air. Because the cart is no longer in contact with the track, friction is eliminated. As a consequence, the cart moves as the law of inertia says it should: once given a shove, it coasts along indefinitely.

The second law: force

Newton's first law of motion describes what happens if no force acts on a body. His second law goes on to describe what happens if there is a force. It states that *if a body is acted on by a force, it accelerates at a rate proportional to the force and inversely proportional to its mass.*

The fact that a body's acceleration is proportional to the force acting upon it means that big forces produce big accelerations. This conforms to everyday experience: if you push very slightly on something it moves only a little, whereas if you push hard it moves a lot. Similarly, the fact that a body's acceleration is inversely proportional to its mass also conforms to everyday experience: if you push on an automobile it moves only slightly, whereas the same force applied to a bicycle sets it moving more rapidly.

It is important to emphasize that a force produces, not just motion in general, but a particular kind of motion: acceleration. Acceleration is *the rate of change of velocity.* That is, the acceleration A equals the change in velocity divided by the time required to make that change:

$$\text{acceleration} = \text{change in velocity}/\text{time}$$
$$A = \Delta v/\Delta t.$$

Imagine that you are in a car waiting at a red light. The light turns green, and you accelerate from zero to 50 miles per hour. What was your acceleration? If you took, say, an entire minute to reach full speed, the acceleration was small. But if you reached full speed in a mere ten seconds, the acceleration was large. Similarly, if your car had reached not 50 but 100 miles per hour in that minute, the acceleration would have been correspondingly larger.

It is also important to emphasize that technically speaking velocity has a direction as well as a magnitude. A velocity of 50 miles per hour due north is different from a velocity of 50 miles per hour due east. If a car travels at a steady 50 miles per hour but makes a turn, in daily life we would say that it has not accelerated. But technically speaking it has. A good way to keep this distinction in mind is to use "speed" to denote the magnitude of a body's velocity without reference to its direction, while "velocity" does take account of direction. Thus 50 miles per hour due north is the same *speed* as 50 miles per hour due east – but it is a different *velocity.*

. . . .

NOW YOU DO IT
Two cars are waiting at a stop light. It turns green and they drive off. One car reaches 60 miles per hour in 5 seconds, while the other reaches 60 miles per hour in 3 seconds. Which car has the greater acceleration?

Two cars are waiting at a stop light. It turns green and they drive off. One car reaches 25 miles per hour in 2 seconds, while the other reaches 50 miles per hour in 3 seconds. Which car has the greater acceleration?

. . . .

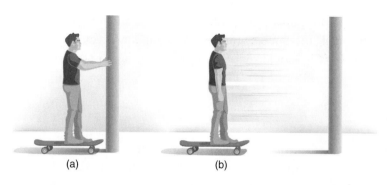

Figure 3.4 **Action and reaction. When you push against a wall, the wall pushes back against you with an equal and opposite force.**

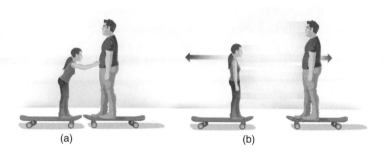

Figure 3.5 **Action, reaction and mass. When you push against a second skateboarder, both of you are set moving. Your force is what accelerates the other person; the reaction to this force is what accelerates you. Since the two forces are equal, each person's acceleration is inversely proportional to that person's mass: the lighter person ends up moving fastest.**

The mathematical statement of Newton's second law of motion is

$$F = MA,$$

where F is the force, M the body's mass and A its resulting acceleration.

The third law: reaction

Newton's third law of motion states that *to every action there is an equal and opposite reaction.* Thus, if one body exerts a certain force on a second body, the second exerts the same force back on the first.

Does this law conform to everyday experience? Imagine that you are standing against a wall and give it a push. You have exerted a certain force on the wall. Newton's third law states that the wall exerts the same force on you. Has it? We can easily illustrate that it has, by supposing that you are standing on a skateboard (Figure 3.4). In such a situation, your push sets you gliding away from the wall. This motion is caused by the reaction force which the wall exerted on you.

Similarly if you were to push, not against a wall, but against a friend also on a skateboard (Figure 3.5), both of you would be set moving. The force that set your friend moving is your push: the force that set you moving is the reaction force, which is opposite to your push.

We can carry this second example a little further, by imagining that your friend weighs a lot more than you. Remember that the reaction force equals the original force. By Newton's second law, then, each person's acceleration is inversely proportional to his mass: the heavier person moves less than the lighter one.

....

NOW YOU DO IT
Suppose you push on a truck that is parked with its engine off. Does the truck push on you?

Suppose you push on a friend who weighs a lot less than you. If both of you are standing on skateboards, which of you moves more?

....

Newton's third law is the principle behind rocket propulsion (Figure 3.6). A rocket engine works by injecting fuel into a combustion chamber. The fuel burns, and expands out through the nozzle. The force that produces this motion comes from the high pressure produced in the gases. By Newton's third law, the same force is exerted on the rocket engine by the gases: this is the force that sets the rocket moving.

Orbital motion

With an understanding in hand of Newton's laws of gravity and motion, we now turn to a study of orbits.

The connection between orbital motion and falling

At first glance orbits seem alien, otherworldly things. But to a great degree this view is mistaken. In reality there is a strong connection between orbits and the common, everyday experience of falling. To begin our study of orbits, therefore, let us begin by thinking about falling.

Take a golf ball and drop it. It falls straight down. But now set the golf ball to moving before you drop it – by rolling it along a table, for instance (Figure 3.7). Once it rolls off the edge of the table, the ball starts falling. But it no longer falls straight down! Rather, it falls in a curving path. As we will shortly see, that is what an orbit is: a fall along a curving path of a very particular type.

As illustrated in Figure 3.7, the golf ball travels a certain distance before striking the ground. The faster you roll it, the greater is this distance. How far would it travel if you were to roll it exceedingly rapidly – at thousands of miles per hour, say? Of course it is not possible to actually perform this experiment, but we can imagine the result. It is illustrated in Figure 3.8. We see that the golf ball travels so far before it lands that the surface of the Earth drops away from it as it flies. The essentially spherical shape of the Earth is beginning to be important.

Throw the ball yet faster. As illustrated in Figure 3.9, it now travels most of the way around the Earth before striking the ground. And finally, if we throw the ball at a sufficiently great speed, it travels in a perfect circle about the Earth, and it returns to its starting point. The ball *never* strikes the ground! Rather, it is now in an orbit about the Earth.

We now understand two things. First, we see that orbits are simply examples of falling. But we also understand why it is that most people do not understand this connection. The connection would be evident only if we could actually perform such an experiment. And of course we cannot; not only would we have to accelerate everyday objects to enormous velocities, but we would also have to clear away all the hills, buildings and trees blocking the path of the flying body. Indeed, we would also have to clear away all the air, which acts to slow the body. In practical terms, it is only possible to perform this experiment in outer space – and this is why people get the erroneous impression that orbits somehow "belong" to outer space.

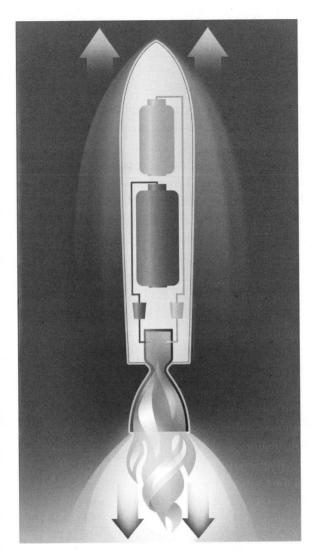

Figure 3.6 **The rocket engine works by Newton's third law. The engine produces a force on the exhaust gases, which are accelerated out through the nozzle. They exert an equal reaction force on the rocket engine, which accelerates it forward.**

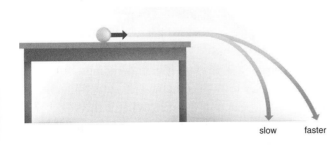

slow   faster

Figure 3.7 **The greater its velocity, the farther an object flies.**

Figure 3.8 **A very fast object travels so far that the Earth's curvature becomes important.**

Figure 3.9 **A sufficiently high velocity will cause the falling object to orbit the Earth.**

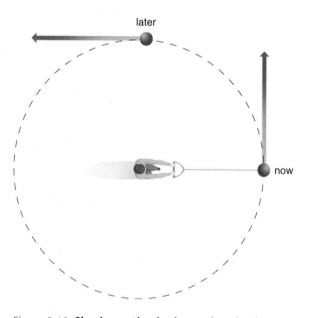

Figure 3.10 **Circular motion is always changing its direction. The resulting acceleration is known as centripetal acceleration.**

Falling in a circle

Because many orbits are nearly perfectly circular, let us think more carefully about circular motion.

Begin with an everyday example. Imagine that you were to take a length of string, tie one end about some object, and whirl the object about in a circle (Figure 3.10). Take care to whirl it about at constant speed, neither speeding up nor slowing down the motion. Is the object accelerating? Your first thought might be that it is not – after all, its speed is never changing. But recall that speed is different from velocity, and that velocity has a direction as well as a magnitude.

Let us therefore ask whether the direction of motion of your whirling object is changing. A glance at Figure 3.10 shows that it is. At the moment labeled "now" in the figure, the object is moving in one direction – but at the moment labeled "later" it has moved along its path, and its direction of motion has changed. Indeed, *an object in circular motion is always accelerating.*

The acceleration of an object moving in a circle with constant speed is known as *centripetal acceleration.* The formula for centripetal acceleration is

$$A_{\text{centripetal}} = V^2/R,$$

where V is the body's velocity and R the radius of the circle in which it moves.

Orbits and Newton's laws: the mathematics of orbits

We are now in a position to apply our knowledge of Newton's laws and of centripetal acceleration to orbits. To make things specific, we will think about the Earth in its orbit about the Sun. But everything we say will be valid for all orbiting bodies.

In Figure 3.11 we illustrate the Earth in its orbit about the Sun. The Earth orbits at a steady 30 kilometers per second. Like the object at the end of the string which we illustrated in Figure 3.10, its direction of motion is always changing even though its speed is constant. Therefore it is undergoing centripetal acceleration. By Newton's second law, if something is accelerating a force must be acting upon it. What is this force? In the case of the object at the end of the string, it was your tug on the string. In the case of the Earth, *the force that bends the Earth's path is gravitation from the Sun.*

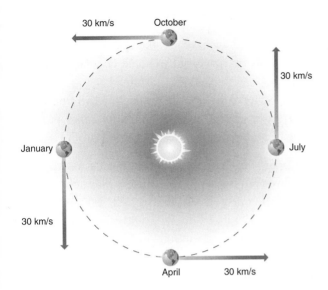

30 km/s October

30 km/s

January July

30 km/s

April 30 km/s

Figure 3.11 **The direction of the Earth's motion is always changing as it orbits the Sun. Therefore the Earth is always accelerating. By Newton's second law, there must always be a force upon it. This force is gravity from the Sun.**

Newton's law of gravity tells us the magnitude of this force:

$$F_{\text{on Earth}} = G\, M_{\text{Sun}}\, m_{\text{Earth}} / R_{\text{Earth}}^2,$$

where R_{Earth} is the distance between the Earth and the Sun. Similarly, Newton's second law tells us the magnitude of the Earth's resulting acceleration A_{Earth}:

$$F_{\text{on Earth}} = m_{\text{Earth}}\, A_{\text{Earth}}.$$

Equating the two:

$$G\, M_{\text{Sun}}\, m_{\text{Earth}} / R_{\text{Earth}}^2 = m_{\text{Earth}}\, A_{\text{Earth}}.$$

Solving for the Earth's acceleration we find:

$$A_{\text{Earth}} = G\, M_{\text{Sun}} / R_{\text{Earth}}^2.$$

Because the Earth's motion is circular, the acceleration A_{Earth} is given by our formula for centripetal acceleration:

$$A_{\text{centripetal}} = V_{\text{Earth}}^2 / R_{\text{Earth}}.$$

Setting $A_{\text{Earth}} = A_{\text{centripetal}}$ we find

$$V_{\text{Earth}}^2 / R_{\text{Earth}} = G\, M_{\text{Sun}} / R_{\text{Earth}}^2,$$

and solving for V we obtain

$$V_{\text{Earth}} = \sqrt{G\, M_{\text{Sun}} / R_{\text{Earth}}}.$$

This formula is the central result of our study of orbits. As we have emphasized, while in deriving it we thought specifically about the Earth, it is valid for every object orbiting in a circle. We can express this more generally.

> The circular orbit formula: if an object moves in a circular orbit of radius R about an object of mass M, its velocity must be $V = \sqrt{GM/R}$.

If we solve our formula for M, we find an equally important formula that allows us to measure the masses of astronomical objects.

A comment on the units of *G*

Before we use these formulas, we must briefly turn aside to discuss the units of G. Previously we stated that G was given by

$$G = 6.67 \times 10^{-11} \text{ newton meter}^2/\text{kilogram}^2.$$

This expression for G is to be used in finding the force of gravity, as we did previously. But if we were to plug it into either the circular orbit or the mass formula, we would find that the units do not balance. This is because the unit of force, the newton, is itself composed of other units. An alternative set of units for G is

$$G = 6.67 \times 10^{-11} \text{ meter}^3/\text{second}^2\text{ kilogram}.$$

In using both the circular orbit formula and the mass formula, we must use these units for G.

> The mass formula: if an object moves in a circular orbit of radius R with velocity V, the object about which it orbits must have mass $M = RV^2/G$.

Using the orbit formulas

Velocity of a low-Earth orbit

With what velocity must we launch a spacecraft in order to place it into a low-Earth orbit? We can use our circular orbit formula to find out.

Detailed calculation

Step 1. Plug the mass of the Earth and the altitude of the spacecraft into the circular orbit formula.

The circular orbit formula is

$$V = \sqrt{GM/R}.$$

In this formula we put M equal to the mass of the Earth, $M = M_{Earth} = 6.0 \times 10^{24}$ kilograms, and for R we use the radius of the spacecraft's orbit, which we have already found to be 6.858×10^6 meters. We then find the spacecraft's velocity in orbit to be

$$V = \sqrt{(6.67 \times 10^{-11} \text{ meter}^3/\text{second}^2 \text{ kilogram}) \times}$$
$$\sqrt{(6.0 \times 10^{24} \text{ kilograms})/6.858 \times 10^6 \text{meters}),}$$

$$V = 7640 \text{ meters/second.}$$

Step 2. The formula gives its result in MKS units: meters per second. Convert this into miles per hour to get an intuitive feel for it.

One meter/second is 2.2 miles/hour. So we calculate

$$V = 7640 \times 2.2 = 17\,000 \text{ miles/hour.}$$

The logic of the calculation

Step 1. Plug the mass of the Earth and the altitude of the spacecraft into the circular orbit formula.

Step 2. The formula gives its result in MKS units: meters per second. Convert this into miles per hour to get an intuitive feel for it.

The result of the detailed calculation tells us why space flight is so expensive. In order to launch the spacecraft, we must accelerate it to 17 000 miles per hour – a very great velocity – and we must also raise it 480 kilometers (300 miles) above the Earth's surface. Since a typical spacecraft weighs thousands of tons, this is a difficult proposition. Thus it is that only governments and large corporations have the resources to engage in space flight.

. . . .

NOW YOU DO IT

What is the circular orbit velocity of a satellite 1000 kilometers above the Earth's surface? Express your answer in miles per hour.

A satellite is orbiting about the Earth at a velocity of 12 000 miles per hour. We want to calculate the radius of its orbit. (A) Describe the logic of the calculation you will use. (B) Carry out the detailed calculation.

. . . .

In contrast, the force of gravity on smaller bodies is weaker, and it is easier to launch satellites about them. As an extreme example, consider an asteroid 30 kilometers in diameter. The orbit velocity about such an asteroid is a mere 25 miles per hour!

Geosynchronous orbits

Communication satellites are often placed in geosynchronous orbits, in which they always remain above the same point on the Earth's surface. Such satellites seem to hover motionlessly, thousands of miles straight up. Why don't they fall down? Figure 3.12 explains the apparent paradox. In fact these satellites are always falling! They are falling in circular orbits, orbits adjusted so that the satellite completes one orbit per day. Since any point on the Earth's surface also completes one revolution per day (due to the Earth's rotation) the satellite always remains above the same point.

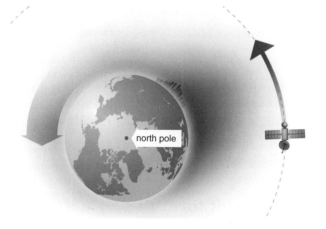

Figure 3.12 **A geosynchronous satellite orbits once a day. Therefore it continually remains above the same point on the Earth.**

Detailed calculation

In our formula $M = R V^2/G$, if we put R equal to the radius of the Earth's orbit and V the Earth's velocity in orbit, the M we find will be the Sun's mass. The radius of the Earth's orbit is just the distance to the Sun, which is $R_{\text{Earth's orbit}}$ $= 1.5 \times 10^{11}$ meters. Our velocity in orbit about the Sun is 30 kilometers per second, or $V_{\text{Earth's orbit}} = 3 \times 10^4$ meters per second. We find then the Sun's mass

$M_{\text{Sun}} = R_{\text{Earth's orbit}} \, (V_{\text{Earth's orbit}})^2/G$

$M_{\text{Sun}} = (1.5 \times 10^{11} \text{ meters})(3 \times 10^4 \text{ meters/second})^2/$
$\quad\quad\quad (6.67 \times 10^{-11} \text{ meter}^3/\text{second}^2 \text{ kilogram})$

$M_{\text{Sun}} = 2 \times 10^{30}$ kilograms.

The mass of the Sun

We can use our mass formula

$$M = RV^2/G$$

to find the masses of astronomical bodies. Let us use it to find the mass of the Sun.

According to our formula, in order to measure the Sun's mass we need to study the orbit of something orbiting about it – a planet. Any planet will do: no matter which one we use, we will get the same answer. As an example, let us use the orbit of the Earth.

The logic of the calculation

Plug the radius of the Earth's orbit, and the Earth's velocity about the Sun, into the formula and find the mass of the Sun.

As you can see from the detailed calculation, the mass of the Sun is 2×10^{30} kilograms.

NOW YOU DO IT

Jupiter's satellite Europa has an orbital radius of 6.7×10^8 meters, and it moves with a velocity of 13 700 meters per second. Use these to find the mass of Jupiter.

Now that you know the mass of Jupiter, use this to predict the orbital velocity of its moon Io (orbital radius 422 000 kilometers).

Detectives on the case

The paradox of weightlessness

All of us have seen movies of astronauts in the Space Station as it circles the Earth. They float weightlessly about. Everything not actually tied down floats loosely in midair. If an astronaut dons a space suit and goes outside the Space Station, she simply hovers beside it. How different from life as we know it here on the Earth!

Why is space so different? Let us create a theory of weightlessness. As before, there is no easy path to finding the right theory. We will simply have to try one hypothesis after another, until eventually we find one that works.

Let us start with the theory that *there is no gravity in space*. According to this idea gravitation simply does not operate above the Earth's atmosphere. Those movies are providing us with a vivid lesson in what it is like to experience the complete absence of gravitation.

But *is* gravitation absent in space? In fact it is not! Visualize the Space Station in its orbit. Notice that the Space Station is not moving in a straight line. It is moving in a circle. As we know from Newton's first law of motion, the principle of inertia demands that if something is not moving in a straight line it must be acted on by a force. That force is gravitation.

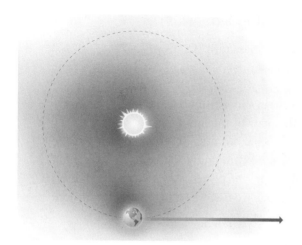

Figure 3.13 **If there were no gravitation in space the Earth would fly away from its orbit about the Sun.**

Furthermore, as we saw in our discussion of orbits, gravitation is what makes the Moon orbit about the Earth – and the Moon is certainly "in space." Similarly, the Earth orbits about the Sun: gravitation from the Sun is reaching across all those enormous reaches of emptiness to draw our planet toward it. So it cannot be true that gravitation is absent in space. If it were, the Earth would detach itself from orbit about the Sun and float off into interplanetary space, where we would all quickly freeze to death (Figure 3.13).

Finally, let us verify that gravitation is not absent in space by calculating *the force of gravity on an astronaut in the Space Station*.

This calculation is precisely analogous to that of the gravitational attraction of the Earth on a 70-kilogram person that we performed above. In that section we found that the force of attraction worked out to 150 pounds. Now let us imagine the same person in the Space Station.

The logic of the calculation

We proceed in exactly the same way as in our calculation of a person's weight, but because the Space Station is high above the Earth we use a larger distance between the astronaut and the Earth. Since R is bigger, we expect to get a smaller force. But we do not expect to get zero for the force!

As you can see from the detailed calculation below, the astronaut, who weighed 150 pounds on Earth, is acted on by a 136-pound force of gravity when in space. She is not free of gravity at all! So how could she possibly float so motionlessly?

Detailed calculation

In our formula for the force on the astronaut from the Earth

$$F = G\, M_{\text{Earth}}\, m_{\text{astronaut}}/R^2$$

everything is the same as in our previous calculation except R, the distance from the astronaut to the center of the Earth. Previously R had been the radius of the Earth. Now it is this plus the height of the Space Station in its orbit about the Earth. This height is 350 kilometers:

$$R = R_{\text{Earth}} + 350 \text{ kilometers}$$
$$= (6378 + 350) \text{ kilometers}$$
$$= 6728 \text{ kilometers}$$
$$= 6.728 \times 10^6 \text{ meters}.$$

We find

$$F = (6.67 \times 10^{-11} \text{ newton meter}^2/\text{kilogram}^2)$$
$$(6 \times 10^{24} \text{ kilograms}) (70 \text{ kilograms})/(6.728 \times 10^6 \text{ meters})^2$$
$$F = 619 \text{ newtons}$$

which, since one newton is 0.22 pounds, is 136 pounds.

NOW YOU DO IT
What is the gravitational force (in pounds) on a 50-kilogram astronaut in a spacecraft 600 kilometers above the Earth?

Our first theory cannot be correct. Let us try a second. Perhaps *the Space Station shields the astronauts inside it from the Earth's gravity.*

Can this theory be true? It cannot. For, after all, think of an astronaut who gets into a space suit and leaves the Space Station. Once this astronaut is outside, she is no longer within the vehicle, and it cannot be shielding her. But she still floats motionlessly beside it!

Our "shielding" theory has failed the first test. Let us test it in a second way. As we saw, the Earth exerts a 136-pound force on the astronaut. If the Space Station were to be counteracting this force with one of its own, the force from the Space Station would also have to equal 136 pounds. Let us use Newton's law of gravitation to see if it does.

Detailed calculation

In our formula for the force on the astronaut from the Space Station,

$$F = G\, M_{station}\, m_{astronaut}/R^2,$$

R is the distance from the astronaut to the Space Station. The Space Station is 20 meters wide and far longer; let us take 20 meters as a rough guess for the distance between the astronaut and its walls. The Space Station's mass is 417 000 kilograms. We have then

$$M_{station} = 4.17 \times 10^5 \text{ kilograms}$$
$$m_{astronaut} = 70 \text{ kilograms}$$
$$R = 20 \text{ meters}$$

$$F = (6.67 \times 10^{-11} \text{ newton meter}^2/\text{kilogram}^2)$$
$$(4.17 \times 10^5 \text{ kilograms})\,(70 \text{ kilograms})/(20 \text{ meters})^2$$
$$F = 4.87 \times 10^{-6} \text{ newtons} = 1.07 \times 10^{-6} \text{ pounds.}$$

The force is *very* much less than 136 pounds!

The logic of the calculation

We proceed as before, using 70 kilograms for m, the mass of the astronaut. For M we use the mass, not of the Earth, but of the Space Station. For R we use the distance between an astronaut and the walls of the Space Station.

NOW YOU DO IT
Suppose an astronaut had a mass of 50 kilograms, and was in a spacecraft as large as the Space Station but half its mass. Calculate the force (in pounds) between the spacecraft and the astronaut.

Do we get 136 pounds? As you can see from the detailed calculation opposite we do not. The force works out to a mere millionth of a pound. We conclude that, while there is indeed a force of gravity from the Space Station on the astronauts within it, the force is too small to make a difference.

Neither of our theories is doing very well. They have failed all their tests. But now let us use these tests in a new way – let us use them as clues. Those tests are telling us that gravitation is very much present in space. Somehow, the phenomenon of weightlessness is not the same thing as "gravity-less-ness." Just the opposite: *weightlessness is an effect of gravitation.*

Return to the astronaut who has donned her space suit and left the Space Station. She is being acted on by gravity from the Earth. Indeed, she can be thought of as a satellite in her own right – a tiny satellite, to be sure, a living one in fact, but a satellite all the same. The astronaut is now in her own separate orbit about the Earth. As a matter of fact, this was also true before she left the Space Station, when she was inside. Every object inside the vehicle is orbiting the Earth.

We have repeatedly emphasized that orbiting is merely a form of falling. So let us think about falling. Can we think of an Earth-bound example of weightlessness? We can! Climb to the top of a high-diving board above a swimming pool. Hold in your hand a video camera, point it at yourself, and jump off. Later on we will watch the video. What we see on the screen is you floating about "weightlessly."

Of course we also see in your video the outside scene rushing by, and the pool rushing up to meet you. These are absent from videos taken in space. So let us refine our Earthly video by imagining a slightly different situation. Imagine that you and your video camera are inside a box – an elevator, perhaps – suspended above a swimming pool (Figure 3.14)

Suddenly, the ropes are cut, and the elevator and its contents plunge downwards into the pool. As it falls, you float about "weightlessly" within it. What we see as we watch your video later is a perfect re-enactment of videos of astronauts inside the Space Station.

Figure 3.14 **A falling elevator mimics the phenomenon of "weightlessness."**

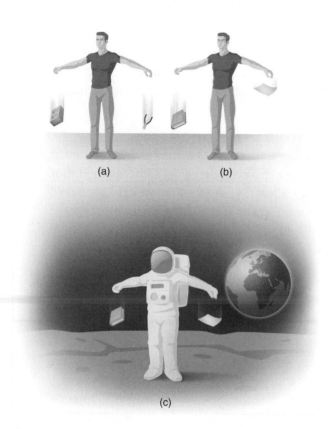

Figure 3.15 **Things fall at the same rate in the absence of air resistance. In (a) we demonstrate this for heavy objects, which can "push aside" the air, in (b) a light object is slowed by air resistance and in (c) in a vacuum a light object falls at the same rate as a heavy one.**

We can also think more mathematically, in terms of our formula for the circular orbit velocity. Notice that this formula does not contain the mass of the orbiting body. This means that the orbit traced out by a small body is the same as that traced out by a large one. So according to our formula for orbit velocity, an astronaut will orbit with the same velocity as the vehicle she rides in. As seen from the Space Station, she floats motionlessly. This is the origin of the phenomenon of "weightlessness."

Is the path of a falling body also independent of its mass? Indeed it is: heavy things fall at the same rate as less heavy ones. You can test this by taking two objects – a cinder block and a pair of pliers, for instance – and dropping them: they will hit the ground at the same moment (Figure 3.15a). (If you were to drop models of the Space Station and an astronaut, the analogy would be perfect.) Of course, you must take care in this experiment not to drop a very light object such as a sheet of paper: air resistance slows the paper's fall (Figure 3.15b). But, were you to perform this second experiment in a vacuum, you would actually see the paper falling at the same rate as everything else (Figure 3.15c).

The importance of quantitative calculations

Notice that we never would have been able to reject our theory had we failed to do a quantitative calculation. The second theory had much to recommend it: it really is true that the Space Station exerts a force of gravity on astronauts within it. Had we not calculated its strength, we would never have realized that this force is so very tiny.

Notice also that our calculation did more than merely help us reject a wrong theory. It actually led us to the right theory. It was this calculation that showed us that gravity is not just present in space – it is nearly as strong as down on the Earth. This led us to the notion that weightlessness has nothing to do with the absence of gravity.

It is never enough in science to have a good idea. In order to be sure that your idea is correct, you need to test it using detailed calculations.

Understanding Kepler's laws

We have already studied properties of orbits in the previous chapter. It can be shown that everything we found in that study follows from Newton's laws. But the proof is too complicated to give here. Rather we will approach an understanding intuitively.

Kepler's first law states that the planets of the Solar System follow elliptical orbits about the Sun. We have seen that an ellipse of zero eccentricity is a circle. So far we have been concentrating on these particularly simple paths. Let us now pass on to elliptical orbits.

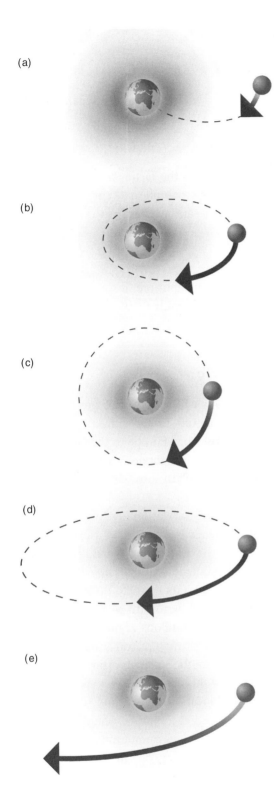

(a)

(b)

(c)

(d)

(e)

Figure 3.16 **Elliptical orbits are produced by launching a satellite at other than circular orbit velocity. The velocity is very small in (a), in (b) it is somewhat larger and in (c) it is just circular orbit velocity. In (d) it is greater than circular orbit velocity, and in (e) it is greater than escape velocity.**

Imagine that we lift a satellite high above the Earth, and propel it sideways with some velocity. It will go into an orbit. If we choose the velocity to be the circular orbit velocity, we will produce a circular path. But if we choose some other velocity, we will get an elliptical orbit. Figure 3.16 illustrates what happens.

If the satellite is launched slowly, it merely falls back to Earth along a curving path (Figure 3.16a). This path is, in fact, a segment of an ellipse: it terminates because it bumps into the Earth. If we now launch our satellite somewhat more rapidly, but still below the circular orbit velocity, it goes into an elliptical orbit (Figure 3.16b). Notice that in this orbit, the satellite is at its greatest distance above the Earth when it is launched. Increasing the launch velocity still more, we reach a perfectly circular orbit (Figure 3.16c). At yet higher velocities we again get an elliptical orbit (Figure 3.16d) but with the difference that now the satellite is at its *least* distance above the Earth at its launch point. And finally, if we launch at the *escape velocity* the satellite flies away from the Earth, never to return (Figure 3.16e). (It can be shown that escape velocity is just $\sqrt{2}$ times circular orbit velocity.)

Kepler's second law states that a line drawn from the Sun to a planet sweeps out equal areas in equal times. As we saw in the previous chapter, this implies that the closer a planet is to the Sun the faster it orbits. We can understand this by returning yet again to our insight that orbiting is just a special kind of falling.

In Figure 3.17 we show a planet in an elliptical orbit. In that figure we have indicated the direction in which gravity from the Sun pulls the planet: this direction is the equivalent of our "down." The planet speeds up as it approaches the Sun because gravity from the Sun is accelerating it. The planet is quite literally falling toward the Sun: as it falls, it picks up speed.

Conversely, once it has passed its point of closest approach, the orbiting planet starts to move away from the Sun. Its motion is now analogous to that of a golf ball thrown upwards. The Sun's gravity now acts to slow it.

Kepler's third law states that the square of the period of a planet in its orbit is proportional to the cube of its semimajor axis. Although we cannot prove this for all orbits, we can do so for a particular case: the circular orbit.

To do so we use our formula for orbit velocity. The semimajor axis of a circular orbit is just the orbit radius R. The period P is the time required for the body to complete one orbit. This time is the distance around the orbit divided by the orbital velocity. For this velocity we use our formula for the velocity in a circular orbit. For the distance around the orbit we use the orbit circumference $2\pi R$.

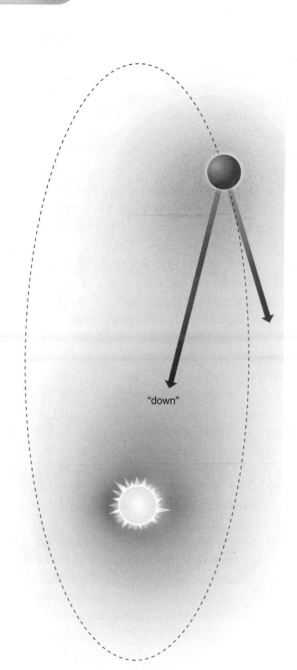

"down"

Figure 3.17 **Falling toward the Sun in an elliptical orbit. The orbital velocity increases as the body nears the Sun.**

Period = circumference/velocity

$$P = 2\pi R/\sqrt{GM/R} = (2\pi R)\sqrt{R/GM}.$$

In the spirit of Kepler's third law, let us find the square of the period:

$$P^2 = (4\pi^2 R^2)(R/GM) = (4\pi^2/GM)R^3.$$

Notice that this formula contains the factor $4\pi^2/GM$, where M is the mass of the Sun. This complicated combination is just a constant; let us call the constant K. Then our formula reads

$$P^2 = KR^3.$$

But this is Kepler's third law! Furthermore, if we were to plug the values of G and M into our formula for K we would find it had the value we found in Chapter 2.

Interplanetary navigation

Centuries from now, historians will write of two great ages of exploration. The first commenced in the fifteenth century, when ocean-going ships set sail from Europe to the far corners of the world. The second is going on right now. Astronauts have walked upon the Moon and regularly orbit the Earth. Space probes ply the Solar System.

Navigation in space is not like navigation here on the Earth. It poses its own special problems. If you were in a canoe on a lake and wished to go home, the best strategy would be to aim your canoe at the dock and paddle in a straight line. But if you wished to send a space probe to Mars it would do no good to send the probe off on a path heading toward Mars. If you did, the probe would miss its mark by millions of miles.

It would miss for two reasons. The first reason is that Mars is a moving target. It is traveling about the Sun in its orbit, and by the time your space probe reached this orbit the planet would have moved away from its original location. The second reason why this strategy would not work is that the probe would not continue traveling in a straight line. Things cannot travel in straight lines in space. Things travel in orbits. *The task of interplanetary navigation is the task of sending space probes on orbits carefully chosen so that they reach their intended destinations.*

There are two requirements for such orbits. On the one hand, the probe's orbit must smoothly join up with that of Mars. If not, the probe would be traveling at an enormous velocity relative to Mars when it got there, and it would simply sail on past the planet and off into space. This means that the elliptical orbit of the probe must be tangent to that of Mars – and, for a similar reason, to the Earth.

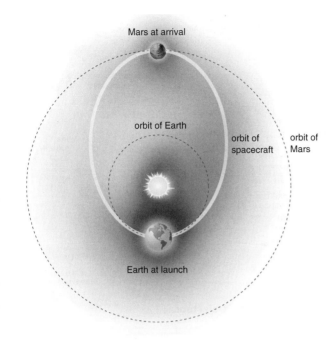

Figure 3.18 **An orbit to Mars is an ellipse tangent to the Earth's orbit at launch, tangent to Mars' orbit at arrival, and timed so that the space probe and Mars reach the same point at the same time.**

The second requirement is that, when the probe arrives at Mars' orbit, Mars must be there. Surely it would do no good were Mars to lie off on the far side of the Sun when the probe arrived! This requirement gives rise to the *launch window* – the brief interval of time during which the probe must be launched. If this launch window is missed, one must wait for many months for the planets to return to their favorable configuration before launching again.

An orbit to Mars satisfying these requirements is illustrated in Figure 3.18.

Navigation to the outer Solar System: gravity-assist orbits

Recall our insight that orbiting is a special kind of falling. Since gravity pulls spacecraft toward the Sun, the direction to the Sun is "down." This means that sending a space probe to an inner planet such as Mercury or Venus is like dropping a stone. But sending a space probe to an outer planet such as Saturn is like throwing a stone upwards. And it is hard to throw a stone upwards; navigation to the outer reaches of the Solar System poses special problems all its own.

Saturn lies 9.5 astronomical units from the Sun. At its closest approach to the Earth, the planet is 790 million miles from us – straight up, in terms of our analogy. In Chapter 8 we will discuss the Cassini mission to this planet; as we mention there, it is the best-instrumented spacecraft ever sent to another planet, weighing more than 6 tons at launch. In terms of our analogy of throwing a stone upwards, sending Cassini to Saturn is like throwing a very heavy boulder very far up into the air. Indeed, we do not possess rocket engines sufficiently powerful to send Cassini to Saturn.

How, then, did the engineers of NASA do it? They employed a wonderful device known as the "gravity-assist orbit."

To understand the gravity-assist orbit, think of hitting a baseball with a bat (Figure 3.19). When a batter hits a baseball, the ball bounces off the bat *with an increased speed*. Some of the energy of motion of the bat has been transferred to the ball.

In the same way, interplanetary navigators can direct their spacecraft toward an oncoming planet (Figure 3.20). The spacecraft does not actually strike the planet: rather it loops around it and flies off in a new direction. The spacecraft ends up with an increased velocity. Just as some of the motion of the bat was transferred to the baseball, so too has some of the motion of the planet been transferred to the spacecraft.

The NASA engineers constructed an extraordinary orbit for Cassini which sent it looping around Venus, then around

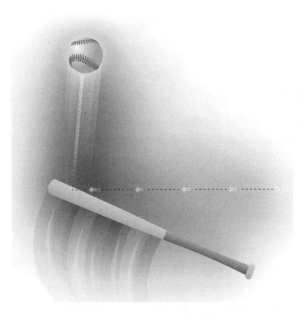

Figure 3.19 **Speeding up a baseball with a bat.**

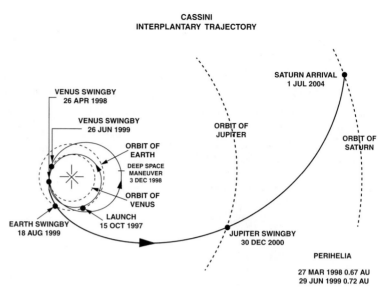

Figure 3.21 **Cassini interplanetary trajectory.**

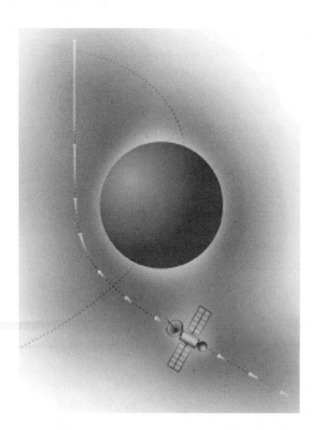

Figure 3.20 **Speeding up a space probe with a planet.**

Venus a second time, then back to the Earth, and finally to Jupiter before heading off to Saturn. In each of these flybys, the spacecraft was sped up by the motion of the planet, and between each of them it was orbiting the Sun in an elliptical orbit designed to carry it to its next close encounter. To appreciate this feat, think of tossing a baseball to a batter, who hits it off with increased speed to a second batter, who hits it with yet greater speed to a third, who hits it to a fourth, who then hits the ball so that it flies farther than any baseball has ever been struck – and such that the baseball lands exactly on the roof of an automobile which is driving by outside the park. This astonishing orbit is illustrated in Figure 3.21. The entire journey of Cassini to Saturn took nearly seven years: it arrived in the summer of 2004 and, at the time of this writing, it is still conducting observations.

THE NATURE OF SCIENCE

CERTAINTY AND UNCERTAINTY IN SCIENCE

Most things in life are uncertain. We are seldom exactly sure of things. Sometimes our uncertainty is over relatively trivial matters – what was the best movie I saw last year; where is the best place to buy a shirt? But often we are unsure about deeply important problems.

Science, on the other hand, seems different. Science reaches conclusions of which we can be certain. The Earth is most definitely not flat, and it most definitely is not the center of the Universe. The answer to a homework problem is either right or wrong: while you might get partial credit for a good try, there is very little room for uncertainty in the matter.

For this reason, science progresses onward. Scientists solve old problems and move on to new ones. They make new discoveries and enlarge their understanding. How different all this is from most areas of life!

Detailed calculation

Step 2. Using a ruler, measure the diameter of a dime. Find the diameter and distance of Mars and plug them into the proportion.

Mars' diameter is 6794 kilometers, or 6.8×10^6 meters. As you can see from Figure 3.18, the distance to Mars at launch is 2.52 AU, or 3.8×10^{11} meters. The diameter of a dime is 3/4 inch, or 0.019 meters. So our proportion reads:

distance to dime/0.019 meters = 3.8×10^{11} meters/6.8×10^6 meters.

Step 3. Solve to find the distance to the dime.
   We find

   distance to dime = 1000 meters.

Step 4. To get an intuitive feeling for the answer, convert it to miles.
   One meter is 6.2×10^{-4} miles, so this is $(1000)(6.2 \times 10^{-4})$ miles = 0.6 miles.

But while science brings certainty, it does so only in the long run. In the short run scientists are no more certain of things than anyone else. Practising researchers, working at the cutting edge of knowledge, are incessantly plagued by uncertainty.

Let us illustrate this by thinking about a very specific example: the problem of sending a mission to Mars. As we discussed above, to do this we must launch our space probe into an elliptical orbit at a tangent to that of the Earth at launch, at a tangent to that of Mars at arrival, and timed so that the space probe and Mars reach the same point at the same time. Surely this is not so very difficult! Surely we must be certain by now of how to do this! Let us see why, even now, it is not so easy to do these things.

To begin with, it is important to understand the magnitude of the problem facing us. Mars is a pretty small target, considering its distance. Let us get a feeling for just how small a target it is. Suppose we think in terms of the analogy of shooting a bullet at a dime. How far away must the dime be to present an equally small bull's eye?

THE LOGIC OF THE CALCULATION

Step 1. Set up a proportion distance to dime/diameter of dime = distance to Mars/diameter of Mars.
Step 2. Using a ruler, measure the diameter of a dime. Find the diameter and distance of Mars and plug them into the proportion.
Step 3. Solve to find the distance to the dime.
Step 4. To get an intuitive feeling for the answer, convert it to miles.

. . . .

NOW YOU DO IT
Suppose we wish to send a spacecraft to Jupiter, and we consider an analogy in which this is like shooting a bullet at a nickel. How distant must the nickel be for this to be a good analogy? (Jupiter's radius is 71 000 kilometers; the radius of its orbit is 5.2 astronomical units.)

. . . .

As we see from the detailed calculation, Mars is a very small target indeed – like a dime more than half a mile off. If we miss a bull's eye in target practice, the consequences are not so bad. But if a space probe misses its target, the consequences are severe. It will either crash into the planet or fly past it and continue on forever through interplanetary space. Furthermore, mistakes are expensive: the Mars Observer mission, for instance, cost just short of a billion dollars.

So the stakes are high and we do not want to make any mistakes in our orbit. Let us ask why it is hard not to make mistakes.

Detailed calculation

Step 1. In order to find the actual and intended distances traveled, multiply the duration of the trip by the actual and intended velocities.

The trip to Mars takes 8.5 months, or 2.2×10^7 seconds. At its correct speed the space probe would have traveled a distance

$$\text{Correct distance} = (33 \text{ kilometers/second})(2.2 \times 10^7 \text{ seconds})$$
$$= 7.26 \times 10^8 \text{ kilometers.}$$

But at its actual speed the distance traveled is
$$\text{Actual distance} = (33.001 \text{ kilometers/second}) \times$$
$$(2.2 \times 10^7 \text{ seconds})$$
$$= 7.2602 \times 10^8 \text{ kilometers.}$$

Step 2. Subtract the results. This will give the error in the distance traveled.

The actual distance traveled is too great, by an amount of
$$\text{error in distance} = (7.2602 \times 10^8) - (7.26 \times 10^8)$$
$$\text{kilometers}$$
$$= 0.0002 \times 10^8 \text{ kilometers}$$
$$= 2 \times 10^4 \text{ kilometers.}$$

Step 3. To understand the significance of this error, compare it to the size of the "bull's eye" – i.e. to the diameter of Mars.

Mars' diameter is (Appendix III) 6794 kilometers. Let us calculate the ratio.

Error in distance/diameter of Mars = 2×10^4 kilometers/ 6794 kilometers = 2.9.

One problem has to do with the rocket engine that powers our space probe. It must send the probe off on its journey with a velocity of 33 kilometers per second. But suppose the engine is just the littlest bit off. Suppose it gives the probe a velocity of 33.001 kilometers per second. That looks like an utterly inconsequential error. But it is not inconsequential: it is fatal.

THE LOGIC OF THE CALCULATION

We want to find out how far the space probe *actually* traveled, and compare this to how far it *should have* traveled during its journey.

Step 1. In order to find the actual and intended distances traveled, multiply the duration of the trip by the actual and intended velocities.

Step 2. Subtract the results. This will give the error in the distance traveled.

Step 3. To understand the significance of this error, compare it to the size of the "bull's eye" – i.e. to the diameter of Mars.

NOW YOU DO IT
Suppose the spacecraft's velocity had been 33.005 kilometers per second. By how much would it have missed Mars? Compare your answer to the diameter of Mars.

The error in distance that we find from the detailed calculation is *nearly three times* the diameter of Mars! The probe would have entirely missed its target. That tiny error on the part of the rocket engine, apparently so trivial, would have doomed the mission.

But the problems do not only have to do with our equipment. They also have to do with our uncertainty in the orbit the space probe will follow, even if it is launched with the right velocity. The probe's orbit is influenced by gravity from the Sun – and we do not know the Sun's gravity with perfect accuracy. This gravity depends on the Sun's mass, and if you were to look up this mass you would not find it given to an infinite number of decimal places!

What can be done? We need to know the Sun's mass to great accuracy. As we have seen in this chapter, to measure the Sun's mass we need to observe the orbits of planets to great accuracy. But the accuracy of our observations is limited.

Even if we did know the Sun's gravity to perfect accuracy, we would still not know the space probe's orbit to perfect accuracy. This is because it is *also* influenced by gravity from the other planets. All nine planets of the Solar System exert forces on the space probe – forces that change as the planets orbit. And, of course, we do not know these forces with perfect accuracy. And lastly, we do not exactly know the orbit of Mars either.

Previously we likened the task of sending a spacecraft to Mars to that of shooting a distant dime. Now we realize that it is like shooting a dime using a slightly erratic gun – and that the bullet, buffeted by winds, follows a slightly erratic path. Furthermore, the dime itself is moving along a slightly erratic path.

All these "slightlys" add up to a hard problem. The laws of nature may be perfect, but their application in the real world is not. Perfection and certainty are no more possible for scientists than for anyone else. The engineers who design spacecraft orbits must learn how to live with uncertainty. They must learn how to make their orbits not perfect, but good enough.

In 1997 this uncertainty culminated in a lawsuit before a federal court seeking to halt the launch of a space probe. At issue was the Cassini mission to Saturn, which was powered by nuclear energy and contained 72 pounds of radioactive plutonium. (Because sunlight is so dim in the outer Solar System, far from the Sun, solar panels could not be used to power the spacecraft.) Antinuclear activists were concerned that, in the event of a disaster, this reactor would pose a terrible health risk.

Two critical stages in the mission were of concern. The first was the launch itself. A rocket explosion during liftoff, similar to the Challenger explosion, would scatter the highly radioactive plutonium across Florida. The second was the near-Earth flyby, during which the spacecraft was to pass 312 miles above the Earth at more than 40 000 miles per hour. The slightest error in its orbit might send the spacecraft with its deadly cargo crashing onto the Earth.

Debate raged. Opponents of the launch challenged NASA's estimate that a disaster was highly unlikely. Many scientists sided with NASA, and argued that the chances of a catastrophe were far too low to justify canceling so important a mission. But not all scientists agreed. Other organizations also weighed in. Dr. Helen Caldicott, founder of Physicians for Social Responsibility, noted that plutonium was one of the most toxic substances known: one millionth of a gram was a carcinogenic dose; and a single pound, if uniformly spread about the world, could hypothetically induce lung cancer in every person on Earth.

What would you have done? Had you the authority, would you have launched Cassini? How about the mission to Mars, which cost a good deal of money but posed no health risk? One approach might be to do nothing until you are sure. But had people never taken any risks in the past, we would not be where we are today. Risk is part of life.

Ultimately the lawsuit was thrown out and Cassini was launched. The rocket did not blow up, and in 1999 it safely swung by the Earth and continued on its way to Saturn. But Cassini was but one episode, and the space program is but one part of the scientific enterprise. The problem of uncertainty in science will never go away. Every citizen has the responsibility of working to establish how we should deal with it.

Summary

Universal gravitation

- Gravitation is the most important factor that determines the structure of the astronomical Universe. There is not a single aspect of astronomy for which gravity fails to play a crucial role.
- It is universal.
- It is a force of attraction between every pair of bodies.
- Its strength is proportional to the product of their masses divided by the square of their distance apart:

  gravitational force $F = GMm/R^2$.

- The gravitational force between everyday objects is small.
- An object's weight is the gravitational force between it and the Earth.

Newton's first law of motion: inertia

- Every body continues what it had been doing unless compelled to change its state of motion by a force.
- In everyday experience we do not witness the effects of this law, since friction and air resistance are forces that act to slow bodies.

Newton's second law: force

- If a body is acted on by a force, it accelerates at a rate proportional to the force and inversely proportional to its mass:

  $F = MA$ so that $A = F/M$.

- Acceleration is rate of change of velocity.
- Velocity has a direction as well as a magnitude, so a change in direction with no change in the number of miler per hour counts as an acceleration.

Newton's third law: reaction

- To every action there is an equal and opposite reaction.
- This is the principle that allows rocket engines to work.

Orbital motion

- Orbital motion is the same thing as falling.
- It is falling in an arcing path.
- An object moving in a circle is always accelerating (centripetal acceleration).
- The magnitude of centripetal acceleration is the velocity squared divided by the radius of the circle in which the body moves:

  centripetal acceleration $= V^2/R$.

The mathematics of orbits

- If an object moves in a circular orbit of radius R about an object of mass M, its velocity must be:

  the circular orbit formula: $V = \sqrt{GM/R}$.

- The circular orbit formula allows us to find the mass of a body by studying the orbit of something about it.

  The mass formula: if an object moves in a circular orbit of radius R with velocity V, the object about which it orbits must have mass
  $M = RV^2/G$.

- This allows us to measure the mass of, for example, the Sun.

Using the orbit formulas

- Velocity of a low-Earth orbit is 17 000 miles per hour.
- A geosynchronous orbit is one for which the orbital period equals one day.
- An object in such an orbit appears to hover motionlessly above the same point on the Earth.
- Communication satellites are placed in geosynchronous orbits.

Understanding Kepler's laws

- These laws can be derived from Newton's theory of orbits.
- A circular orbit arises if the orbiting object is given a certain critical velocity. If it is given a different velocity it follows an elliptical orbit.
- Kepler's second law implies that, as an object nears the Sun, it speeds up. This arises because orbiting is a form of falling: the object accelerates as it falls.
- It is easy to prove Kepler's third law from our orbit formula, which applies to circular orbits.
- We can also prove Kepler's third law for elliptical orbits, but it is harder.

Interplanetary navigation

- When we send a spacecraft to a distant planet, it moves in an orbit about the Sun.
- This orbit is designed to intersect that of the target planet.
- A gravity-assist orbit sends a spacecraft close by one planet in order to send it on to another.

DETECTIVES ON THE CASE

Weightlessness

We considered a variety of theories of weightlessness. The first, that there is no gravity in space, was shown to be untenable. The second, that spacecraft shield the astronauts inside from gravity, was also shown to be untenable. We finally realized that weightlessness is an effect of gravitation: it arises because everything falls at the same rate.

THE NATURE OF SCIENCE

Uncertainty in science

Scientists are continually dealing with uncertainty. As an example, we considered the task of sending a spacecraft to a planet.

• The "target size" is exceedingly small.

• A tiny error in the velocity of our spacecraft would cause it to miss its target.
• Similarly, a tiny error in the spacecraft's orbit would cause it to miss.
• This error can be caused by errors in our knowledge of the masses and orbits of other bodies in the Solar System.
• Occasionally this uncertainty can have policy consequences, for an error in a spacecraft's orbit as it approaches the Earth could lead to a devastating accident.

Problems

(1) Find the force of gravity in pounds between two 100-kilogram people 2 meters apart.

(2) (A) Find the weight of a 100-kilogram person on the Earth. (B) Now suppose this person traveled to a planet with the same mass as the Earth, but a larger diameter. Would the person's *weight* be the same, more or less than on the Earth? Would the person's *mass* be the same, more or less than on the Earth? Explain your answers. (C) Now suppose this person traveled to a planet with the same diameter as the Earth, but a smaller mass. Would the person's *weight* be the same, more or less than on the Earth? Would the person's *mass* be the same, more or less than on the Earth? Explain your answers.

(3) (A) Find the weight of a 100-kilogram person on Mars. (B) Find the weight of the Mars Rover Spirit as it stands on Mars. (Mars' radius is 3397 kilometers; its mass is 0.11 that of the Earth. The mass of Spirit is 180 kilograms.)

(4) Suppose you stepped on the bathroom scale, and it told you that you weighed 170 pounds. (A) Describe the logic of the calculation you would use to determine from this your mass. (B) Carry through the detailed calculation and find your mass.

(5) Suppose there were a planet with the mass of the Earth, but a different radius. Suppose that you went there and found that you weighed 60 pounds.
(A) Describe the logic of the calculation you would use to determine from this the planet's radius.
(B) Carry through the detailed calculation and find this radius.

(6) Suppose there were a force between two objects – not gravity! – which obeyed the law $F = MmR$. Explain why this force would be small for objects close to one another, but large if they were far apart.

(7) Suppose there were a force between two objects – also not gravity! – which obeyed the law $F = Mm/R^4$. Explain why this force would be exceedingly large for objects close to one another, and exceedingly small if they were far apart.

(8) Suppose Newton's constant of gravitation G were big. Explain why you would feel a physical force of attraction to other people.

(9) Two cars are waiting at a stop light. It turns green and they drive off. One car reaches 25 miles per hour in 2 seconds, while the other reaches 25 miles per hour in 3 seconds. Which car has the greater acceleration?

(10) Two cars are waiting at a stop light. It turns green and they drive off. One car reaches 25 miles per hour in 2 seconds, while the other reaches 50 miles per hour in 4 seconds. Which car has the greater acceleration?

(11) A car is driving down the road at 50 miles per hour. The driver does a U-turn, and drives back from where she came at 50 miles per hour. Did the car change its speed? Did it change its velocity?

(12) Two skaters are standing on the ice in a skating rink. One pushes on the other. Does only one start moving, or do both start moving?

(13) Suppose we were to receive a message from an extraterrestrial civilization, and suppose that the message told us that orbital velocity around this civilization's home planet was very low. Explain why we know that this civilization's home planet must be of very low mass.

(14) (A) Find the velocity of a low orbit about our Moon (mass 7.4×10^{22} kilograms; radius 1740 kilometers). (B) Suppose one wished to launch a satellite into a high orbit about the Moon – namely an orbit lying 2000 kilometers above its surface. Would the required velocity be the same as, greater than or less than the

velocity you calculated in part (A)? Explain your answer.

(15) In "The nature of science. The importance of skepticism" in Chapter 1 we emphasized that skepticism is the most cherished scientific virtue, and that every scientific theory must be tested over and over again. Let us therefore test Newton's theory. Your task is to do this by using Newton's circular orbit formula to *predict* the orbital velocity of Mars about the Sun, and then *compare your result* with Mars' actual velocity. (A) Describe the logic of the calculation you would use to find the predicted value for the orbital velocity. (B) Carry out the detailed calculation. (C) Does the predicted velocity agree with the measured velocity? (The radius of Mars' orbit is 1.52 AU and its measured orbital velocity is 24.1 kilometers per second.)

(16) In "The nature of science. The importance of skepticism" in Chapter 1 we emphasized that skepticism is the most cherished scientific virtue, and that every scientific theory must be tested over and over again. Let us therefore conduct a second test of Newton's theory. One prediction of his theory is that, when we find the mass of the Sun by using his mass formula, it makes no difference which planet's orbit we use. In the text we used the orbit of the Earth and found the Sun's mass. Your task is *to select any other planet and use its orbit to find the Sun's mass*. Do you get the same answer? (Orbital data for the planets are given in Appendix III.)

(17) Can you think of a third test of Newton's theory? (A) Describe it in detail and (B) carry out the calculation.

(18) What is the gravitational force (in pounds) from the Earth on an 80-kilogram astronaut in a spacecraft 900 kilometers above the Earth?

(19) What is the gravitational force (in pounds) from the spacecraft on that 80-kilogram astronaut 10 meters from a 1-million-kilogram spacecraft?

(20) At the time of this writing, the New Horizons spacecraft is on its way to Pluto (radius 1200 kilometers, orbital radius 39 astronomical units). Consider an analogy in which this is like shooting a bullet at a dime. How distant must the dime be for this to be a good analogy?

(21) Suppose we wish to send a spacecraft to Mars, and by mistake we sent it off with a velocity of 29.003 kilometers per second. By how much would it miss its target? Compare your answer to the diameter of Mars.

You must decide

(1) You are the judge who must decide whether to allow the launch of the Cassini mission to Saturn. You have heard testimony from both sides of the debate. Write a detailed memorandum in which you describe
   • the arguments each side presented,
   • the rebuttal to each side's argument that the opposing side gave,
   • the decision that you reached,
   and (this is the most important of all!)
   • your reasons for your decision.

(2) You are a program officer at the National Aeronautics and Space Administration (NASA) charged with supporting NASA's ability to conduct space science missions. NASA is considering two programs.
   • One would develop technology for a new generation of solar panels, which would be able to capture solar energy with greater efficiency than is currently available. As a consequence, missions to the outer Solar System would no longer need to be powered by nuclear reactors.
   • The other would send a small satellite into orbit about the Sun, and concurrently would develop technology to track the orbit of this satellite (and any other as well) with unprecedented accuracy. This would enable scientists to determine more precisely the masses of the Sun and of all the planets of the Solar System. This, in turn, would allow NASA to navigate more accurately its missions throughout the Solar System.
   You must decide which option to pursue. Write a detailed memorandum in which you describe
   • the arguments each side presented,
   • the rebuttal to each side's argument that the opposing side gave,
   • the decision that you reached,
   and (this is the most important of all!)
   • your reasons for your decision.

4

Light

Astronomy faces a difficulty not shared by other sciences: we cannot get our hands on what we study. The geologist can pick up a rock and examine it; the biologist can dissect an animal. But among all the objects in the Universe, only the Moon, Venus, Mars, the satellite Titan of Saturn and a few comets and asteroids have actually been touched by spacecraft. The rest of the cosmos we are forced to study from afar.

Luckily, the Universe is continually broadcasting information to us, coded into light – or, more generally, electromagnetic radiation. It is by studying this light that we gain information about the cosmos. Indeed, until the advent of the space program and its robotic landers, this was the *only* means we had of gathering information about the cosmos.

In this chapter we will begin our study of light by discussing the concept of brightness, and then move on to the wave nature of light and the Doppler effect. Next we will consider the spectrum of light, both the continuous spectrum and spectral lines. It is truly amazing how much we can learn simply by gazing carefully at things.

Brightness: the inverse square law

We begin our study of light by discussing the concept of brightness. It is obvious that some lights are bright and some are dim. But a little thought shows that there are two kinds of brightness.

Two kinds of brightness

To appreciate this fact, consider the following conundrum. A 100-watt light bulb is brighter than a 40-watt light bulb. But suppose the 100-watt bulb is miles away, and the 40-watt bulb is pressed up close to your face. Then which is brighter?

When we ask this question, we are thinking of "brightness" in two different ways: an *absolute brightness*, which measures how bright a source of light really is; and an *apparent brightness*, which measures how bright you see it to be.

- The *absolute brightness* of a source of light is the quantity of light it *emits*. In the case of light bulbs, this is measured by the number of watts emitted by the bulb. Astronomers customarily refer to absolute brightness as *luminosity, L.* The luminosity of a 100-watt light bulb is simply $L = 100$ watts. For comparison, the luminosity of the Sun is a gigantic $L = 4 \times 10^{26}$ watts.

- The *apparent brightness* of that source is the quantity of light *that reaches us.* Apparent brightness is measured by the number of watts that hits a detector of unit surface area. For instance, if the detector is a telescope, and if the lens of the telescope happens to have a surface area of one square meter, then the apparent brightness would measure how much light enters the telescope. Astronomers customarily refer to apparent brightness as *flux, f.*

Clearly the 100-watt bulb remains a 100-watt bulb even when it is very far away: *luminosity (absolute brightness) does not depend on distance.* But equally clearly, the apparent brightness of the bulb grows fainter as it is carried farther away: *apparent brightness does depend on distance.* So the answer to our conundrum is that the absolute brightness of the faraway 100-watt bulb still exceeds that of the nearby 40-watt bulb, but the apparent brightness is far less.

Light carries energy: the brightness of a light beam is a measure of how much energy it carries. Specifically, *the absolute brightness of a source of light is the amount of light energy it emits per second.* The watt, a familiar unit of luminosity, is then a measure of energy emitted per second. Similarly, *the apparent brightness is the amount of light energy entering a detector of unit surface area per second.*

NOW YOU DO IT

(A) Which has the greater *luminosity*: a searchlight 100 miles away, or a candle one inch away? (B) Which has the greater *flux*: a searchlight 100 miles away, or a candle one inch away? (C) Which has the greater *apparent brightness*: a searchlight 100 miles away, or a candle one inch away? (D) Which has the greater *absolute brightness*: a searchlight 100 miles away, or a candle one inch away?

The inverse square law

The formula describing how apparent brightness depends on distance is the inverse square law of the propagation of light.

To understand this law, consider Figure 4.1, which illustrates the light flying away from a source in all directions. The farther the light travels, the more spread out it has become. We can think of the light as being diluted by distance. Figure 4.1 also illustrates a telescope, which is gathering the light. If the telescope has a gathering area that is one meter square, then the flux is the quantity of light the telescope "catches." As you can see, if the telescope is close to the source, it "catches" a great deal of the light emitted by the source. But if the telescope is far away, it "catches" only a small amount of the emitted light. This is why a light source appears to grow dimmer as we move farther and farther away from it.

Recall that the apparent brightness measures the amount of light hitting a detector of unit surface area a distance R from a source. In Figure 4.2 we illustrate an imaginary sphere surrounding the light source. The *total* amount of light

Figure 4.1 **The inverse square law. Light flying away from a source in all directions becomes diluted by distance. The farther a telescope is from the light source, the less light the telescope catches.**

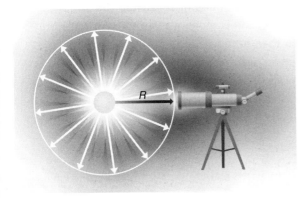

Figure 4.2 **Understanding the inverse square law.**

passing through this sphere is just the total amount of light the source emits, which is its absolute brightness or luminosity L. We need to find how much of this actually reaches the detector. The fraction that reaches the detector is just the fraction of that imaginary sphere occupied by the detector. This fraction is the area of the detector divided by the area of the sphere. The area of the detector is 1, and the area of the imaginary sphere is 4π times its radius squared, or $4\pi R^2$. Thus the fraction of light that reaches the detector is $1/4\pi R^2$. Therefore the apparent brightness of a source equals its absolute brightness divided by $4\pi R^2$. In symbols,

$$f = L/4\pi R^2$$

where

- f is the flux, i.e. the apparent brightness of the source,
- L is the luminosity, i.e. the absolute brightness of the source,
- R is the distance from the source to the detector.

This is the inverse square law of the propagation of light.

Using the inverse square law

Let us give some examples of the use of this law.

(1) Suppose you are a certain distance from a light bulb – and then move twice as far away. Your distance R has doubled. According to the inverse square law, f is divided by 4: moving twice as far away makes a light 4 times fainter.

(2) Similarly, if you move till you are half as close to that light bulb, R is divided by 2, and f is multiplied by 4: halving the distance to a light source makes it 4 times brighter.

(3) Suppose you are one meter away – this is roughly arm's length – from a 100-watt light bulb. What apparent brightness f do you receive from it? Using the inverse square law,

$$f = (100 \text{ watts})/4\pi (1 \text{ meter})^2$$

$$f = 8 \text{ watts/meter}^2.$$

(4) Now let the source of light be the Sun. And let us turn the calculation around: what we can measure is the apparent brightness we receive from the Sun, and the Sun's distance – so let us use the inverse square law to *find* the Sun's luminosity!

The logic of the calculation

Step 1. Solve the inverse square law for the luminosity.
Step 2. Plug in the measured flux from the Sun and its distance from us.

Detailed calculation

Step 1. Solve the inverse square law for the luminosity.
We find

$$L = 4\pi R^2 f.$$

Step 2. Plug in the measured flux from the Sun and its distance from us.

Measurements reveal that the apparent brightness of the Sun is

$$f_{Sun} = 1.34 \times 10^3 \text{ watts/meter}^2$$

and the Sun's distance is the astronomical unit, or 1.5×10^{11} meters. So we find

$$L_{Sun} = 4\pi \, (1.5 \times 10^{11} \text{ meters})^2 \, (1.34 \times 10^3 \text{ watts/meter}^2)$$
$$L_{Sun} = 3.8 \times 10^{26} \text{ watts.}$$

As you can see from the detailed calculation, the answer we get is $L_{Sun} = 3.8 \times 10^{26}$ watts. This is, of course, an enormous brightness: the Sun produces energy at a prodigious rate. We will return to how the Sun manages to do this in Chapter 14.

NOW YOU DO IT

You are looking at two 100-watt light bulbs. One seems 25 times brighter than the other. How much closer to you is the brighter one than the dimmer one?

You are looking at two light bulbs, both the same distance from you. One seems 25 times brighter than the other. How much more luminous is the brighter one than the dimmer one?

Saturn is 9.5 times farther from the Sun than we are. How much dimmer is sunlight on Saturn than here?

Suppose we receive the same flux from a 100-watt light bulb as from the Sun. How far away is that light bulb?

Electromagnetic waves

So far we have been discussing the energy carried by light without mentioning what light is. In fact *light is an electromagnetic wave.*

Wave motion

Before discussing electromagnetic waves, let us consider something more familiar: water waves.

Recall what happens when you wiggle your hand to and fro in a bathtub. As your hand moves, it creates water waves that move outward away from it. Recall in particular the pattern these waves make where they meet the bathtub wall. This pattern is studied in Figure 4.3, where we graph the water level along the wall versus position.

Figure 4.3 illustrates the configuration at a certain definite moment: we can think of it as a snapshot of the wave. We could also study what the wave does as time passes. Let us suppose that we have glued to our bathtub wall a ruler running vertically up and down. We will use it to study how the height of the water level changes as a wave passes by. We measure the height of the water against the ruler – first at one instant, then a moment later, then still later, and so forth. Then we graph the results. Figure 4.4 shows the result.

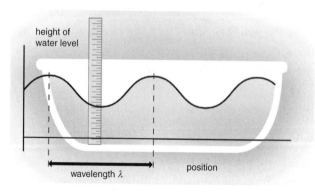

Figure 4.3 **A water wave.**

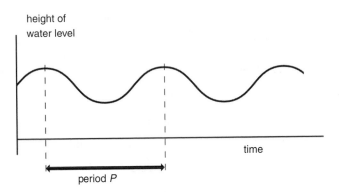

Figure 4.4 **The water level goes up and down as a wave passes.**

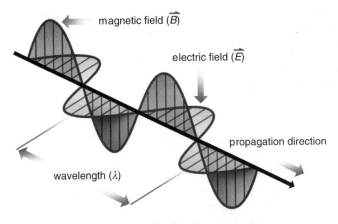

Figure 4.5 **An electromagnetic wave. The electric and magnetic fields oscillate in concert, and remain perpendicular to one another.**

Notice how similar Figure 4.4 is to Figure 4.3! Both show the characteristic "wavy" pattern we associate with a wave. This teaches us that this pattern means two quite different things. It describes how the amplitude of a wave varies

- from one *place* to another at a certain moment in time,
- from one *time* to another at a certain location in space.

Water waves are not the only types of waves one finds in nature. Sound is also a wave: specifically, it is a variation in the density and pressure of air. Similarly, *an electromagnetic wave is a variation in the strengths of electric and magnetic fields.* These fields are perpendicular to one another, and both undergo oscillations as illustrated in Figure 4.5. They are perfectly analogous to the wave illustrated in Figures 4.3 and 4.4.

Measuring waves

In Figure 4.3 we mapped out a wave's variation from one place to another, and indicated the distance between wave crests. This distance is *the wavelength,* λ (the Greek letter "lambda"). Similarly, in Figure 4.4 we mapped out the variation from one time to another, and indicated the time interval between the passage of successive wave crests. This time is *the period, P.* We can also study how many cycles of the wave pass by in one second. This is *the frequency, f.* The frequency is measured in cycles per second, or hertz. (Because a cycle has no units, the actual units of frequency are 1/seconds.) And finally, there is the *amplitude* of the wave – the wave height.

A little thought shows that frequency and period are related. If half a second is required for one cycle, then in one second two cycles will be completed. Similarly, if $1/10$th of a second is required for one cycle, then in one second we will have ten cycles. Clearly, the frequency is one divided by the period:

$$f = 1/P.$$

We can also find a relationship between wavelength and period. Imagine that we were to make a movie of our water wave passing the ruler of Figure 4.3. This movie would show

(a) ripples moving along – with speed "*c*," let us say,
(b) the water level at the ruler oscillating up and down.

It is easy to see that these are connected: the water level at the ruler is oscillating *because* the wave is passing the ruler. In one period the water level at the ruler goes through one cycle of oscillation, and the wave moves forward a distance just equal to its wavelength. But this distance also equals the wave's speed times the period. So the wavelength must equal the speed times the period:

$$\lambda = cP.$$

Combining this result with the above formula for frequency, we find the basic equation relating wavelength, frequency and speed:

$$\lambda f = c.$$

Wavelength and perception: the electromagnetic spectrum

Water, sound and light waves differ in many ways. The velocity of water waves, for example, is a few miles per hour, while that of sound is a healthy 740 miles per hour. The speed of light, in contrast, is a remarkable 186 000 miles per second, or 3×10^8 meters per second.

Water waves have wavelengths ranging from less than an inch (characteristic of small ripples) to many yards (characteristic of ocean waves). The wavelengths of sound waves are, in fact, somewhat similar: our ears are sensitive to sound wavelengths in the range 8 centimeters to 17 meters. The human auditory system is so constructed that our sensation of musical pitch is related to the wavelength of the sound waves striking our ears: a sound wave whose wavelength is 1 meter is almost exactly the note E above middle C. A shorter wave produces a higher pitch, and a longer wave a lower. A wave of high amplitude we perceive as louder, a wave of smaller amplitude as softer.

. . . .

NOW YOU DO IT

Suppose you are standing waist-deep in a pond. (a) Ripples on its surface make the water level move up and down against your tummy with a frequency of 7 cycles per second. What is the period of these waves? (b) Other waves come by that have a period of 4 seconds. With what frequency do these waves make the water level move up and down against your tummy? In each case, (A) describe the logic of your calculation, and (B) carry out the detailed calculation.

A loudspeaker is broadcasting a musical note with a frequency of 400 cycles per second. (A) What is the period of this wave? (B) Now you turn up the volume on the loudspeaker: in what way does this change the period of the wave?

The ordinary AM radio broadcast band runs from a frequency of 540 kilocycles to 1600 kilocycles. (A kilocycle is a thousand cycles per second.) What are the wavelengths corresponding to these two frequencies? Express your answers in miles in order to get a feeling for them. In each case, (A) describe the logic of your calculation, and (B) carry out the detailed calculation.

. . . .

In contrast, the light to which our eyes are sensitive has astonishingly small wavelengths. Green light, for example, has a wavelength of 5×10^{-7} meters. Just as the wavelength of sound waves influences the pitch we hear, so does that of a light wave influence the color we see: red light has a longer wavelength λ, and violet a shorter λ. In Figure 4.6 we illustrate the spectrum of visible light.

But these are not the only electromagnetic waves that exist: they are just the ones that correspond to visible light. Electromagnetic waves of longer and shorter wavelength surround us every moment of our lives, but we cannot directly perceive them.

Figure 4.6 **The spectrum of visible light. One nm is 10^{-9} meters. 👁 (Also see color plate section.)**

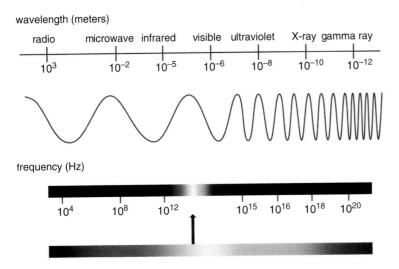

Figure 4.7 **The full electromagnetic spectrum.**

Rather we build devices that detect them. For instance, radio waves have wavelengths ranging from meters to millimeters, whereas the wavelengths of X-rays range from 10^{-11} meters to 10^{-8} meters. The full electromagnetic spectrum is illustrated in Figure 4.7.

The Doppler effect

The Doppler effect is one of the most important tools the astronomer possesses. It allows us to measure the velocity with which an astronomical object is moving.

While you may never have *heard of* the Doppler effect, you have already *heard* it. Recall what happens when you stand beside a highway and listen to the sound of a car as it drives by. As the car approaches and then passes, this sound grows louder and then softer. But something else happens to the sound as well: at the moment the car passes, it shifts from a higher to a lower pitch. This shift in pitch is due to the Doppler effect. As we have just seen, pitch is related to the wavelength of sound waves. The Doppler effect describes *the shift in wavelength due to the state of motion of the source emitting the waves.*

When we listen to the sound of a passing car, we are detecting sound waves, which cannot travel through the vacuum of space. But light and other forms of electromagnetic radiation do, and the Doppler effect applies to them as well:

- the light emitted by an object moving away from us will be shifted toward the red in color (*redshift*),
- the light emitted by an object moving toward us will be shifted toward the blue in color (*blueshift*).

The usefulness of this to astronomy is as follows. If you think back to the sound of a car driving by, you will recall that the faster the car goes, the larger is the shift in pitch of its sound. This is also true of light: the faster an astronomical object is moving, the more altered is the wavelength of the light we receive from it. Astronomers turn this logic around: the greater the shift in wavelength, the faster the source must be moving. *By measuring the magnitude of the Doppler shift, we can infer the velocity of the object emitting the wave.* The Doppler effect provides us with a "speedometer."

There are two important limitations to this "speedometer."

First, it *only measures relative motion.* So far we have discussed the Doppler effect if the source is moving and the observer is stationary. But there is also a Doppler effect if the source is holding still and it is the observer that moves. If you are standing on a sidewalk and a car with its horn honking drives past, you hear a shift. But you also hear a shift if you are in a moving car and the horn is honking in a parked car. The Doppler effect provides us with no means of deciding whether it is

Figure 4.8 **The Doppler shift measures only relative motion. In both cases, the shift is the same.**

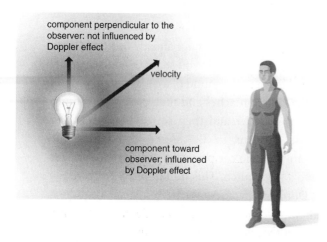

component perpendicular to the observer: not influenced by Doppler effect

velocity

component toward observer: influenced by Doppler effect

Figure 4.9 **Components of motion. No matter how an object moves, its velocity can be split into a component toward or away from an observer, and a second component crossways to the observer's line of sight. Only the component toward or away from the observer is affected by the Doppler effect. (The same separation of velocity into components can be done if it is the observer that moves.)**

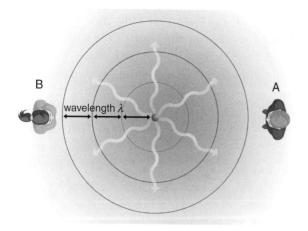

B

A

wavelength λ

Figure 4.10 **Waves from a stationary source. The wavelength is the same everywhere.**

the source or the observer that is moving: all that it tells us about is their relative motion (Figure 4.8).

Second, *it only measures motion of an object toward or away from us: it tells us nothing about crossways motion.* So far we have concentrated our attention on motion toward or away from us. What happens if the motion is sideways? In this case there is no shift in wavelength. We therefore conclude that the Doppler effect allows us to measure the *component of an object's velocity toward or away from us.* It cannot be used to detect the component of velocity across the line of sight. Figure 4.9 shows an object moving toward an observer, and illustrates these two components.

Origin of the Doppler effect

Figure 4.10 illustrates the pattern of wave crests expanding away from some source of waves. As illustrated in that figure, the wavelength is the same everywhere about the pattern.

But Figure 4.10 only applies if the source of the waves is holding still. Figure 4.11 illustrates the case of a moving source. Notice that in this case the wavelength is different in various parts of the diagram. This is the Doppler effect.

In Figure 4.11 the source is moving toward the observer A. Notice that the waves emitted in that direction – the waves that will reach observer A – are compressed by the motion of the source. They have a shorter wavelength, which corresponds to a higher pitch. This is just what the example of standing beside a roadway teaches: if a car is moving toward you, the sound you hear has a higher pitch. Similarly, in Figure 4.11 the source is moving away from observer B, and the waves reaching this observer are longer (lower pitch).

Finally, notice that the faster the source is moving, the greater the distortion of the wavelengths produced by its motion. Thus, by measuring the distortion we can infer the velocity of the source. This is how we measure the velocities of astronomical objects.

Mathematics of the Doppler effect

From the mathematics of wave motion, a formula giving the magnitude of the Doppler effect can be derived:

$$\frac{\text{change in wavelength}}{\text{true wavelength}} = \frac{\text{velocity}}{\text{velocity of wave}}$$

or, in symbols,

$$\Delta\lambda/\lambda = V/C,$$

where

- $\Delta\lambda$ is the change in wavelength,
- λ is the true wavelength,
- V is the component along the line of sight of the relative velocity between the source and the observer,
- C is the velocity of the wave.

In using the Doppler effect to measure the velocity of an astronomical object, we measure the observed wavelength of its light and compare this with the true wavelength λ: this allows us to find $\Delta\lambda$. We then use the above formula to find V. But in order to do all this we need to know the light's true wavelength. This information is provided by a study of spectral lines, to which we will turn below.

As an example, suppose a car is emitting sound of wavelength 1 meter, and is driving away from you at 74 miles per hour. What is the wavelength of the sound you hear?

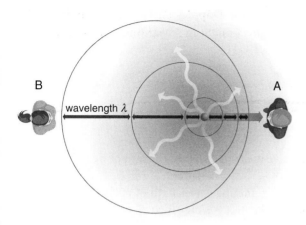

Figure 4.11 **Waves from a moving source. The wavelength is no longer the same everywhere. Rather, it is altered by the motion of the source.**

The logic of the calculation

Step 1. Solve the Doppler effect formula for the shift in wavelength.

Step 2. Plug in the true wavelength, the velocity of the car and the speed of sound. This allows us to find the shift in wavelength.

Step 3. Since the motion is away from you, the Doppler effect leads to a *larger* wavelength. So to find the observed wavelength, we *add* this shift to the true wavelength.

As you can see from the detailed calculation, the answer is that the Doppler effect stretches the wavelength from 1 to 1.1 meters.

Conversely, had the car been driving toward you at the same speed, $\Delta\lambda$ would have been the same, but we would have subtracted it from λ to find the observed wavelength. And had the car been driving crossways to your line of sight, even though it was moving there would have been no change in wavelength.

Detailed calculation

Step 1. Solve the Doppler effect formula for the shift in wavelength. We get

$\Delta\lambda/\lambda = V/C.$

Step 2. Plug in the true wavelength, the velocity of the car and the speed of sound. This allows us to find the shift in wavelength.
The speed of sound is $C_{sound} = 740$ miles per hour. The car's speed (74 miles per hour) is $1/10$th of this. The true wavelength is 1 meter. Thus

shift in wavelength $= (1/10)(1 \text{ meter}) = 0.1$ meter.

Step 3. Since the motion is away from you, the Doppler effect leads to a larger wavelength. So to find the observed wavelength, we add this shift to the true wavelength.
We get

observed wavelength = emitted wavelength + change in wavelength
= 1 meter + 0.1 meter = 1.1 meter.

NOW YOU DO IT
Suppose a car is driving toward you at twice the speed of the example we have just worked. What would be the wavelength of the sound we hear?

Suppose an airplane emits a sound of wavelength 2 meters. How fast does the airplane have to travel for its observed wavelength to be 3 meters? In what direction must it travel? (A) describe the logic of your calculation, and (B) carry out the detailed calculation.

As a *second example*, let us consider what the Doppler effect does to starlight. Astronomers customarily measure wavelengths of light in angstroms: one angstrom is 10^{-10} meters.

$$\text{One angstrom} = 10^{-10} \text{ meters.}$$

Let us explore how, once astronomers have measured the wavelength of light from a star, they can use this to find the velocity.

Suppose we know that a star is emitting light at a wavelength of 6000 angstroms, but that when we measure the wavelength we find it to be 5990 angstroms. What is the velocity?

Detailed calculation

Step 1. Solve the Doppler effect formula for the velocity.
  We find $V = C\,\Delta\lambda/\lambda$.

Step 2. From the observed λ and true λ, find $\Delta\lambda$.
  The true wavelength is 6000 angstroms, the observed wavelength is 5990 angstroms. So the shift in wavelength $\Delta\lambda$ is (6000 angstroms − 5990 angstroms) = 10 angstroms.

Step 3. Plug in this $\Delta\lambda$ and the speed of light C_{light} to find V.
  The speed of light C_{light} is 3×10^8 meters/second. So we get

  $V = (3 \times 10^8 \text{ meters/second}) \, (10 \text{ angstroms}/6000$ angstroms$) = 5 \times 10^5$ meters/second.

Step 4. Is the velocity toward or away from us?
  Since the wavelength is decreased, the velocity is toward us.

The logic of the calculation

Step 1. Solve the Doppler effect formula for the velocity.
Step 2. From the observed λ and true λ, find $\Delta\lambda$.
Step 3. Plug in this $\Delta\lambda$ and the speed of light C_{light} to find V.
Step 4. Is the velocity toward or away from us?

As you can see from the detailed calculation, the answer is 5×10^5 meters per second = 500 kilometers per second toward us. Remember that this does not mean that the star is moving this fast, but only that this is the relative velocity between the star and us: we could be moving, or the star could be moving, or both could be moving. Furthermore, this is only the component along the line of sight of this relative motion.

Astronomical uses of the Doppler effect

As we have discussed, in using the Doppler effect in astronomy, we measure $\Delta\lambda/\lambda$ and use the formula to find the velocity of the astronomical object. In spite of its limitations, there are many uses of this "speedometer." Here are a few: we will encounter many others.

(a) *Rotation of the Sun.* The light from the left-hand edge of the Sun is found to be shifted in wavelength as compared with that from its right-hand edge. This tells us that the Sun is rotating on its axis (Figure 4.12).

(b) *Discovering planets around other stars.* As we will see in Chapter 10, we can find planets orbiting distant stars by studying the motions they induce in their suns.

(c) *Tracking of interplanetary probes.* Engineers with NASA regularly receive radio messages from interplanetary space probes as they orbit through the Solar System. The received messages are found to be at the "wrong wavelength" – i.e. not the wavelengths at which the onboard transmitters emit their signals. This is because of the Doppler effect. The shifts in wavelength can be used to determine the spacecrafts' velocities, and so track their orbits.

(d) *Radar astronomy.* Just as the light *emitted* by an object is subject to the Doppler effect, so too with light (or other forms of electromagnetic radiation) *reflected*

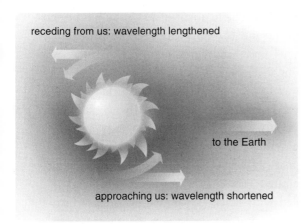

receding from us: wavelength lengthened

to the Earth

approaching us: wavelength shortened

Figure 4.12 **Rotation of the Sun shifts the wavelength of light from its edges as seen from the Earth.**

off an object. This is how police "radar guns" operate. Astronomers use the same methods to track the orbits of planets. Radar signals of one wavelength are bounced off planets: the return echoes are found to be shifted in wavelength. Using the shift, we can measure the relative velocities of planets and so determine their orbits.

(e) *Motions of stars.* To the naked eye, stars seem motionless. But as we have seen, studies show that the light they emit is shifted in wavelength. They must be moving! Their velocities turn out to be roughly 10 kilometers per second, or about 20 000 miles per hour.

(f) *Motions within interstellar clouds.* Atoms and molecules within interstellar clouds emit radio waves that are found to be at the "wrong" wavelengths. By measuring the shifts in wavelength, we can measure the velocities of "winds" within the clouds.

(g) *Expansion of the Universe.* The wavelength of light from distant galaxies is observed to be shifted to longer values. This tells us that these galaxies are all moving away from us. This is the expansion of the Universe. (Chapter 18 studies this issue, but it is important to emphasize that, because the expansion of the Universe involves relativity, the actual situation is more complicated than this.)

(h) *Dark matter.* The Doppler effect allows us to study the orbits of stars in galaxies. As we will discuss in Chapter 16, such studies have revealed the existence of vast amounts of unseen matter in them.

NOW YOU DO IT
A space probe emits a radio signal whose frequency is 700 thousand cycles per second. (A) What is the wavelength of this radiation? (B) Suppose we observe its wavelength to be 428.9 meters. How fast is it moving? In what direction? In each case, (A) describe the logic of your calculation, and (B) carry out the detailed calculation.

Blackbody radiation

Just as you have already heard the Doppler effect, so too you have already seen blackbody radiation. Sunlight is blackbody radiation. So is the light from an incandescent light bulb, from the burner of an electric stove, and from every star.

Let us begin our study of blackbody radiation with a familiar example. Wait until it is night, and then go into a kitchen with an electric stove. Turn out all the lights, and then turn on a burner on the stove. Watch it carefully.

At first you see nothing. But soon the burner has heated up somewhat, and it begins to glow a dull red. As time passes the burner grows yet hotter, and two things happen: the light it emits grows brighter, and it grows yellower.

This light emitted by the burner is *blackbody radiation.* (The term "blackbody" comes from the fact that the only objects that emit perfect blackbody radiation are those that absorb all light incident upon them, i.e. that are black.) Blackbody radiation is *the electromagnetic radiation emitted by a hot object.*

Every hot object emits blackbody radiation. The filament of a light bulb is heated by an electric current passing through it, and so emits visible light. The Sun is heated by nuclear reactions deep in its core, and does so too. But blackbody radiation does not have to be visible light. Depending on the temperature of the emitting body, it can exist at other wavelengths. While the Sun's surface is sufficiently hot to emit visible light, its core is far hotter, and is filled with blackbody radiation in the X-ray region of the spectrum. Conversely, your own body, heated by metabolic processes to 98.6 degrees Fahrenheit, emits infrared radiation – too long to be seen by the naked eye, although it can be detected by various instruments. Indeed, the entire Earth glows at infrared wavelengths. Cosmologists have discovered that the very Universe is filled with blackbody radiation – the Cosmic Microwave Background radiation, discussed in Chapter 18 – at radio wavelengths.

As the example of the electric stove makes clear, daily experience teaches that as the temperature of a body increases

(A) it grows brighter – that is, the blackbody radiation grows more intense: this is the *Stefan–Boltzmann law*;
(B) its color shifts from red to yellow – that is, the blackbody radiation shifts to shorter wavelengths: this is the *Wien law*.

These two laws have been quantitatively investigated in the laboratory. We now discuss each.

The Stefan–Boltzmann law

The Stefan–Boltzmann law describes fact (A), that the hotter a blackbody grows the more intense its emission grows. It states that *a blackbody radiates energy at a rate proportional to its surface area multiplied by the fourth power of its temperature.* Specifically, the luminosity L of a blackbody is

$$L = 0.57 \times 10^{-7} A T^4 \text{ watts,}$$

where A is the surface area of the hot object in square meters and T its temperature in kelvin (degrees above absolute zero).

Let us explore the consequences of this relation.

(1) Suppose the big burner on your stove has twice the surface area as the little burner. If they have the same temperature, the big one puts out twice as much light as the little one.
(2) Suppose we find a star with the same temperature as the Sun, but half its surface area. Then the smaller star's luminosity must be half that of the Sun.
(3) Suppose a burner on your stove doubles its temperature. Then T^4 is multiplied by $2^4 = 16$, and it puts out 16 times as much light.
(4) Suppose we find a star with the same surface area as the Sun, but twice the temperature. Then the hotter star has a luminosity 16 times greater than the Sun.

....

NOW YOU DO IT
Suppose the front burner on your stove has a diameter of 0.1 meter, the back burner has a diameter of 0.2 meters. If they have the same temperature, how much more light does the big burner emit than the small one?

Suppose the temperature of the burner on your stove is divided by 2. By what factor does its light emission change?

We now pass on to a more complicated calculation. Because your body is warm, *you* emit blackbody radiation! Let us calculate how much.

Detailed calculation

Step 1. We need to know your surface area and temperature.
Your body temperature is 98.6 degrees Fahrenheit, or 310 kelvin, and a typical person's surface area is roughly 1.5 square meters.

Step 2. Plug these into the Stefan–Boltzmann law.
We get

$$L_{\text{you}} = (0.57 \times 10^{-7})\,(1.5)\,(310)^4 \text{ watts} = 790 \text{ watts}.$$

Step 3. To get an intuitive feel for the answer, calculate how many 100 watt light bulbs this corresponds to.
790 watts corresponds to 7.9, or a bit less than eight, 100-watt light bulbs.

The logic of the calculation

Step 1. We need to know your surface area and temperature.

Step 2. Plug these into the Stefan–Boltzmann law.

Step 3. To get an intuitive feel for the answer, calculate how many 100 watt light bulbs this corresponds to.

As you can see from the detailed calculation, you shine with the intensity of nearly eight light bulbs! Why are you not aware of this emission? Because it is invisible to the naked eye; as we will see when we turn to the Wien law, it lies in the infrared region of the spectrum.

NOW YOU DO IT
Measure the diameter of a basketball, and then calculate its area using the formula $A = 4\pi R^2$. If its temperature is 70 degrees Fahrenheit (°F) = 294 kelvin (K), how many watts does it emit? (A) Describe the logic of your calculation, and (B) carry out the detailed calculation. (C) Why don't you see it glowing?

Finally, notice that the Stefan–Boltzmann law provides us with a thermometer: we can use it to take the temperature of an astronomical body if we know its luminosity and surface area. Let us illustrate this by finding the temperature of the Sun.

The logic of the calculation

Above we found the Sun's luminosity. If we knew the Sun's surface area we would know all the factors in the Stefan–Boltzmann law except for the temperature. We could then use the law to *find* the temperature.

Step 1. Find the surface area of the Sun.

Step 2. Solve the Stefan–Boltzmann law for the temperature. It will depend on the Sun's luminosity and surface area.

Step 3. Plug the Sun's area and luminosity into the result to find the Sun's temperature.

As you can see from the detailed calculation overleaf, the Sun's temperature is a bit less than 6000 kelvin.

Detailed calculation

Step 1. Find the surface area of the Sun.

The Sun is spherical, and we know the surface area of a sphere is $A = 4\pi R^2$, where R is its radius. Measurements reveal that the radius of the Sun is $R_{Sun} = 6.96 \times 10^8$ meters.

We then find

$$A_{Sun} = 4\pi R_{Sun}^2 = 4\pi \, (6.96 \times 10^8 \text{ meters})^2 = 6.2 \times 10^{18}$$

square meters.

Step 2. Solve the Stefan–Boltzmann law for the temperature. It will depend on the Sun's luminosity and surface area.

The Stefan–Boltzmann law is

$$L = 0.57 \times 10^{-7} \, A \, T^4 \text{ watts},$$

so that

$$T^4 = L/0.57 \times 10^{-7} \, A$$

and

$$T = [L/0.57 \times 10^{-7} \, A]^{1/4}.$$

(The temperature we find here will be in kelvin.)

Step 3. Plug the Sun's area and luminosity into the result to find the Sun's temperature.

We know the Sun's luminosity to be 3.8×10^{26} watts. We then calculate

$$T = [3.8 \times 10^{26}/(0.57 \times 10^{-7} \times 6.2 \times 10^{18})]^{1/4}$$
$$T = 5750 \text{ K}.$$

NOW YOU DO IT

Suppose a star had the same temperature as the Sun, but a luminosity of 3×10^{26} watts. What would be its surface temperature? (A) Describe the logic of your calculation, and (B) carry out the detailed calculation.

The Wien law

The Wien law describes fact (B): that the hotter a blackbody grows, the shorter is the wavelength it emits. To be precise, it describes the wavelength of most of the radiation – for, as we will shortly see, a blackbody emits energy at many wavelengths. The Wien law states that the bulk of the emission occurs at a wavelength $\lambda_{\text{most emission}}$ given by

$$\lambda_{\text{most emission}} = 0.003/T \text{ meters},$$

where again T is the temperature in kelvin.

Notice that this law agrees with our experience of the electric stove in the kitchen. The lower T is, the bigger is $\lambda_{\text{most emission}}$ as predicted by the Wien law. And long wavelengths correspond to red light. Similarly, as time passed and the burner heated, T increased and, by Wien's law, $\lambda_{\text{most emission}}$ got less, making for a whiter light.

Let us also explore the consequences of this relation.

(1) Above we saw that, when the burner on a stove grows twice as hot, it emits 16 times as much light. The Wien law tells us that this light is concentrated at half the original wavelength. The same is true of a star twice as hot as the Sun.

(2) Let us return to the question of the blackbody emission from your body. In the previous section we found that you emit nearly as much light as eight light bulbs. But why do we not see this emission? We are now able to answer this question. Using a body temperature of 98.6 degrees Fahrenheit (310 kelvin) in the Wien law, we find that most of your blackbody emission is at a wavelength $\lambda_{\text{most emission}} = 0.003/310$ meters, or about 10^{-5} meters. This lies in the infrared region of the spectrum. This is why your own blackbody radiation is invisible to you: you shine by infrared light.

(3) Most of the emission from the Sun is at a wavelength $\lambda_{\text{most emission}} = 0.003/5750$ meters $= 5.2 \times 10^{-7}$ meters. This is in the visible region of the spectrum. Thus we can detect blackbody radiation from the Sun with our naked eyes – it is sunlight.

NOW YOU DO IT

Return to that 70 °F basketball. At what wavelength does most of its emission lie? (A) Describe the logic of your calculation, and (B) carry out the detailed calculation.

At what wavelength is most of the emission from a blackbody whose temperature is 3000 kelvin? (A) Describe the logic of your calculation, and (B) carry out the detailed calculation.

. . . .

Spectra

As we have mentioned, a hot object does not emit all its blackbody radiation at the wavelength $\lambda_{\text{most emission}}$ given by the Wien law. Rather it emits more strongly at this wavelength than at any other. During the latter part of the nineteenth century, a series of experiments mapped out the detailed spectrum of this radiation. Before describing their results, let us clarify the concept of a spectrum.

The *spectrum* (plural: spectra) of light is the analysis of this light into its constituent colors, or wavelengths. An analogy with music might be helpful. Suppose you strike a single key on a piano – middle C, say. All the waves emanating from the piano will have the wavelength of middle C. But suppose now you strike a pair of notes – C and E, perhaps. In this case, the waves from the piano are a mixture of the two wavelengths corresponding to C and E. Figure 4.13 shows the spectra of these two cases.

In a similar way, the spectrum of light reveals its constituent wavelengths, i.e. colors. You might think that such a spectrum would be an abstract, unfamiliar thing. But in reality all of us are familiar with one from daily experience: every time you see a rainbow you are seeing the spectrum of the Sun.

Sunlight, in fact, consists of a "chord" containing not two, but a wide range of wavelengths. The various colors of the rainbow are simply those colors spread out in an arc. And, like the spectrum of sound of Figure 4.13, the intensity of each color in a rainbow reflects the intensity in sunlight of the radiation at the corresponding wavelength. In Figure 4.14 the Sun's spectrum is displayed.

Other stars emit different "chords" from the Sun. Rainbows on any planets orbiting these stars would be different from ours. We say that the spectra of these stars differ from that of our Sun.

Notice the dark "lanes" running through the spectrum in Figure 4.14(a), and the corresponding dips in Figure 4.14(b). These are spectral lines, to which we turn below.

How are spectra produced in the laboratory? One common method is to employ a glass prism. Figure 4.15 illustrates how a prism works (see also Chapter 5 on telescopes). A beam of light is bent as it passes from air into glass. Furthermore, different wavelengths are bent by different amounts. So light of each color follows its own separate path, and what exits the prism is a "spray" of differing colors. This is, in fact, how rainbows are formed: in this case the "prisms" are raindrops.

The spectrum of blackbody radiation

We turn now to the observed spectrum of blackbody radiation. Figure 4.16 illustrates the spectrum of emission from a blackbody at a variety of temperatures.

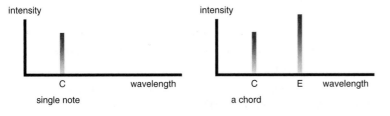

Figure 4.13 **Spectra of sound waves from a piano. In the second figure the note E was struck with greater force than C.**

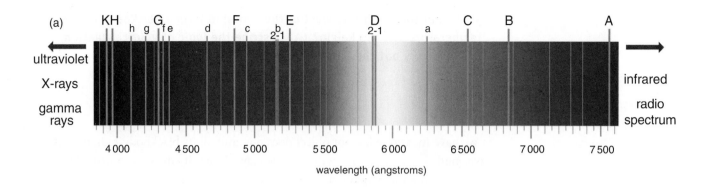

(a)

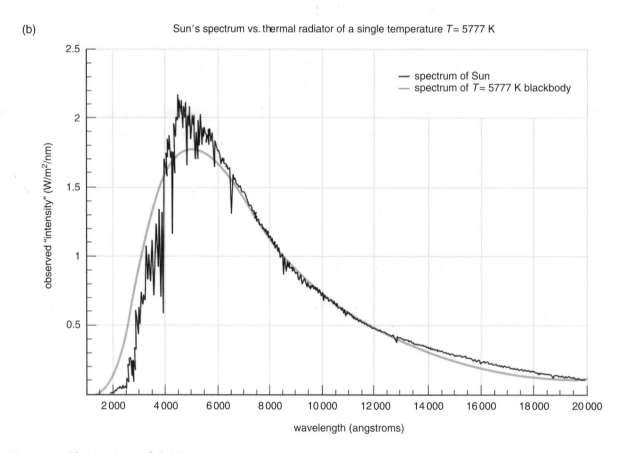

(b)

Figure 4.14 **The spectrum of the Sun.**

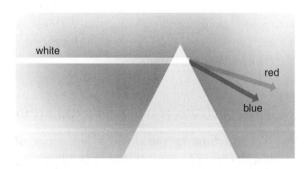

Figure 4.15: **A prism splits light into its constituent colors.**

Note five things about this figure.

- The emission from the blackbody consists of a "chord" of many wavelengths.
- Some of this radiation lies in the visible region of the spectrum. But some does not. The emission also extends into the ultraviolet and infrared regions of the spectrum.
- The emission is strong at some wavelengths, and weak at others. Indeed, the peaks of the emission in Figure 4.16 occur at the wavelengths $\lambda_{\text{most emission}}$ given by the Wien law.

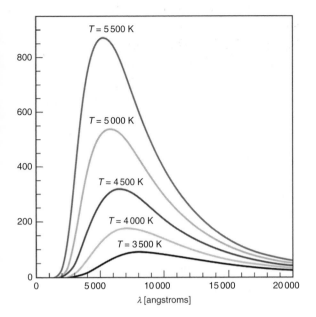

Figure 4.16 **The spectrum of blackbody radiation.**

NOW YOU DO IT

Figure 4.16 shows three curves corresponding to three different temperatures. For each of these temperatures, predict the wavelengths at which the curves should peak. Now measure these wavelengths on the graph. How accurate is the graph?

- The hotter the blackbody, the shorter is the wavelength corresponding to the peak of each curve. This too is simply the Wien law.
- The hotter the blackbody, the more intense is the emission. This is simply the Stefan–Boltzmann law.

Spectral lines

If you compare the actual spectrum of the Sun (Figure 4.14) with those of Figure 4.16, you will see that the Sun is nearly a perfect blackbody. But there is one striking difference: unlike a perfect blackbody, the Sun's spectrum is interrupted at various wavelengths. These dark bands in the solar spectrum are known for obvious reasons as *spectral lines*. They are found in the spectrum, not just of the Sun, but of every astronomical source.

Two experiments

We can learn more about spectral lines by means of a pair of experiments.

For the *first experiment*, we will need a source that produces light at all wavelengths. Although this is not essential, let us agree that this source will be a blackbody. In our first experiment, we let the emission from this blackbody pass through a cool gas, as shown in Figure 4.17. The result is a series of dark lines in the spectrum, similar to those found in sunlight.

These lines mark wavelengths at which the light is being absorbed by the gas. We term them *absorption lines.* We conclude that a dilute gas is not capable of absorbing light of all wavelengths: rather it absorbs light of a discrete set of wavelengths.

We now realize that the dark lines in the Sun's spectrum (Figure 4.14) do not mark wavelengths at which the Sun fails to emit, but rather wavelengths at which its emission is being absorbed. It turns out that the gas responsible for this absorption lies within the Sun: it is the cool outer layers of the Sun, where the gas is dilute enough and cool enough to form absorption lines.

In our *second experiment*, we dispense with the source of light, but rather heat the gas (Figure 4.18). From our example of heating the burner of an electric stove – and from everything we learned in the previous section – we would expect the gas to emit blackbody radiation. But this is not what turns out to happen! Rather we

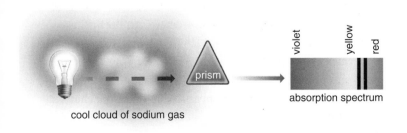

Figure 4.17 **Formation of absorption lines.**

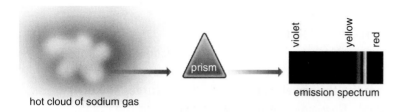

hot cloud of sodium gas

emission spectrum

Figure 4.18 **Formation of emission lines.**

get emission confined to a set of discrete wave-lengths known as emission lines.

This experiment teaches us that a dilute gas is not capable of emitting blackbody radiation. Such a gas, rather than emitting light of all wavelengths, only emits light of a discrete set of wavelengths. It turns out that density is the key here: *blackbody radiation is emitted by dense objects: spectral lines are emitted by rarified gases.*

Furthermore, it turns out that the wavelengths of these emitted lines are exactly the same as the wavelengths of the dark lines we saw in Figure 4.17. *If a rarified gas is heated it emits a certain set of wavelengths: alternatively if light passes through it, it absorbs exactly the same set of wavelengths.*

Finally, in Figure 4.19 we add one more element to our study of spectral lines. In this figure, the spectra of a number of different gases are compared. We see that *each element possesses its own unique spectrum,* in the same way that each person possesses a unique fingerprint. Hydrogen has one spectrum, helium another, and so forth.

Astronomical uses of spectral lines

Spectroscopy is the study of spectral lines. Using spectroscopy we can learn many things.

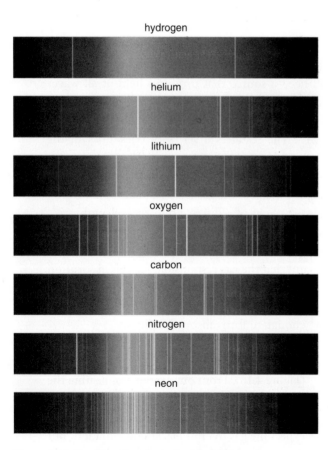

hydrogen

helium

lithium

oxygen

carbon

nitrogen

neon

Figure 4.19 **Spectra of various chemical elements.**

(1) In our discussion of the Doppler effect, we commented that we need to know the true wavelength λ of emission in order to use the Doppler formula to find a star's velocity. We do this by using spectral lines. The lines appearing in Figure 4.19 occur at known wavelengths. If we see that the spectrum of a certain object looks like Figure 4.19, but that the lines occur at "wrong" wavelengths, we know that the Doppler effect is operating. Each line's true wavelength is known, so that the shift $\Delta\lambda$ can be found.

(2) The lines are influenced by the temperature of the gas: as the temperature is raised, some spectral lines grow stronger while others grow weaker. By means of such studies, spectral lines allow us to take the temperatures of the stars. As we saw in the previous section, the Stefan–Boltzmann and Wien laws allow us to measure the temperatures of astronomical objects: we now see that spectroscopy is yet another way to do this.

(3) Because they function as "fingerprints" for chemical elements, spectra allow us to deduce which elements are present in a gas. Through analysis of the spectra of stars, astronomers have measured in detail the chemical compositions of the stars.

(4) The lines are influenced by any magnetic fields that might be present in the emitting gases. By means of

such analyses, we have learned that sunspots are regions of intense magnetism on the surface of the Sun.

(5) The lines are influenced by the pressure and density of the emitting gas. By means of such studies, we can learn about conditions in the atmospheres of stars.

Atomic structure and the origin of spectral lines

We have already emphasized that blackbody radiation is emitted by dense objects, whereas spectral lines arise only in a dilute gas. This is because blackbody radiation is caused by large numbers of atoms acting together, whereas *spectral lines arise from the emission and absorption of light by individual atoms* in the gas. So to understand spectral lines we need to understand atoms. We now turn to a study of these atoms, and how they emit and absorb light.

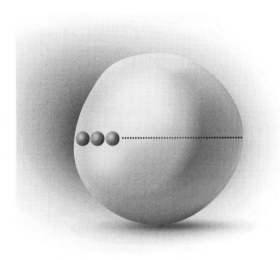

Figure 4.20 **Sixty million atoms of hydrogen would be required to cross a pea.**

Atoms

Size

An atom is extraordinarily small. As an example, the atom of the element hydrogen, the simplest and smallest of all atoms, has a diameter of 10^{-10} meters. To get a feeling for this, imagine that you were to lay down a series of these atoms in a line: how many would you require for the line to stretch across a pea (Figure 4.20)? If a pea is ¼ inch across, or 0.006 meters, we calculate

$$\text{number of atoms} = 0.006 \text{ meters}/10^{-10} \text{ meters} = 60 \text{ million.}$$

For comparison, if we were to lay this number of *peas* down in a line, they would stretch a distance

$$(60 \text{ million})(0.006 \text{ meters}) = 3.6 \times 10^5 \text{ meters} = 220 \text{ miles!}$$

NOW YOU DO IT
Put a billion billion (10^{18}) hydrogen atoms in a line. How long is that line? (A) Describe the logic of your calculation, and (B) carry out the detailed calculation.

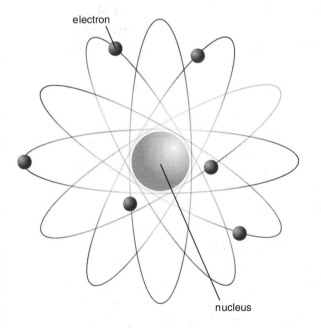

Figure 4.21 **An atom consists of a nucleus surrounded by a cloud of electrons.**

Structure

As illustrated in Figure 4.21, an atom consists of a *nucleus* surrounded by a cloud of *electrons*. While an atom is small, its nucleus is smaller still. The nucleus of hydrogen is 1.6×10^{-15} meters in diameter. Comparing this to that of the hydrogen atom as a whole:

$$D_{\text{H atom}}/D_{\text{H nucleus}} = -10^{-10} \text{meters}/1.6 \times 10^{-15} \text{meters} = 63\,000.$$

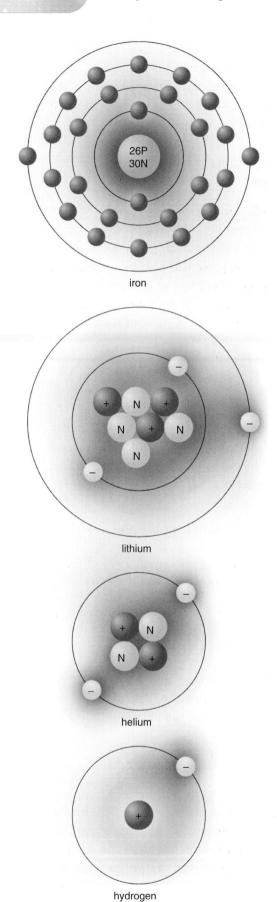

iron

lithium

helium

hydrogen

Figure 4.22 **Atomic structure. Protons are labeled +, electrons −.**

Actually an atom is mostly empty space; if we were to inflate the nucleus of a hydrogen atom to be an inch across, the atom itself would be a mile across.

NOW YOU DO IT
Verify the above statement. (A) Describe the logic of your calculation, and (B) carry out the detailed calculation.

The atomic nucleus is composed of *neutrons* and *protons*. The bigger the atom, the more neutrons and protons its nucleus contains. An atom contains an electron for every proton, so that the number of electrons equals the number of protons. Similarly, the number of neutrons in a nucleus roughly equals the number of protons, although they are not exactly equal in all cases. As illustrated in Figure 4.22, the farther up the periodic table of elements we go, the more particles the atom contains.

Mass
While the atomic nucleus is very small, it contains most of the atom's mass. The masses of the neutron and proton are 1836 times that of the electron. So, while most of the atom's space is taken up by the electrons, they contribute very little to its mass.

NOW YOU DO IT
Suppose you have a one ounce grape. What sort of object has 1836 times as much mass? (A) Describe the logic of your calculation, and (B) carry out the detailed calculation.

Charge
Electrons possess a negative charge, protons an exactly equal positive charge and neutrons zero charge. Since the number of electrons equals the number of protons, their net charge adds to zero. So an atom possesses zero net charge. (If one or more electrons are removed from an atom, we get an ion.)

Emission and absorption of light
It is the electrons in an atom that are responsible for the emission and absorption of light, and the production of spectral lines. An electron moving about its nucleus has a certain energy, and if its state of motion changes this energy changes. If, for instance, an electron suddenly lowers its energy, the lost energy will be radiated away as a burst of

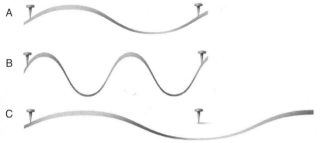

Figure 4.23 **A string can vibrate only in certain allowed modes. Modes A and B are allowed, but C is not.**

light and we get an emission line. Similarly, light falling upon an atom can be absorbed: the increased energy goes into an increased energy of the electron and we get an absorption line. These processes are governed by quantum mechanics, to which we now turn.

Quantum mechanics

In many ways an atom looks like a miniature Solar System. But the laws that govern the two are utterly different. In the previous chapter we built up an understanding of the orbits of the planets in terms of Newton's laws of gravitation and motion. But these laws do not work for atoms. It is quantum mechanics that allows us to understand the motion of electrons in an atom.

Matter waves

A central principle of quantum mechanics is *wave–particle duality*. This is the principle that *to every wave there is an associated particle, and to every particle an associated wave*.

Let us apply this principle to an electron. It has an associated wave, called a *matter wave*. An analogy might be the waves on a string stretched tight between two nails. If such a string is plucked, it will vibrate. But as illustrated in Figure 4.23, it can vibrate only in certain ways.

One mode of vibration is illustrated in Figure 4.23(A). This pattern of vibration of the string has a crest and a trough. A second allowed mode is illustrated in Figure 4.23(B), where there is a crest, a trough, and then a second crest and trough. But in Figure 4.23(C) we see a mode that is not allowed. This mode is forbidden because such a pattern of vibration would require the ends of the string to be moving. But they cannot move – they are nailed down!

....

NOW YOU DO IT
In Figure 4.23 we sketched some allowed and forbidden modes of vibration of a string stretched between two nails. Sketch several more allowed and forbidden modes.

    Our discussion applies to a guitar string just as well as an ordinary string. Explain why a guitar player changes the note she is playing by pressing her finger on the string at a new place. Suppose she moves her finger up along the fret board: does the string vibrate with a longer or shorter wavelength? Does it emit a higher or lower pitch?

....

Figure 4.24 **Matter waves in an atom can vibrate only in certain allowed modes. Mode A is allowed, mode B is not.**

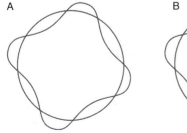
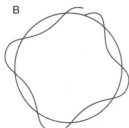

How does this reasoning apply to the waves in an atom? Figure 4.24 illustrates the principle. In Figure 4.24(A) we show the matter wave associated with an electron in an atom. This is an allowed mode of vibration. But in Figure 4.24(B) we sketch a mode of vibration that is forbidden. It is forbidden not because there are nails that prevent the ends of the wave from moving, but because as the wave wraps around in a circle its ends do not match up. These two ends must join smoothly together. But in this forbidden mode of vibration they do not.

....

NOW YOU DO IT
In Figure 4.24 we sketched some allowed and forbidden modes of vibration of the matter waves in an atom. Sketch several more allowed and forbidden modes.

....

Thus quantum mechanics requires that the waves within an atom can vibrate only in certain allowed patterns. But the more rapidly a wave vibrates, the higher its energy! *To each mode of vibration in an atom, there is a corresponding energy.* So we see that quantum mechanics predicts that *an electron in an atom may only possess certain allowed energies.* It can have such-and-such an energy, or such-and-such another energy – but not an energy in between. This is totally unlike normal circumstances, since usually things can move with any energy at all. It is only submicroscopic objects such as electrons and atoms whose energies are restricted by quantum mechanics.

The emission and absorption of light are caused by jumps between these allowed energies. If an atom jumps from a higher to a lower energy, it emits a burst of light that carries away the difference in energy. Similarly, if it absorbs light, the atom moves from a lower to a higher energy state. Let us now turn our attention to this light, and how it carries energy.

Photons and the Planck formula

Recall the central principle of quantum mechanics: to every wave there is an associated particle, and to every particle an associated wave. We have applied it to electrons. Now let us apply it to light.

As we have seen, light is a wave. The particle associated with this wave is the *photon.* Just as a moving electron has an energy, so too with the photon. A photon's energy is given by the Planck formula

$$E = hf,$$

if the light wave is vibrating with frequency f. Here h is a new constant of nature, known as Planck's constant. In the MKS system, its value is

$$h = 6.6 \times 10^{-34} \text{ MKS units.}$$

We have already emphasized that, when an atom jumps from one of its allowed energies to another, light is emitted or absorbed, which can have only certain allowed energies. Planck's formula now tells us that this light can have only certain allowed frequencies. These are the frequencies of the atom's spectral lines.

Energy-level diagrams and the origin of spectral lines

It is convenient to represent the allowed energy states of an atom by means of an energy-level diagram. Such a diagram indicates the various energies the atom may possess. In Figure 4.25 we illustrate the energy-level diagram of the simplest of all elements, hydrogen.

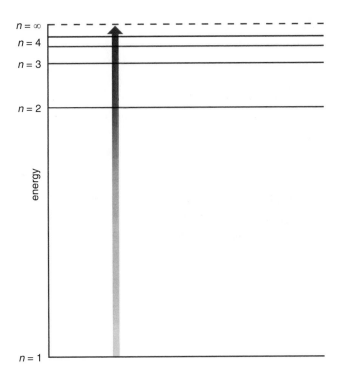

Figure 4.25 **Allowed energy levels of hydrogen, corresponding to its allowed modes.**

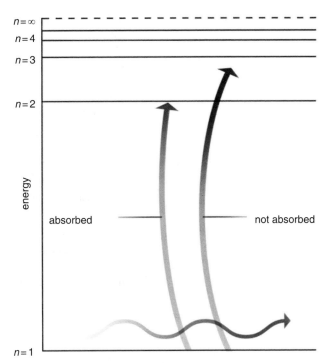

Figure 4.26 **Absorption lines arise when the photon hitting an atom has precisely the energy required to move it from one energy state to another. Photons of other energies are not absorbed, because the atom refuses to enter into the energy state their absorption would produce.**

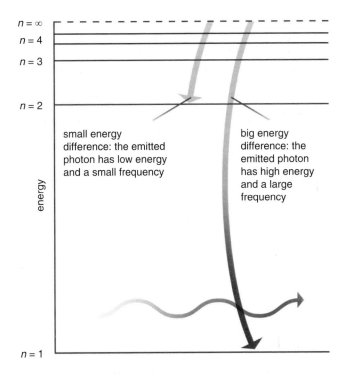

Figure 4.27 **Emission lines arise when an atom reduces its energy, dropping to a lower energy state and emitting photons whose energies correspond to the energy differences between the two states.**

How can such diagrams be used to understand emission and absorption lines? Imagine a beam of light shining upon hydrogen atoms in a gas. If the light has a spectrum consisting of many colors, the waves have many frequencies and the photons many energies. Let us first concentrate our attention on those photons whose energies precisely equal the *difference* in energy between two allowed energy levels of an atom (Figure 4.26). These photons have exactly the energy required to raise the energies of the atoms. When they strike the atoms, some photons will be absorbed by them. Once this has occurred,

- some of the atoms will now have higher energies,
- some of the photons in the light beam will no longer exist, since they have been absorbed by the atoms.

Thus some of the light of precisely the wavelength corresponding to these photons has been absorbed.

Now let us move on to concentrate our attention on photons that *do not* have precisely the energy corresponding to a difference in energy levels of the atom. These photons are *not* absorbed, because the atom refuses to enter into the energy state their absorption would produce. We conclude that the hydrogen atoms absorb only light of a discrete set of colors – those colors whose photons possess just the energies required to raise the atoms' energies from one energy level to another. *In this way we account for the absorption lines.*

Similarly (Figure 4.27), t*he emission lines* arise when an atom reduces its energy, dropping to a lower energy state and emitting photons whose energies correspond to the energy differences between the two states.

NOW YOU DO IT

In Figure 4.26 we indicated some photons that can be absorbed by an atom, and some others that cannot. Sketch several more allowed and forbidden absorptions.

In Figure 4.27, indicate a jump in energy in which an atom emits (a) a photon of exceedingly large frequency, and (b) a photon of exceedingly small frequency.

Summary

Inverse square law
- There are two kinds of brightness:
  - absolute brightness
    - how bright something really is
    - doesn't depend on distance
    - the amount of light energy given off per second
    - measured in watts
    - known as luminosity L,
  - apparent brightness
    - how bright something appears to be
    - depends on distance
    - the amount of light energy reaching each square meter per second
    - measured in watts per square meter
    - known as flux f.
- They are related by the inverse square law: $f = L/4\pi R^2$ (R is the distance).
- We can use this to measure the luminosity of the Sun (and other bodies).

Electromagnetic waves
- Waves have a wavelength λ, frequency f and speed c.
- These are related by $\lambda f = c$.
- The speed of light is very fast: 186 000 miles per second, or 3×10^8 meters per second.
- Light is an electromagnetic wave.
- Radio waves and infrared have longer wavelengths than light: ultraviolet, X-rays and gamma rays have shorter wavelengths.

Doppler effect
- If the source moves away from the observer, wavelengths are increased.
- If the source moves toward the observer, wavelengths are decreased.
- Provides us with a "speedometer": measure the wavelength and deduce the velocity.
- The "speedometer" only measures relative motion: either the source or the observer (or both) can be moving.
- The "speedometer" only measures motion toward or away from the observer: it does not measure sideways motion.
- Formula: $\Delta\lambda/\lambda = V/C$, where $\Delta\lambda/\lambda$ is the fractional change in wavelength, V is the relative speed between source and observer and C the velocity of the wave.

Blackbody radiation
- Is the electromagnetic radiation emitted by a hot body.
- Stefan–Boltzmann law:
  - tells us the total amount of energy given off per second,
  - grows larger as the body grows hotter,
  - formula: $L = 0.57 \times 10^{-7} A T^4$ watts.
- Wien law:
  - tells us the wavelength at which most radiation is emitted,
  - this wavelength grows shorter as the body grows hotter,
  - formula: $\lambda_{most\ emission} = 0.003/T$ meters.

Spectra
- Are studied by breaking light into its component colors (wavelengths).

- Blackbody radiation:
  - is emitted by dense objects,
  - the spectrum consists of many wavelengths,
  - not all are in the visible portion of the spectrum.
- Spectral lines:
  - pertain to rarefied gases,
  - emission lines are individual wavelengths at which radiation is absorbed,
  - absorption lines are individual wavelengths at which radiation is absorbed,
  - if a rarified gas is heated it emits a certain set of wavelengths: alternatively if light passes through it, it absorbs exactly the same set of wavelengths,
  - each element possesses its own unique spectrum, so that a spectrum can be used as a "fingerprint" identifying the element.

Atomic structure

- An atom is very small.
- An atom consists of a nucleus surrounded by electrons.
- Nucleus:
  - much smaller than the atom,
  - most of the mass of the atom,
  - composed of protons (positive charge) and neutrons (no charge).

- Electrons:
  - move about the nucleus,
  - have very small mass,
  - negative charge.
- If an electron lowers its energy of motion, radiation is emitted carrying away the lost energy.
- If an electron increases its energy of motion, this is because it absorbed radiation giving it the gained energy.

Quantum mechanics

- Governs the behavior of tiny things.
- Tells us matter has wave properties and waves have particle properties.
- Predicts that electrons in an atom can only have certain energies.
- Predicts that the particles associated with light ("photons") have energy $E = hf$ (where h is Planck's constant).
- Spectral lines arise when electrons jump from one allowed energy state to another:
  - if the jump is to lower energy, a photon is emitted and an emission line is formed,
  - if the jump is to higher energy, this is because a photon was absorbed and an absorption line was formed.

Problems

(1) (A) Which has the greater *luminosity*: a distant star, or a candle one inch away? (B) Which has the greater *flux*: a distant star, or a candle one inch away? (C) Which has the greater *apparent brightness*: a distant star, or a candle one inch away? (D) Which has the greater *absolute brightness*: a distant star, or a candle one inch away?

(2) Just as we do for light, we can think about the *apparent loudness* and *absolute loudness* of a sound. Suppose the sound we are thinking about is that of a loudspeaker playing some music. Discuss these two concepts with regard to this example.

(3) (A) Suppose you move three times farther from a light bulb. By what factor does its brightness change? (B) How about if you move three times closer to it?

(4) You are looking at two 100-watt light bulbs. One seems 16 times brighter than the other. How much closer to you is the brighter one than the dimmer one?

(5) You are looking at two light bulbs, both the same distance from you. One seems 16 times brighter than

the other. How much more luminous is the brighter one than the dimmer one?

(6) (A) You are looking at two light bulbs. One has four times the luminosity of the other, but it is twice as far away. Do the two appear to you to be equally bright? If not, which is brighter? (B) You are looking at two light bulbs. One has four times the luminosity of the other but it is three times farther away. Do the two appear to you to be equally bright? If not, which is brighter?

(7) Neptune is 30 times farther from the Sun than we are. How much dimmer is sunlight on Neptune than here?

(8) We want to find the flux we receive from a 40-watt light bulb 1 meter away. (A) Describe the logic of the calculation you will use to do this. (B) Carry out the calculation.

(9) We want to find the flux we receive from a 40-watt light bulb 3 meters away. (A) Before doing any calculation predict how much fainter this flux will be than the answer you got in the previous problem. (B) Carry out the calculation. (C) Compare your result to the answer you got in

Problem (3) and check to see if your prediction had been correct.

(10) Suppose we receive a flux of 20 watts per meter^2 from a light bulb 10 meters away, and we want to find the light bulb's luminosity. (A) Describe the logic of the calculation you will use to do this. (B) Carry out the calculation.

(11) Suppose we receive a flux of 3 watts per meter^2 from a 100-watt light bulb, and we want to find out how far away the bulb is. (A) Describe the logic of the calculation you will use to do this. (B) Carry out the calculation.

(12) Suppose you are standing waist-deep in a pond. (A) Ripples on its surface make the water level move up and down against your tummy with a frequency of 3 cycles per second. What is the period of these waves? (B) Other waves come by that have a period of 5 seconds. With what frequency do these waves make the water level move up and down against your tummy?

(13) Suppose you are standing waist-deep in a pond. Ripples on its surface make the water level move up and down against your tummy by an inch, with a frequency of 3 cycles per second. Now some much larger waves come by that move the water level up and down across your tummy by 5 inches, but still with a frequency of 3 cycles per second. Do these larger waves have the same period as the smaller ones?

(14) The note D above middle C has a frequency of 293 cycles per second, and the note E above that has a frequency of 330 cycles per second. (A) Find the periods of the waves corresponding to these notes. (B) Find the wavelengths of these waves. Express your answers in inches and feet to get an intuitive feel for them.

(15) In each of the situations illustrated in Figure 4.28, indicate whether the observer sees a redshift, a blueshift, or no shift in the wavelength of light from the source.

(16) The orbital velocity of the Earth around the Sun is 30 kilometers per second. Consider a star that emits light of wavelength 6000 angstroms and that is not moving. Suppose that the star lies directly in front of us in our orbit, so that we are heading directly toward it. We want to find the wavelength of the light we receive from it. (A) Describe the logic of the calculation you will use to do this. (B) Carry out the calculation. (C) Suppose we wait six months. Draw a diagram illustrating why the Earth is now moving directly

Figure 4.28 **Relative motion and the Doppler effect.**

away from the star. (D) Describe the logic of the calculation you will use to find the wavelength of the light we receive from the star in this configuration. (E) Carry out the detailed calculation. (F) Notice that during this six-month period the wavelength changed from being *less than* to *more than* 6000 angstroms. So there must have been some time when the wavelength just *equalled* 6000 angstroms! Draw a diagram illustrating where the Earth is in its orbit at this time.

(17) Suppose a star is moving at an unknown velocity, and it is emitting light of a true wavelength 6000 angstroms. But when we measure the wavelength of that light, we find the observed wavelength to be 6001 angstroms. We want to find the component along the line of sight of our relative velocity. (A) Describe the logic of the calculation you will use to do this. (B) Carry out the calculation.

(18) Suppose the burner on an electric stove triples its temperature. By what factor does its blackbody emission increase?

(19) Suppose one person has half the surface area of another. How much more intense is the emission of infrared radiation from the bigger person than the smaller?

(20) Imagine two stars: A and B. Star A has half the surface area of star B, and twice its temperature. (A) Do they have the same luminosity? (B) If not, which is brighter? (C) By what factor?

(21) Suppose you wanted to use the Stefan–Boltzmann law to calculate the total amount of infrared emission put out by the Earth. What would you need to know about the Earth in order to do this?

(22) Suppose you wanted to use the Stefan–Boltzmann law to calculate the size of a star. What would you have to know about this star in order to do this?

(23) Consider a baked potato in an oven at 400 °F = 478 kelvin. (A) If its surface area is 0.02 square meters, how many watts does it emit? (B) At what wavelength does most of this radiation lie? Why can't you see it? (C) Describe the logic of your calculation, and (D) carry out the detailed calculation.

(24) Suppose a star had half the surface area of our Sun, and twice its luminosity. What would be its surface temperature? (A) Describe the logic of your calculation, and (B) carry out the detailed calculation.

(25) The entire Universe is filled with the so-called Cosmic Background Radiation, a faint glow left over from the Big Bang (Chapter 18). Most of this radiation is at a wavelength close to 0.001 meters. What is its temperature?

(26) (A) Suppose you have taken the spectrum of a star, and you see it consists of blackbody radiation together with a set of absorption lines. Describe two ways that you can find the star's temperature from the spectrum. (B) Suppose now that you find a way to measure the star's distance and apparent brightness (i.e. flux). Describe how you can now find the star's luminosity. (C) Describe how you can now determine the star's surface area and therefore its radius.

(27) (A) How many He nuclei does it take to cross a pea? (B) If you lined up that many peas end-to-end, how far would they stretch?

(28) Suppose you had a one ton truck. What sort of object has 1/1836 of its mass?

(29) Consider two strings on a harp: a long one and a short one. Which emits the higher pitch? Which vibrates at a longer wavelength?

5

The astronomers' tools: telescopes and space probes

In the early days, the only way to study the Universe was to look at it with the naked eye. As we have seen, Galileo's use of the telescope revolutionized astronomy. But his was not the only revolution. More recently new instruments such as radio telescopes and X-ray telescopes have been constructed, capable of "looking" at the sky in entirely new wavelengths. Some of these telescopes sit on the ground: others orbit in space. Each has changed the way we do astronomy.

In recent decades the space program has made it possible to study the Moon and planets by actually visiting them. Astronauts have walked on the Moon, and robotic space probes have visited every planet of the Solar System save Pluto. This entirely new way of studying the Solar System has brought back a wealth of information.

The telescope is the very symbol of astronomy, the most important instrument at our disposal. Why do we need it? What does a telescope do for us that our naked eyes cannot?

- On the one hand, telescopes *magnify* the heavens. They allow us to discern things too small to be seen with the naked eye. It was this ability to expand our view that allowed Galileo to discover spots on the Sun, mountains on the Moon and the satellites of Jupiter.
- On the other hand, telescopes *gather more light* than the naked eye. They allow us to see objects too faint to be seen otherwise. As an example, consider the beautifully detailed view of the nearby Andromeda Galaxy (Figure 5.1). This view could only have been obtained through a telescope – not because this galaxy is too small to be seen with the naked eye (it is actually larger in the sky than the Moon!) but because it is so faint that to the naked eye it appears merely as a diffuse, vague glow.
- Finally, telescopes *extend the range of wavelengths* accessible to observation. The human eye is sensitive only to electromagnetic radiation with wavelengths in a very narrow range. But with telescopes we can observe across the full span of the electromagnetic spectrum.

Figure 5.1 **The Andromeda Galaxy.** 👁 **(Also see color plate section.)**

The pinhole telescope

How does a telescope work? We can gain an important insight into this by studying an amazingly simple device: the pinhole telescope (alternatively known as the pinhole camera). While it is impractical for actually observing astronomical objects, such a telescope is so simple that it is easy to see how it functions. In later sections we will use the insights we have gained from it as we move on to more complex actual telescopes.

We begin with two experiments.

In the *first experiment* we simply take a detector of light and stick it in one end of a cylinder. Other than this, our cylinder is completely empty: it has no lens at the other end, nor any other fancy optical equipment. This first device could be so simple as an empty cardboard cylinder with a light detector at one end (Figure 5.2).

We point this device at a scene. Does our light detector show an image of the scene? We find that it does not. Rather, it records a uniform distribution of light (Figure 5.3).

Our first attempt at building a telescope was a failure. In the *second experiment* we simply cover up the left-hand end of our first attempt, and then punch a tiny hole in the cover with a pin (Figure 5.4).

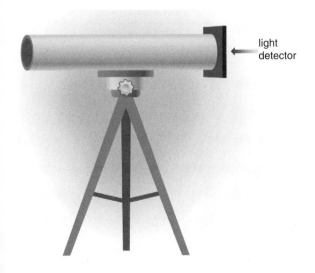

light detector

Figure 5.2 **First attempt at a telescope.**

Figure 5.3 **Our first attempt at building a telescope fails. It produces only a uniform wash of light across the detector.**

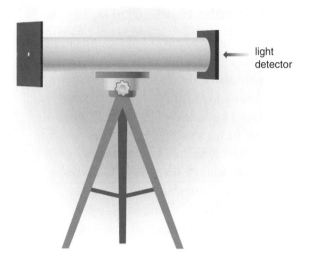

Figure 5.4 **The pinhole telescope makes an image.**

Figure 5.5 **The image produced by a pinhole telescope of a single light bulb on a Christmas tree.**

Now we find to our astonishment that an image is produced! We have succeeded in building a crude telescope, known as a *pinhole telescope.*

How does so simple a device work – and why did our first attempt not work? We need to think about rays of light from the scene at which our telescope is pointing, and how these rays get to the light detector. Let us imagine that our scene is a Christmas tree at nighttime festooned with lights. We will begin with a particularly simple scene in which only a single light is shining: other than this one point of light, the Christmas tree is dark. Figure 5.5 illustrates what we find on the light detector if we point our pinhole telescope at such a scene. A single point on the detector is receiving light: this is the image of our single light bulb.

Figure 5.6 illustrates how the telescope produced this image. Light rays stream outwards in every direction away from the bulb. But only one of them reaches the detector – and it arrives at only one point on it. The other rays in Figure 5.6 would have reached other points on the detector, but they are blocked by the opaque screen on the telescope's left-hand end.

In our first attempt at building a telescope this screen was absent, so these other rays *did* reach the detector, and they were recorded at other points (Figure 5.7). This is why the entire detector had recorded light in our first attempt at building a telescope: the light from the bulb is spread out across it.

We have reached an important insight.

> An image is formed when the rays from a source of light reach *a single point* on the light detector. If the rays from this source reach many points on the detector, no image is formed.

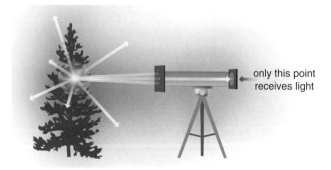

Figure 5.6 **How a pinhole telescope produces an image. Light from the light bulb reaches only one point on the detector.**

Figure 5.7 **The entire light detector receives light from the light bulb in the absence of a pinhole cover.**

If the rays reach many points, we do not get an image. This is why our first attempt at building a telescope failed to produced an image: the rays from a single bulb arrive at every point on the detector.

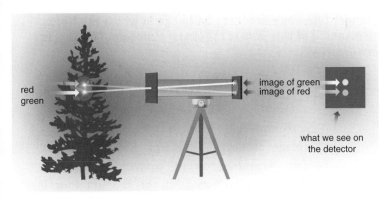

Figure 5.8 **A pinhole telescope produces an upside-down image.**

....

NOW YOU DO IT
Suppose you punch *two* tiny holes in your pinhole telescope. Draw a picture illustrating the image this produces of a single light bulb.

....

Notice that the image of the Christmas tree in Figure 5.5 is upside down. Figure 5.8 illustrates the reason. It shows rays of light emanating from *two* light bulbs on our Christmas tree: a red one and a green one. As you can see, while in reality the red bulb lies above the green one, in the image it lies below it.

....

NOW YOU DO IT
In this problem we study the way a pinhole telescope (with one hole) magnifies things. Imagine that the scene the telescope is looking at has two light bulbs separated by a certain distance:

If a pinhole telescope points at such a scene, it will produce an image like this:

Your task in the problem is to imagine *a longer telescope*: one with a pinhole the same size and the cylinder just as wide – but in which the cylinder is longer, so that the detector is farther away from the pinhole. Draw pictures that show that the image produced by this longer telescope looks like this:

As you can see, the longer telescope produces an image in which the lights are farther apart. The meaning of this is that *the longer a pinhole telescope is, the greater is its magnification.*

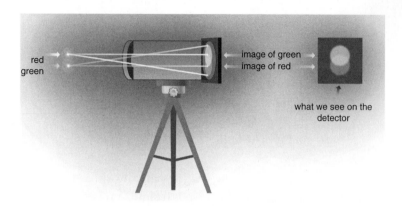

Figure 5.9 **A wide pinhole produces a blurred image.**

While it is simple and easy to understand, the pinhole telescope is no good for studying astronomical objects. This is because the image produced by such a telescope is very faint – and if you are studying objects that are themselves very faint, this is a fatal flaw. It is not hard to see why the pinhole telescope suffers from this flaw. Most rays from the source are blocked by the telescope's opaque screen: only a few make it through the tiny pinhole onto the detector. Furthermore, if we widen the pinhole to accommodate more rays and so make the image brighter, the image blurs. Figure 5.9 illustrates why.

The problem with the pinhole telescope is that it admits only one light ray, and so produces a faint image. *All the complex optical equipment within a modern telescope is designed to cure this one fatal flaw.* Their function is to gather *lots* of light rays, and bring them *to a single point* on the light detector. The pinhole could not do this: mirrors and lenses can.

Reflecting telescopes

Whether it be an optical observatory in Hawaii, a radio observatory in Puerto Rico or an X-ray satellite in space, the basic principle behind a telescope is always the same: the heart of every modern telescope is the *reflector*, which gathers incoming electromagnetic radiation over a large area and reflects it to a focus. Let us study how a mirror can do this.

Begin with a flat mirror. It intercepts many light rays, which is what we need it to do. But remember our basic rule: an image is formed when the rays from a source of light reach *a single point* on the light detector. If the rays from this source reach many points on the detector, no image is formed. As illustrated in Figure 5.10, a flat mirror does not focus rays to a single point. So it does not produce an image.

In order to bring all the light rays to a single point and so produce an image, we need to bend the mirror into a curve (Figure 5.11).

Figure 5.11 illustrates light rays reaching the telescope from a single point, such as a star. Figure 5.12 illustrates how the telescope forms an image of a larger object, in which we can actually see detail. Just as the pinhole telescope, the image formed is upside down. Remarkably, this is also true of the human eye: the light rays striking the retina form an inverted

Figure 5.10 **A flat mirror fails to focus light rays onto a single point.**

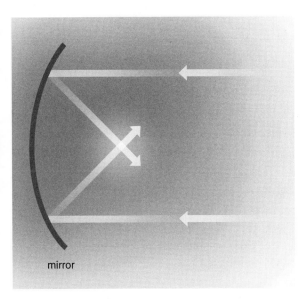

Figure 5.11 **A curved mirror collects incoming light rays and brings them to a focus.**

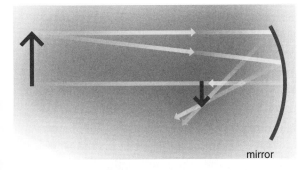

Figure 5.12 **Images are inverted by a reflector.**

view of the world! This inversion is reversed somewhere within the brain, so that what we experience appears right side up. Binoculars and home telescopes employ special optics to accomplish the reversal. Astronomical telescopes, on the other hand, are able to dispense with these optics since there is no "up" or "down" in the sky.

Notice an awkward element to Figure 5.12: the focus point lies in front of the reflector, directly in the path of the incoming light. Any equipment we use to study the image must sit at this focus point, as in the so-called *prime focus* design of Figure 5.13(a). If the observing equipment is not too bulky, this does not pose much of a problem. But bulky equipment will block much of the light, and the telescope's usefulness will be degraded.

Several designs have been developed to circumvent this problem (Figures 5.13b–d). In the *Newtonian* system, a small mirror is placed just short of the focus point, which directs the converging beam to a focus to the side of the telescope. In the *Cassegrain* system, the mirror reflects the converging beam straight back toward the reflector, where it passes through a small hole in the reflector's center and is brought to a focus beneath it. (One might think that the hole in the mirror would produce a corresponding hole in the image, but this is not true: the image is formed by all the light rays falling on all of the reflector, and even though part of the reflector is missing a complete image is formed.) Finally, the preferred arrangement if the observing equipment is particularly bulky is the *coudé* arrangement (from the French for "elbow"), which sends the beam to a focus located in a room entirely separated from the telescope.

Refracting telescopes

So far we have been considering *reflecting telescopes* – those with mirrors. Historically speaking, *refracting telescopes* – those with lenses – were developed first, and even today they remain the instrument of choice for many amateur astronomers. But they are hardly ever used for astronomical research.

The basic principle of the lens is illustrated in Figure 5.14. A ray of light passing from air into glass is bent, or *refracted*. If the glass is formed into the shape of a lens, the rays striking it are refracted to a focus, as in Figure 5.15. Just as a mirror brings light rays to a point, so too with the lens. Figure 5.16(a) illustrates a refracting telescope using the

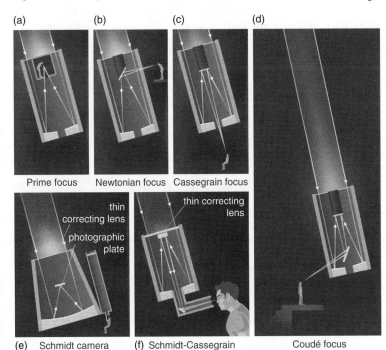

(a) Prime focus (b) Newtonian focus (c) Cassegrain focus

(e) Schmidt camera (f) Schmidt-Cassegrain Coudé focus

thin correcting lens
photographic plate
thin correcting lens

Figure 5.13 **Six basic telescope designs of reflectors.**

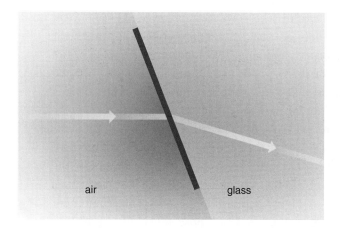

Figure 5.14 **Light passing from air into glass is refracted, i.e. its path is bent.**

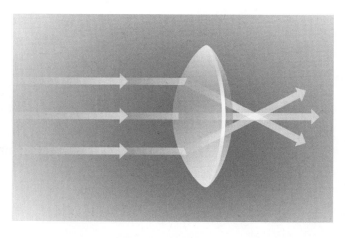

Figure 5.15 **Light passing through a lens is refracted so that it is focused to a point.**

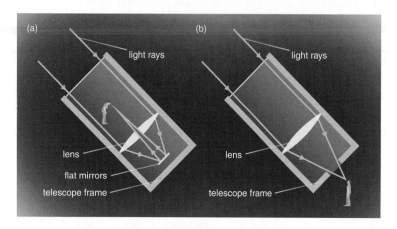

Figure 5.16 **Two telescope designs of refractors. (a) Cassegrain is never used, (b) prime focus is common.**

Cassegrain design. As you can see, the incoming light rays are partially blocked by the observer. For this reason refracting telescopes never use this design, but instead employ the prime focus design of Fig. 5.16(b).

Why are refracting telescopes not employed for astronomical research? Because they suffer from a flaw. Figure 5.14, illustrating the bending of light passing from air into glass, concentrated on a single wavelength. As we mentioned in Chapter 4, however, the bending depends on the wavelength. The shorter the wavelength, the greater the bending (Figure 5.17). The consequence is that light of different wavelengths – different colors – is focused to different points, as illustrated in Figure 5.18. Since the light from an astronomical object contains many colors, the image is sharp for only one color, and is blurred for all the rest. This distortion is known as *chromatic aberration*. Reflecting telescopes do not suffer from this problem, so that every color is focused to the same point. With refractors, sophisticated multiple lenses have been designed that alleviate the problem, but only partially; for research it has proved best to employ only reflectors.

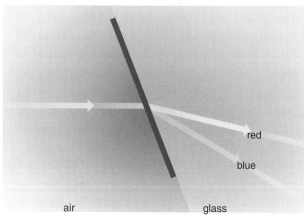

Figure 5.17 **Differing wavelengths are refracted by differing amounts, with short wavelengths bent more than long.**

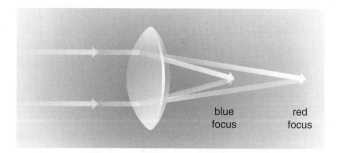

Figure 5.18 **Chromatic aberration: different wavelengths are focused to different points by a lens.**

Telescope mountings

⇐ **Looking backward**
We studied the apparent rotation of the sky in Chapter 1.

Amateur telescopes are sometimes mounted on a tripod. Anyone who has used such a mounting immediately discovers its shortcomings. If the telescope is trained at an astronomical object, the object almost immediately drifts out of the field of view.

The telescope did not slip. Rather the sky rotated overhead. This is caused by the Earth's eastward rotation: as we saw in Chapter 1, our rotation causes not just the Sun but everything to rise, drift across the sky, and set. To compensate for this, the telescope must be steadily swung westward, tracking the apparent motion of the sky. This is accomplished by means of a *sidereal drive* ("star" drive), sometimes known as a "clock drive."

(a)

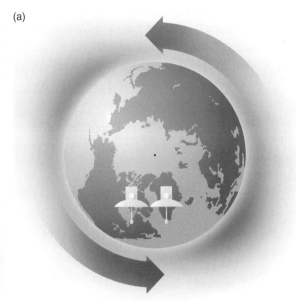

(b)

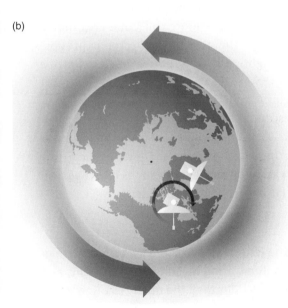

Figure 5.19 **A sidereal drive swings the Canadian telescope to compensate for the rotation of the Earth. The telescope in Greenland does not have such a drive.**

NOW YOU DO IT
Imagine that you were to set up a telescope on a planet that rotated more rapidly than the Earth. Would objects seen through that telescope remain in its field of view for a longer or shorter time than they do here?

Figure 5.19 shows two telescopes: one in Northern Canada and the other in Greenland. The Canadian telescope is equipped with a sidereal drive while the one in Greenland is not. Figure 5.19(a) shows the configuration at first: both telescopes are pointing at a distant star (lying straight down in the figure). Figure 5.19(b) shows the configuration after time has passed and the Earth has rotated: as you can see, the Canadian telescope is still pointing at the star but the Greenland one is not.

Notice in Figure 5.19 that the axis around which the telescope has rotated is pointing directly at you – which is to say that it is pointing parallel to the Earth's rotation axis. This is known as a *polar mount*. With the alternative *altitude-azimuth mount*, the telescope is mounted on an axis pointing straight up – that is, directly away from the center of the Earth. This is simpler to construct, and in the case of a large, heavy telescope it is preferred. The downside is that in this case the sidereal drive must swing the telescope about two axes simultaneously. This can be accomplished with a computer-controlled drive, but it makes the polar mount preferred by amateurs.

It is vital that the telescope mounting follow the rotation of the sky with great accuracy, and protect the telescope against the slightest vibration. For this reason, the cost of the mounting can exceed that of the optical system.

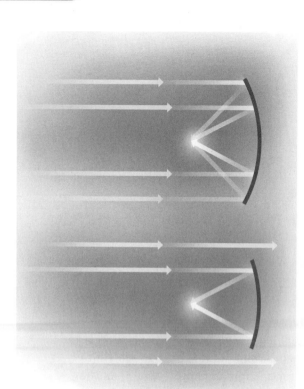

Figure 5.20 **A large mirror "catches" more light rays than a small one.**

The powers of a telescope

As we discussed in the introduction to this chapter, most astronomical objects are too small and/or too faint to be seen with the naked eye. We need telescopes in order to (a) gather more light than the naked eye, and (b) magnify the heavens. We now consider each of these powers in turn.

Light-gathering power

We have emphasized that pinhole telescopes produce images that are too faint because they admit only a small amount of light. Reflecting and refracting telescopes, on the other hand, "catch" a great deal of light, and so produce brighter images. As illustrated in Figure 5.20, the larger the mirror the more light is caught. So a telescope with a large mirror can see fainter objects, and more distant objects, than a small one. (Everything we say in this section applies to refracting telescopes as well as reflectors.) This is one reason why astronomers persist in building ever-larger telescopes.

We can also investigate this quantitatively. Recall that the apparent brightness of a distant astronomical object – its flux – is the quantity of light energy we get from it per second *per area*. So the total quantity of light gathered by a telescope is the light per unit area f reaching it times the area A of the telescope's mirror:

$$\text{Total light "caught" by the telescope} = fA$$

> **⇐ Looking backward**
> We studied the inverse square law in Chapter 4.

As you can see, the total light "caught" by the telescope is proportional to the surface area of its mirror. Since a big telescope has more collecting area than a small one, it gathers more light. And since the area of a reflector is proportional to its diameter squared, the light-gathering power of a telescope depends on the square of its diameter. So a telescope twice as big does not gather twice as much light – it gathers four times as much light.

Let us compare the capability of a telescope to that of the naked eye. The diameter of the pupil of your eye is very roughly ¼ cm. If we choose 10 cm as the diameter of even a modest telescope, we see that this telescope gathers $[(10 \text{ cm})/(¼ \text{ cm})]^2 = 40^2 = 1600$ times as much light as your eye. So by looking though the telescope, you will be able to detect light fainter by a factor of 1600 than otherwise.

....

NOW YOU DO IT
Suppose you have two telescopes: the lens of one is three times bigger than that of the other. (A) Draw diagrams illustrating incoming rays of light, and showing why the bigger one catches more light than the other. (B) How much more light does it catch than the other?

The biggest telescope we have ever built is the Keck Observatory: its mirror is 10 meters in diameter. The Hubble Space Telescope, on the other hand, has a mirror 2.4 meters in diameter. How much more light does the Keck telescope gather than the Hubble? (A) Describe the logic of the calculation you will use. (B) Carry out the detailed calculation.

....

We can also calculate *how much farther* you can see with such a telescope than with your naked eye. With the naked eye you can see stars out to a certain distance. But if you were to use that 10-cm telescope, you would be able to see stars with a flux fainter by a factor of 1600. Since the flux we receive from the star follows the inverse square law,

$$f = L/4\pi R^2$$

(where R is the star's distance), a star $\sqrt{1600} = 40$ times farther away is a factor of 1600 fainter. We conclude that, looking through our telescope, we would be capable of seeing a star 40 times farther away than with the naked eye.

We emphasize, furthermore, that a 10-cm telescope is tiny by modern standards. Consider the Keck Observatory, with its 10-meter mirror. Since 10 meters is 1000 cm, the area of this telescope is $[(1000 \text{ cm})/(1/4 \text{ cm})]^2 = 16$ million times that of the human eye.

NOW YOU DO IT
How much farther can the Keck see than the naked eye? (A) Describe the logic of the calculation you will use. (B) Carry out the detailed calculation.

Modern detectors such as the charge-coupled device (CCD) render telescopes such as the Keck even more powerful than this.

Magnification

For observing the heavens, we want our telescope to magnify as much as possible. The *resolving power* of a telescope measures how well it can discern fine detail. For example, looking through a telescope with a good resolving power you might be able to read the writing on a faraway billboard, while using a telescope of poor resolving power you could not. The letters on that billboard subtend a certain angle: the resolving power is the smallest value this angle could have which the telescope can distinguish (Figure 5.21). (Be careful here: this is a little confusing, since the *smaller* this angle the *better* the resolving power.)

Unfortunately there are limits to the resolving power a telescope can achieve. Let us study these limits.

To improve the resolving power we need to increase a telescope's magnification. We will start with a very simple kind of magnification. Suppose we have a photograph taken through a telescope, and we merely zoom in on the photograph. This will allow us to make out finer and finer details in it, details that originally were invisible. But will we be able to enlarge our photo indefinitely, searching for ever-finer details? It turns out that we cannot.

Figure 5.22 illustrates what happens if we zoom in on a photograph too much. The original photograph appears quite smooth. But in succeeding enlargements, we begin to see that a CCD image is actually composed of pixels. Normally these pixels are so small that we do not notice

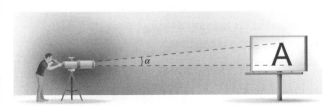

Figure 5.21 **The resolving power of a telescope measures the smallest value of α it can distinguish.**

zoom 32x

zoom 16x

zoom 8x

zoom 4x

original image:
no zoom

Figure 5.22 **Enlarging a CCD image too much reveals its pixel structure.**

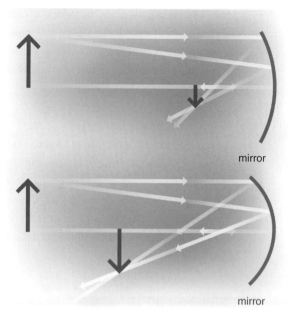

Figure 5.23 **A larger image is produced by a reflector with a longer focal length.**

them. But as we zoom in, they eventually become noticeable. And finally, we have zoomed in so much that nothing can be made out.

We have reached a limit imposed by the pixelated nature of a light detector. The surface of every light detector is peppered with a very large number of individual tiny pixels. If the photograph is enlarged too much, eventually we will merely be seeing the pixels themselves, rather than resolving finer and finer details.

NOW YOU DO IT
Try this for yourself!

We conclude that a light detector cannot produce a perfect image: enlarging the image indefinitely will not allow us to see ever-finer details within it. Let us therefore try a different method. Let us try to make our original image so large that we don't *need* to enlarge it in order to see fine details! Figure 5.23 shows how we can accomplish this. The distance from a mirror to its focus point is its *focal length*: as illustrated, the greater the focal length, the larger the image. So to obtain a great magnification of the heavens, we build a telescope with a long focal length. (We saw earlier in the chapter that this is also true of a pinhole telescope: the longer such a telescope, the greater its magnification.)

Unfortunately, there are limits to this method as well. Figure 5.24 shows what happens if we observe the same object with telescopes of ever-longer focal length. As you can see, succeeding images are larger and show more detail. Furthermore, the early images appear quite sharp. But as we go to larger and larger magnifications, they begin to become blurred. And finally, the last magnification is so great that the blurring has become overwhelming: nothing can be made out.

Here, too, we are encountering a limitation in how fine a detail can be observed – and this limitation has nothing to do with some

zoom 64x

zoom 32x

zoom 16x

zoom 4x

original image:
no zoom

Figure 5.24 **Large images are blurred, both by atmospheric turbulence (the "seeing limit") and by the wave nature of light (the "diffraction limit").**

imperfection in the telescope. The blurring we are seeing here is intrinsic to the light arriving at the detector. It happens for two separate reasons: (A) a blurring caused by turbulence in the Earth's atmosphere, and (B) a blurring caused by the wave nature of light.

We turn to each of these in turn.

(A) Atmospheric turbulence: "seeing"

If you look at a stone lying on the bottom of a swimming pool, you will notice that it appears to waver to and fro. The image of the stone dances about, constantly shifting its position and shape, sometimes even splitting in two. This phenomenon is caused by the irregular refraction of light from the stone as it passes through the ripples and waves on the pool's surface.

A similar thing happens to the light from a distant star as it passes through the Earth's atmosphere. The atmosphere is never static. Rather, warm and cool cells of air shift about overhead, refracting the light rays from the star this way and that. As a consequence, the image of the star continuously shifts about. This shifting blurs the image of every astronomical object. For instance, if you look at a star through a telescope you will not perceive it to be the tiny, point-like object it actually should appear to be: rather its image will be an extended, fuzzy blob that churns and fluctuates incessantly.

This fluctuation, known as "seeing," is what causes stars to twinkle. It is the bane of ground-based astronomy, and it blurs every astronomical image made from the ground. The only way to avoid it is to lift the telescope above the atmosphere, and launch it into space.

(B) The wave nature of light: the "diffraction limit" on a telescope's resolving power

A telescope focuses light perfectly only so long as light rays travel in straight lines through the telescope. As we have seen in Chapter 4, light (and all other forms of electromagnetic radiation) is actually composed of waves. These waves do not travel in exactly straight lines. Rather, as they travel through the telescope, they are subtly bent, a phenomenon known as "diffraction." As a consequence, they do not converge to a perfect focus.

It can be shown that diffraction blurs the image produced by a telescope by an angle ϕ given by

$$\phi = 2.1 \times 10^5 \lambda / D \text{ seconds of arc,}$$

where λ is the wavelength of the electromagnetic radiation the telescope is observing, and D is the diameter of its mirror (or lens). (One second of arc is 1/60th of a minute of arc, which itself is 1/60th of a degree.)

Return to Figure 5.21 illustrating a telescope focused on a distant billboard. Suppose the angle subtended by the letters on the billboard is α. But diffraction smears out everything seen through our telescope by the angle ϕ. If ϕ is less than α, the letters will be legible. But if ϕ is greater than α, the blurring is so great as to render them illegible. So the angle ϕ given by the above formula sets the resolving power of our telescope.

This formula tells us the best possible resolving power a telescope could achieve, even if all its optics and detectors were perfect and if there were no atmospheric "seeing." We see that this resolving power depends on both the size of our telescope (D) and the wavelength of electromagnetic radiation (λ) we are observing. Consider the first variable. We see from the formula that the bigger is D the smaller is ϕ: *bigger telescopes have better resolving power than small ones* (remember: small ϕ means a good resolving power).

This formula is valid for a telescope observing electromagnetic radiation of any wavelength whatsoever, whether it be visible light, radio waves or gamma rays. Let us apply it to an optical telescope, and choose the observing wavelength λ to be a typical visible wavelength of 5500 angstroms, or 5.5×10^{-7} meters. We obtain *the maximum possible resolving power of an optical telescope:*

$\phi = 0.116/D$ seconds of arc if D is measured in meters.

Let us get some practice using this formula. Let us return to our 10-cm telescope. In the previous section we studied its light-gathering power. Now let us study its resolving power.

The logic of the calculation

Step 1. Convert the diameter D into meters.
Step 2. Calculate ϕ.

Detailed calculation

Step 1. Convert the diameter D into meters.
  10 centimeters is $1/10$th of a meter.

Step 2. Calculate ϕ.
  We have

  $\phi = 0.116/D = 0.116/0.1 = 1.16$ seconds of arc.

As you can see from the detailed calculation, such a telescope has a resolving power of 1.16 seconds of arc.

NOW YOU DO IT
We want to calculate the resolving power of a 1-meter telescope. (A) Describe the logic of the calculation you will use. (B) Carry out the detailed calculation.

Is such a telescope capable of reading the writing on a distant billboard? Let us suppose that each letter is a foot high, and the billboard is a mile away.

The logic of the calculation

Step 1. Find the angle α subtended by something a foot high a mile away.
Step 2. If α is greater than 1.16 seconds of arc the telescope can resolve the writing: if α is less than 1.16 seconds the writing will be illegible through the telescope.

Detailed calculation

Step 1. Find the angle α subtended by something a foot high a mile away.

The small-angle formula helps us here (Appendix I):

$$\alpha = [d/2\pi D]\ 360\ \text{degrees},$$

where d, the height of the letters, is 1 foot and D, their distance away, is 1 mile. Plugging in we find

$$\begin{aligned}\alpha &= [1\ \text{foot}/(2\pi)\ (1\ \text{mile})]\ 360\ \text{degrees}\\ &= [1\ \text{foot}/(2\pi)\ (5280\ \text{feet})]\ 360\ \text{degrees}\\ &= 1.1 \times 10^{-2}\ \text{degrees} = 40\ \text{seconds of arc}.\end{aligned}$$

Step 2. If α is greater than 1.16 seconds of arc the telescope can resolve the writing: if α is less than 1.16 seconds the writing will be illegible through the telescope.

Because 40 seconds of arc is a far bigger angle than 1.16 seconds of arc, the writing will be legible through the telescope.

As you can see from the detailed calculation, such a telescope is more than capable of reading the writing on the billboard.

NOW YOU DO IT
We want to find out if a 1-meter telescope is capable of resolving the writing on a distant billboard. Suppose the letters are 1 foot high, and the billboard is 25 miles away. (A) Describe the logic of the calculation you will use. (B) Carry out the detailed calculation.

Now let us ask a more astronomically interesting question: is such a telescope capable of seeing a crater on the Moon? We will consider a crater 100 meters in diameter – this is about the size of a city block.

The logic of the calculation

Step 1. Find the distance D to the Moon.

Step 2. Set $d = 100$ meters and use the small-angle formula to calculate α.

Step 3. If α is greater than 1.16 seconds of arc the telescope can resolve the crater: if α is less than 1.16 seconds the crater is too small to be seen through the telescope.

As you can see from the detailed calculation, such a telescope will not be capable of resolving a 100 meter crater on the Moon.

NOW YOU DO IT
Is our 1-meter telescope big enough to resolve a 100-meter crater on the Moon? (A) Describe the logic of the calculation you will use. (B) Carry out the detailed calculation.

Next-generation telescopes

Modern telescopes employ two new technologies that greatly enhance their capabilities: active optics and adaptive optics.

Active optics

The shape of a telescope's mirror must be perfect in order accurately to focus the incoming light. But several factors act to deform a mirror.

One such factor is its weight, which causes it to sag. This sag depends on the orientation of the telescope: when it points straight up the mirror

Detailed calculation

Step 1. Find the distance D to the Moon.

The Moon's distance is 384 000 kilometers = 3.84×10^8 meters.

Step 2. Set d = 100 meters and use the small-angle formula to calculate α.

The small-angle formula is

$$\begin{aligned}\alpha &= [d/2\pi D]\ 360\ \text{degrees}\\ &= [100\ \text{meters}/(2\pi)(3.84 \times 10^8\ \text{meters})]\ 360\ \text{degrees}\\ &= 1.5 \times 10^{-5}\ \text{degrees} = 0.05\ \text{seconds of arc}.\end{aligned}$$

Step 3. If α is greater than 1.16 seconds of arc the telescope can resolve the crater: if α is less than 1.16 seconds the crater is too small to be seen through the telescope

Because 0.05 seconds of arc is far smaller than 1.16 seconds of arc, the telescope is not capable of making out such a crater on the Moon.

lies flat, but if the telescope points at a star near the horizon the mirror is vertical. Another factor is its temperature, which drops during the course of a night's observing. Since glass contracts as it cools, the mirror's shape changes accordingly. Minute variations over its surface of temperature and bearing strength introduce yet further irregular deviations in the reflector's shape.

In an active optical system, sensors continually track the shape of a mirror, and large numbers of small motors act continually to change that shape. For example, the reflectors of the Keck Observatory are actually composed of a great many individual small mirrors, whose orientation is continually monitored and corrected in this way.

Adaptive optics

Adaptive optics go further, and attempt to correct the mirror's shape for the effects of seeing. In one technique a bright "guide star" is identified, lying close to the fainter star one wishes to observe. The guide star is carefully monitored: deviations from the expected point-like image contain information about atmospheric seeing, and can be used to correct for this seeing.

A difficulty with this method is that it only works for regions of the sky relatively close to the guide star, which must be quite bright: since there are not that many bright stars, this technique cannot be used everywhere. A highly experimental technique currently under development uses an intense brief pulse from a laser, pointed at the region of interest (Figure 5.25). This causes atmospheric atoms to emit light and so creates a brief "artificial star" on the sky, by means of which we can probe the state of the atmosphere in that direction. Here, too, information about fluctuations in the atmosphere are used continually to alter the shape of the mirror.

Figure 5.25 **Laser guide star. The Keck telescope in Hawaii projects a laser beam into the night sky. The galactic plane of the Milky Way is visible in the sky to the right of the image. The stars are trailed in this time exposure due to the rotation of the Earth. 👁 (Also see color plate section.)**

Site selection

Once a telescope is constructed and provided with the finest of mountings and optical systems – where should it be put? Many factors govern the siting of a telescope.

On the one hand, it should be far from civilization, so that the sky is as dark as possible. Many historic telescopes have been severely compromised by light pollution from expanding cities and suburban sprawl. For instance, the famous 100-inch telescope atop Mount Wilson in California, with which Edwin Hubble discovered the expansion of the Universe (Chapter 18), looks down upon Los Angeles. The city lights make for a pretty sight at night, but in recent decades they have rendered the telescope relatively useless for front-line research.

In recent years astronomers have worked with nearby cities to enact policies to reduce their light pollution. A conspicuous success story is Tucson, not far from a suite of major observatories. Indeed, among all the states in the USA, Arizona has taken the most steps to limit light pollution.

The blurring of a telescope's view of the sky caused by atmospheric turbulence ("seeing") can

Figure 5.26 **The observatories on Mauna Kea.**

be reduced by situating a telescope atop a mountain. Because much of the atmosphere then lies beneath the telescope, the seeing is consequently reduced. It can be reduced yet further by choosing a mountain above which the atmosphere is particularly stable, in the same way that the wavering of the image of a stone on the bottom of a swimming pool depends on how still is the pool's surface. When a large telescope is being planned, a site selection team conducts detailed observations of the stability of the atmosphere in various locations, searching for the best possible site.

Such studies have identified a number of optimal locations. One is on the summit of Mauna Kea, a 13 800-foot dormant volcano on Hawaii (Figure 5.26). The seeing there is unusually good since the summit is surrounded by thousands of miles of relatively constant-temperature ocean, and has no nearby other mountains to roil the upper atmosphere. Few city lights spoil the viewing. For most of the year, the atmosphere above the summit is calm and dry. Because of these advantages, a dozen world-class telescopes occupy the summit's 11 600-acre science reserve, representing a capital investment of more than half a billion dollars.

Astronomical detectors

Once a telescope has brought the light from a distant astronomical object to a focus, the light must be studied. On the one hand, the *charge-coupled device* (CCD) is used to make images; on the other hand, the *spectrograph* is used to analyze its colors.

The CCD

The charge-coupled device, or CCD, has made possible a dramatic improvement in how astronomers make images of astronomical objects. Previous ages have seen similar advances. As we saw in Chapter 2, the early astronomers had only their naked eyes to study the heavens, and Galileo's invention of the astronomical telescope expanded our view of the sky immeasurably, revealing spots on the Sun, mountains on the Moon and satellites of Jupiter. The invention of photography in the mid 1800s instituted a comparable increase in information-gathering power: in a single night of observing, numerous astronomical photographs could be obtained, each containing more data than could be seen through the telescope eyepiece.

But photographic film is not a perfect device for gathering light, and for many purposes it has been supplanted by CCDs. A CCD is an array of tiny silicon diodes, each of which is extraordinarily sensitive to light. They are arranged in a square about the size of a postage stamp, consisting of several million "picture elements," or *pixels*. When light falls upon a pixel a tiny bit of electric charge is liberated. As the exposure progresses more light arrives, and more charge accumulates. Finally, when the exposure is finished, the quantity of charge on each diode is measured. The net result is a square array of numbers, each one measuring how much light fell on a pixel.

The CCD is a far superior detector of light than photographic film. Fully 90% of the light striking a CCD is actually captured and used to create the image, in contrast to roughly 10% of that striking photographic film. A further advantage is that the output of a CCD is an array of numbers, which can be directly fed into a computer for image analysis. Most commercial camcorders and digital cameras use them as well.

The spectrograph

As we saw in Chapter 4, the spectrum of light contains a wealth of information. To obtain a spectrum, we need to split the light into its constituent colors. This is accomplished by means of a spectrograph.

A prism is a simple spectrograph (Figure 5.27). As we noted in our discussion of refracting telescopes, when light passes from air into glass the degree of bending depends on the wavelength. Thus the various wavelengths present in the light are arranged along a band, and form the spectrum of the light. So chromatic aberration, which makes lenses so unsuitable for astronomical research, here is put to good use.

A defect of the prism is that some of the light is absorbed in passing through the glass. Thus modern spectrographs employ *diffraction gratings*, in which thousands of parallel lines are etched on the surface of a reflecting surface. Light of different wavelengths reflects off the grating at different angles, so producing a spectrum. The iridescent colors one observes in light reflected from a CD arise from a similar effect.

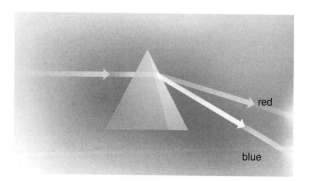

red

blue

Figure 5.27 **A prism splits white light into its constituent colors.**

Radio telescopes

Radio astronomy began in 1933 at Bell Labs in New Jersey, where the engineer Karl Jansky (Figure 5.28a) was given the task of investigating sources of shortwave radio interference. He constructed an antenna on the lawn outside the lab (Figure 5.28b), and found that some of the radio interference was coming from the center of the Milky Way.

Jansky's first radio telescope bore little resemblance to the traditional telescope: it was an array of antennas, each employing the same principle as the radio antenna sticking up out of an

(a)

(b)

(c)

Figure 5.28 **Early days of radio astronomy. (a) Karl Jansky. (b) A reconstruction of Jansky's original radio telescope. (c) Grote Reber.**

automobile. The next important step was taken by Grote Reber (Figure 5.28c), who built a 9.4-meter reflecting dish following the same basic design as an optical reflector (Figure 5.11). He set it up in his mother's back yard in Wheaton, Illinois, and in 1944 produced a radio map of the sky. With the close of the Second World War, the new technology of radar which had been developed for military purposes was pressed into service, and radio astronomy began in earnest.

Radio signals are electromagnetic waves whose wavelengths range from many meters to millimeters. The radio waves emitted by astronomical objects are no different in kind from those you pick up with a home radio set. The only difference is that the astronomical sources are so faint that your set is not capable of detecting them.

But if you could transport your home set into space, and journey with it close to a source of astronomical radio emission, what you would hear over your radio would be a hiss of static. With few exceptions, this hiss would be utterly steady in time, never varying in intensity or pitch. On Earth, each radio station emits its signal at a precisely defined wavelength, so that as you tune your set from one wavelength to the next you receive first one, and then another radio station. But an astronomical source emits at all wavelengths of the radio band, so you would hear its hiss no matter how you tuned your set.

Radio telescopes are always reflectors, the reflecting surface being not a glass mirror but a metallic sheet or mesh. They are often of the prime focus design of Figure 5.13(a), with the detector situated above the reflector at its focus. As opposed to optical telescopes, in which the detector is usually a CCD, the detector of radio signals is an antenna in conjunction with an amplifier. These amplifiers are in principle no different from those in a home radio, but they are immeasurably more sensitive. The needs of science being what they are, a radio astronomer does not bother to feed the amplifier's output into a loudspeaker, but rather sends it straight to a computer for analysis.

Resolving power of a radio telescope

Above we discussed the resolving power of a telescope. Our formula for the resolving power

$$\phi = 2.1 \times 10^5 \lambda/D \text{ seconds of arc}$$

is valid for every region of the electromagnetic spectrum. Let us use it to find the resolving power of a radio telescope.

According to this formula, big values of λ imply big values of ϕ; *the longer the wavelength we observe, the poorer is our resolving power* (remember that large values of ϕ imply poor resolving power). This means that radio telescopes, which observe long wavelengths, have inherently poorer resolving power than optical telescopes that observe shorter wavelengths. For this reason radio astronomers are forced to build particularly large telescopes. They increase the diameters D of their telescopes, so as to compensate for the very large λs at which they operate. The radio telescope in Arecibo, Puerto Rico, for instance, is the largest telescope of any kind in the world, with a diameter of 300 meters – about three city blocks (Figure 5.29).

Hydrogen gas, one of the prime constituents of interstellar clouds, emits a spectral line with a wavelength of 0.21 meters. Suppose we use the Arecibo radio telescope to observe this line. What is its resolving power? In our formula for resolving power we take $\lambda = 0.21$ meters and $D = 300$ meters, and we find $\phi = 150$ seconds of arc, or a bit more than 2 minutes of arc. This is very much poorer than even a small optical telescope: for all its huge size Arecibo, the largest telescope in the world, has very poor resolving power.

Figure 5.29 **The Arecibo radio telescope.** 👁 **(Also see color plate section.)**

NOW YOU DO IT
Suppose we used the Arecibo telescope to observe at a wavelength of one meter. What would be its resolving power? (A) Describe the logic of the calculation you will use. (B) Carry out the detailed calculation.

Interferometry

To compensate for their poor resolving power, radio astronomers have pioneered the development of a technique known as interferometry. Interferometry uses the wave nature of electromagnetic radiation to achieve a resolving power greater than that allowed by normal means. The outputs of two (or more) telescopes are combined, with careful attention being paid to maintaining their wave natures in the process

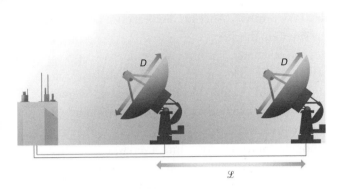

Figure 5.30 **Radio interferometer: the outputs of two radio telescopes are combined, yielding a resolving power depending on their separation $\mathcal{L}$.**

Figure 5.31 **The Very Large Array.**

Figure 5.32 **ALMA, the Atacama Large Millimeter Array (artist's rendering of the completed telescope).**

of combination (Figure 5.30). The result has the resolving power of a telescope whose diameter is not D, and not twice D, but rather the distance $\mathcal{L}$ between the two individual telescopes! Since they can be quite far apart, the resulting resolving power is very much greater than that of each acting alone.

This is the principle behind the Very Large Array, an interferometer consisting of 27 radio telescopes located on the plains of New Mexico (Figure 5.31). They are arranged in a Y-shaped pattern along railroad tracks, each track 21 km long. Each of these telescopes is 25 meters in diameter, itself a respectable size. The total collecting area of all these telescopes added together is that of a single instrument 130 meters across. But by using the techniques of interferometry, the resolving power of the VLA is that of a telescope nearly 40 *kilometers* across.

Even higher resolving powers have been attained by the Very Long Baseline Array, a network of ten radio telescopes spread from the Caribbean to Hawaii. Using such a large baseline, the resulting interferometer has achieved an angular resolution of 0.0002 seconds of arc, which is 500 times better than even the Hubble Space Telescope has been able to achieve.

ALMA, the Atacama Large Millimeter Array, currently under construction in the Atacama Desert of Chile, will be the largest telescope in existence when completed (Figure 5.32). This array of radio telescopes will operate at millimeter and submillimeter wavelengths, and will have an angular resolution of 0.01 seconds of arc, the highest of any telescope in the world – five times better than the Hubble Space Telescope. It will consist of 66 radio telescopes of sizes 12 meters and 7 meters, and is being built by an international partnership between Europe, North America, East Asia and the Republic of Chile. It is scheduled to be operational by the end of 2012.

A new window on the Universe

Jansky's observation that the center of our Milky Way Galaxy was a source of radio emission came as a surprise. We knew very little about the galactic center in those days, because our view of it was obscured by interstellar clouds of gas and dust. But based on what little we did know, nothing prepared us for his discovery. The galactic center was thought to be merely the point in space about

(a)

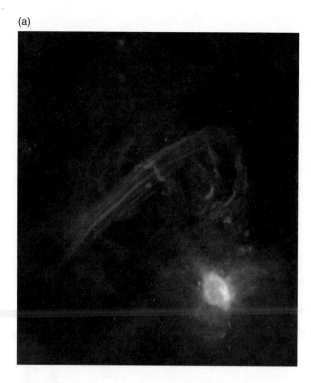

(b)

Figure 5.33 **Two views of the center of our Galaxy.
(a) In radio waves. (b) In X-rays.** 👁 (Also see color
plate section.)

which the Milky Way revolved. After his discovery, we knew there was more to the situation than that.

This story has two outcomes. The first is that radio telescopes can penetrate to where optical telescopes cannot. Interstellar clouds, which block optical light, cannot block radio waves. So radio astronomers can observe places hidden to optical astronomers.

The second outcomes is that the Universe is more interesting than we had thought. Since Jansky's time, everywhere we have looked we have found surprises. Our Sun, relatively nearby and certainly easy to observe, turns out to have some tricks up its sleeve. Its visible light is blackbody emission, which at the Sun's temperature contains no observable emission at radio wavelengths. Nevertheless, Reber found it to be a strong source of radio signals.

Radio telescopes allow us directly to observe clouds of interstellar hydrogen. Such clouds emit no optical light, and they are very difficult to detect by optical means. We see one only if by chance it lies sufficiently close to a star to shine by reflected light, or if it blocks the light of background stars. But by observing its 0.21-meter spectral line, radio astronomers have mapped hydrogen's distribution throughout the Galaxy, and have found it in quantities never before suspected.

More generally, every time we observe the cosmos at a new wavelength, we have been surprised at the result. Figure 5.33 dramatizes this point. It shows two images of the center of our Milky Way Galaxy, the first (Figure 5.33a) in radio wavelengths, and the second (Figure 5.33b) in X-rays. As you can see, the two views of the cosmos are utterly different.

Perhaps it is not so surprising that telescopes utilizing new wavelengths reveal new and unexpected aspects to the Universe. As an analogy, suppose that you were to observe the transmitting antenna of a radio station using two different wavelengths. Let us say that one wavelength is that of visible light: you will merely look at the antenna with your naked eye. The other wavelength is that of radio waves: you will "observe" these by means of a radio in your hands. Looking at the antenna, you can learn that it is metallic, with such-and-such a shape, and so-and-so many feet high. To your eyes it would appear to be utterly quiescent. But now turn on your radio, and listen to the transmitter's broadcasts. Nothing that your visual inspection has revealed would prepare you for what you hear. The converse is also true: listening to a radio tells you little about what the transmitting antenna looks like.

⟵ **Looking backward**
We studied spectra in Chapter 4.

....

NOW YOU DO IT
Recall that a spectrum is a graph illustrating which wavelengths are observed from a body. Sketch the spectrum of emission from a radio antenna (A) when it is broadcasting, and (B) when it is not broadcasting.

....

Astronomy from space

In recent decades much astronomical research has been done from space. There are two reasons for this transformation.

- As we have commented, all ground-based telescopes are subject to "seeing," whereby their view of the sky is blurred by atmospheric turbulence. While siting a telescope on a mountain can reduce this blurring, it cannot eliminate it. Only in space can a telescope reach its full resolving power.
- The advent of radio astronomy has taught us that it would be wise to extend our capabilities yet further, to other regions of the electromagnetic spectrum. But with the exception of the near infrared, the atmosphere is opaque at these wavelengths. Trying to observe them, even from atop a mountain, is like trying to see the sky on a cloudy night.

For these reasons, numerous telescopes have been launched into space. NASA's Great Observatories, an ambitious set of four orbiting telescopes intended to cover the full range of the electromagnetic spectrum, are complemented by a wide variety of smaller space observatories. Voyaging yet farther into space, astronauts have walked upon the Moon, and unmanned space probes have visited every planet of the Solar System.

We now turn to these new forms of astronomy. Surprises have greeted us at every turn.

Telescopes in space

The Hubble Space Telescope

At a cost of $2.2 billion dollars and a weight of nearly 13 tons, the Hubble Space Telescope (Figure 5.34) cost about $4000 an ounce – more then ten times the value of gold. Named in honor of Edwin Hubble, the discoverer of the expansion of the Universe (Chapter 18), it was launched in 1990 from the Space Shuttle Discovery. The Hubble is in a low-Earth orbit, 612 km above the Earth's surface.

The heart of the telescope is its 2.4-meter primary mirror. The telescope is of the Cassegrain design, with the light reflected from a secondary 0.3-meter mirror back through a hole in the primary to a focus. At the focus point a variety of cameras and spectrographs can be inserted, depending on the needs of the observation being conducted.

The Hubble's mirrors were to be ground to a shape deviating no more than 1/800 000 th of an inch from perfect. Unfortunately, after the telescope was launched it was discovered that, while perfectly smooth, there was an error in the shape to which the primary had been ground. Three years after launch, a Space Shuttle servicing mission succeeded in installing corrective optics which compensated for the problem.

It was the requirement that the telescope be accessible to the Shuttle for such servicing missions that dictated the choice of a low-Earth orbit. Many components of the telescope are of modular design, so that they can be removed by an astronaut and replaced. The initial suite of scientific instruments has been upgraded and replaced several times since the telescope's launch.

Figure 5.34 **The Hubble Space Telescope.** 👁 **(Also see color plate section.)**

Our formula for the maximum possible resolving power of an optical telescope,

$$\phi = 0.116/D \text{ seconds of arc,}$$

tells us that the Hubble, whose mirror diameter $D = 2.4$ meters, has a maximum resolving power $\phi = 0.05$ seconds of arc. Since the telescope is above the Earth's atmosphere, seeing does not blur its images, making it more powerful than larger ground-based telescopes. But in order to achieve this acuity of vision, the telescope must remain pointed at its target with extreme accuracy. This is relatively easy to accomplish on the solid ground, but it is far more difficult in space. The Hubble employs gyroscopes and a set of fine-guidance sensors to maintain pointing. The sensors, in fact, are themselves state-of-the-art telescopes. They point at a set of reference stars whose positions have been measured to high accuracy, and they determine the Hubble's pointing direction relative to them. So accurate is the complete system that the Hubble comes close to its maximum resolving power, regularly achieving 0.2 seconds of arc.

Operation of a telescope in space has other difficulties not encountered on the ground. As the telescope passes from the dark to the sunlit sides of the Earth, it is exposed to the direct heat of the Sun unfiltered by our atmosphere. As a consequence, it undergoes fluctuations in temperature exceeding 100 degrees

Fahrenheit in each 90 minute orbit. Because materials expand upon heating, the entire spacecraft flexes in response. The optical system, however, is so well isolated from the spacecraft that it maintains its shape while this is going on.

Minute-to-minute operation of the telescope and spacecraft is robotic. Much as the sensation of pain warns us of danger, so too a set of sensors continually monitor the state of the instrument and protect it from threatening situations (pointing directly at the Sun, for instance). The swinging of the telescope from one astronomical source to the next is automatic, as is the transmission of data back to the ground. The entire telescope plus spacecraft is powered by two solar panels that provide 2400 watts, about the power needed to light a home.

Over the long term, the Hubble is controlled by the Flight Operations Team at the Space Telescope Science Institute, on the campus of the Johns Hopkins University in Baltimore. Because the telescope is public property any scientist can use it. Scientists wishing to do so submit a proposal to a review committee of astronomers, which chooses the most worthy and schedules observations long in advance. As you can imagine, competition to use the Hubble is intense.

The James Webb Space Telescope

Named for the NASA administrator who played a critical role in the Apollo Moon landings, the James Webb Space Telescope is planned to be an orbiting telescope with a huge mirror 6.5 meters in diameter (Figure 5.35). Because it operates in both the visible and infrared regions of the electromagnetic spectrum, it is envisaged as the successor to both NASA's Hubble Space Telescope and the Spitzer Infrared Space Telescope discussed below. The JWST has been planned since 1997 although its launch date is currently very much in doubt. Indeed, as discussed below, the project is currently threatened with cancellation.

Figure 5.35 **The James Webb Space Telescope. A full-scale model displayed at the American Astronomical Society Meeting in Seattle, January 2007.** 👁 **(Also see color plate section.)**

NOW YOU DO IT
The Hubble Space Telescope has a mirror 2.4 meters in diameter. How much more light will the James Webb Space Telescope telescope gather than the Hubble? (A) Describe the logic of the calculation you will use. (B) Carry out the detailed calculation.

Infrared astronomy

In 1800 the British astronomer William Herschel undertook a study to measure the amount of energy emitted by the Sun at various wavelengths. He used a prism to spread the Sun's light into its constituent colors, and he placed a thermometer at each color to see how warm the thermometer got. One day he made a mistake, and left the thermometer lying where no light was arriving – just beyond the long-wavelength end of the spectrum. To his surprise he found that the thermometer still got warm. He had discovered infrared radiation.

Infrared radiation is an electromagnetic wave of slightly longer wavelength than the visible region of the spectrum. Infrared wavelengths are customarily measured in *microns*, where the micron is a millionth of a meter, or equivalently 10 000 angstroms. The midrange of the visible spectrum is at about half a micron: the infrared spectrum begins at about 1 micron and extends to 100 microns.

Our atmosphere is partially transparent to wavelengths ranging from 1 to 40 microns. The primary absorbers of this so-called "near infrared" are water and carbon dioxide molecules. Because these are concentrated at low altitudes, infrared astronomy must be conducted from high elevations. Beyond 40 microns lies the so-called "far infrared," which can only be observed from space.

Near infrared can be detected by CCDs and other light detectors. For longer wavelengths we must use an extension of Herschel's method, and seek to observe the warming effects of the radiation. Such a detector is called a *bolometer*. A common bolometer is a tiny germanium chip a mere few millimeters across. When warmed by infrared radiation, the bolometer's electrical resistance changes: by measuring this resistance, we can infer the infrared flux reaching the bolometer.

We can gain great insight into infrared astronomy by recalling what we know of blackbody radiation. Recall that the Wien law tells us that most emission from a blackbody of temperature T is at a wavelength $\lambda_{\text{most emission}}$ given by

$$\lambda_{\text{most emission}} = 0.003/T \text{ meters,}$$

> ⇐ **Looking backward**
> We studied the Wien law in Chapter 4.

where T is given in kelvin. Let us use this to ask what sorts of things emit infrared radiation. If we set $\lambda_{\text{most emission}}$ equal to a typical near infrared wavelength of 10 microns (10^{-5} meters), we can solve this to find $T = 300$ kelvin, or just over 60 degrees Fahrenheit.

....

NOW YOU DO IT
For what temperature does an object emit radio signals with a wavelength of one meter? (A) Describe the logic of the calculation you will use. (B) Carry out the detailed calculation.

....

But this is room temperature. It is the temperature of the infrared telescope and the dome that houses it – even of the bolometer we use to detect infrared radiation. Each of these is constantly emitting its own infrared radiation! How can we possibly observe the emission from a far-off astronomical source in such circumstances? It would be like trying to see a faint star through a telescope that itself glows brightly.

Infrared astronomers are forced to use special techniques, not required for any other kind of astronomy. No part of an infrared telescope or its housing is visible from the detector, so that their emission simply does not reach it. The emission from the detector itself is reduced by lowering its temperature. This is accomplished by cooling it to close to absolute zero, for instance by immersing it in a bath of ultra-cold liquid helium.

This requirement of extreme cold places severe limits on what infrared astronomers can accomplish. The Infrared Astronomy Satellite (IRAS), launched in 1983, was the world's first orbiting infrared telescope. It conducted extensive observations over a ten-month period but then ceased operations – not because anything on board had failed, but because it had run out of liquid helium coolant.

Our knowledge of blackbody radiation also allows us to learn something about the astronomical sources of infrared. We know from Chapter 4 that the cooler an object is, the longer is the wavelength of the bulk of its blackbody emission. The converse is also true: infrared radiation is emitted by cool objects. Stars, with surface temperatures of many thousands of degrees, emit visible light. The sources of infrared emission, on the other hand, must be far cooler than stars. We conclude that infrared telescopes are particularly useful for studying cool objects such as interstellar clouds. As we will see in Chapter 13, these are the sites of star formation, so that infrared astronomy is the natural way to study the birth of stars and planets.

Infrared is also a good wavelength band with which to study the early history of the Universe. As we will see in Chapter 18, the Universe is expanding: the farther away an object, the faster it is moving. This motion causes redshifts, which grow larger the farther away we look – or equivalently, the farther back in time we look. These can be sufficient to shift the visible-light emission of objects early in the history of the cosmos into the infrared band.

As we have noted, near infrared radiation can be observed from the ground. A number of infrared telescopes are sited atop the summit of Mauna Kea in Hawaii. Starting in 1974, NASA built and flew a far-infrared observatory on board a C-141 aircraft. Named the Kuiper Airborne Observatory after an early pioneer of infrared astronomy, the aircraft flew at altitudes of up to 13 kilometers, above 99% of the absorbing atmosphere, and carried a 0.9-meter telescope.

The first orbiting infrared observatory was IRAS. Although its telescope was relatively small, because the satellite was in space it made revolutionary advances, cataloging 250 000 separate sources of infrared emission. NASA's fourth Great Observatory, the Spitzer Infrared Space Telescope (Figure 5.36), carried a 0.85-meter mirror cooled by 95 gallons of liquid helium to 5.5 degrees above absolute zero. It was launched in 2003 and its supply of liquid helium lasted until 2009. Currently out of coolant, it is now able to conduct only limited observations.

X-ray astronomy

The X-ray region of the electromagnetic spectrum has wavelengths ranging from 10^{-11} meters to 10^{-8} meters, or equivalently 0.1 angstroms to 100 angstroms. Since visible light has wavelengths running from 4000 to 7000 angstroms, we see that X-rays have wavelengths far shorter than light – just the opposite of infrared radiation, which has longer wavelengths.

Because they work with radiation of such short wavelengths, X-ray telescopes are quite unlike their optical counterparts. Reflection from a mirror, the basic principle of telescope design, does not work at these wavelengths. Rather than being reflected, X-rays falling upon a mirror penetrate straight into it and are absorbed (Figure 5.37). Nor do lenses work at X-ray wavelengths.

Figure 5.36 **The Spitzer Infrared Space Telescope prior to its launch.**

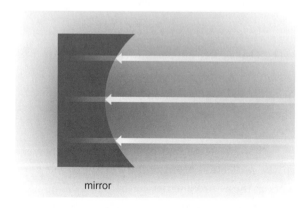

mirror

Figure 5.37 **X-rays are not reflected from a mirror, but penetrate into it and are absorbed.**

Early telescopes therefore had no means of gathering X-rays over a wide area and focusing them to a point. How could one possibly do astronomy in such a situation? An X-ray pinhole telescope would produce too faint an image to be useful. As an analogy, imagine what photography would be like if there was no such thing as a lens.

Here's a crude way to deal with this situation: take a long tube and point it at the source of emission (Figure 5.38). With such an arrangement, the only light reaching the indicated point on the detector is from the general vicinity of the source: the detector "looks down the tube." No image is formed: the detector will merely record the existence of X-rays — but we do know where they came from. Furthermore, by swinging our tube up and down along the source we can get crude information about it. Such a device is known as a *collimator*. It was the principle of operation of early X-ray telescopes.

The first observation of celestial X-rays was made in 1962. A small rocket was launched from White Sands, New Mexico. Too small to achieve orbit, it remained above the atmosphere for a mere six minutes. During that brief time, it observed X-rays and radioed the information back to Earth before plummeting back downwards. So crude was the collimator employed by this first instrument that it was only able to say that the detected X-rays came from outside the Solar System.

NOW YOU DO IT
How do you think scientists knew that the source of the X-rays was not a planet within our own Solar System, if the only thing they knew was the direction to the source? (Hint: recall our discussion of the ecliptic and the motion of the planets in Chapter 2.)

For years thereafter the fledgling field relied on small rockets, none of which remained above the atmosphere for more than a matter of minutes. In 1970, however, the first X-ray satellite was launched from a platform off the coast of Kenya. Named "Uhuru" (Swahili for "freedom"), it remained in orbit for three years and identified several hundred sources of X-ray emission, making a host of exciting discoveries.

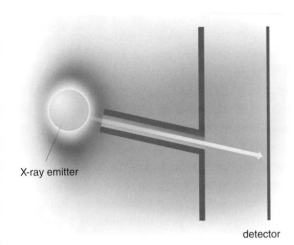

X-ray emitter

detector

Figure 5.38 **The principle behind a collimator, which provides crude directional information.**

In those years researchers were severely hampered by their inability to make images of the sky at X-ray wavelengths, and it was often difficult or impossible to tell which astronomical object was emitting the X-radiation they observed. Much later, however, a means was found to build mirrors for X-ray astronomy. These rely on the fact that X-rays incident upon a surface at very shallow angles are partially reflected.

Such "grazing incidence" reflection is the cornerstone of the Chandra X-ray Observatory, which was launched in 1999, and which actually produces images of X-ray emission from the sky. The Chandra mission was named for the Nobel-prize winning Indian astrophysicist Subrahmanyan Chandrasekhar, who discovered the maximum mass of white-dwarf stars (Chapter 15). Chandra was the third of NASA's Great Observatories.

As with celestial infrared, we can gain insight into the nature of celestial X-ray sources by using our knowledge of blackbody radiation. As we emphasized, the shorter the wavelength of the observed radiation, the hotter the body emitting it. X-rays have such short wavelengths that the required temperatures are millions of degrees or more. This is enormously greater than the surface temperature of any normal star.

....

NOW YOU DO IT
For what temperature does $\lambda_{most\ emission}$ given by the Wien law equal the X-ray wavelength 10^{-10} meters? (A) Describe the logic of the calculation you will use. (B) Carry out the detailed calculation.

....

Many processes other than blackbody radiation are known to produce celestial X-rays, but the principle remains the same: while infrared astronomy is the study of cool objects, X ray astronomy is the study of hot, highly energetic objects. Sources of celestial X-rays include superheated regions above the Sun's surface, neutron stars heated to great temperatures by catastrophic collapse and hot intergalactic gas.

Gamma-ray astronomy

Beyond the X-ray region of the electromagnetic spectrum lie gamma rays, whose wavelengths are less than 10^{-11} meters, or 0.1 angstroms. Such ultra-short wavelengths correspond to ultra-high energies: gamma-ray astronomy is the study of the hottest, most violent processes in the Universe.

Gamma rays are produced, among other means, by the decay of radioactive nuclei. So high are the energies of these gamma rays that they are very dangerous: gamma radiation is one of the three components of radioactivity, the other two being high-energy electrons and nuclei.

Gamma rays cannot be detected by photographic film, CCDs or any of the other traditional methods of astronomy. Rather, special devices are employed, which emit minute flashes of light when struck by them. So powerful are gamma rays that no known substance reflects or refracts them, so that no imaging gamma-ray telescope has ever been built. Rather, crude directional information is provided by measuring the direction of the light flash produced by the gamma ray.

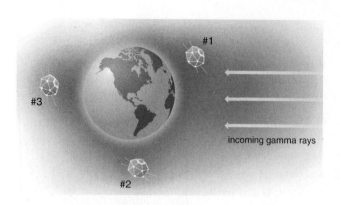

incoming gamma rays

Figure 5.39 **The Vela satellite system finds the direction to a gamma-ray source by a timing method.**

A second major difficulty facing gamma-ray astronomers is the fact that the Universe produces very few gammas. People used to joke that this form of astronomy relied on such a small number of detections that each detected gamma could be given its own name.

The first detection of celestial gamma rays was inadvertent. The Vela satellites had been placed in orbit in order to monitor clandestine tests of thermonuclear weapons (a hydrogen bomb emits this radiation when detonated). They determined the direction to a source by the timing method illustrated in Figure 5.39. Gammas arriving from the indicated direction would trigger satellite number 1 first, then number 2 and finally number 3. Precise measurement of the time delays between triggerings of successive satellites allowed a determination of the direction of arriving radiation. (For example, in the indicated configuration the incoming gamma rays arrive at satellite number 1 first, and then shortly afterwards at satellite number 2. Finally, a good deal of time after this, they arrive at satellite number 3). In 1967 the system detected brief bursts of gamma radiation whose timing indicated that they could not have originated on the Earth.

....

NOW YOU DO IT
Suppose satellite number 1 had detected gamma rays and then, somewhat later, satellites number 2 and number 3 had simultaneously detected them. What is the direction to the source of gamma rays?

Suppose alternatively that all three satellites had detected gamma rays at the same instant. What is the direction to the source of these gamma rays? Actually there are *two* answers to this question: can you find them both?

....

Gamma-ray astronomy made a giant stride forward with the launch in 1991 of the Compton Gamma Ray Observatory. Named in honor of the Nobel-prize winning physicist Arthur Holly Compton, it was the second of NASA's Great Observatories. Compton had four instruments that covered an unprecedented six decades (factors of ten) of the electromagnetic spectrum: each of them achieved an improvement in sensitivity of better than a factor of ten over previous missions. Because the number of gammas detected depends on a detector's mass, and because the Universe produces so few gammas, the telescope had to be unusually massive: it weighed 17 tons, making it at the time the heaviest payload that had ever been lifted into orbit. The mission lasted for nine years before the observatory was deorbited and reentered the atmosphere in 2000.

NASA's current gamma-ray observatory is Fermi: launched in 2006, it is far more powerful than Compton.

Space probes

The Apollo missions to the Moon

The Moon is the only celestial body to have been visited by humans. After a long series of test flights, the Apollo program culminated on July 20, 1969, when

Neil Armstrong stepped out on the lunar surface with the historic words "that's one small step for [a] man, one giant leap for mankind" (Figure 5.40).

The 45 000-kilogram Apollo 11 spacecraft had required some 2½ days to journey from an orbit about the Earth to one about the Moon. There, while Michael Collins remained on board the orbiting "command module," Armstrong and Edwin Aldrin descended to the Moon's surface in the "lunar module." They remained there for about a day, deploying an experiment to study the solar wind, another to study moonquakes and in a third a laser reflector with which the distance from Earth to Moon could be carefully monitored. They collected samples of the Moon's soil, planted a flag and took photographs. All the while a television camera they had set up some distance away relayed images of them moving about the Moon to television sets throughout the world.

Armstrong and Aldrin then returned to the lunar module, the upper portion of which – a spacecraft in its own right – lifted off and rejoined the command module. They then departed for the Earth, hitting the atmosphere along a precisely defined reentry corridor and splashing down in the Pacific Ocean, where they were transferred by helicopter to a nearby ship. Because of the conceivable, although exceedingly unlikely, threat of contamination by lunar organisms, the three astronauts were quarantined for some time after their historic journey.

Six other Apollo missions took place over succeeding years, all of which were successful, with the exception of Apollo 13 which suffered an explosion on the way to the Moon (the three astronauts made it back to Earth safely). The last Apollo mission took place in 1972. Since then no manned spacecraft has set forth for any celestial body.

Figure 5.40 **An Apollo astronaut stepping onto the Moon.**

Figure 5.41 **The Voyager spacecraft.**

The Voyager missions to the outer Solar System

Five years after the last Apollo flight, the unmanned Voyager missions (Figure 5.41) were launched to study the two greatest planets of the Solar

System, Jupiter and Saturn. The two identical spacecraft were originally designed to survive for five years, but they proved so durable that their missions were extended for an extraordinary twelve years. After studying their initial objectives, the spacecraft went on to study the two other gas giant planets, Uranus and Neptune, and a full 48 moons. Even now, decades after their launch, the spacecraft are still functional: having left the Solar System, they are now returning data about interstellar space.

This remarkable triumph was made possible by a rare alignment of planets. The Voyager spacecraft employed the gravity-assist technique (Chapter 3) in which they would journey to a target, complete their studies on the fly as they zoomed by, and were then accelerated by the planet's gravity on to the next target. Only once in about every 175 years do the planets align in such a way as to make such a series of "bounces" through the outer Solar System possible.

The Voyager missions were launched in 1977 and required two years to reach their first objective, Jupiter. Accelerating on from there to Saturn, Voyager 1 passed close to its large moon Titan – the only moon known to possess an atmosphere – and behind its beautiful rings. There the spacecraft's trajectory was bent out of the ecliptic plane, and so out of the Solar System. Voyager 2, in contrast, was sent along a trajectory that brought it first to Jupiter, then to Saturn, next to Uranus and finally, in 1989, to Neptune.

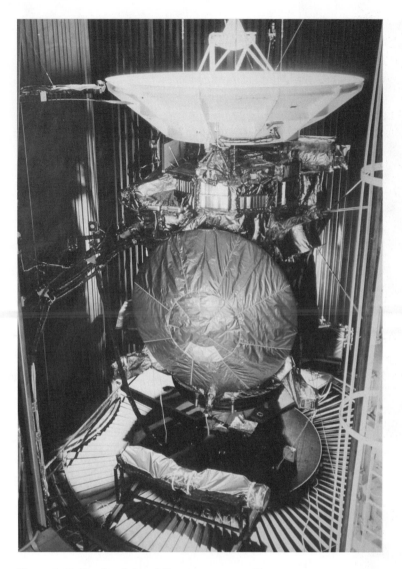

Figure 5.42 **The Cassini and Huygens spacecrafts.**

Both spacecraft are now on their way to the stars and will never return. On board each is a message, designed to be read by any extraterrestrial civilization that happens to chance upon it at some unimaginably distant time in the future. Because of the absence of erosion in space, these two frail machines will still be in pristine condition long after the great pyramids have been worn away to dust.

The Cassini mission to Saturn

The Cassini mission (Figure 5.42), launched in 1997, arrived at Saturn in 2004. It was an ambitious mission to the second-largest planet of the Solar System. Cassini was the best-instrumented spacecraft ever sent to another planet, standing 22 feet high and weighing more than 6 tons at launch. Its objective was to study Saturn's internal structure and atmosphere, the composition and structure of its beautiful rings, the surfaces and internal structures of its many moons, Saturn's magnetosphere and its interaction with the solar wind.

On board was a second spacecraft, the Huygens probe, which detached from Cassini and dropped onto Saturn's moon Titan. Titan is the largest moon in the Solar System – indeed, it is larger than Pluto – and is the only satellite known to possess an atmosphere. Once in orbit about Saturn, Cassini released Huygens, which freely fell toward Titan. Entering its atmosphere, Huygens deployed a parachute and robotically landed on the satellite's surface. Both during its descent and after landing, Huygens' robotic laboratory studied the structure and composition of Titan's atmosphere, its wind and weather patterns, energy flux, and the moon's surface. The laboratory operated for only a few hours, transmitting its measurements to Cassini which relayed them on to Earth. Afterwards Cassini embarked on its own four-year tour of the Saturnian system, studying both its rings and moons.

The Discovery program

Apollo, Voyager and Cassini were hugely ambitious missions. Each involved large teams of people, many scientific instruments, great expense and a process of planning and development extending over many years. The Discovery program represents NASA's implementation of an entirely different approach to space science. Embodying the philosophy of "Faster, Better, Cheaper," Discovery missions are intended to return important scientific data in a relatively short amount of time for a cost of no more than $299 million each. In contrast, Cassini's budget was ten times bigger.

In the past, NASA would announce its plan to study a certain planet and solicit separate bids for (1) construction of the spacecraft and (2) its scientific instruments and for (3) operation of the mission. In contrast, the Discovery program solicits proposals for an entire mission, put together by a team composed of people from industry, small businesses, government laboratories and universities. Strict cost caps are maintained, and rapid schedules imposed. The philosophy is for each mission to be tightly focused in scope with a short development time. NASA's goal is to launch a Discovery mission every one to two years.

One such mission was the Mars Pathfinder probe to Mars. The probe arrived on July 4, 1997, slowed its descent by a system of parachutes and retro-rockets, and then fell freely for the last few hundred feet. The entire spacecraft was encased in a set of inflated airbags – a sort of "bubble-wrap" – which allowed it to bounce harmlessly when it slammed into the surface. After coming to rest the airbags deflated, a set of "petals" were unfolded, and a small robotic vehicle, entirely free ranging, then drove down one of the petals and explored the Martian surface.

The Mars Science Laboratory

Launched in November 2011 and arriving at Mars in 2012, the Mars Science Laboratory (Figure 5.43) is the most ambitious rover to date to land on the planet. Its primary mission is to determine if Mars ever possessed an environment suitable for life. It will do this by searching for organic compounds and possible metabolic products of ancient organisms, and by studying the rocks for details about the past climate in which they formed.

The laboratory is twice as long and four times as heavy as the Mars Exploration Rovers. Within it are ten science instruments. They will analyze up to 70 samples of rock and soil delivered to them by a robotic arm, which will scoop up the samples

Figure 5.43 **The Mars Science Laboratory rover inside NASA's Spacecraft Assembly Facility.**

and deliver them to the spacecraft. Instruments will identify organic compounds and measure the isotopic ratios of chemical elements important to life. Other instruments will measure the elemental composition of rocks and soils, allowing scientists to infer their mineral composition. On the robotic arm is a camera to examine the surfaces of rocks and soil. On a mast is a stereo color movie camera, and a laser that can zap rocks from up to 10 meters away, and then examine the atoms excited by the laser beam.

Another set of instruments will study the current environment around the rover. One will observe the radiation levels, a measurement critical to the planning of future human exploration of Mars. A second will function as a miniature weather station, observing pressure, temperature, humidity, wind speeds and ultraviolet radiation. A third will measure the abundance of hydrogen just below the surface, which might identify water.

In 2009 a public poll was conducted on the NASA website to name the Laboratory: the winning name ("Curiosity") had been submitted by a sixth-grader, Clara Ma, from Kansas.

Space-based versus ground-based astronomy

Should we do astronomy from space? Are the advantages it brings worth the cost? The question would be academic if space-based astronomy were cheap. But in reality it is enormously difficult, and therefore enormously expensive. Is this money well spent?

To put the question in perspective, compare the cost of the Hubble Space Telescope, $2.2 billion, to the $140 million price tag of the world's largest telescope, the Keck. For the cost of the Hubble Space Telescope, 16 Kecks could have been built.

Note also that the Hubble is, by modern standards, a fairly small instrument. Its mirror diameter of 2.4 meters is far less than the 10-meter diameter of the Keck. Since, as we have seen, the light-gathering power of a telescope depends on its collecting area, which is proportional to the square of its diameter, we see that the Keck is capable of detecting objects fainter by a factor of $(10/2.4)^2 = 17.4$ than the Hubble. This means that the Keck can peer farther into the Universe than Hubble. Recall from the inverse square law that the apparent brightness of a galaxy depends on its distance squared; a galaxy just barely visible to the Hubble would be visible to Keck $(10/2.4) = 4.2$ times farther away.

But even so, the Hubble can do many things that ground-based telescopes cannot. It can observe at wavelengths to which the atmosphere is opaque. Its freedom from the blurring effects of seeing give it a wonderful angular resolution. The Hubble Deep Field, one of the most famous astronomical images ever made, could never have been taken without its extraordinary powers. For work requiring wide wavelength coverage and/or high angular resolution, the Hubble is pre-eminent.

But in some respects ground-based telescopes are catching up to their space-based counterparts. A further factor to be considered in evaluating the trade-offs between space- and ground-based astronomy is that ground-based astronomy can take advantage of the latest technological breakthroughs. Space missions, on the other hand, employ the technology available when they were built – and since the design of a space mission is established long before its launch, these missions are technologically obsolete by the time they *begin* operations. Recent work with adaptive optics has allowed ground-based telescopes partially to cancel out the effects of seeing. For instance, the Gemini South telescope, situated in Chile, in 2002 achieved a resolution of 0.4 seconds of arc – close to Hubble's 0.2 seconds of arc.

Space probes are subject to breakdowns that would be trivial on the ground, but in space are catastrophic. The Mars Observer space probe, launched in 1992, voyaged safely to Mars and then abruptly ceased communications. All contact with the spacecraft was lost: no one ever figured out what had happened. On the ground the malfunction could have been diagnosed and fixed, but in space this was impossible. The mission, costing $980 million, was irretrievably lost.

This is why the Hubble was placed into a low-Earth orbit, accessible to the Space Shuttle so that it could be repaired by astronauts. Many other missions, on the other hand, are in orbits not accessible to the Shuttle: to them, the most "trivial" of failures would be disastrous. The Shuttle program was retired in 2011.

Ground-based astronomy is the method of choice in certain circumstances, and space-based astronomy in others. Progress in the field depends on a balance between the two approaches. It is always a delicate matter to achieve this balance.

THE NATURE OF SCIENCE

BIG SCIENCE

Modern science is in the throes of a transformation. Two anecdotes – both true! – illustrate the nature of this transformation.

The first anecdote concerns a young physicist who wished to consult with Albert Einstein about some research he was doing. They agreed to meet on a certain street corner at a certain time. Unfortunately, the young scientist was late. When he finally got there, he rushed up to the great man and began apologizing profusely. But Einstein brushed the apologies aside. "There was no difficulty," he explained. "I can do my work anywhere." Einstein had continued thinking as he stood there on the street corner.

The second anecdote concerns another young physicist who wished to speak with Carlo Rubbia, leader of a research group in which she worked. But Rubbia's schedule was crowded, and it was hard to find a time to meet. One day the phone rang in her office. It was Rubbia, saying he had a few minutes free. Slamming down the phone she dashed over to his office, only to find it locked. Rubbia had been calling from the airport, where he was about to board a plane. The research group he directed was in Switzerland – but he taught at Harvard. Every week he would fly back and forth between the two.

It is probably fair to say that most people think of Einstein as the model of what it is like to be a scientist: the solitary individual, working alone. That is "small science." But more and more nowadays, "big science" is becoming the norm. This is science done by large teams of people.

A good way to see this transformation is to look at research papers published in scientific journals. These are how scientists formally announce the results of their work. In 1985 about a quarter of the research articles in a particular issue of the primary astronomical journal of research had one author. Conversely, no article in that issue had more than five authors. But by 2002 the percentage of single-author articles in a particular issue was only half what it had been in 1985 – and one article had 36 different authors from 12 different institutions ranging from Japan to Maryland.

WHAT IT IS LIKE TO DO BIG SCIENCE?

Research in a large group is fundamentally different from research done alone. Big groups have so many members that the junior people – and these are trained scientists – often have little say in the overall direction of the work. Most important decisions are made by the senior group leaders, and most day-to-day decisions are made by middle-level managers. Sometimes the junior scientists can feel a little left out.

Work on a big project can take literally decades. Furthermore, if the project is a space probe, the probe will take many years to reach its destination: the Cassini mission, which arrived at Saturn in 2004, was launched in 1997. Many of the scientists who did the early work on this mission had died or retired by the time it arrived at its destination and began making observations.

The experience of using a big telescope is radically different from that of using a small one. The astronomer using a small telescope uses it all by herself. She has complete use of the instrument all night, or for many nights, and she operates it alone – or at most with the help of a single technical assistant. While working with the instrument, it is she who makes the important scientific decisions.

How different this is from the experience of using a big telescope! The astronomer wishing to use the Hubble Space Telescope is required to submit an observing proposal. (Competition for the use of the Hubble is intense: for its first 1200 hours of observing time, proposals adding up to 11 000 hours were submitted.) If her proposal is accepted, it is not she who will actually conduct the observation. Indeed, she may not even know just when it is happening, for the actual operation of the telescope is done by a team of engineers familiar with its complexities. If the research calls for using several instruments – a camera and a spectrometer, say – it might be conducted in parts spread over several weeks, for the Hubble does not switch instruments very often. Ultimately, the results of the observation are telemetered back down to Earth and emailed to the astronomer.

There is a saying: "more is different."

THE FUNDING OF BIG SCIENCE

Little science is expensive. But big science is very expensive in comparison to small science. The two are financially supported in very different ways in the USA.

Small science is largely funded by two government agencies, the National Science Foundation and NASA. These agencies receive research proposals from practising scientists, and they decide which to support. In reaching this decision, they employ carefully crafted procedures. They begin by formulating policies designed to ensure integrated support of an entire field of science. Officials might decide that too little attention has been paid to stellar astrophysics recently, and that they should accordingly increase their support for this area. Because these decisions are made by career officials who are usually themselves scientists, they are informed by an extensive knowledge of the field.

They are also largely insulated from political pressures. And when it comes down to deciding specifically which proposals to fund, the insulation is complete. These decisions are made on the basis of recommendations from practising astronomers who have no ties to the government whatsoever. They are astronomers gathered from universities and colleges around the country. They meet as panels, read all the proposals submitted in their area of expertise and make their recommendations.

Big science, however, is funded directly by Congress. These projects are too expensive to be supported by the normal process. Funding for the single most expensive scientific instrument ever proposed, the famous Superconducting Super Collider, was tacked on as an amendment to an Energy and Water Development Appropriations bill. The people who made the decision to fund this accelerator were not professional scientists but politicians, and their decisions were not made on the basis of careful study of the needs of the field. Is this a good way to support such enormously expensive and ambitious projects?

The following quotation from Wikipedia (August 2, 2011) illustrates the nature of funding big science projects:

> On 6 July 2011, the United States House of Representatives' appropriations committee on Commerce, Justice, and Science moved to cancel the James Webb Space Telescope project by proposing an FY2012 budget that removed $1.9 billion from NASA's overall budget, of which roughly one quarter was for JWST. This budget proposal was approved by subcommittee vote the following day, but it remains to be seen how the rest of the US House of Representatives and Senate will weigh in on this issue during the ongoing federal budget negotiations...
>
> The committee charged that the project was 'billions of dollars over budget and plagued by poor management'. The telescope was originally estimated to cost $1.6 billion but the cost estimate grew throughout the early development reaching about $5 billion by the time the mission was formally confirmed for construction start in 2008. In summer 2010, the mission passed its Critical Design Review with excellent grades on all technical matters, but schedule and cost slips at that time prompted US Senator Barbara Mikulski to call for an independent review of the project. The Independent Comprehensive Review Panel ... found that the earliest launch date was in late 2015 at an extra cost of $1.5 billion (for a total of $6.5 billion). They also pointed out that this would have required extra funding in FY2011 and FY2012 and that any later launch date would lead to a higher total cost. Because the runaway budget diverted funding from other research, the science journal *Nature* described the James Webb as 'the telescope that ate astronomy'. However, termination of the project as proposed by the House appropriation committee does not provide funding to other missions as the JWST line is simply terminated with the funding simply leaving astrophysics (and leaving the NASA budget) entirely.
>
> The American Astronomical Society has issued a statement in support of JWST, as did US Senator Barbara Mikulski. A grassroots movement to save the telescope has been started and a number of editorials supporting JWST have appeared in the international press.

DOES SCIENCE COST TOO MUCH?

The Hubble Space Telescope cost $2.2 billion. Is that too much money? Because it is the federal government that paid for the Hubble, and because the government draws its money from taxpayers, only the public can decide.

One approach is to ask how much each American citizen spent on the Hubble. Let us calculate this. Since there were about 280 million Americans when the Hubble was being built, each person spent $2.2 billion/280 million, or $786. Would you say that they have gotten their money's worth from it? And does the fact that they spent this money not all at once, but over the many years it took to build the Hubble, lead you to change your mind?

A second approach is to ask what this sum of money would have bought, had we spent it on something else. $2.2 billion is enough to fund the entire public education system of Los Angeles county for three years. It is five times the endowment of one of the nation's premier artistic institutions, the Museum of Modern Art. Would you say that the money spent on the Hubble was well spent?

WILL SCIENCE COST TOO MUCH IN THE FUTURE?

The most expensive scientific instrument ever proposed, the Superconducting Super Collider, was never built. The project was terminated after its projected cost had escalated to a staggering $13 billion. While the design studies that were conducted had some peripheral relevance to other machines, and while a few technological spinoffs did result from the project, they are hardly worth the $2 billion that had been expended by that point. That enormous sum of money – nearly enough to build a second Hubble Space Telescope – was for all intents and purposes wasted.

There is reason to worry that this debacle is a harbinger of things to come. This is because science necessarily grows more and more expensive as time passes. Perhaps it will ultimately grow too expensive for society to support. We built the Hubble, and we are building other giant scientific instruments. But perhaps we will not be able to continue this forever.

Let us see why science keeps growing ever more costly. First, let us look at it historically. We will focus on three different periods – early in the twentieth century, the mid twentieth century and now. Table 5.1 lists the biggest telescope in the world at each of these points in time, and how much each cost.

Clearly, our telescopes have been costing more and more. The last column tells why: they have been growing bigger and bigger.

Furthermore, our telescopes *must* grow bigger and bigger. If they did not, astronomy would stagnate. Whenever a new telescope is built, one bigger and better than all the others, a whole new field opens up. What used to be hard is now easy. Every time this bigger and better instrument is trained on the sky, it finds something exciting. New discoveries are made rapidly. But with the passage of time the pace slows! All the easy things have been done. Now our new telescope has to work harder to discover something new. It does not make the headlines quite as often as it used to. And in the long run, the telescope will simply have finished doing everything it is capable of. Rather than spend the money to keep it operating, astronomers shut it down.

But science cannot be conducted without scientific instruments. Thus it is that astronomers perpetually seek to build bigger and bigger telescopes. And thus it is that astronomy grows steadily more expensive.

Will it eventually grow *too* expensive? What is the biggest telescope we will ever build? Surely we will never build one bigger than the Earth itself! Probably we will never spend more than the Gross National Product to build one. When we have reached these limits, will it no longer be possible to build new telescopes? Will astronomy come to an end?

Table 5.1. **The biggest telescopes in the world.**

Date	Name	Cost when built	Cost in 2002 dollars	Mirror diameter (meters)
Early twentieth century	100-inch Mt. Wilson telescope	$645 000	$6.7 million	2.5
Mid twentieth century	200-inch Mt. Palomar telescope	$6 million	$63 million	5
Now	Keck telescopes	$140 million	$140 million	10

Summary

- Telescopes magnify the heavens, gather more light than the naked eye and observe in wavelengths other than those the naked eye can see.

Pinhole telescope

- Constructed by punching a tiny hole in an opaque screen and putting it at the end of a cylinder.
- An image is formed when the rays from a point of light reach a single point on the film: the pinhole telescope does this.
- It produces an upside-down image.
- The image is very faint. Making the pinhole bigger makes the image less faint, but also blurs it: lenses and mirrors correct this defect.

Reflecting telescopes

- A flat mirror directs rays of light from a single point to many points: therefore it does not produce an image.
- A curved mirror directs rays of light from a single point to a single point: therefore it does produce an image.
- The image is upside down.
- The four basic designs of reflecting telescopes: the prime focus, Newtonian, Cassegrain and Coudé.

Refracting telescopes

- Refraction: a ray of light, passing from air into glass and then out again, is bent.
- If the glass is in the shape of a lens, rays from a single point are bent to fall on a single point: thus the lens produces an image.
- The image is upside down.
- Refracting telescopes use the prime focus design.
- Refracting telescopes suffer from a flaw: they focus different wavelengths to different points. Thus refracting telescopes are not used in astronomy.

Telescope mountings

- The rotation of the Earth causes the sky to appear to rotate overhead. Telescopes must be swung to follow this motion.
- Two kinds of telescope mountings: polar and altitude-azimuth.

The powers of a telescope

- Light-gathering power.
  - Bigger telescopes produce brighter images.
  - Total light "caught" by a telescope is proportional to the mirror's surface area.
  - Big telescopes can "see" farther than small ones.
  - This is why we build large telescopes.
- Magnification (resolving power).
  - Resolving power: measures the smallest angle a telescope can distinguish.
  - Large resolving power means the telescope can resolve fine detail (i.e. distinguish small angles).
  - The longer a telescope the larger is its resolving power.
  - Resolving power cannot be made better than a limit caused by turbulence in the Earth's atmosphere. For this reason we set telescopes atop mountains or launch them into space.
  - Resolving power cannot be made better than a limit caused by the wave nature of light. The bigger our telescope, the better is the limit on its resolving power set by this phenomenon. This is another reason why we build large telescopes.
  - The resolving power of a telescope also depends on the wavelength of the electromagnetic radiation it observes. Radio telescopes (long wavelengths) have lower resolving power than optical ones. Thus we need our radio telescopes to be *very* large.

Next-generation telescopes

- Active optics: the mirror is deformed to achieve its optimal shape.
- Adaptive optics: a system senses the state of the atmosphere above the telescope and adjusts the shape of the telescope's mirror accordingly.
- Site selection: telescopes must be far from sources of light pollution (cities, etc.) and at high altitude (to minimize the effects of atmospheric turbulence).

Astronomical detectors

- The CCD (charge-coupled device): superior to photographic film.
- The spectrograph: splits light into its component wavelengths (colors).

Radio telescopes

- The first radio telescope was an array of antennas.
- Modern radio telescopes employ reflecting curved surfaces made of metal.
- Interferometry can greatly increase the resolving power of a radio telescope.
- Observing in new wavelengths reveals new information about the Universe.

Astronomy from space

- The Earth's atmosphere degrades the performance of a telescope in two ways.
  - Turbulence in the atmosphere blurs the images produced by a telescope.
  - The atmosphere is opaque to many wavelengths: the only way to observe at these wavelengths is from space.
- The Hubble Space Telescope is the most famous of the orbiting observatories.
  - $2.2 billion.
  - Orbits 612 kilometers up.
  - 2.4-meter mirror: this is not large as telescopes go, but since it is above the atmosphere Hubble is one of the finest telescopes that we have.
- Infrared astronomy.
  - Infrared radiation has longer wavelength than optical radiation.
  - The most common detection method employs a bolometer: the infrared radiation warms it and we measure this warming.
  - Infrared radiation is emitted by room-temperature objects: i.e. objects far cooler than stars.
  - The telescopes themselves must be cooled so that they are prevented from emitting their own infrared radiation, which would obscure the radiation we are trying to observe.
  - Most famous observatory: the Spitzer observatory in space.
- X-ray astronomy.
  - X-rays have shorter wavelengths, i.e. higher energy, than visible light.
  - Most mirrors do not work at such high energies: early telescopes used collimators and modern ones use "grazing incidence" reflection.
  - Celestial X-rays are emitted by exceedingly hot, energetic processes.
  - Most famous X-ray observatory: the Chandra observatory in space.
- Gamma-ray astronomy.
  - Gamma rays have the shortest wavelengths and the highest energies that we observe.
  - They are produced by the hottest, most violent processes in the Universe.
  - Most famous gamma-ray telescope: the Compton gamma-ray observatory in space.

Manned space probes

- The Moon is the only celestial object to be visited by humans.
- First landing: Apollo 11 in 1969. Neil Armstrong and Edwin Aldrin walked on the Moon for about a day.
- Six more missions were launched.
- All were successful except Apollo 13. Although the Apollo 13 astronauts did not walk on the Moon, they returned to Earth safely.

Robotic space probes

- Voyager missions.
  - Twin spacecraft that flew by Jupiter, Saturn, Uranus and Neptune.
  - They used the "gravity assist" technique.
  - They carry a message to any alien civilization that might come upon them.
- Cassini/Huygens mission.
  - Studied Saturn and its moons.
  - Huygens landed on Titan, the only satellite possessing an atmosphere.
- The Discovery program: "faster, better, cheaper."
- Designed to be relatively cheap and rapid.

Ground-based versus space-based telescopes

- Astronomy from space is far more expensive than astronomy from the ground.
- Space-based telescopes take so long to build that they are technologically obsolete the day they are launched, unless (expensive) servicing missions regularly visit them to install up-to-date equipment.
- Ground-based telescopes are far larger than space-based ones.
- Nevertheless, practising astronomy from space brings advantages that ground-based astronomy does not.

The nature of science. Big science

- Much contemporary science is done by large teams of people.
- Working in such a team is very different from working alone.
- "More is different."
- Funding of science (USA).
  - Small science – i.e. science done by individuals or small groups – is funded by government agencies such as the National Science Foundation or NASA.
  - Decisions on what projects to fund are based upon recommendations by practising scientists and are independent of political considerations.
  - Big science is funded by acts of Congress, and depends heavily on political considerations.
- The costs of science.
  - Science is expensive.
  - It is sure to grow more expensive, since astronomers need better and better telescopes and better telescopes cost more money.

Problems

(1) Suppose you punch *three* tiny holes in your pinhole telescope. Draw a picture illustrating the image this produces of a single light bulb.

(2) In this problem we want to study the blurring produced by a wide pinhole. We will do this by imagining that we are pointing our pinhole telescope at a particularly simple scene: a wall, half of which is painted red, and the other half green (Figure 5.44): We want our pinhole to produce a good image of this. In particular, we want the junction between the red and green images to be sharp.

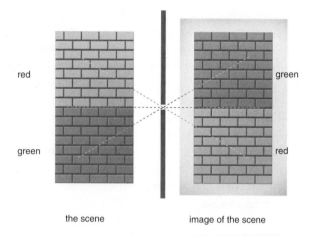

Figure 5.45 **A small pinhole produces a sharp junction between the red and green portions of the scene.**

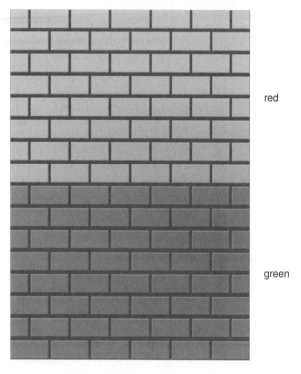

Figure 5.44 **The scene we want our pinhole telescope to image.**

First, let's study a regular pinhole. Figure 5.45 maps out how the red and green rays pass through the pinhole and reach the detector. As you can see, the junction between the red and green images is sharp.

But now let us look at this with a wide pinhole (Figure 5.46).

(A) As you can see from the figure, certain locations on the light detector receive only red rays, certain other locations receive only green rays – *but some special locations receive both red and green rays.* These "special locations" are the blurred image of the sharp junction between red and green that we see in the scene.

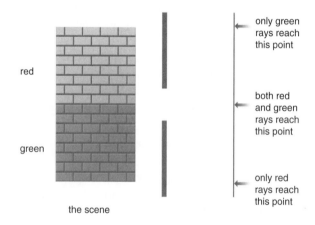

Figure 5.46 **A wide pinhole blurs the image of a sharp junction.**

(B) Your task in the problem is to figure out how wide this blurred region is. You can do this by drawing red and green rays as they pass through the wide pinhole and reach the detector. In particular, draw a series of pictures that illustrate that *if the pinhole is slightly wide, the image is only slightly blurred, but if the pinhole is very wide, the image is very blurred.*

(3) Does a telescope situated at the north pole need a clock drive? How about one situated at the south pole?

(4) Imagine that at some time in the future we set up an observatory on the Moon. Recall that the Moon keeps the same side always facing toward the Earth. Would

the clock drive in our lunar observatory need to swing its telescope more or less rapidly than those on Earth?

(5) Suppose you had two telescopes: the lens of one is two times bigger than that of the other. (A) Draw diagrams illustrating incoming rays of light, and showing why the bigger one catches more light than the other. (B) How much more light does it catch than the other?

(6) Suppose you had two telescopes: one catches just four times as much light as the other. (A) Draw diagrams illustrating the incoming rays of light, and showing why the one that catches more light must be the bigger of the two. (B) How much bigger is this one than the other?

(7) How much farther can a 20-centimeter telescope see a star than the naked eye? (A) Describe the logic of the calculation you will use. (B) Carry out the detailed calculation.

(8) The mirror of the Hubble Space Telescope has a diameter of 2.4 meters. We want to compare the quantity of light gathered by this telescope to that gathered by the pupil of the naked eye. (A) Describe the logic of the calculation you will use. (B) Carry out the detailed calculation

(9) We want to calculate how much farther the Hubble can see a star than the naked eye. (A) Describe the logic of the calculation you will use. (B) Carry out the detailed calculation.

(10) We want to calculate the resolving power of a 15-meter telescope. (A) Describe the logic of the calculation you will use. (B) Carry out the detailed calculation.

(11) We want to find out if a 15-meter telescope is capable of resolving the writing on a distant billboard. Suppose the letters are an inch high, and the billboard is 11 miles away. (A) Describe the logic of the calculation you will use. (B) Carry out the detailed calculation.

(12) How big a telescope do you need to resolve a 100-meter crater on the Moon? (A) Describe the logic of the calculation you will use. (B) Carry out the detailed calculation.

(13) Google Earth shows images of the Earth obtained by orbiting satellites. In some of these images one can make out individual automobiles. Suppose the satellite that obtained this image were in an orbit 200 miles up. (A) Explain how you would be able to calculate from this the size of the orbiting telescope. (B) Carry out the detailed calculation.

(14) How big a radio telescope operating at a wavelength of 0.21 meters would you need to get a resolution of one second of arc? (A) Describe the logic of the calculation you will use. (B) Carry out the detailed calculation.

(15) For what temperature does $\lambda_{\text{most emission}}$ given by the Wien law equal the gamma-ray wavelength 10^{-11} meters? (A) Describe the logic of the calculation you will use. (B) Carry out the detailed calculation.

You must decide

(1) Suppose you have been appointed Presidential Science Advisor. The President wishes to allocate half a billion dollars to support the development of new instruments for astronomy. There are two ways to do this.
   - One option would provide increased funding for space missions devoted to science.
   - The other option would set the money aside for the development of ground-based telescopes.

The first option would require channeling the increased funding to NASA, which is the sole agency supporting missions in space. The second would channel the increased funding to the NSF, which supports ground-based astronomy.

Your task is to write a memorandum analyzing these two options, and to recommend which course of action the President should take. Your memorandum should include a *detailed* account of
   - the arguments for each option,
   - the arguments against each option,
   - the decision that you reached,
   and (this is the most important of all!)
   - your reasons for your decision.

(2) You are a young astronomer just beginning your professional career. You have received two job offers.
   - One offer from NASA will enable you to join a large team which is already at work designing instrumentation that will be deployed on an ambitious space mission to search for life elsewhere in the Universe.

- The other offer from a prestigious university will enable you to conduct research of your own, by yourself, on any topic you desire.
- Unfortunately, your parents are worried about you, and they have a tendency to micro-manage your career. Your task is to write them a letter analyzing

  - the arguments for each option,
  - the arguments against each option,
  - the decision that you reached,

  and (this is the most important of all!)
  - your reasons for your decision.

Part II

The Solar System

The Solar System is our first step outward into the cosmos. It consists of the Sun and the family of bodies orbiting about it. Largest among this family are the planets. Until recently there were nine in all, but in 2006 Pluto was redefined to be no longer a planet. So now there are eight. Mercury is the innermost, and Neptune the outermost planet: our Earth is the third.

Solar System

- Sun
- Mercury
- Venus
- Earth
- Mars
- (Asteroid Belt)
- Jupiter
- Saturn
- Uranus
- Neptune
- (Pluto)

Orbiting about the planets are the moons. We have one, some planets have none – Mercury and Venus, for example – while others have many – Saturn had more than 60 at last count.

Lying between the orbits of Mars and Jupiter are the asteroids: small, rocky bodies ranging in size from tiny bits of rubble to that of a small moon. Orbiting through the Solar System in highly inclined, highly elliptical orbits are the comets, "dirty snowballs" that vaporize when near the Sun, the vapor streaming away in graceful tails. Also orbiting through the Solar System are the meteoroids, small bodies that occasionally slam into us producing fiery trails in the sky. Some are sufficiently large to survive their passage through the atmosphere and fall to Earth, there to dig craters. When we find them strewn about the ground we have been granted free samples of the cosmos.

Should we also include in our list the space probes? They have voyaged to every planet of the Solar System. Even those that have fulfilled their missions and been shut down still exist: tiny artificial satellites, orbiting the Sun or some planet. Quiescent now, they too are part of the Sun's family.

6

Introducing the Solar System

People used to think that the Solar System was essentially the entire Universe, and that beyond its bounds lay little more than "lots of stars." We now know this is not so, and that the full astronomical Universe is far richer than that. Our Solar System is one among many, one small part of the whole. I like to think of it as our astronomical home base: the "house" in which we live.

We will begin to study our "house" by considering its most general features. We will ask how big it is and how massive; and what the orbits are of the bodies within it. The Solar System turns out to have a strikingly regular shape, and the planets within it divide naturally into two groups: inner and outer. In subsequent chapters we will focus more closely on its individual members.

Measuring the Solar System

In Part I of this book we prepared ourselves for our study of astronomy. In particular, we amassed a set of tools that we can use to measure various properties of the Universe. Let us now use these tools to find:

(1) the size of the Solar System,
(2) the sizes of the Sun and planets,
(3) the masses of the Sun and planets.

(1) The size of the Solar System

To measure the size of the Solar System we need to measure the sizes of orbits within it. We will begin by measuring the size of our orbit, and then move on to that of the planets.

The distance to the Sun
The radius of the circle the Earth moves in (recall that the orbit of the Earth is very nearly a perfect circle) is the distance to the Sun. We measure this distance by finding our orbit's circumference C and using the formula $C = 2\pi R$ to find its radius R. The circumference, in turn, is the distance we travel in one year: we find this distance by measuring our velocity in orbit and multiplying this velocity by one year.

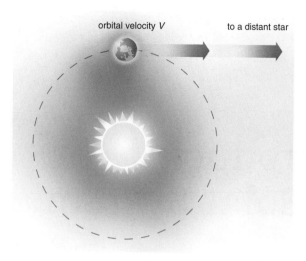

orbital velocity *V*   to a distant star

Figure 6.1 **Measuring Earth's velocity in orbit by observing the blueshift of a distant star's spectral lines.**

To find our orbital velocity we use the Doppler effect. Recall that this effect allows us to measure the relative velocity between us and a source of light. This source of light will be a distant star: if that star is stationary, the result we get will be *our* orbital velocity, which is what we need to measure.

So we arrive at the paradoxical conclusion that, to find the distance to the Sun, we don't need to observe the Sun. We need to observe some other star! Figure 6.1 illustrates the principle of the measurement. In the illustrated configuration we are approaching the distant star. When we measure the wavelengths of lines in its spectrum, we find that these wavelengths are shorter than they should be. The difference between the observed and expected wavelengths is the Doppler shift $\Delta\lambda$. By using the Doppler effect formula

$$\Delta\lambda/\lambda = V/c$$

we can then find the Earth's orbital velocity V.

⇐ **Looking backward**
We studied the Doppler effect in Chapter 4.

NOW YOU DO IT
In the configuration illustrated in Figure 6.1, suppose that we observe a spectral line from the distant star that ought to be at a wavelength λ = 5000 angstroms – but in fact we find it to have a wavelength $\lambda_{observed}$ = 4999.5 angstroms. We want to find from this the Earth's orbital velocity about the Sun. (A) Describe the logic of the calculation you will use to do this. (B) Carry out the detailed calculation. (The speed of light is 3×10^8 meters per second.)

The result you get from the above exercise is V = 30 kilometers per second. From this we can calculate the distance to the Sun as follows.

The logic of the calculation

Step 1. Find the distance we travel in one year by multiplying our velocity by one year. This distance is the circumference C of our orbit.

Step 2. Find the radius R of our orbit using $R = C/2\pi$. This is the distance to the Sun.

NOW YOU DO IT
Suppose you were an astronomer living on Venus, and you performed the same observations. You would have found that your planet's orbital velocity about the Sun was 35 kilometers per second, and its year was 225 Earth days long.

Detailed calculation

Step 1. Multiply the Earth's velocity by one year to find the circumference of our orbit.

One year = (365 days/year)(24 hours/day)(60 minutes/hour)(60 seconds/minute) = 3.15×10^7 seconds.

We then calculate

C = (30 kilometers/second)(3.15×10^7 seconds) = 9.45×10^8 kilometers = 9.45×10^{11} meters.

Step 2. Find the radius R of our orbit using $R = C/2\pi$.
We calculate

$R = 9.45 \times 10^{11}$ meters$/2\pi = 1.5 \times 10^{11}$ meters.

A more accurate measurement gives the result 1.496×10^{11} meters.

The astronomical unit: our distance from the Sun

The distance from the Earth to the Sun is commonly referred to as the *astronomical unit* or AU.

1 AU = 1.496×10^{11} meters or about 93 million miles.

From this, you want to find Venus' distance from the Sun. (A) Describe the logic of the calculation you will use to do this. (B) Carry out the detailed calculation.

. . . .

Radii of the orbits of the planets

In the above exercise we imagined what an astronomer on Venus would observe: this allows us to find the radius of Venus' orbit. Of course we cannot really do this if we stay right here on Earth. Here is a method that we can use.

If you observe the planet Venus over the course of many months you will see that it never gets very far from the Sun. During certain seasons it can be seen in the evening shortly after sunset: in these configurations you will find it to be fairly close to the horizon – which is to say, fairly close to the Sun. During other seasons it can be seen in the morning shortly before sunrise – and here too, you will always find it close to the Sun.

Why should this be so? Because the orbit of Venus is smaller than that of the Earth. Figure 6.2 illustrates the configuration. Venus, of course, always lies somewhere along its orbit; as illustrated in Figure 6.2, the angle between it and the Sun is never very great.

Careful measurements reveal that the observed angle between Venus and the Sun is never greater than 46.3 degrees. We can use this to find the *radius of the orbit of Venus* as follows.

When Venus appears farthest from the Sun, the configuration is as illustrated in Figure 6.2. In this configuration, the line of sight leading from us to Venus is tangent to Venus' orbit. This means that it forms a right angle with the line from the Sun to Venus (Figure 6.3). The triangle illustrated in Figure 6.3 is therefore a right triangle. Furthermore, two other quantities in this triangle are known. One is the length of the line leading from the Earth to the Sun: this is 1 AU. The other is the maximum angle between Venus and the Sun: 46.3 degrees.

We can use this to find the length of the side X, the distance from Venus to the Sun. There are two ways to do this.

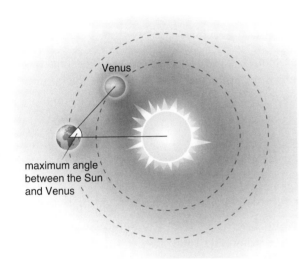

Figure 6.2 **Venus always lies close to the Sun as seen from the Earth.**

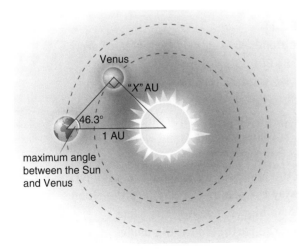

Figure 6.3 **The radius of Venus' orbit X can be found from this construction.**

(1) One way is to draw on a piece of paper a right triangle, the hypotenuse of which is 1 foot (0.3 meters) long and the indicated angle of which is 46.3 degrees. We can then simply measure with a ruler the other leg! If we do this we find that its length is 0.723 feet (0.217 meters). This demonstrates that Venus' orbit has a radius of 0.723 AU.

(2) If you are familiar with trigonometry, you will recall that the sine of an angle is the opposite over the hypotenuse. So we have sine (46.3°) = (X AU)/(1 AU). Using a calculator we find that sine (46.3°) = 0.723, showing that X = 0.723 AU.

. . . .

NOW YOU DO IT
If you observe Mercury you will find that it never gets more than 22.8 degrees from the Sun. We want to use this to find the distance between Mercury and the Sun. (A) Describe how you plan to do this. (B) Carry out your work and find the radius of Mercury's orbit.

. . . .

The orbital radii of other planets can be found by similar geometrical constructions.

(2) The sizes of the Sun and planets

Now that we know how far away the Sun is, we can easily measure how big it is. Figure 6.4 illustrates the method. If we measure the angle that the Sun subtends we can calculate its diameter d_{Sun} by using the small-angle formula (Appendix I):

$$d_{Sun} = 2\pi D[\theta/360°]$$

where D is the Sun's distance, which we have just measured.

The Sun is observed to subtend an angle $\theta = \frac{1}{2}$ degree. Combining this with the measured value of the astronomical unit, we calculate the Sun's diameter to be 1.39×10^9 meters, or 1.39 million kilometers. Similarly, we can measure the angles subtended by the various planets of the Solar System, and so find their sizes as well.

. . . .

NOW YOU DO IT
The Moon subtends an angle of ½ degree of arc, and it lies 384 000 kilometers from the Earth. From these, we want to find the diameter of the Moon. (A) Describe the logic of the calculation you will use to do this. (B) Carry out the detailed calculation. (C) Suppose the Moon were to move farther from the Earth, while keeping the same size. Would the angle it subtends become bigger, smaller, or stay the same? (D) Suppose the Moon were to become bigger, keeping the same distance from the Earth. Would the angle it subtends become bigger, smaller, or stay the same?

. . . .

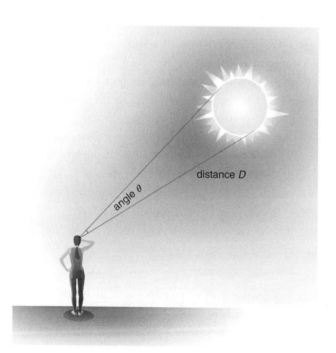

distance D

angle θ

Figure 6.4 **Diameter of the Sun can be found by measuring the angle θ it subtends and its distance D.**

(3)   The masses of the Sun and planets

In Chapter 3 we developed the *mass formula*, which allows us to measure the mass M of any body by observing the orbit of some other body about it. If the orbiting body moves with velocity V in an orbit of radius R, then

$$M = RV^2/G,$$

where G is Newton's constant of gravitation. In Chapter 3 we used the orbit of the Earth to measure the mass of the Sun: using the measured values of R and V for the Earth's orbit, we found the Sun's mass to be $M_{Sun} = 2 \times 10^{30}$ kilograms. We can use the same method to find the masses of other bodies in the Solar System: by observing the orbit of a satellite of Mars, for instance, we can measure Mars' mass.

. . . .

NOW YOU DO IT
Mars lies 1.52 AU from the Sun, and it orbits once every 1.88 years.
(A) What celestial body's mass can be determined from this? Explain your answer. **(B)** Describe the logic of the calculation you will use to find this body's mass. **(C)** Carry out the detailed calculation.

Deimos moves in an orbit of radius 23 500 kilometers about Mars, and it orbits that planet once every 1.26 days. **(A)** What celestial body's mass can be determined from this? Explain your answer. **(B)** Describe the logic of the calculation you will use to find this body's mass. **(C)** Carry out the detailed calculation.

. . . .

The measured Solar System

The results of all these measurements are given in Table 6.1. (We include in this table Pluto, ignoring for now the question of whether it really is a planet.)

Let us study this table carefully, trying to make sense of all those numbers. Our goal is to build up an intuitive understanding of what it is telling us.

Table 6.1. **The Solar System.**

	Mass (Earth=1)	Radius (kilometers)	Mean distance from Sun (AU)	Orbital eccentricity	Orbital inclination to ecliptic (degrees)
Mercury	0.055 8	2 439.7	0.387	0.21	7.0
Venus	0.815	6 051.8	0.723	0.006 8	3.4
Earth	1	6 378.1	1	0.017	0.0
Mars	0.107 4	3 397	1.524	0.093	1.8
Jupiter	317.9	71 492	5.203	0.048	1.3
Saturn	95.15	60 268	9.539	0.056	2.5
Uranus	14.54	25 559	19.18	0.047	0.77
Neptune	17.23	24 764	30.06	0.008 6	1.77
(Pluto	0.002 2	1151	39.44	0.25	17.1)

A scale model of the Solar System

Sizes of the planets

Begin by focusing on the sizes of the Sun and planets – their radii. In understanding these data it might be helpful to think in terms of a scale model.

Suppose we shrink everything till the planet Earth is the size of a grapefruit. Into this small fruit every continent, every mountain and desert and ocean of our great world is now compressed. On this scale, the Sun is immense: a full 12 yards (10.97 meters) across. Indeed, if we were to somehow put together *every* planet in the Solar System, and add to the pile all the moons, asteroids, comets and meteors, we would still end up with a body far smaller than the Sun! With this simple mental exercise we have reached an important understanding: in a very real sense, the Solar System consists of nothing but the Sun. In terms of sheer quantity of matter, everything else – our own Earth included – is tiny.

Two comments may be made here.

- If we think in terms of the emission of energy this conclusion grows yet stronger. Among all the bodies of the Solar System, only the Sun pours out immense quantities of it. This energy, generated by nuclear reactions deep within its interior, is the source of all the light and heat in the Solar System. If by some terrible mischance the Sun were to cease generating energy, eternal night would fall, and the Earth and every body in the Solar System would quickly cool down to a few degrees above absolute zero – 459 degrees below zero Fahrenheit.
- This is why the task of searching for planets orbiting about other stars in the Universe is so difficult. The planets for which we are searching are both tiny and dim compared with the stars about which they orbit. We will discuss this further in Chapter 10.

After the Sun, the second biggest object in the Solar System is Jupiter, whose diameter is about 1/10th that of the Sun. In terms of our scale model in which the Earth is the size of a grapefruit, Jupiter is nearly 4 feet (1.22 meters) in diameter. Indeed, Jupiter is not just the biggest of the planets: it is bigger than all the rest of them combined. So if we wished to render our previous picture more accurate, we would say that the Solar System consists not just of the Sun, but of the Sun and Jupiter.

Jupiter is in certain ways "almost a star" in its own right. The larger moons orbiting about it form a miniature solar system of their own, and its composition is strikingly similar to that of the Sun. While Jupiter does not generate energy by nuclear reactions in its core, it would if it were somewhat more massive.

Four of the planets in the Solar System are larger than the Earth (Jupiter, Saturn, Uranus and Neptune). The remainder all have sizes somewhat smaller than that of our world. Mercury, the closest planet to the Sun, is 1.5 inches (3.8 centimeters) across in our scale model: Pluto, ¾ of an inch (1.9 centimeters).

Sizes of the orbits of the planets

Let us now move on to consider the distances of the various planets from the Sun (which are customarily expressed in astronomical units). Within our scale model, in which the Earth is represented by a grapefruit, we lie ¾ of a mile (1.207 kilometers)

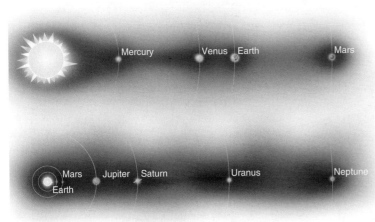

Figure 6.5 **Relative sizes of the orbits of the planets.**

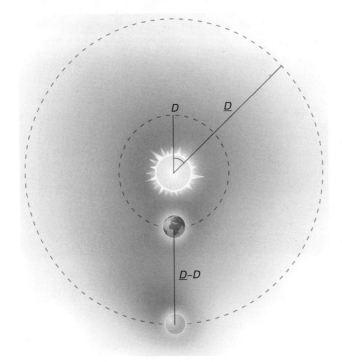

Figure 6.6 **In the most favorable configuration for studying a planet its distance from us is its orbital radius minus that of the Earth.**

from the Sun, while Pluto's orbit has a semi-major axis of fully 29 miles (nearly 47 kilometers). Conversely, if we wanted to make our model of the Solar System a more manageable size, we would be forced to make the planets almost too small to see. If, for example, we think of a model in which the Earth is a millimeter in diameter – a bit more than the size of the period at the end of this sentence – the outermost planet is a bit more than a quarter of a mile from the Sun. Figure 6.5 shows the orbits of the planets to scale.

As you can see from this figure, the outer planets are very widely spaced compared to the inner ones. Saturn, the next planet out after Jupiter, lies nearly twice as far from the Sun as Jupiter. Similarly, Uranus lies twice as far from the Sun as Saturn.

This makes it far harder to study the outer planets of the Solar System than the inner ones. Consider the following.

(1) Imagine that we wish to study a planet by telescope. We will get the best view when it is closest to us. As you can see from Figure 6.6, this distance of closest approach is the radius of the planet's orbit minus that of ours. As an example, let us compare the difficulty of observing Uranus with that of observing Mars. From Table 6.1 we calculate that Uranus' distance of closest approach is 18.2 AU, while Mars' is 0.52 AU. So in their most favorable viewing configurations, Uranus is 18.2/0.52 = 35 times farther away from us than Mars!

(2) The outer planets are more faintly illuminated by the Sun than the inner ones. Recalling the inverse square law from Chapter 4, we see that the intensity of sunlight on a planet decreases as the inverse square of its distance from the Sun. Thus the flux of sunlight falling on Uranus, as compared to the flux falling on Mars is

$$\frac{\text{Flux of sunlight on Uranus}}{\text{Flux of sunlight on Mars}} = \left[\frac{\text{Mars' distance from the Sun}}{\text{Uranus' distance from the Sun}}\right]^2$$
$$= [1.52\,\text{AU}/19.2\,\text{AU}]^2$$
$$= 0.0063.$$

The flux of sunlight illuminating Uranus is 0.63% that illuminating Mars. Daylight is dim out there!

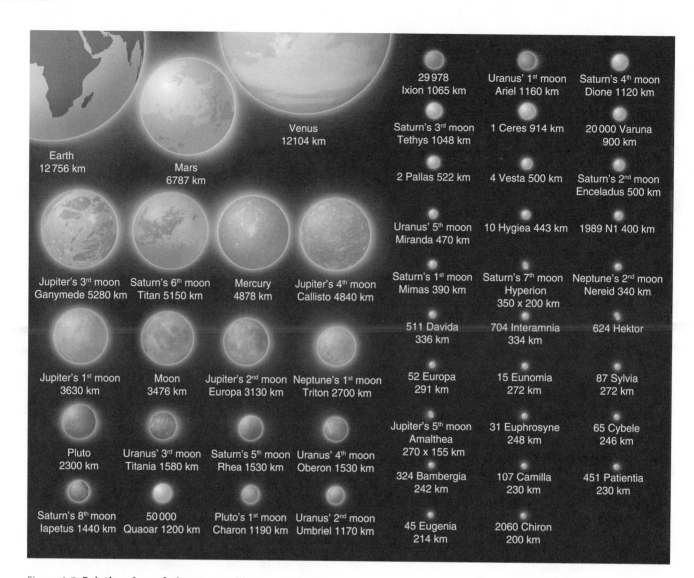

Figure 6.7 **Relative sizes of planets, satellites and asteroids.**

(3) When we study a planet by telescope, we are observing the sunlight it reflects back to us. As it travels from the planet to us, this reflected light itself obeys the inverse square law. So we are hurt doubly: sunlight out there is dim, and very little of that reflected sunlight reaches us! This makes a distant planet far dimmer as it appears in our telescopes than a nearer one.

(4) Suppose alternatively that we wish to send a spacecraft to study our target planet. The distance this spacecraft must travel is far greater to an outer planet than to an inner one. Mars Pathfinder took seven months to reach that planet – but Voyager took eight *years* to reach Uranus. Recall also from Chapter 3 that gravitation from the Sun tugs a space probe toward it, so that sending a probe to an outer planet is like throwing a boulder upwards, itself a difficult task. And finally, since the intensity of sunlight is so faint in the outer Solar System, spacecraft out there cannot use solar power and must carry onboard nuclear reactors for power. Recall from Chapter 3 the danger of radioactive contamination from these reactors, and the furor that erupted over the launch of the Cassini mission to Saturn.

....

NOW YOU DO IT

Suppose we wish to observe the planet Neptune. We want to know how far away it is when it lies closest to the Earth, as compared with how far away Mars is at closest approach. (A) Draw a diagram, similar to Figure 6.6, in which you illustrate this configuration. (B) Carry out the calculation to find (distance to Neptune at closest)/(distance to Mars at closest).

We wish to compare the flux of sunlight on Neptune with that on Mars. (A) Describe the logic of the calculation you will use to do this. (B) Carry out the detailed calculation.

....

Structure of the Solar System

Shapes of orbits

Recall two essential characteristics of the Solar System that we learned in Chapter 2.

(A) The Solar System is a disk: the orbits of the planets lie nearly in a plane.
(B) These orbits are very nearly circular.

We can quantify (A) by measuring the angle of inclination between each planet's orbit and the ecliptic, which as we recall is the plane of the Earth's orbit. Similarly, we can quantify (B) by measuring each planet's orbit's eccentricity. These are listed in Table 6.1. As we can see, both the inclinations and eccentricities are small.

All the planets orbit in the same direction. This direction is that of the Sun's rotation on its axis: indeed, the ecliptic plane nearly coincides with the Sun's equator. Similarly, many planets rotate in the same sense as does the Sun. The orbits of most of the larger moons also lie in the ecliptic plane. Finally, most moons orbit in the same direction as their planets.

Planets and satellites

Many people believe that planets and moons are totally different. Planets, for instance, are big but their moons are small. There is much truth to this: certainly *our* Moon is smaller than the Earth. But the full situation is more interesting than this.

Figure 6.7 shows the relative sizes of selected planets, moons and asteroids. We see that the distinction between these three classes of object is not so clear as one might think. Ganymede, one of the moons of Jupiter, is bigger than both Mercury and Pluto. Similarly, our own Moon is bigger than Pluto. Ceres, the largest asteroid, is bigger than many of the smaller moons.

Similarly, one of the moons of Saturn (Titan) is known to have a thick atmosphere – indeed, its pressure is greater than air pressure on the Earth! And finally, while many moons are geologically inactive, others show signs of geological reworking of their surfaces.

Thus, the more we learn the more fuzzy does the distinction between moon and planet become.

Inner planets and outer planets

Table 6.1 reveals a striking regularity in the Solar System: the inner planets are a lot smaller than the outer ones. In terms of our scale model in which the Earth is

Detailed calculation

Step 1. Average density is defined to be mass divided by volume.

Density = M/V, where M is the Earth's mass

Step 2. The volume is given by $V = (4/3)\pi R^3$, where R is the Earth's radius.

Step 3: Plug in values of the mass and radius from Table 6.1.

For the Earth, Table 6.1 shows

mass = 5.98×10^{24} kilograms
radius = 6378 km = 6.378×10^6 meters.

We find the Earth's volume to be

$V = (4/3) \pi (6.378 \times 10^6 \text{ meters})^3 = 1.09 \times 10^{21}$ cubic meters,

and its density:

average density of Earth = $(5.98 \times 10^{24}$ kg$)/(1.09 \times 10^{21}$
cubic meters)
= 5490 kilograms/cubic meter.

represented by a grapefruit, the four inner planets (Mercury through Mars) have diameters measured in inches. But the four outer planets (Jupiter through Neptune) have diameters measured in feet. Only tiny Pluto fails to conform to this pattern.

We can also think in terms of the planets' average densities. Here, too, we find a major difference between the inner and outer planets. These are important because density is a clue to a planet's composition. For instance, an object made of iron has a large density, while an object made of foam has a low density.

Let us find the Earth's average density.

The logic of the calculation

Step 1. Average density is defined to be mass divided by volume.

Step 2. The volume is given by $V = (4/3)\pi R^3$, where R is the planet's radius.

Step 3. Plug in values of the mass and radius from Table 6.1.

We find the Earth's average density to be 5490 kilograms per cubic meter. On the other hand, if we carry out the same calculation for an outer planet we get a far lower result. Jupiter's average density, for instance, works out to 1246 kilograms per cubic meter. This is a mere one-quarter the density of the Earth.

NOW YOU DO IT
Calculate Saturn's average density.

What does this tell us about the planets' compositions? The density we obtained for the Earth is characteristic of rocks and metals – and, indeed, that is what the Earth is made of: solid rock and an iron core. The far lower density of the outer planets, on the other hand, tells us that they cannot possibly be made of rock! Spectroscopic analysis confirms this conclusion. Spectra of the inner planets show they are composed of rocky substances, but spectra of the outer planets reveal the presence of gases. The outer planets are known as the *gas giant* planets.

Smaller bodies of the Solar System

Between the inner and the outer planets lies the *asteroid belt*, a broad region of space populated by innumerable small rocky bodies. Ceres, the largest, is 950 km across. That's about the distance between Washington DC and Detroit. Most asteroids are far smaller – boulder-sized or pebble-sized. Space probes have sent back images of a few: they are irregular in shape, and covered with craters. From time to time two asteroids collide, shattering into pieces: these pieces appear to be the *meteoroids* that occasionally fall to Earth.

When a meteoroid strikes the Earth's atmosphere it is moving so rapidly that it is heated to incandescent temperatures by friction. It glows with a brilliant light, which we on the ground see as a "shooting star," or *meteor*. Most meteoroids are so small that they are entirely burned up in our atmosphere, but a few of the larger ones survive to land on the ground – only the very biggest dig craters. When we find them, many look to the untrained eye like ordinary rocks, although detailed analysis reveals important differences.

There is another source of these "shooting stars" – the comets. Comets have been described as "dirty snowballs." They are loose, fluffy things, composed of ices mixed with dust and tiny grains. In contrast to every other object in the Solar System, the comets move in highly elliptical orbits. Most of the time they lie far from the Sun, where their ices are frozen in the deep cold of space. But occasionally their orbits carry them closer in, where they are warmed by sunlight and their ices vaporize. The emitted gases are combed outward away from the Sun by the solar wind, where they form the beautiful comet tails. These tails are immense, and can be as much as a hundred million miles long.

How do we know that some meteors are bits of comets? Because the Earth, in its orbit around the Sun, occasionally passes through the orbit of a known comet – and when it does, a meteor shower results. The Perseid meteor shower, for instance, which occurs in mid August, is associated with comet Swift–Tuttle. Apparently the emitted gases from a comet blow off the dust and grains from its surface. These continue to travel about the Sun in the same path as the parent comet, tiny orbiters in their own right. When we plow through them we get a meteor shower.

Summary

Measuring the Solar System
- To measure the radius of the Earth's orbit (the astronomical unit, AU):
  - use the Doppler effect to measure our velocity about the Sun;
  - this velocity equals the circumference of our orbit, which is $2\pi R$, divided by one year: solve this for R.
- Use trigonometry to measure the radii of the orbits of the other planets in terms of the AU.
- Measure the size of the Sun (or any planet) by measuring the angle it subtends, and use the known value of its distance and the small-angle formula.
- Measure the mass of the Sun by observing the orbit of a planet about it and using Newton's mass formula.
- We can do the same for any planet with moons. For a planet without moons, we must send a space probe past it and observe the probe's orbit.

A scale model of the Solar System
- If we imagine the Earth to be the size of a grapefruit, the Sun would be 12 yards (11 meters) across.

- The Sun is the largest body in the Solar System. Indeed, it is bigger than everything else put together.
- Next in size is Jupiter, which is "almost a star."
- The orbits of the planets are very large: in the above scale model:
  - the Sun would be ¾ miles (1.2 kilometers) away,
  - the semimajor axis of Pluto's orbit would be 29 miles (46.67 kilometers).
- The Solar System is mostly empty space.
- The Solar System is disk-shaped: planetary orbits are nearly circular, and they lie nearly in the same plane.
- Some moons are bigger than planets. The largest asteroid is bigger than many moons.
- Inner versus outer planets:
  - inner planets: low mass, high density, composed of rock,
  - outer planets: high mass, low density, mostly or entirely composed of gas.
- Asteroids:
  - very small (hundreds of kilometers to pebble-sized),
  - rocky,
  - most lie between Mars and Jupiter,
  - moderately elliptical orbits.

- Comets:
  - "dirty snowballs,"
  - when they near the Sun they warm and sprout tails of vaporized gases,
  - move in highly elliptical orbits.
- Meteors come from collisions between asteroids and comets.

Problems

(1) Imagine that you are an astronomer living on Jupiter, and you observe a spectral line from a distant star toward which Jupiter is orbiting. The spectral line from this star, which ought to be at a wavelength $\lambda = 5000$ angstroms, turns out to have a wavelength $\lambda_{observed} = 4999.78$ angstroms. You want to find from this Jupiter's orbital velocity about the Sun. (A) Describe the logic of the calculation you will use to do this. (B) Carry out the detailed calculation. Jupiter's year is 11.87 Earth years long: you want to find from this the distance between Jupiter and the Sun. (C) Describe the logic of the calculation you will use to do this. (D) Carry out the detailed calculation.

(2) Let us consider a distant star at four different times over the course of a year: in January, March, June and September. As you can see, in March we are heading directly away from that star, and in September we are heading directly towards it (Figure 6.8).

So in March the wavelength of light from that star will be lengthened by the Doppler effect, and in September it will be shortened. (A) Your first task is to figure out what the Doppler effect does to the light's wavelength at other times of year. In particular, draw a rough sketch in which you complete Figure 6.9.

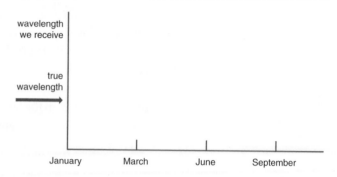

Figure 6.9 **Wavelength of light we receive from a distant star depends on the season.**

In drawing this diagram, remember that in Chapter 4 we emphasized that the Doppler effect does not measure the full velocity of an object: it only measures the component of its velocity along the line of sight leading from the object to the observer. (B) You have now made a series of theoretical predictions. As we have repeatedly emphasized, scientific predictions need to be tested by observations. Can *all* of your predictions be so tested – or only some of them? Explain your answer.

(3) In finding the Earth's velocity about the Sun, we used the Doppler effect. But this effect only allows us to measure the *relative* velocity between us and the distant stars – and the stars are moving too! Furthermore, their velocities are themselves unknown. How can we disentangle our motion from that of the stars?

Here's a method: observe a star not once but twice. We can choose any star at all. We first observe it on a night when the Earth is heading directly toward it – and we next observe it when the Earth is heading directly away from it.

Here's a diagram of the Earth in its orbit about the Sun, and a distant star (Figure 6.10).

- We want the first observation to take place on a night when the Earth is heading directly *toward* the star. By using Figure 6.10, explain why the direction to the star on that night makes a *right angle* with the direction to the Sun.
- We want the second observation to take place on a night when the Earth is heading directly away from the star.

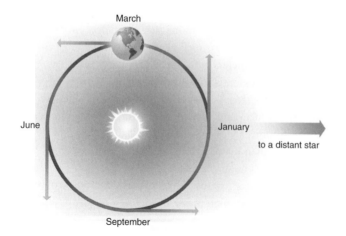

Figure 6.8 **Our velocity relative to a distant star depends on the season.**

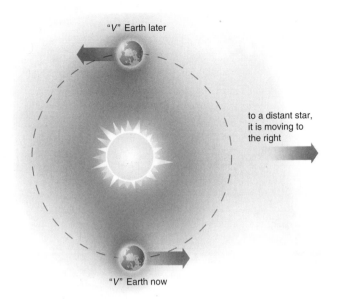

"V" Earth later

to a distant star,
it is moving to
the right

"V" Earth now

Figure 6.10 **Doppler shift measurements of a moving star.**

By using Figure 6.10, explain why the second observation must take place exactly six months after the first.

- Let us call V_1 the relative velocity between us and the star which we find from the first observation. Similarly, let V_2 be the relative velocity we find from the second observation. Explain why

$$V_1 = V_{Earth} - V_{star}$$
$$V_2 = V_{Earth} + V_{star.}$$

- Here are the results of these two observations:

$$V_1 = 20.9 \text{ km/second}$$
$$V_2 = 40.1 \text{ km/second.}$$

From these, we want to find both our velocity about the Sun, and the star's velocity. (A) Describe the logic of the calculation you will use to do this. (B) Carry out the detailed calculation.

(4) Imagine a hypothetical planet that never gets more than 15 degrees from the Sun. We want to use this to find the distance between this planet and the Sun. (A) Describe how you plan to do this. (B) Carry out your work and find the radius of its orbit.

(5) If you observe Saturn you will find that it can be at any angle from the Sun. Draw a diagram, analogous to Figure 6.2, illustrating that the radius of Saturn's orbit must be greater than 1 AU.

(6) When it is closest to the Earth, Jupiter subtends an angle of 0.013 degrees of arc, and it lies 4.2 AU from us. From these, we want to find the diameter of Jupiter. (A) Describe the logic of the calculation you will use to do this. (B) Carry out the detailed calculation.

(7) Uranus lies 19.2 AU from the Sun, and it orbits once every 84 years. (A) What celestial body's mass can be determined from this? Explain your answer. (B) Describe the logic of the calculation you will use to find this body's mass. (C) Carry out the detailed calculation.

(8) The satellite Oberon lies 583 000 kilometers from Uranus, and it orbits Uranus once every 13 days. (A) What celestial body's mass can be determined from this? Explain your answer. (B) Describe the logic of the calculation you will use to find this body's mass. (C) Carry out the detailed calculation.

(9) Suppose we wish to observe the planet Uranus. We want to know how far away it is when it lies closest to the Earth, as compared with how far away Mars is at closest approach. (A) Draw a diagram, similar to Figure 6.6, in which you illustrate this configuration. (B) Carry out the calculation to find (distance to Uranus at closest)/(distance to Mars at closest).

(10) We wish to compare the flux of sunlight on Uranus with that on Mars. (A) Describe the logic of the calculation you will use to do this. (B) Carry out the detailed calculation.

(11) We want to find the average density of Venus, using the data in Table 6.1. (A) Describe the logic of the calculation you will use to do this. (B) Carry out the detailed calculation. (C) Suppose Venus were to grow bigger, while keeping the same mass. Would its average density become larger, smaller, or stay the same? (D) Suppose Venus' mass were to increase, while keeping the same size. Would its average density become larger, smaller, or stay the same?

The inner Solar System

We now turn to a more detailed tour of the Solar System.

In this chapter we will visit our Moon, and then tour those planets lying closest to the Sun. Our own home world is one of them: the others are Mercury, Venus and Mars. These planets constitute what might be called our immediate vicinity, the portions of the Solar System that lie closest to us. But the real reason to treat all these bodies at once is that, taken together, they constitute the inner Solar System, all of whose planets are relatively small and dense and are composed of rock. In the next chapter we will encounter the outer planets, which are entirely different: far larger, and composed primarily of gas.

The farther out we will look, the more alien will the worlds become: in this chapter we'll start with something more or less familiar. But even these, our nearest-neighbor worlds, will turn out to be full of surprises. And perhaps this is the most important lesson of all: the Universe continually surprises us, and it is endlessly fascinating.

The Moon

Observing the Moon

The first stop on our journey will be our own Moon (Figure 7.1). To the naked eye the Moon consists of light areas and irregular dark splotches. These dark regions were termed *maria* ("MAH-ria") by the ancients, from the Latin *mare* ("MAH-ray") for "sea." Although we now know they are ancient lava flows, not oceans, the name has stuck. The view through even a small telescope or pair of binoculars is far richer. In lighter areas can be seen mountains together with innumerable craters, from the largest of which great white rays are seen to emanate. The maria too have craters, though far fewer of them.

Prior to the space program, no one had ever seen the Moon's back side, for it forever keeps the same face toward the Earth. Not until the Apollo missions circled about the Moon did we get a detailed picture of its far side (Figure 7.2). Remarkably, this side is composed almost exclusively of the lighter, heavily cratered terrain: there are far fewer maria back there.

We can learn more about the topography of the Moon by choosing carefully which portion of it to observe. How long is the shadow cast by a lunar object?

Figure 7.1 **The Moon.**

Figure 7.2 **The far side of the Moon.**

Figure 7.3 illustrates two of them. As we can see, man "A" casts no shadow while the shadow from man "B" is very long. This tells us that we should concentrate on the line dividing the Moon's illuminated side from its dark side: shadows there are highly accentuated, making the topography easy to make out. We can see this behavior in Figure 7.4.

Detailed studies of the lengths of shadows on the Moon allow us to measure the heights of the Moon's mountains. They turn out to be about as high as Earthly ones.

If you look closely at Figure 7.1 you will notice a striking regularity: the lighter areas of the Moon have lots of craters and mountains, while the darker areas do not. And the lunar maria are not just mostly free of craters: they are also quite flat. But why should this be so? Let us keep this question in mind as we continue our exploration of the Moon.

A major difference between the Moon and the Earth is that the Moon is peppered with craters, while the Earth has only a few. What accounts for this striking dissimilarity? Until recently we might have asked why the Moon has so many craters. But recent advances have allowed us to get good, close-up views of many planets, satellites and asteroids: these views have taught us that craters are common. Furthermore, as we will see in Chapter 13, the final stage of the formation of the Solar System was marked by a continual hail of impacts. So the real question is not why the Moon has so many craters – it is why has the Earth so few?

There are two answers to this question:

- The surface of the Earth is continually being re-shaped by geological processes that "erase" its craters. (We will study these processes later in this chapter.) But the Moon is geologically inactive.
- Earthly craters are continually being eroded away by wind and rain. But the Moon has no atmosphere and no water, so this erosion is absent.

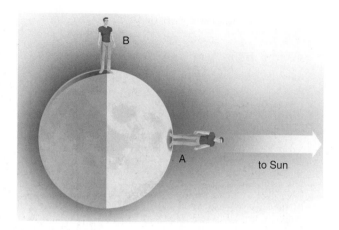

Figure 7.3 **Shadows cast by objects on the Moon are accentuated along the dividing line between its light and dark halves.**

Air on the Moon?

The Moon has no atmosphere. How do we know this? One line of evidence comes from the *absence* of a phenomenon that we would expect to happen if the Moon did have an atmosphere.

This phenomenon concerns the occultation of a star by the Moon. As it orbits the Earth, from time to time the Moon passes in front of – "occults" – a distant star (Figure 7.5a). We will now show that, if the Moon had an atmosphere, we would expect the star to grow gradually dimmer just prior to the occultation. Conversely, if the Moon does not have an atmosphere, we would expect it to abruptly vanish as the Moon passes in front of it (Figure 7.5b).

No atmosphere is perfectly transparent. Tiny particles within it scatter and absorb light. On Earth these particles produce the haze that so often obscures distant objects. Similarly, when we look at a star just prior to occultation by the Moon, its light rays are passing through any lunar atmosphere – and its haze – that might exist.

As illustrated in Figure 7.6, the closer the star is to the Moon's edge the more atmosphere its rays would pass through. Thus, the closer the star is to the moment of occultation, the more its light would be obscured by haze. So it is that, had the Moon an atmosphere, we would expect the star's light to be gradually dimmed: initially by only a little, but then by more and more until finally the body of the Moon itself passes in front of the star.

Observations reveal that this does not happen. Stars are found to simply wink out, without dimming, upon being occulted. Therefore the Moon has no atmosphere.

This tells us that the Moon is an eerily silent and motionless place. Because it has no air, there are no sounds on the Moon. Nor are there any clouds, wind or rain. Untold ages can pass before anything *happens* there. With nothing but the occasional impacts of tiny micrometeoroids to erode the footsteps left by the Apollo astronauts, they are eternal in human terms. Millions of years from now, when the human race has possibly gone extinct, those faint footsteps in the lunar dust will still be there.

Figure 7.4 **Relief is particularly visible along the edge of the illuminated Moon. This image was taken from Apollo 12 as it orbited the Moon.**

NOW YOU DO IT
Suppose an astronaut is standing on the Moon and looking at the Earth. Does this astronaut ever see the Earth occult a distant star? If so, does the star suddenly wink out or gradually fade as this happens?

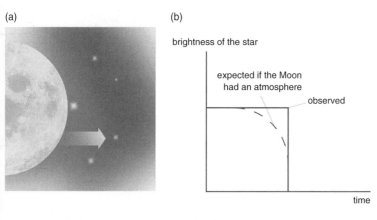

(a) (b)

brightness of the star

expected if the Moon
had an atmosphere

observed

time

Figure 7.5 **(a) Occultation of a distant star by the Moon. (b) Evidence for a lunar atmosphere would be a gradual dimming of a star just before the occultation.**

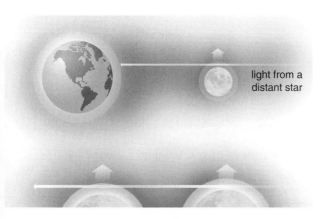

light from a
distant star

Figure 7.6 **A light ray from a distant star would pass through more and more haze as it approaches the moment of occultation by the Moon.**

Lunar cratering rates

The Moon, like the other bodies of the Solar System, was formed by a series of catastrophic impacts of smaller bodies (we will study this process in Chapter 13). On the surface of the Moon are the scars of the final stages of this process.

These scars are the Moon's craters. They were formed by the impacts. The ages of these craters tell us something about the history of the impacts. Studies show that there are lots of old craters and fewer young ones: this tells us that there were lots of impacts long ago, and fewer recently.

Let us study this in detail. To make things easier, we will begin with an analogy. In our analogy we will replace the Moon by a town, and the Moon's craters by people living in that town. In our analogy the formation of a crater – the impact of a body – will be replaced by the birth of a baby. Craters, of course, never leave the Moon, so to make our analogy complete we will have to imagine that nobody ever leaves our town, nobody dies and nobody ever enters it. Let us call our town "Mooney."

Suppose we conduct a census of Mooney, and we find there are a thousand newborn babies but a greater number of elementary-school kids. This means that this year there were a thousand births, but more than that in previous years. So our census tells us that in Mooney the birth rate has been decreasing.

Imagine that our census is taken in the year 2010, and it yields the data listed in Table 7.1.

From this we can figure out Mooney's birth rate over the past 60 years. After all, those 1 year olds were born in 2009, and the 10 year olds had been born in the year 2000. Simply by subtracting, we can find how Mooney's birth rate has been changing over time (Table 7.2).

If we read this table from bottom to top we will get the history of Mooney's birth rate shown in Figure 7.7.

Let us apply the very same method to craters on the Moon. By measuring their ages, we can compile data similar to those in Table 7.1. In Table 7.3 we give results for a representative portion of the Moon, chosen to contain on average one young crater. It turns out that this region contains on average slightly more than one 2-billion-year-old crater – but a huge number of 4-billion-year-old craters!

Clearly, the "birth rate" of craters was enormous far back in the past compared to now. Just as we did for the town of Mooney, we can find the birth rate on the Moon simply by subtracting. For example, two billion years ago it was

Table 7.1. **A census of Mooney (2010).**

Age	Number of people
1	1 000
10	1 100
20	1 300
30	1 300
40	1 400
50	1 600
60	2 300

Table 7.2. **History of the birth rate in Mooney.**

Age	Number of people	Year born
1	1 000	2009
10	1 100	2000
20	1 300	1990
30	1 300	1980
40	1 400	1970
50	1 600	1960
60	2 300	1950

Table 7.3. **A census of craters on the Moon.**

Age of crater	Relative number of craters
Young	1
2 billion years	1.3
4 billion years	100

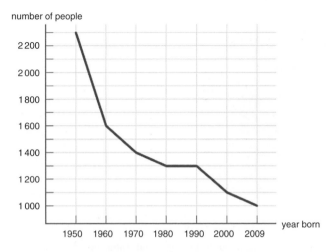

Figure 7.7 **The birth rate in Mooney as determined from a census. The birth rate has been declining during the past 60 years.**

1.3 times greater than it is now. The result we obtain is shown in Figure 7.8, where we plot the impact rate so determined over the entire history of the Moon.

....

NOW YOU DO IT

(A) **Suppose that in the year 2015 a census of the town of "Looney" yields the following results.**

Age	Number of people
1	1 000
10	1 000
20	900
30	800
40	700

Compile a table, similar to Table 7.2, giving the history of the birth rate in Looney over the past 40 years.

(B) **Suppose that in the town of "Steady" there have been one thousand births each year over the past 50 years. Compile a table, similar to Table 7.1, showing the results of a census of Steady.**

....

History of the lunar surface

Figure 7.8 tells us that very little has changed on the Moon during the past three billion years. In contrast, the surface of the Earth is very young indeed. Even an ancient mountain range such as the Allegenys is only one-tenth as old as lunar craters, and the Himalayas are very much younger than that. Compared with the Earth, the Moon is an ancient, quiescent place.

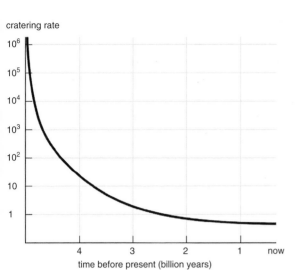

cratering rate

time before present (billion years)

Figure 7.8 **The cratering rate on the Moon was very great early in the Solar System's history. It has decreased dramatically since then.**

Billions of years ago, the Solar System was a hellish place, in which moons and planets were incessantly pummeled by cataclysmic impacts. This phase of the Moon's history appears to have begun about 4.5 billion years ago, when the Solar System formed, and it terminated about 3 billion years ago, when the rate of impacts tapered off.

The lunar maria were also formed during this period. They are the remnants of particularly gigantic impacts, in which bodies of enormous size slammed into the Moon. These immense collisions dug out vast areas of the lunar surface to form *impact basins*. Beneath these impact basins, the shattered crust of the Moon was threaded by huge fissures. Over subsequent ages the deep interior of the Moon slowly heated. This warming was caused by the gradual decay of radioactive elements such as uranium. The heat released by these decays slowly accumulated until it was sufficient to melt inner portions of the Moon. Molten lava flooded upwards into the impact basins through the fissures lying below them. Craters and mountains within the basins were obliterated. Thus were formed the maria.

We are now in a position to understand the striking regularity that we noticed earlier concerning the flatness of the lunar maria. The maria are flat because they are lava flows. It is not as if no meteoroids had ever landed there. They did land – but the craters they formed now lie buried beneath the lava.

This stage in the history of the Moon appears to have extended from 4.2 to 3.1 billion years ago. Since that time little has happened on the Moon save for the occasional impact by the few remaining meteoroids.

Tides and the Moon

As we all know, tides in our oceans are caused by the Moon. But how do we know this? Let us follow along the path of discovery that leads us to this insight.

Detectives on the case

What causes tides?

Let us try to understand what might cause tides. Our first attempt will turn out to conflict with observation. Looking carefully at the nature of the discrepancy will give us a clue of how to modify our theory until it works.

Our theory will be that tides arise from the gravitational influence of the Moon on the Earth's oceans. Recall from Newton's law that the force of gravitation is universal: everything attracts everything else. Thus, even though you are not aware of it, the Moon exerts a gravitational force on you. It also exerts a gravitational force on the oceans.

Perhaps variations in this force from place to place give rise to the tides. After all, different parts of the ocean are at different distances from the Moon. Since the force of gravitation depends on distance (recall that gravity obeys an inverse square law), it

⇐ **Looking backward**
We studied gravitation in Chapter 3.

Figure 7.9 **A wrong theory of the tides. The hypothesis is that the portion of the ocean closest to the Moon is attracted most strongly toward it, and therefore bulges upwards toward it. This theory predicts one tide per day, as opposed to the two per day that we actually observe.**

is those portions of the ocean that are closest to the Moon that must be attracted with the greatest force.

Our first try at a theory will be as follows: the portion of the ocean closest to the Moon is attracted most strongly toward it, and therefore bulges upwards toward it.

This theory is illustrated in Figure 7.9. Is there anything wrong with it? Let us ask: how many tides does it predict per day? Notice that, according to this theory, the ocean bulge is always located on that part of the Earth closest to the Moon. But the Earth is rotating! Figure 7.9 shows a man standing on the beach. As the Earth rotates, he is carried into and then out the other end of the bulge. He will experience high tide when he is in the bulge. And according to Figure 7.9 this happens *once* a day.

But this prediction conflicts with observation. In reality there is not one high tide per day. There are two.

Our theory has failed. Let us try to fix it. As usual, the Universe itself gives us clues as to how to do so. Observation reveals that there are two tides per day, and we can use this as a guide. It is telling us that there are two bulges in the ocean. Furthermore, since successive high tides come at 12-hour intervals, the bulges must be on opposite sides of the Earth.

Figure 7.10 illustrates what must be the correct configuration. But how mysterious it is! In that figure the Moon is tugging the oceans to the right. Why should one of the bulges point to the left? Furthermore, notice that the right-hand bulge in Figure 7.10 occurs where the force from the Moon is greatest – but the left-hand bulge occurs where the force is *weakest*. Why should the point of least force experience a bulge?

So far we have been thinking about the points of greatest and least forces. Let us broaden our thinking. Where is the force intermediate in strength? Midway between the two bulges. And this intermediate point is not located in an ocean at all! It is located within the body of the Earth itself (Figure 7.11).

We now realize what had been wrong with our initial theory. We had been thinking only about the oceans, but we should have been thinking about the Earth as a whole.

Figure 7.10 **Observation tells us that there are two bulges in the ocean.**

- That portion of the ocean closest to the Moon is pulled most strongly toward it.
- That portion of the ocean farthest from the Moon is pulled least strongly toward it.
- The main body of the Earth is pulled with an intermediate force toward the Moon.

With this recognition, we have finally created the correct theory of tides. It is these *differences in gravitational forces that produce the tides*.

- The *right*-hand bulge in the ocean is produced because the ocean is pulled more strongly than the Earth, so that *the ocean is pulled away from the Earth*.
- The *left*-hand bulge in the ocean is produced because the Earth is pulled more strongly than the ocean, so that *the Earth is pulled away from the ocean*.

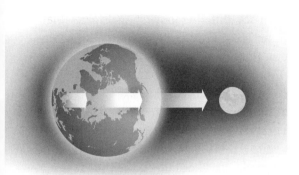

It is important to emphasize that we have given here only a simplified theory of the origin of tides. In fact both the Sun and the Moon influence them. So too do details of the coastline topography and the depth of the seabed. So complicated is the actual situation that mariners must rely on tide tables, which give the times of high and low tides for every day of the year.

Figure 7.11 **Tides are caused by differing gravitational forces from the Moon. The side of the Earth closest to the Moon experiences a greater force than the main body of the Earth, which in turn experiences a greater force than that on the opposite side.**

NOW YOU DO IT
We know that the gravitational attraction between two bodies grows weaker as their distance increases. But suppose it did not. Suppose the force of gravitation were the same no matter how far apart the two bodies were. In this case, would there be tides? Explain your answer.

Tides and the rotation of the Moon

The Moon keeps the same face toward the Earth. An astronaut standing on its near side would see the Earth forever hanging at the same point in the sky. Conversely, an astronaut standing on its far side would *never* see the Earth. As seen from the Moon, the Sun rises and sets, the stars rise and set – but the Earth does not. We say that the Moon undergoes *synchronous rotation*, in which its period of rotation is the same as its period of revolution about the Earth (Figure 7.12). This is in sharp contrast to the Earth, whose period of rotation – one day – is very different from its period of revolution – one year.

Is the Moon's strange rotation simply a coincidence? We might be tempted to think that it is, were it not for a striking fact: most other satellites in the Solar System do the same thing. They also undergo synchronous rotation. They all have different orbital periods, but they all rotate in synchronism with their orbits, so as to keep the same face toward the planet. This striking regularity shows that there must be some explanation for this peculiar behavior.

Figure 7.12 **Synchronous rotation.**

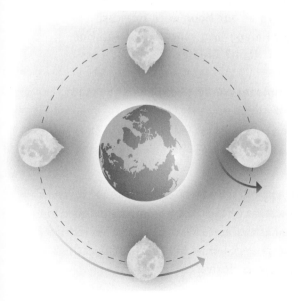

A clue is provided by the only moon in the Solar System that does *not* rotate synchronously: a satellite of Saturn known as Phoebe. And Phoebe is Saturn's *outermost* satellite.

Recall our discussion of "Detectives on the case": that observations give us clues as we create our theories. In this case the clue is telling us that whatever the process is that enforces synchronous rotation, it must grow weaker with a satellite's distance from its planet. Gravitation, of course, grows weaker with distance. So let us guess that the responsible agency is *the very same difference in gravitational forces that produces tides on the Earth*.

This leads us to guess that we should turn our attention away from tides on the Earth, and think of tides on the Moon. At first glance this seems foolish, for tides occur in oceans, and the Moon has no water. But as we have just seen, to understand the tides we must consider the entire body of the Earth. And over geological ages, even solid rock flows smoothly.

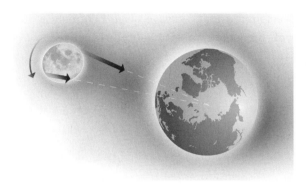

Figure 7.13 **Tides on the Moon enforce its synchronous rotation. The rotation of the Moon twists its tidal bulge away from the Earth–Moon line. Differing gravitational forces from the Earth on the two bulges act to slow the Moon's rotation.**

Consider, therefore, the differing gravitational forces *from* the Earth *on* the Moon – and in particular, on the entire body of the Moon. The near side of the Moon is tugged by a big force, the far side by a little force. In response to these differences, the Moon develops two bulges – just like the oceans of the Earth. Far back in the past the Moon presumably rotated more rapidly, so that it did not rotate synchronously. It is easy to see that, in this situation, the Earth's gravitational force on the Moon's bulges would have steadily slowed its rotation until it became synchronous.

Figure 7.13 illustrates the situation far back in the past, when the Moon was rotating more rapidly. As on the Earth today, the Moon's rotation carried each part of it into and out of the bulges. But time was required for rock to flow, and internal friction opposed this flow. The result was that *the bulges lagged behind the force*. As illustrated in Figure 7.13, the Moon's bulges were somewhat twisted away from the line joining the Moon to the Earth.

In this configuration the differing forces from the Earth on the two bulges acted to slow the Moon's rotation. In Figure 7.13 the force on the bulge closer to the Earth twists the Moon clockwise. The force on the far bulge twists the Moon in the opposite direction, but it is weaker. So the net result is a force acting against the Moon's rotation. This force, acting over astronomical ages, slowly dragged the Moon into synchronous rotation.

If the Earth has slowed the Moon's rotation, why hasn't the Moon slowed the Earth's? The answer is that it has, but because the Earth is so much more massive than the Moon the process is slower. As the ages pass, our rotation rate is growing slower and our days are growing longer. Ages hence the Earth too will rotate synchronously. This has already happened with Pluto and its moon Charon: each keeps the same face toward the other.

NOW YOU DO IT
Saturn is a lot more massive than the Earth, and its moon Helene lies roughly the same distance from Saturn as our Moon does from us.
Would you expect the force twisting Helene into synchronous rotation to be the same as that acting on our Moon? Greater than this? Less than this?
   Explain why the moons close to a planet would be expected to rotate synchronously, but those far from it might not.

Mercury

The closest planet to the Sun is Mercury. Before the advent of the space age we knew very little about it. Furthermore, some of what we did "know" ultimately turned out to be wrong. This is a remarkable state of affairs, since Mercury's orbit lies relatively close to ours: it is a mere 0.6 AU away, as opposed to the 80 AU diameter of the Solar System as a whole. Why was Mercury so elusive?

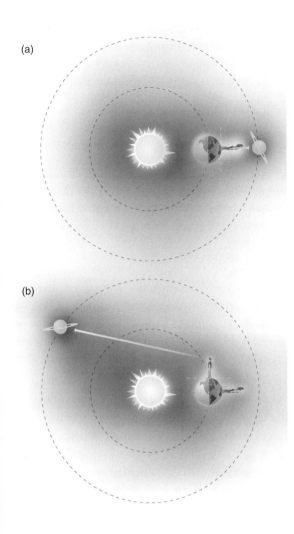

(a)

(b)

Figure 7.14 **An outer planet can be observed in the middle of the night at certain points in its orbit. (a) When the planet is at the indicated position, a person experiencing midnight sees it directly overhead. That is a good observing configuration. But when the planet has moved to the far side of the Sun (b), it can be seen only by an observer close to the sunrise line, who sees it close to the horizon. In such a configuration, observations are hampered by haze and the glare of the Sun.**

Observing inner planets

Mercury was elusive because of its proximity to the Sun. This proximity renders it hard to see from here. Not until we were able to send space probes to observe it from other vantage points were we able to get a good look at Mercury.

A planet can be seen only at nighttime and, depending on where it lies in its orbit, this can impose severe restrictions on our ability to observe it. Let us begin by thinking about the problem of observing not Mercury, but a planet whose orbit lies *farther* from the Sun than our own – Saturn perhaps. In Figure 7.14(a) we illustrate a configuration in which it is easy to observe such a planet. For the person shown – let's call her Alice – Saturn is situated on the part of the Earth directly away from the Sun: she is experiencing midnight. Furthermore, Saturn is directly overhead Alice. This is a good state of affairs for astronomers: the object they wish to study lies high in the sky during most of the night.

In contrast, Figure 7.14(b) illustrates the opposite situation, in which it would be hard to observe Saturn. This is because it has moved to a different point in its orbit, one lying behind the Sun as seen from the Earth. Now Alice can no longer see it: it lies below her horizon. Only for Bob does Saturn lie above the horizon. But notice that Bob is situated close to the line dividing day from night. And this means that he would have a hard time observing Saturn – for two reasons.

- Bob is experiencing twilight. The sky is not entirely dark, but is faintly illuminated by the Sun.
- Saturn lies close to Bob's horizon, where it might be obscured by clouds, haze and the like.

We have learned an important lesson: when observing a planet lying farther from the Sun than we, astronomers need to choose their observing times carefully. Some times are good and some are bad.

But in observing a planet lying closer to the Sun than we, all times are bad. As illustrated in Figure 7.15, no matter where such a planet is located in its orbit, it is close to the Sun as seen from the Earth. And so close does Mercury lie to the Sun that, throughout all history, Earth-bound astronomers found it almost impossible to gather information about it.

Radar astronomy and the rotation of Mercury

This difficulty once led to an important error.

During the 1880s the Italian astronomer Giovanni Schiaparelli (1835–1910), peering with his small telescope through obscuring haze in the few short minutes of twilight, thought he could see faint markings on Mercury. By timing their passage across its visible disk he attempted to measure Mercury's rotation rate. He concluded that it spun about on its axis once every 88 days.

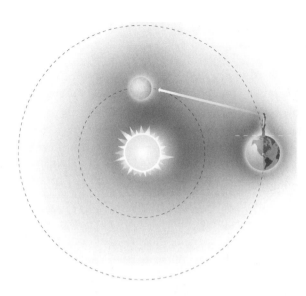

Figure 7.15 **An inner planet always lies close to the Sun as seen from Earth, no matter where the planet is in its orbit.**

This was a remarkable discovery, since Mercury also took 88 days to orbit the Sun. Schiaparelli had discovered that Mercury, like the Moon, was undergoing synchronous rotation. It made sense, since tidal forces from the Sun could well have enforced this kind of motion.

....

NOW YOU DO IT
Why did Schiaparelli feel that tidal forces could have forced Mercury into synchronous rotation, when these forces clearly have not been able to force the Earth into synchronous rotation?

....

In 1965, however, a new kind of technique was brought to bear on Mercury, and it showed Schiaparelli's "discovery" to have been false: Mercury does not rotate synchronously. This new technique was *radar astronomy*.

Radar works by beaming a burst of radio waves toward a target. They are reflected by the target, and their return "echo" is detected. Astronomers used the giant radio telescope in Arecibo, Puerto Rico, as their "radar set." They arranged for it to emit a brief and very powerful burst of radio signals toward Mercury. Then they waited. So vast are interplanetary distances that fully ten minutes were required for the waves, moving at the speed of light, to reach Mercury, bounce back, and return. When they did, they were detected by the telescope.

The emitted radio waves had been of a single wavelength. Remarkably, however, the astronomers found that the return echo was not. Rather it was composed of a variety of wavelengths.

Why? Figure 7.16 shows the radar waves as they impinged upon Mercury. Because the planet was rotating, part of it was moving toward the waves, and part away from them. Under these circumstances the Doppler effect came into play. Just as the Doppler effect alters the waves *emitted* by a moving body, so too does it alter those *reflected* by a moving body. The waves reflected from the part of Mercury moving toward them were blueshifted. Those reflected from the part of Mercury moving away from the Earth were redshifted. By measuring these shifts in wavelength, the astronomers deduced the speed of rotation of Mercury. They found that it rotated not once every 88 days, but once every 58.6 days.

Figure 7.16 **Reflected waves from a rotating body suffer red- and blueshifts.**

The radar observations proved Schiaparelli to have been mistaken: despite its proximity to the Sun, Mercury does not undergo synchronous rotation. There is, however, a remarkable connection between Mercury's rotation and revolution. Its observed period of rotation is exactly ⅔rd its period of revolution. Mercury rotates three times for every two orbits about the Sun.

This strange state of affairs appears to have been produced by tidal forces from the Sun acting on Mercury's elliptical orbit. Were its orbit circular, they would have locked it into synchronous rotation. But Mercury's

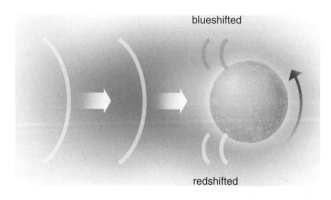

blueshifted

redshifted

relatively elliptical orbit make the tidal forces alternately stronger and weaker, thus leading to the observed ⅔rd rotational pattern.

Spacecraft observations of Mercury

In 1974 astronomers got their first good look at Mercury as the Mariner 10 spacecraft swung by it. The space probe went into an orbit about the Sun, which periodically brought it back to the vicinity of the planet. Unfortunately, the supply of maneuvering gas was depleted after only three encounters, then radio transmissions were turned off.

⇐ **Looking backward**
We studied orbits, particularly gravity-assist orbits, in Chapter 3.

After that, no space mission visited Mercury for more than 30 years. Why the long wait? The answer is that an orbital path taking a spacecraft from the Earth to Mercury and then into an orbit around the planet appeared to be prohibitively difficult to achieve.

Since Mercury lies closer to the Sun than Earth, during a voyage to Mercury a spacecraft would be accelerated by the gravity of the Sun. In terms of the Sun's gravitational pull acting upon it as it traveled, the spacecraft would be traveling "downwards." Recall our analogy in Chapter 3 that orbiting is like falling – and recall that, as a body falls downwards, it picks up speed. This means that, once it arrived at Mercury, the spacecraft would be traveling exceedingly rapidly. In order to put itself into orbit about the planet, such a spacecraft would have to slow down when it arrived. And to slow down it would have to carry fuel – lots of fuel, too much to be practicable.

Ultimately it was realized that the same sort of "gravity-assist" maneuver (Chapter 3) used to *accelerate* a spacecraft on a voyage to the outer Solar System could be used to *decelerate* one on a trip to Mercury. Such an orbit requires a rare alignment of planets. A mission dubbed MESSENGER (MErcury Surface, Space ENvironment, GEochemistry and Ranging) was launched in August of 2004. One year later it performed a gravity-assist maneuver as it passed close by the Earth, and then two more in 2006 and 2007 as it passed by Venus. MESSENGER reached Mercury in January of 2008 and flew by it – and then again later that year and a third time in 2009. Each encounter altered its path. Finally, in March of 2011, the spacecraft had slowed sufficiently to enter an orbit about Mercury.

This orbit is highly elliptical, taking it within 200 kilometers (120 miles) of Mercury's surface and then 15 000 km (9300 miles) away from it every 12 hours. This was chosen to shield the probe from the heat radiated by Mercury's hot surface. Only a small portion of each orbit is at low altitude where the spacecraft is subjected to heating from the hot side of the planet.

Figure 7.17 shows a photomosaic of the surface of Mercury. Mercury looks a lot like the Moon. Spacecraft observations have revealed the same craters, impact basins and absence of atmosphere as we find on our own satellite. Subtle differences can be accounted for on the basis of Mercury's greater mass. For example, the rays radiating away from craters on Mercury are shorter than those on the Moon because ejecta from an impact travel a shorter distance on Mercury, owing to its stronger gravity.

Perhaps the most striking feature on Mercury is the so-called Caloris impact basin. As its name attests, it is one of the hottest places on the planet ("caloris" is Latin for "heat"). This is because every other orbit the basin directly faces the Sun

Figure 7.17 **Mercury as revealed by the MESSENGER space probe.**

when Mercury lies closest to it in its elliptical orbit. The impact that created it must have occurred early in the planet's history, because the basin's vast lava plains are studded with numerous craters.

Owing to its proximity to the Sun, and the great length of its day, the surface of Mercury is ferociously hot. Temperatures there can exceed 700 kelvin – 800 degrees Fahrenheit (about 430 degrees Celsius). That is hot enough to melt certain metals. Conversely, during the long Mercurian nights, the temperature can plummet to 100 kelvin – nearly 280 degrees Fahrenheit below zero (173 degrees Celsius below zero).

The escape of planetary atmospheres

Mariner 10 found that Mercury, like the Moon, has no air. Why not? Is it only an accident that these two worlds, so alike in many ways, are also alike in this?

It is not an accident. Both the Moon and Mercury have relatively low masses, and Mercury is not just low-mass but hot. It turns out that *bodies that are low-mass and/ or hot cannot possess atmospheres.* Let us see why.

Air is a gas, and a gas is composed of atoms and molecules in motion. Each particle flies about randomly, perpetually colliding with others. If you could see these particles they might look like insects in a swarm, incessantly darting this way and that. The motion is quite rapid. Within your lungs, for instance, oxygen molecules are zipping about at more than a thousand miles per hour.

What is the consequence of this motion? Gravitation attracts the particles in an atmosphere, just as it attracts everything else. But if a particle moves fast enough, it can overcome gravitation. If its velocity exceeds the so-called *escape velocity,* it will leave the planet. On Earth, escape velocity is 7 miles per second. Anything moving slower than this will remain on the Earth. But anything moving faster will fly off into space.

In reality, essentially no particles in our atmosphere are moving this fast. Thus the Earth is able to keep its air. But consider now the situation on a low-mass body such as the Moon. Because its mass is lower than the Earth's, gravity there is weaker. This means that escape velocity from the Moon is less than that from Earth. It is so low that the Moon has lost all its air.

Mercury's mass is greater than the Moon's, so escape velocity there is greater (though less than on Earth). But let us remind ourselves of an important property of all gases.

> **Ideal gases**
>
> • The hotter the gas, the more rapidly do particles within it move.
> • The lighter the particle in a gas, the more rapidly it moves.

So any particles in a Mercurian atmosphere would be moving more rapidly than those on the Moon. It turns out that they would be moving at greater than escape velocity, so Mercury, too, cannot have air.

These ideas can also be applied to other bodies in the Solar System. The asteroids for instance, with their exceedingly low masses, possess no air at all. Mars, whose mass is intermediate between the Earth's and the Moon's, has a thin atmosphere. And the giant planets, far more massive than the Earth, have immense atmospheres.

NOW YOU DO IT
Venus lies between Mercury and the Earth, and its mass is nearly the same as the Earth's. Would you expect it to have more or less atmosphere than we do? (As we will shortly see, both of these expectations are wrong, owing to a "runaway greenhouse effect" on Venus.)

Venus

Let us move outward away from the Sun in our tour of the Solar System. The next planet we encounter is Venus. Like Mercury, before the space age we knew very little about it. But unlike Mercury, this was not because of its proximity to the Sun. Venus lies sufficiently far from the Sun that the view we get of it from Earth is fairly good. But Venus is perpetually covered by clouds, clouds so thick and impenetrable as to entirely mask its surface. Not once have they ever parted to give us a view.

Venus has been called Earth's twin. It is nearly the same size (its diameter is 95% that of the Earth) and mass (its mass is 81% that of the Earth). It lies only slightly closer to the Sun than we do (the semimajor axis of its orbit is 72% that of the Earth). This led astronomers to speculate that conditions there would be similar to conditions here. Assuming the clouds to be made of water, people imagined a humid, hot world not so very different from terrestrial jungles.

But in the early 1960s two US spacecraft flew by Venus, and in 1970 a Soviet space probe landed on its surface and broadcast data for nearly half an hour. In 1978 two Soviet spacecraft landed on the planet and broadcast pictures of its surface (Figure 7.18). Since then, numerous space missions have dropped balloons into Venus' atmosphere and mapped its surface from orbit. The reality they found has turned out to be very different from what people had expected.

The rotation of Venus

Figure 7.18 **The surface of Venus as revealed by the Soviet Venera space probe.**

Even before the space age, radar observations of Venus revealed that it rotates backward. On Venus, the Sun rises in the west and sets in the east. Furthermore, it does so very slowly: Venus' day lasts 243 Earth days. Since its orbital period is 225 Earth days, this implies that Venus does not rotate synchronously. We are not sure how this state of affairs arose. There may be some sort of gravitational influence from the Earth, for Venus presents nearly the same face to us every time it comes closest to us. However, the match is not perfect, and another theory is that Venus' abnormally slow rotation was imparted by a giant impact during its formation.

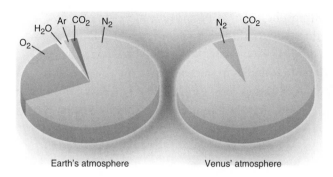

Figure 7.19 **Composition of Venus' atmosphere as compared with that of Earth.**

The atmosphere of Venus

Astronomers were astonished to find that the atmosphere of Venus is hideously hot and enormously massive. The temperature on the planet's surface is high enough to melt lead – 730 kelvin, or 855 degrees Fahrenheit. That makes Venus the hottest place in the Solar System – hotter even than Mercury. Air pressure on the surface is 90 times greater than on Earth: that is the pressure on a submarine 3000 feet below the surface of an Earthly ocean. So efficiently does this massive, thick atmosphere transport heat that night is just as hot as day there, and the poles just as hot as the equator. There is no place an astronaut might go to cool off, and the planet has no seasons.

So thick are Venus' clouds that they filter out 98% of the sunlight reaching the planet, so that sunlight on the surface is oppressively dim. Owing to the clouds' colors, the sky on Venus is orange.

Venus' clouds are composed of droplets, not of water like terrestrial clouds, but of sulfuric acid and particles of crystalline sulfur. When the acid droplets grow sufficiently large they fall as acid rain, but so hot is it there that they evaporate before reaching the ground.

The upper clouds, visible from Earth and space, are observed to rotate in the same direction as the planet, but far more rapidly. The entire atmosphere is rotating relative to the planet at speeds of 360 kilometers per hour.

Venus' atmosphere is far thicker than Earth's. Our clouds extend upwards to an altitude of roughly 6 kilometers. Venus' extend to about 70 kilometers.

The composition of Venus' atmosphere is also utterly unlike ours (Figure 7.19). It is almost entirely composed of carbon dioxide (CO_2). In contrast, our air is 78% nitrogen (N_2) and 21% oxygen (O_2), with only trace quantities (0.03%) of CO_2.

The greenhouse effect and the atmosphere of Venus

How can we understand Venus' extraordinary atmosphere? It defies all our expectations. Farther from the Sun than Mercury, Venus ought to be cooler than it. Less massive and hotter than the Earth, it ought to have less air. It turns out that the atmosphere of Venus results from *the greenhouse effect* run amok.

Before exploring how this effect has wrought such havoc on Venus, let us explore how it works in greenhouses. The remarkable thing about a greenhouse is that it can stay warm even without a heating system. It does so by using solar energy combined with a striking property of glass: glass is transparent to visible light, but not to infrared.

Figure 7.20 shows how the greenhouse effect works. Sunlight enters through the glass roof of the greenhouse, and warms what's inside. As we know from our study of blackbody radiation, every warm body emits electromagnetic radiation. So the greenhouse's interior emits. This radiation carries energy and would cool the interior *if it could leave*. But the Wien law tells us that, at the greenhouse's temperature, this radiation falls in the infrared region of the spectrum. Because

⇐ **Looking backward**
We studied blackbody radiation and the Wien law in Chapter 4.

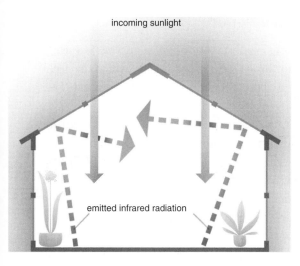

incoming sunlight

emitted infrared radiation

Figure 7.20 **The greenhouse effect. The glass roof is transparent to incoming sunlight, but it traps the infrared radiation by which the greenhouse would have cooled.**

Figure 7.21 **Cloudy nights are warm. Via the greenhouse effect, water droplets in clouds trap infrared radiation emitted by the ground.**

glass does not transmit infrared, the radiation is trapped. Thus the greenhouse stays warm.

It is not only glass that is opaque to infrared. In the Earth's atmosphere, water droplets in clouds block this radiation as well. In this way the greenhouse effect influences our weather. You may have noticed that clear nights are usually colder than cloudy nights. This is because the water droplets of which our clouds are composed trap the infrared radiation emitted by the ground (Figure 7.21).

Carbon dioxide is also a "greenhouse gas," which traps infrared radiation. And because there is so much CO_2 in the atmosphere of Venus, it has drastically raised the planet's temperature.

But why is there so much CO_2 on Venus? Since the Earth is Venus' "twin," why don't we have as much carbon dioxide as Venus? Remarkably, we do! It's just not in our atmosphere. It is in rocks, particularly limestone, in the Earth's crust. If all the CO_2 in our crust were liberated as a gas, we would have an atmosphere as thick as Venus' – and, via the greenhouse effect, nearly as high a temperature.

Why is carbon dioxide present as a gas on Venus, but as a solid on Earth? Because Venus has no water. For two reasons, the presence of water on Earth is responsible for removing CO_2 from our atmosphere.

- CO_2 is soluble in water. It is dissolved in our oceans.
- There is a chemical reaction that combines gaseous CO_2 with certain minerals to form limestone. This reaction proceeds more efficiently in the presence of water.

Both water and CO_2 are emitted by Earthly volcanoes. Presumably the same was true of Venus early in its history, when it had no massive atmosphere to give it such a big greenhouse effect. But even without this effect, because Venus was relatively close to the Sun it was quite warm. This would have steamed away a little water from its oceans into its atmosphere. But as we have seen, water vapor is a greenhouse gas! So this produced a mild greenhouse effect, which heated the planet yet further. This evaporated a little more liquid water into the atmosphere, which produced yet more heating, and so on in a vicious circle. Ultimately there developed a *runaway greenhouse effect*, which produced the superheated, massive atmosphere we find today.

Where did all Venus' water go? Venus has no ozone layer to protect it from the Sun's ultraviolet radiation. These ultraviolet photons would have then disassociated H_2O into hydrogen and oxygen. The hydrogen, being a lighter particle, would then have escaped into space, via the same mechanism we discussed in our treatment of the escape of planetary atmospheres.

Remarkably, the very same runaway would have happened on Earth, had we been even slightly closer to the Sun. It is a sobering thought: had the Earth been formed in a slightly different orbit, life could never have originated here. Our existence seems to be due to a lucky chance.

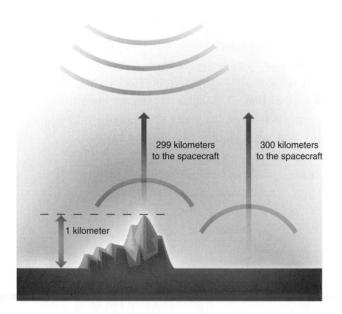

299 kilometers to the spacecraft

300 kilometers to the spacecraft

1 kilometer

Figure 7.22 **Radar echo from a mountain travels a shorter distance than from the surrounding terrain.**

The surface of Venus

The most extensive observations of Venus' surface were conducted by the Magellan spacecraft, which in 1990 went into orbit about the planet and surveyed its surface with radar.

This radar, which could penetrate the obscuring clouds, was used to map the surface topography. It did this by timing the return radar echo. Regularly, the spacecraft beamed a brief burst of radio signals downwards onto the planet. It then measured the time required for the return echo to arrive. A time interval shorter than expected meant that the surface of Venus was somewhat closer to the spacecraft than normal, indicating a mountain. A longer than expected time interval meant the spacecraft was orbiting over a canyon.

Let us pause a moment to consider the extraordinary accuracy that must be achieved in order to make such a measurement. The Magellan spacecraft's orbit carried it to within 300 kilometers of the surface of Venus. Imagine that at this point it sent out a radar pulse. That pulse had to travel the 300 kilometers down to the surface, and the return echo had to travel another 300 kilometers back to the spacecraft, for a total of 600 kilometers = 6×10^5 meters. The radar pulse, being electromagnetic radiation, traveled at the speed of light, which is $c = 3 \times 10^8$ meters per second. The total time elapsed between sending and receiving the pulse was then

$$\text{elapsed time} = \text{distance/time} = (6 \times 10^5 \text{ meters})/(3 \times 10^8 \text{ meters/second})$$
$$= 2 \times 10^{-3} \text{ seconds} = 0.002 \text{ seconds}.$$

Now imagine that the spacecraft had been passing over a mountain when it sent out its pulse (Figure 7.22). In this case it would have received *two* echoes: first an echo from the mountain top, and next a second echo from the surrounding, lower-level terrain. We already know when the second echo arrived: let us now calculate when the earlier echo from the mountain top did. If the mountain is just one kilometer high (3281 feet), the distance from the spacecraft to its top was not 300 kilometers but 299 kilometers, and the total distance the radar pulse had to travel down to its top and then back up was not twice 300 kilometers but twice 299 kilometers, or 598 kilometers = 5.98×10^5 meters. So the time between sending and receiving the first echo is

$$\text{elapsed time} = \text{distance/time} = (5.98 \times 10^5 \text{ meters})/(3 \times 10^8 \text{ meters/second})$$
$$= 1.9933 \times 10^{-3} \text{ seconds} = 0.001\,9933 \text{ seconds}.$$

Now let us find the time *delay* between the arrival of the first echo (from the mountain top) and the second (from the surrounding terrain). This delay is the difference between the two time intervals we have just calculated:

$$0.002\,0000$$
$$-0.001\,9933$$
$$\overline{0.000\,0067 \text{ seconds.}}$$

In order to even "know" the mountain was there, the "timers" on board the spacecraft had to be capable of distinguishing the two return echoes separated by a mere 0.000 0067 seconds.

Figure 7.23 **Map of Venus produced by the Magellan spacecraft.**

NOW YOU DO IT
Suppose the spacecraft had received a *triple* echo: one sooner than expected, and one later. What topography would this have revealed?
 Suppose the spacecraft had been 500 kilometers above the surface of Venus, and it had passed over a canyon one kilometer deep. (A) How much time would have elapsed between sending out the pulse and receiving an echo from the surrounding terrain? (B) Would it have received an echo from the canyon before or after this moment? (C) What is the time delay between the two echoes?

As a result of its radar observations, Magellan compiled maps of the surface of Venus. They show that Venus is an exceedingly flat planet (Figure 7.23). Its topography is primarily rolling plains, with a relatively small fraction of the planet rising high above, or dropping far down in canyons.

The Earth

Watery planet

Continuing on our journey away from the Sun, we reach the third planet, our home. There is something wrong with the name we chose for it. The word "earth" means "ground" or "dirt." The name makes sense, for the ground is where we live. But in a view of the Earth from space (Figure 7.24), the solid ground is nearly overwhelmed by the blue of oceans and the white of clouds. And it is important to realize that both oceans and clouds are made of the same thing: H_2O – liquid in the oceans, vapor in the clouds. We should have named our planet "water."

In our tour of the Solar System, the Earth is the first body we have encountered that possesses water. But it is not unique in the Solar System as a whole. Indeed, as we continue on our journey outward away from the Sun, we will find water playing an increasingly important role. The inner regions of the Solar System will turn out to be the exception rather than the rule. The only reason they are dry is that they are so hot: any water they once contained has boiled into gaseous H_2O, and the H_2O molecules have escaped.

The Earth *is* unique in possessing *liquid* water. Water can exist as a liquid only within a relatively narrow range of temperatures and pressures. Hotter than this at a given pressure and it is a gas: colder and it is a solid. Since a planet's temperature is determined by its distance from the Sun, this means that it must lie within a certain range of distances from the Sun to have liquid water. As we have seen from our study of Venus' greenhouse effect, this range is very narrow indeed. The Earth is just the right distance from the Sun to contain liquid water.

And liquid water is essential to life as we know it. Our own bodies are roughly 60% water. It is similarly ubiquitous within the bodies of every other living organism,

Figure 7.24 **The Earth photographed from the Apollo 17 mission as it was on its way to the Moon. Note how much H_2O is visible in this image.** ◉ **(Also see color plate section.)**

ranging from trees to birds to bacteria. Furthermore, liquid water has a number of properties that make it particularly useful in the metabolism of living things. It is for this reason that astronomers are excited by the discovery of water's prevalence in the outer regions of the Solar System. As we will see, there is some chance that the satellite Europa contains liquid water and that Mars did in the past. This makes these worlds particularly exciting to scientists interested in searching for life elsewhere in the Solar System. We will return to this issue later.

(a)

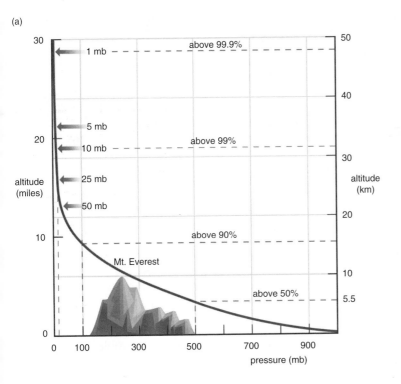

(b)

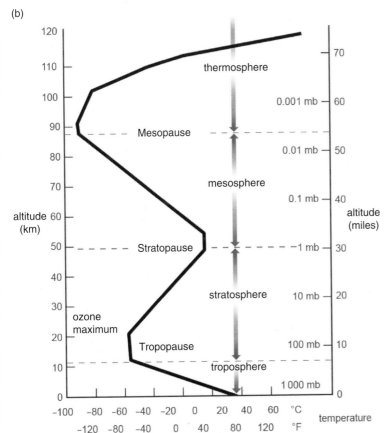

Figure 7.25 **Pressure and temperature of our atmosphere has a complicated dependence on altitude.**

The atmosphere

Atmospheric pressure is customarily reckoned in "millibars," with pressure at ground level being close to 1000 millibars. Figure 7.25(a) indicates how air pressure drops as we gain altitude. At an altitude of only 5.5 kilometers (3.5 miles) the pressure has dropped to half that at sea level. Similarly, 99.9% of all air lies below 50 kilometers.

NOW YOU DO IT
What is air pressure on the summit of Mount Everest? (That's why Himalayan mountaineers often carry their own oxygen!)
What is air pressure 10 miles up?

While 50 kilometers seems like a pretty high altitude, it is very small compared to the 6378-kilometer radius of the Earth. On the scale of the Earth as a whole, air is confined to an exceedingly thin layer. Notice also that there is no "top" to the atmosphere, no altitude at which it suddenly stops. Rather, the atmosphere simply grows more and more rarified the higher we go. Even in what we customarily regard as outer space faint traces remain, and they exert a slight drag upon orbiting satellites.

NOW YOU DO IT
To get a feeling for how thin our atmosphere is, let us a construct a scale model in which the Earth is represented by a sphere one meter across. On this scale, how far above the sphere's surface would be the point at which atmospheric pressure has dropped to half that at sea level? To get an even better intuitive feeling for this, convert the sphere's diameter into feet and your answer into inches.

We are all familiar with the fact that the atmosphere grows colder as we gain altitude: that's why mountain tops are covered with snow even in summer. Remarkably, however, this behavior does not continue as we go to yet greater heights. As shown in Figure 7.25(b), at an altitude of about 10 kilometers (6 miles), the cooling trend ceases, and the temperature stays almost constant. Above this the temperature actually starts to increase with increasing altitude. Yet higher the temperature goes through a second cycle of first cooling and then warming with height.

. . . .

NOW YOU DO IT
There are two points on this curve above which the atmosphere actually starts to grow warmer with increasing altitude. Where are these points?
    You have already figured out the air pressure 10 miles up. What is the temperature there?
    At what altitude is it coldest? What is the temperature (in °F) there?

. . . .

How can we understand this strange behavior? Our atmosphere is heated by sunlight. But sunlight is composed of many different wavelengths, and it turns out that each layer of our atmosphere is warmed by light of a different wavelength. The outermost layer is heated by the absorption of X-rays from the Sun. These X-rays strip electrons from atoms, producing the ionosphere which, as we see from Figure 7.25(b), is hotter than ground level. Lower down, at an altitude of about 50 kilometers, is the ozone layer, which absorbs solar ultraviolet radiation and is heated by it. Finally, the base of the atmosphere, where we live, is warmed by visible light.

As we have mentioned, our air is 78% nitrogen (N_2) and 21% oxygen (O_2), with only trace quantities of other gases.

Human alteration of the atmosphere

In recent years we have realized that humanity is exerting a profound influence on the Earth's atmosphere. We are depleting the ozone layer, and we are emitting greenhouse gases, which lead to global warming. Both of these arise from the effects of modern industrial civilization, and both have happened in a mere blink of the eye in geological terms.

Ozone is a molecule consisting of three oxygen atoms, in contrast to the normal oxygen molecule, which has two. It is the primary component of photochemical smog, the pollution encountered in major cities. Ozone has an unpleasant odor and irritates eyes and noses. Nevertheless, it is supremely vital to our health. This is because ozone exists in the Earth's upper atmosphere, where it is instrumental in absorbing ultraviolet radiation from the Sun. This ultraviolet can be very dangerous to living things. It is known to cause skin cancer, and to break apart DNA.

Stratospheric ozone is concentrated in a layer, about 10 kilometers thick, 25 kilometers up. Even within this layer, the concentration of ozone is very low: there are only about 12 ozone molecules per million air molecules. Nevertheless, so efficient is ozone in absorbing solar ultraviolet that it is all that protects us from a far greater prevalence of cancers, eye lesions such as cataracts, intense sunburns and a suppressed immune system – both in humans and in other animals. (This absorption of solar ultraviolet in this layer is partly responsible for the striking increase of temperature with altitude documented in Figure 7.25(b).)

Ozone is produced in a two-step process. First, a solar ultraviolet photon breaks apart a normal oxygen molecule into its component atoms,

$$\text{ultraviolet photon} + O_2 \rightarrow O + O,$$

and then one of these atoms combines with a normal oxygen molecule to form ozone:

$$O + O_2 \rightarrow O_3.$$

In 1974 the chemists F. Sherwood Rowland and Mario J. Molina argued that certain chemicals known as chlorofluorocarbons (CFCs) can break apart ozone, so depriving us of our protective shield against solar ultraviolet. Furthermore, the CFC itself is not destroyed in the process, so that a single one of these molecules can destroy a lot of ozone molecules.

CFC molecules contain (among other atoms) atoms of the element chlorine. It is this chlorine that is primarily responsible for destroying ozone. Once a molecule of CFC has wafted into the upper atmosphere, solar ultraviolet radiation breaks it apart, releasing the chlorine atom. The chlorine then combines with ozone as follows:

Step 1. $Cl + O_3 \rightarrow ClO + O_2.$

The ozone molecule has been destroyed. Of course, the chlorine has combined with an oxygen to form the harmless molecule ClO – but now, in a second step, the ClO is broken apart to release the chlorine back into the atmosphere:

Step 2. $ClO + O \rightarrow Cl + O_2.$

So the chlorine is now free to seek out another ozone molecule and repeat the process. Thus the sequence *Step 1* followed by *Step 2* cycles over and over again. It is estimated that a single chlorine atom can destroy as many as 100 000 ozone atoms in this way.

Until scientists realized how dangerous CFCs were, they were widely used in our modern industrial society. It is estimated that as many as 5 billion kilograms of CFCs have already been emitted into the atmosphere. Thus, even if industrial production were to entirely cease, their effect would not go away. In a 1991 study it was found that over the previous 12 years the ozone layer had thinned by roughly 3% over the northern hemisphere.

The ozone hole
Furthermore, over Antarctica the drop in ozone levels has been far greater. Figure 7.26 documents how the total amount of ozone over Antarctica has decreased over time. In Figure 7.26(a) each point represents the average over the month of October for that year: the data have been taken over a 40-year time span. As you can see, prior to 1975 the ozone level was steady, but since then it has dropped.

NOW YOU DO IT
By what factor did the amount of ozone over Antarctica decrease between 1975 and 1990?

Figure 7.26(b) shows observations of the ozone hole from space. As you can see, the hole is larger than the entire continent of Antarctica.

NOW YOU DO IT
According to Figure 7.26(b), the ozone hole is variable. In that figure, in what year was there the least amount of ozone above Antarctica? In what year was there the greatest?

Ozone depletion over Antarctica is extreme because of the special circumstances that obtain there. On the one hand, Antarctica is exceedingly cold. On the other

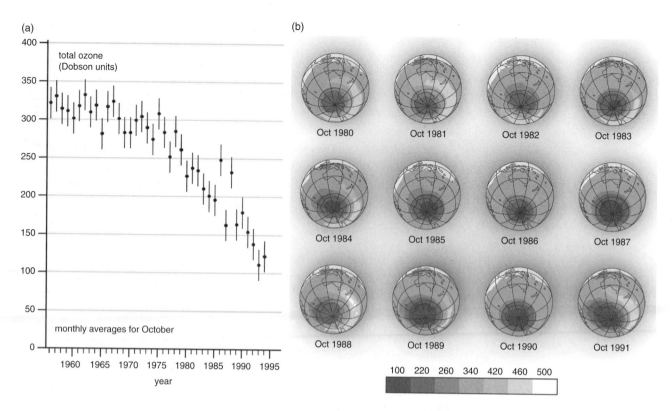

Figure 7.26 **Ozone depletion in the Antarctic. (a) Each data point represents the average over the month of October for that year. (b) Ozone concentrations as measured by the Total Ozone Mapping Spectrometer Earth-orbiting satellite.**

hand, during the winter a so-called "polar vortex" forms, a pattern of air currents that circle about the continent. These circulating currents prevent warmer air from mid latitudes from mixing with the ultra-cold Antarctic air. The net result is that atmospheric temperatures over the pole are even lower than we would expect: about 80 °C (more than 110 °F) below zero. Under these frigid circumstances so-called "polar stratospheric clouds" form. They are composed not of water droplets, but of ices mixed with numerous other chemical species. Large molecules react on the surfaces of these droplets to release their ozone-destroying chlorine atoms into the air. These reactions are extremely fast, which is why ozone is so strongly depleted over Antarctica, and why the growth of the hole has been so rapid.

In 1987 an international agreement known as the Montreal Protocol was signed, in an effort to reduce drastically the emission of CFCs by industrial societies. Revisions of this agreement were later made in the light of increased scientific knowledge. The agreements appear to be bearing fruit: in 1996 measurements revealed that, for the first time, the amount of chlorine in the atmosphere had decreased.

Global warming

Our study of the superheated atmosphere of Venus has taught us how important the greenhouse effect can be in altering an atmosphere's temperature. Because CO_2 is so prevalent in Venus' air, it has had a huge effect there. Because it is so rare in our air, we might think it has little effect on Earth.

But contemporary society is changing the situation – in two ways.

- Industrial emissions inject a good deal of carbon dioxide into our atmosphere. It is the burning of fossil fuels that is the main culprit here. In the combustion process, atmospheric oxygen combines with carbon in the fossil fuel to produce CO_2.

Mauna Loa Observatory, Hawaii
monthly average carbon dioxide concentration

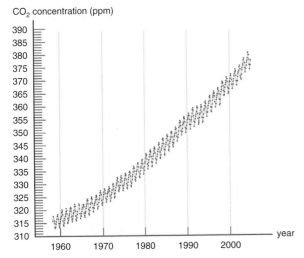

Figure 7.27 **Measured concentration of CO_2 has steadily increased. The small up-and-down fluctuations are seasonal variations.**

- During photosynthesis, plants absorb CO_2 and so remove it from the atmosphere. But the deforestation of large tracts of land, particularly the Amazon rain forest, is removing these trees.

Figure 7.27 shows the measured concentration of CO_2 since 1960. This graph shows two things. On the one hand, there are small "wiggles." If you study this graph carefully, you will see that they occur exactly once per year.

NOW YOU DO IT
Try verifying this for yourself by counting the number of wiggles between any two dates.

So these wiggles must be related to the seasons; and indeed they are caused, among other things, by seasonal variation in the uptake of CO_2 by growing plants. But apart from these seasonal variations, the concentration of CO_2 in the atmosphere also shows a steady increase.

NOW YOU DO IT
(A) By what factor did the concentration of CO_2 in the atmosphere increase between 1958 and 2004? (B) If the same behavior persists unchanged, what will be the concentration of CO_2 in the atmosphere in 2050?

This increase has persisted for decades. Furthermore, in geological terms it has occurred in a mere blink of an eye.

NOW YOU DO IT
Figure 7.27 shows the concentration of CO_2 in the atmosphere between 1958 and 2004. (A) What fraction is this of the length of time during which the human race has existed as a species (about a million years)? (B) To put this in perspective, consider an analogy in which the duration of our species is represented by one day: how long in this analogy is the period of time from 1958 to 2004?

There is a widespread consensus among scientists that this increase in carbon dioxide levels has already begun to trigger global warming. There is a very serious debate about how costly it will be to reverse the trend. But there is no debate that on the scale of geologic time we are causing a spectacularly rapid injection of a potentially dangerous gas into our atmosphere.

Continental drift and plate tectonics

A glance at a map is enough to show that many of the continents have symmetrical shapes. North America fits well into Europe and the northern portions of Africa. South America fits into the central and southern coast of Africa. It is almost as if the continents are like pieces of a jigsaw puzzle, which has been broken apart and scattered over the face of the globe.

The theory of *continental drift* was first proposed in the early twentieth century by Alfred Wegener. Initially most scientists did not take it seriously, for nobody could see how the continents could move. After all, the Earth is made of solid rock.

Recently we have learned, however, that rock is not so very solid as all that. Over long periods of time it can flow. Perhaps you have experimented with Silly Putty, a peculiar substance that shatters like a solid if given a sudden jerk – but that stretches like taffy if pulled slowly. Rock is like that.

We now can directly observe the sliding about of the continents. Distances between continents can be measured to an accuracy of about a centimeter. These studies reveal that the surface of the Earth is like a collection of rafts slowly drifting about on a lake. North America, for instance, is receding from Europe at the rate of nearly 2 meters per century, or about an inch per year. Many years ago, we must have been part of a single super-continent.

How long ago was this? We can find out if we suppose that our motion has remained steady over past geological ages. Of course we do not know if this has been so, but the assumption will serve to give us a rough idea of the true situation.

Today North America is about 3000 miles from Europe. Traveling away from it at 2 meters per century, how long was required to cover this distance? We calculate:

time = distance/velocity
distance = 3000 miles = 4800 kilometers = 4.8×10^6 meters
time = 4.8×10^6 meters/2 meters per century
   = 2.4×10^6 centuries or 240 million years.

As we emphasized, this is a rough calculation: nevertheless, it tells that, over time spans of several hundred million years, continental drift entirely redraws the map of the Earth.

. . . .

NOW YOU DO IT
How much narrower was the Atlantic Ocean when Columbus sailed across it?
. . . .

In Figure 7.28 we show the map of the Earth in past ages, as deduced from detailed studies. Our simple calculation turns out to have been roughly correct: 225 million years ago, all continents were united into a single land mass. This super-continent is called Pangaea. By 200 million years ago Pangaea had broken up into two smaller masses known as Laurasia and Gondwanaland. North America and Europe were still joined together 65 million years ago, although by this time South America and Africa had parted to form separate continents. Only relatively recently did the map we know today come into being.

Plate tectonics

What causes the continents to move about? An important piece of the answer was obtained when new techniques allowed us to map the ocean floors. These techniques revealed a topography utterly unlike that of the continents (Figure 7.29). Snaking along the entire length of the Atlantic Ocean is an enormous chain of

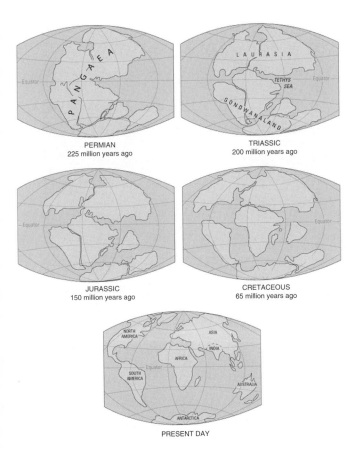

PERMIAN
225 million years ago

TRIASSIC
200 million years ago

JURASSIC
150 million years ago

CRETACEOUS
65 million years ago

PRESENT DAY

Figure 7.28 **Continental drift has remade the map of the world.**

Figure 7.29 **The ocean floors have a topography unlike that of the continents.**

mountains. This chain, known as a "midocean ridge," is longer than any mountain range previously known. Splitting the middle of this ridge is an immense rift. When a series of samples of the ocean floor about it were obtained, a remarkable pattern was revealed: the closer the floor was to the rise, the younger it was.

The midocean ridges are sites where the crust of the Earth is being created (Figure 7.30). Along them, hot magma rises upwards from the Earth's deep interior. It spreads laterally outward, cooling and solidifying to form vast "plates." The plates' motion is what we know of as continental drift: it is driven by the up- and down-welling motions deep within the Earth.

These plates are illustrated in Figure 7.31. Notice in that figure that the plates are larger than the continents, and do not coincide with them. The South American plate, for instance, also encompasses a good deal of the Atlantic Ocean.

These plates are continually encountering one another as they move. When two plates slide sideways relative to one another, such as the Pacific and North American plates, earthquakes and volcanoes result. When they directly collide, "subduction zones" result (Figure 7.32) in which one plate slides beneath the other. These zones, in which the crust is being forced downwards into the Earth's interior, are marked by ocean trenches many miles deep. Finally, if the plates' colliding motion rams two continents together, mountain ranges are thrust up: the Himalayas is the most spectacular example.

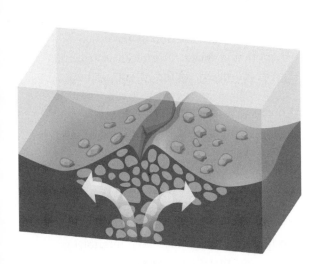

Figure 7.30 **New crust is formed as currents of rock well upwards along the midocean ridge and then spread laterally outward. Their motion drives continental drift.**

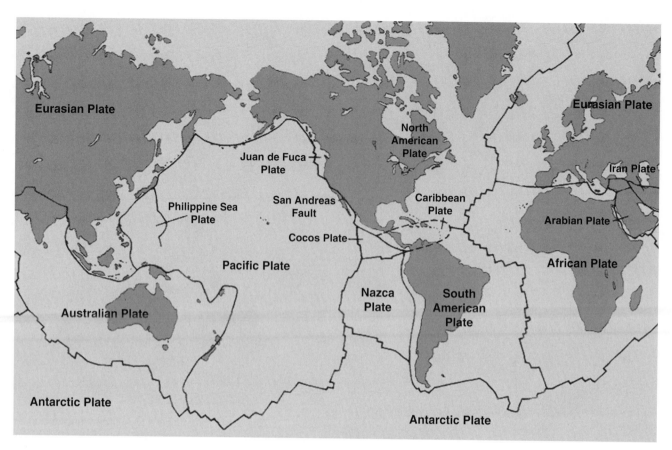

Figure 7.31 **Plates.**

The ultimate energy source for all this motion is the upwelling currents of magma, driven by the great heat of the Earth's interior. We now turn to a study of this interior.

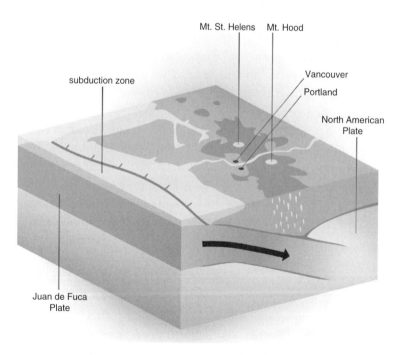

Figure 7.32 **When plates collide subduction zones result.**

The interior of the Earth

Although it is very close to us in astronomical terms, the interior of the Earth is a remote and inaccessible place. Astronauts have journeyed to the Moon, and space probes have journeyed to remote reaches of the Solar System. But no person, and no robotic probe, has ever penetrated into the Earth's interior. The deepest mine we have ever dug has reached a mere few kilometers below the Earth's surface – a tiny fraction of its 6378-kilometer radius. Nevertheless, we have found ways to indirectly probe our planet's deep interior.

As our first step in this exploration, let us consider the average density of the Earth. As we commented in Chapter 6, this provides us with a clue to its makeup.

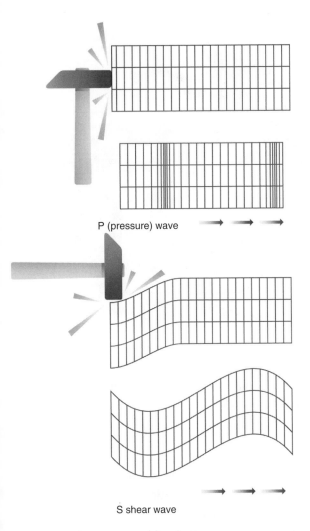

P (pressure) wave

S shear wave

Figure 7.33 **Seismic waves are of two types:**
P (pressure) waves and S (shear) waves.

In that chapter we found Earth's density to be 5490 kilograms per cubic meter. But now bend down and pick up a stone and measure *its* density. You will find it to be far less: about 3000 kilograms per cubic meter. The Earth as a whole is a good deal denser than the rocks on its surface.

This tells us that lying deep within the Earth is a region of ultra-high density. Because iron has been found to be very common in meteorites, and because it is a high-density material, we have good reason to hypothesize that iron might be the responsible culprit. As we will now see, the study of seismology has born out the hypothesis that the Earth possesses an iron core.

Seismic waves provide us with a highly detailed way to explore the Earth's interior. Seismic waves, which are waves in rock, are produced by earthquakes ("seismos" is Greek for "to shake"). When an earthquake occurs, these waves are generated in the same way that striking a bell with a hammer causes it to ring. Seismometers scattered about the Earth record these vibrations. By analyzing them we can "look" into the Earth's interior, in much the same way that a doctor "looks" into the body of a patient using ultrasound (also a wave).

Seismic waves are of two types: P waves (for "pressure") and S waves (for "shear"). They are illustrated in Figure 7.33. P waves are analogous to the pressure waves in air that we experience as sound: an earthquake gives the rock a push, raising its pressure and density to produce the wave. S waves, on the other hand, are produced when an earthquake gives the rock a sideways shear, and they have no analog in sound.

Observations by seismographs scattered about the Earth have revealed a striking anomaly about these waves. When an earthquake occurs, the P waves it generates are detected everywhere across the face of the globe. But the S waves are not! They are detected everywhere *except throughout a broad region on the opposite side of the Earth from the quake* (Figure 7.34). Let us call this region, in which P waves are detected but S waves are not, the "zone of avoidance."

Why should this be? A clue is provided by the following facts.

- P waves can travel through a liquid. (This is not surprising: as we have seen, sound is a P wave, and sounds can be heard under water.)
- S waves *cannot* travel through a liquid.

The absence of S waves in a region of avoidance is therefore telling us that something liquid lies between it and the quake. Because the region lies on the opposite side of the Earth from the quake, this liquid must lie at the Earth's center. We have discovered that *the Earth has a liquid core.* We now understand the region of avoidance, within which S waves are never found. It is the "shadow" of the liquid core.

....

NOW YOU DO IT
All we can really say from Figure 7.34 is that the liquid region lies *somewhere*
along the line joining the earthquake to the zone of avoidance. So there are

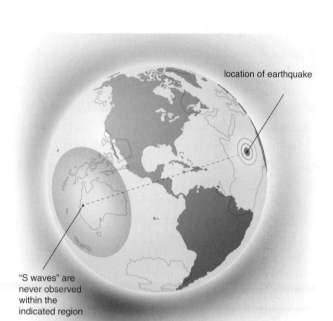

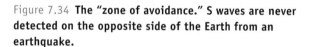

Figure 7.34 **The "zone of avoidance." S waves are never detected on the opposite side of the Earth from an earthquake.**

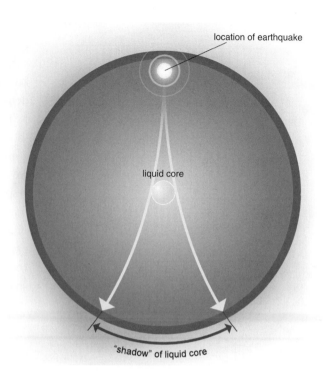

Figure 7.35 **Size of the "shadow" of S waves allows us to measure the size of the Earth's liquid core.**

many possible locations for the liquid region! How can we narrow down the range of possibilities to conclude that the liquid actually lies in the Earth's core? (A) Explain how repeated observations of *many* earthquakes allow us to do this. (B) Can you think of a medical procedure that uses the same logic?

Suppose that there was a liquid region 100 miles in diameter lying 100 miles beneath the Earth's surface, and suppose that an earthquake occurred in Kansas. Where would S and P waves be detected? Consider two possibilities: (A) the liquid region lay just beneath the location of the earthquake, and (B) it lay just beneath Sydney, Australia.

We can even use seismic observations to measure this core's diameter. The core lies halfway between the earthquake and its "shadow" on the opposite surface of the Earth. Therefore, the diameter of the core can be inferred from the observed diameter of the shadow.

NOW YOU DO IT
Suppose seismic waves traveled in straight lines. Explain why the diameter of the liquid core would be just half the diameter of the zone of avoidance.

As indicated in Figure 7.35, however, seismic waves do not travel in straight lines. This is due to the phenomenon of refraction, which affects every type of wave. Just as a light ray is bent when passing from one medium to another (see Figure 5.14), so the path of a seismic wave bends when passing from one region of the Earth's interior to another. Similarly, just as a light ray is reflected when striking a mirror, so a

⇐ **Looking backward**
We studied refraction in Chapter 5.

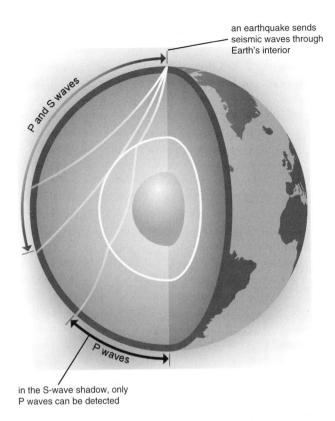

in the S-wave shadow, only
P waves can be detected

Figure 7.36 **Reflection and refraction of seismic waves as they travel through the Earth.**

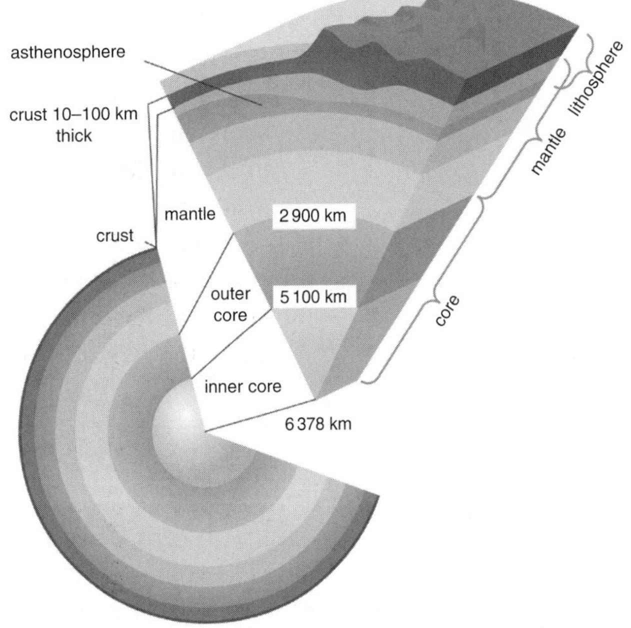

Figure 7.37 **Interior of the Earth as inferred from seismic observations.**

seismic wave is reflected from a sharp junction between one region of the Earth and another (Figure 7.36).

NOW YOU DO IT
You have just shown that, if seismic waves traveled in straight lines, the diameter of the liquid core would be half the diameter of the zone of avoidance. Given that the waves actually bend as shown in Figure 7.36, is the actual diameter of the liquid core greater or less than this?

So far we have made inferences about the Earth's interior by using two lines of evidence:

- the overall density of the Earth has provided evidence for an iron core,
- the absence of S waves in certain regions has provided evidence for a liquid core.

These are important as far as they go. But by studying the complex patterns of seismic waves arriving at observing stations scattered over the face of the Earth, seismologists have been able to piece together a far more detailed picture of our world's interior. This picture is summarized in Figure 7.37.

Deep in the Earth's interior is a *solid inner core*, whose radius is about 20% that of the Earth as a whole. Theoretical calculations indicate that it is indeed composed of iron, with significant amounts of nickel as well. Its temperature appears to be roughly 6000 kelvin – hotter than the surface of the Sun! At such high temperatures we would normally expect iron to melt, but the seismic evidence indicates that it is solid: apparently the enormous pressure there keeps the material solid.

Wrapped about the inner core is a *liquid outer core*, whose radius is about half that of the Earth as a whole. It appears to be composed of similar minerals to the inner core, and its temperature is similar. But because the weight of the overlying rock is less, the pressure is less and the material is molten. This is the liquid core we hypothesized earlier.

Wrapped about the two cores is the *mantle*, which extends nearly to the Earth's surface. Cooler than the core, it is solid – but only solid in the sense that we discussed above, and it is capable of smoothly flowing over geological ages. The flow of heat out of the Earth's deep interior drives slow *convection currents* in the mantle, much as a sauce simmering on a burner slowly churns. These currents are what drive continental drift.

Finally we reach the only portion of the Earth that we can directly experience: *the crust*. The crust's density is lower than that of the underlying material, so that it buoyantly floats upon it like a sheet of plywood floats on a lake. The crust is thicker under the continents (60 kilometers) than under the oceans (10 kilometers).

Perhaps we should also include in our picture the *biosphere*, an ultra-thin layer atop the crust. Only within this fragile film does life exist on Earth.

Mars

Mars is the last of the inner planets we encounter on our journey away from the Sun. Smaller than the Earth and Venus, larger than Mercury and the Moon, it is intermediate among the bodies of the inner Solar System.

⇐ **Looking backward**
We studied "seeing" in Chapter 5.

The length of the Martian day is nearly the same as ours – 24 hours and 37 minutes. Its year is 1.9 Earth years long. Since the tilt of its spin axis is almost the same as ours, we see that Mars' seasons are quite like ours, but each is nearly twice as long.

Mars illustrates yet another way in which our knowledge of the Solar System has been revolutionized by the space program. While in the old days Mercury was hard to study owing to its proximity to the Sun, and Venus was hard to study owing to its atmosphere, Mars was hard to study owing to *our* atmosphere. Figure 7.38 shows a photograph of Mars taken by a ground-based telescope. Notice how blurred the image is. You might think that the telescope used to produce this photograph was pretty poor, but it was not. In fact the image is blurred because of "seeing" – the distortion in the paths of light rays as they pass through our turbulent atmosphere. Figure 7.38 represents about the best view astronomers could get prior to the advent of the space age.

In 1877 the Italian astronomer Giovanni Schiaparelli announced that he

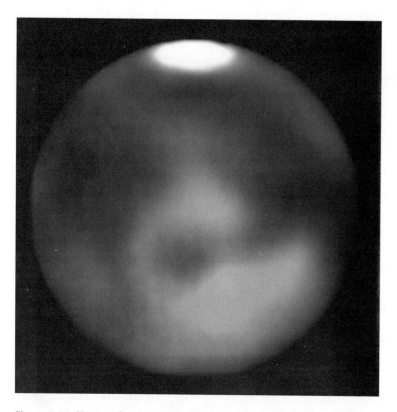

Figure 7.38 **Mars as photographed with a ground-based telescope. The image is blurred by atmospheric "seeing."** 👁 **(Also see color plate section.)**

had observed faint straight lines on the surface of Mars. The word he used to describe them was "canali," which means "channels" in Italian. But it was translated into English as "canals." Over the years, the idea developed that Schiaparelli had discovered a gigantic feat of engineering produced by a race of intelligent beings.

One astronomer who was particularly taken with this idea was Percival Lowell, who built an observatory in Arizona specifically dedicated to the study of the canals of Mars. Peering through the distorting atmosphere of the Earth, Lowell succeeded in mapping out what he thought was an extensive network of canals crisscrossing the face of the planet (Figure 7.39).

Eventually the idea became widespread that Mars was inhabited by a superior civilization. In 1938 Orson Welles' famous radio broadcast "The War of the Worlds" capitalized on this. Welles' dramatization of an invasion from Mars took the form of a series of alarming news bulletins interrupting a regular programming schedule. What began as an innocent Halloween prank ended up as a panic, as thousands of listeners, unaware that the "news broadcasts" were fictitious, fled from their homes in terror.

Many astronomers felt that the evidence for canals on Mars was sketchy, owing to the extreme difficulty of making them out using ground-based telescopes. Many other astronomers were convinced of the canals' existence. Not until 1965 was the matter settled. Mariner 4, the first space probe to visit the planet, returned close-up images of its surface, none of which showed canals. They had been an illusion.

The Martian atmosphere

Mars' atmosphere, like Venus', is mostly carbon dioxide. Oxygen is present, but only in trace quantities. But unlike Venus, the Martian atmosphere is very thin:

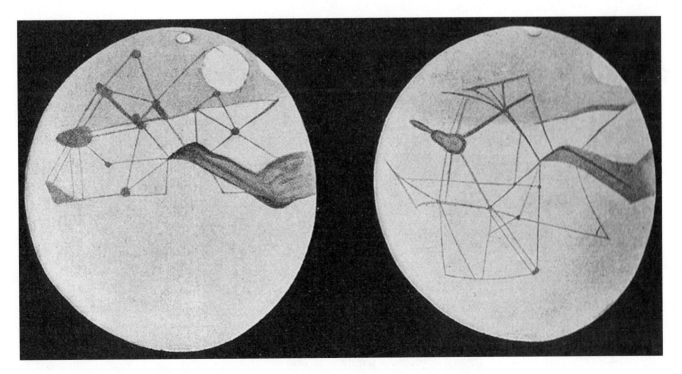

Figure 7.39 **Canals on Mars as mapped in the early twentieth century by Percival Lowell.**

atmospheric pressure there is a mere 0.6% of ours. An astronaut on Mars would need a space suit.

Wind is an enormous presence on Mars. It can whip up immense dust storms. Indeed, when Mariner 9 arrived at the planet in 1971 and went into orbit about it, the entire world was obscured by dust. The storm took weeks to slowly clear.

This is all the more surprising because the air on Mars is so thin. To pick up dust, such a rarified wind must blow very hard. Winds of 180 kilometers per hour – 112 miles per hour – are required to produce the vast dust storms we observe covering the planet. Because the Martian soil is red, the dust particles that perpetually fill the air make the sky on Mars pink, not blue.

Mars is a cold place. Viking 1, which landed fairly close to the equator, recorded a maximum summertime temperature of minus 33 degrees Celsius, or 27 degrees Fahrenheit below zero. It gets this cold here on Earth at many locations in winter – but this was the *hottest* the Viking Lander ever recorded. At the very same location, just before dawn, things had cooled down to minus 83 degrees Celsius, or 117 degrees Fahrenheit below zero! And a second Viking spacecraft, which landed at 48 degrees north latitude, recorded a temperature of minus 100 degrees Celsius, or minus 148 degrees Fahrenheit.

The Viking Landers also observed a curious alteration in Mars' air pressure. As the Martian seasons progressed, its barometric pressure changed. Furthermore, it changed in a regular fashion: it was lowest in Mars' northern hemisphere summer, and highest in the northern hemisphere winter. Our own barometric pressure, of course, also undergoes alterations, but they are irregular. Furthermore, they are tiny in comparison with those on Mars.

What could possibly cause such a striking alteration? It clearly has something to do with the seasons – and seasons, of course, have something to do with temperature. In understanding this phenomenon, we already have some facts at our disposal:

· Mars' air is made of CO_2,
· Mars is *cold*,
· CO_2 freezes at sufficiently low temperatures.

How cold does it have to be for CO_2 to freeze? Experiment shows that a temperature of minus 123 degrees Celsius, or minus 190 degrees Fahrenheit, is required. This is colder than any temperature recorded by the Viking Landers. But these spacecraft did not land in the coldest part of the planet! One landed close to the equator, and the other at an intermediate latitude.

Like the Earth, the coldest parts of Mars are its poles. Remote observations of the temperatures there show that, in winter, the poles of Mars are cold enough to freeze carbon dioxide.

We have made a remarkable discovery: every winter, the poles of Mars are so cold that *the air there freezes*. Since a good deal of gas has been removed from the atmosphere by this freezing, air pressure drops across the entire planet. When the weather warms, the solid CO_2 steams away back into the atmosphere, and the air pressure rises. This is the explanation for the seasonal alteration in pressure thousands of miles away, as recorded by the Viking Landers. (However, *be warned*: there is something wrong with our description of this theory. You will be asked to give a correct description at the end of this chapter.)

Figure 7.40 **The surface of Mars as revealed by the Hubble Space Telescope.** 👁 **(Also see color plate section.)**

One question remains: why doesn't this happen here? After all, just like Mars, our poles are coldest in winter. The answer is that our polar caps are made of frozen water – but Mars' polar caps are made of frozen Martian air.

The surface of Mars

Mars has more craters than the Earth, but fewer than Mercury or the Moon. This is not surprising, since wind erodes its craters over geologic ages, and Mars has less air than Earth and more than Mercury or the Moon. There is a striking asymmetry: Mars' southern hemisphere is more heavily cratered than its northern hemisphere. Furthermore, the southern hemisphere is somewhat elevated compared to the northern. The relative absence of craters in the north must indicate that the land there is younger. And indeed, observations show that the northern hemisphere craters are less eroded than those in the south. An image of Mars is shown in Figure 7.40.

Mars has volcanoes. Indeed its largest, Olympus Mons, is the biggest known volcano in the Solar System. It is roughly 600 kilometers – 370 miles – in diameter: that is the distance from Los Angeles to Phoenix. Its summit is 25 kilometers (82 000 feet) high, as compared to Mount Everest's 29 028 foot elevation. We do not know if the eruption that formed it was relatively recent, or far back in the planet's history.

Many things on Mars are outsized. Valles Marineris ("Valley of the Mariners") is an immense canyon stretching nearly half way across the visible face of the planet. Its length is greater than the distance across the United States. Six hundred kilometers wide at the widest, it is more than six kilometers deep – four times deeper than the Grand Canyon. Indeed, the Grand Canyon would fit into one of its side canyons.

Water on Mars

There is much evidence for water on Mars. Images of the surface sent back by Landers commonly show small patches of frost in the shadows cast by rocks just after dawn. Orbiters have returned numerous images showing clouds ringing the bases of volcanoes, or within low-lying canyons. Measurements have revealed the presence of water vapor in the air above the polar caps, indicating that they contain great quantities of water ice beneath their frozen CO_2: in summer the CO_2 vaporizes, but the water ice remains.

All this does not add up to a lot of water, though. Even if you added up all the water in the Martian atmosphere, you would not get very much. But images of Mars' surface made from space give us reason to think that the planet once

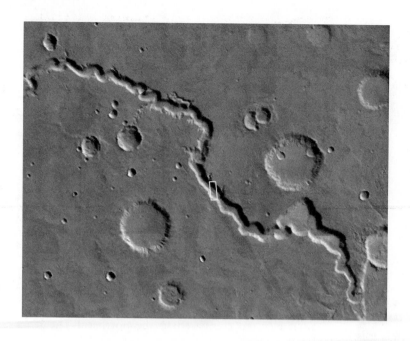

Figure 7.41 **A sinuous channel.**

contained lots of water. There are many lines of evidence. Among them are the following.

- *Sinuous channels* (Figure 7.41) are long, meandering channels in the Martian surface. They are dry now, but they were clearly carved by slowly flowing water. We might think of them as dry river washes.
- *Outflow channels* (Figure 7.42) are broad regions showing the effects of torrents of flowing water. These too are currently dry, but they appear to be the results of cata-strophic floods. The amount of water involved in these floods is tens of thousands of times as much as the yearly discharge of a great river such as the Amazon or the Mississippi.
- *Sludge* (Figure 7.43): certain Martian impact craters are surrounded by a curious terrain not seen about craters on other worlds. It seems to have been produced by a sudden melting of great amounts of permafrost lying beneath the surface. Presumably it was the energy of the impact that suddenly heated the permafrost. The liquid water thus produced mixed with the soil to produce a sludge, which flowed downwards in a "wet avalanche" before solidifying.

It is likely that the entire surface of Mars is permafrost (soil mixed with frozen water), possibly extending to as much as 1 kilometer below the surface. This adds up to a lot of water – enough to cover the planet with an ocean 10 to 100 meters deep.

The mystery of liquid water on Mars

Figure 7.42 **An outflow channel.**

It is not so surprising that Mars has water. As we will see, water is common in the Solar System. But it is very surprising that Mars once had *liquid* water. The problem is that it is quite impossible for liquid water to exist on Mars as it is today. If the planet once contained liquid water, it must have been a very different place then than now.

Why is it impossible for Mars to contain liquid water today? There are two reasons. One is easy: it's too cold! Even at the warmest, temperatures on Mars are far below freezing. So if Mars once had liquid water, it once was warmer.

The second reason involves the fact that air pressure on Mars is a mere 0.6% of ours. That is almost a vacuum. To understand what this means, let us build a "Mars chamber," and use

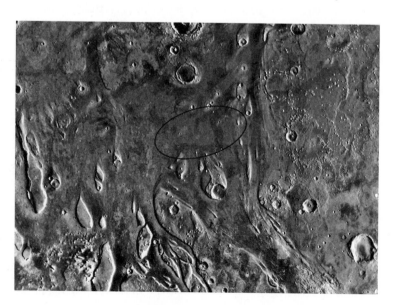

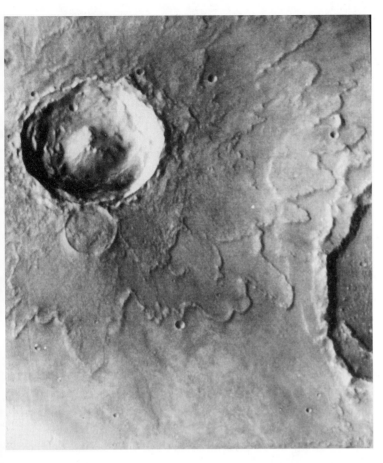

Figure 7.43 **Sludge (soil wetted by melted permafrost) appears to have formed the curious "splash" pattern surrounding Crater Yuty.**

it to duplicate in the laboratory conditions on Mars. Our chamber will do two things. On the one hand it will be a refrigerator, which can mimic Mars' frigid environment. On the other hand it will be a vacuum chamber, which can mimic Mars' ultra-low air pressure. We will watch what it does to liquid water.

Let us begin with our "Mars chamber" at room temperature. Put a glass of water in it, and shut the door. Now pump out most of the air, till the pressure inside duplicates that on Mars. What happens to the water in the glass? Observation shows that *it boils*. Even though we have not heated the water – indeed, it is still at room temperature – at low pressures it turns into steam.

Of course we have not yet succeeded in mimicking the Martian environment with our chamber. To do so we need to cool it down, keeping the pressure low as we do so. What happens when we do this? Does the water vapor condense into water droplets? It does not! Observation shows that *it snows*. The water vapor turns into solid water.

In our daily experience here on Earth, water can exist in three forms. If we begin with a high temperature, it is a gas – water vapor. If we then cool the gas, it first turns into a liquid (the vapor condenses into rain) and then into a solid (the water freezes). But our "Mars chamber" has taught us that at low pressures the intermediate form of liquid water simply does not exist. Rather, if we begin with a high temperature, it is a gas – and if we cool the gas it freezes. *At low pressures liquid water cannot exist.*

This sounds like an astonishing state of affairs, but in fact it is not so strange as all that. Dry ice – frozen carbon dioxide – does this on Earth. If you have ever seen a cake of dry ice removed from its refrigerator, you may have noticed that the stuff does not melt – it steams. Just as liquid water cannot exist in the ultra-low pressures found on Mars, so liquid carbon dioxide cannot exist in the pressures found on Earth.

The evidence for liquid water on Mars far back in its past is therefore evidence for an astonishing fact: *Mars was not just warm long ago. It also had more air.* Mars is now a cold, icy world with very little atmosphere. But this was not always so. Some planet-wide transformation must have occurred, which brought it from an early hospitable state into its present harsh state.

What produced this extraordinary transformation? There are two theories.

On the one hand, we believe that early in every planet's history the emission of gases from volcanoes led to a thick atmosphere. Conceivably, the greenhouse effect might have warmed the original Martian atmosphere enough to allow water to melt. But since Mars is a relatively low-mass planet, it cannot have held onto all this air for long. As we described in our discussion of Mercury, it would have ultimately been lost into space.

Alternatively, other theories posit that both the tilt angle of Mars' rotation axis, and the eccentricity of its orbit, change in a repetitive pattern. This would have led to a repetitive change in the amount of sunlight reaching its polar caps, and therefore in the amount of CO_2 steamed off from them in summer. Perhaps the variation is enough to have produced a regular alternation between a warm, thick atmosphere and the cold, thin atmosphere we observe today.

Notice that these two theories make very different predictions as to how long ago this warm, thick atmosphere existed. According to the first theory, it was early in the planet's history – billions of years ago. According to the second, it happened many times, the most recent being not so very long ago at all in geological terms. In order to decide between these two theories, therefore, we need to measure the ages of its sinuous channels, outflow channels and sludge formations. But we cannot do this without actually getting our hands on rock samples from them. For this reason, there is great interest in mounting a "sample return" mission, which would visit each of these formations, scoop up rock samples, and return them to our laboratories. The technical difficulties in conducting such a mission, however, are enormous.

The Earth's ice ages are evidence that worldwide alterations in climate can occur. But the changes that produced our ice ages are tiny in comparison to those that must have occurred on Mars. When these happened, whether they happened only once or over and over again, whether they have anything to do with our ice ages – these are questions that we cannot answer at present. We are left with a tantalizing mystery.

Summary

The Moon
- Lighter areas are covered with craters.
- Darker areas ("maria") are vast lava flows that have obliterated the craters.
- The Earth has few craters because here they have been "erased" by geological processes and erosion: both these are absent on the Moon.
- The Moon has no atmosphere.
- The Moon is geologically inactive: little has happened there in many billions of years.
- The Moon's craters:
  - were formed by impacts,
  - most of these impacts happened early in the Solar System's history,
  - the maria were produced by particularly huge impacts, which melted vast portions of the lunar surface.

- The Moon keeps the same face toward the Earth ("synchronous rotation") because of tides raised by the Earth in the Moon's solid body.

Mercury
- Very difficult to observe because it lies so close to the Sun.
- This led to a mistake: for years it had been thought that Mercury exhibited synchronous rotation. Ultimately, radar astronomy showed this to be false.
- Spacecraft observations show that Mercury looks much like the Moon.
- Because it is so close to the Sun, Mercury is very hot.
- Because it is so hot, and has such a low mass, Mercury has no atmosphere.

Venus
- Permanently covered by clouds: we study it using radar astronomy.

- Rotation:
  - rotates backward: the Sun rises in the west and sets in the east on Venus,
  - rotates very slowly: one Venus day is 243 Earth days.
- Atmosphere:
  - enormously hot (855 °F),
  - very dense (air pressure 90 atmospheres),
  - mostly CO_2,
  - these extraordinary conditions are the result of a runaway greenhouse effect.
- Surface is very flat.

Earth

- The only planet known to contain liquid water (which is essential to life).
- Atmosphere:
  - forms a very thin layer (for example, 99.9% lies below 50 kilometers),
  - different layers of the atmosphere are warmed by sunlight of different wavelengths; this leads to the temperature falling, rising, falling and then rising as we go upwards,
  - most abundant constituent: nitrogen; second most: oxygen.
- Ozone (O_3):
  - the ozone layer absorbs ultraviolet radiation from the Sun, which is dangerous to life,
  - chlorofluorocarbons (CFCs) can break apart ozone,
  - the ozone layer is known to have thinned over the entire northern hemisphere: over Antarctica it has dramatically thinned (the "ozone hole").
- Global warming:
  - caused by the greenhouse effect (involves CO_2),
  - industrial emissions inject a good deal of carbon dioxide into our atmosphere,
  - the deforestation of large tracts of land is removing trees (which absorb CO_2).

- Continental drift and plate tectonics:
  - over long time scales, rock can flow,
  - continents move at roughly an inch per year,
  - driven by hot magma rising from the Earth's deep interior and spreading outward at midocean ridges,
  - where plates collide, earthquakes and volcanoes result.
- Earth's interior:
  - studied via seismic waves (produced by earthquakes),
  - such studies reveal that Earth has a solid inner core (iron) surrounded by a liquid outer core, then the solid mantle and finally the crust.

Mars

- Prior to the space program we knew very little about Mars, and some of what we "knew" (i.e. canals) was wrong.
- Atmosphere: very thin and very cold.
- There is much ice on Mars (permafrost).
- There is much evidence that liquid water once existed in great quantities on Mars.
- This tells us that Mars used to be warmer and had more air.

DETECTIVES ON THE CASE

What causes tides?

We built a theory of tides by thinking about the gravitational force of the Moon on the oceans.
- Our first theory was that the Moon attracted the oceans toward it.
- But this theory predicts one tide per day, while observations show there are two.
- Eventually we realized that tides are caused by variations in the Moon's gravitational pull from one side of the Earth to the other.
- In reality, tides are also affected by the Sun, and the local coastline.

Problems

(1) Draw a diagram, analogous to Figure 7.6, showing the Earth and its atmosphere. Indicate on your diagram rays of light from the Sun (a) at noon, and (b) at sunset. Use your diagram to explain why it is possible to look directly at the Sun at sunset, while you would be blinded if you looked at it at noon.

(2) Draw a diagram, analogous to Figure 7.5, showing how the brightness of a distant star would vary as the Moon passed in front of it, for two cases: (A) the Moon had a

very tenuous atmosphere, and (B) the Moon had a very thick atmosphere.

(3) Draw a diagram, analogous to Figure 7.5, showing how the brightness of a distant star would vary as the Moon passed in front of it: (A) if the Moon orbited the Earth far more rapidly than it actually does, and (B) if it orbited more slowly.

(4) Suppose that in the year 2025 a census of a town yields the following results:

Age	Number of people
5	11 000
15	12 000
25	10 000
35	9 000
45	9 500

Compile a table, similar to Table 7.2, giving the history of the birth rate in this town over the past 45 years.

(5) (A) Suppose an asteroid is discovered on which there are the same number of old craters as young ones. Draw a graph, similar to Figure 7.8, showing the history of the rate of impacts on this asteroid. (B) Now do the same for another asteroid, on which there are lots of young craters and few old ones.

(6) We know that the gravitational attraction between two bodies grows weaker as their distance increases. But suppose it did not. Suppose the force of gravitation were just the opposite, and grew *stronger* as their distance increased. In this case, would there be tides? Would they be just like they are in reality, or different? Explain your answer.

(7) (A) Draw diagrams analogous to Figures 7.14 and 7.15 illustrating why it is easier for ground-based telescopes to observe Venus than Mercury. (B) Suppose you lived on a space ship that was voyaging outwards through the Solar System, drawing steadily farther and farther away from the Sun. Draw diagrams analogous to Figures 7.14 and 7.15 illustrating why it would grow progressively more difficult for you to observe the various planets of the Solar System as your voyage continued.

(8) The Sun is composed of gas, and it is of course very hot – hotter than any planet. Nevertheless, the gases in the Sun have not escaped. Why not?

(9) Suppose the Magellan spacecraft is one thousand kilometers above the surface of Venus, and it sends out a radar pulse. The return echo consists of a double echo, with a time delay indicating that it is passing over a two-kilometer mountain. (A) What is the time interval between sending out the pulse and receiving the echo from the surrounding terrain? (B) What is the time delay between the two return echoes? Now suppose that on its next passage over the same mountain, the spacecraft has lowered its orbit, so that it now is 500 kilometers above the surrounding terrain. (C) Now answer questions (A) and (B) again.

(10) Suppose the Magellan spacecraft sends out a pulse and receives a return echo 0.003 seconds later from the surrounding terrain – and then, 0.000 008 seconds after that, it receives a second echo. (A) How far is it from the surface of Venus? (B) How deep is the canyon it is passing over?

(11) Calculate how much narrower the Atlantic Ocean was when humans first emerged as a separate species (very roughly one million years ago).

(12) Suppose that wrapped entirely about the Earth there was a liquid region lying 100 miles beneath the surface, and suppose that an earthquake occurred in Kansas. Where would S and P waves be detected?

(13) Recently you considered a scale model in which the Earth was a sphere one meter across. In terms of this scale model, how big would be the Earth's (a) solid inner core, (b) liquid outer core and (c) mantle?

(14) If you could ascend to enormous altitudes in our atmosphere, you would eventually reach a point at which air pressure equaled Mars' surface pressure. How high is this altitude? What is the temperature there, and how does it compare to the surface temperature on Mars?

(15) In our discussion of the mystery of liquid water on Mars, we presented a theory involving the escape of gases from its surface. Suppose Venus were suddenly to become a lot less massive. (Of course this is completely impossible, but we are merely playing with ideas here!) What might happen to the temperature on Venus' surface?

· ·

WHAT DO YOU THINK?

(1) Throughout the Solar System we almost always find that it is the moons, and not the planets, that rotate synchronously. For example, our Moon keeps the same face towards the Earth – but the Earth does *not* keep the same face toward the Moon. If we survey the Solar System we find this is also true for the other planets and their moons – with one exception: Pluto. This one body is found to rotate synchronously with its moon Charon: Pluto keeps the same face toward its moon, and its moon keeps the same face toward it.

Recall our discussion in "Detectives on the case. Observations give us clues as we create our theories." In this case the observations you need are the data contained in Table 6.1 (in Chapter 6). See if you can use those data to find a theory of why it should be only Pluto that rotates synchronously with its moon.

(2) In discussing the Viking Landers we presented a theory to account for the seasonal alteration of air pressure on Mars. But there is something wrong with that theory! For suppose it is summer in Mars' northern hemisphere: then it is winter in the southern hemisphere. What does our theory predict that air pressure should be at these times of year?

(3) Now focus on the fact that in reality the air pressure is observed to be greatest only once per Martian year, namely in the northern hemisphere's summer! Recall our discussion in "Detectives on the case" of how observations give us clues as we create our theories. In this case the clue is that *the south polar cap of Mars is much bigger than the north polar cap*. See if you can use this clue to correct our error.

You must decide

Suppose you are a program officer at the National Science Foundation charged with supporting the development of our understanding of the Solar System. You have received two grant proposals.

(1) One requests support for a program of research aimed at learning more about the history of Venus' atmosphere, with the goal of understanding how its runaway greenhouse effect occurred.

(2) The other requests support for a robotic sample return mission, which would land in a sinuous channel on Mars and send rocks from the channel back to Earth. The purpose is to conduct age dating tests on these rocks in the laboratory, in order to determine when Mars possessed liquid water.

Unfortunately, you only have enough money to fund one of these proposals. Your task is to write a memorandum analyzing these two options, and to recommend which course of action the NSF should take. Your memorandum should include a *detailed* account of

• the arguments for each option,

• the arguments against each option,

• the decision that you reached,

and (this is the most important of all!)

• your reasons for your decision.

8

The outer Solar System

In the previous chapter we studied the inner Solar System, whose planets are relatively small and dense, and are composed of rock. Now, proceeding farther outward away from the Sun, we skip over the asteroid belt to reach the outer Solar System. This is the domain of the mighty Jupiter, Saturn, Uranus and Neptune; and then tiny Pluto. (As we have mentioned, it was recently decided that Pluto is not really a planet: we will defer this question to Chapter 9, and for now we will simply call it not a planet but a "planet.") The giant planets are immensely larger than our own, and are composed almost entirely of gas. They are utterly unlike the Earth: indeed one of them, Jupiter, can be thought of as almost a star. They are graced with lovely rings.

Like stars, these outer planets have miniature "solar systems" of their own: great numbers of moons revolving about them, just as planets revolve about the Sun. Before the advent of the space age, Earth-bound telescopes were unable to resolve these satellites, and we knew them as little more than tiny points of light. But recent space missions have revealed these satellites as being worlds in their own right, utterly unlike our own Moon – and indeed utterly unlike anything we have seen before.

Jupiter

As we already commented, Jupiter is the biggest planet: indeed it is bigger than all the other planets combined. Jupiter's radius is 11 times the Earth's, and its mass is 318 Earth masses. Jupiter lies 5.2 AU from the Sun, and its year is 11.86 Earth years. It rotates on its axis once every 9 hours, 55 minutes, 29 seconds.

The *visible surface* of Jupiter is not solid: it is the top of its atmosphere. Indeed, everything we see on Jupiter can be termed "Jupiter weather." Meteorologists accustomed to studying our own weather have had a field day trying to understand that on Jupiter: it is utterly different from our own.

As we see from Figure 8.1, Jupiter's surface is marked by bands of alternating light and dark colors, running parallel to its rotational equator. We are not certain what gives the bands their distinctive color. Studies utilizing the Doppler effect show that the light-colored bands are regions of rising currents of gas, while in the dark bands the gas is descending.

Figure 8.1 **Jupiter.** 👁 (Also see color plate section.)

Figure 8.2 **Jupiter's Great Red Spot is the size of two Earths.**
👁 (Also see color plate section.)

NOW YOU DO IT
What do you think the observation was that led us to this conclusion?

Easily visible with even a small telescope is Jupiter's *Great Red Spot*. It is the size of two Earths. First seen through a telescope in 1831 and possibly earlier (reports are vague), it has persisted ever since. Remarkably, the spot is not fixed on the surface of the planet, but erratically and slowly migrates east and west – but never north or south! Astronomers thought it was unique until spacecraft got to Jupiter and returned close-up images showing that spots are actually common – the Great Red Spot is only the biggest.

Such images (see Figure 8.2) show that the spots are giant whirlpools in the atmosphere, somewhat similar to our hurricanes but vastly bigger. Some rotate clockwise, some counter-clockwise. (The Great Red Spot rotates counter-clockwise once every six days.) Smaller spots are seen to merge, and roll about each other like ball bearings.

Temperature

Because Jupiter is so far from the Sun, it is very cold. The temperature of Jupiter's visible surface is minus 159 degrees Celsius, or 254 degrees Fahrenheit below zero. But as we descend beneath the visible surface, the temperature increases. About 60 kilometers below the cloud tops, temperatures are similar to room temperature.

Jupiter's atmosphere is composed of compounds lethal to us, and indeed to every organism that exists on Earth today. But it is not necessarily lethal to all forms of life. Indeed, as we mentioned in the previous chapter the Earth's primitive atmosphere, in which life originated, would also be lethal to contemporary life forms. Jupiter does not have any solid surface for living creatures to live on, but perhaps it is conceivable that some form of life floats permanently within this narrow warm zone among the clouds.

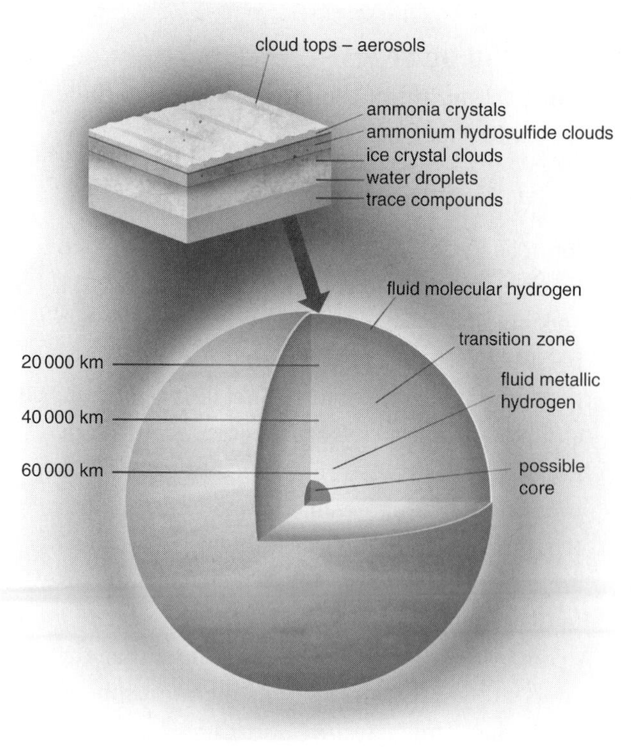

cloud tops – aerosols

ammonia crystals
ammonium hydrosulfide clouds
ice crystal clouds
water droplets
trace compounds

fluid molecular hydrogen

transition zone

fluid metallic
hydrogen

20 000 km

40 000 km

60 000 km

possible
core

Figure 8.3 **Jupiter's interior.**

Composition

In Chapter 6 we calculated Jupiter's average density, and concluded that this density is characteristic of gases. Spectroscopic studies have confirmed this. Jupiter is composed almost exclusively of gaseous hydrogen (79%) and helium (19%), with only trace quantities of other elements. (It may have a solid core.) This composition is utterly different from that of the inner planets.

Remarkably, however, Jupiter's composition is almost identical to that of the Sun! How can we understand this? Surely it is no accident that the composition of the second most massive body in the Solar System – Jupiter – matches that of the most massive body – the Sun.

Let us return to what we know of the escape of planetary atmospheres (Chapter 7). We already understand that, the more massive a body, the less liable it is to lose its atmosphere. So this explains why Jupiter, the most massive planet, has so much gas in it. But why this particular combination of gases? In our discussion of the escape of planetary atmospheres, we also noted that the less massive the particles making up the gas, the faster it is moving. Since hydrogen and helium are the lightest elements, they move the most rapidly. So these two elements will be preferentially lost from a planet's atmosphere.

This allows us to understand Jupiter's composition. Only massive bodies like Jupiter and the Sun have sufficient gravity to hold on to these two slippery elements. Other less-massive planets like the Earth may once have had great quantities of hydrogen and helium, but they were rapidly lost into space.

Interior

At high pressures, hydrogen, the primary constituent of Jupiter, is known to become a liquid: these pressures are reached perhaps 1000 kilometers beneath the planet's surface. At still higher pressures, the electrons in the hydrogen atoms leave their atoms and wander freely about, making the liquid a good conductor of electricity. Because metals are good electrical conductors, this state is known as metallic hydrogen. Studies of the way in which Jupiter's rotation make it bulge at its equator suggest that the planet might have a solid core, slightly bigger than the Earth, containing a mass of perhaps 15 Earth masses. Figure 8.3 sketches our present understanding of the interior of Jupiter.

Jupiter's energy emission

We turn now to a surprising fact about Jupiter: it emits more energy than it receives from the Sun. Why is this surprising? Because planets are not supposed to do such a thing. Among the bodies in the Solar System, only the Sun emits light. We used to think that all the other bodies – moons, planets and so forth – are simply illuminated by the Sun, much as the tables and chairs in a room are illuminated by a light bulb. But recently we have learned that, while Jupiter is indeed illuminated by

⇐ **Looking backward**
We studied blackbody radiation in Chapter 4, and infrared telescopes in Chapter 5.

sunlight, it emits more energy than this illumination could provide. The only possible conclusion is that Jupiter is a "light bulb" in its own right.

This discovery came about as a result of the development of infrared astronomy. As we know, a warm body emits blackbody radiation, and at relatively cool temperatures this radiation falls in the infrared portion of the spectrum. Once telescopes were constructed capable of detecting this infrared, it became possible to measure the total quantity of heat radiation emitted by Jupiter. Astronomers were surprised to find that it emitted more than it should.

Let us illustrate this, by calculating (A) how much energy Jupiter receives from the Sun, and comparing this to (B) how much energy it emits. We will find that (B) is more than (A). What is the source of this "excess" energy emission? The only thing possible is that it is somehow generated within the planet.

We'll start with (A).

Intuitive mathematics

Interstellar space is very cold, and the Sun is very hot. As an analogy we can think of the Sun as being like a campfire on a winter night. It warms everything in its vicinity. Similarly, planets are warmed by the Sun. And since they are warm, they emit infrared radiation.

The closer you are to a campfire, the more it warms you. Furthermore, as you stand by a campfire, the larger you are the more of its rays you will "catch" from it. So by analogy, we expect our formula for how much energy Jupiter receives from the Sun to predict that

- a planet close to the Sun must receive more energy from it than one far away,
- a big planet catches more rays than a small one, and so must receive more energy from the Sun than a small one.

The logic of the calculation

⇐ **Looking backward**
We studied the inverse square law in Chapter 4.

The inverse square law is

$$\text{flux} = L_{\text{Sun}} / 4\pi D^2,$$

where L_{Sun} is the Sun's luminosity and D its distance from Jupiter. This gives us the flux of light Jupiter receives from the Sun – the rate at which energy falls on each unit area. So the steps in our calculation are as follows.

Step 1. Find the luminosity of the Sun and Jupiter's distance from the Sun.
Step 2. Plug them into the inverse square law to find the flux of energy reaching Jupiter from the Sun.

What we want to know is the *total* energy reaching Jupiter from the Sun. To find this:

Step 3. Multiply the result of step 2 by Jupiter's area.

As you can see from the detailed calculation, Jupiter receives 8×10^{17} watts from the Sun.

NOW YOU DO IT
What is the flux of sunlight reaching Neptune? What is the total number of watts reaching Neptune? (You can look up Neptune's radius and distance from the Sun in Appendix III.)

Detailed calculation

Step 1. Find the luminosity of the Sun and Jupiter's distance from the Sun.

Jupiter is 5.2 astronomical units from the Sun:

$D = 5.2$ AU $= (5.2$ AU$)$ $(1.496 \times 10^{11}$ meters/AU$) = 7.78 \times 10^{11}$ meters and the Sun's rate of emission of energy – its luminosity – is

$L_{Sun} = 3.8 \times 10^{26}$ watts.

Step 2. Plug them into the inverse square law to find the flux of energy reaching Jupiter from the Sun.

The inverse square law is

flux $= L_{Sun}/4 \pi D^2$,

where D is Jupiter's distance from the Sun, and L_{Sun} is the Sun's rate of emission of energy – its luminosity. So we calculate:

$$\text{flux} = \frac{(3.8 \times 10^{26} \text{ watts})}{4\pi(7.78 \times 10^{11} \text{ meters})^2}$$
$$= 50 \text{ watts/meter}^2.$$

Notice that this result conforms to *our first intuitive math* result. Had Jupiter been closer to the Sun, we would have been dividing by a smaller number and gotten a bigger answer for the flux.

Step 3. Multiply the result of step 2 by Jupiter's area.

Our first thought might be that we should multiply the result of step 2 by Jupiter's surface area. But this cannot be correct, since for one half of Jupiter it is night and there is no sunlight there. The easiest way to think of this is to think about the *shadow* cast by Jupiter (Figure 8.4). The energy intercepted by Jupiter is just the energy that would have reached the shadow, had Jupiter not been there! Since the shadow is a disk whose radius is that of Jupiter, the energy intercepted by Jupiter is given by the flux times the shadow's area. This area is πR^2, where R is Jupiter's radius.

Since Jupiter's radius is

$R = 71\,400$ kilometers $= 7.14 \times 10^7$ meters

we can find the total amount of energy from the Sun reaching Jupiter:

rate of energy reaching Jupiter $=$ (flux) (πR^2)
$= (50 \text{ watts/meter}^2)$ (π) $(7.14 \times 10^7 \text{ meters})^2$
$= 8 \times 10^{17}$ watts.

This is what we set out to calculate in (A).

Notice that this result conforms to *our second intuitive math* result. Had Jupiter been bigger, we would have been multiplying by a bigger number and gotten a bigger answer for the flux.

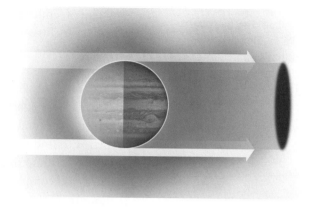

Figure 8.4 **The energy intercepted by Jupiter is the energy that would have reached the shadowed region had Jupiter not been there.**

(B) Now let us do the second part of our calculation, and find the energy leaving Jupiter. The situation is a little complicated, since Jupiter both emits its own light and reflects sunlight. To make things simple we will concentrate on the first process, and calculate the energy carried by its emitted infrared radiation.

Intuitive mathematics

As we discussed in Chapter 4, this energy is *blackbody radiation*. So also is the heat energy emitted by the burner of an electric stove. As we know from experience,

- a hot burner emits more energy than a warm one,
- a big burner emits more energy than a small one.

So we expect our calculation of the energy emitted by a planet to be greater for a warm planet than a cool one, and greater for a big planet than a small one.

The logic of the calculation

Recall the Stefan–Boltzmann law from our study of blackbody radiation in Chapter 4:

$$L = 0.57 \times 10^{-7} A T^4 \text{ watts.}$$

Here L is the rate of emission of energy from a blackbody, A is the blackbody's surface area and T is its temperature. So the steps in our calculation are as follows.

Step 1. Find the radius and temperature of Jupiter.
Step 2. Plug them into the Stefan–Boltzmann law to calculate the rate at which Jupiter emits energy.

Detailed calculation

Step 1. Find the area and temperature of Jupiter.

Jupiter's measured temperature is 145 kelvin. To find Jupiter's surface area, we use the formula $A = 4\pi R^2$, where R is Jupiter's radius. As we saw above, Jupiter's radius is $R = 71\,400$ kilometers $= 7.14 \times 10^7$ meters. We calculate

$$\text{Jupiter's surface area} = 4\pi R^2 = 4\pi(7.14 \times 10^7 \text{ meters})^2$$
$$= 6.41 \times 10^{16} \text{ meters}^2.$$

Step 2. Plug them into the Stefan–Boltzmann law to calculate the rate at which Jupiter emits energy.

We know from the Stefan–Boltzmann law (recall our study of blackbody radiation in Chapter 4) that the rate of emission of energy from a blackbody – its luminosity – is given by:

$$\text{rate energy leaves Jupiter} = L_{\text{Jupiter}} = 0.57 \times 10^{-7}\, A\, T^4 \text{ watts}$$

$$L_{\text{Jupiter}} = (0.57 \times 10^{-7})\,(6.41 \times 10^{16})\,(145)^4 \text{ watts}$$
$$L_{\text{Jupiter}} = 1.6 \times 10^{18} \text{ watts}.$$

This is what we set out to calculate in (B).

Notice that our result conforms to our *intuitive math* results. Had Jupiter been hotter we would have been multiplying by a bigger number and gotten a bigger answer. Similarly, had it been bigger we would have been multiplying by a bigger number and gotten a bigger answer.

We have now completed our two calculations. They have been a little complicated. Let us pause and take a breath, and then look at our two answers. We found for the energy reaching Jupiter from the Sun:

$$\text{energy reaching Jupiter} = 8 \times 10^{17} \text{ watts,}$$

and for the energy leaving Jupiter via its blackbody radiation:

$$\text{energy leaving Jupiter} = 1.6 \times 10^{18} \text{ watts.}$$

Of course even more energy is leaving Jupiter than this, since it is not just emitting blackbody radiation: it is also reflecting sunlight. So our conclusion is clear. *More energy is leaving Jupiter than it gets from the Sun.* Indeed, if we compare the two we find that Jupiter is emitting energy at twice the rate it is receiving.

....

NOW YOU DO IT

It is very difficult to measure the surface temperature of Neptune, but a rough value is 50 kelvin. How many watts is Neptune emitting? Do you conclude that it is generating energy in its interior?

....

What could be the source of Jupiter's "extra" energy? One possibility is that it is generating energy in its interior by some sort of chemical reactions, much as the Sun generates energy by nuclear reactions (we will study this process in Chapter 14). A second possibility is that Jupiter was formed as an extended body and has been steadily contracting ever since, generating energy by this contraction. Astronomers at present feel that the second theory is the more likely of the two.

What is the difference between a planet and a star?

At first glance this sounds like a silly question. After all, the Earth and the Sun are utterly different from one another. But our study of Jupiter has shown that the actual situation is more complicated than we had initially thought. Consider the following.

(1) One might think that planets are small and stars are big. But Jupiter's size lies midway between that of the Earth and the Sun, as shown in Figure 8.5.

(2) One might think that planets are solid and stars are not. But Jupiter is almost entirely a gas surrounding a liquid.

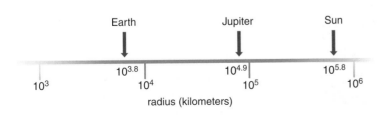

Figure 8.5 **Jupiter's radius is midway between the Earth's and the Sun's.**

(3) One might think that stars are composed of different chemical elements from planets. But Jupiter is more similar to the Sun than it is to the Earth in terms of composition. Indeed, as we have noted above, Jupiter's composition is almost identical to that of the Sun.

(4) One might think that stars emit light and planets do not. But Jupiter emits its own (infrared) light.

All this shows that the distinction between planet and star is not so clear as we had originally thought. The point is not that there is no difference between the two: rather the point is that there is no well-defined demarcation separating them. It is a little like thinking about the difference between a pond and an ocean. Oceans are different from ponds, but nature gives us no hard and fast line separating the two. In thinking about stars and planets we must think in terms of a range of properties. At one end of this range it is clear that we are talking about planets, and at the other end it is clear that we are talking about stars. But in between there is a gray area in which the situation is not so clear.

One way to resolve this ambiguity is to adopt a definition, with the understanding that every definition involves a certain degree of arbitrariness. For instance, we might define a pond to be a body of water containing no more than so-and-so many gallons. Similarly, astronomers customarily define a star to be something that generates its own energy *by nuclear reactions*. According to this definition, a star must have a mass greater than 0.08 times the mass of the Sun, which is 80 times the mass of Jupiter. According to this definition, Jupiter is a planet – or at any rate, it is not a star.

NOW YOU DO IT
Where would you place the dividing line between a pond and an ocean – and how do you justify your answer?

Figure 8.6 **Io.** 👁 **(Also see color plate section.)**

Jupiter's Galilean satellites

The four largest satellites of Jupiter, discovered by Galileo, are graced with the lovely names Io, Europa, Ganymede and Callisto. They are so interesting that we will discuss them in a section of their own.

Before the Voyager spacecraft reached Jupiter we knew almost nothing about these worlds. So distant is Jupiter and its family of satellites that even our best ground-based telescopes could make out little of their surfaces. The images Voyager returned were therefore the first decent look anyone had ever had of these satellites.

Figure 8.7 **Io's volcanoes.**

Io

Io, Jupiter's innermost moon, is about the size of our own Moon. It lies 422 000 kilometers from Jupiter, and moves in an orbit of eccentricity 0.004.

A representative view of Io is shown in Figure 8.6. When astronomers first saw it they found it hard to believe their eyes. Io looked like nothing they had ever seen. What could account for its amazing features? To resolve the mystery consider Figure 8.7. The plume shown in that image is a volcanic eruption.

Astronomers realized that the entire surface of Io consists of ejecta from an extensive system of volcanoes. Io is the most volcanically active body in the Solar System. When Voyager visited it, eight separate volcanoes were violently spewing out sulfur compounds. We estimate that these volcanoes eject something like one ton of material per second. The entire surface of Io is continually being buried by this ejecta, obliterating craters as fast as they form. So explosive are these eruptions that some of their ejecta is actually propelled entirely away from Io and into space.

What powers this extraordinary activity? Volcanoes, we know, are driven by underground heat. Why is Io's interior so hot?

Detectives on the case

Why is Io so hot?

Let us continue exploring how scientists create theories, and consider the mystery of Io's spectacular volcanism. We want to think up our own explanation for it, and – this is the most important of all – we want to pay attention to how the observational facts help us in our task.

The question is "why is the interior of Io so hot?" In searching for an answer, the first step is to assemble what we know about Io.

- It is the closest satellite to Jupiter.
- Its orbit is not exactly circular: the eccentricity of its orbit is small but not zero.
- Jupiter is the biggest planet.
- Jupiter emits more energy than it receives from the Sun.

Somewhere in this list of facts is a clue that will lead us to the solution of our mystery.

There is no hard and fast rule for inventing a new theory. A scientist simply proceeds by trial and error. So let us do the same. Let us think up a possible theory, and then test it to see if it fits the facts.

Suppose we begin with the fact that Jupiter emits its own energy. Does this lead us to a possible theory? Perhaps we might think of Jupiter as being a little like the Sun in this regard. If so, we should then think of Io as being like a planet warmed by this "sun." Which planet? Well, since Io is the closest satellite to Jupiter, we should think of Io as being like Mercury, the planet closest to the Sun. Mercury, of course, is very hot because it is so close to this heat source. So let us propose *our first theory: Io's great heat comes about because it is warmed by Jupiter.*

As we emphasized, it is never enough to think up a theory. A scientist must test the theory, by comparing its predictions with observation. Here's one such prediction: since our theory is proposing that Io is warmed by Jupiter, most of its volcanoes should be on the side facing the planet (since Io is in synchronous rotation, the same side always faces Jupiter). Are they? Observations shows that they are not: volcanoes are found all over Io.

Our proposed theory has failed one test. Can we think of yet other tests? A second one would be to calculate how much energy Io receives from Jupiter, to see if it is enough to heat Io significantly. Let us do this calculation.

Detailed calculation

Step 1. Find D and $L_{Jupiter}$.

Io's distance from Jupiter is $D = 4.2 \times 10^8$ meters. We found the rate of emission of energy from Jupiter, its luminosity, to be $L_{Jupiter} = 1.6 \times 10^{18}$ watts.

Step 2. Plug them into the inverse square law.

$$\begin{aligned} \text{Flux at Io from Jupiter} &= L_{Jupiter}/4\pi D^2 \\ &= (1.6 \times 10^{18} \text{ watts})/4\pi(4.2 \times 10^8 \text{ meters})^2 \\ &= 0.72 \text{ watts/meter}^2. \end{aligned}$$

Step 3. Decide whether the result we get from step 2 is enough to heat Io significantly

Is this a lot or a little? Is it enough to heat Io to great temperatures? In order to decide, we need to compare it to some flux which we know is *not* very great. For instance, we could compare it to the flux you receive from a fireplace: this is certainly a noticeable flux, but it does not heat you up to super-high temperatures. Alternatively, we could compare it to the flux of energy we on the Earth get from the Sun, which is enough to heat the Earth to a comfortable temperature, but no higher – certainly not enough to lead to volcanoes over our planet's entire surface!

Let us do the second, and compare Io's flux from Jupiter to the flux we on Earth get from the Sun. To find this, we just repeat the above calculation, using $D = 1$ AU, and the luminosity of the Sun for L. (Alternatively, we could simply step outdoors and measure this flux.) When we do so we get

flux at Earth from Sun = 1350 watts/meter ^{2}.

Notice that this is far more than the flux arriving at Io. Our conclusion is that *the flux arriving at Io from Jupiter is not enough to heat it greatly.* It is only enough to heat it moderately.

Intuitive mathematics

This calculation is just like the one we did before, when we found the energy that Jupiter receives from the Sun: now we are finding how much energy Io receives from Jupiter. As before, the inverse square law tells us that the flux of energy at Io from Jupiter is given by

$$\text{flux} = L_{Jupiter}/4\pi D^2,$$

where $L_{Jupiter}$ is the rate of emission of energy from Jupiter – its luminosity – and D is Io's distance from Jupiter. We need to do the following.

Step 1. Find D and $L_{Jupiter}$.
Step 2. Plug them into the inverse square law.
Step 3. Decide whether the result we get from step 2 is enough to heat Io significantly.

As illustrated by the detailed calculation, our proposed theory is not doing very well. The flux of energy Io gets from Jupiter is not very great at all. Perhaps we should abandon our first theory, and seek another. What should we choose for our second theory?

NOW YOU DO IT
Suppose our calculation had shown that the flux reaching Io from Jupiter was 1350 watts per meter^2. Would you have said that this was a possible theory of Io's intense volcanism?

A clue is provided by the facts that Io is Jupiter's closest satellite, and Jupiter is the most massive planet. This tells us that Io is experiencing a particularly large force from Jupiter. To see why, let us remind ourselves of Newton's law of gravitation:

$$F = GMm/D^2,$$

where M is the planet's mass, m is the satellite's mass and D the satellite's distance from its planet. This formula is telling us that the force of gravitation from a planet on its moon grows stronger with (a) the mass of the planet, and (b) the proximity of the moon.

Is this helping us answer our question? Why should a particularly large gravitational force make Io hot? There is no reason at all. Nobody has been able to think of a reason why this should explain Io's volcanoes.

But let us not abandon the idea of gravitation so quickly. Our study of tides has shown us that *differences* in gravitational forces on a satellite's near and far sides can also be important. Because Io is experiencing large gravitational forces, these differences (the "tidal forces") are also large. And as a consequence, the tidal bulge on Io is particularly large.

> ⇐ **Looking backward**
> We studied tidal forces in Chapter 7.

Is *this* helping us answer our question? Why should a particularly large tidal bulge make Io hot? Here, too, nobody can see why it should. So now let us integrate into our thinking yet another fact about Io: it is not in a perfectly circular orbit. Io's orbit is elliptical. As it moves about Jupiter, Io alternately approaches and recedes from it. And as it does so, its tidal bulge grows larger and smaller. *Io's tidal bulge is constantly changing.*

Would this make Io hot? Let us do an experiment and see. Take something and distort its shape – and then keep on changing the distortion. One way to do this is to take an elastic ball – a tennis ball, say – and rapidly flex it in your hand. If you do so *you will find the ball growing warm*. Alternatively, you might try rapidly stretching and loosening a rubber band: you will find the band grow warm. Continually changing the shape of an elastic body releases heat within it.

We have at last succeeded in finding a second theory of Io's great heat: it arises from the continual variation in its tidal bulge as it orbits Jupiter.

As before, we must now test our theory. Where on Io does it predict volcanoes should lie? Because flexing liberates heat throughout the entire body of Io, our theory predicts that volcanoes should be found everywhere upon its surface. As we have seen they are, so the theory has passed the first test.

The second test is to calculate how much heat is released in Io by flexing. This calculation is complicated, so we will not do it here. But it turns out that the theory passes this test as well: enough heat can be generated within Io to account for its numerous volcanoes.

Europa

The outer three Galilean satellites all possess rocky cores, but their surfaces are entirely covered by ice.

A general view of Europa is shown in Figure 8.8(a). Europa has the flattest terrain of any body in the Solar System, and it is quite bright. A close-up view of Europa's surface (Figure 8.8b) reveals vast ice rafts imbedded in a frozen sea. It looks much like our own Arctic Ocean.

Europa has only a few craters, indicating that its surface is relatively young in geological terms. Whereas Io's surface is perpetually renewed by volcanic eruptions,

(a)

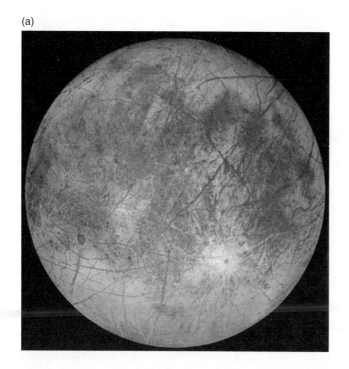

(b)

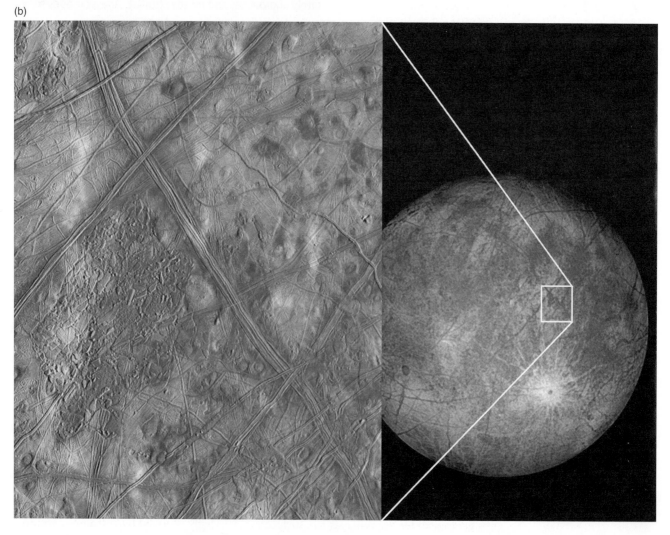

Figure 8.8 **Europa. (a) A distant view. (b) A close-up image.**

(a)

(b)

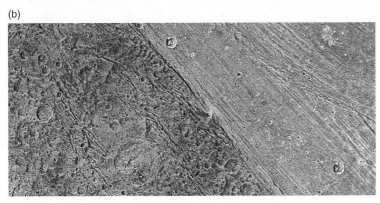

Figure 8.9 **Ganymede. (a) A distant view. (b) A close-up image.**

that of Europa is renewed by liquid water flowing outward from its interior. Europa's interior seems to be warm enough to melt ice. Indeed, the satellite is thought to possess an ocean, several hundred kilometers deep, lying just below its frozen surface. The dark veins covering the surface are probably fissures through which this water has upwelled: the fact that they are dark indicates that this water is somewhat "dirty."

Conceivably, life might exist within these underground oceans. Although the technical problems in doing so are enormous, NASA has considered the possibility of sending a space probe to the surface, and of attempting to drill through the ice floes to reach the ocean beneath.

Ganymede

Ganymede is the third satellite out from Jupiter (Figure 8.9a). It is the largest moon in the Solar System, bigger than Mercury and Pluto. Its surface consists of light and dark areas, the light areas marked by an amazing pattern of winding parallel grooves (Figure 8.9b). The origin of these grooves is not well understood.

Ganymede is the first of Jupiter's moons we have encountered to possess craters, indicating that its surface is older than Io's and Europa's. Notice in Figure 8.9 that these craters are white! This indicates that the icy surface of Ganymede has been darkened, and that the impacting meteoroids have punched through the dark layer to reveal the pristine white ice beneath.

What has darkened Ganymede's surface? The responsible agent is probably the same incessant hail of micrometeoroids that strikes the Earth. On Earth these tiny bits of dust and gravel burn up in the upper atmosphere, producing "shooting stars." Since Ganymede has no atmosphere, they survive to reach its surface, thereby darkening it.

Callisto

The farthest of the Galilean satellites from Jupiter, Callisto is shown in Figure 8.10. Its surface is quite dark, indicating that it has been exposed for a relatively long time to the "dirtying" effects of impacting micrometeoroids.

Callisto's surface is entirely covered by craters. Shown in Figure 8.10(b) is Valhalla, an enormous set of concentric rings, produced by some immense impact in the distant past. The fact that no crater is found at the rings' center indicates that the

(a)

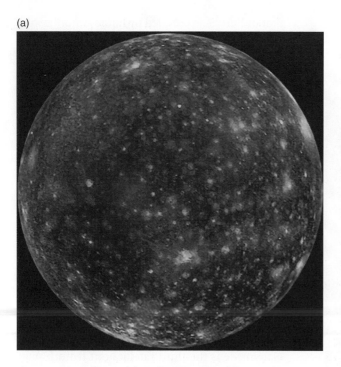

(b)

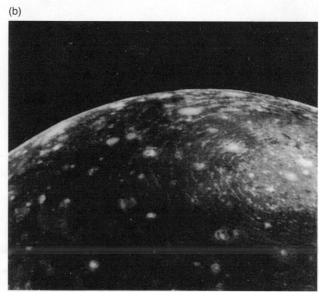

Figure 8.10 **Callisto. (a) A distant view. (b) A close-up image.**

impacting object penetrated deep into Callisto's interior. Callisto appears to be an ancient world, and its surface has been but slightly altered by geological processes.

Detectives on the case

Craters on the moons of Jupiter

As we look over the moons of Jupiter, a remarkable pattern emerges: *the farther a moon is from Jupiter, the more craters it is found to contain*. The innermost moon has none, the outermost moon has the most, and there is a smooth progression in between.

Why should this be so? Here are two possible theories.

(1) There are lots of impacting meteoroids far from Jupiter, and only a few close to it.
(2) Geological processes have been more effective in erasing craters on the inner satellites than on the outer ones.

Let us test each theory. We will start with the first.

One way to test this theory is to point a telescope at Jupiter and count how many meteoroids there are far from the planet as opposed to close in. But this is impossibly difficult. Meteoroids are so small that we cannot detect them around Jupiter through even our biggest telescopes. Even a telescope mounted on a spacecraft orbiting Jupiter would not be capable of detecting any but the very closest meteoroids. We will have to think of another way to test this proposal.

Let us test it by using our scientific understanding. This is what we did when we tested our initial theory of Io's volcanoes: we did a theoretical calculation of how much heat Io receives from Jupiter. Here we can do a similar thing: we can ask if our proposal makes scientific sense.

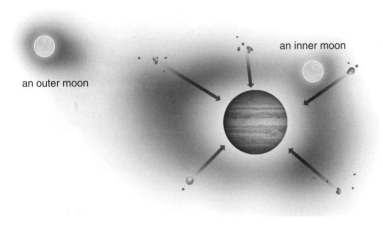

Figure 8.11 **Jupiter attracts meteoroids toward it. So an inner moon is exposed to more impacts than an outer one.**

Once we start thinking this way, we realize that it does not. Indeed, our first theory *conflicts* with what we know about gravity. Because Jupiter is so massive, its force of gravitation is large, and it attracts meteoroids to it (Figure 8.11). The meteoroids are "focused" toward the planet. This exposes an inner satellite to *more* impacts than an outer one – just the opposite of what the first theory proposes!

Let us therefore pass on to the second theory. How would we test it? Here's a method. If we are hypothesizing that craters on Jupiter's moons are being continually re-worked by geological processes, then the moons with few craters will be those with the most rapid re-working, while those with many craters will be those with the least rapid re-working. This leads to a *prediction of this theory*: that the surfaces of Jupiter's innermost moons are *younger* than those of the outermost moons.

Is there a way to see whether this prediction agrees with observation? There is. Recall our discussion of the darkening of a satellite's surface by impacting dust and gravel. The older the surface, the more this dark material will be covering up the underlying fresh white ice. So the second theory then predicts that outer satellites ought to be darker than inner ones. And they are! As we have noted, outer Ganymede and Callisto are far darker than inner Europa. Our second theory has passed this test.

We could go on and propose other tests. But as before, our main goal in this section is to explore the nature of science. Recall another comment we made concerning the nature of science: that a scientist never stops asking questions. Now we must ask *why* geological processes are more active on the inner moons. Scientists believe that the explanation has to do with the same process that we invoked to understand Io's many volcanoes: the liberation of heat by tidal squeezing. As we commented, tidal heating is strongest close to Jupiter, and it grows weaker as we progress away from it. So Io's interior is most strongly heated, and Callisto's interior is heated hardly at all. This heating leads to a continual re-working of the satellites' surfaces by upwelling water: the closer a satellite is to Jupiter, the more efficient is this re-working.

Saturn

Saturn (Figure 8.12), sixth planet from the Sun, is the most distant planet visible with the naked eye. Before the advent of telescopes, it marked the outer edge of the known Solar System.

Saturn is the second largest planet, with a diameter nearly ten times that of the Earth. It orbits 9.5 AU from the Sun with an orbital period of 29.5 years. Its period of rotation is 10 hours, 39 minutes and 22 seconds. It is marked with the same bands parallel to its equator as Jupiter, but they are less distinct.

Figure 8.12 **Saturn.** 👁 **(Also see color plate section.)**

Saturn's rings

By far the most remarkable features of Saturn are its rings. First seen by Galileo in 1610, his telescope was so small that they appeared to him to be faint blurred "handles" on the planet's two sides. Not until 1655 did the 26-year-old astronomer Christiaan Huygens realize the true nature of these "handles." Huygens chose an unusual way to report his discovery – an anagram:

> aaaaaaacccccdeeeee
> ghiiiiiiilllllmmnnnnnnnnnn
> ooooppqrrstttttuuuuu

Not for four years did Huygens reveal the secret of his anagram. In 1659, in a treatise on Saturn, he explained that it could be unscrambled to read (in Latin) "Saturn is girdled by a thin flat ring, nowhere touching it, and inclined to the ecliptic."

We now know that Saturn is girdled by an entire system of rings. The outermost, or "A" ring, has an outer radius of 136 500 kilometers. Within it is a "B" ring, within which is a gap known as the "Cassini division" after the astronomer who discovered it. Inside this is the "C" ring, with an inner radius of 73 000 kilometers. Because these enormous rings are only a number of meters thick, this makes them the flattest objects we have ever discovered.

Detectives on the case

What are Saturn's rings?

How can we understand these amazing and beautiful structures? An important clue is provided by the Doppler effect. Spectroscopic observations show that the light from them is shifted in wavelength. Light from one side of the rings is shifted to shorter wavelengths: light from the other side is shifted to longer wavelengths (Figure 8.13). Furthermore, the shifts are symmetrical, with the magnitude of one side's blueshift being the same as that of the opposite side's redshift.

These observations tell us that *the rings are circling about Saturn*. Furthermore, the velocity of this circular motion depends on position. Figure 8.14 shows the results of more detailed observations, in which Doppler shifts were obtained at a number of locations within the rings. As you can see, the inner portions are circling most rapidly and the outer portions least rapidly.

All this should be ringing a bell within your mind. It sounds exactly like the motion of planets about the Sun – as we know from our study of Kepler's laws, an inner planet orbits more rapidly than an outer one. We are led to consider the possibility that the rings are not "circling" about Saturn: they are orbiting about it! We are proposing that *the rings of Saturn are composed of innumerable small bodies, each circling about Saturn in its own*

Figure 8.13 **Observed Doppler shifts in Saturn's rings.**

Figure 8.14 **Rotational velocities in Saturn's rings as inferred from Doppler shifts.**

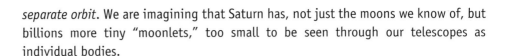

separate orbit. We are imagining that Saturn has, not just the moons we know of, but billions more tiny "moonlets," too small to be seen through our telescopes as individual bodies.

Test the proposed theory

How can we test this theory? We must check to see if its predictions conform with observations. These purported orbits must obey Newton's laws. We know from Chapter 3 that the velocity of an object in orbit is given by the circular orbit formula:

$$V = \sqrt{GM/R},$$

where M is the mass of the body about which it orbits, and R the radius of its orbit. As we emphasized in that chapter, this formula applies to *every* orbiting object: planets orbiting the Sun, artificial satellites orbiting the Earth – even microscopic "moonlets" orbiting Saturn. So let us apply it to Saturn's rings.

Doppler shift observations have been taken at selected distances from Saturn. Their results are shown in Table 8.1.

We need to use Newton's formula to predict the velocity of motion of a body orbiting about Saturn at these same distances. If these predictions match the observed velocities, we will have reason to accept our theory.

> ⇐ **Looking backward**
> We studied the orbit formula in Chapter 3.

Table 8.1. **Saturn's rings.**

Distance from Saturn (millions of meters)	Velocity measured using the Doppler effect (kilometers per second)
72.4	22.5
90.5	20.0
121	17.8

The logic of the calculation

Step 1. Select any one of the rows in this table as the data point to be checked.

Step 2. Insert the selected value for R into the circular orbit formula, using as M the measured mass of Saturn. Use the formula to calculate the *predicted* orbital velocity at this distance from Saturn.

Step 3. Compare this with the *observed* value given in Table 8.1.

As you can see from the detailed calculation, the agreement is excellent. Problem (6) at the end of this chapter asks you also to check the other two data points. They, too, turn out to conform with the observations. Figure 8.15 shows the theory compared with observation.

Detailed calculation

Step 1. Select any one of the rows in the table as the data point to be checked.

Let us check the first data point.

Step 2. Insert the selected value for R into the circular orbit formula, using as M the measured mass of Saturn. Use the formula to calculate the predicted orbital velocity at this distance from Saturn.

Saturn's mass is 5.69×10^{26} kilograms, and Newton's constant G is 6.67×10^{-11} meter³/second² kilogram. The first data point lies at $R = 72.4$ million meters, or 7.24×10^{6} meters. Using the circular orbit formula

$$V = \sqrt{GM/R},$$

we calculate the predicted velocity at this location in the rings:

$$V_{predicted} = \sqrt{(6.67 \times 10^{-11})(5.69 \times 10^{26})/7.24 \times 10^{6})} \text{ meters/second,}$$

$V_{predicted} = 2.25 \times 10^{4}$ meters/second = 22.5 kilometers/second.

Step 3. Compare this with the observed value given in the table.

The predicted value agrees with the observed value. We conclude that the first data point conforms to our theory.

Confirmation of the proposed theory

Every time we test a theory one of two things happens: either it fails or it passes. Up to now we have focused attention on how failing a test leads us to reject a hypothesis. Let us now pay attention to the process of *confirmation* when a theory passes.

Recall our analogy (Chapter 2) of testing a theory as being like shaking a ladder. The point is not to knock the ladder down: the point is to make sure that nothing else can knock it down. We want to make sure the ladder

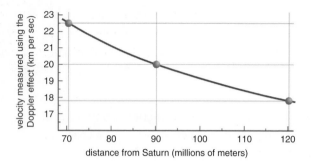

Figure 8.15 **Theory compared with observation. The dots are measured velocities at various points within the rings. The solid line is the prediction of the theory that the rings are composed of orbiting particles. Agreement is excellent.**

is secure. Similarly, we want to "shake" our theories in order to make sure that we can trust them.

The more ways we can shake a ladder – to the right, to the left, forward and backward – the more confident we can be of it. Similarly, we always want to confirm a theory in as many different ways as possible. Our theory of Saturn's rings has made three separate predictions – the predicted velocity at three separate locations. All three of them turned out to agree with observation. It would be good to predict the velocity at yet other locations, and make observations to test these predictions.

Best of all would be to find some completely different *kind* of prediction that we could test. Here is one: if the rings are composed of many tiny bodies, there must be spaces between these bodies. So it must be possible to peer between them, right through the spaces. And it turns out that we can. The rings are transparent: stars can sometimes be seen shining through them.

NOW YOU DO IT
Here is yet another validation of our theory. Observation reveals that the velocity of particles making up the rings on one side of the planet is equal and opposite to that on the other side. For example, whereas the rings that lie 72.4 million kilometers from Saturn on one side are *receding* from us at 22.5 kilometers per second, those that lie 72.4 million kilometers from Saturn on the other side are *approaching* us at 22.5 kilometers per second. Why do we say that this confirms our theory?

Spacecraft observations of Saturn's rings

Figure 8.16 is a close-up image of Saturn's rings. Extending from 7000 kilometers out to 80 000 kilometers from Saturn and with a thickness of only a meter, they form the flattest structure known. As you can see, the rings are actually thousands of tiny ringlets. Each ringlet, in turn, is composed of innumerable tiny bodies ranging from fractions of a centimeter to a few meters in size: they are almost entirely composed of water ice and grains of dust. Their total mass is similar to that of a moon of Saturn.

Origin of Saturn's rings

Saturn's rings lie inside Saturn's *Roche limit* – the critical distance within which a satellite is pulled apart by tides. To understand the Roche limit, return to our previous discussion of tides. We know that the closer a satellite is to a planet, the bigger is the bulge raised on that satellite by tidal forces. If the satellite gets too close to the planet, the bulge can be so great as to tear the satellite apart!

This leads us to formulate three hypotheses for the origin of Saturn's rings.

(1) Ages past, a satellite of Saturn might have wandered inside the Roche limit. If so, it would have been torn into innumerable tiny moonlets to form Saturn's rings.

Figure 8.16 **Spacecraft images of Saturn's rings. The upper image was obtained by the Cassini spacecraft in visible light, the lower is a simulation based on radio occultation studies.**

(2) In Chapter 13 we will show that planets and moons were formed by the amalgamation of innumerable tiny bodies. But this amalgamation cannot happen within the Roche limit. Perhaps the particles making up Saturn's rings are the particles that failed to coalesce into a moon.

(3) Possibly a large body struck one of Saturn's moons, disrupting it.

We are not sure at present which of these hypotheses is correct.

Planetary rings

Observations reveal that Saturn is not the only planet to possess rings. Indeed, rings have also been discovered around Jupiter, Uranus and Neptune. Saturn's rings are by far the most extensive, however.

Rings are often shaped by so-called "shepherd moons," tiny bodies found orbiting just outside a ring, or within a gap between two rings. Gravity from these moons helps to give the ring a sharp edge: any ring particle that wanders too close to the shepherd moon is either sent back into the ring, propelled away from the ring or lands on the moon itself.

Enceladus

Saturn's moon Enceladus is quite small – somewhat smaller than Great Britain. Nevertheless, it is quite remarkable owing to the presence of liquid water within it.

The spacecraft Cassini, which is currently orbiting Saturn, first passed close to Enceladus in 2005. Remarkably, Cassini's instruments detected signs of water molecules coming from the moon. Further observations made during the flyby revealed geysers of water – "volcanoes of water" – on its south pole. Three years later, Cassini's orbit carried it to within 50 kilometers of the moon's surface, and it actually flew right through the plume of a geyser. Onboard instruments detected molecules of H_2O, CO_2 and various hydrocarbons.

Enceladus orbits within Saturn's so-called E ring, and indeed it is situated within the ring's most dense region. This has led scientists to propose that the E ring in fact consists of particles from the moon, some propelled by its geysers and others by meteoric impacts upon it.

The surface of Enceladus consists mostly of clean ice, making it highly reflective. Cassini revealed that many regions of Enceladus possess very few craters, lending support to the idea that it is geologically active. Cassini has discovered numerous winding canyons, rifts and fractures within older, heavily cratered terrain, also suggesting that the moon is undergoing geologic activity.

Cassini's infrared spectrometer mapped the distribution of heat on the surface of Enceladus, and revealed a network of warm regions beneath its south pole. This has led to a model in which the geysers on Enceladus are fueled by underground

(a)

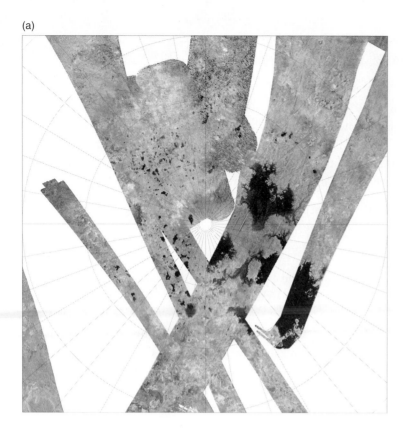

(b)

Figure 8.17 **Titan. (a) A radar image of its polar regions obtained by the Cassini space probe. (b) Titan's surface as imaged by the Huygens lander.**

pressurized pockets of liquid water. Indeed, these pockets might be as warm as minus 3 degrees Celsius.

For all these reasons, in 2011 NASA scientists at an Enceladus Focus Group Conference reported that Enceladus could be an even more hospitable abode than Mars for extraterrestrial life.

Titan and its atmosphere

Titan, Saturn's largest moon, is the only satellite known to possess a substantial atmosphere. In December 2004 the Huygens space probe detached itself from the Cassini spacecraft, which was in orbit about Saturn. After a 20-day coast, the probe entered Titan's atmosphere. Parachutes opened to slow its descent. As Huygens drifted downwards, instruments measured the temperature, pressure and chemical composition of the atmosphere. The data were radioed to the Cassini orbiter, which then transferred them to Earth.

Figure 8.17(a) is a mosaic composed of radar observations of Titan by Cassini. In some ways Titan resembles the Earth, in particular because of the clear evidence of flowing liquid on Titan. The image shows erosion, lakes and riverbeds cut by flowing liquid of some kind. Titan is too cold for this liquid to be water: possibly it is methane. The largest lake shown in this image is slightly larger than Lake Superior.

NOW YOU DO IT

If the largest lake in this image is slightly larger than Lake Superior, how large is the *second* largest lake in this image?

Two hours and 27 minutes after it entered Titan's atmosphere, the Huygens probe landed on a soft sandy riverbed. It survived the landing, and maintained communication with Cassini for several more minutes, sending the image of Titan's surface shown in Figure 8.16(b). This moon is now the most distant celestial body upon which a human-made object has landed.

Titan is particularly interesting because the chemistry of its atmosphere and surface involves organic compounds. Despite their name, these compounds are not solely produced by living organisms. Nevertheless, Titan's organic chemistry may be similar to that of the primitive Earth. Studies of Titan may shed light on our world's early evolution, and the chemical processes that preceded the origin of life here.

The discovery of Uranus, Neptune and Pluto

As we have noted, Saturn is the most distant planet visible to the naked eye. Prior to the invention of the telescope, it marked the outer boundary of the Solar System. No one ever "discovered" Mercury, Venus, Mars, Jupiter or Saturn: they are visible to the naked eye to anyone who looks carefully at the nighttime sky. But in terms of distance from the Sun, they constitute only the inner ¼ of the Solar System! The remaining planets were discovered, and the story of how this came about is a fascinating account of hard work, accident and error.

Uranus

This planet was discovered by accident on the night of March 13, 1781, by William Herschel (Figure 8.18), a then-unknown astronomer who earned his living as a musician. As we have noted in Chapter 5 the stars are so far away that, even through a telescope, they appear to be simply points of light. On that fateful night Herschel found a "star" that had a visible disk.

Initially Herschel thought he had discovered a comet, but it was eventually shown that his new object was a planet. It lay twice as far from the Sun as Saturn. The first new world in history had been discovered, the size of the Solar System had doubled. Herschel became instantly famous.

Neptune

As the years passed, it became clear that Uranus was not moving as it should. There were slight discrepancies between its orbit as observed and its orbit as predicted by Newton's laws. Sixty years after Herschel's discovery, the English astronomer John Couch Adams (Figure 8.19a) became interested in these discrepancies. He considered the possibility that they arose because Uranus was being gravitationally attracted by some unknown planet. He set about the task of using the observed discrepancies to predict where this new planet should lie.

In 1845 Adams, who was young and not well known, sent his results to the Astronomer Royal, George Biddell Airy. Unfortunately, Airy paid no attention to Adams' work. At the same time, however, and entirely unknown to Adams, a second astronomer was doing the very same calculation: the Frenchman Urbain Jean Joseph Le Verrier (Figure 8.19b). When Le Verrier's results reached Airy he compared them to those

Figure 8.18 **William Herschel.**

(a)

(b)

Figure 8.19 **(a) John Couch Adams. (b) Urbain Jean Joseph Le Verrier.**

of Adams, and found them to be essentially identical. Finally convinced that the matter had to be taken seriously, Airy contacted astronomers at Cambridge University, who began searching for the planet.

Meanwhile, Le Verrier too had found it difficult to persuade any astronomer with a large telescope to take his work seriously. Eventually he wrote to Johann Galle at the Berlin Observatory. Galle received Le Verrier's letter on September 23, 1846: that very night he found Neptune lying close to its predicted position.

Pluto

In the years following Neptune's discovery, astronomers decided that it did not exactly explain all the discrepancies in the orbit of Uranus. Even after accounting for Neptune's gravitational influence upon it, tiny deviations still remained. Astronomers became convinced that yet another unknown planet existed, lying beyond the orbit of Neptune.

Because Neptune orbits so slowly about the Sun, and because it is so very distant, it was impossible to measure its orbit with great accuracy. So it was impossible to use deviations in its orbit to search for this new world. Accordingly, astronomers attempted to use the deviations in Uranus' orbit to predict its location. One of the astronomers who did so was Percival Lowell, whose work on the "canals" of Mars we have recounted in the previous chapter. Lowell concluded that the new planet most probably lay in the constellation of Gemini, and he searched for it there from 1906 till his death a decade later.

Years later the young astronomer Clyde Tombaugh (Figure 8.20) was hired to continue Lowell's work. Tombaugh worked with astronomical photographs of regions of the sky thought likely to contain the new planet. It was exhausting work, for each such photograph contained hundreds of thousands of stars. Eventually, on the night of February 18, 1930, Tombaugh found an object that was moving against the backdrop of the fixed stars. He had discovered Pluto.

Sadly, in recent years we have realized that Tombaugh's triumph was something of an accident. Recent measurements have shown that Pluto's mass is too small for it significantly to influence the orbits of Uranus and Neptune. Furthermore, we now know that the supposed discrepancies between Uranus' predicted and observed orbits are probably not real, but are the result of measurement errors. A wrong analysis had led to a wrong prediction, which nevertheless happened by accident to guide Tombaugh to his great discovery.

THE NATURE OF SCIENCE

THE ROLE OF LUCK IN SCIENTIFIC DISCOVERY

What role does chance play in the progress of science? Can a great scientific discovery be made by sheer good fortune? Can an incompetent, lazy researcher make a profound contribution to knowledge? Should the Nobel Prize be awarded for luck?

The scientists who discovered Neptune and Pluto had set out to find their worlds. But the discovery of Uranus is different. Herschel had not been looking for any new planets. He made his discovery one night out in his backyard while looking through a telescope. So was his discovery nothing but luck?

To put this issue in perspective, imagine a basketball player who is truly incompetent. He drifts aimlessly about the court, pretty much getting in his teammates' way. If the ball ends up in his hands, he tosses it away more or less at random. But imagine that he is lucky and, purely by chance one day, he happens to throw the ball into the hoop. He makes the point. This is unlikely, but it is certainly possible.

A good player can make points over and over again: that is what distinguishes experts from the merely lucky. But imagine that our incompetent player is *very* lucky. Imagine that, purely by chance, he keeps on managing to get the ball through the hoop. This is not just unlikely – it is extremely unlikely. But it is not completely impossible. His team wins the game: had he not been on it, they would have lost.

In contrast, consider now the following description of Herschel's way of doing astronomy.

Herschel developed an avid interest in astronomy and this became his first love. He rented a small reflecting telescope, but this only whetted his appetite to own his own, larger telescope. Since he didn't have enough money to buy one, he contrived to build it with the help of [his brother] Alexander and [sister] Caroline... Because of his great zeal and methodical nature, William Herschel became one of the most notable observers in the history of astronomy. His systematic survey of the sky was one of the most important of his accomplishments.

Figure 8.20 **Clyde Tombaugh.**

On March 13, 1781, while scanning the skies with a 7-inch reflecting telescope, he observed an unusual object; it presented an extended disk-like shape. Herschel thought he had discovered a comet. He continued his observations, and calculations for months, discovering the orbit lay well beyond the orbit of Saturn and was fairly circular. Herschel's 'comet' was in fact a planet ....

The instruments Herschel had lacked clock drives to keep them trained on the moving sky, so the method he used was to direct his telescope to a point on the meridian and watch what crossed the field of view. Since Herschel had to stand on a ladder to do his observing, he would call out descriptions of whatever he saw of interest to his sister Caroline at the foot of the ladder. She would then record the information and time. By using this method he was able to observe objects in a thin east–west strip of sky. As the nights progressed, he would change the position of the telescope to an elevation higher or lower than the previous night. This enabled him to observe another strip of sky. They eventually were able to observe all the sky visible in Great Britain.[1]

Is our incompetent basketball player a good analogy to Herschel? He is not. There is no comparison between our player's aimless wanderings and the focused, dedicated work of Herschel. Herschel's credit for discovering the first new planet is deserved. Our basketball player, on the other hand, would not have deserved credit for helping his teammates win the game.

There is a saying: "Chance favors the prepared mind."

SERENDIPITY: DIFFERENT FROM LUCK.

Herschel had not known of his new planet's existence: therefore, he had not gone looking for it. Furthermore, the planet might not have existed. Or it might not have been visible from the location of his telescope. Or it might have been too small for his telescope to spot. Finally, someone else might have discovered it prior to Herschel. For all these reasons, his discovery was fortuitous.

Nevertheless, we must not speak of luck. We must speak of *serendipity*. Serendipity is not the same thing as luck: our basketball player made his points by luck, whereas Herschel made his discovery serendipitously.

---

[1] Peggy Taylor & Sara Saey at http://www.astroleague.org/al/obsclubs/herschel/fwhershs.html

Many discoveries are made through serendipity. Merely in this one chapter there are three.

- The first spacecraft to visit Saturn discovered that its rings are braided. We had not expected this. As recounted above, we had known they were composed of many little particles, but this is not the same thing as saying the particles are arranged in braids.
- It was a surprise to learn that Jupiter radiates more energy than it receives from the Sun. Nothing we had known about this planet predicted this strange fact.
- We will shortly find a third example of serendipity when we describe the discovery of Pluto's moon.

Indeed, discoveries through serendipity are common in science. Every time a new telescope is built, one capable of seeing farther than earlier ones, that telescope discovers unexpected things. Every time we send a space probe to a previously unvisited world, that probe tells us things we never knew before.

There is another saying: "Expect the unexpected." It is the motto of every practising scientist. It is why we go exploring.

....

NOW YOU DO IT
Can you think of an experience of your own that illustrates the distinction between luck and serendipity?

....

Uranus

Uranus (Figure 8.21) lies 19 AU from the Sun, taking 84 years to complete one orbit. Four times the diameter of the Earth, it has 15 times its mass. Uranus' day is 17.24 hours long.

....

NOW YOU DO IT
How many "Uranus years" have passed (that is, how many times has Uranus circled about the Sun) since we discovered it?

....

Seasons on Uranus

Uranus is the only planet of the Solar System whose rotation axis lies in the plane of its orbit (Figure 8.22) (the precise angle is 98 degrees). This makes for remarkable seasons.

Consider first the configuration (labeled "A" in Figure 8.22) in which the planet's spin axis points at the Sun. An observer on the north pole would see the Sun directly overhead, never

Figure 8.21 **Uranus as imaged by the Voyager spacecraft.**

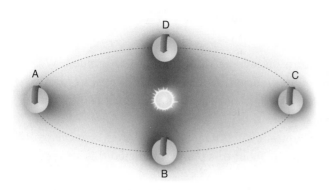

Figure 8.22 **Uranus' rotation axis lies in the plane of its orbit.**

moving as Uranus' day passes. An observer on the planet's equator would see the Sun skimming along the horizon, making one complete circuit every Uranus day. An observer located anywhere in the planet's southern hemisphere would *never* see the Sun, and would be in perpetual night.

Consider next the configuration ¼ of a Uranus year, or 21 Earth years, later. By this time the planet has moved to position "B" in Figure 8.22. Now an observer on the north pole would see what the equatorial observer had seen before: the Sun skimming along the horizon, completing a circuit every Uranus day. An observer on the equator, on the other hand, would see the Sun rise, pass directly overhead, and then set. At this point, the observer on the south pole is just beginning to see the Sun, which lies on the horizon and skims about it in the opposite direction from that as seen from the north pole. Similarly, as Uranus proceeds to points "C" and "D" on its orbit, its seasons are just the reverse of those at "A" and "B."

Astronomers are not sure why Uranus' rotation axis is tilted so strongly relative to its orbit plane. Remarkably, the magnetic axis on Uranus is also unusual, in that it is tilted by a full 55 degrees from its spin axis. (For comparison, the Earth's magnetic axis tilts only 12 degrees from its spin axis.) Thus compass needles on Uranus would point nowhere near north. This may be a clue to the origin of Uranus' strange rotation. We will see in Chapter 13 that planets were formed by the catastrophic collision of large bodies: perhaps one of these collisions was at a glancing angle, so twisting its rotation axis about but failing to twist the magnetic axis.

Uranus' atmosphere

The atmosphere of Uranus consists primarily of hydrogen and helium, as are those of Jupiter and Saturn, with significant admixtures of methane, ammonia and water vapor. So distant is the planet that prior to the space age we had no idea what it looked like. But in 1986 Voyager 2 flew by Uranus, and returned numerous images of it and its moons. The image of Uranus returned by Voyager is of a featureless blue/green ball, with none of the bands characteristic of Jupiter and Saturn. Deep within its interior the atmospheric gases are crushed by the enormous pressure to form a liquid interior, while at the very center lies a rocky core roughly the size of the Earth.

Rings of Uranus

In Chapter 7 we noted that from time to time the Moon, in its orbit about the Earth, passes in front of a distant star – an occultation. Planets too can occult stars. Such events are very rare, since planets are no more than tiny points of light as they appear in the sky. Nevertheless, they do occur from time to time, and when they do they are always interesting to astronomers, for they provide unique opportunities to study the planets' atmospheres. Accordingly, astronomers were particularly interested in a predicted occultation of a star by Uranus on March 10, 1977.

Because the precise time of the event was somewhat uncertain, one team of astronomers began taking data 45 minutes before the predicted moment of occultation. They measured the brightness of the distant star as Uranus passed in front

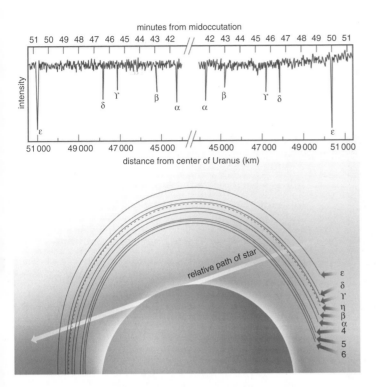

Figure 8.23 Occultation of a distant star by Uranus showed that it has rings. Rings 4, 5, and 6 were discovered subsequently.

Figure 8.24 **Voyager images of the rings of Uranus.**

of it. The data they obtained are shown in Figure 8.23. Notice in that figure that, long before the actual occultation, the star suddenly and briefly dimmed – and not once, but five times. Then, after the occultation, the star repeated this pattern of sudden brief dimmings.

What could cause such a strange phenomenon? Something must have briefly obscured the star. Because the dimmings after the occultation just mirrored those before it, we know that the responsible agents are distributed on both sides of the planet. Because the dimmings are brief, we know the agents are thin. And because the star did not dim to invisibility, we know the agents are partially transparent. The occultations had revealed the presence of rings about Uranus.

When the Voyager spacecraft reached Uranus, it confirmed the presence of the rings and discovered many more (Figure 8.24). Unlike Saturn's rings, those of Uranus are exceedingly thin. Also unlike Saturn, Uranus' rings are composed of exceedingly dark particles.

Moons of Uranus

As before, we can get a clue as to the composition of Uranus' moons by calculating their average densities. These densities are slightly greater than that of water, suggesting that they consist of rocky cores surrounded by thick layers of ice.

All of Uranus' moons are quite small. Titania, the largest, is 790 kilometers in radius. Contrast this with our own Moon's 1738-kilometer radius. As a consequence, astronomers thought that these moons would be more like stones than geologically active worlds. They were surprised when Voyager returned images of their surfaces, showing many signs of geological processes.

A network of canyons winds across the surface of Ariel (Figure 8.25). Rather than being V-shaped, these canyons are remarkable in having broad flat floors, with sharply rising sidewalls.

Ariel has very few craters, suggesting that its surface has been renewed. We speculate that this may have been caused by ice welling upward along the canyons, much as on Earth the mid-ocean ridges mark locations where interior magma flows out from the interior.

Miranda (Figure 8.26) is the most striking of all Uranus' moons. Containing many craters, its surface is like none other in the Solar System. Certain regions show mountains, enormous

Figure 8.25 **Ariel.**

cliffs many kilometers high, and broad undulating terrain. Other, oval-shaped regions are marked by an astonishing pattern of parallel grooves. The surface of Miranda appears to have been re-worked by volcanoes of ice.

Neptune

Neptune lies 30.1 AU from the Sun, requiring a full 165 Earth years to complete one orbit. Not until the year 2011 did a single "Neptune year" pass since its discovery.

Neptune's day is 16 hours and 3 minutes long. It has been called Uranus' twin, being of similar mass and size. Like Uranus, Neptune has a set of thin rings. Both planets have a rocky core larger than the Earth, an inner mantle consisting of liquid water, methane and ammonia and a thick gaseous atmosphere.

In 1989 Voyager 2 swept past Neptune and gave us our first good look at it (Figure 8.27). The planet is a beautiful blue in color. While we may think this is appropriate – Neptune is the god of the ocean – unfortunately the color does not arise from water. Neptune's visible surface is the top of its atmosphere, and its color arises from methane mixed with its hydrogen-rich gases.

Visible in Figure 8.27 is Neptune's Great Dark Spot, named in analogy to Jupiter's Red Spot. The Dark Spot is about the size of the Earth, and appears to be the site of upwelling currents from the planet's deep interior. Associated with the Spot are clouds of a brilliant white, composed of frozen methane crystals. Several other dark spots and white clouds are also visible on the planet's surface.

The Voyager spacecraft gave us an unprecedented close-up look at Neptune, but it was a one-shot affair: after passing Neptune the spacecraft continued on its journey outward away from the Solar System. But a year after the Voyager encounter the Hubble Space Telescope was launched, and several years later it commenced a regular program of observations of Neptune. The images Hubble returns to us are not as detailed as those of Voyager, but Hubble can "return" to Neptune as often as we wish. As a consequence of Hubble's observations, we now know that the features on Neptune's surface come and go. The Great Dark Spot, for instance, no longer exists. Neptune is the site of dynamically evolving weather patterns.

The mystery of Neptune's moons

Prior to the Voyager encounter we knew of two moons of Neptune. Voyager added six more. All the moons are black as soot, and most are very small and irregular in shape, like lumps of coal. Two of Neptune's moons, Triton and Nereid, have very unusual orbits (Figure 8.28).

Triton, the largest of Neptune's moons, is only slightly smaller than our own Moon. Its orbit is

- retrograde – in the opposite sense to the rotation of Neptune,
- quite circular,
- not in Neptune's equatorial plane.

Why is this unusual? After all, many satellites are known whose orbits are retrograde, and/or do not lie in their planet's equatorial plane. But until now every such retrograde satellite we have encountered has been

- low mass,
- in an elliptical orbit.

Triton does not fit this pattern.

Satellites in the Solar System fall into two groups. On the one hand there is the "regular" group, consisting of higher-mass satellites in forward-moving circular orbits lying in their planets' equatorial planes. On the other hand is the "irregular" group, consisting of lower-mass satellites in elliptical, retrograde and highly inclined orbits. We think that moons in the "regular" group were formed along with their planets, while those in the "irregular" group are bodies that wandered too close and were captured.

The problem with Triton is that it does not fall into either group. If it had been formed with Neptune it would not be in a retrograde orbit so highly inclined relative to the planet's equator. If it had been captured it would not be in a circular orbit.

Nereid is the other satellite of Neptune with an unusual orbit. It lies quite far from Neptune, traveling with a period of nearly an Earth year. And this orbit is by far the most eccentric of any moon in the Solar System.

There is no agreement among astronomers about how to understand these strange orbits. One possible clue is that the orbit of Pluto is highly elliptical, and actually intersects that of Neptune. Therefore, once in a while Pluto and Neptune pass very close to one another. Since Pluto itself has a very low mass, it was once speculated that Pluto used to be a satellite of Neptune. Possibly some process ejected Pluto

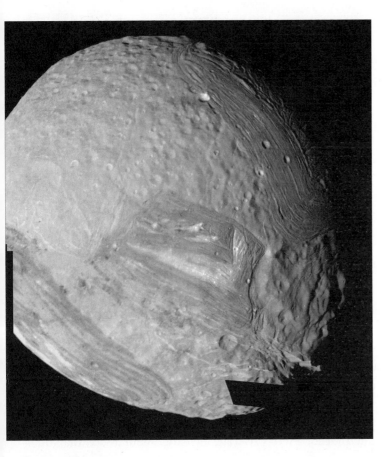

Figure 8.26 **Miranda.**

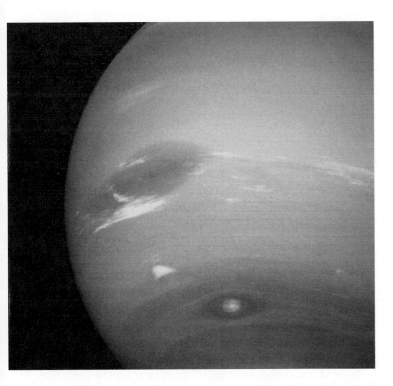

Figure 8.27 **Neptune.** 👁 **(Also see color plate section.)**

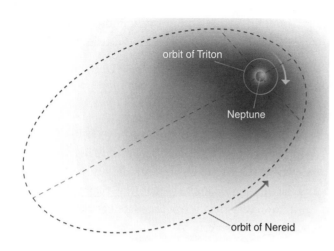

Figure 8.28 **Orbits of Triton and Nereid.**

and simultaneously altered the orbits of the other moons. A second possibility involves the idea that both Pluto and the moons of Neptune are remnants of a new class of objects known as plutoids or Kuiper Belt Objects. We will turn to this possibility in the next chapter.

Pluto

Pluto, once considered the most distant planet in the Solar System, also has the most elliptical orbit, with an eccentricity of 0.25. Its mean distance from the Sun – its semimajor axis – is 39.5 AU, but its actual distance from the Sun ranges from 30 AU to 49 AU. Similarly, it is the only "planet" whose orbit lies far out of the ecliptic (its angle of inclination is 17 degrees). So eccentric is Pluto's orbit that it crosses that of Neptune: indeed, from 1979 to 1999 Pluto lay closer to the Sun than Neptune. One Pluto year lasts 248 Earth years.

Pluto is smaller than our Moon, with a radius of 1150 kilometers. Its mass is a mere 0.002 Earth masses, and it rotates on its axis every 6.4 days.

Not surprisingly Pluto is the coldest "planet" in the Solar System, with a surface temperature of 233 degrees Celsius below zero, or 387 degrees Fahrenheit below zero. Unlike the other outer planets – and like the inner planets – the surface of Pluto is solid. Indeed, Pluto's average density, about twice that of water, indicates that it is predominantly rocky in composition. So cold is Pluto that its surface is covered by a "snow" of frozen methane. So dim is sunlight out there that stars can be seen in broad daylight, and the Sun looks merely like an unusually bright star.

To get a feeling for Pluto's enormous distance from the Sun, let us ask what the Sun would look like were we to go to Pluto. Let us find out how large the Sun would appear to be from that far-distant vantage point. To calculate the angle θ subtended by the Sun as seen from Pluto, we can use the small-angle formula (Appendix I):

$$\theta = 360 \left[d/2\pi D \right] \text{ degrees,}$$

where d is the Sun's diameter, and D is its distance from Pluto.

The logic of the calculation

Step 1. Insert the Sun's diameter and its distance from Pluto into the small-angle formula.

As usual with any scientific calculation, it is important to get an intuitive feel for the answer. To do so, let us ask how far away a dime would need to be in order to look as small – that is, to subtend this same angle. We can use the small-angle formula in reverse.

Step 2. Insert the calculated value for θ, and the diameter of a dime, into the small-angle formula. Solve it for D, the distance we must be from the dime for it to appear as small as the Sun does from Pluto.

Detailed calculation

Step 1. Insert the Sun's diameter and its distance from Pluto into the small-angle formula.

$\theta = 360\,[d_{Sun}/2\pi D_{Sun}]$ degrees.

The diameter of the Sun is

$d_{Sun} = 1.4 \times 10^9$ meters

and its distance from Pluto is 39.5 AU, or

$D_{Sun} = (39.5\ \text{AU})\,(1.496 \times 10^{11}\ \text{meters/AU}) = 5.91 \times 10^{12}$ meters.

So we calculate:

$\theta = 360\,[1.4 \times 10^9/2\pi\,(5.91 \times 10^{12})]$ degrees
$\theta = 1.35 \times 10^{-2}$ degrees = 0.8 minutes of arc.

An angle of 0.8 minutes of arc is certainly small, but we do not yet really comprehend just how small it is. To do so, proceed to step 2.

Step 2. Insert the calculated value for θ, and the diameter of a dime, into the small-angle formula. Solve it for D, the distance we must be from the dime for it to appear as small as the Sun does from Pluto.

1.35×10^{-2} degrees $= 360\,[d_{dime}/2\pi D_{dime}]$.

Taking the dime's diameter to be about 2 centimeters, or 0.02 meters, we find

$D_{dime} = (360)(0.02)/(2\pi)(1.35 \times 10^{-2})$ meters
$D_{dime} = 85$ meters,

which is 93 yards – roughly the length of a city block.

As you can see from the detailed calculation, the Sun as seen from Pluto looks like a dime seen from a city block away. This gives us a good "seat of the pants" feel for our answer.

Pluto's moon Charon

Pluto's moon, Charon, was discovered by serendipity in 1978. A team of astronomers was conducting ultra-precise observations of Pluto in order more accurately to measure its orbit. They found that the "planet" – so small and so distant that it appeared a mere fuzzy blob through their telescope – appeared elongated (Figure 8.29a). Furthermore, the elongation kept changing, occasionally disappearing entirely. The team hypothesized that the "blob" was a satellite, not resolved through their telescope. Observations with a better telescope (Figure 8.29b) finally succeeded in verifying this hypothesis. The new moon, named Charon, orbits about Pluto once every six days.

Recall from Chapter 3 that we can measure the mass of a planet by observing the orbit of a moon about it. Indeed, this is the *only* means we have of finding the mass of a planet. Since prior to 1978 no moons of Pluto were known, astronomers did not know Pluto's mass. The discovery of Charon enabled us finally to measure it.

Charon itself is an unusually massive moon. Indeed, its mass is not much less than Pluto's. In Chapter 12 we will discuss binary stars, which revolve around each other. Similarly, Pluto and Charon form what might be called a "binary planet." As we have mentioned, each rotates synchronously so that each keeps the same face toward the other. Figure 8.29(b) also shows two of the three smaller moons orbiting Pluto.

The mystery of Pluto

Pluto does not fit into the categories of inner versus outer planets that we described in Chapter 6. Recall that the inner planets are

- small mass,
- rocky in composition,
- high density;

whereas the outer planets are

- large mass,
- gaseous/liquid in composition with small rocky cores,
- low density.

(a)

(b)

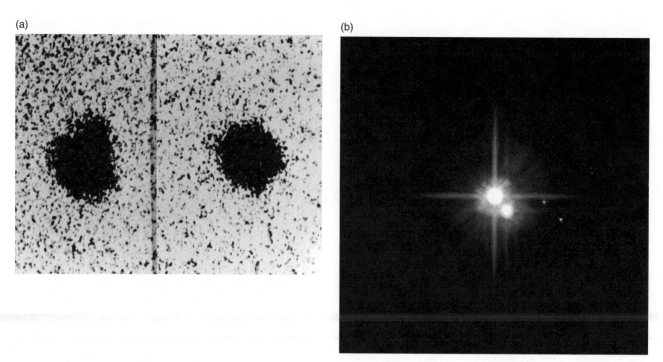

Figure 8.29 **Pluto's moons. (a) In the initial observation the resolution was so poor that Pluto and Charon were melded into one. (b) In a finer resolution we see Charon and two of the three smaller moons: Nix and Hydra. The third smaller moon, S/2011 P1 is not pictured.**

But Pluto does not fit into this scheme. Indeed, it is just like the inner planets – except that it is situated among the outer ones.

Furthermore, as we will see in Chapter 13 on the origin of the Sun and planets, there are good theoretical reasons for this division between inner and outer planets. So Pluto does not just fail to fit into a progression: it also fails to fit our theories. Recall that in 2006 it was decided that Pluto should not be classified as a planet. We will have more to say about this in the next chapter.

Summary

Jupiter

- Jupiter is bigger than all the other planets combined.
- Its visible surface, and much of its interior, is gaseous.
- Its many vividly beautiful markings such as the Great Red Spot are immense weather systems.
- Composition: similar to that of the Sun.
- The outer planets' great gravitational pull allows them to retain the lighter, more volatile elements.
- Jupiter emits more energy than it receives from the Sun. We are not sure why this is so.

Satellites of Jupiter

- Innermost: Io.
  - Covered with volcanoes.
- Next: Europa.
  - Surface: solid ice.
  - Beneath this ice: liquid water (underground oceans).
  - Occasionally this water erupts through the ice to form "water volcanoes."
  - The underground oceans might conceivably support life.

- Next: Ganymede.
  - Largest moon in the Solar System, bigger than Mercury and Pluto.
  - Surface (solid ice) covered by a network of parallel grooves, and craters.
  - Ganymede's surface has been darkened and the craters reveal the whiter surface beneath.
- Next: Callisto.
  - Surface: solid ice.
  - Has more craters than even Ganymede.

Saturn
- Before the advent of telescopes, Saturn marked the outer edge of the Solar System.
- Second largest planet.
- Like Jupiter, Saturn's surface and much of its interior is gaseous.
- Rings.
  - Saturn is encircled by an entire system of rings.
  - A mere few meters thick and hundreds of thousands of kilometers across, the rings form the flattest structure ever discovered.
  - Spacecraft observations show that the rings are composed of orbiting particles arranged in many slender braids.
  - They lie so close to Saturn that the planet's tidal forces would pull apart any satellite that wandered too close: this may explain their origin.
- Saturn's moon Titan.
  - Only moon with an atmosphere.
  - The space probe Huygens landed on Titan and returned a few images.
  - It has lakes and rivers (not composed of water: it is too cold for water be liquid there).
  - The chemistry of Titan's atmosphere and surface may be similar to that of the primitive Earth.

The discovery of Uranus, Neptune and Pluto
- Uranus is the first planet to have been discovered: i.e. the first not visible to the naked eye.
- Found "by accident" by William Herschel in 1781.
- Anomalies in the orbit of Uranus led John Couch Adams and Urbain Jean Joseph Le Verrier to the idea that gravity from a previously unknown planet was disturbing its orbit.
- They calculated where this new planet should be: it was found there in 1846 and named Neptune.
- In subsequent years astronomers decided that Neptune could not explain some of the anomalies in the orbit of Uranus: they postulated yet another planet disturbing its orbit.
- In 1930 Clyde Tombaugh found Pluto at the predicted position of this object.

- Since then we have realized that this discovery was something of a mistake: there are no unexplained anomalies in Uranus' orbit, and Pluto's mass is far too small to influence it anyway.

Uranus
- The rotation axis of Uranus lies in its orbital plane, which gives it very strange seasons.
- Uranus has very thin rings, discovered fortuitously when a star passed behind them.
- Like Jupiter and Saturn, Uranus is mostly gaseous with liquid and solid inner regions.
- Moons: even though they are very small, they show signs of significant geologic activity.

Neptune
- Also mostly gaseous with liquid and solid inner regions.
- Surface is a beautiful vivid blue, arising from methane.
- Two of Neptune's moons have very anomalous (and unexplained) orbits.

Pluto
- Orbit is very elliptical for a planet.
- Unlike the outer planets, Pluto is very small and rocky.
- From Pluto, the Sun is merely a very bright star.
- Pluto's moon Charon is not much smaller than Pluto.

DETECTIVES ON THE CASE

Why is Io so hot?
- We first tried out the theory that Io is heated by its proximity to Jupiter, which emits energy.
- But we then calculated this heating to be too small to make a difference.
- Eventually we adopted the theory that Io is heated by variations in its tidal bulge produced by Jupiter.
- This theory passed several tests.

DETECTIVES ON THE CASE

Craters on the moons of Jupiter
The farther a moon is from Jupiter, the more craters it contains. Why?
- We first tried out the theory that there are more impacting bodies far from Jupiter than close in.
- But we realized that our understanding of gravity predicts just the opposite.
- We then tried the theory that geological processes that erase craters are stronger on the inner moons.
- This theory passes several tests.

DETECTIVES ON THE CASE

What are Saturn's rings?

• Observations employing the Doppler effect show that the rings are rotating about Saturn.
• The rotation velocity decreases as we move outwards.
• This led us to propose that the rings are composed of innumerable small bodies orbiting Saturn.
• We tested the theory in two ways.

THE NATURE OF SCIENCE

The role of luck in scientific discovery

• Herschel had not gone looking for Uranus: in this sense his discovery was fortuitous.
• Nevertheless, Herschel deserves credit for his discovery, whereas a discovery made by pure luck would deserve none.
• Many scientific discoveries come about in a similar serendipitous fashion.
• "Chance favors the prepared mind."
• "Expect the unexpected."

Problems

(1) Figure 8.30(a) shows a solar power panel illuminated by sunlight. It generates a certain amount of power. But then (Figure 8.30b) someone sticks a spherical object half a meter across in front of the panel. That object casts a shadow on the solar panel. Your task is to figure out *how many watts less power the panel now generates.* (A) Describe the logic of the calculation you will perform. (B) Carry out the detailed calculation. (You will need to know that the flux of light from the Sun is 1400 watts per square meter.) (C) Suppose alternatively that someone stuck, not a sphere, but a square piece of cardboard ½ meter on a side in front of the panel. How much less power would it generate? (D) Suppose that we now double the size of the solar panel, but we keep the size of the spherical object the same. How would this change your answer to (A)?

(2) Suppose we were to completely cover the entire surface of the Earth with solar panels. We want to calculate how much power could be generated by these panels. (A) Describe the logic of the calculation you will perform. (B) Carry out

the detailed calculation. (C) For comparison, the total amount of solar energy generated worldwide nowadays is about 3 billion watts. Given your result in (B) comment on the future of the solar power industry.

(3) Suppose we knew a planet's radius and total infrared luminosity. How could we find its temperature?

(4) Where would you place the dividing line between day and night? In your discussion make sure that you take into account the fact that the Sun does not set in an instant, since it has a certain size (Figure 8.31). Furthermore, things do not grow dark the moment after the Sun has finished setting.

(5) What is the difference between a planet and a moon? Where would you place the dividing line between them? Justify your answer.

(6) In our discussion of Saturn's rings, we formulated the theory that they are composed of orbiting particles. We tested this theory by comparing one particle's predicted orbital velocity with the observed velocity.

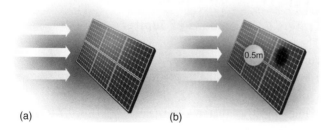

(a)   (b)

Figure 8.30 **If a shadow falls on a solar panel it produces less power.**

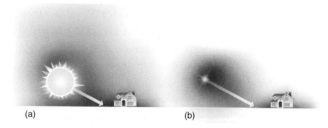

(a)   (b)

Figure 8.31 **Because the Sun is extended (a), it does not set in an instant. But a point-like star (b) does set in an instant.**

But we carried out the test for only one data point. You should now test the other two points. (A) Describe the logic of the calculation you will perform. (B) Carry out the detailed calculation.

(7) Consider (A) the passage of one day and (B) the passage of the seasons on (i) the north pole of Uranus and (ii) at our north pole. What are their similarities and differences?

(8) (A) Suppose a planet were to possess rings that were elliptical in shape, rather than circular. Draw a figure, analogous to Figure 8.23, illustrating the brightness of a distant star as it is being occulted by that planet. Now re-draw your figure supposing that one of these rings were (B) far thicker than the others, or alternatively (C) as thin as the others but far more opaque than they.

(9) Suppose you went to the planet Mercury. (i) How big would the Sun appear – that is, what angle would it subtend in the sky? (ii) How far away would you have to hold a dime in order for it to subtend the same angle? For each question (A) describe the logic of the calculation you will perform, and (B) carry out the detailed calculation.

(10) Table 6.1 lists the radii of the orbits of the known planets. (A) Use these data to take a guess as to how far from the Sun a tenth planet might lie. (B) How confident are you in the reliability of this guess?

WHAT DO YOU THINK?

(1) In our discussion of Io, we formulated the theory that its volcanism results from the periodic squeezing of the moon by tidal forces from Jupiter. We formulated two tests of this theory. But it would be better to find other ways to check it – after all, two is not very many. Can you think of a new test? Describe in detail the observations you would need to conduct in order to carry it out.

(2) So far we have performed a number of *kinds* of tests of our theory of Saturn's rings: we tested the predicted orbital velocity, and the prediction that they should be transparent. Can you think of still other kinds of tests? Be careful to describe in detail how you would carry them out.

You must decide

THE NOBEL PRIZE AND NEPTUNE

Let us imagine that the Nobel Prize existed at the time of the discovery of Neptune (in reality it didn't). Suppose you had lived back then and had been charged with deciding who was to receive the Prize for the discovery of this planet. Three candidates have been nominated: Adams from England, Le Verrier from France and Galle in Germany. Which of them would you choose?

You are, of course, free to choose one, two or all three. But you are under great pressure from all sides. Intensely patriotic scientists and politicians in all three nations are up in arms over the coming nomination. Whoever you choose, you will be attacked.

Who do you choose – and what will you say to those who attack you for your choice?

9

Smaller bodies in the Solar System

Planets are not the only bodies orbiting the Sun. The Solar System also contains innumerable smaller bodies: meteoroids, asteroids and comets. We now turn to a study of these.

We have already commented on how common craters are in the Solar System. Our Moon is covered with them, as are many other moons and planets. Indeed, the Earth is unusual in having so very few craters. As we have already mentioned, this is not because the Earth never had craters: it is because they have been "erased" by erosion and other geological processes.

All those craters were caused by impacts. Indeed, small objects are continually striking everything in the Solar System. The Solar System is filled with these bodies – meteoroids, asteroids and comets – flying about every which way. Encounters with them are common: roughly 40 000 tons of extraterrestrial material fall upon the Earth each year.

Once in a long while these impacts are catastrophic. Indeed, we believe it was just such an impact that led to the extinction of the dinosaurs. Only highly sensitive and highly accurate observations are capable of telling us whether an incoming body poses a similar threat today. Efforts are under way to study the feasibility of deflecting any such potentially threatening objects that we might discover.

Meteors

Meteor showers

Several times each year a meteor shower occurs. Perhaps you have seen one. They can be very beautiful as, utterly soundlessly, meteors streak across the sky. Sometimes a particularly large one will leave a faint glowing trail. The meteoroids that produce these showers are tiny – about the size of snowflakes. Because they are so small they entirely burn up in their fiery passage through the atmosphere: none of them survives to reach the ground. During a particularly intense shower you might see one a minute, although occasionally even greater rates are seen.

If you look at a meteor shower for any length of time, you will quickly notice a striking fact: the "shooting stars" all appear to be coming from the same place. As illustrated in Figure 9.1(a), each track points directly away from the same point on the sky. This point is termed the "radiant."

Definitions of terms

Three different terms need to be kept in mind.

- A *meteoroid* is a small body as it flies through space, before it has encountered the Earth.
- A *meteor* is that body as it flies through our atmosphere. They are sometimes termed "shooting stars," but the term is a misnomer: they are not rapidly moving stars, but tiny bits of extraterrestrial matter colliding with the Earth. Because they travel at from 10 to 70 kilometers per second (20 thousand to 160 thousand miles per hour), they are heated to incandescence by air friction and glow brilliantly.
- A *meteorite* is the body after it has landed on the Earth.

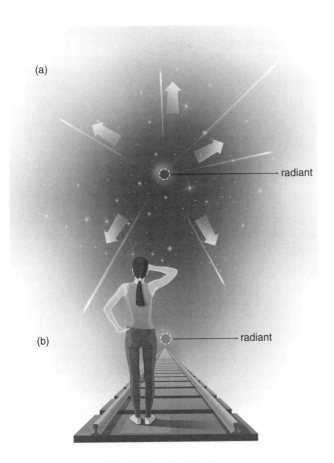

(a)

(b)

— radiant

— radiant

Figure 9.1 **(a) is the radiant of a meteor shower (b) is the radiant of the railway track – not where the meteors come from.**

As illustrated in Figure 9.1(b), this is a trick of perspective. It arises because the meteors are heading straight toward us. Their trails appear to diverge for the same reason that railroad tracks appear to diverge to an observer standing between them. In fact the railroad tracks are parallel, and so too are the paths of the meteors as they streak through the air.

Just as the "radiant" of the railroad tracks marks the distant point from which they come, so too does a meteor shower's radiant tell us which part of the sky the meteoroids came from. We can think of the meteoroids as a swarm, consisting of tiny bits of extraterrestrial material, traveling on its own orbit about the Sun. The radiant tells us the part of the sky from which this swarm was traveling. Each meteor shower has a name, describing where its radiant lies. The Perseid shower's radiant is in the constellation of Perseus, the Leonid's in the constellation of Leo.

Another striking fact about meteor showers is they always occur at the same time of year. The Perseid shower comes in mid August, the Leonid shower comes in mid November. Since the Earth is regularly orbiting the Sun, this means that each meteor shower comes when the Earth reaches a certain definite location in its orbit. And it turns out that these locations mark the points at which our orbit intersects that of a comet.

We have gained an important insight: the meteoroids that give rise to meteor showers are associated with comets.

⇒ **Looking forward**
In our study of comets we will see why comets are accompanied by swarms of debris.

As illustrated in Figure 9.2, a comet is accompanied by a swarm of tiny debris that travels about the Sun in the same orbit as it does. When we turn to our study of comets we will understand why this is so: for now, the important lesson we have learned is that whenever the Earth in its orbit about the Sun plows through this swarm, we get a meteor shower.

Table 9.1. **Meteor showers.**

Name	Dates	Hourly rate	Associated comet
Quadrantids	Jan. 2–4	30	?
Eta Aquarids	May 2–7	10	Halley
Perseids	Aug. 10–14	40	Swift–Tuttle
Orionids	Oct. 18–23	15	Halley
Leonids	Nov. 14–19	6	Tempel–Tuttle
Geminids	Dec. 10–13	50	3200 Phaeton

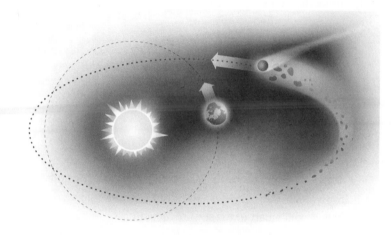

Figure 9.2 **When the Earth passes through cometary debris we get a meteor shower.**

Halley's comet is perhaps the most famous of them all. This comet, in fact, gives rise to *two* meteor showers: the Eta Aquarids when we pass through its orbit in May, and the Orionids when we pass through the other side of its orbit in October. Some meteor showers are associated with "defunct comets": those that have entirely evaporated away and no longer exist. Table 9.1 lists the most notable meteor showers and their associated comets, if known.

Meteorites

Among all the 40 000 tons of meteoroids that hit the Earth each year, most are so small that they burn up in the atmosphere. Only about 200 tons of these are large enough to survive and reach the ground to become meteorites.

You might think that meteorites would be red-hot when they land. In fact they turn out to be quite cool! How can this be, since they were heated to incandescence by friction against the atmosphere?

The solution to this puzzle can be found by calculating how long they were heated. As we saw in Chapter 6 most of our atmosphere lies within 10 kilometers of the Earth's surface. That is where most of the meteor's heating takes place. Furthermore, as we have seen meteors move at between 10 and 70 kilometers per second. How long, then, does it take the meteor to traverse the atmosphere?

The logic of the calculation

Step 1. Choose a representative value for the meteor's velocity.
Step 2. Divide this into the thickness of the Earth's atmosphere.

As you can see from the detailed calculation, it takes less than a second for a meteor traveling straight down to traverse the atmosphere (and somewhat longer for one on a slanting path). This is such a short interval of time that the heat generated by friction cannot penetrate very deeply into its body. Only the very outermost layer is heated, and this cools rapidly once the meteorite has come to rest.

Detailed calculation

Step 1. Choose a representative value for the meteor's velocity.
Let's choose a value somewhere between the two extremes of velocity: say, 30 kilometers per second.

Step 2. Divide this into the thickness of the Earth's atmosphere.
We calculate:

time = distance/velocity = 10 kilometers/30 kilometers per second = ⅓rd of a second.

NOW YOU DO IT
Suppose a meteoroid traveling at 50 kilometers per second strikes the Earth. For how long is it heated as it travels through the atmosphere?

Figure 9.3 **Barringer Crater.**

Meteor craters

As it plunges through the atmosphere, a meteor is not just heated by friction. It is also slowed. The degree to which it is slowed depends on its size. A small one – baseball sized, say – has very little inertia, and it is so decelerated by atmospheric friction that it merely hits the ground with a thump. But a large meteor is hardly affected by friction at all. You can verify this for yourself by dropping a piece of paper and a book. Atmospheric drag so slows the paper that it gently flutters down – but the book lands with a good deal of force. Similarly, a large meteor slams into the ground with cataclysmic force. Figure 9.3 shows the result, Barringer Crater in Arizona.

Such a meteor hits the ground at speeds of 10 to 70 kilometers per second, or 20 thousand to 160 thousand miles per hour. This enormous velocity is so great that the rock cannot get out of the meteor's way as it plunges downward. An immense pressure wave through the solid rock builds up in front of it – the seismic equivalent of a sonic boom. The wave expands outward in all directions, and produces an immense explosion. Indeed, the result is very similar to what would happen were a bomb to go off underground. Theoretical computer models show that the resulting crater will be roughly ten times bigger than the impacting body that made it.

Barringer Crater in Arizona (Figure 9.3) is an example of a young impact crater. About 1 kilometer in diameter, it is 50 000 years old. The body that produced it must have been about 50 meters in diameter, which is about the size of a large apartment building, and it released the energy equivalent of a 3-megaton hydrogen bomb.

But this is an unusually young crater. Most are far older. Many craters are known whose ages are hundreds of millions of years. These have been so eroded by wind and rain that they are very difficult to see from the ground. Figure 9.4 is an example. Indeed, it is possible to live within an ancient, eroded crater and not even know it!

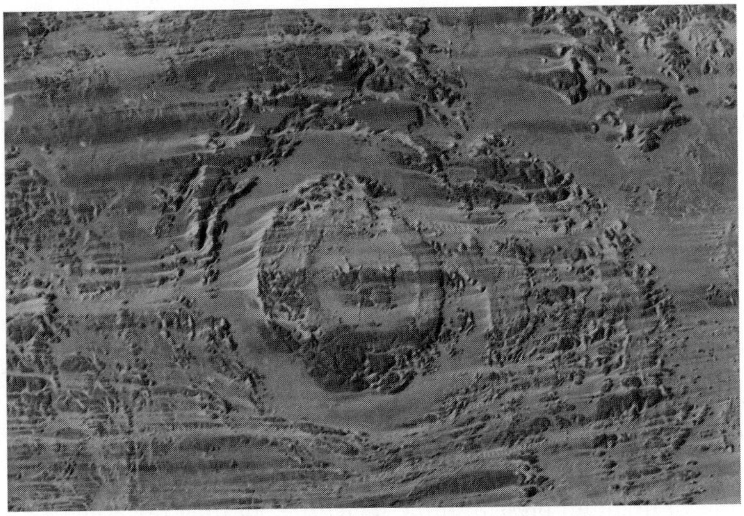

Mineralogy of meteorites

Because most meteorites look like ordinary rocks to the untrained eye, not that many are found each year. The rest just lie around undiscovered. You may have picked up a meteorite at some point in your life without knowing it.

Those meteorites that we do find are particularly valuable. Long before the advent of the space program, astronomers had them for study and observation. They were free samples of the cosmos.

Such studies have shown that 95% of all meteorites are stony. They are composed of silicates and metallic nickel–iron, and have densities of 3500 to 3800 kilograms per cubic meter. For

Figure 9.4 **Aorounga Crater, a 250-million-year-old impact crater located in Chad.**

comparison, terrestrial stony rocks are composed primarily of silicates and have somewhat lower densities: 3100 to 3300 kilograms per cubic meter.

Among the remaining meteorites, 4% are composed of nickel and iron. They are very much denser than the stony meteorites, with densities lying between 7600 and 7700 kilograms per cubic meter. The remaining 1% have properties in between those of the stony and iron meteorites.

Meteorite ages

The ages of meteorites can be found by two different means: radioactive age dating, and cosmic ray exposure ages.

Radioactive age dating

Radioactive age dating is a common method that is employed to find the ages of things. It uses the decay of an unstable atomic nucleus as a "clock."

To understand the principle of this method, let us study the radioactive decay of uranium. Nuclei of the isotope uranium-235 are found to decay into nuclei of lead and helium as follows:

$$^{235}\text{U} \rightarrow {}^{207}\text{Pb} + 7\,{}^{4}\text{He}.$$

Each decaying uranium nucleus fissions into eight pieces: one lead nucleus and seven helium nuclei.

While the decay of any individual nucleus is random (i.e. we cannot predict when it will split), the behavior of large numbers of them is predictable. The basic principle of radioactive decay involves the concept of a *half-life*, which is the interval of time in which half the uranium nuclei decay. When the above decay is studied in the laboratory, it is found that it has a half-life of 700 million years. If we begin with a certain number of uranium nuclei, during the first 700 million years half of them decay: during the second 700 million years half of the remaining nuclei decay, and so on.

Let us use this to build a "clock" with which we can measure the age of a meteorite. Suppose we start with a meteorite containing some number of uranium nuclei – say, one million of them. If we then wait 700 million years, we will find that half of them have decayed, so that it now contains only half its initial number of uraniums – i.e. it contains 500 000 of them.

Now let us wait another 700 million years! At the beginning of this second interval of time, there were 500 000 uraniums present. By the basic principle of radioactive decay, half of these decay in one half-life. So at the end of this second 700 million year interval 250 000 uraniums have decayed, leaving only 250 000 left. This is only ¼ of the initial amount of uranium. At this point, our meteorite is two half-lives, or 1.4 billion years, old.

If we now wait a third half-life, how much uranium is left? At the beginning of this interval of time there were 250 000 uranium nuclei; during the interval half of them decayed, leaving 125 000 by its end. At this point, our meteorite is three half-lives, or 2.1 billion years, old.

The more time passes, the less uranium remains in the meteorite. In Figure 9.5 we collect our results. We can use this figure to find the age of a meteorite. We do this by measuring the amount of uranium present in the meteorite, locating this value on the vertical axis of Figure 9.5 and reading off the age on its horizontal axis. For example, if a meteorite has half its initial amount of uranium in it, it must be one half-life old, i.e. 700 million years old.

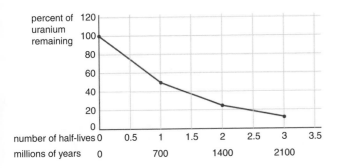

Figure 9.5 **Radioactive decay. The more time passes, the less ^{235}U remains. By measuring the amount of uranium remaining, we can measure the age of the sample.**

NOW YOU DO IT
Suppose we find a meteorite with 35% of its initial ^{235}U present. How many half-lives have passed since it formed? How many years?

So far we have concentrated on the isotope ^{235}U of uranium. A different isotope, ^{238}U, decays with a half-life of 4.5 billion years. (A) Carry through the analysis of how much ^{238}U is present after one, two and three half-lives. (B) Draw a graph, similar to Figure 9.5, showing how much ^{238}U is present after 4.5 billion years, 9 billion years and 13.5 billion years. (C) Suppose we find a meteorite with 25% of its initial ^{238}U present. How many half-lives have passed since it formed? How many years?

Solidification ages

Unfortunately there is a complication we must address before this method can be used in practice. Notice that the method consists of comparing the amount of uranium in a sample with the amount that was originally present. The complication is that usually we have no way of knowing this initial amount. Did our rock begin its days as pure uranium? Or did it contain uranium mixed with other elements?

Fortunately there is a way around this difficulty. It is to measure not only the amount of uranium in the meteorite, but also how much helium it contains. Notice from the basic decay scheme that every time a uranium nucleus has decayed, seven helium nuclei have appeared. Helium, of course, is a gas. If a meteoroid is solid, this gas will remain within it. But suppose the meteoroid had melted at some point in the past! In this case the gaseous helium would have bubbled away. This has the effect of "resetting our clock to zero": once the meteoroid has re-frozen, we know that it contains no helium. Any helium that rock *does* contain must have come from subsequent radioactive decay.

This gives us a more practical means of determining the age of our meteorite.

> In this case, however, what we mean by "age" is something different from "the time since the meteorite was created." Rather it is the time since it last melted and then solidified. For this reason the age so found is termed the "solidification age."

Let us therefore go back to our analysis of the decay of uranium, and think not just about how much uranium is present, but also how much helium. We will say that at some time our rock melted, liberating any helium it might have contained. Then, at time = zero, the rock froze. Thereafter, we have the following.

At time = 0 suppose there are N uranium nuclei in the rock, so that half this number (i.e. *N/2*) of decays take place in the next half-life.

Table 9.2. **Radioactive decay.**

Time	Number of uranium nuclei in the sample at the beginning of this period of time	Number of helium nuclei in the sample at the beginning of this period of time	Number of He/number of U in the sample at the beginning of this period of time	Number of decays that will take place in the next half-life
0	N	0	0	$N/2$
1 half-life	$N/2$	$7N/2$	7	$N/4$
2 half-lives	$N/4$	$21N/4$	21	$N/8$
3 half-lives	$N/8$	$49N/8$	49	$N/16$

At time = one half-life. In the previous half-life, $N/2$ of the uranium nuclei decayed, so $N/2$ are left. Furthermore, each decay produced seven helium nuclei, so the number of helium nuclei after one half-life is $(7)(N/2)$. The number of decays that will take place in the next half-life is half the number of uraniums present at one half-life, namely half of $N/2$, or $N/4$.

At time = two half-lives. After these $N/4$ decays take place, the number of uranium nuclei has been reduced from $N/2$ to $N/4$. Each decay produced seven helium nuclei, so the number of helium nuclei created during the past one half-life is $(7)(N/4)$. This is added to the number of heliums present at the beginning of this period $(7 N/2)$ to get $21 N/4$ helium nuclei. The number of decays that will take place in the next half-life is half the number of uraniums present at time = two half-lives, namely half of $N/4$, or $N/8$.

At time = three half-lives. After these $N/8$ decays take place, the number of uranium nuclei has been reduced from $N/4$ to $N/8$. Each decay produced seven helium nuclei, so the number of helium nuclei created during the past one half-life is $(7)(N/8)$. This is added to the number of heliums present at the beginning of this period $(21 N/4)$ to get $49 N/8$ helium nuclei.

Let us pause and collect our results in Table 9.2 and Figure 9.6. They allow us to determine the solidification ages of meteorites. For instance, a meteorite with seven times as many helium as uranium nuclei must have solidified one half-life ago.

....

NOW YOU DO IT

Suppose a rock sample is found that contains 20 times as many helium nuclei as ^{235}U nuclei. How many half-lives have passed since it last melted and then solidified? How many years?

Return to our isotope ^{238}U, which decays with a half-life of 4.5 billion years. (A) Carry through the analysis of how many helium and ^{238}U nuclei are present after one, two and three half-lives. (B) Draw a graph, similar to Figure 9.6, showing the ratio of the number of helium to ^{238}U nuclei that are present after 4.5 billion years, 9 billion years and 13.5 billion years. (C) Suppose we find a meteorite with 25 times as many helium as ^{238}U nuclei. How many half-lives have passed since it formed? How many years?

....

When such methods are applied to meteorites, it is found that they last were melted between 4.0 and 4.5 billion years ago. Because the age of the Solar System as

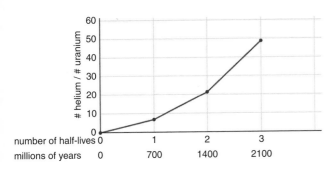

Figure 9.6 **Helium to uranium ratios can be used to determine solidification ages without knowing the initial composition of a rock sample.**

a whole is 4.5 billion years, this probably means that the solidification age is the true age of the meteorites: they solidified when they were formed, and they have not been melted since then.

Cosmic ray exposure ages

We now turn to the second way to find the ages of meteorites. We can determine how long they have been exposed to cosmic rays.

Cosmic rays fill all of space. They are rapidly moving subatomic particles: electrons, atomic nuclei and gamma rays. When they strike a meteoroid they penetrate into it, and they leave telltale signs of their passage.

- They can strike nuclei of atoms within the meteoroid, transmuting them into decaying nuclei.
- They can leave microscopic tracks in the meteoroids.

In both cases we can measure how long the meteoroid has been exposed to the cosmic rays: we can apply the usual techniques of radioactive age dating to the decaying nuclei, and we can count how many microscopic tracks the meteorite contains.

Cosmic ray exposure ages are different from solidification ages. This is because cosmic rays cannot penetrate very far into solid rock. In fact they penetrate only about a meter. So if a meteoroid is large, most of its interior will be shielded from them.

When these methods are applied to meteorites, it is found that they have been exposed to cosmic rays for times ranging from 5 million to 1 billion years.

Interpretation of meteorite ages

We have discussed two methods of determining the ages of meteorites – and the answers we get are different (Table 9.3)! What does this tell us?

Let us first focus attention on the fact that cosmic ray exposure ages are much less than solidification ages. The only way to understand this is to postulate that meteors come from deep inside large *parent bodies*, and that for most of their histories they were shielded from cosmic rays. At some point in the past these parent bodies must have been shattered into smaller pieces, and it is these that were exposed to cosmic rays. They flew through space and eventually hit the Earth.

> Cosmic ray exposure ages tell us not when a body was formed, but rather how long ago the shattering of their parent bodies took place.

Now let us turn our attention to the fact that cosmic ray exposure ages differ from one meteorite to another. Some are relatively short (5 million years). Others are far larger (1 billion years). This tells us that there were many such shattering events. It is not the case that there was one big

Table 9.3. **Observed meteorite ages.**

Solidification ages	4.0 billion–4.5 billion years
Cosmic ray exposure ages	5 million–1 billion years

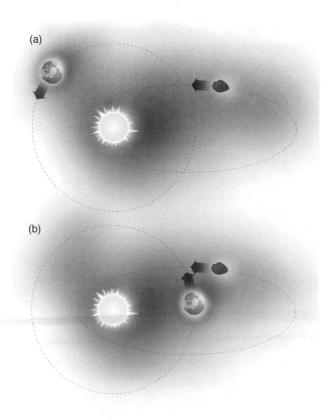

Figure 9.7 **No threat of collision arises in the configuration (a), since when the meteoroid crosses the orbit of the Earth, the Earth is elsewhere. Only in configuration (b) does a threat arise.**

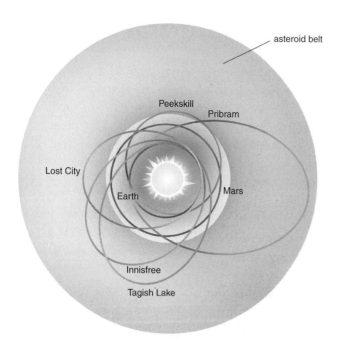

Figure 9.8 **Orbits of meteors originate in the asteroid belt.**

cataclysm: there were lots of little ones. Different meteorites come from different parent bodies that were broken apart at different times.

NOW YOU DO IT
Suppose you found a meteorite whose solidification age was 100 million years, and whose cosmic ray exposure age was also 100 million years. Speculate on the history of this object.

Where do meteorites come from?

What were these parent bodies? We have already seen that meteor showers are associated with comets. But recall that the meteoroids that give rise to meteor showers are very small. The larger meteoroids that reach the Earth to become meteorites are another matter, for they do not recur at the same date each year and they are not associated with comets. Where do they come from?

Here is a clue. A meteoroid, before it hit the Earth, must have been moving in an orbit about the Sun that intersected our own. Every time that meteoroid reached this intersection point, there was a danger of collision. As illustrated in Figure 9.7 however, most of the time no collision ensued, and the meteoroid survived. We can ask: how long could it have survived, before being annihilated by its collision with the Earth?

The answer is that a meteoroid in an Earth-crossing orbit could survive only for a short time in cosmic terms. So there must be some "reservoir" somewhere, from which new meteoroids are continually being produced and sent on their orbits toward the Earth. Presumably these "production events" are the shatterings we invoked in the previous section, in which large parent bodies were broken into smaller pieces.

Two lines of evidence have been found which indicate that these parent bodies are the asteroids:

• Reflectance spectra of meteorites are similar to those of asteroids.
• Some meteors have been observed with such care that it has been actually possible to determine their orbits. These orbits are found to originate within the asteroid belt (Figure 9.8).

We therefore now turn to a study of the asteroids.

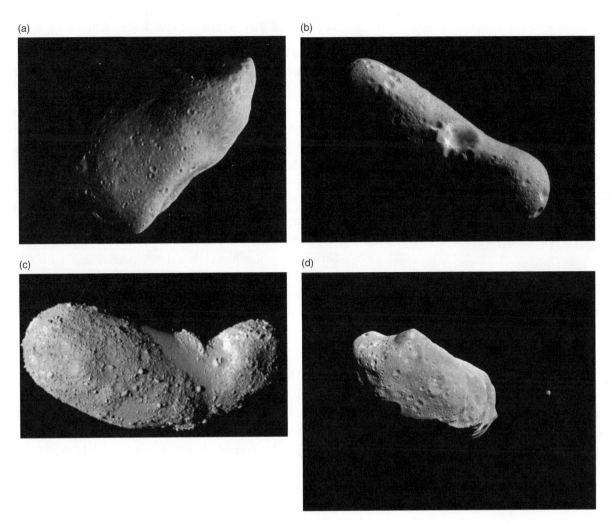

Figure 9.9 **Asteroids.**

Asteroids

The first asteroid (the name means "star like") was discovered by Giuseppi Piazzi, a Sicilian monk, on the night of January 1, 1801. Later named Ceres, after the Roman god of the harvest, it is a mere 900 kilometers in diameter – quite small as Solar System bodies go, although Ceres is in fact the largest of the asteroids.

Figure 9.9 shows a collection of asteroid images obtained by spacecraft flybys. They are irregular in shape, and pockmarked with craters. Note that one asteroid – Ida, a mere 52 kilometers long – has its own moon, Dactyl (Figure 9.9d).

Almost all asteroids lie in a broad belt between the orbits of Mars and Jupiter, with semimajor axes ranging from 2.2 AU to 3.3 AU. Their periods of orbit range from 3 to 6 years. In contrast to the planets, their orbits are fairly elliptical, with eccentricities of 0.1 or 0.2 being common. Contrast this with the planets, whose orbital eccentricities are far smaller: that of the Earth is 0.02, that of Jupiter 0.05.

> ⇐ **Looking backward**
> Recall from Chapter 2 that orbits of small eccentricity are nearly circular, whereas those of large eccentricity are highly "squashed."

> We see that planets move in nearly circular orbits, but asteroids' orbits are nowhere near circular.

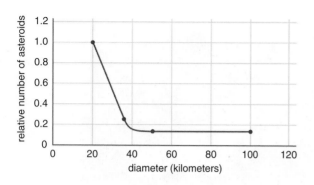

Figure 9.10 **Distribution of asteroid sizes. There are few large asteroids, and many small ones.**

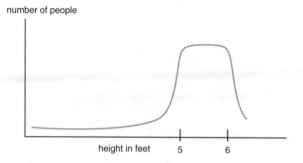

Figure 9.11 **Distribution of heights of humans is nothing like the distribution of sizes of asteroid.**

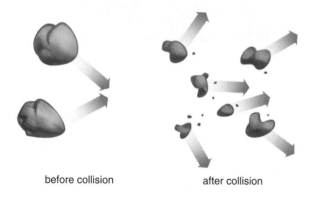

Figure 9.12 **Fragmentation of asteroids upon collision gives rise to many small ones.**

Numbers and sizes of the asteroids

Hundreds of thousands of asteroids are known. Orbits have been determined for many of them. Observations show that there are more small asteroids than big ones. Only three are known whose diameters exceed 500 kilometers – but several hundred have diameters exceeding 100 kilometers, and there may be half a million asteroids larger than 1 kilometer in size. Figure 9.10 graphs this distribution of sizes of asteroids.

It is worthwhile to pause and consider how unusual a distribution this is. To appreciate this, let us consider the distribution of sizes of some other object – human beings, say. Most adults are more or less the same size – they are between 5 and 6 feet (1.5 and 1.8 meters) in height. Children have heights ranging from about a foot (infants) on up – but there are fewer children than adults. So the distribution of heights of humans looks like that shown in Figure 9.11. Note how completely unlike Figure 9.10 this is.

....

STOP AND THINK
Can you think of familiar, everyday objects whose size distribution is similar to that of asteroids?
   Can you think of objects whose size distribution is just the opposite: lots of big ones, but few little ones?
....

Why would small asteroids be so common? One reason is *collisions* and *fragmentation* among the asteroids. If two of them collide with sufficient velocity, they will shatter upon impact. What started out as two relatively large bodies ends up as a lot of smaller ones (Figure 9.12). As we have already seen, these collisions are probably the events that give rise to the meteoroids.

Let us analyze this more carefully. Suppose that a long time ago there were ten asteroids, each of which had a mass of one million kilograms. Suppose that after 100 000 years half of them had collided, and that each colliding asteroid had split in half. Then, at the end of this 100 000 year period, we would find the following.

• Half of the initial million-kilogram asteroids – i.e. five of them – have escaped collision. These are still around.
• The other half – i.e. five asteroids – have collided. Each of these split in two, yielding ten asteroids. Each of them had half the mass of their "parents," i.e. 500 000 kilograms.

Figure 9.13 collects these results. Notice an interesting thing about this figure. The initial distribution looks nothing like the observed distribution of asteroids – but the later distribution does: it shows more small asteroids than big ones! This agreement between the observations and the prediction of our "fragmentation

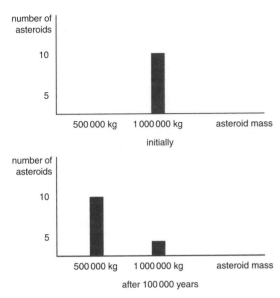

Figure 9.13 **Collision and fragmentation alter the distribution of asteroids, producing more small ones than big ones.**

theory" is one line of evidence for the correctness of the theory. Of course, as with all scientific theories, the more evidence the better.

NOW YOU DO IT

Suppose that another 50 000 years now passes, during which half of the asteroids collide. By this we mean that half of the large asteroids, which had not yet suffered a collision, now do so; and that half of the smaller ones, which had suffered a collision, now suffer a second one. Suppose further that in every such collision, each object splits in two. Draw a sketch, analogous to Figure 9.13, showing the distribution after the end of this additional 50 000-year period.

Suppose rather that each of the colliding asteroids had split into *three equal pieces*. (A) What would be the distribution of asteroid masses at the end of the first 100 000-year span? (B) Draw a sketch, analogous to Figure 9.13, showing this distribution.

It is very difficult to measure how many asteroids there are in total. We believe that we have found all the large ones, but it is hard to find the little ones – and since there are so many little ones, they are the important ones when it comes to counting the total number. We believe there may be something like one billion asteroids. While this is a large number, so small are they that it does not add up to a great amount of mass. The total amount of mass in the asteroid belt is far less than that of the Moon.

While there are a great many asteroids, it would be wrong to think that the asteroid belt is crowded. Because interplanetary space is so vast, even a billion of them scattered about the belt turn out to be pretty thinly spread. If you were to be suddenly transported out to the asteroid belt, you might not be able to see a single one of them with your naked eye! Similarly, every space probe that has ever traveled to the outer Solar System passed through the belt – and not a single one of them has ever been struck by an asteroid.

Rotation of asteroids

Asteroids are so small that, even through our largest ground-based telescopes, they cannot be clearly resolved. Rather they appear to be merely points or vaguely extended smudges of light (recall our discussion in Chapter 5 of the resolving power of a telescope). Prior to the flybys that returned images such as Figure 9.9 we had only a rough idea of what an asteroid looked like. Nevertheless, long before the space age we knew that asteroids rotated, and indeed we were able to measure how rapidly they rotated.

This accomplishment was made possible by observations of the apparent brightnesses of asteroids. It was found that they varied in brightness, growing alternately brighter and dimmer as illustrated in Figure 9.14.

Careful observations showed that this variation was repetitive, with the pattern repeating itself every 5 to 10 hours. We can understand this by recognizing that an

⇐ **Looking backward**
Recall our discussion in Chapter 5 of the resolving power of a telescope.

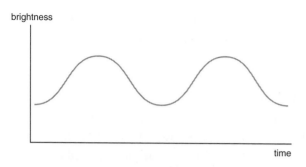

Figure 9.14 **Brightness variation of an asteroid.**

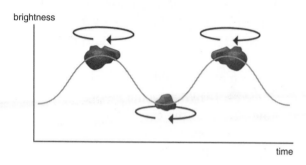

Figure 9.15 **Brightness variation of an asteroid arises from its rotation.**

⇐ **Looking backward**
Recall from Chapter 2 that Kepler's third law states that the period of an orbit depends on its semimajor axis.

asteroid has an irregular shape. As an irregularly shaped body rotates, it alternately presents its broad side and its narrow side to us, and it appears alternately brighter and dimmer (Figure 9.15).

Such observations reveal that asteroids as a group rotate more rapidly than planets, with periods of rotation typically ranging between 5 and 10 hours. For comparison, the most rapidly rotating planet (Jupiter) has a rotational period of 9 hours 55 minutes.

Furthermore, by observing an asteroid at different points in our orbit, we can track its rotation as seen from different vantage points, and gain information about the direction of its spin axis. Recall that planetary rotation axes all point roughly perpendicularly to the plane of the ecliptic. But this is not true of the asteroids: their spin axes point every which way. Presumably this too is due to collisions among asteroids, whereby they deliver each other glancing blows and twist their rotation axes about randomly.

Kirkwood gaps

In 1866 the American astronomer Daniel Kirkwood, who was studying the orbits of the asteroids, assembled data on their semimajor axes. We have already seen that they orbit with semimajor axes lying between 2.2 AU and 3.3 AU. Kirkwood was interested in studying how these semimajor axes were distributed within this range. Were they distributed uniformly throughout it? Or were there more asteroids with certain semimajor axes than others?

Figure 9.16 graphs the data. It shows the number of known asteroids as a function of their orbital semimajor axis a. If the asteroids were uniformly distributed, we would find the same number for every value of a, so the data would form a straight line. But as you can see from Figure 9.16, they do not. Certain values of a seem to be avoided: there are no asteroids at all with these special values of the semimajor axis. These gaps in the distribution of semimajor axes are known as the *Kirkwood gaps*.

Recall from Kepler's third law (Chapter 2) that the period of an orbit depends on its semimajor axis. This gives us an alternative way to think about the Kirkwood gaps: they are gaps in the orbital periods of asteroids. For some reason, there are no asteroids with certain periods.

Kirkwood's discovery was that these "missing" orbital periods are related to the orbital period of Jupiter. Consider as an example the gap at $a = 3.28$ AU. Kepler's third law tells us that its orbital period P (in years) is

$$P^2 = a^3,$$

where a is in AU. With $a = 3.28$, this gives $P = 5.9$ years. *And this is just half of Jupiter's orbital period.* The other Kirkwood gaps also correspond to orbital periods that are simple multiples of that of Jupiter. The gap at 2.5 AU has an orbital period $\frac{1}{3}$rd that of Jupiter: that at 2.82 AU has a period that is $\frac{2}{5}$th of Jupiter's.

How can we understand this strange state of affairs? Jupiter, of course, is the most massive planet, and it exerts the biggest force of gravity. Let us therefore think about the force of gravity from Jupiter on an asteroid. Because gravity follows an

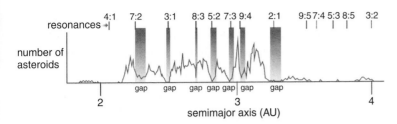

Figure 9.16 **Kirkwood gaps in the distribution of asteroids.**

inverse square law, this force is strongest when the asteroid is closest to Jupiter. As a simple approximation, we can imagine Jupiter giving the asteroid a tug every time it passes by.

Let us return to the gap at $a = 3.28$ AU. Why are there no asteroids with this semimajor axis? To answer this question, let us imagine what would happen if there were an asteroid with this semimajor axis. Its period of orbit is just half that of Jupiter, so that it completes two revolutions during the time required for Jupiter to complete one. Figure 9.17 illustrates the situation. In this diagram, the numbers beside the orbits represent certain selected moments of time. At the time "1" Jupiter and the asteroid are closest, and there is a strong tug (upwards in the diagram) on the asteroid.

Thereafter, the asteroid draws away from Jupiter. At the time "3," the asteroid has completed half of its orbit and Jupiter one quarter of its own orbit. The two are now fairly far apart, and the gravitational tug is a good deal smaller.

Eventually, at time "5," the asteroid has returned to its initial position. Is the gravitational tug again large? It is not! At this time, Jupiter has only completed half its orbit, and is quite far from the asteroid. Not until time "9" have *both* Jupiter and the asteroid returned to their original configuration, and the force is greatest.

So the gravitational tug on the asteroid reaches its greatest value after the asteroid has orbited twice. And this happens at the same point in the asteroid's orbit – the point lying at the top of Figure 9.17.

To appreciate the significance of this, let us now consider a second asteroid – one *not* in a Kirkwood gap. This configuration is shown in Figure 9.18.

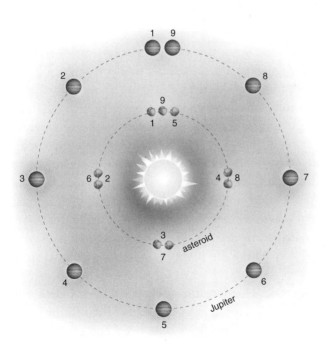

Figure 9.17 **Jupiter's gravitational tug on an asteroid is greatest when they are closest. For the indicated asteroid, this always occurs at the same point in its orbit.**

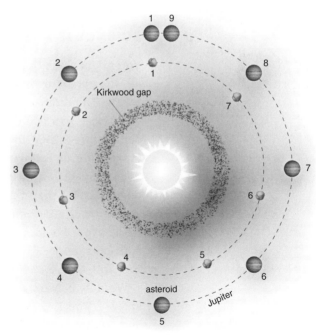

Figure 9.18 **An asteroid not in a Kirkwood gap experiences the greatest gravitational tugs from Jupiter at varying points in its orbit.**

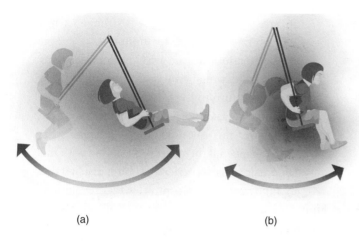

(a)   (b)

Figure 9.19 **Pumping a swing is more efficient if the child pumps synchronously (a) with the motion of the swing, rather than asynchronously (b).**

Notice that now the gravitational tug reaches its greatest value at different points in the asteroid's orbit. At the first time of greatest tug (labeled "1") the asteroid is at the top of the diagram and the force is up – but at the next time of greatest tug ("7" for the asteroid) it is elsewhere, and the force is up and to the right.

We have reached an important understanding. The gravitational tugs on an asteroid in a Kirkwood gap are synchronized. But those on an asteroid not in a gap are not synchronized.

Figure 9.19 illustrates an analogous situation. If the child on a swing synchronizes her pumping motion with that of the swing, she will build up a large motion (a). That is like an asteroid in a Kirkwood gap. If, however, the child pumps out of synchrony with the swing's motion (b), she will never build up a large motion. That is like an asteroid outside of a Kirkwood gap.

An asteroid in a Kirkwood gap therefore builds up a large motion, which carries it out of the gap. Detailed studies show that this motion eventually becomes chaotic, and the object is ejected from the asteroid belt.

THE NATURE OF SCIENCE

SCIENCE IS ABSTRACT

You might think that the Kirkwood gaps are something that you can see. After all, they are gaps in the distribution of asteroids. But it turns out that this is wrong. The Kirkwood gaps are not something you can see.

Suppose you could travel far into space and look down upon the asteroid belt. Figure 9.20(a) diagrams what you might naïvely expect: asteroids scattered more or less randomly about, but with a gap at 3.28 AU. Naïvely, you might expect to find plenty of asteroids closer to the Sun than 3.28 AU, and plenty farther than this distance – but none whose distance from the Sun just equaled 3.28 AU. Remarkably, however, this is not what is actually observed. Rather you would see the distribution illustrated in Figure 9.20(b): asteroids both outside the gap *and* inside it. There are plenty of asteroids 3.28 AU from the Sun!

What was wrong with our naïve expectation? Figure 9.21 supplies the answer. The Kirkwood gaps are not gaps in the asteroids' *distances from the Sun*. They are gaps in their *orbital semimajor axes*. The orbit shown in Figure 9.21 has a semimajor axis greater that 3.28 AU – and yet the asteroid lies 3.28 AU from the Sun, placing it within the Kirkwood gap. A body in an elliptical orbit can be found anywhere along its orbit: it can lie close to the

(a) (b)

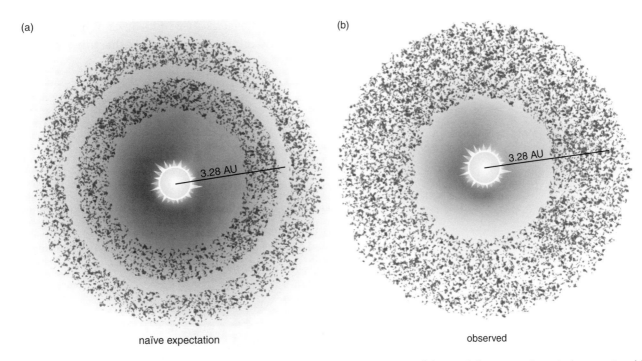

3.28 AU

3.28 AU

naïve expectation

observed

Figure 9.20 **Kirkwood gaps are not gaps in the distribution of asteroids. Naïvely (a) we might expect there to be no asteroids found in the gap at 3.28 AU – but (b) there are!**

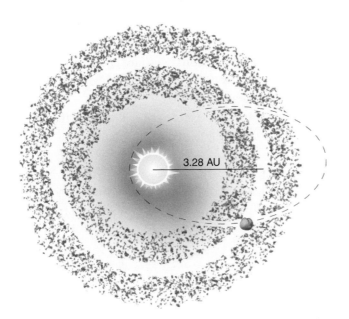

3.28 AU

Figure 9.21 **An elliptical orbit crosses the Kirkwood gap – and an asteroid can lie anywhere along this orbit, even inside the gap!**

Sun or far from it. Only if asteroid orbits were circular would their distances from the Sun equal the semimajor axes of their orbits.

The important point here is that *the Kirkwood gaps are abstract*. They are gaps in the distribution of a mathematical entity known as a semimajor axis, and this is not something that can be directly seen. This turns out to be true of much of science: science is often concerned with abstract, unfamiliar entities.

Energy is another example of a purely abstract quantity of great importance in science. A speeding car has energy of motion. You might think that you can actually see this energy, but that is mistaken: what you see is the car, and the car's motion – but you are not seeing the energy carried by that motion. Similarly, gasoline carries the chemical energy that powers the car. But here too, this energy cannot be seen: what you can see is gasoline, but you can't see its energy. Scientists find that in some important sense the energy is just as real as the motion and the gasoline: it can be changed from one form to another, but its sum total is conserved.

Scientists concern themselves with these strange, abstract entities not out of some perverse satisfaction, but because they need to. We often find that studying concrete entities does not get us very far. Had Kirkwood concentrated attention on something less abstract, such as asteroids' distances from the Sun, he

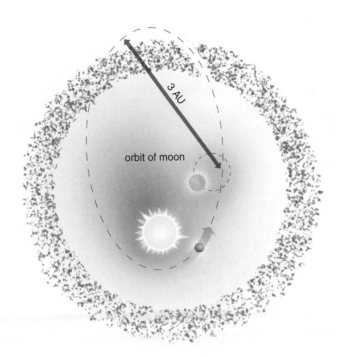

Figure 9.22 **The 1991 asteroid, which passed within the orbit of the Moon, had gone 3 AU and hit an exceedingly small "bull's eye."**

never would have discovered his gaps. Whether we like it or not, it often turns out that the truly important things in science are abstract. It is these that are significant.

We will encounter the abstractness of science at many points throughout the book.

Collision with an asteroid?

While most remain within the belt, beyond the orbit of Mars, some asteroids have orbits extending into the inner Solar System. Some of these orbits even intersect that of the Earth. These are called the *Earth-crossing asteroids*. Do they pose a threat to us? Could they collide with the Earth?

A body whose orbit crosses that of the Earth does not automatically pose a threat to us. As we already saw, the question is whether it ever gets close to the Earth itself. As illustrated in Figure 9.7(a), it is entirely possible that when it crosses the orbit of the Earth, we lie safely far away. But this configuration does not always obtain. In Figure 9.7(b) we illustrate a configuration in which there is a serious danger of a collision.

In 1991 this configuration actually came to pass, and an asteroid passed within the Moon's orbit. You might think that this was a pretty distant encounter, since the Moon lies a full 384 thousand kilometers from us. But on the astronomical scale, 384 thousand kilometers is a very narrow miss indeed.

Let us show this by means of an analogy. Figure 9.22 illustrates the Earth-crossing asteroid on its way toward us. Also indicated in this figure is the Moon's orbit.

We can think of this orbit as a "bull's eye": while the asteroid missed the Earth, it did manage to hit this "target." As an analogy, we can compare this to the task of shooting a bullet at a target. Let us study how accurately we would need to aim the bullet in order to hit the bull's eye.

The logic of the calculation

Step 1. Set up a proportion:

$$\frac{\text{diameter of Moon's orbit}}{\text{distance asteroid traveled}} = \frac{\text{diameter of bull's eye}}{\text{distance bullet traveled}}.$$

Step 2. Solve the proportion for the diameter of the bull's eye.

Step 3. Suppose the target is 100 feet away. Plug in the diameter of the Moon's orbit and the distance the asteroid traveled to find the diameter of the bull's eye.

As you can see from the detailed calculation, the asteroid's passing this close to the Earth was like hitting a two-inch target at a distance of one hundred feet – a very good aim indeed! Of course that asteroid had not been actually aimed at the Earth.

Detailed calculation

We have our proportion.

Step 2. Solve the proportion for the diameter of the bull's eye.
We get

diameter = (distance bullet traveled) (diameter of Moon's orbit)/(distance asteroid traveled).

Step 3. Suppose the target is 100 feet away. Plug in the diameter of the Moon's orbit and the distance the asteroid traveled to find the diameter of the bull's eye.

The Moon lies 384 thousand kilometers from the Earth, so that the diameter of its orbit is 7.68×10^5 kilometers. Let us suppose that the asteroid came from the central portions of the asteroid belt, lying perhaps 3 AU from us. Then the distance the asteroid traveled was (3 AU) (1.496×10^8 kilometers/AU), or 4.49×10^8 kilometers. Then this proportion tells us that the diameter of the bull's eye is:

diameter of bull's eye = (100 feet) (7.68×10^5 kilometers)/
(4.49×10^8 kilometers)
= 0.17 feet,

which is two inches (5 centimeters).

⇐ **Looking backward**
Recall our discussion in Chapter 3 of Kepler's second law.

Its path was entirely random. Nevertheless, purely by bad luck, it had come close to striking us.

NOW YOU DO IT
Suppose an asteroid in an elliptical orbit passes within 100 000 kilometers of the Earth. Consider an analogy of shooting a bullet at a target and hitting a bull's eye 2 inches (5 centimeters) across. We want to calculate how far away the target would need to be to make this a good analogy. (A) Describe the logic of the calculation you will use. (B) Carry out the calculation.

Suppose an asteroid were actually to strike the Earth, and that it came from 2.5 AU away. Consider an analogy of a mother bird dropping a worm into a baby bird's mouth which is 1 inch (2.5 centimeters) across. We want to calculate how far away from the baby the mother would need to be to make this a good analogy. (A) Describe the logic of the calculation you will use. (B) Carry out the calculation.

How devastating would collision with an asteroid be? It depends on how rapidly the asteroid is moving, and on its mass. Recall our discussion in Chapter 3 of the fact that an orbiting body picks up speed as it moves inward toward the Sun. By the time it reaches our orbit, an asteroid that has traveled in from the asteroid belt will be moving at about 42 kilometers per second. Of course, we too are moving. The Earth's orbital velocity about the Sun is 30 kilometers per second.

What matters is the *relative velocity* between us and the asteroid. This depends on the direction each is moving. In Figure 9.23(a), in which the asteroid is "chasing" the Earth, its velocity relative to us is 12 kilometers per second. But in Figure 9.23(b), in which Earth and asteroid are heading directly at each other, the relative velocity is far higher: 72 kilometers per second. Other orientations, such as the "crossing" configuration of Figure 9.7, yield relative velocities in between these values.

We emphasize that these are enormous speeds. Even the smallest, 12 kilometers per second, is nearly 27 thousand miles per hour. Detailed calculations show that collision with an asteroid one kilometer in diameter would release the energy of one thousand 100-megaton bombs. Furthermore, one kilometer is not very big as asteroids go. Many have even larger diameters, and would release even greater energy in a collision.

NOW YOU DO IT
Suppose that an asteroid moving at 42 kilometers per second were to approach Mars. Recall that the farther a planet is from the Sun the slower it orbits. Would you expect the collision between the asteroid and Mars to be

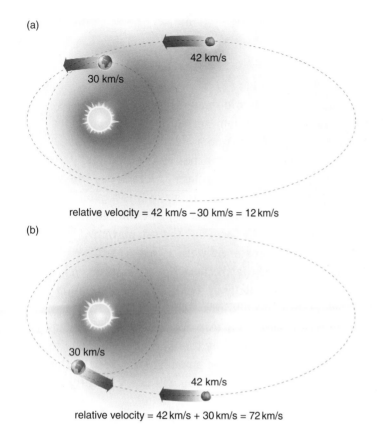

(a)

42 km/s

30 km/s

relative velocity = 42 km/s − 30 km/s = 12 km/s

(b)

30 km/s

42 km/s

relative velocity = 42 km/s + 30 km/s = 72 km/s

Figure 9.23 **The velocity of approach of an asteroid depends on which direction it is going.**

more catastrophic, less catastrophic or as catastrophic as a collision with the Earth? Suppose an asteroid moving at 35 kilometers per second were to approach the Earth. What would be the greatest and least relative velocity between us and the asteroid?

The extinction of the dinosaurs

Collision with a small asteroid would have little effect on the Earth as a whole. But collision with a large one would be catastrophic.

The dying out of the dinosaurs 65 million years ago was one of the most dramatic extinction events known. Within a relatively short amount of time most dinosaur species went extinct. Furthermore, the dinosaurs were not the only group to suffer this fate. A significant fraction of all the species that existed at the time appears to have also died out. For this reason, paleontologists speak of this as something entirely beyond the normal pattern of extinctions: *a mass extinction.*

For many years the cause of this extraordinary event was unknown. But in 1980 evidence was found linking it to a giant impact. This evidence was an excess of the element iridium in the geological layer marking the epoch of the mass extinction. In itself iridium is not particularly dangerous in any way. Its significance for the extinction is that it is rare on the Earth – but it is common in asteroids and comets.

If we hypothesize that this iridium was brought to Earth by collision with an extraterrestrial body, we can calculate the body's size from the amount of iridium in the layer. The size calculated in this way works out to about 10 kilometers. That is the diameter of an asteroid or comet.

Geologists have found abundant evidence that such a cataclysm actually did occur. Excess iridium has been found in geological strata marking the extinction event everywhere across the globe, and in undersea sediments as well. A type of mineral known as shocked quartz, which is only produced under enormous pressures, has also been found in these strata. Fossilized deposits indicative of giant tsunamis have been found, and enormous quantities of soot indicating gigantic worldwide forest fires.

The magnitude of such an impact is hard to imagine. As we have seen, the velocity of the collision was enormous. The energy released in the impact was 100 billion times greater than the energy released in a giant earthquake, which itself causes untold devastation. So large was the impacting object that its top would have been above most of the atmosphere at the moment its base struck the ground. The crater it dug, now buried by sediments, is known as Chicxulub, for the Mexican village of that name within it. It is over 150 kilometers across (Figure 9.24).

(a)

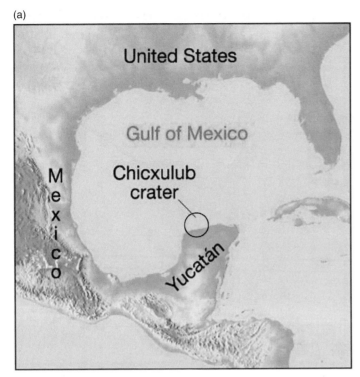

(b)

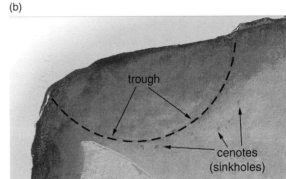

Figure 9.24 **The Chicxulub crater.**

THE NATURE OF SCIENCE

SCIENCE AND PUBLIC POLICY

HOW GREAT IS THE THREAT OF COLLISION?

Clearly such a collision with an asteroid or a comet would be a terrible, catastrophic event. Should we be concerned about this? What are the chances that it will occur? Figure 9.25 shows the rate of impacts of bodies of various sizes. We see from this graph that a 10-kilometer object of the sort that killed the dinosaurs collides with the Earth once every 100 million years.

. . . .

NOW YOU DO IT
(A) What size meteoroids collide with the Earth once a month? (B) What size meteoroid is likely to collide at some point between now and the year 2100?

. . . .

This impact, we know, occurred 65 million years ago. Should we conclude that the Earth is safe until the full 100-million-year interval is up – i.e. for another 35 million years? Unfortunately, we cannot draw this conclusion. Impacts are not *regular* processes, they are *random* processes. The data in Figure 9.25 does not mean that impacts happen on schedule every 100 million years. Rather it means that, in any 100-million-year interval, the chances are good that an impact will occur. So we do not know that we have 35 million years before the next collision. It might happen sooner.

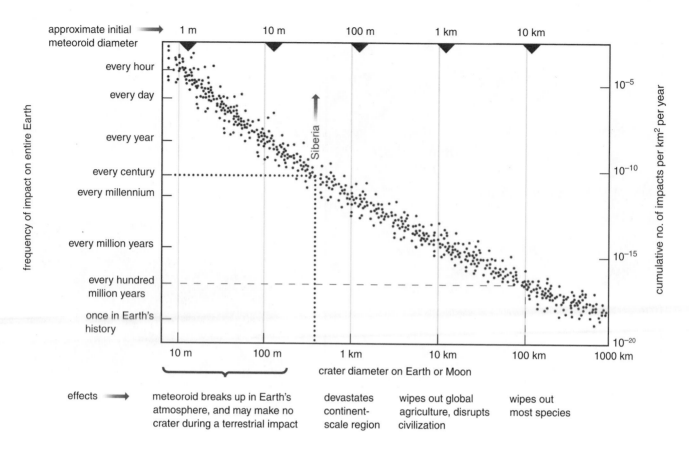

approximate initial meteoroid diameter →

effects →

| meteoroid breaks up in Earth's atmosphere, and may make no crater during a terrestrial impact | devastates continent-scale region | wipes out global agriculture, disrupts civilization | wipes out most species |

Figure 9.25 **Impact rates of bodies of various sizes with the Earth.**

How much sooner? Specifically, what are the chances that this catastrophe will happen *to you* – i.e. sometime during your life? To find out, we need to know how much longer you will live. While of course we don't know this, we can make a rough estimate as follows: if you are a college student you are probably about 20 years old right now, and your average life expectancy is close to 80 years. Therefore, if you are an average college student you will live for about another 60 years. Let us compare this with the average time between impacts:

60 years/100 million years = 1/17 million.

The span of your future life is only a 17 millionth of the time between impacts. Clearly, there is only the tiniest of chances that you will experience this terrible catastrophe.

NOW YOU DO IT
(A) Carry out the same calculation for a household pet. (B) Carry it out for an animal that lives a century.

So should you worry about it? Because our calculated probability is so low, one's initial thought might be to forget about the whole thing. But if an impact does occur, it is not just you who will die. Nearly everybody else will too! We should not be thinking about the 1 in 17 million chance that you are facing. We should be thinking of the 1 in 17 million chance that the entire Earth is facing: the chance of a terrible, planet-wide catastrophe, one in which billions of people die, industrial civilization comes to an end and many species go extinct.

The situation is similar to that of taking out an insurance policy. The chances are not very great that your house will burn down. But if this does happen the consequences will be terrible. So it behooves you to take out fire insurance.

CONGRESSIONAL HEARINGS

A series of congressional hearings has been held, during which testimony was heard from scientists and military concerned with the threat of impacts. Part of the testimony at one of these hearings referred to a distinctly non-scientific issue: the danger that a thermonuclear war might be triggered by a collision with an asteroid or comet. A military official testified as follows.

> Two and a half months ago, Pakistan and India were at full alert and poised for a large-scale war, which both sides appeared ready to escalate into nuclear war. The situation has defused – for now. Most of the world knew about this situation and watched and worried. But few know of an event over the Mediterranean on June 6th of this year that could have had a serious bearing on that outcome. US early warning satellites detected a flash that indicated an energy release comparable to the Hiroshima burst. We see about 30 such bursts per year, but this one was one of the largest we have ever seen. The event was caused by the impact of a small asteroid, probably about 5–10 meters in diameter, on the Earth's atmosphere. Had you been situated on a vessel directly underneath, the intensely bright flash would have been followed by a shock wave that would have rattled the entire ship, and possibly caused minor damage.
>
> The event of this June received little or no notice as far as we can tell. However, if it had occurred at the same latitude just a few hours earlier, the result on human affairs might have been much worse. Imagine that the bright flash accompanied by a damaging shock wave had occurred over India or Pakistan. To our knowledge, neither of those nations have the sophisticated sensors that can determine the difference between a natural NEO [Near Earth Object] impact and a nuclear detonation. The resulting panic in the nuclear-armed and hair-triggered opposing forces could have been the spark that ignited a nuclear horror we have avoided for over a half century.[1]

As a result of these hearings the House of Representatives, in its NASA Multiyear Authorization Act of 1990, directed NASA to convene a series of workshops evaluating means to find approaching objects and to divert them from their paths. Since these workshops, a variety of programs have started up designed to identify all asteroids whose orbits carry them close to the Earth. Since the late 1990s, these search programs have been discovering somewhat less than 100 large bodies – those more than 1 kilometer across – per year.

ASTEROID DEFLECTION: THE DILEMMA

A terrible dilemma faces us in thinking about how to deal with the threat of collision with an asteroid or a comet. Suppose one of these search programs discovers an object on a collision course with the Earth. What should we do?

---

[1] Testimony of Brigadier General Simon Worden, United States Air Force, before the Subcommittee on Space and Aeronautics, Committee on Science, US House of Representatives, October 3, 2002.

The obvious answer is that we should send a space mission out to this body, and somehow deflect it from its path. Perhaps the mission would explode a bomb alongside it, pushing it sideways a bit. Alternatively, rockets might be attached to the approaching object, which would steadily propel it sideways. Other schemes involve shining an intense laser beam on it, heating it so that it evaporates: the evaporating material would function as does exhaust from a rocket, also gently pressing it sideways.

This is clearly what we should do once we have discovered a threatening body. But at the current time, no asteroid or comet is known whose orbit poses a threat to our planet. As we have seen, there are more small asteroids than big ones. It is very unlikely that a watch program will discover an incoming 10-kilometer body that could cause catastrophic damage to Earth. However, a small asteroid could still cause damage and once we have spotted a threatening asteroid there will not be enough time to develop the technology required to deflect the object. So should we be investing time, money and resources to develop technology as a precaution for an event that is very unlikely but potentially devastating?

A further consideration to the debate was raised by Carl Sagan, an American astronomer, in the 1990s. Sagan noted that if we develop technology that is capable of deflecting asteroids, there is a risk that this technology may fall into the wrong hands and be misused to direct asteroids towards Earth. In creating technology to protect us from the unlikely but potentially catastrophic consequences of asteroid collision, we may inadvertently create a terrible new weapon.

SCIENCE AND PUBLIC POLICY

It is not scientists who will decide whether or not to launch a program to defend ourselves from an approaching asteroid. It is the entire country – say, the United States. Furthermore, the cost of launching this program is going to be gigantic, so that the only way to deal with the threat is to mobilize the resources of the entire nation – or indeed the entire world. And so, suddenly and without warning, scientists who study asteroids and comets have found themselves thrust willy-nilly into the public arena. What started out as pure, abstract scientific research into asteroidal orbits has moved into a larger sphere.

People used to regard scientists as other-worldly individuals, lost in their own research and far removed from the rough-and-tumble of the "outside" world. This popular perception might have been accurate once – but no longer. It was the terrible experience of the Second World War and the bombing of Hiroshima and Nagasaki that decisively marked this transition. As the historian of science Gerald Holton has written:

> a secret army of scientists, quartered in secret cities, was suddenly revealed to have found a way of reproducing at will the Biblical destruction of cities and of anticipating the apocalyptic end of man that has always haunted his thoughts. That one August day in 1945 changed the imagination of mankind as a whole... The traumatic experience of one brief, cataclysmic event on a given day can reverberate in the spirit for as long as the individual exists, perhaps as long as

the race exists. Hiroshima, the flight of Sputnik and of Gagarin – these were such mythopoeic events. Every child will know hereafter that "science" prepared these happenings. This knowledge is now embedded in dreams no less than in waking thoughts; and just as a society cannot do what its members do not dream of, it cannot cease doing that which is part of its dreams.[2]

Whether they like it or not, scientists are now regularly involved in public policy issues. They are called upon to testify before Congress, and they are employed by multinational corporations and the military along with universities and colleges. Science used to be thought of as an ivory-tower profession. No longer.

THE NATURE OF SCIENCE

CERTAINTY AND UNCERTAINTY IN SCIENCE

In our discussion of "Certainty and Uncertainty in Science" (in Chapter 3, "Newton's laws: gravity and orbits") we emphasized that while science brings certainty, it does so only in the long run. In the short run, scientists are no more sure of things than anyone else. Practising researchers, working at the cutting edge of knowledge, are incessantly plagued by uncertainty. In that discussion we were concerned with the difficulty of sending a spacecraft to Mars. Now we wish to return to the topic, this time with regard to the difficulty of deciding whether an approaching asteroid poses a threat of collision with the Earth.

You may have seen news reports in which astronomers announce that an asteroid has been found coming toward the Earth, and that there is such-and-such a probability that it will hit us. Why, you may have wondered, do they seem to be hedging their bets? Why don't they simply come out and say whether it will collide with us or not? Let us explore this question.

Suppose that you have discovered an asteroid that seems to be heading toward us. Is it exactly heading toward us? Is it going to collide with our planet, producing a worldwide catastrophe? Or will it zoom by in a harmless near miss?

To put this problem in context, let us imagine that technology has been developed capable of deflecting such an incoming asteroid from its path. Perhaps this technology involves launching a spacecraft toward the asteroid: once it arrives there it will shove the asteroid sideways, causing it to miss us. Of course, that spacecraft is going to take some time to reach its target. So it behooves us to launch it soon, giving it enough time to reach the asteroid. And this means that you, the astronomer who discovered the asteroid, have very little time to decide whether it poses a threat. Furthermore the mission to deflect the asteroid will cost many billions of dollars, so your prediction had better be right.

Figure 9.26 illustrates the configuration. Both the Earth and the asteroid are moving: projecting their paths forward in time, you conclude that they will come very close to colliding.

But notice that your conclusion depends on knowing how far the asteroid is from us. As illustrated in Figure 9.27, if the asteroid is actually farther away than you think, it will miss the Earth.

---

[2] Gerald Holton, "Scientific research and scholarship: notes towards the design of proper scales," *Deadalus* Spring 1962.

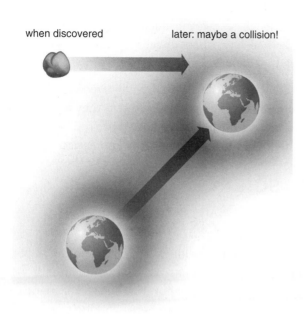

Figure 9.26 **Earth and asteroid, possibly on a collision course.**

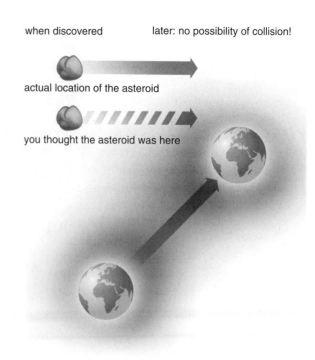

Figure 9.27 **Earth and asteroid not on a collision course.**

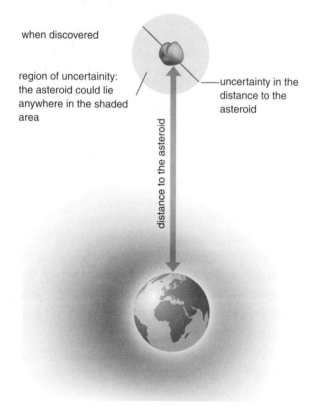

Figure 9.28 **You cannot find out exactly where the asteroid lies. All you can find out is that it lies somewhere within the shaded region.**

This means that, in order to decide whether it poses a threat, you must measure the distance to your asteroid. Suppose that you have done this. Of course, no measurement is perfect, so there is always going to be some uncertainty in your measurement. This means that there is some uncertainty in the location of the asteroid on the night that you discovered it (Figure 9.28).

Depending on where the asteroid really is, it will follow different paths. Figure 9.29 illustrates a configuration in which none of these possible paths intersects the Earth's. So although you cannot predict just how closely the asteroid will pass us by, you know for sure that there will be no catastrophe.

But notice that the region of uncertainty in Figure 9.29 is small. Suppose, however, that it is not. Suppose it is big. In this case you cannot be sure that the asteroid will miss the Earth. Maybe it will – but maybe it won't! Figure 9.30 illustrates a situation in which we cannot reach any definite conclusion at all.

Figure 9.29 illustrates a situation in which you can decide with certainty whether the asteroid poses a threat: it applies if the region of uncertainty is far smaller than the size of the Earth. Figure 9.30 illustrates the opposite

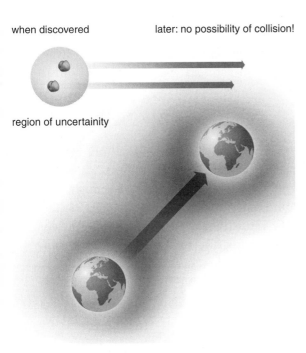

Figure 9.29 If the region of uncertainty is small, we can be sure the asteroid will not strike the Earth.

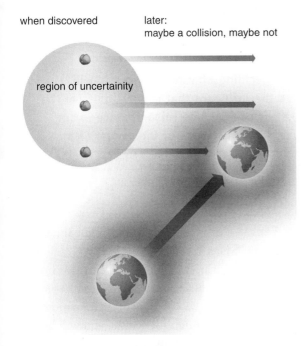

Figure 9.30 If the region of uncertainty is big, we cannot be sure whether the asteroid will strike the Earth or not.

situation, in which you cannot decide with certainty whether the asteroid poses a threat: it applies when the region of uncertainty is far larger than the Earth. This tells us that, in order to be sure of your conclusion, you need to be able to pinpoint the asteroid's location to within an accuracy far smaller than the size of the Earth.

To pinpoint the location of the asteroid, in turn, you need to measure its distance from us. Let us call the answer you get D. Because your measurement cannot be perfect, you will only be able to measure D to a certain accuracy. Let us call this accuracy ΔD. So the true distance to the asteroid could be $D + \Delta D$, or $D - \Delta D$, or anything in between. This tells us that the radius of the region of uncertainty is ΔD. Then, in order for you to be sure of your conclusion, ΔD must be a lot smaller than the size of the Earth, so that $\Delta D/D$ must be much less than the size of the Earth divided by the asteroid's distance.

Suppose you got $D = 2$ astronomical units when you measured your asteroid's distance. Then, in order to decide whether it poses a threat, you need $\Delta D/D$ to be much less than

(2)(radius of Earth)/(2)(one astronomical unit)
  = (radius of Earth)/(one astronomical unit)
  = 6.4×10^6 meters/1.5×10^{11} meters
  = 4.27×10^{-5}.

Let us get an intuitive feel for this by considering the analogy of an automobile collision. Suppose you are driving down a road and you spot a car heading down a cross street on what might be a collision course (Figure 9.31).

Suppose that car is about a mile away. To decide whether it is truly on a collision course you would need to find its distance to an accuracy of a mere few inches:

If $\Delta D/D$ is much less than 4.27×10^{-5}
then ΔD is much less than $(4.27 \times 10^{-5})(D)$
  = $(4.27 \times 10^{-5})(1 \text{ mile} = 5280 \text{ feet}) = 0.23$ feet
  = 2.7 inches (6.9 centimeters).

Furthermore, you need to do this quickly, while both you and the other car are moving, and while the other car is still quite far away.

In our discussion we have left out a lot of complicating factors. On the one hand, we have assumed that we knew the asteroid's velocity perfectly. In reality we do not – we need to measure it, and that measurement will be subject to its own uncertainty. Furthermore, neither the asteroid nor the Earth is traveling in a straight line: both are orbiting the Sun, so both are moving in ellipses. And finally, as we discussed in Certainty and Uncertainty in Science (in Chapter 3, "Newton's laws: Gravity and orbits"), both these orbits are slightly "wiggly" as they are tugged this way and that by gravity from all the planets of the Solar System.

Figure 9.31 **An approaching car.**

All these uncertainties combine to make it very difficult to decide whether an approaching asteroid actually poses a serious threat. And finally, as we have emphasized, if we ever do discover such an asteroid we will have very little time to figure out what to do about it.

THE IMPORTANCE OF ACCURACY

Scientists are perpetually striving for greater and greater accuracy. Suppose that astronomers spot a new object in the sky. You might think that would be good enough, and they would now turn their attention to other matters. But no – in reality they return again and again to the newly discovered object, observing it in greater and greater detail. Why this endless push for higher accuracy?

Our work in this section has answered that question. What we wanted to know was whether the approaching asteroid posed the threat of a terrible catastrophe. And we have seen that a low-accuracy measurement of the asteroid's distance told us nothing at all about this. All it could do is tell us that there might be a catastrophe – which is something we already knew! Only a high-accuracy measurement could answer our question.

We will return to the subject of the need for accuracy in the very next chapter, where we will realize that only super-high-accuracy observations are capable of telling us whether the distant stars have planets.

··

NOW YOU DO IT
In our discussion of the difficulty of deciding whether an asteroid will collide with the Earth, we neglected the influence that gravitation from the Earth has on the asteroid's path. Let us now include this influence in our thinking. Will this increase or decrease our estimate of the danger of a collision? Explain your reasoning.

··

Comets

Comets (Figure 9.32) move along highly elongated orbits that extend from the inner portions far into the outer reaches of the Solar System.

Only during the relatively brief period of time during which they are close to the Sun do they become visible. With their long, luminous tails comets – the name comes from Greek *aster kometes* which means *long-haired star* – are one of the most beautiful of all astronomical sights. If you are lucky enough to see one, you will be looking at the largest object you have ever seen: comet tails often exceed 1 AU (93 million miles, about 1.5×10^8 kilometers) in length.

But comets are rare. Unless you are a dedicated comet-hunter, you are likely to see only a few in your lifetime. While several comets appear each year, most require a telescope to be seen. Among the most recent truly spectacular comets was Hale–Bopp, which appeared in 1997. It was so bright that it could easily be seen with the naked eye. Indeed, one night I saw it while getting out of my car in a brightly illuminated downtown parking lot. The so-called "great comet of 2007," the

(a)

(b)

(c)

(d)

Figure 9.32 **Comets.**

brightest in over 40 years, was actually visible in broad daylight to observers in the southern hemisphere.

In 1910 the Earth passed through the tail of Halley's comet. Since spectroscopic observations had shown that certain poisonous gases were present in the tail, there was some fear that all life on Earth might be threatened. But so tenuous are the gases of a comet's tail that nothing of the sort happened. Indeed, a comet's tail is mostly empty space.

Like planets and meteors, comets move across the backdrop of the constellations. In contrast to meteors, though, they do so very slowly. A comet does not flash across the sky: it hovers in the same location for weeks, only slowly changing its position. In this regard a comet is like a planet. But in every other regard, cometary motions are unlike those of planets. While the motion of a planet is regular, most comets

Figure 9.33 **Edmond Halley.**

come at utterly unpredictable times. While planets orbit within the plane of the Solar System – the ecliptic – cometary orbits lie at any angle to the ecliptic. While planetary orbits are nearly circular, those of comets are highly elongated ellipses. And while all planetary orbits are prograde – in the same direction – there are just as many retrograde comets as prograde ones.

For all these reasons, comets stand in sharp contrast to the extreme regularity that is so characteristic of the heavens. In consequence, they were in past times often regarded as harbingers of ill fortune. The destruction of Jerusalem in 70 CE was widely thought to have been "predicted" by the appearance of a comet four years earlier (we now know it to be Halley's comet). In 1066 the Norman invasion of England coincided with a return of Halley's comet, while a threatened Turkish invasion of Europe coincided with yet another return in 1456.

For centuries the hypothesis was entertained that comets might be some rare form of atmospheric phenomenon. Not until 1577 was it shown that they lie far beyond the atmosphere. In that year Tycho Brahe observed a comet from his great observatory at Hven (Chapter 2) outside Copenhagen. Tycho later learned that an astronomer in Prague had also observed this comet – and had found it to be in the same position against the stars. But had the comet been in the atmosphere it would have been relatively nearby, and would have been at a different position against the stars as seen from Prague than from Copenhagen – just as a nearby tree appears to change its position against the horizon when you move your head from side to side. (The effect of changing perspective in astronomy is known as parallax: it is studied in Chapter 12.) Tycho concluded that comets lie very far from the Earth, and are celestial phenomena.

Halley's comet

In 1704 Edmond Halley (Figure 9.33) was computing the orbits of various historical comets. Four of them particularly struck his attention. All had retrograde orbits, together with other orbital similarities as well.

Here are the comets that struck Halley's attention:

- the Great Comet of 1456,
- the comet of 1531, observed by Petrus Apianus,
- the comet of 1607, observed by Kepler,
- the recent comet of 1682.

Halley noticed a striking fact about these comets: they occurred regularly. We can see this by computing the intervals between their dates.

Date	Time interval (years)
1456	
	75
1531	
	76
1607	
	75
1682	

Halley guessed that these were actually all the same comet, a comet that swung about the Sun with a period of roughly 75 years. Based on this hypothesis, he predicted when it would next appear.

Sure enough, the comet was indeed sighted on schedule – on Christmas night of 1758 by the German farmer and amateur astronomer George Palitzch, using a homemade telescope. Halley was by this time long dead, but the comet was posthumously named for him in his honor.

The return of Halley's comet at its predicted time provided a decisive confirmation of Newton's laws. Nowadays we might not find this very interesting, but at the time it was very important. After all, in those days Newton's theories were brand new – when Halley made his prediction, Newton's *Principia* was a mere 17 years old. As we have seen, it is essential that every new scientific theory be tested over and over again before we are to accept it as valid.

⇐ **Looking backward**
Recall "The nature of science. The importance of skepticism" in Chapter 2.

Furthermore, it is important to recognize that at the time Newton's theories had been tested only in a limited set of circumstances. We have already emphasized how different the orbits of the comets are from those of the planets. Indeed, the differences are greater than that, for whereas planets are relatively quiescent bodies, comets are active: they have tails that grow and shrink and change their orientation. It was entirely possible that they might respond to gravitation differently from planets. After all, wood responds differently from metal to a magnet. The fact that comets obeyed Newton's laws was decisive confirmation of the universality of gravitation. Wood and metal, comets and planets – all respond to gravity in the same way.

The physical nature of comets

We can learn much about the physical nature of comets from three observations.

(1) There is a striking relationship between the length of a comet's tail and its distance from the Sun. This is particularly easy to observe if we can see the comet and the Sun at nearly the same time – i.e. just after sunset, or just before sunrise. For instance, a comet might lie close to the western horizon shortly after sunset (Figure 9.34).

A single observation proves nothing, for there is nothing to indicate the direction of motion of the comet. If you were to see a comet in such a configuration, you would not be able to see it moving. Only as the weeks roll by would its motion become evident. In Figure 9.34 we illustrate a series of "snapshots" of the comet, taken over several weeks. In this series the comet's motion is evident. In Figure 9.34 the comet is approaching the Sun and, as we can see, its tail is growing longer as time passes. Similarly, after the comet has passed the Sun and is flying away from it. The tail grows smaller as time passes!

Figure 9.34 **The closer a comet is to the Sun, the longer is its tail.**

The closer a comet is to the Sun, the longer is its tail.

> ⇐ **Looking backward**
> Recall our discussion of
> spectroscopy in Chapter 4.

Table 9.4. **Composition of cometary tails**

Molecule	Abundance
H_2O	80%
CO	10%
CO_2	3.5%
Other molecules	A few %

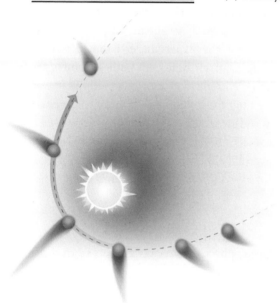

Figure 9.35 **Comet and Sun.**

(2) We can find the composition of a comet's tail by means of spectroscopic observations. It is found that there are two components to the spectrum.

- A set of spectral lines indicating the presence of gases in the tail. The composition of the gases deduced from these observations is shown in Table 9.4.

As you can see, a comet's tail is mostly composed of water vapor, together with a small admixture of other volatile compounds.

- There is a second set of spectral lines that are merely the reflected spectrum of the Sun. We interpret these as arising from some component of the tail that is reflecting sunlight to us. More detailed observations reveal that this component is a swarm of dust and small pebbles.

(3) Finally, there is a striking relationship between the direction a comet tail points and the direction to the Sun. As before, this is particularly easy to observe if we can see the comet and the Sun at the same time – i.e. just after sunset, or just before sunrise. We can see this by returning to our previous diagrams of comet tails. Figure 9.34 illustrated the configuration of a comet close to the western horizon shortly after sunset. Notice that in this figure the tail is pointing away from the Sun. This also turns out to be true in other configurations as well.

> Comet tails always point away from the Sun.

We might think of the comet's tail as streaming out behind it. But this turns out to be true only some of the time. It is true in Figure 9.34, in which the comet is approching the Sun. But as it recedes from the Sun, the tail is streaming out in front! Sometimes the tail follows the comet in its travels. Other times it leads it. But the tail always points away from the Sun.

Observations (1) and (3) are summarized in Figure 9.35.

Detectives on the case

What are the comets?

We are now ready to use these three observations to build an understanding of the physical nature of comets. Let us try to create in our minds a model that can explain the data. This is a process we have gone through many times before, and by now we are getting used to it. We will go through a similar process here.

Begin by remembering the clues that observations have given us. We know that a comet's tail is made of gas and tiny particles. The tail is not static. It keeps changing, in both length and direction. The image comes to mind of the gas and particles being emitted by something – something that we have not yet seen. What is this "something"? We will call it the *nucleus* of the comet. Our task is to understand what it is.

Whatever the nucleus is, it emits more material as it gets closer to the Sun. Why should this be true? It would be true if it emitted *because* it was closer to the Sun – i.e. because it was warmed by the Sun.

Because the tail of a comet is mostly H_2O, our nucleus must be primarily composed of H_2O as well. When a comet is far from the Sun it is very cold, so the H_2O must be in the form of ice. And as it approaches the Sun, the ice warms.

Does it melt? As we learned in Chapter 7 when we were considering the mystery of liquid water on Mars, water cannot exist as a liquid at low pressures. In the vacuum of interplanetary space, H_2O can exist only as a solid or a gas. So as the nucleus approaches the Sun, its ice does not melt: it steams away as water vapor.

Of course H_2O is not the only component of comet tails. Other substances such as CO and CO_2 are found there as well, together with minute solid particles. So the nucleus must be made of ice, frozen CO and CO_2 (dry ice) and dust and pebbles.

We have succeeded in creating a model: *the nucleus of a comet is a dirty snowball*. When the snowball is far from the Sun, it is so small as to be invisible to us. But as it approaches the Sun in its elongated orbit it is heated, and gases and solid matter stream away from it to form the tail.

Our model nicely explains most of our observations of comets. One observation is left: the fact that comet tails always point away from the Sun. Why should this be so?

If we consider a comet falling inward toward the Sun, we might interpret this as telling us that the gas and particles are simply left behind the nucleus as it travels. They might be like bits of paper tossed from a speeding car. But if we now consider our observations of a comet after it has passed the Sun, we see that this cannot be correct. Comet tails sometimes follow behind the comet – but sometimes they lead it. In contrast, stuff tossed from a speeding car always streams out behind it.

We need a different analogy. Think of bits of paper tossed from a *slowly moving car on a windy day*. Suppose in particular that the wind is blowing faster than the car is moving! Then as illustrated in Figure 9.35, sometimes the trail of paper follows behind the car, and sometimes it streams out ahead.

Observation (3) is therefore telling us that there is a wind in space. This is the solar wind, which steadily blows outward away from the Sun (we will study it further in Chapter 11). It combs the debris emitted by the cometary nucleus out into a long tail pointing away from the Sun. A further influence on comet tails is the pressure exerted by sunlight from the Sun, which has the same effect.

Let us pause now, and look back at the path we followed as we developed our theory. As before, many elements were present.

⇒ **Looking forward**
We will study the solar wind in Chapter 11.

- We used clues provided by observations as helpful hints as we developed our theory.
- We thought in terms of analogies: we thought of a comet as being like a car, and its tail as like bits of paper.
- There was no clear path toward our theory: we tried one idea and then another, and finally found one that seemed to work.

Now we need to test our theory – and as usual, we need to test it in as many ways as possible.

Confirmation of the theory

There are four separate lines of evidence that serve to confirm our theory.

(1) The theory predicts that comets eject tiny bits of solid matter. We noted in our discussion of meteors that meteor showers occur when the Earth in its motion crosses the orbit of a comet. We now understand why this is so: the meteoroids which produce the showers are the material ejected by the comet.

(2) Very precise measurements have shown that the orbits of comets are not exactly elliptical. There are tiny deviations, deviations that grow bigger as the comets approach the Sun. These deviations can be explained by remembering that our theory predicts that comets eject gases, and they do so more strongly as they approach the Sun. As they emit these gases the comets experience a reaction force via Newton's third law, which pushes them sideways. The comets are behaving like rockets!

(3) Our theory predicts that there is a wind in space, blowing comet tails away from the Sun. As we mentioned, this is the solar wind: it has been directly observed.

(4) These three confirmations constituted very strong evidence in favor of the theory. Years afterward, a fourth came from spacecraft observations, which directly observed the nucleus of a comet and verified the theory in detail. This, of course, was the most decisive validation of them all. It is important to note, however, that the first three confirmations were so compelling that very few scientists doubted the overall validity of the theory at the time of the space missions.

Anatomy of a comet

> **⇐ Looking backward**
> Recall from Chapter 2 that Kepler's second law shows that bodies in elliptical orbits move most slowly when farthest from the Sun.

When far from the Sun, the nucleus is all there is to a comet. It is a "dirty snowball," consisting of various ices mixed with dust and gravel. It can be between 0.1 and 100 kilometers in diameter. Since Kepler's second law (Chapter 2) shows that bodies in elliptical orbits move most slowly when farthest from the Sun, the comet spends most of the time in this state.

But as it falls inward toward the Sun, the nucleus picks up speed – and it warms. When it is a few AU from the Sun the ices begin to vaporize. This process can be somewhat violent, and the gas spews outward with a high velocity, carrying the dust and gravel with it. This ejected material forms a cloud termed the *coma*, which surrounds the nucleus: it can be up to 0.01 AU in diameter. Beyond this lies the tail, combed outwards by the solar wind and the pressure of sunlight: it can exceed 1 AU in length (Figure 9.36).

When the comet swings by the Sun it is moving most rapidly, and (other things being equal) it is ejecting the most material. Thereafter the comet flies outward away from the Sun, its tail preceding it and steadily shrinking – back into the deep freeze from which it came.

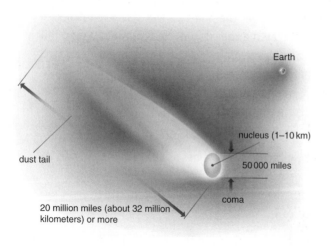

Earth

nucleus (1–10 km)

50 000 miles

coma

dust tail

20 million miles (about 32 million kilometers) or more

Figure 9.36 **Anatomy of a comet.**

Spacecraft observations of Halley's comet

All the information we have summarized above concerning cometary nuclei was obtained indirectly. Indeed, prior

Figure 9.37 **The nucleus of Halley's comet as observed by the Giotto spacecraft.**

to 1986 no one had ever actually seen the nucleus of a comet. But the situation changed when Halley's comet passed through the inner Solar System.

Four different spacecraft paid it a visit. Two Soviet probes conducted a detailed study of its tail. A Japanese probe studied ultraviolet light from its coma and tail. A European craft named Giotto (for the renaissance painter who depicted a comet in a painting) was sent deep into the coma, and observed the nucleus itself. Giotto's mission was somewhat dangerous, for the spacecraft could have been damaged or destroyed by "sandblasting" from the dust and gravel streaming away from the nucleus. Luckily the probe survived, and it returned the first close-up look humanity ever had of the nucleus of a comet.

Figure 9.37 is the image it obtained. Giotto found Halley's nucleus to be surprisingly dark. In fact it is one of the darkest objects in the Solar System – as black as coal.

In this image we see the black nucleus spewing forth gas, which itself is brightly illuminated by the Sun. Several "active regions" can be seen from which the gas is being emitted: these turn on and off as the rotation of the nucleus carries them into and out of direct sunlight. Halley's nucleus turned out to be irregular in shape, measuring 16 kilometers by 8 kilometers by 7 kilometers. Its density is surprisingly low: about one quarter that of water. So we cannot think of the nucleus as a solid chunk of ices: it is more of a loose, fluffy agglomeration.

Deep Impact

The Deep Impact mission, launched in 2005, is the only man-made object ever to have touched a comet. It consisted of a space probe that, upon reaching comet Tempel-1, hovered at a safe distance from it and then "fired a bullet" at the nucleus, and studied the debris from the impact.

The "bullet" was a 370-kilogram Smart Impactor – a cylinder 1 meter in diameter and 0.8 meters tall. It struck the comet's nucleus at 37 000 kilometers per hour. It contained an Impactor Targeting Sensor whose purpose was to sense the Impactor's trajectory, which could then be adjusted up to four times between release and impact, and to image the comet from close range. As the Impactor neared the comet's surface, this camera took high-resolution pictures of the nucleus that were transmitted in real time to the main spacecraft before it and the Impactor were destroyed. The final image taken by the Impactor was snapped only 3.7 seconds before impact.

Just minutes after the impact, the flyby probe passed by the nucleus at a distance of 500 km, taking pictures of the crater position, the debris from the impact, and the entire nucleus. The results showed that the comet contained more dust, and less ice, than expected: scientists compared it to talcum powder. The comet continued outgassing from the impact for 13 days, with a peak 5 days after impact. A total of 5 million kilograms of water and between 10 and 25 million kilograms of dust were blown away by the impact. The crater from the impact is about 150 meters in diameter.

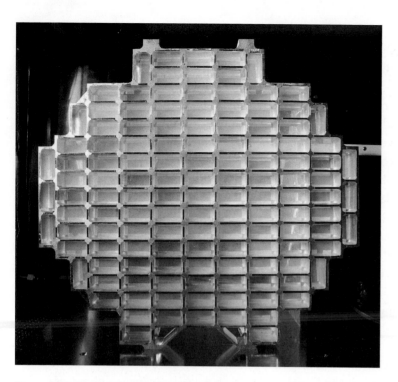

Figure 9.38 **Stardust's sample collector.**

The Stardust and Hayabusa missions

In 2004 NASA's "Stardust mission" flew past a comet, gathered some of the dust particles expelled from its nucleus and returned them to Earth. This was the first time material from a comet had ever been brought back to Earth. In 2005 the Japanese mission Hayabusa ("peregrine falcon") landed on an asteroid, collected tiny grains and returned them to Earth – also a first. We will focus on the Stardust mission, in order to illustrate the extraordinary nature of these undertakings.

Launched in 1999, the spacecraft was put into an orbit about the Sun that would bring it back past Earth for a gravity-assist maneuver in 2001. It then passed by asteroid Annefrank in 2002, and ultimately reached its target, comet Wild 2, two years later.

On January 2, 2004, Stardust encountered the comet at a distance of 237 kilometers. The orbit of Stardust was designed such that, as the spacecraft passed through the comet's coma, it was traveling at the relatively low velocity of 6.1 kilometers per second so as not to damage the dust grains as they were captured. The relative velocity between the comet and the spacecraft was such that the comet actually overtook the spacecraft from behind as they traveled around the Sun. During the encounter, the spacecraft was on the sunlit side of the nucleus.

Once in the vicinity of the comet, the spacecraft deployed a "dust collector" (Figure 9.38). The collector, about the size of a tennis racket, consisted of 90 blocks of aerogel, a porous, sponge-like material in which 99.8% of the volume is empty space. Aerogel is 1000 times less dense than glass. When a dust particle from the comet hit the aerogel, it became buried in the material, creating a long track. The aerogel collector stuck up out of a Sample Return Capsule, which itself looked somewhat like a giant clamshell. After passing through the comet's coma, the collector was folded into the capsule, which then clamped shut. The spacecraft then performed a deep-space maneuver that would allow it to pass by Earth a second time in 2006, to release the Sample Return Capsule for a landing in Utah.

On January 16, 2006, the Sample Return Capsule successfully separated from Stardust as it flew by the Earth, and re-entered the atmosphere at a velocity of 12.9 kilometers per second, the fastest re-entry speed ever achieved by a man-made object. The capsule then parachuted to the ground (Figure 9.39), finally landing at the Utah Test and Training Range. It was then transported to the Johnson Space Center's "clean room," which is 100 times cleaner than a hospital operating room. Only there was it opened, so as to ensure the collected grains would not be contaminated by Earthly dust.

Preliminary estimations suggested at least a million microscopic specks of dust were embedded in the aerogel collector. Remarkably, it also appears that about

Figure 9.39 **Stardust's return. The Sample Return Capsule, containing dust captured from a comet, parachutes to Earth (artist's conception).**

45 interstellar grains had also been captured. Among the many outcomes of the mission was an announcement by scientists from the University of Arizona that they had found evidence for the presence of liquid water in the comet. The discovery shatters the existing paradigm that comets never get warm enough to melt their icy bulk.

A subsequent mission extension was approved in 2007 to bring the spacecraft back to full operation for a flyby of comet Tempel-1 in 2011. This allowed it to visit this comet after it had been hit by Deep Impact, and to take an image of the impact crater, which had actually been rendered invisible to Deep Impact by the debris from the impact.

How long does a comet last?

Each time it passes close to the Sun in its orbit, a bit more of the nucleus of a comet is vaporized and blown away. Eventually it will be entirely vaporized, and nothing will be left beyond a loose collection of gravel. At this point, the comet will have become inactive, and will no longer be capable of generating a tail.

> We have reached an interesting conclusion: comets are ephemeral. They do not last forever.

How long do they last? How long does it take for a comet to be exhausted? We could find out if we knew:

(A) the mass of the nucleus,
(B) the rate at which it is losing mass.

If we divide (A) by (B) we will find how long the comet can survive. And spacecraft observations of Halley's comet have sent back data that will allow us to calculate both these quantities.

(A) Mass

The Giotto space probe measured the nucleus of Halley's comet to be 16 kilometers by 8 kilometers by 7 kilometers, and its density to be about one quarter that of water, i.e. about 250 kilograms per cubic meter.

The logic of the calculation

The mass of the nucleus is its density times its volume, and we know the density of the nucleus.

Step 1. To find the volume, multiply its length by its depth by its height.

Step 2. Multiply the volume by the density to find the mass.

As you can see from the detailed calculation, the comet's mass is about 2.3×10^{14} kilograms.

Detailed calculation

Step 1. To find the volume, multiply the length by the depth by the height of the nucleus.

We need to work in MKS units, so we need to convert kilometers to meters: there are 10^3 meters in one kilometer. So:

length = 16 kilometers = 1.6×10^4 meters,
width = 8 kilometers = 8×10^3 meters,
height = 7 kilometers = 7×10^3 meters.

Then the volume is

$V = (1.6 \times 10^4 \text{ meters}) (8 \times 10^3 \text{ meters}) (7 \times 10^3 \text{ meters})$
$= 9.0 \times 10^{11} \text{ meters}^3.$

Step 2. Multiply the volume by the density to find the mass.

Since the density is 250 kilograms/meter^3 we find

mass = $(9.0 \times 10^{11} \text{ meters}^3) (250 \text{ kilograms/meter}^3)$
$= 2.3 \times 10^{14} \text{ kilograms.}$

NOW YOU DO IT
(A) What is the mass of a cometary nucleus 12 kilometers by 6 kilometers by 3 kilometers if its density is 250 kilograms per cubic meter? (B) What is the mass if the density is 150 kilograms per cubic meter? In both cases (C) describe the logic of the calculation you will use. (D) Carry out the calculation.

(B) Rate of loss of mass

The Giotto spacecraft observed that Halley's comet was spewing out water vapor at the rate of 25 tons per second, together with between 5 and 10 tons per second of dust. Let us take the total rate of emission to be about 30 tons per second, or 2.7×10^4 kilograms per second.

We are now ready to calculate how long a comet can survive before it has blown away all its mass.

The logic of the calculation

We find the time a comet can survive by dividing its mass by its rate of loss of mass. But the calculation is more complicated than this, because comets do not emit material all the time. As we have seen, a comet spends most of its time far from the Sun, where it is quiescent. Only when it approaches the Sun does it begin to steam away. So we proceed as follows.

Step 1. Figure out how much time in each orbit the comet spends sufficiently close to the Sun to emit gases.

Step 2. Multiply the rate of emitting mass by the result of step 1 to find the total mass lost in each orbit.

Step 3. What fraction of the total mass of the comet is this?

Step 4. One divided by this fraction is the number of orbits about the Sun that the comet can make before it has entirely dissipated.

Step 5. Multiply the result of step 4 by the orbital period of the comet to find the total amount of time it can survive before dissipating away.

As you can see from the detailed calculation, a typical comet can survive for about 40 000 years.

Detailed calculation

Step 1. Figure out how much time in each orbit the comet spends sufficiently close to the Sun to emit gases.

   Comets vary: some emit for longer times than others. But observations reveal that a typical comet emits gases for very roughly half a year, or half a year = $(365/2)(24)(60)(60)$ seconds = 1.6×10^7 seconds.

Step 2. Multiply the rate of emitting mass by the result of step 1 to find the total mass lost in each orbit.

$$\text{Mass lost} = (\text{rate of mass loss})(\text{time})$$
$$= (2.7 \times 10^4 \text{ kilograms per second})(1.6 \times 10^7 \text{ seconds})$$
$$= 4.3 \times 10^{11} \text{ kilograms.}$$

Step 3. What fraction of the total mass of the comet is this?

   Since the mass of Halley's comet is 2.3×10^{14} kilograms, we see that in each passage by the Sun, the comet loses (4.3×10^{11} kilograms)/(2.3×10^{14} kilograms) = 1/530 of its mass.

Step 4. One divided by this fraction is the number of orbits about the Sun that the comet can make before it has entirely dissipated.

   It can survive 530 such passages before it has entirely evaporated away.

Step 5. Multiply the result of step 4 by the orbital period of the comet to find the total amount of time it can survive before dissipating away.

   Since the orbital period of Halley's comet is 76 years, it can survive only for about $(76)(530) = 40\,000$ years.

NOW YOU DO IT
Suppose a comet has a mass of 10^{14} kilograms, that it orbits the Sun once every 60 years and

that during the 3 months it is closest to the Sun it emits gases at a rate 10^3 kilograms per second. We want to calculate how long it can survive before being entirely dissipated away. **(A)** Describe the logic of the calculation you will use. **(B)** Carry out the calculation.

Consider two comets. They have the same mass, and take the same amount of time to complete an orbit. But one ("Comet A") remains sufficiently close to the Sun to emit gas for longer than the other ("Comet B"). Which one lasts longer?

Consider two comets. They have the same mass, and take the same amount of time to complete an orbit. But Comet A remains sufficiently close to the Sun to emit gas for three months each orbit, while Comet B does so for 4 months. If Comet A lasts 75 000 years, how long does Comet B last?

The Oort Cloud

Since our calculation has been approximate, we should not take the precise figure we have obtained too literally. But we can conclude with confidence that the active phase of a comet is at most a few tens of thousands of years. The durations of these active phases are utterly insignificant in terms of the history of the Solar System as a whole, which is 4.5 billion years old. Our calculation has been crude but our general insight is certain all the same: in astronomical terms, every comet is a very recent newcomer to the Solar System.

Suddenly we are faced with a mystery: if comets are so short-lived, why are they still around? There must be some reservoir somewhere, which continually supplies new comets. Otherwise they would all have become exhausted a few tens of thousands of years after the origin of the Solar System, and there would be none of them left.

Where is this reservoir? It is known as the *Oort Cloud* of comets, named after the Dutch astronomer who first hypothesized its existence. We can learn more about the Oort Cloud by studying the orbits of comets.

These observations reveal the following.

- Comets come in from far out in space.
- Their orbits have exceedingly long major axes: about 50 000 AU.
- Comets come in from every direction in space. There are just as many comets with orbits in one direction as in other directions. This is in contrast to planets, all of which orbit in the same direction.

From these we conclude that the Oort Cloud of comets is an enormous sphere surrounding the Solar System, with a radius of about 50 000 AU (Figure 9.40).

Our picture of the Solar System has suddenly expanded. Everything we have been studying up to now – the Sun and planets, the

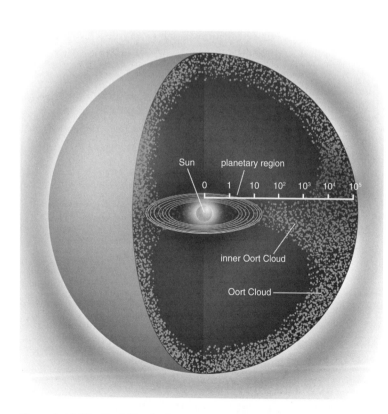

Figure 9.40 **The Oort Cloud of comets surrounds the Sun at enormous distances.**

moons and asteroids and meteors – all this shrinks to insignificance compared to the immense reaches of space spanned by this swarm of comets. The radius of the zone of planets – 40 AU is the radius of Pluto's orbit – is tiny compared with that of this huge cloud. Indeed, the Oort Cloud extends far out into interstellar space: ⅕th of the way to the nearest star.

The Oort Cloud contains a great number of comets. From observations of the rate at which new comets are discovered, we can calculate that it contains hundreds of billions of them. Remarkably, however, the total mass of the cloud is not very large. Since the mass of a single comet is 1.3×10^{14} kilograms, the total mass of all the comets in the Oort Cloud is:

$$\text{total mass} = (1.3 \times 10^{14} \text{ kilograms/comet})(10^{11} \text{ comets})$$
$$= 1.3 \times 10^{25} \text{ kilograms},$$

which is only a few times the mass of the Earth.

. . . .

NOW YOU DO IT
Suppose the Oort Cloud contained 10 billion comets, each of mass 5×10^{14} kilograms. What is the total mass of all these comets? Do they add up to more or less than the mass of the Earth? (A) Describe the logic of the calculation you will perform. (B) Carry out the detailed calculation.

Suppose each comet had a mass of 10^{13} kilograms. We want to know how many of them there would have to be to add up to the mass of the Moon. (A) Describe the logic of the calculation you will perform. (B) Carry out the detailed calculation.

. . . .

We can easily calculate the orbital period of a comet in the Oort Cloud. As we saw in Chapter 2, if the semimajor axis a of an elliptical orbit is measured in AU and the period P in years, Kepler's third law states

$$P^2 = a^3.$$

Since the *major* axis of a typical orbit is 50 000 AU, the semimajor axis is half this, or $a = 25\,000$ AU. Then we find

$$P = a^{3/2} \text{ years} = (25\,000)^{3/2} \text{ years} = 4 \text{ million years}.$$

. . . .

NOW YOU DO IT
Suppose their orbital major axes were 90 000 AU: what would be their orbital periods?

. . . .

⇒ **Looking forward**
We will see in Chapter 12 that stars move randomly about. We will look at the Milky Way in Chapter 16.

If you could be suddenly transported out to the Oort Cloud, you would be so far from the Sun that it would appear to be just another star. Indeed, it might be difficult to decide at first glance which among all the myriad stars in the sky is the one about which the cloud's comets are orbiting. So distant are these comets from the Sun that they are exceedingly cold: a mere few degrees above absolute zero. So weak is the Sun's gravity at such great distances that comets orbit very slowly, requiring millions of years to complete one circuit. Even if you had the Hubble Space Telescope with you out there, you would be so far from the zone of planets that you would be hard pressed to see them through it.

The cometary nuclei making up the cloud remain for untold ages in this state of "suspended animation." But once in a very long while some outside agency disturbs their orbits. Perhaps this agency is gravity from a passing star (as we will see in Chapter 12, stars move randomly about). Perhaps it is the passage of the Solar System together with its swarm of comets through the plane of our Milky Way Galaxy (Chapter 16). In any event, since gravity from the Sun is so weak out there, it does not take much of a force to disturb the comets' orbits.

As a result of this disturbance, whatever it might be, comets are scattered about. One might be directed, purely by chance, toward the far-distant Sun. Millions of years later, it enters the realm of the planets. Warmed by the Sun, it starts to evaporate. The evaporated material is combed outward away from the Sun by the solar wind to form a graceful tail. But the nucleus continues inexorably on its elliptical orbit, passes the Sun, and continues on its journey – back out to the frigid reaches of the Oort Cloud from which it came.

Long-period and short-period comets

You may have been struck by a curious discrepancy. We have just calculated that a comet in the Oort Cloud has an orbital period of 4 million years. But observations of Halley's comet show that it has a period of a mere 76 years.

In fact there are two classes of comets: long period and short period. Most comets (84%) are long period. The remainder (16%) are short period.

Kepler's third law

$$P^2 = a^3$$

tells us that while a long-period comet has a large semimajor axis, a short-period comet has a much smaller one. We have already seen that a period of 4 million years corresponds to a semimajor axis of 25 000 AU. Similarly, we can calculate the semimajor axis of the orbit of Halley's comet to be

$$a = P^{2/3}\text{AU} = (76)^{2/3}\text{AU} = 18 \text{ AU}.$$

For comparison, Uranus' orbital radius is 19 AU. So the path of Halley's comet carries it from our vicinity out to the realm of the outer planets – a long way, but nowhere near the Oort Cloud.

....

NOW YOU DO IT
What is the semimajor axis of a comet with an orbital period of a century?
What planet has a similar semimajor axis?

....

We believe that short-period comets are captured long-period comets. When a long-period comet from the Oort Cloud enters the realm of the planets, it usually passes through with no appreciable change in its motion. If so, it continues on its elliptical orbit back out to the cloud. But occasionally, and purely by chance, a comet will pass close by one of the planets. If so, gravity from the planet can distort its orbit. Once in a while the distortion is so great as to capture the comet into a

Figure 9.41 **A continuous range of compositions. Asteroids have little or no water ice: comets have lots.**

short-period orbit. (Because Jupiter has the most gravity, most of the time it is an encounter with Jupiter that results in capture.)

What is the difference between a comet and an asteroid?

In many ways it is hard to tell the difference between the nucleus of a comet and an asteroid. As we have noted, cometary nuclei range in diameter up to 100 kilometers. While this is smaller than the largest asteroid (914 kilometers), many asteroids are of just this size. If the orbit of a comet happens to lie in the plane of the Solar System, when far from the Sun it might be very difficult to distinguish from an asteroid.

This difficulty was dramatically emphasized when it was found that Asteroid 4015, which had been discovered in 1979, had the same orbit as Comet Wilson–Harrington, discovered in 1949. The two were the same body!

It may be that the most fruitful way to think is to order the smaller bodies of the Solar System in terms of the quantity of water ice they contain. Perhaps we should think of a continuous spectrum, as depicted in Figure 9.41.

"Asteroid" and "comet" would then merely be labels we apply to the two ends of the spectrum. Asteroids would be defined as bodies with little or no ice, and comets as bodies with lots of ice. In this view a burned-out comet might even *be* an asteroid.

One difference between asteroids and comets is that they have very different orbits – asteroids lie in the asteroid belt, and comets have highly elliptical orbits ranging throughout the Solar System and out into the Oort Cloud. But the more we observe, the more we discover "asteroids" lying outside the main asteroid belt. A striking example is a body known as Chiron. Discovered in 1977, Chiron was found to have a spectrum matching that of the asteroids, so that is what it was initially thought to be. Its orbit, however, was quite elliptical, and lay far from the asteroid belt, beyond the orbit of Saturn. Furthermore, as the years passed and it moved toward the Sun along its orbit, it began to emit gases. The "asteroid" was turning into a "comet"!

Recently this has led to the creation of a new category: that of the main-belt comet. This is a comet with a nearly circular orbit lying within the asteroid belt. We may also think of them as icy asteroids.

Kuiper Belt Objects: is Pluto a planet?

⇐ **Looking backward**
In the previous chapter we saw that in many ways Pluto is quite unusual.

Kuiper Belt Objects (named after the Dutch-American astronomer who first hypothesized their existence) are a group of small bodies orbiting between 40 and 400 AU from the Sun. Thus they occupy the outer fringes of the realm of the planets. They appear to be physically similar to the nuclei of comets in that they are composed of ices mixed with pebbles and dust. A further similarity between the Kuiper Belt Objects and comets are their quite elliptical orbits. The only differences are that the orbits of these objects lie relatively close to the plane of the Solar System, and far closer to the Sun than the Oort Cloud.

In contrast to the outer planets, which are large, low density, gaseous bodies, Pluto is small, high density and composed of rock and ices. It moves along an elliptical orbit, in contrast to the planets, which move in highly circular orbits.

All in all, Pluto looks quite unlike the outer planets. *But it looks a lot like a Kuiper Belt Object.* Many astronomers feel that Pluto is not really the outermost planet at all. They feel that it is the first Kuiper Belt Object to have been discovered. This is one of the reasons that led astronomers to remove Pluto from the list of planets in 2006. It is currently classified as a dwarf planet, a recently invented category.

The New Horizons mission to Pluto

The New Horizons mission was originally planned as a voyage to what was then the only unexplored planet in the Solar System. Indeed, when the spacecraft was launched, Pluto was still classified as a planet. Pluto's newly discovered satellites, Nix and Hydra, also have a connection with the spacecraft: the first letters of their names, N and H, are the initials of "New Horizons." The moons' discoverers chose these names for this reason, in addition to Nix and Hydra's relationship to the mythological Pluto.

Recall our analogy that launching a spacecraft is like throwing a stone. In particular, launching one in a direction away from the Sun is like throwing a stone upwards – and, since Pluto is so distant from the Sun, our "stone" must be thrown upwards with great force. Indeed, New Horizons is one of the fastest objects ever launched. The spacecraft took only nine hours to reach the Moon's orbit – contrast this with the three days the Apollo astronauts had taken to travel this far.

New Horizons was launched on January 19, 2006: it passed the orbit of Saturn in 2008, and that of Uranus in 2011. It is projected to reach its destination on July 14, 2015, where it will conduct observations of both Pluto and its moon Charon.

After passing by Pluto, New Horizons will continue farther into the Kuiper Belt. Mission planners are now searching for one or more additional Kuiper Belt Objects for similar flybys. Since the spacecraft will have very little fuel left for maneuvering, this phase of the mission is contingent on finding suitable objects close to its flight path. Sadly, this rules out any possibility for a flyby of Eris, a Kuiper Belt Object comparable in size to Pluto. Once the craft is more than 55 astronomical units from the Sun, the communication link with it will become too weak to be used.

To commemorate the discovery of Pluto, 1 ounce (28.35 grams) of the ashes of Pluto's discoverer Clyde Tombaugh are aboard the spacecraft. Also, one of the science packages onboard – a dust counter built by students at the University of Colorado – is named after Venetia Burney, who at the age of 11 named Pluto after having learned of its discovery from her grandfather.

Origin of comets

In Chapter 13 we will study the origin of the Solar System. An important conclusion of that chapter is that the initial stage in the formation of planets was the condensation, out of a hot rotating disk of gas, of objects known as *planetesimals*: small clusters of frozen H_2O, CO and CO_2 mixed with dust and pebbles. These planetesimals collided with one another to form the planets. We can think of them as being the building blocks out of which the Solar System was formed.

> ⇒ **Looking forward**
> In Chapter 13 we will study the origin of the Solar System.

> ⇐ **Looking backward**
> We studied orbits in Chapter 3.

There is a striking similarity between the physical makeup of these planetesimals and the nuclei of comets.

> Indeed, we believe that comets are leftover planetesimals – planetesimals that did not amalgamate together to form moons and planets. They are like unused bricks left lying around a construction site when the job is done.

How can we understand the fact that the moons and planets lie in a plane, but the comets of the Oort Cloud do not? In the final stages of the formation of the planets, numerous planetesimals must have passed near them without suffering actual collisions. During these encounters the planets' strong gravitational pulls would have gripped the planetesimals and thrown them outward, far from the Solar System, there to orbit in what we know today as the Oort Cloud. The Kuiper Belt Objects are merely those planetesimals lying in the outermost fringes of the "construction site," where there were no planets available to fling them outward.

If this is true, the Oort Cloud and Kuiper Belt are cosmic time capsules, in which the original building blocks of the Solar System have been preserved unchanged for billions of years. For this reason, we are particularly interested in sending space probes to study these comets. Because these objects are so enormously far from us, this will be a technically difficult task. We have, of course, already sent a probe to Halley's comet. This, however, is a nearby comet, one that has passed close to the Sun numerous times and suffered much expulsion of material. The truly pristine bodies, which have been completely unchanged for cosmic ages, are far more distant and difficult to study.

Summary

Meteors
- Meteors often come in meteor showers, which are associated with comets.
- Even though they are heated to incandescence by their passage through the atmosphere, meteors are quite cold when they land.
- Smaller meteors burn up in the atmosphere: big ones dig immense craters.
- We find their ages via radioactive age dating and cosmic ray exposure age dating.
- These ages tell us much about the past history of meteors.
- Meteors originate in the asteroid belt.

Asteroids
- Asteroids swing about the Sun in highly elliptical orbits lying between Mars and Jupiter.
- There are many small asteroids, and few large ones. This is evidence for collisions among asteroids, leading to their fragmentation.
- Brightness variations allow us to infer their rotation rates.
- The so-called "Kirkwood gaps" are gaps in the distribution of asteroid semimajor axes: they arise from synchronized gravitational tugs from planets.
- There is danger of a collision with an asteroid. In the past such collisions have led to mass extinctions.

Comets
- Comets appear without warning: in contrast to planets, they seemed to early astronomers to be unpredictable.
- However, Edmond Halley found that in fact they are predictable: they move in highly elliptical orbits leading far into the outer reaches of the Solar System. This leads to a test of Newton's laws.
- Comet tails grow longer as they approach the Sun, they always point away from the Sun, and they contain volatile compounds such as H_2O.
- Comets are ephemeral: they come from the Oort Cloud.
- Short-period comets are captured from this cloud.

- Comets and asteroids may really be similar, differing only in the amount of water they contain.
- The Kuiper Belt lies beyond the orbit of Pluto: indeed Pluto may be not a planet but a Kuiper Belt Object.
- Comets are thought to be remnant planetesimals, "building blocks" left over from the formation of the Solar System.

The nature of science
- Many scientific concepts are abstract, and have no counterpart in daily life. The Kirkwood gaps are an example: energy is another.

- Often there is an intimate connection between science and public policy. Scientists make discoveries that raise important public issues.
- In order to make a prediction, we often need highly accurate observations. Lacking them, scientists are forced to live in a state of uncertainty.

DETECTIVES ON THE CASE

- Observations of comet tails give us clues that help us formulate a model of a comet as a "dirty snowball." This is confirmed by spacecraft observations.

Problems

(1) Suppose a meteoroid traveling at 70 kilometers per second strikes the Earth. For how long is it heated as it travels through the atmosphere?

(2) Explain in words why the answer to problem (1) depends on whether the meteor was traveling on a slanting path through the atmosphere, as opposed to straight down.

(3) Suppose we find a meteorite with 75% of its initial ^{235}U present. How many half-lives have passed since it formed? How many years?

(4) In our discussion of radioactive age dating, we concentrated on the isotope ^{235}U of uranium. A different isotope, ^{238}U, decays with a half-life of 4.5 billion years. (A) Carry through the analysis of how much ^{238}U is present after one, two and three half-lives. (B) Draw a graph, similar to Figure 9.5, showing how much ^{238}U is present after 4.5 billion years, 9 billion years and 13.5 billion years. (C) Suppose we find a meteorite with 75% of its initial ^{238}U present. How many half-lives have passed since it formed? How many years?

(5) Suppose a rock sample is found that contains 30 times as many helium nuclei as ^{235}U nuclei. How many half-lives have passed since it last melted and then solidified? How many years?

(6) Return to our isotope ^{238}U, which decays with a half-life of 4.5 billion years. Suppose we find a meteorite with four times as many helium as ^{238}U nuclei. How many half-lives have passed since it formed? How many years?

(7) What is the ratio between the number of helium nuclei and the number of ^{235}U nuclei after *four* half-lives since the solidification of ^{235}U?

(8) Suppose there were an element – we will call it X – that decays into *six* helium nuclei, rather than into seven:

$$X \rightarrow Pb + 6He$$

with a half-life of 500 million years. (A) Carry through the analysis of how many helium and X nuclei are present after one, two and three half-lives. (B) Draw a graph, similar to Figure 9.5, showing the ratio of the number of helium to X nuclei present after 500 million, one billion and 1.5 billion years. (C) Suppose we find a meteorite with 13 times as many helium as X nuclei. How many half-lives have passed since it formed? How many years?

(9) Suppose you found a meteorite whose solidification age was 100 million years, and whose cosmic ray exposure age was 250 million years. Speculate on the history of this object.

(10) We have seen three examples of *distributions* in this chapter: the distribution of asteroid sizes (Figure 9.10), the distribution of sizes of people (Figure 9.11) and the distribution of the semimajor axes of asteroids (Figure 9.16). In this problem we get some practice of our own with distributions.

(A) Suppose you surveyed a great number of
- basketball players,
- thoroughbred racing jockeys,
and you measured their heights. Draw two sketches, each similar to Figure 9.11, in which your results are displayed.

(B) Supposed you surveyed a great number of people in
- a large city,
- a college campus,
and you measured their ages. Draw two sketches, each similar to Figure 9.11, in which your results are displayed.

(11) Draw diagrams similar to Figure 9.11 showing the distribution of sizes of human beings in (A) an elementary school, (B) a basketball team.

(12) In our discussion of asteroid collision and fragmentation, we analyzed what would happen if after 100 000 years half the asteroids had collided and split in two. Suppose rather that each of these colliding asteroids had split into *four equal pieces*. (A) What would be the distribution of asteroid masses at the end of this 100 000-year span? (B) Draw a sketch, analogous to Figure 9.13, showing this distribution.

(13) Suppose an asteroid in an elliptical orbit passes within 150 000 kilometers of the Earth. Consider an analogy in which this is like shooting a bullet at a target and hitting a bull's eye three inches across. We want to calculate how far away the target would need to be to make this a good analogy. (A) Describe the logic of the calculation you will use. (B) Carry out the calculation.

(14) Suppose an asteroid were to strike the Moon, and that it came from 2.5 AU away. Consider an analogy of shooting a bullet at a target and hitting a bull's eye two inches across. We want to calculate how far away the target would need to be to make this a good analogy. (A) Describe the logic of the calculation you will use. (B) Carry out the calculation.

(15) Figure 9.23 shows an asteroid moving at 42 kilometers per second approaching the Earth (velocity 30 kilometers per second). (A) Suppose alternatively that the asteroid were moving at 60 kilometers per second. What would be the maximum and minimum relative velocities between asteroid and Earth in this case? (B) Suppose alternatively that an asteroid moving at 42 kilometers per second were to approach Venus. Recall that the closer a planet is to the Sun the faster it orbits. Would you expect the collision between asteroid and Venus to be more catastrophic, less catastrophic, or as catastrophic as a collision with the Earth?

(16) (A) What size meteoroids collide with the Earth once a year? (B) What size meteoroid is likely to collide at some point between now and the day you die? (See Figure 9.25.)

(17) Suppose we ask not whether an incoming asteroid will strike the Earth, but whether it will strike the Moon. (i) How small does $\Delta D/D$ have to be for us to decide whether it poses this threat? (ii) Consider an analogy in which the asteroid is a car roughly two miles away. How accurately would we need to measure the car's distance in order to decide if it actually poses a threat? Suppose that your measured distance to the asteroid is 3 astronomical units. For both questions (A) describe

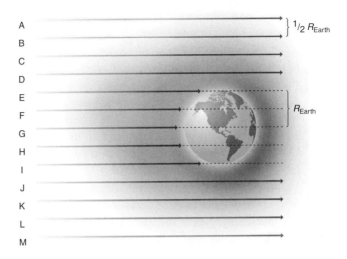

Figure 9.42 **Possible paths of an incoming asteroid.**

the logic of the calculation you will use, and (B) carry out the calculation.

(18) Which planet is in greater danger of a collision with an asteroid: the Earth or Jupiter? Explain your reasoning.

(19) Figure 9.42 illustrates a number of paths that an incoming asteroid might follow. They lie a distance $R_{Earth}/2$ from each other. As you can see, path G leads directly through the center of the Earth, paths F and H lead halfway from the Earth's center to its edge and paths E and I just graze the Earth's surface. The remaining paths miss the Earth.

If the region of uncertainty were very small, we would be confident that a collision with the Earth would occur for only three of these possible paths, namely F, G and H. Similarly, we would be sure that there would be no collision for eight paths, namely A–D and J–M. Finally two paths (E and I) just graze the Earth's surface – a "sort-of" catastrophe.

But suppose the region of uncertainty were not small. Then there can be a possibility of collision for some of the paths A–D and J–M.

Your task is to decide *which paths allow for at least the possibility of collision if the region of uncertainty is large*. In your work, study two cases:
• the radius of the region of uncertainty equals half the radius of the Earth;
• the radius of the region of uncertainty equals twice the radius of the Earth.

(20) Suppose the nucleus of a comet has dimensions 20 kilometers by 11 kilometers by 9 kilometers and its density is 125 kilograms per cubic meter. We want to find its mass. (A) Describe the logic of the calculation you will use. (B) Carry out the calculation.

(21) Suppose a comet's nucleus is a sphere 10 kilometers in diameter, with a density of 400 kilograms per cubic meter. We want to find its mass. (A) Describe the logic of the calculation you will use. (B) Carry out the calculation.

(22) A spacecraft measures the mass of the nucleus of a comet to be 1.5×10^{14} kilograms. It is spherical, with a radius of 4 kilometers. We want to find its density. (A) Describe the logic of the calculation you will use. (B) Carry out the calculation.

(23) Suppose a comet has a mass 10^{14} kilograms, that it orbits the Sun once every 50 years and that during the 4 months it is closest to the Sun it emits gases at a rate 10^5 kilograms per second. We want to calculate how long it can survive before being entirely dissipated away. (A) Describe the logic of the calculation you will use. (B) Carry out the calculation.

(24) Suppose you sleep for 8 hours each night. We want to find out how much time elapses before you have slept for a total of one year. (A) Describe the logic of the calculation you will use. (B) Carry out the calculation. (C) What does this have to do with how long comets last? (D) You will probably spend four years going to college. During your time in college, how much time will you have been asleep?

(25) Consider two comets. They have the same orbits, but one ("Comet A") is twice as massive as the other ("Comet B"). Suppose Comet A lasts 100 million years. How long will Comet B last?

(26) Consider two comets. They have the same orbits, but Comet A lasts three times longer than Comet B. If the mass of Comet A is 10^{14} kilograms, what is the mass of Comet B?

(27) Suppose the Oort Cloud contains 12 billion comets, each of mass 3×10^{14} kilograms. What is the total mass of all these comets? Do they add up to more or less than the mass of the Earth? (A) Describe the logic of the calculation you will perform. (B) Carry out the detailed calculation.

(28) Suppose each comet has a mass of 10^{14} kilograms. We want to know how many of them there would have to be to add up to the mass of the Earth. (A) Describe the logic of the calculation you will perform. (B) Carry out the detailed calculation.

(29) Suppose the comets' orbital major axes are 100 000 AU: what would be their orbital periods?

(30) What is the semimajor axis of a comet with an orbital period of 35 years?

You must decide

(1) Suppose you have been appointed Presidential Science Advisor. The President is trying to decide whether to initiate a crash program aimed at developing technology capable of deflecting an incoming asteroid or comet. She is haunted by two scenarios:

In the *first scenario* no attempt is made to develop this technology. However, by a terrible and unlikely chance a 10-kilometer asteroid is discovered heading toward us. Frantic efforts are made to develop a means of shoving it aside, but there is not enough time. The asteroid collides with the Earth, causing worldwide destruction and loss of life.

In the *second scenario* enormous expense is incurred in developing a means of deflecting an incoming asteroid or comet. (This expense is diverted from programs designed to provide affordable health care for all.) No incoming asteroid is discovered. But in spite of the fact that this program has been classified top secret, a spy obtains the details of the technology and sells

them to an enemy nation. This nation uses the technology to divert a 1-kilometer asteroid toward the Earth. It lands within the continental United States, causing terrible destruction and loss of life.

Your task is to write a memorandum analyzing these two scenarios, and to recommend which course of action the President should take. Your memorandum should include a *detailed* account of your reasons for choosing this course of action – reasons which the President will be able to use in explaining her choice to the American people during the upcoming election.

(2) Suppose you are charged with deciding whether Clyde Tombaugh should receive the Nobel Prize posthumously for having discovered Pluto. Considering the discussion both in this chapter and the previous one, what is your decision and how will you defend it against critics?

(3) **The "Doom Theory" of comets**   At many places throughout this book[3] we have described the

---

[3] See "The nature of science. Lessons from history" in Chapter 2, and "Detectives on the case" in this and the previous chapters.

characteristics a theory must have in order for it to be a legitimate scientific theory. Also, in this chapter we discussed a theory of the nature of comets. We also briefly mentioned a different theory. To quote from our earlier discussion –

[Comets] were in past times often regarded as harbingers of ill fortune. The destruction of Jerusalem in 70 CE was widely thought to have been "predicted" by the appearance of a comet four years earlier (we now know it to be Halley's comet). In 1066 the Norman invasion of England coincided with a return of Halley's comet, while a threatened Turkish invasion of Europe coincided with yet another return in 1456.

The "Doom Theory" is that *comets are harbingers of ill fortune.* Your task is to write an essay in which you discuss
• whether the "Doom Theory" is a legitimate scientific theory
and
• if it isn't, then what is it?

10

Planets beyond the Solar System

As we saw in Chapter 8, finding new planets is hard. Uranus, the first planet not visible to the naked eye, was discovered in 1781, Neptune in 1846 and Pluto in 1930. That works out to only one new planet in each of the last three centuries! But in recent years the pace of discovery has accelerated spectacularly. A new planet was discovered in 1995; by now more than a thousand potential candidates have been found. Plans are under way for yet more dramatic advances in technology, which should greatly improve our ability to find these new worlds.

Most remarkable of all is that none of these new planets orbits our own Sun. They are orbiting the distant stars.

These discoveries have completely changed our view of our place in the Universe. Until the first of these new worlds was discovered, for all we knew our own Solar System might have been unique. If so, we would have been utterly alone in the cosmos. But by now we know that planets are common. This has dramatic implications for the search for life elsewhere in the Universe.

> ⇒ **Looking forward**
> We will study the search for life in the Universe in Chapter 19.

Direct detection of extrasolar planets

How have these far-distant worlds been found? You might think that they were discovered in just the same way that planets in our own Solar System were: just by looking for them through a telescope. Remarkably, however, this was not the case. In almost every case, even though we know of their existence, we have never seen these new worlds.

There are two reasons.

(1) They lie very close to their "parent stars" as we see them through a telescope. This is illustrated in Figure 10.1, which shows two configurations. In (a) we are observing a planet within our own Solar System, whereas in (b) we are observing a hypothetical planet orbiting a distant star. As you can see, the angle between the extrasolar planet and its star is tiny: far smaller than that between a Solar System planet and our Sun. Indeed, as we will see, this angle is below the ultimate resolving power of a telescope for all but the closest stars. So we cannot distinguish between the planet and its star: the two appear to be merged into a single fuzzy blob.

(a)

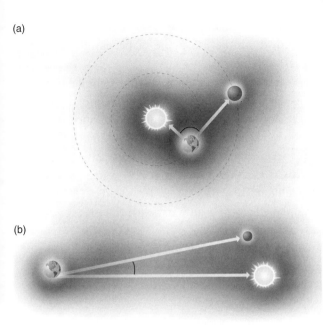

(b)

Figure 10.1 **Extrasolar planets lie close to their "parent stars" as seen from the Earth. From our vantage point, (a) the angle between a Solar System planet and the Sun is large, but (b) the angle between an extrasolar planet and its star is small.**

⇐ **Looking backward**
We studied the resolving power of telescopes in Chapter 5.

Detailed calculation

Step 1. The formula we need is the small-angle formula.

$$\theta = \left[\frac{d}{2\pi D} \right] 360 \text{ degrees},$$

where D is the distance from us to the star, and d is the distance between the star and its planet.

Step 2. Plug in the distance from us to α Centauri, and the distance between α Centauri and its hypothetical planet.

For our hypothetical Jupiter-like planet, $d = 5.2$ AU $= 7.8 \times 10^{11}$ meters. For α Centauri $D = 4.2$ light years $= (4.2 \text{ light years})(9.46 \times 10^{15} \text{ meters/light year}) = 3.97 \times 10^{16}$ meters. We calculate

$$\theta = \left[\frac{7.8 \times 10^{11} \text{ meters}}{2\pi(3.97 \times 10^{16} \text{ meters})} \right] 360 \text{ degrees}$$

$$= 1.13 \times 10^{-3} \text{ degrees} = 4 \text{ seconds of arc}.$$

Step 3. To see whether the planet can be resolved, compare the result with the resolving power of our telescopes.

As we saw in Chapter 5, this is very small: atmospheric "seeing" and the diffraction limit on a telescope's resolving power make this very difficult to resolve.

(2) They are very faint. In particular, they are faint compared to the stars about which they orbit. The result is that looking for a planet orbiting a distant star is like looking for a firefly buzzing close to a searchlight. We are effectively blinded by light from the parent star when attempting to observe the planet.

To appreciate the magnitude of this problem, let us calculate each of the factors contributing to it. To be specific, let us imagine that there might be a planet just like Jupiter in orbit about the closest star, α Centauri, and let us see how hard it would be for us to detect it. We will start with (1) the angle between the planet and the star, and then move on to (2) the brightness of the planet as compared to that of the star.

(1) First we look at the angle between an extrasolar planet and its star.

The logic of the calculation

Step 1. The formula we need is the small-angle formula (Appendix I).

Step 2. Plug in the distance from us to α Centauri, and the distance between α Centauri and its hypothetical planet.

Step 3. To see whether the planet can be resolved, compare the result with the resolving power of our telescopes.

As you can see from the detailed calculation the angle is close to the resolving power of even our biggest telescopes. Furthermore, had we done the same calculation for a planet about a more distant star, our result would have been even smaller, and the hypothetical planet would have been impossible to resolve.

NOW YOU DO IT
Suppose you were an astronomer living on a planet orbiting α Centauri, and suppose you possessed a telescope capable of resolving an angle of one second of arc. If you pointed this telescope at *our* Solar System, which planets could you resolve and which could you not resolve?

Suppose a star 10 light years away has a planet just like Mars orbiting about it – i.e. in an orbit of radius 1.52 astronomical units from its star. As seen from Earth, what is the angle between the planet and the star?

(2) We now turn to a calculation of *the brightness of this planet, as compared to that of its star.* Since a planet shines by reflected light, we need to find the

total amount of light falling upon it from its star. The amount it reflects can certainly be no greater than this! Indeed, it will be less if the planet is not very reflective – i.e. if it is predominantly dark in color.

Let us be generous, and imagine that the planet reflects all the light falling upon it. We already went through just this calculation in Chapter 8, when we calculated the total amount of light falling on Jupiter from our Sun. Since the light emitted by α Centauri happens to equal that of our Sun, the total amount of light falling on our hypothetical planet from its star equals that falling on Jupiter from our Sun. We calculated this in Chapter 8 to be 8×10^{17} watts.

But the luminosity of the star α Centauri is a full 3.8×10^{26} watts. This is *475 million* times that reflected by the planet! Put it a different way: the planet we are searching for is one 475 millionth as bright as its star.

We now understand why the task of directly searching for such a planet about its star is so extraordinarily difficult. We must detect an exceedingly faint point of light, lying exceedingly close to a far brighter star. This task is so technically difficult that it has hardly ever been accomplished.

And yet, extrasolar planets have been found! If we cannot see them, how have we done this? The answer is that they have been found by indirect methods.

Indirect detection of extrasolar planets

The method involves so-called "center-of-mass motion." We now turn to a description of this method. It involves looking closely not at the planet, but at the star about which it orbits. The planet's motion causes the star to move: we detect this motion, and so infer the existence of the planet.

Center of mass

Center-of-mass motion is a tiny movement of a star, caused by the movement of a planet about it. The star orbits in a small circle as the planet orbits in a big circle. While we cannot directly see the planet, *we have been able to detect the star's movements and so infer the planet's existence.* Furthermore, measuring the magnitude of the star's movements allows us to infer the mass and orbit of the planet.

The center of mass is a purely imaginary point. But it has an important physical significance. In our discussion of orbits up to now, we have spoken of one body orbiting another. The Moon, we said, orbits the Earth. Similarly, we said that the Earth orbits the Sun. In reality, however, these are not quite true. In fact, *two bodies orbit about their mutual center of mass.* We can infer the existence of one body by observing the motion of the other.

This mutual motion is analogous to that of two ballroom dancers. As they dance, neither holds still. Rather, each moves in a circle. If the two dancers have equal masses (Figure 10.2a), they move in equally sized circles. Alternately, if one dancer is far more massive than the other – an adult dancing with a child, say – the adult

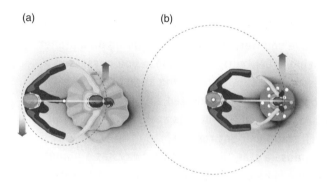

(a)   (b)

Figure 10.2 **Center of mass of two dancers lies on the line joining them. The dancers swing about this center of mass, so that each moves in a circle. If the dancers have equal masses (a), their circles are of equal radius. If, however, one dancer's mass greatly exceeds that of the other (b), the low-mass dancer swings in a large circle while the high-mass dancer hardly moves at all.**

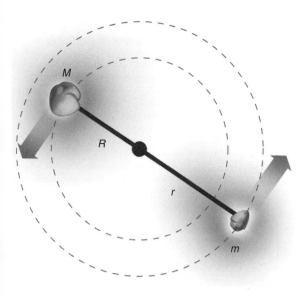

Figure 10.3 **The center of mass of two orbiting bodies lies on the line joining them. The bodies orbit about this center of mass. The higher the mass of the body, the closer it lies to the center of mass, and the smaller the orbit it moves in.**

moves hardly at all, while the child is swung about in a large circle (Figure 10.2b).

While Figure 10.2(b) looks very different from Figure 10.2(a), in reality they are both examples of the same thing. In both cases, the dancers are moving about their common center of mass. The only difference is the location of this center of mass. It turns out that the center of mass always lies

- on the line joining the two bodies,
- closer to the higher-mass dancer (if their masses are unequal),
- midway between them (if their masses are equal).

Since each dancer moves about the center of mass, we can see that *the higher the mass of a dancer, the smaller the circle in which that dancer moves.*

Similarly, two astronomical bodies orbit about their common center of mass. Figure 10.3 illustrates the configuration: M and m are the bodies' masses, and R and r are their distances from the center of mass. Each body moves in an orbit about the center of mass. If the orbits are circular, the larger mass M moves in a circle of radius R, while the smaller mass m moves in a circle of radius r.

The location of the center of mass can be found from the formula

$$R/r = m/M.$$

Let us get some practice using this formula. Suppose first that the two bodies have the same mass. (We have not yet encountered this situation, but we will in Chapter 12 when we study binary stars, which are two stars orbiting one another.) Then $m = M$, and the formula tells us that $R = r$: the radii of the two orbits are the same. Thus two equal-mass bodies move in equal-sized orbits, as in Figure 10.4.

Suppose alternatively that one body is far less massive than the other – which is the case for a planet and a star! In this case m is very much less than M, and our formula tells us that r is very much bigger than R. Since r is the size of the planet's orbit, and R that of the star's, we see that (Figure 10.5):

- the planet moves in a big circle,
- the star moves in a small circle.

Furthermore, since the two bodies require the same amount of time to complete one orbit (this is certainly true for our two dancers!), we see that

- the planet moves rapidly,
- the star moves slowly.

This small movement of the star is its center-of-mass motion. It is so small that we were justified in ignoring it in Chapter 3's discussion of orbits. Nevertheless, *it is the detection of this motion that has allowed us to infer the existence of extrasolar planets.*

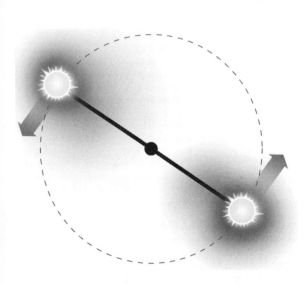

Figure 10.4 **Two equal-mass bodies orbit in equal-sized orbits, with equal velocities. Their center of mass lies midway between them.**

Figure 10.5 **Low- and high-mass bodies. The low-mass body moves in a big circle with high velocity, while the high-mass body moves in a small circle with low velocity. The center of mass lies close to the high-mass body.**

Let us plug some representative numbers into our formula to gain an appreciation of how hard it is to detect the motion caused by extrasolar planets. We will return to the hypothetical Jupiter-like planet about the star α Centauri which we analyzed above. What is the center-of-mass motion of the star caused by this planet? In particular, *what is the radius of the orbit of the star induced by the planet?*

The logic of the calculation

The center-of-mass formula

$$R/r = m/M$$

tells us the ratio between the orbital radii of the star and planet. We know the planet's orbital radius r: we are assuming it is the orbital radius of Jupiter in its motion about our Sun. And since the planet is so much less massive than the star, r is essentially the distance between planet and star. We also know the masses of both bodies. So the only unknown is R, the radius of the orbit the star moves in – and this is what we want to know.

Step 1. Solve the center-of-mass formula for R.

Step 2. Plug in values for r, M and m and calculate R.

Step 3. To get an intuitive feeling for the answer, compare it with some known quantity.

As you can see from the detailed calculation, α Centauri moves in a circle only slightly larger than the star itself.

Detailed calculation

Step 1. Solve the center-of-mass formula for R.
We get

$R = rm/M.$

Step 2. Plug in values for r, M and m and calculate R.
The hypothetical planet's mass is that of Jupiter, which is $m = 1.90 \times 10^{27}$ kilograms. The mass of α Centauri happens to equal that of the Sun, which is $M = 2.0 \times 10^{30}$ kilograms. Since the planet's orbital radius is by hypothesis that of Jupiter, which is $r = 5.2$ AU $= 7.8 \times 10^{11}$ meters, we see that the radius of the orbit in which α Centauri moves is

$R = (7.8 \times 10^{11}$ meters$) (1.90 \times 10^{27}$ kg$)/(2.0 \times 10^{30}$ kg$)$
$= 7.41 \times 10^{8}$ meters.

Step 3. To get an intuitive feeling for the answer, compare it with some known quantity.
For comparison, the radius of the star itself is that of the Sun, which is 6.96×10^{8} meters. We see that α Centauri moves in a circle only slightly larger than the star itself.

NOW YOU DO IT
Consider two planets orbiting a star:

- **a low-mass planet close to its star,**
- **a low-mass planet far from its star.**

Which makes the star move in the bigger circle?

Consider two other planets orbiting a star:

- **a low-mass planet at a certain distance from its star,**
- **a high-mass planet the same distance from its star.**

Which makes the star move in the bigger circle?

Above you began analyzing the problem of finding a Mars-like planet orbiting a star 10 light years away. Now continue that analysis. Suppose the star's mass is that of our Sun. What is the size of the circle the star moves in? If the star's radius equals that of our Sun, is the star's orbit larger or smaller than the size of the star?

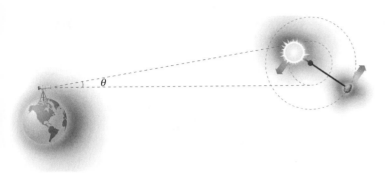

Figure 10.6 **Directly observing the center-of-mass motion of a star is too difficult. The angle θ is too small.**

Directly searching for the center-of-mass motion induced by extrasolar planets

We have already seen that directly searching for extrasolar planets is too hard for us to do right now. We will now show that directly searching for the center-of-mass motion they induce is *also* technically too hard right now. Our conclusion will be that we need to find a means of *indirectly* searching for this motion: direct detection is beyond our capabilities at present.

Figure 10.6 shows an astronomer on the Earth observing a star undergoing center-of-mass motion. In order to know that the star is moving, as opposed to holding still, we need to be able to resolve the angle θ. But θ is too small to be observed. We can show this as follows.

The logic of the calculation

We need to calculate θ and compare it to the resolving power of a telescope. We will continue thinking about the effects of a hypothetical Jupiter-like planet orbiting α Centauri.

Step 1. The formula we need is the small-angle formula (Appendix I).

Step 2. Plug in the distance from us to α Centauri, and the diameter of the circle that α Centauri moves in.

Step 3. To see whether the star's motion can be resolved, compare the result with the resolving power of our telescopes.

As you can see from the detailed calculation the angle is very much less than the resolving power of even our biggest telescopes. Furthermore, this calculation was done for the closest star: had we done the calculation for a more distant star, we would have gotten an even smaller answer. We conclude that *directly searching for the center-of-mass motion of a star induced by a planet orbiting about it is impossible.*

Detailed calculation

Step 1. The formula we need is the small-angle formula.

$$\theta = \left[\frac{d}{2\pi D}\right] 360 \text{ degrees,}$$

where D is the distance from us to the star, and d is now the diameter of the circle the star moves in.

Step 2. Plug in the distance from us to α Centauri, and the diameter of the circle that α Centauri moves in.

For α Centauri, $D = 4.2$ light years = (4.2 light years)(9.46 × 10^{15} meters/light year) = 3.97×10^{16} meters. The diameter of the circle it moves in is twice its orbital radius, which we calculated to be 7.41×10^{8} meters. So $d = 1.48 \times 10^{9}$ meters. We then find

$$\theta = \left[\frac{1.48 \times 10^{9}}{2\pi 3.97 \times 10^{16}}\right] 360 \text{ degrees}$$
$$= 2.14 \times 10^{-6} \text{ degrees} = 0.008 \text{ seconds of arc.}$$

Step 3. To see whether the star's motion can be resolved, compare the result with the resolving power of our telescopes.

This is very much smaller than the resolving power of any telescope we possess.

NOW YOU DO IT

You have been analyzing the problem of finding a Mars-like planet orbiting a star 10 light years away, supposing that the star's mass is that of our Sun. You have already calculated the size of the circle the star moves in. What is the angle θ subtended by this circle?

Indirectly searching for the center-of-mass motion induced by extrasolar planets

We now turn to an indirect technique that *has* been successful. This technique relies on the fact that, if a planet orbits a star, both the planet and the star are moving. The star is undergoing center-of-mass motion and therefore has a certain velocity of its own. The technique is to search for tiny Doppler shifts in the spectral lines of stars, caused by the small velocities of the stars in their orbits.

Let us calculate the magnitude of the Doppler effect caused by such a center-of-mass motion. To do this we need to know (A) the velocity v of the star, and (B) the Doppler shift induced by this velocity. We will consider each in turn.

(A)   The velocity of the star

The logic of the calculation

To find the star's velocity, we divide the distance it moves by the time it takes to move this distance. Since the star is moving in a circle, the distance it moves is the circumference of that circle.

>   Step 1. Find the circumference of a circle whose radius is the orbital radius of the star.

To find the time the star takes to move this distance, we use the fact that the orbital period P of the star equals that of the planet (recall our two dancers: each circles about in the same time as the other).

>   Step 2. P equals the period of Jupiter's orbit.
>   Step 3. Divide the result of step 1 by the result of step 2 to find the star's velocity.

As you can see from the detailed calculation, the answer works out to a mere 28 miles per hour (48 kilometers per hour).

Detailed calculation

Step 1. Find the circumference of a circle whose radius is the orbital radius of the star.
    This circumference $C = 2\pi R = 2\pi(7.41 \times 10^8 \text{ meters}) = 4.65 \times 10^9$ meters.

Step 2. P equals the period of Jupiter's orbit.
    The period of Jupiter's orbit is 11.87 years or 3.74×10^8 seconds.

Step 3. Divide the result of step 1 by the result of step 2 to find the star's velocity.
    We find

$$v = 4.65 \times 10^9 \text{ meters}/3.74 \times 10^8 \text{ seconds} = 12.4 \text{ meters/second} = 28 \text{ miles per hour (48 kilometers per hour)}.$$

NOW YOU DO IT
You have been analyzing the problem of finding a Mars-like planet orbiting a star 10 light years away, supposing that the star's mass is that of our Sun. You have already calculated the size of the circle the star moves in. What is the star's velocity? (The orbital period of Mars is 1.88 Earth years.)

This is tiny by astronomical standards. Nevertheless, it is not too tiny to be detected. Let us pass on to calculating (B).

(B)   The Doppler shift induced by the star's velocity

The logic of the calculation

The Doppler shift can be found from the Doppler formula

Detailed calculation

We have

$$\Delta\lambda/\lambda = v/c,$$

where c is the velocity of light (3×10^8 meters/second) and $v = 12.4$ meters/second. We calculate

$$\Delta\lambda/\lambda = (12.4 \text{ meters/second})/(3 \times 10^8 \text{ meters/second})$$
$$= 4.17 \times 10^{-8}.$$

$$\Delta\lambda/\lambda = v/c,$$

where c is the velocity of light.

Step 1. Plug in the star's velocity and that of light.

As you can see from the detailed calculation, $\Delta\lambda/\lambda$ is extraordinarily small. Nevertheless, using ultra-sensitive modern techniques, astronomers have succeeded in actually detecting it in distant stars. Once they have found it, they can calculate the mass and orbit of the planets responsible.

....

NOW YOU DO IT

What is the Doppler shift $\Delta\lambda/\lambda$ induced by the velocity you calculated for our Mars-like planet orbiting a distant star?

....

We have *finally* figured out how to find planets orbiting about distant stars: observe spectral lines emitted by the stars, and search for tiny periodic shifts in their wavelengths from long to short and back again.

THE NATURE OF SCIENCE

THE DESIGN OF OBSERVATIONS

Many people think that the way astronomers make a discovery is simply by pointing a telescope at something and looking at it. Sometimes this will work: that is how Galileo discovered mountains on the Moon, for instance, and spots on the Sun. But as we have just seen, often this will not work.

Think back over the path we have followed in this chapter. The most obvious way to search for extrasolar planets is simply to look for them through our telescopes. But we quickly realized that, were we to point a telescope at a star, we would not be able to see through it any planets the star possesses. We concluded that an indirect means was needed, and we fastened on a means involving the motion of the star induced by a planet. So the next most obvious way would be to search for this motion. But here too, we realized that the motion could not be directly seen through a telescope, and that an even more indirect means was required.

The lesson we take from this is that astronomers do not merely look through their telescopes and record what they see. Great care is required in designing observations. We will see this several more times throughout this book.[1]

---

[1] In Chapter 12 we will discuss "Representative samples and observational selection," and in Chapter 18 we will return to "The design of observations."

Observed extrasolar planets

Figure 10.7 gives an example of our method for finding extrasolar planets in operation. It shows observations of the orbital velocity, inferred from Doppler shifts in its spectral lines, of the star HD 209458 ("HD" stands for a catalog of stars known as the Henry Draper catalog). As you can see, the star is orbiting in a circle with a velocity 87.1 meters per second with an orbital period of 3.53 days. The inferred mass of the planet causing this motion is 0.63 Jupiter masses, and it is 0.045 AU from its star.

Figure 10.8 lists all the extrasolar planets that had been discovered by this technique, as of September 2003. Since that time many new planets have been discovered. Nevertheless, the broad outlines of the situation were already clear by that time.

The most important lesson to be learned from this figure is that *planets are common*. A full 110 new worlds are shown on this figure. Compare this with the eight known planets of our Solar System! Furthermore, as we will see in the next section, subsequent searches have been far more sensitive, and have revealed many other new planets.

A closer look at the data in Figure 10.8 reveals a number of interesting peculiarities about these new worlds.

In the first case, they are all very massive. The least massive planet to be found in Figure 10.8 is that orbiting the star HD 49674. As you can see from the figure, it has a mass 0.11 times that of Jupiter, or 35 times that of the Earth. Indeed, most entries on the figure are even more massive than Jupiter, until a few short years ago the most massive planet known. These new worlds are gas giants, not small rocky planets like the Earth.

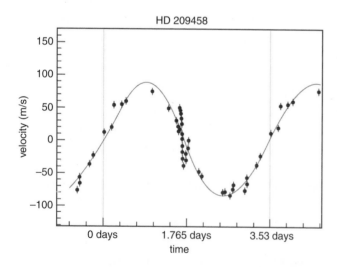

Figure 10.7 **Center-of-mass motion of the star HD 209458 inferred from shifts in its spectral lines. The odd behavior in the velocities near time = 1.7 days is due to the rotation of the star, as the planet blocks the approaching and receding limbs of the star as it transits across it.**

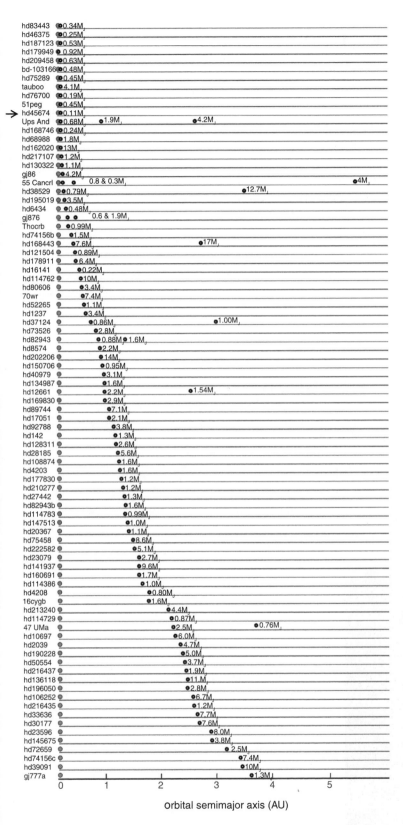

Figure 10.8 **New worlds. The extrasolar planets known as of September 2003 are arranged in order of increasing distance from their star. Each planet's mass is listed as a multiple of Jupiter's mass M_J.**

Furthermore, these gas giants lie very close to their stars. More than half of them lie closer to their stars than the Earth does to the Sun. Only a single one lies as far from its star as Jupiter does from the Sun.

NOW YOU DO IT

What is the *most* massive planet listed in Figure 10.8?

Which planet in this figure is closest to its star? How far away is it from its star?

How many planets in this figure lie closer to their star than Mercury does to our Sun?

Which planet in this figure is farthest from its star? How many planets in our Solar System are farther from the Sun than this planet is from its star?

The state of affairs revealed by Figure 10.8 is very surprising. And it is surprising for two different reasons.

- It is radically different from the situation here. As we saw in Chapter 6, our Solar System is neatly divided into the inner and outer planets. Those close to the Sun are all of relatively low mass, while those far from the Sun are of high mass. In our Solar System, the gas giant planets invariably lie far from the Sun, not close to it. Why should the extrasolar planets be so different?

- There are good theoretical reasons for the configuration we find in our Solar System. As we will see in Chapter 13, those planets that form close to a star *must* be of low mass. Conversely, the high-mass planets can only have been formed far from a star. The extrasolar planets therefore pose a serious problem for our understanding of how planets form.

To resolve these dilemmas, theories are being developed whereby the extrasolar giant planets initially formed far from their stars, but then migrated inwards by a combination of gravitational forces. These ideas are currently very new, and it is certain that as time passes our understanding will rapidly evolve.

THE NATURE OF SCIENCE

THE IMPORTANCE OF ACCURACY

We emphasized in the previous chapter (in "The nature of science. Certainty and uncertainty in science") that scientists are perpetually striving for greater and greater accuracy in their observations. We have seen in this chapter more examples of why this is so. We have seen that only an extremely high accuracy measurement of the wavelength of light from a distant star is capable of detecting the tiny Doppler shift that is the signal of a planet orbiting about it. Had our measurement of the light's wavelength been of lower accuracy, we could only have concluded that the star might or might not possess a planet – which is no news at all.

Notice also that even this great accuracy suffices to detect large, Jupiter-sized planets – but, as shown in Problems 3 and 5, even *greater* accuracy will be required to detect any smaller, Earth-sized planets that might exist.

THE NATURE OF SCIENCE

INDIRECT EVIDENCE

As we have discussed, obtaining an actual image of an extrasolar planet is exceedingly difficult. Furthermore, this is not how extrasolar planets have been discovered. Nevertheless, a very few images actually have been obtained. In 2008 direct evidence for a planet orbiting the star Fomalhaut was found (Figure 10.9). Fomalhaut is 25 light years away, and is surrounded by a disk of gas and dust: this image was obtained by blocking out the light from the star, allowing us to observe the planet orbiting about it.

We emphasize, however, that such actual images of extrasolar planets are rare. Indeed, in almost every case *we have no direct evidence at all for the existence of these extrasolar planets.* We had actual images of Uranus, Neptune and Pluto: we have nothing of the sort for almost all of these newer worlds. Indeed, the only evidence we have for their existence is indirect: a periodic shifting to and fro in the wavelengths of spectral lines emitted by certain stars. It is a long way from this to new worlds orbiting distant suns! Are we really so sure that they exist? Or is it possible that we are mistaken?

When the evidence we have is indirect, there are two things we can do.

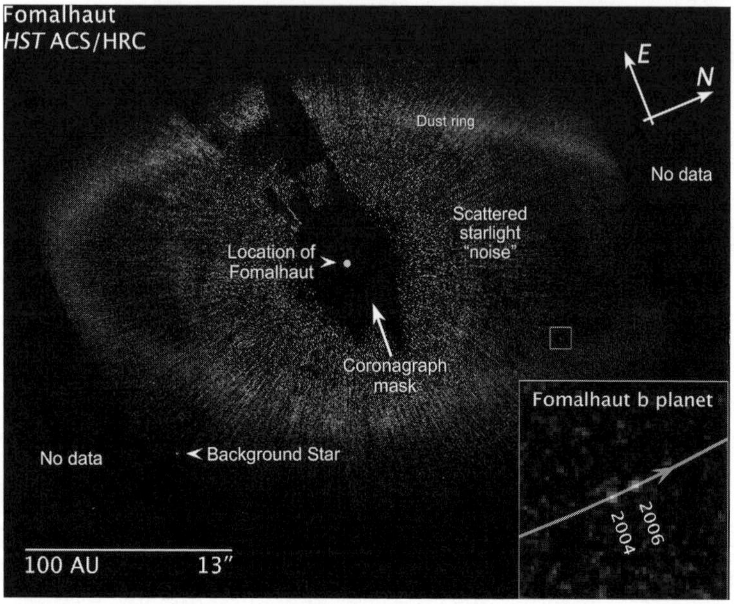

Figure 10.9 **Image of an extrasolar planet orbiting the star Fomalhaut.**

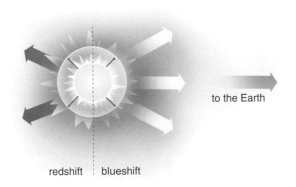

to the Earth

redshift ⋮ blueshift

Figure 10.10 **A vibrating star shows Doppler shifts. We receive light only from the near side of the star. At the illustrated point in the star's vibrational cycle, this side is moving toward us, so its light shows a blueshift. When the star starts contracting, it will show a redshift.**

(1) We can *search for alternative explanations for the data*. If we cannot find any such alternatives, we have reason to believe that our interpretation of the evidence is correct.

(2) We can *search for confirming evidence*. If we find it, even if it too is indirect, we have further reason to believe that our interpretation of the evidence is correct.

We consider each of these in turn.

(1) *Alternative explanations for the data*. We have observed a periodic shifting to and fro of the wavelengths of spectral lines in a star, and we interpreted this as being caused by center-of-mass motion induced by an orbiting planet. But could the shifts be caused by *something else*?

Here is a possible "something else": a *vibration of the star*.

As we will see in Chapter 15, the so-called Cepheid variable stars are known to vibrate. They grow larger and then smaller. In Chapter 15 we will study why this happens. For now we are interested in the fact that, when this happens, their light exhibits just the sort of Doppler shifts that have been observed.

Figure 10.10 illustrates why. A vibrating star grows periodically larger and then smaller. It "breathes" in and out. Figure 10.10 is a "snapshot" in which the star happens to be expanding: at this moment the near edge of the star is moving toward us. All the light we receive from this edge is therefore Doppler shifted to shorter wavelengths. (Of course the far edge of the star is moving away from the Earth, so that its light is shifted to longer wavelengths – but we do not receive this light: it is hidden by the opaque bulk of the star.) Later the star will cease expanding and begin to contract. At this point the light we receive will be coming from a surface that is moving away from us, and it will be redshifted. So the net result is that the light from the star periodically shifts from longer to shorter wavelengths and then back again. This, of course, is exactly what had been observed.

Is this a possible alternate explanation for the observations? There are two reasons for arguing that the answer was No. Studies of vibrating stars reveal that as a star expands and then contracts

- it changes its luminosity, growing brighter and then dimmer during the cycle,
- it changes its temperature, growing hotter and then cooler during the cycle.

Neither of these has been observed in any of the stars listed in Figure 10.8. These stars show periodic red- and blueshifts *without* accompanying changes in luminosity or temperature. For this reason, the vibrating-star scenario is untenable.

Furthermore, *no* tenable alternate explanation has ever been found to account for the observations. This is good reason to believe that ours was the correct explanation, and that we really had found extrasolar planets.

(2) Let us pass on to a second means of checking to see whether we actually have discovered extrasolar planets. It is to *search for confirming evidence*. Such evidence has been found: a phenomenon known as a *transit*.

(a)

(b)

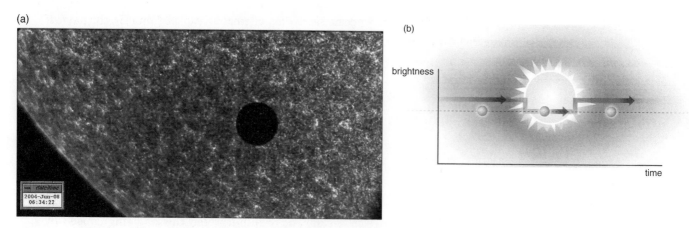

Figure 10.11 **(a) A transit of Venus across the face of the Sun. In (b) we illustrate the fact that this transit causes the Sun to dim briefly.**

Transits occur in our own Solar System. Figure 10.11 illustrates a transit of a planet across the face our Sun. A planet lying closer to the Sun than us passes from time to time across its face as viewed from the Earth. When this happens, a small portion of the solar surface is obscured, and the total amount of light we receive from the Sun decreases.

Similarly, from time to time an extrasolar planet could be expected to pass in its orbit between us and its star. When this happens the star would briefly dim. Because a planet is so much smaller than a star, the amount of decrease should be very small. Nevertheless it should be measurable.

Such a phenomenon was in fact observed. As we already saw (Figure 10.7) the star HD 209458 shows periodic Doppler shifts. Figure 10.12 shows the same star's brightness over the span of several hours. As you can see, the brightness undergoes a brief decrease.

The magnitude of the decrease is just what would be expected were a planet to transit across the face of the star. Furthermore, the *timing* of the decrease is also what we would expect. This is illustrated in Figure 10.13, which shows that a transit is expected to occur midway between the times of maximum redshift and maximum blueshift. This is just when the transit shown in Figure 10.12 happened. This is strong confirming evidence that a planet actually does orbit about HD 209458.

These two lines of argument, taken together, give us strong reason to believe that we really have discovered planets orbiting about distant stars. As we have discussed, many years after their discovery direct evidence was found, in the form of the actual image of a tiny point of light beside a distant star (Figure 10.9). But so compelling was the indirect evidence that, by the time this image was obtained, most astronomers had accepted the indirect evidence as highly persuasive.

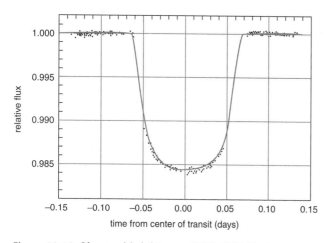

Figure 10.12 **Observed brightness of HD 209458 shows a sudden decrease, presumably caused by the transit of its planet across its face.**

This kind of indirect evidence is widespread in science. Often it is all that we have. We will see further examples of it time and again throughout this book.

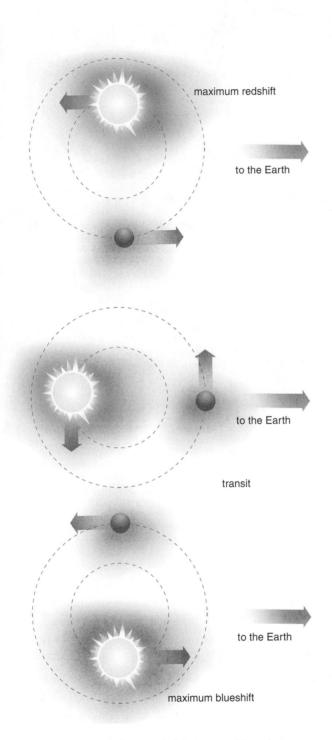

Figure 10.13 A transit is expected to occur midway between the times of maximum redshift and maximum blueshift.

NOW YOU DO IT
Explain why a big planet makes its star dim a lot as it transits, whereas a small planet makes it dim only a little.

Suppose it were possible to observe spectral lines coming from not just a star, but also the planet orbiting that star. Draw a diagram similar to Figure 10.11, illustrating the variation in brightness of the star as the planet transits, and including also (similarly to Figure 10.7) variations in the Doppler shift in the velocity of *both* the planet and the star.

The Kepler mission

The Kepler mission (Figures 10.14 and 10.15), launched in 2009, is a small orbiting telescope devoted to the study of

Figure 10.14 **Launch of the Kepler mission.**

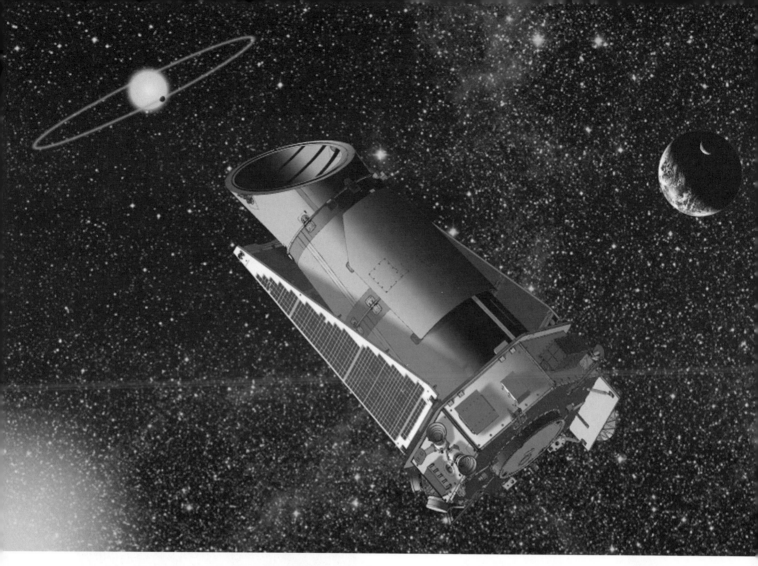

Figure 10.15 **The Kepler spacecraft in orbit (artist's conception).**

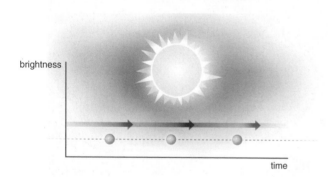

Figure 10.16 **Transits are rare.**

planets transiting across their stars. Kepler discovers extrasolar planets by observing the minute diminutions in the brightness of a star caused by the transit of a planet in front of it.

At the time of its launch, a few such transits had already been found. But the designers of the Kepler mission knew that they would be very rare. In our own Solar System, Mercury and Venus, the only planets lying closer to the Sun than us, usually "miss" the Sun as illustrated in Figure 10.16. This happens because the Earth does not orbit in exactly the same plane as Mercury or Venus. As illustrated in Figure 10.17, only if the two orbital planes coincide very closely will a transit occur.

Similarly, only if an extrasolar planet lies nearly on the line of sight leading from its star to the Earth will a transit occur (Figure 10.18). Because this usually does not occur, the Kepler satellite was designed to observe many thousands of stars in order to find transits. It was pointed at a region of the sky rich in stars, and it recorded the brightness of each over many months.

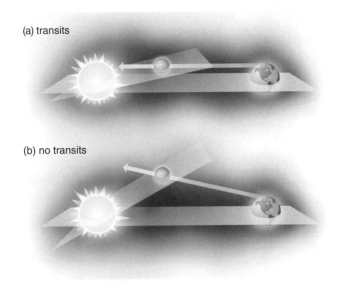

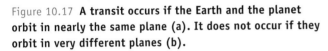

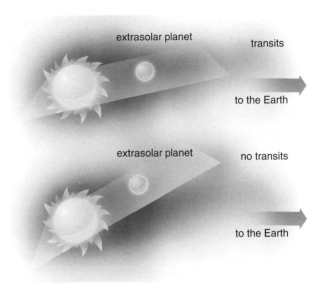

Figure 10.17 **A transit occurs if the Earth and the planet orbit in nearly the same plane (a). It does not occur if they orbit in very different planes (b).**

Figure 10.18 **A transit of a star by its planet will occur only if the line of sight to the Earth nearly coincides with the orbital plane of the planet.**

NOW YOU DO IT
Draw diagrams illustrating why (A) a planet orbiting a big star is more likely to transit across it than one orbiting a small star, and why (B) a planet lying close to its star is more likely to transit across it than a planet lying far from its star.

⇒ **Looking forward**
We will study habitable zones in Chapter 19.

The results of this mission have been spectacular. A year after its launch, the Kepler mission released the results of an analysis of its first 4 months of observations. It had discovered an extraordinary 1235 possible new planets orbiting about 997 different stars. Most remarkable of all, fully 54 of these new worlds lie in so-called "habitable zones" – regions at the correct distance from their stars to be warm enough to sustain life. The Kepler scientists were careful to describe their discoveries as "candidates" rather than outright detections of new worlds, emphasizing that follow-up studies need to be conducted. But even at this early date it is clear from their results that planets are common in the Universe.

The Terrestrial Planet Finder

NASA's Terrestrial Planet Finder is planned to be a mission dedicated to actually observing extrasolar planets. Just as we now can directly image planets in our Solar System, the goal of this mission is directly to image planets orbiting the distant stars. So sensitive will it be that it should be capable of producing images of small planets like the Earth.

As we discussed above, to accomplish this goal the mission must be capable of spotting the tiny speck of light that is a planet in the glare of light from its star. To do this, some means of blocking out the star's light must be devised. How this will be accomplished has not yet been finally determined. Two general techniques are currently under study.

- "Coronagraph": a large optical telescope, with a mirror three to four times bigger than the Hubble's, and at least ten times more accurate. Special optics would block out the star's light, thus allowing us to see the fainter planet.
- "Infrared Interferometer": many small telescopes, perhaps flying in formation or perhaps on a fixed structure, would be combined to produce an interferometer. Waves from the star would be combined in such a way as to reduce their intensity, thus allowing us to see the fainter planet.

Currently the Terrestrial Planet Finder is not a well-defined project: it is a long-term goal. Precise specifications for the mission have not yet been established. NASA currently hopes to finalize plans for it in the near future, pending progress in the relevant science and technology.

Summary

Direct detection of extrasolar planets

- We know of great numbers of planets orbiting other stars.
- With only a few exceptions, they have not been detected directly. There are two reasons.
  - They lie exceedingly close to their "parent stars."
  - They are very faint.

Indirect detection of extrasolar planets

- Relies on center-of-mass motion: the planet induces a small orbit in its "parent star."
- Directly detecting this motion of the star is technically impossible at present.
- The velocity of this motion leads to a Doppler shift, which *can* be detected.

Observed extrasolar planets

- Extrasolar planets are common.
- All known extrasolar planets are very massive – far more massive than the Earth.
- They lie very close to their "parent stars."
- This is a problem:
  - it is utterly unlike our own Solar System,
  - it is forbidden by our theory of the origin of the Solar System,
  - we do not yet know how to solve this problem.

THE NATURE OF SCIENCE

The design of observations

Many people think that, to make a discovery, all you need do is look through a telescope. Nothing could be further from the truth. In order to find these extrasolar planets we needed to think long and hard about *which* observation to conduct!

THE NATURE OF SCIENCE

Indirect evidence and the role of accuracy in science

For many years we had no *direct* evidence for the existence of these planets beyond the Solar System. In such a case we can do the following.

- Search for alternative explanations for the data.
  - But no such alternative explanation has ever been found.
- Search for confirming evidence.
  - Such evidence has been found: transits of extrasolar planets.
- Neither of these is conclusive proof, but they do provide reasons to accept our interpretation of the data.

Notice that we detect these new worlds only via extremely high accuracy observations. This is but one example of the fact that progress in science usually depends on extreme accuracy.

The Kepler mission detects dimmings caused by the transit of an extrasolar planet across the face of its star. It has detected many possible planets.

The Terrestrial Planet Finder (on the drawing board): a highly ambitious mission to build a telescope capable of directly observing Earth-sized planets about nearby stars.

Problems

(1) In calculating the size of the orbit α Centauri moves in due to a hypothetical Jupiter-sized planet, we stated "we know the planet's orbital radius r: since the planet is so much less massive than the star, r is essentially the distance between planet and star." Explain the logic behind this statement (A) in intuitive terms, and (B) mathematically, by looking at how we used the center-of-mass formula.

(2) When we analyzed whether it was possible to directly observe the center-of-mass motion of a star induced by an orbiting planet, we wrote that "this calculation was done for the closest star: had we done the calculation for a more distant star, we would have gotten an even smaller answer." Explain the logic of this statement (A) in intuitive terms, and (B) mathematically, by looking at how we used the small-angle formula.

(3) In this chapter we have analyzed the problem of searching for a hypothetical Jupiter-like planet orbiting the closest star, α Centauri. Your task now is to analyze the problem of searching for a hypothetical Earth-like planet orbiting α Centauri. (A) Before doing any calculations, think about whether this task will be easier, harder or no different from the task of finding a Jupiter-like planet. (B) Now calculate the following.
- The angle between such a planet and the star.
- The brightness of the planet, compared to that of the star. (The total amount of sunlight falling on the Earth is 1.72×10^{17} watts: this will be the amount of light from α Centauri falling on this planet since α Centauri has the same luminosity as our Sun, and our hypothetical planet is just 1 AU from its star.) What is the ratio between the brightness of the star and the brightness of the planet?
- The radius of the orbit of the star induced by the planet. Compare this radius to that of the star itself.
- The angle subtended at the Earth by the diameter of the star's orbit.
- The velocity of the star as it moves in its orbit.
- The Doppler shift $\Delta\lambda/\lambda$ induced by this velocity.

Do the results of these calculations bear out or contradict the intuitive conclusions you reached in (A)? If there are contradictions, what was wrong about your intuitive thinking?

(4) Suppose you are an astronomer getting ready to search for planets orbiting about distant stars. You are going to use the technique of searching for Doppler shifts in the spectral lines of stars. But you know that this is going to be a very difficult task. Accordingly, you are putting a good deal of thought into deciding *which stars to study*.
You have two options:
- study nearby stars,
- study faraway stars.
You want to choose the option that yields *the biggest Doppler shifts*. Which option do you choose? Why?

(5) Explain why it is that a high-mass planet induces a bigger Doppler shift in its star's spectrum than a low-mass planet. Begin with (A) an intuitive explanation, but then pass on to (B) an explanation in terms of the formulas we have been using.

(6) (A) Explain why it is that we would have to observe a star for a long time to discover any planet that exists orbiting very far from that star, but would only have to observe for a short time to discover any planet that exists orbiting close to it. (B) Suppose that aliens living on a planet orbiting α Centauri had a telescope similar to ours but far more accurate, so that they could detect a motion of our Sun no matter how small it was. But suppose that they could observe us with that telescope only for 7 years. Which planets of our Solar System would be easiest for them to discover? Which would be the hardest?

(7) Suppose a star dims by 1% due to a transiting planet. Which of the following is true:
- the diameter of the planet is 1% that of the star,
- the area of the planet is 1% that of the star,
- the volume of the planet is 1% that of the star,
- the mass of the planet is 1% that of the star?

WHAT DO YOU THINK?

In our discussion of the results presented in Figure 10.8, we claimed that they pose a severe problem for our understanding. But in Problems 5 and 6 you have demonstrated that *it is easier to detect high-mass planets than low-mass ones*, and *it is easier to detect planets lying close to their stars than ones far away*. With these facts in mind, study carefully the observational results given in Figure 10.8. What conclusions can *really* be drawn from them? Is it possible that the problem facing us is really not so very severe as we had thought?

You must decide

Your kid sister is 8 years old: in 10 years she will be old enough to enter college. But your parents have little money, and they will not be able to pay for her education. One day you meet a person named Mephisto, who proposes a bet.

• If 10 years from now most astronomers still think that we have discovered extrasolar planets, Mephisto will pay for your sister's education.

• If 10 years from now most astronomers think that we were mistaken and that in fact we have not discovered these planets, you will turn over to Mephisto half of all your earnings for the rest of your life.

Do you take the bet? Why, or why not?

Part III

Introducing stars

To the naked eye, it is stars that make up the astronomical Universe. They are what we see when we look up at the sky at night. It is sad that so few people nowadays have the experience of seeing them in all their beauty. Our forebears, who lived before the advent of the electric light, were more privileged than we in this regard. If you have never done so, you owe it to yourself to see the nighttime sky in all its glory on a clear night somewhere far from city lights. Then you will know the beauty of the astronomical Universe.

Stars scatter across the sky in enormous multitudes. About 6000 are bright enough to be seen with the naked eye. Through even a small pair of binoculars many more are visible, and a telescope reveals yet more (Figure III.1). Our own Galaxy, the vast disk-shaped assemblage of stars in which we live, contains 200–400 billion stars.

Almost without exception, stars in our vicinity are scattered randomly across the sky. There is no overall pattern to their distribution. In particular, the constellations are not real, physical groupings. Rather they are the sort of pattern that arises by accident when things are distributed by chance. If you were to toss a handful of pebbles on the ground, and then bend down to study how they fell, you would similarly see a few striking groupings. Here you might find several pebbles in a line, there a few that seem to form a square.

There are exceptions to this general rule. Some groups of stars are physically meaningful. An example is the Pleiades, a so-called open cluster of stars (Figure III.2), recently formed from an interstellar cloud of gas and dust. Figure III.3 shows another type of grouping, a so-called globular cluster of stars. Globular clusters can contain as many as a hundred thousand stars, all gravitationally bound together into a sphere and orbiting about a common center of mass.

To us short-lived humans, the constellations seem eternal. But in fact they are not. If you could live for geological ages, and gaze upwards with a vision in which thousands of years were compressed into seconds, you would see the constellations smoothly distorting (Figure III.4).

They are distorting because the stars are moving. This motion is random. One star might be moving up and to the

Figure III.1 **A universe of stars.** 👁 **(Also see color plate section.)**

Figure III.2 **The Pleiades.** 👁 **(Also see color plate section.)**

Figure III.3 **A globular cluster of stars, Messier 80.**

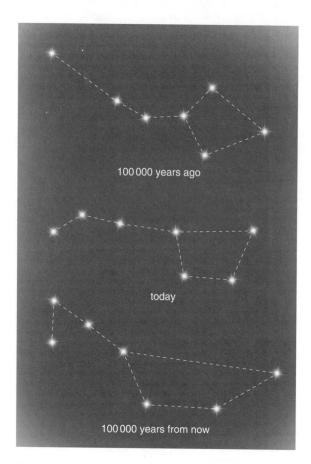

100 000 years ago

today

100 000 years from now

Figure III.4 **Distorting constellations. Stars move randomly, smoothly altering the constellations over historical ages.**

right, the one next to it downwards and somewhat toward us. The motion is quite rapid in human terms: some 10 kilometers per second, or 22 thousand miles per hour. But so gigantic are the distances between stars that ages are required for them to move an appreciable distance.

Indeed, the distances between stars are so great as to defy the imagination. A ray of light, traveling at 186 000 miles per second (almost 300 000 kilometers per second), requires 8 minutes to reach us from the Sun – but it requires 4 *years* to reach us from the next closest star. Many visible to the naked eye are hundreds of times more distant than that. Deneb, a bright star in the constellation of Cygnus, is so far away that the light by which we see it was emitted at about the time of the fall of Rome.

Our own Sun is a typical star – the only reason it looks so different from the others is that we are very close to it in astronomical terms. Some stars are bigger than our Sun and some smaller, some more luminous and some less so, some more massive and some less massive. But as a general class, stars are Suns in their own right. Indeed, as we have already seen, many have planets, just like our Sun.

Stars and the planets that accompany them are formed when the cores of interstellar clouds commence contracting, pulled together by gravitation. They are very massive. Our own Sun contains more matter than all the planets of the Solar System combined, and the same is true of the other stars and their planets. Similarly, stars are very big. Our Sun, for instance, is more than a hundred times bigger than the entire Earth. They are very hot – hot enough to vaporize solid substances, so a star is nothing more than a big ball of superheated gas. And stars emit energy at a prodigious rate.

This energy comes from nuclear reactions proceeding in their deep interiors. A star is a gigantic, naturally occurring thermo-nuclear reactor. These reactors pass through various stages of nuclear burning. In the initial stages these reactions convert hydrogen into helium: our Sun is in this phase of its evolution. Other stars, in later stages of their life cycles, are converting helium into carbon, or carbon into yet heavier elements. As a star passes through these various stages it alters its structure, changing in color, growing brighter and dimmer, larger and smaller.

Ultimately the thermonuclear fuel within a star is exhausted. Deprived of its source of energy, the star comes to the end of its evolution. Depending on the details, various ends are possible, some gradual, some violent and spectacular. But in the very long run, every star will eventually cease shining. Many billions of years hence the stars will have gone out, and the Universe will be a place of unrelieved darkness.

Our Sun

We will begin our work on stars by studying the one we know best: our Sun. The Sun is the closest star. Unlike every other star, it is so close that we can minutely observe its surface and surrounding layers.

We owe our existence to the Sun. Its great emissions warm us, and protect us from the unimaginable cold of interstellar space. Indeed most of the Universe, far from the Sun or other stars, is just a few degrees above absolute zero, which is 459 degrees below zero Fahrenheit. We lie huddled close to a warming campfire in a frigid wasteland.

As we have noted, the Sun is gigantic – more than a hundred times the size of the Earth – and immensely massive – far more massive than all the planets of the Solar System combined. And it is so hot as to vaporize every known substance. We might say that the Sun is nothing more than a ball of superheated gas. But perhaps we should put quotation marks around that "nothing more." As we will see, this blazing sphere of gas exhibits a rich and complex behavior.

In this chapter we will discuss the outer regions of the Sun: the Sun that we can see. In Chapter 14 we will turn our attention to the Sun that we cannot see: its deep interior.

Measuring the Sun

The Sun is the most obvious of all astronomical objects. You might think that it would be easy, simply by looking at the Sun, to find out a number of things about it. Certainly we do this every day as we go about our lives: we look at a man and estimate how heavy he is, and we look at a car and estimate how far away it is. But nothing in the visual appearance of the Sun helps us do this. For example, the *apparent* size of the Sun is about half that of a dime (a dime has a diameter of about ¾ of an inch or 1.9 centimeters) held at arm's length. But what is the Sun's *true* size? Is it large and distant, or smaller and closer? Nothing in its naked-eye appearance helps us to answer this question.

Here a number of things that we would like to know about the Sun.

(1) How far away is it?
(2) How big is it?
(3) How massive is it?
(4) How bright is it?

(5) How hot is it?

(6) Is it turning on its axis?

(7) What is it made of?

Naked-eye observations cannot help us answer these questions. But in Part I of this book we collected a set of tools that we can use to measure various properties of the Universe. Let us use them to answer our questions about the Sun.

(1)   The distance to the Sun

⟸ **Looking backward**
We measured the distance to the Sun and its size in Chapter 6.

As we discussed in Chapter 6, the Sun's distance is the radius of the Earth's orbit. This radius can be measured by finding the circumference of the Earth's orbit, and dividing by 2π. The circumference, in turn, is found by measuring the Earth's orbital velocity via the Doppler effect, and then multiplying by the length of a year to find the distance we travel in a year.

The result we obtained is known as the astronomical unit or AU.

> Distance to the Sun = 1 astronomical unit = 1.496×10^{11} meters.

NOW YOU DO IT

Suppose you lived on a planet orbiting the Sun at a speed of 24.1 kilometers per second, and that the length of this planet's year is 1.88 Earth years. We want to find the distance of this planet from the Sun – both in kilometers and in astronomical units. (A) Describe the logic of the calculation you will do. (B) Carry out the detailed calculation.

Jupiter's moon Io moves in an orbit of radius 422 000 kilometers, and it orbits once every 1.77 days. We want to know how fast it is moving. (A) Describe the logic of the calculation you will do. (B) Carry out the detailed calculation.

(2)   The size of the Sun

This, too, we found in Chapter 6. As we discussed, the Sun's size can be measured by finding the angle θ it subtends. The small-angle formula (Appendix I) then tells us that the diameter of the Sun, d_{Sun}, is given by

$$d_{Sun} = 2\pi D[\theta/360°],$$

where D is the Sun's distance from us, which we have just measured.

The result we obtained is as follows.

> Diameter of the Sun = 1.39×10^9 meters = 1.39 million kilometers.

NOW YOU DO IT
Suppose you lived on a planet 1.52 AU from the Sun, from which the Sun subtended an angle of 0.35 degrees. We want to use this to calculate the diameter of the Sun. **(A)** Describe the logic of the calculation you will do. **(B)** Carry out the detailed calculation.

Suppose you lived on a planet 0.39 AU from the Sun, and that the Sun's radius was 6.96×10^8 meters. We want to calculate the angle the Sun would subtend from this planet. **(A)** Describe the logic of the calculation you will do. **(B)** Carry out the detailed calculation.

(3)   The mass of the Sun

As we discussed in Chapter 3, the Sun's mass can be measured by observing the orbits of planets about it. If a planet moves with velocity V in an orbit of radius R, then the Sun's mass is given by

$$M_{Sun} = RV^2/G,$$

← Looking backward
We studied orbits in Chapter 3.

where G is Newton's constant of gravitation.

The result we obtained is as follows.

Mass of the Sun $= 1.989 \times 10^{30}$ kilograms.

NOW YOU DO IT
Suppose you lived on a planet 1.52 AU from the Sun which orbited at a velocity of 24.1 kilometers per second. We want to use this to find the mass of the Sun. **(A)** Describe the logic of the calculation you will do. **(B)** Carry out the detailed calculation.

Suppose we placed an artificial satellite in orbit 3 AU from the Sun. We want to predict that satellite's velocity. **(A)** Describe the logic of the calculation you will do. **(B)** Carry out the detailed calculation.

(4)   The luminosity of the Sun

← Looking backward
We studied light in Chapter 4.

We discussed this in Chapter 4. Now that we know how far away the Sun is, we can easily find its luminosity by measuring the flux of light we receive from it. This flux f is measured to be $f = 1400$ watts per square meter. Using the inverse square law

$$f = L_{Sun}/4\pi D^2$$

(where D is the Sun's distance) we obtained the following result.

> Luminosity of the Sun $= 3.8 \times 10^{26}$ watts.

NOW YOU DO IT

Suppose you lived on a planet 0.72 AU from the Sun, from which the Sun's flux was 2700 watts per square meter. We want to use this to find the luminosity of the Sun. (A) Describe the logic of the calculation you will do. (B) Carry out the detailed calculation.

What would be the flux of sunlight you would observe if you lived on Mars (1.52 AU from the Sun)? (A) Describe the logic of the calculation you will do. (B) Carry out the detailed calculation.

(5)   The temperature of the Sun

The Sun's temperature can be measured by using the properties of the light it emits, which is blackbody radiation. There are two ways to do this.

Spectrum

According to the Wien law (Chapter 4), the bulk of the emission from a blackbody occurs at a wavelength $\lambda_{\text{most emission}}$ given by

$$\lambda_{\text{most emission}} = 0.003/T \text{ meters},$$

where T is the blackbody's temperature in kelvin. The Sun's emission is found to be greatest at a wavelength of 5190 angstroms, or 5.19×10^{-7} meters. We can insert this in the Wien law and solve to find the temperature.

> ⇐ **Looking backward**
> We discussed blackbody radiation and spectral lines in Chapter 4.

NOW YOU DO IT

Suppose a certain star's emission is found to be greatest at a wavelength of 4900 angstroms. We want to find its temperature. (A) Describe the logic of the calculation you will do. (B) Carry out the detailed calculation.

Predict the wavelength at which the emission from a star at 7000 kelvin is greatest. (A) Describe the logic of the calculation you will do. (B) Carry out the detailed calculation.

Luminosity

We used this technique in Chapter 4. According to the Stefan–Boltzmann law the Sun's luminosity is related to its temperature by

$$L_{\text{Sun}} = 0.57 \times 10^{-7} A T^4 \text{ watts}.$$

Here A is the Sun's surface area, given by $A = 4\pi R^2$ (R is the Sun's radius). We inserted the measured values of the Sun's luminosity and radius into the Stefan–Boltzmann law and solved to find its temperature.

Both methods yield the same result for the Sun's surface temperature.

> Temperature of the Sun = 5780 kelvin.

. . . .

NOW YOU DO IT

Suppose we observe a star to have a luminosity of 2×10^{26} watts and a radius of 6.1×10^8 meters. We want to calculate its temperature. (A) Describe the logic of the calculation you will do. (B) Carry out the detailed calculation.

Predict the luminosity of a star whose radius is 5×10^8 meters and whose temperature is 5000 kelvin.

. . . .

(6) The rotation of the Sun

Figure 11.1 shows a series of images of the Sun obtained over many days. In these images a number of sunspots can be seen. Soon we will discuss the origin of these sunspots; for now we are interested in the fact that they can be seen to drift across its face. This is because they are fixed on the Sun's surface, and the Sun is rotating. By timing the motion of the sunspots, we can determine how the Sun rotates.

The images in Figure 11.1 cover a certain time span. In this period of time, a sunspot that began at the Sun's left-hand edge has been carried almost to its

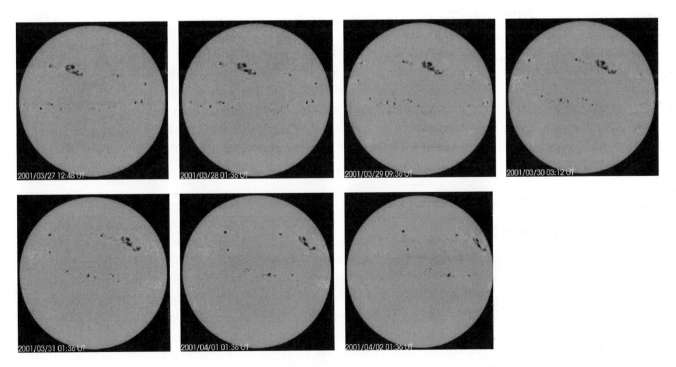

Figure 11.1 **Rotation of the Sun can be observed by tracking the paths of sunspots across its face.**

right-hand edge. So the Sun has rotated nearly halfway around in this period. In this way we calculate that the Sun rotates once every 24 days.

More careful observations show that the Sun does not rotate uniformly. The equator turns out to rotate more rapidly than the poles. While the Sun's equator rotates once every 24 days, at the poles the rotational period is longer. Such a motion is called *differential rotation*.

(7)   The composition of the Sun

The composition of the Sun's surface can be determined from its spectrum. The result is as follows.

> The Sun is about 73% hydrogen, about 25% helium, and only small amounts of all the other elements.

The distant stars also turn out to have similar compositions.

(8)   The Sun in general

We have succeeded in answering all of our questions about the Sun. Let us now step back, and build up an intuitive understanding of what our results are telling us.

The Sun is very far away

Although it does not look particularly distant to the naked eye, we have found the Sun to be very far away indeed. It is difficult to intuitively comprehend just how distant it is. A commercial jetliner, which can fly entirely around the Earth in a few days, would require 21 years to reach the Sun!

· ·

NOW YOU DO IT
Imagine a space ship that traveled at 1000 miles per hour. We want to calculate how long it would take to get to the Sun. (A) Describe the logic of the calculation you will do. (B) Carry out the detailed calculation.

Imagine a space ship that reached Venus (0.28 AU from us at closest approach) in a month. We want to calculate its speed. (A) Describe the logic of the calculation you will do. (B) Carry out the detailed calculation.

· ·

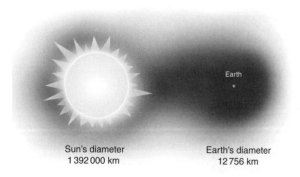

Sun's diameter
1 392 000 km

Earth's diameter
12 756 km

Earth

Figure 11.2 **The diameters of the Sun and Earth (drawing not to scale).**

The Sun is very big

Our result for the Sun's diameter translates into 865 thousand miles (1 392 000 kilometers). This is more than one hundred times the diameter of the Earth (Figure 11.2)! Our entire world would be no more than a speck, if we could view it alongside the Sun. In fact, many sunspots are larger than the Earth.

The Sun is not just big – it is also massive. The mass of the Sun is enormously greater than that of the Earth. It is greater than that of Jupiter, the largest planet in the Solar System. Indeed, if we were somehow to put together every planet of the Solar System, and add to them all the asteroids, comets and meteors, we still would get a mass far less than that of the Sun. In terms of the sheer quantity of matter it contains, the Solar System *is* the Sun: everything else – including our own Earth – is insignificant.

The Sun is very bright

Our result for the luminosity of the Sun represents an enormous rate of energy output. To get an intuitive sense of this, let us compare it to the total power requirements of our modern industrial civilization. If we consider the power needed to light up every lamp on Earth, and add to this that required to power every speeding automobile, every airplane and ocean-going ship, and add to these the power consumption of every factory in the world – if we were to form this sum, we would get a net rate of energy usage of our entire planet of roughly 10^{13} watts. This is a lot of power – but the Sun's rate of output of energy is 38 000 000 000 000 times greater than this! Indeed, in a few ten-thousandths of a second the Sun emits enough energy to supply the total quantity of energy that our worldwide civilization has consumed since the industrial revolution.

. . . .

NOW YOU DO IT

At our current rate of energy usage, how long would it take us to use up the energy that the Sun emits in a second? **(A)** Describe the logic of the calculation you will do. **(B)** Carry out the detailed calculation.

At our current rate of energy usage, how much time does the Sun take to emit enough power to fuel our civilization for a thousand years? **(A)** Describe the logic of the calculation you will do. **(B)** Carry out the detailed calculation.

. . . .

The Sun is a ball of gas

The Sun is too hot to be solid or liquid: it must be gaseous. Its surface temperature, 5780 kelvin, is high enough to vaporize many solids. Furthermore, this is only the Sun's *surface* temperature. Its interior is even hotter: the temperature of the Sun's core is an astonishing 15 million kelvin. No solid, no liquid, could survive such temperatures.

Even had we not known the Sun's temperature, we would still have known it cannot be solid by observations of its differential rotation. A solid body rotates all at once: every point upon it requires the same amount of time to complete one rotation. But as we have seen, different regions of the Sun rotate at different rates. No solid body could do such a thing.

Because the Sun is not solid but gaseous, it can undergo differential rotation. But friction tends to oppose this peculiar state of rotation: we do not fully understand what it is that overcomes the opposing friction force and maintains it.

We now turn to more detailed observations of the Sun.

Observing the Sun

The Sun is so bright that *you should never look directly at it* – even through dark glasses. Serious damage to your eyes can result.

There are a number of means you can adopt to observe the Sun.

- Special ultra-dark eyeglasses for viewing the Sun can be purchased from any scientific supply store.
- A telescope can be pointed at the Sun. *Don't look through it*: instead, put a sheet of white paper behind the eyepiece (Figure 11.3) and adjust its distance from the eyepiece till the Sun is in focus. Then look at the image of the Sun formed on the paper.

Astronomers observe the Sun in a number of ways.

A number of orbiting satellites have been launched that study the Sun. The most prominent is SOHO, the *Solar and Heliospheric Observatory*. It is a two-ton, billion-dollar spacecraft developed by the European Space Agency and NASA. It has been observing the Sun 24 hours per day since it was launched in 1995. The mission was initially planned to last two years. The spacecraft has performed so well, however, that the mission lifetime was advanced: indeed, at the time of writing (2011) it is still functioning.

The engineers who designed SOHO chose an ingenious orbit for it. As illustrated in Figure 11.4, any spacecraft orbiting the Earth has a limited view of the sky. Had the engineers chosen the orbit illustrated in Figure 11.4, SOHO's view of the Sun would have been obscured for half of the time.

One solution would have been to place SOHO into a polar orbit. As illustrated in Figure 11.5, this would have provided an uninterrupted view of the Sun – but only in the configuration illustrated in Figure 11.5(a). Unfortunately, because the orientation of the spacecraft's orbit does not change as the Earth swings about the Sun, one quarter of an orbit later the Earth would again be in an unsuitable configuration (Figure 11.5b).

Figure 11.3 **Observing the Sun safely.**

view of the Sun is blocked by the Earth from over here

orbit of SOHO

Figure 11.4 **View of the Sun is obscured by the Earth for half of a spacecraft's orbit in the illustrated configuration.**

Figure 11.5 **An unblocked view of the Sun cannot be achieved by a polar orbit. The view is unblocked in the configuration (a), but because the orientation of a spacecraft's orbit remains constant as the Earth swings about the Sun, one quarter of an orbit later (b) the Sun is again blocked for half of each orbit.**

(b)

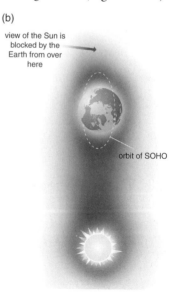

view of the Sun is blocked by the Earth from over here

orbit of SOHO

(a)

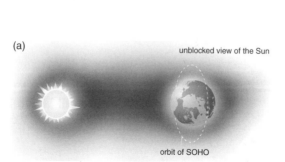

unblocked view of the Sun

orbit of SOHO

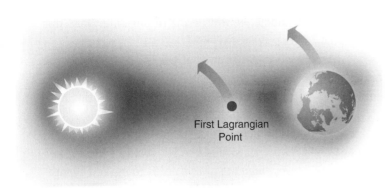

Figure 11.6 **At the First Lagrangian Point the combined gravity from the Sun and Earth keep a spacecraft fixed along the Sun–Earth line, allowing for uninterrupted viewing of the Sun.**

First Lagrangian Point

The engineers decided to adopt an entirely different strategy. They placed SOHO at a special point, known as the First Lagrangian Point, where the gravitational attraction of the Earth on SOHO just balances that from the Sun. This would keep SOHO fixed on the line leading from the Earth to the Sun. As illustrated in Figure 11.6, the advantage of this choice was that, because its view of the Sun would never be obscured by the Earth, SOHO could continuously monitor the Sun. This Lagrangian point is situated some 1.5 million kilometers away from the Earth – about four times the distance to the Moon.

⇐ Looking backward
We studied telescopes and adaptive optics in Chapter 5.

The *Solar Dynamics Observatory* is a space-based telescope operated by NASA. Launched in 2010, it is designed to observe the Sun with far greater angular resolution than SOHO. The primary mission of this observatory is to further our understanding of the ways in which the Sun influences the Earth: it is designed to operate for 5 to 10 years.

The *National Solar Observatory* operates a number of telescopes devoted to observing the Sun from the ground. The Advanced Technology Solar Telescope, currently being considered, will be a 4-meter telescope employing the most sophisticated techniques of adaptive optics, and is designed to achieve an angular resolution a full ten times better than any existing telescope.

Detectives on the case

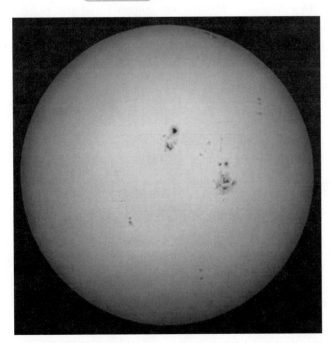

Figure 11.7 **The surface of the Sun as photographed in visible light. Note that the center appears brighter than the limb, a phenomenon known as limb darkening.**

Limb darkening

A photograph of the Sun is shown in Figure 11.7. Notice that in this photograph the center is brighter than the edges. This is termed *limb darkening* (the limb of the Sun is its outer perimeter as seen from the Earth). Why does it happen? Let us see if we can create a theory.

For our first theory let us try the notion that *part of the Sun's surface is brighter than the rest*. We illustrate the configuration in Figure 11.8. The whole Sun glows: we are postulating that part of it glows with an extra brightness. In Figure 11.8 this extra-bright region is pointing toward us. So from our vantage point it will appear in the middle of the Sun's visible disk – which is what we see in Figure 11.7.

This theory makes a prediction. Since we are orbiting, a few months later we will have moved in our orbit, and we will be viewing the Sun from a new vantage point. Therefore, the Sun's bright region will no longer appear to

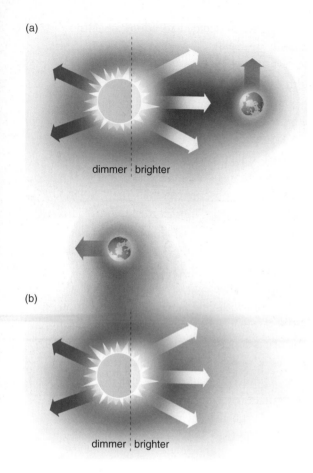

(a)

dimmer : brighter

(b)

dimmer : brighter

Figure 11.8 **First theory of limb darkening is that part of the Sun glows more brightly than the rest: (a) now, (b) 3 months later.**

be situated in its middle. It will be off to one side. Does this theory match the observations?

It does not. Observations show that the bright region on the Sun's visible disk is *always* situated in its center, no matter when we observe it. No matter what the season, it is always the Sun's center as seen from the Earth that is darker than its limb.

Can we patch up our theory to account for this? Suppose we add an additional hypothesis: that *the Sun rotates so as to keep its bright portion always pointing toward us*. Of course our hypothesis is a little awkward: why should the Sun's rotation be so perfectly tied to our orbit? Nevertheless, it is clear that this would fix the problem.

Does our new theory agree with observations? Unfortunately it does not. As we saw at the beginning of this chapter, the Sun does indeed rotate. But it does not rotate at the rate our assumption requires. The Earth takes a year to go around the Sun, so if we want the Sun's bright region to keep pointing toward us we need the Sun to rotate once a year too. But, as we saw in our study of the Sun's rotation, it does not: it spins about on its axis once every 24 days.

How can we meet this difficulty? Here's a thought. The portion of the Sun facing us is also closest to us. Perhaps the Earth *heats* those portions. After all, we know from our study of light that a hotter object glows more brightly than a less hot one.

The difficulty with this idea is that we cannot think of any process that would make it happen. For example, suppose we imagine that the Earth is reflecting sunlight back onto the Sun's surface. A detailed calculation shows that the amount of reflected sunlight is far too small to make a difference. Alternatively, suppose that gravity from the Earth is somehow altering the structure of the Sun's outer layers. Calculation shows that the force of gravity from the Earth is far too small to have any effect.

Let us give up on our theory and look elsewhere. Go back to the beginning, and suppose that the Sun is equally bright everywhere across its surface. This is certainly what we would expect, since the Sun is nothing more than a brilliantly shining sphere.

Think, then, of a glowing sphere – a superhot cannonball, perhaps. Figure 11.9 illustrates our lines of sight to it. These lines of sight lead from our eyes to its surface. What is wrong with this picture? What is it about the real Sun that this figure does not capture?

Here's what is wrong. The most important thing we know about the Sun is that it is hot. It is so hot that it cannot be solid: it is a gas. *And a gas does not have a surface.*

Perhaps we had better think more carefully about what it means to speak about the surface of the Sun. In fact what appears in Figure 11.9 as a sharp edge is actually somewhat fuzzy. Figure 11.10 illustrates the blurred nature of the "edge" of another gas: the Earth's atmosphere. As you can see, our atmosphere does not have a sharp outer surface: rather it grows more and more tenuous and eventually fades away entirely.

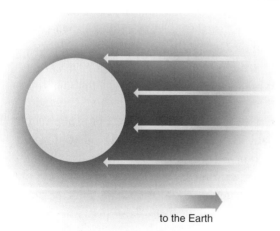

to the Earth

Figure 11.9 **Lines of sight to a sphere terminate at its edge.**

This must be true of the Sun's "surface." Does this lead us to a better theory? As an analogy to looking at the Sun and its dark limb, we can think about looking at another gas: our atmosphere. Because the gases making up the Sun are only partially transparent – the Sun is somewhat "hazy" – we will think about air with haze in our analogy. Of course, in reality we do not look *at* our atmosphere: we look *through* it. And when we do this, we can only see a certain distance before our view is obscured.

In Figure 11.11(a) we illustrate the view of a distant scene on a hazy day. Nearby objects we see clearly, but more distant ones only vaguely. Indeed, truly distant objects, those on the horizon, are completely invisible in this image. In Figure 11.11(b) we illustrate why this is so. Here we show how lines of sight manage to reach nearby objects, allowing us to see them, while those to more distant objects grow more and more tenuous, until they are completely obscured by the haze. Finally, in Figure 11.11(c) we apply the same principles to our view of the Sun. This figure shows that our lines of sight to the Sun do not really end at some sharp edge, as shown in Figure 11.9. Rather they penetrate a certain distance into the Sun and then fade away – just like in Figure 11.11(b).

But notice that in this figure some lines penetrate more deeply than others! The line leading toward the middle of the Sun as seen from the Earth penetrates relatively deeply into its interior. But the line leading toward the Sun's limb as seen from the Earth does not! Throughout its whole length, this particular line stays in the Sun's outer regions. We conclude that *when we look at the center of the Sun we are seeing into deeper layers than when we look at the outer limb.*

What then does it mean that the center appears brighter than the edges? It can only mean that the Sun's interior is brighter than its outer regions. And this means that the Sun's interior is hotter than the outer regions.

Limb darkening has taught us an important lesson: *the Sun grows hotter as we move inward from its surface.*

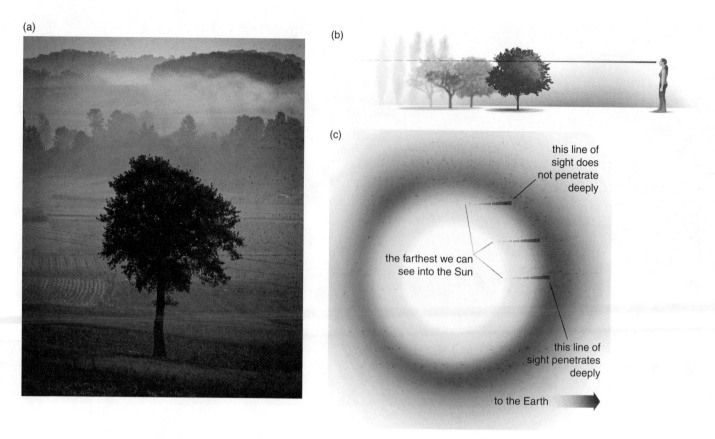

Figure 11.11 **Haze allows us to see only a limited distance. (a) A hazy day. (b) Lines of sight penetrate only a limited distance before haze obscures them. (c) Similarly, lines of sight to a gaseous sphere penetrate only a short distance before being obscured.**

Testing our theory

As before, it is imperative to test our new theory. How can we do this?

Our theory predicts that the temperature of the center of the visible disk of the Sun as seen from the Earth is greater than that of its limb. Recall Chapter 4's discussion of blackbody radiation and spectral lines. These provide us with a means to "take the temperature" of any body.

⇐ **Looking backward**
We discussed blackbody radiation and spectral lines in Chapter 4.

- The intensity of blackbody radiation can be used to determine the body's temperature, via the Stefan–Boltzmann law.
- The spectrum of blackbody radiation can also be used to determine a body's temperature, via the Wien law.
- Spectral lines can give us information about the temperature of the gases in which they form.

Using any of these methods, we can test our theory. It passes all the tests.

The Sun's "surface": the photosphere

When we look at the Sun, whether through so simple an instrument as special dark glasses or through more complex telescopes, what we see is its *photosphere*. The photosphere is the Sun's visible surface. In this section we study its properties.

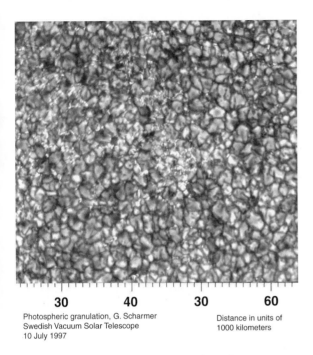

Photospheric granulation, G. Scharmer
Swedish Vacuum Solar Telescope
10 July 1997

Distance in units of
1000 kilometers

Figure 11.12 **Granulation of the Sun's surface.**

⇐ **Looking backward**
We studied the Doppler
shift in Chapter 4.

Granulation and supergranulation

Figure 11.12 shows a close-up view of the Sun's surface. We see it has a mottled quality, consisting of bright, irregularly shaped patches surrounded by dark lanes. These bright patches are known as *granules*.

Granules are more than 1000 km across – about the size of Texas. They are not permanent features. Rather they appear and disappear, persisting for perhaps 20 minutes. Doppler-shift observations show that the bright regions consist of upward-moving gases, while in the dark lanes the gas is moving downwards.

These Doppler shifts tell us what is happening: granules are giant plumes of brightly glowing, superheated gases welling up from the Sun's interior. The granules' dark edges mark locations at which the gases have cooled and are sinking back downwards. The upward velocity of these plumes is quite large: up to 3 kilometers per second, or 6000 miles per hour. The Sun's photosphere is a chaotic, violently churning place.

More careful measurements employing the Doppler effect also reveal a slower upwelling motion persisting over far greater distances, known as *supergranulation*. Supergranules can be several times larger than the Earth. They are evidence of much larger convection currents deep within the Sun's interior.

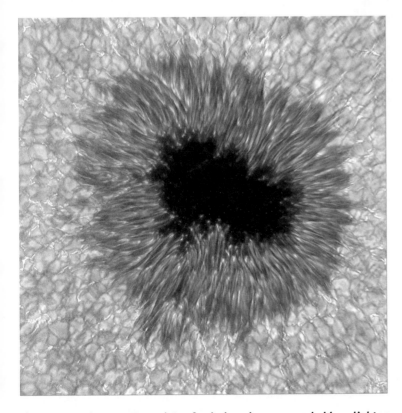

Figure 11.13 **A sunspot consists of a dark umbra surrounded by a lighter penumbra.**

Sunspots

Mottling the Sun, dark irregular patches known as sunspots can often be seen. Figure 11.13 shows one. Sunspots can be larger than the Earth.

Like granules, sunspots are not permanent features of the Sun. A spot will initially appear as a small, dark speck on the Sun's surface, only gradually growing to full size. While most disappear within a day, some persist for months. Although in Figure 11.13 they appear dark, this appearance is an illusion. In reality they emit a great deal of light – but less light than their surroundings, making them appear dark by comparison. Just as the Sun shines because it is hot, sunspots shine less than their surroundings because they are less hot. *Sunspots are regions where the Sun's surface is cooler than usual.* Temperature differences between a sunspot and the surrounding photosphere can amount to 1500 kelvin.

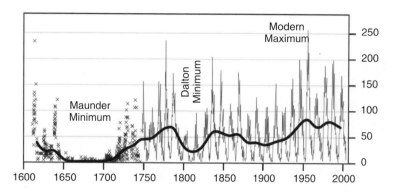

Figure 11.14 **Sunspot counts over several centuries. Notice the absence of spots from 1645 to 1715.**

Sunspot counts: the solar cycle

The number of sunspots visible on the Sun is always fluctuating. Sometimes there are many, while at other times there are few. Remarkably, this fluctuation turns out to be periodic, so that there is a regular, cyclic variation in the number of sunspots. Figure 11.14 shows sunspot counts as observed over the past several centuries. Their number is found to vary over an 11-year cycle. This regular pattern is known as the "solar cycle," or alternatively the "sunspot cycle."

As you can see from Figure 11.14, during the period 1645 to 1715 sunspots were almost entirely non-existent. Indeed, except for two small groups seen in 1705, there are no recorded sightings of sunspots during this entire stretch of time. This striking absence is known as the *Maunder Minimum* after the astronomer who drew attention to it. It does not seem to be a result of faulty observations, since telescopes were routinely being used by astronomers during this period, and we do possess records of sunspots prior to 1645.

Most striking of all is that this minimum seems to have coincided with a time of unusually cold winters on Earth. The most severe portion of the "little ice age," a period of abnormal cold throughout Europe and North America, occurred during the Maunder Minimum (Figure 11.15).

Any given sunspot remains fixed in location on the Sun. But as old spots fade and new ones form, the new ones preferentially form in new locations. Early in the sunspot cycle, spots are found at relatively high solar latitudes. But as the cycle proceeds, spots are found to form closer to the Sun's equator. Figure 11.16, illustrating this phenomenon, is known for obvious reasons as a "butterfly diagram."

Figure 11.15 **A Frost Fair on the Thames at Temple Stairs, London, 1684 (oil on canvas), Hondius, Abraham Danielsz (c. 1625–95).**

Magnetism of sunspots

As we mentioned in Chapter 4, spectroscopic observations can be used to detect magnetic fields. Such observations have shown that, like the Earth, the Sun possesses an overall magnetic field. This field is quite weak – but within a sunspot it is very strong. The typical field within a sunspot can be as much as 4000 gauss, some 1000 times stronger than the Sun's average field. (For comparison, the Earth's overall magnetic field is only half a gauss.) You would not be strong enough to hold onto anything iron in so strong a field: it would be torn from your hand no matter how hard you gripped it. We have reached an important insight: *sunspots are regions of intense magnetism.*

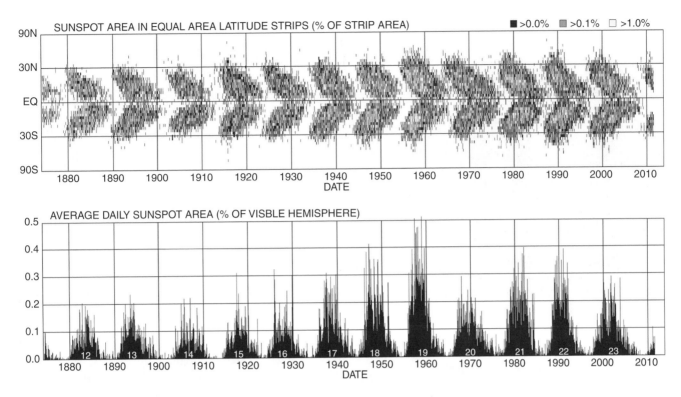

Figure 11.16 **Butterfly diagram showing how the locations of sunspots systematically vary over the sunspot cycle.**

Just as a magnet has two poles, north and south, so too is a sunspot polarized either north or south. Sunspots often come in pairs, oriented roughly east–west. The two members of such a pair always have opposite polarities. If one member of the pair is north, the other invariably turns out to be south.

Furthermore, this pattern of polarities exhibits a large-scale order. If the eastern member of one pair is north, *all* the other eastern members throughout that entire hemisphere of the Sun turn out to be north as well. Furthermore, in the opposite hemisphere the pattern is reversed: it is the western members of the pairs that turn out to be north. Finally, this pattern persists only throughout one 11-year period of the sunspot cycle: during the next 11-year period the pattern reverses, so that it is the western member of the pair that turns out to be north in the northern hemisphere.

Theory of solar magnetism

In Table 11.1 we summarize the observed properties of sunspots.

How can we understand this strange table? What accounts for all these peculiar properties of sunspots? It turns out that it is their

Table 11.1. **Observed properties of sunspots.**

(1) Sunspots are cooler than their surroundings.
(2) Their number increases and decreases regularly over an 11-year sunspot cycle.
(3) Their location on the Sun systematically shifts during the sunspot cycle (the butterfly diagram).
(4) Sometimes they vanish entirely (as in the Maunder Minimum).
(5) They come in pairs, oriented east–west.
(6) The two members of a pair have opposite magnetic polarities: one is north and the other south.
(7) The pattern of magnetic polarities of a pair is reversed between the northern and southern hemispheres of the Sun.
(8) The pattern of magnetic polarities of a pair is reversed between one sunspot cycle and the next.

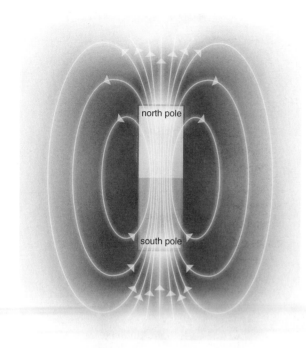

Figure 11.17 **Magnetic field lines in a bar magnet.**

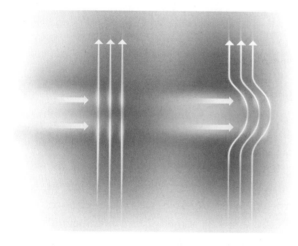

Figure 11.18 **A flowing plasma distorts magnetic field lines.**

magnetism that is the critical feature of sunspots. By thinking about how magnetism works in the Sun, we will be able to reach an understanding of many of the strange properties of sunspots.

In thinking about magnetism it is helpful to think in terms of *magnetic field lines* (Figure 11.17). They indicate the direction of the magnetic field. As indicated, these field lines emanate outward from a north magnetic pole, wrap about the magnet and lead into a south pole.

We are all familiar with how magnets exert forces on things, and how things exert forces on magnets. One pole of a magnet attracts or repels the pole of another, for instance. Similarly, a magnet is attracted to a refrigerator door and so sticks to it. But magnets do not exert forces on everything. A magnet exerts no force on a strip of cloth, for instance. Nor is there a force acting to attract a magnet to a plank of wood.

In particular, there is no force between a magnet and air. But this turns out to be because air is a gas, as opposed to a *plasma*. A plasma is a gas heated to high temperatures, so that its constituent atoms ionize. The gas making up the Sun, of course, is heated to very high temperatures, and so is ionized. And such a plasma *is* affected by a magnetic field.

Two facts will help us understand how a plasma interacts with a magnetic field.

> (A) It is difficult for a plasma to move *across* magnetic field lines. But it can easily move *along* the field lines.
>
> (B) These magnetic field lines are elastic. They can be stretched.

Here is an analogy that might be helpful. We can think of magnetic field lines in a plasma as acting like *elastic pipes* that channel the flow of the plasma. The plasma flows easily through the pipes, but if some force pushes the plasma sideways across the pipes, they stretch and are deformed by this flow, as in Figure 11.18. Flow along the field lines, however, does not distort them at all.

Let us now use these facts to build a theory of sunspots. As we do so, we must keep in mind Table 11.1, which gives a summary of the observed properties of sunspots. Our theory should be able to explain all of these properties.

Figure 11.19(a) illustrates the Sun's overall magnetic field at the start of a sunspot cycle. The field lines look very much like those of Figure 11.17: they lead upwards through the Sun's interior from the south to the north magnetic pole, and then arc outside it from north to south pole.

What does the rotation of the Sun do to this pattern? If the Sun were rotating uniformly, the answer would be simple: the magnetic field lines, because they are

(a)

magnetic field line

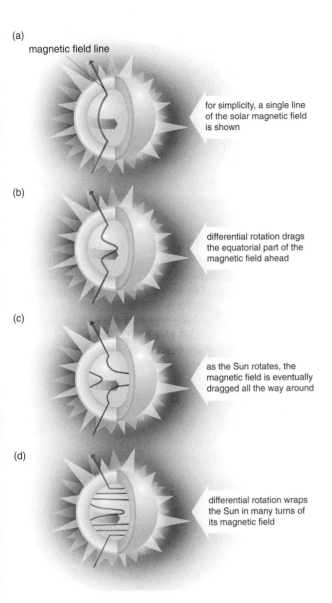

for simplicity, a single line of the solar magnetic field is shown

(b)

differential rotation drags the equatorial part of the magnetic field ahead

(c)

as the Sun rotates, the magnetic field is eventually dragged all the way around

(d)

differential rotation wraps the Sun in many turns of its magnetic field

Figure 11.19 **Differential rotation of the Sun stretches its magnetic field lines. In (a) the field is uniform. But after a while (b), differential rotation has stretched and wrapped the lines. Ultimately (c) and (d), the wrapping is very tight.**

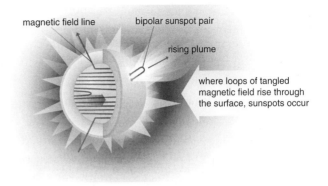

magnetic field line · bipolar sunspot pair

rising plume

where loops of tangled magnetic field rise through the surface, sunspots occur

Figure 11.20 **Loop of magnetic field lines is formed by an upward-moving plume of plasma.**

attached to the plasma making up the Sun, would rotate with it. But the Sun does not rotate uniformly! It rotates differentially, such that the equator spins about more rapidly than the poles. This differential motion stretches the magnetic lines. As illustrated in Figure 11.19(b), it wraps them up. And the more time passes, the more tightly wrapped do they become (Figure 11.19c and d).

Let us now add another element to this picture: the fact that the outer layers of the Sun are regions of violent, chaotic motion. Upward-moving plumes of plasma drag the field lines upward: downward-moving plumes drag them downward. This churning motion scrambles the magnetic field lines, and in some regions it compresses them, making the field stronger. Tight bundles of field lines are formed, tubes of amplified magnetic field.

Figure 11.20 illustrates what a rising plume does to such a tube of field lines.

As you can see, it stretches them upwards to form a loop. *Sunspots mark the points at which these loops cross the solar surface.*

Let us now return to the observed properties of sunspots that we listed in Table 11.1. Notice that a number of these facts can now be understood.

Fact (1) Sunspots are cooler than their surroundings. Explanation: Recall that the surface of the Sun is heated by hot plasma rising upward from its deep interior. Recall also that the magnetic field of sunspots is far greater than the general solar field. Since a plasma has difficulty crossing magnetic field lines, this ultra-strong field inhibits the upward motion of the plasma. Thus we now understand Fact (1): sunspots are cooler than their surroundings because they are not heated so effectively.

Fact (5) Sunspots come in pairs, oriented east–west. Explanation: As we see from Figure 11.19(c), differential rotation in the Sun stretches the magnetic field lines to a configuration in which they are parallel to the Sun's equator. A loop of field lines will be oriented in the same way, such that it runs east–west.

Fact (6) The two members of a pair have opposite magnetic polarities: one is north and the other south. Explanation: Recall that lines of force lead from one magnetic pole to the other. So the point at which a loop exits the Sun forms a sunspot of one polarity, while the point at which it returns is a sunspot of the other polarity. So we now understand Fact (6): a loop that goes up eventually comes down.

Fact (7) The pattern of magnetic polarities of a pair is reversed between the northern and southern hemispheres. Explanation: As we can see from Figure 11.19(c), field lines in the Sun's northern hemisphere run in the opposite direction to those in its southern hemisphere.

The theory we have developed so far accounts naturally for four of our eight observed properties of sunspots. No more progress is possible without a further development of the theory. Detailed studies have shown that a pattern in which the magnetic field lines point as illustrated in Figure 11.19(a) represents the configuration at the start of the sunspot cycle, in which there are few spots. It turns out that in this state the formation of loops happens preferentially at high latitudes. Later on in the cycle, however, differential rotation has wrapped the field lines to the configurations of Figure 11.19(b) and c, and in these configurations loops form preferentially at lower latitudes. In this way we can understand Fact (3), the butterfly diagram.

Eventually the wrapping by differential rotation, and the churning by convection, become so great that the Sun's field dies out. It is then reconstructed by internal motions in the Sun's core. Because of the details of this process, the new field points in the opposite direction, thus leading to the reversal in magnetic polarities from one sunspot cycle to the next (Fact (8)).

THE NATURE OF SCIENCE

THE UNDERSTANDING THAT SCIENCE BRINGS

The summary of the observed properties of sunspots in Table 11.1 is a strange list indeed. It appears to be no more than a confused jumble of unrelated facts. Learning these facts might seem just as difficult, and just as meaningless, as memorizing a list of random numbers.

But notice how our theory of sunspots has brought coherence to this list. It has given us something that we can understand, something that sets the facts of the list in a new light. The situation is a little like that in a murder mystery. At the beginning of the mystery the various clues add up to nothing more than a series of disconnected facts – but once the detective has solved the mystery, the clues all make sense, and they have come together to form a comprehensible story. In the same way, a scientific theory is a story we tell ourselves which gives observed facts meaning. This is one of the most important functions of a theory: it replaces *facts* with *understanding*.

Notice, however, that this understanding is not couched in terms of familiar concepts. Our theory of sunspots does indeed give us understanding – but only in terms of an abstract, exotic subject, that of magnetic field lines and how they interact with a plasma. This is true of much of the understanding that science gives. We have seen this in our study of Kirkwood gaps.[1] Many others of its concepts also have no place in day-to-day life: planetary orbits, for instance, and electrons.

Nevertheless, we scientists do indeed think intuitively about our research. We continually think in terms of analogies. I often visualize magnetic field lines as being like rubber bands, attached somehow to the material of the Sun. I usually visualize orbits as being like the paths of falling bodies (we used this analogy in Chapter 3) and flying electrons as being like tiny bullets.

Nor should it be so surprising that a person can think intuitively about fundamentally abstract entities. You, yourself, do this all the time. For instance, you are probably quite comfortable thinking in terms of negative numbers. Indeed, they might seem just as natural to you as positive numbers. But in fact they are not! For after all, we know what two apples look like – but what do *minus* two apples look like?

Finally, notice that we have not reached an understanding of *all* of Table 11.1. Our theory has not explained every fact about sunspots. There is no generally accepted explanation for the Maunder Minimum, for instance, and its strange connection with climate here on Earth. This is a general aspect of all scientific research. The understanding that it brings is always incomplete. We never know everything. There is always room for more research.

Beyond the Sun's surface: the chromosphere, corona and solar wind

Under normal circumstances, what we see of the Sun is only its photosphere. But during a solar eclipse, this visible surface is blocked by the Moon. Under these circumstances, we are briefly able to see two other layers whose light is normally overwhelmed by the far more intense glare of light from the photosphere: the *chromosphere* and the *corona*. Beyond the corona is the *solar wind*, which extends all the way to the Earth and beyond.

Until relatively recently, everything we knew of the chromosphere and corona was by necessity gathered in the brief few minutes of a total solar eclipse. Nowadays, however, special telescopes have been devised that contain disks to produce "artificial eclipses" by blocking out the photosphere, and baffles to reduce reflected light within the telescope. Satellites and interplanetary space probes have also directly sampled the solar wind as it flows by.

The chromosphere

The chromosphere lies just above the photosphere, and extends upwards several thousand kilometers from it. It has a reddish tinge, which gives rise to its name ("chroma" is Greek for "color"). The chromosphere is the setting of the two most spectacular phenomena the Sun has to offer, *prominences* and *flares*.

---

[1] See "The nature of science: Science is abstract" in Chapter 9.

Figure 11.21 **Prominences.** 👁 (Also see color plate section.)

Prominences

These are immense arches of reddish chromospheric gas extending upwards for tens of thousands of kilometers (Figure 11.21). Sometimes the gas remains stationary within the arch, while other times it flows upwards or downwards along it in graceful streamers. As for the arch itself, it can remain quiescent, oscillate back and forth, or in some spectacular events surge explosively upwards. The greatest prominences are far larger than even Jupiter.

The two "feet" of the arch coincide with sunspots, so that a prominence constitutes a loop leading from one sunspot to another. Indeed, a prominence *is* the magnetic loop to which we referred in our discussion of the theory of the sunspot cycle. Since a plasma cannot move across magnetic field lines of force, the magnetic field lines act like a pipe, through which the plasma flows. We conclude that prominences directly map out the loops of magnetic field lines, as they arch high above the solar surface.

Flares

Flares (Figure 11.22) are enormous explosions on the solar surface, in which great quantities of energy are violently released. They are accompanied by brief but brilliant bursts of light.

Emission at many other wavelengths, ranging from radio waves to ultraviolet, X-rays and gamma rays, have also been detected from them. A flare will brighten over a time span of a few minutes, and then slowly fade to invisibility over several hours. At greatest brilliancy, a flare is a stupendous event – it can actually outshine the entire Sun in ultraviolet light.

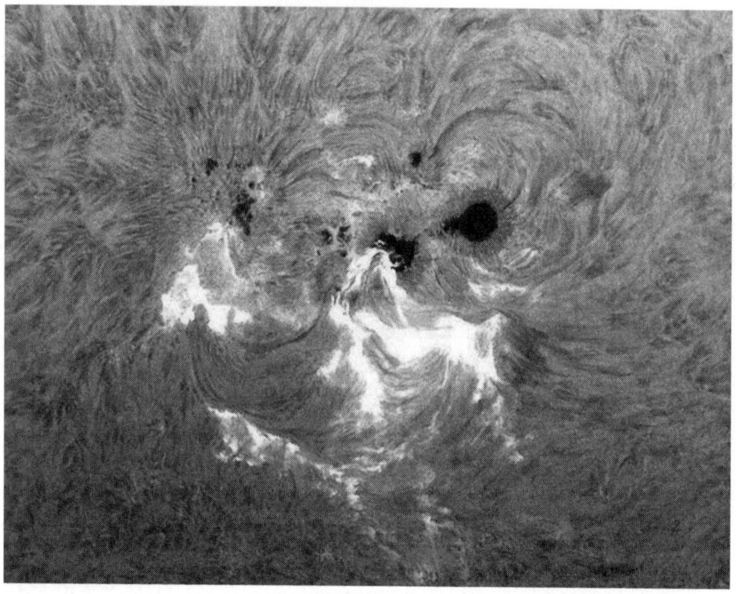

Flares are fairly common: small ones occur several times per day, and the biggest several times per month. Often a flare will recur several times at the same location. The affected region can range from a few thousand to a few tens of thousand kilometers – the biggest flares are bigger than the entire Earth!

The X-ray radiation from a flare has a blackbody spectrum, showing that it arises from matter heated to high temperatures. Recall from the Wien law that the hotter the gas, the shorter the wavelength of the blackbody radiation it emits. In order to emit X-rays, which have a very short wavelength, the flare must have very high

Figure 11.22 **A solar flare (white region) often takes place above a sunspot.** 👁 (Also see color plate section.)

Figure 11.23 **An aurora.** 👁 **(Also see color plate section.)**

temperatures: close to 20 million degrees. This is far hotter than the solar surface, and indeed is comparable to the temperature of the Sun's core.

Particles – electrons, protons and atomic nuclei – are exploded outward by a solar flare at great velocities. Although these explosions are associated with flares, the actual region from which the particles are emitted seems to lie far above the flare itself. The most rapidly moving such particles reach the Earth in half an hour and can pose a serious radiation hazard to astronauts. The slower particles can require several days to reach the Earth.

When they do so, they interact with Earth's upper atmosphere to cause auroras (Figure 11.23). In their motion these particles act like a plasma: they are guided by magnetic field lines (recall our analogy of these lines as "pipes" through which a plasma flows). In this case, it is the Earth's field along which the particles move. These lines guide particles toward the Earth's north and south magnetic poles, which is why auroras occur close to the poles, and are seldom observed at temperate latitudes.

Both the particles and the ultraviolet and X-ray radiation emitted from flares can disrupt the Earth's ionosphere, the region in our upper atmosphere in which atoms are ionized. It normally reflects shortwave radio transmissions back down to the Earth, thus facilitating radio communication. But when the ionosphere is disrupted we experience radio "fadeouts." On March 6, 1989, the satellite Solar Max detected the brightest solar flare in two decades. One week later, an intense burst of particles from the Sun ignited widespread auroras, disrupted radio communications and produced such intense surges of power in electrical grids that millions of residents in Quebec suffered a blackout.

Figure 11.24 **The solar corona.**

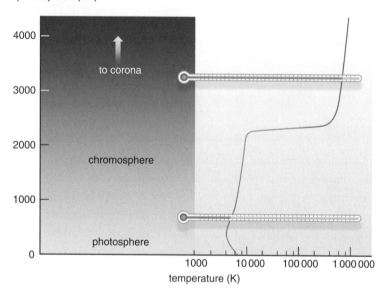

Figure 11.25 **The temperature of the chromosphere and corona is far higher than that of the photosphere. The temperature undergoes a dramatic increase several thousand kilometers above the solar surface.**

Flares are more prevalent near sunspots, and they are more prevalent at sunspot maximum than sunspot minimum (auroras, therefore, are most common during sunspot maximum). We conclude that flares are some kind of magnetic phenomenon. Indeed, they are thought to arise when the magnetic field suddenly changes its state, releasing great quantities of energy.

The corona

Anyone who has seen a total eclipse of the Sun will testify that its most lovely aspect is the solar corona (Figure 11.24). The corona glows with an unearthly silver light, and it surrounds the Sun like a halo. It can extend out to several times the Sun's radius. ("Corona" is the Greek word for "crown.")

The corona can be about as bright as a full Moon: its light is actually sunlight, reflected by electrons and dust particles in the space surrounding the Sun. Its shape is constantly changing. Furthermore, its shape is related to the sunspot cycle: at sunspot maximum the corona is bigger and roughly symmetric, while at sunspot minimum it is smaller and bulges outward at the equator. Note the streamers visible in Figure 11.24. They can extend fully 20 solar radii away from the Sun.

The density of the corona is very low: about one ten-billionth of that of our atmosphere. But its temperature is very high: roughly one million kelvin. Figure 11.25 shows the temperatures of the chromosphere and corona. Notice that they are far hotter than the photosphere. Note also the dramatic increase in temperature that occurs several thousand kilometers above the solar surface.

THE NATURE OF SCIENCE

SCIENTISTS CHANGE THEIR MINDS

The high temperature of the corona is very surprising. Normally we would expect the temperature to grow cooler as we progress outwards from the Sun's deep interior. The Sun's heat is generated by nuclear reactions within its core: we might think of it as being like a furnace. And just as the interior of a furnace is hot and its surface cool, so too should this be the case with the Sun. To some degree, this expectation is borne out: recall that limb darkening demonstrates the Sun's surface to be cooler than its interior. Continuing this trend, we expect the corona to be cooler still. But it isn't – it is hotter!

It is very hard to see how this could be. Indeed, at first glance the corona's great temperature seems to violate a law of physics. The second law of thermodynamics guarantees that heat cannot flow from a cooler body (the Sun's surface) to a hotter one (the corona).

Here are quotations from two textbooks that have addressed this question. The first was written in 1975.

> Hannes Alfvén and Ludwig Biermann independently pointed out that the corona is heated by the dissipation of friction generated by convective motions below the visible surface of the sun. The temperature of the corona has nothing to do with the flow of heat from the visible surface, and so there is no thermodynamic limitation on the temperature of the corona. The phenomenon is analogous to what happens when a man, with a temperature of 37 degrees Celsius, rubs sticks together to achieve temperatures of several hundred degrees Celsius to light a fire.[2]

The second was written in 2000.

> Coronal heating is usually greatest where the magnetic fields are strongest. The Skylab, Yohkoh and SOHO spacecraft have demonstrated that the hottest, densest material in the low corona… is concentrated within thin, long and strongly magnetized loops… Magnetic concentrations merge together and cancel all the time and all over the sun, providing a plausible explanation for heating the low corona.[3]

We are not concerned here with the details of these explanations. What we *are* concerned with is that they are completely different! At some point between 1975 and 2000, scientists changed their minds.

In fact this is a common feature of scientific research. Scientists engaged in discovery are constantly changing their minds. Our goal in this section is to begin our exploration of why and how this happens.

What led scientists to abandon the first theory? There were two main reasons.

- The first textbook is saying that the corona's great temperature has to do with how the energy of motion of upwelling plumes of convection is dissipated by friction. In the early years, calculations had been made of how much heat is dissipated in this way. These calculations indicated that the first theory might be able to account for the corona's great temperature. But then more detailed calculations were made: they revealed that most heat would be dissipated *below* the corona, rather than within it.

[2] E. N. Parker, "The Sun" in *The Solar System: A Scientific American Book*, copyright 1975 by Scientific American Inc.

[3] *The Sun from Space* by Kenneth R. Lang. Springer, 2000.

- Observations of distant stars showed their coronas to be hot, too. But some of these stars did not possess the upwelling plumes of convection that were theorized to heat coronas.

Taken together, these two developments spelled the death of the first theory.

There are four points to be made about this example of how scientists change their minds.

(1) The first theory had much to recommend it: there are indeed enormous plumes of upwelling gas on the surface of the Sun, and their energy of motion is indeed very great. Nevertheless, scientists eventually decided it was incorrect. The lesson we should take from this is that it is not enough that a theory have much to recommend it. It is not even enough that it agree fairly well with observations. More research is always needed. Scientists are forever trying to do their work better. They are constantly re-doing their calculations, adding more and more fine points. They are constantly repeating their observations, studying phenomena in greater and greater depth. You might think it unnecessary to be so obsessive about what appear to be mere details. But in this case, those "details" turned out to be critical. *Nearly enough is not good enough.*

(2) A further lesson we should take from this is that it is *hard for scientists to know when their work is finished.* The first theory survived for years before people realized that it was incorrect. Similarly, are we so very sure that our modern theory of the corona's great temperature is correct? The research that has gone into verifying it is very detailed – but it is not perfectly detailed.

Well-accepted scientific theories have a huge weight of evidence behind them, and there is no reasonable doubt as to their validity. Newton's laws, the age of the Earth and the theory of quantum mechanics – these are examples of scientific knowledge that we have every reason to trust, and no reason at all to distrust. But this is not true of theories at the cutting edge of research.

(3) Notice that the second textbook never even mentioned the theory given by the first textbook. In a textbook there isn't room to show what theories were accepted in the past, only to give currently accepted theories. How a theory evolved is of interest to some scientists, but they also focus on more recent findings to help in their work. Research and theories from tens of years ago tend to be of greater interest to historians than scientists. *Once we know the truth, the story of how we found it may be interesting, but it is not essential.*

(4) One of the originators of the early theory is identified in the first quote as being one Hannes Alfvén. This is a name that probably means little to you – but every astronomer will recognize it, for Alfvén was one of the most famous scientists of his time. Indeed he received the Nobel Prize for his work in the physics of plasmas. Notice that all this respect is accorded him in spite of his being the author of a wrong theory!

This is true of science in general. The scientist engaged in research does not have to be right all the time. It is only important to do the best one can, and make one's theories as true as possible. Indeed, the very nature of cutting-edge research is such that one is often wrong. The scientist who never makes a mistake is a scientist who never takes risks, and will probably never achieve anything very important. *Nothing is wrong with being wrong.*

The solar wind

Extending beyond the corona is the solar wind, in which the outer regions of the Sun flow steadily outward.

Close to the surface of the Sun the wind's velocity is relatively low. But it accelerates as it flows. By the time it reaches the Earth the solar wind is blowing at 400 kilometers per second, or a bit less than a million miles per hour. Furthermore the wind gusts, occasionally reaching velocities almost twice as great.

What generates the solar wind? After all, the force of gravity from the Sun, which attracts the matter of the corona to it, is quite large. The answer involves the high temperature of the corona. Recall that a gas exerts a pressure, and the hotter the gas the higher the pressure. The fact that the corona is very hot means that its pressure is quite large – large enough to overcome gravity and so accelerate the wind.

An ultraviolet image of the solar surface is shown in Figure 11.26. The dark regions visible in this image are termed "coronal holes." Studies have shown that the density of the corona is low in coronal holes, and that within these holes magnetic field lines point away from the Sun like spokes on a wheel. Conversely, in the brighter regions the field lines loop backward to the solar surface. Since a plasma cannot flow across field lines, it is from the coronal holes that the solar wind escapes.

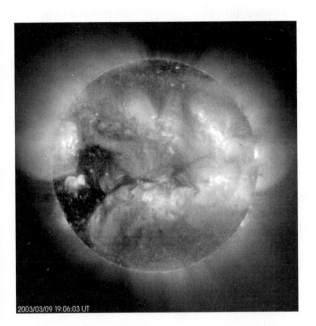

2003/03/09 19:06:03 UT

Figure 11.26 **An ultraviolet image of the Sun. The dark regions mark coronal holes, where the corona is cooler and less dense, and magnetic field lines point radially outward. The solar wind is thought to flow from coronal holes.**

Because magnetic field lines are attached to a plasma, the flowing plasma that is the solar wind combs the Sun's magnetic field lines outward. Figure 11.27 illustrates the resulting configuration. This figure also illustrates what happens to the field lines from the Sun when they encounter the Earth's magnetic field. At the point at which the solar wind plows into our field, a bow shock forms. Particles of the solar wind are trapped in this region by the Earth's magnetic field, there to form the Van Allen radiation belts. These belts can pose significant dangers to astronauts and satellites. Beyond the Earth, the Earth's and the Sun's fields merge into a long tail streaming outward. Our own magnetic field, and our atmosphere, serve to protect us from particles emitted by solar flares.

Figure 11.27 **The magnetic field of the Sun, combed outward by the solar wind, encounters that of the Earth. The regions containing trapped radiation are the Van Allen belts. Where the incoming particles strike our atmosphere, auroras result.**

Summary

- Our Sun is the closest star.
- It is very far away, very big and very bright.
- It is a ball of gas.
- We observe the Sun with both space-based and ground-based observatories.

Photosphere

- The photosphere is the Sun's visible surface.
- Granules and supergranules are the tops of rising plumes of gas.
- Sunspots are regions where the Sun is cooler than normal.
- The number of sunspots fluctuates over an 11-year cycle.
- This cycle is irregular: sometimes sunspots simply disappear for long periods of time.
- Sunspots are regions of intense magnetism.
- Our theory of sunspots involves the stretching of magnetic lines of force by the Sun's plasma.

Chromosphere

- Prominences are plasma trapped in magnetic field loops extending above the solar surface.

- Flares are explosions on the solar surface: they are associated with sudden rearrangements of the Sun's magnetic field.

Corona

- The corona is sunlight reflected off particles expelled from the Sun.
- It is hotter than the Sun's surface.
- The solar wind blows away from the Sun.

The nature of science

- Scientific theories replace lists of unrelated and meaningless facts with understanding.
- Theories are changed when required by new observations or calculations.

DETECTIVES ON THE CASE

Limb darkening

We initially theorized that limb darkening arises because different parts of the Sun's surface are brighter than others, but we soon realized that it arises because its interior is hotter than its visible surface. Observational test confirmed this theory.

Problems

(1) Suppose you are an astronomer who lives on Neptune. You measure your planet's orbital velocity to be 5.46 kilometers per second, and the length of its year to be 164.8 Earth years. The angle the Sun subtends from your home planet is 0.018 degrees, and the flux of sunlight on Neptune from the Sun is 1.5 watts/meter^2. From these calculate
- the distance from Neptune to the Sun,
- the diameter of the Sun,
- the mass of the Sun,
- the luminosity of the Sun.

For each of these, (A) describe the logic of the calculation you will use, (B) carry out the detailed calculation.

(2) Mars' moon Deimos travels in an orbit of radius 23 500 kilometers, and it orbits once every 1.26 days. We want to calculate how fast it is moving. (A) Describe the logic of the calculation you will use. (B) Carry out the detailed calculation.

(3) Suppose you went to the Moon, which lies 384 000 kilometers from the Earth, and you looked up at the Earth, whose radius is 6378 kilometers. We want to calculate the angle the Earth would subtend from your vantage point. (A) Describe the logic of the calculation you will do. (B) Carry out the detailed calculation.

(4) Suppose you observed a planet 9.54 AU from the Sun, and found that it orbited the Sun with a velocity of 9.64 kilometers per second. We want to use this to calculate the mass of the Sun. (A) Describe the logic of the calculation you will do. (B) Carry out the detailed calculation.

(5) Suppose you placed an artificial satellite in an orbit of radius 10 000 kilometers about Venus (whose mass is 0.82 times that of the Earth). We want to calculate the satellite's velocity. (A) Describe the logic of the calculation you will do. (B) Carry out the detailed calculation.

(6) Suppose you lived on an asteroid 2 AU from the Sun, and you observed that the flux from the Sun were 350 watts per square meter. We want to use this to calculate the Sun's luminosity. (A) Describe the logic of the calculation you will do. (B) Carry out the detailed calculation.

(7) Suppose you lived on Pluto (39.4 AU from the Sun). We want to calculate the flux of sunlight you would observe. (A) Describe the logic of the calculation you will do. (B) Carry out the detailed calculation.

(8) A certain star has a blackbody spectrum for which the bulk of the emission occurs at a wavelength of 4000 angstroms. What is its surface temperature? (A) Describe the logic of the calculation you will use. (B) Carry out the detailed calculation.

(9) Suppose the star of the previous problem grows cooler without changing its size. (A) Will the wavelength for which the bulk of the emission occurs increase, decrease or stay the same? (B) How will its color change? (C) Will its luminosity increase, decrease or stay the same?

(10) Imagine a star with a blackbody spectrum with a temperature of 8000 kelvin. We want to find the wavelength for which its spectrum reaches a maximum. (A) Describe the logic of the calculation you will use. (B) Carry out the detailed calculation.

(11) Imagine a star with the same luminosity as the Sun, but three times its radius. We want to calculate its temperature. (A) Describe the logic of the calculation you will use. (B) Carry out the detailed calculation.

(12) Suppose we observe a star to have a luminosity of 7×10^{26} watts and a radius of 6.8×10^8 meters. We want to calculate its temperature. (A) Describe the logic of the calculation you will do. (B) Carry out the detailed calculation.

(13) Predict the luminosity of a star whose radius is 5×10^7 meters and whose temperature is 9000 kelvin.

(14) We mentioned a hypothetical airplane that requires 21 years to reach the Sun. We want to calculate
• its speed,
• and how long it would require to fly around the Sun.
For each of these, (A) describe the logic of the calculation you will use and (B) carry out the detailed calculation.

(15) Imagine a space ship that traveled at 900 miles per hour. We want to calculate how long it would take to get to the Sun. (A) Describe the logic of the calculation you will do. (B) Carry out the detailed calculation.

(16) Imagine a space ship that got to Mars (0.52 AU from us at closest approach) in a month. We want to

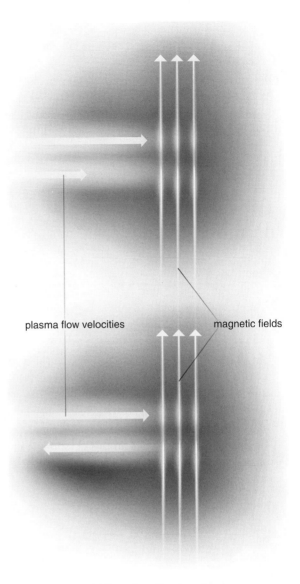

Figure 11.28 **Field lines in a flowing plasma.**

calculate its speed. (A) Describe the logic of the calculation you will do. (B) Carry out the detailed calculation.

(17) Suppose our worldwide industrial civilization used 5×10^{12} watts. How long would we take to use up the energy the Sun emits in $1/100$th of a second? (A) Describe the logic of the calculation you will do. (B) Carry out the detailed calculation.

(18) Suppose our worldwide industrial civilization used 5×10^{12} watts. How long would it take the Sun to emit the amount of energy we have used up in the past century? (A) Describe the logic of the calculation you will do. (B) Carry out the detailed calculation.

(19) (A) Suppose that the Sun were *more transparent* than it actually is. Would this affect the phenomenon of limb darkening? In what way?

(B) Suppose that the Sun's interior were *cooler* than its surface. Would this affect the phenomenon of limb darkening? In what way?

(20) One of our theories of limb darkening postulated that the Sun rotates in such a way as to keep the same side facing the Earth at all times. We want to calculate the velocity of rotation the Sun would have if this were true. (A) Describe the logic of the calculation you will use. (B) Carry out the detailed calculation. (Observations using the Doppler effect can measure the Sun's rotational velocity: it turns out to be far greater than this!)

(21) Suppose that magnetic field lines in a plasma point vertically upwards, but the plasma is flowing sideways (Figure 11.28). Draw how these plasma flows distort the magnetic field lines.

You must decide

(1) You work at a university specifically devoted to science, technology and engineering. Graduates of this university almost invariably go on to jobs in high-tech industrial or government labs, or basic research in science.

The university has recently received a large gift from an alumnus. Now the university president is trying to decide what to do with the money. She is considering two options: (A) create a new department of aeronautical engineering, or (B) create a new department of the history of science. The president has asked you for your advice on this question. Write a memorandum in which you address
  • the arguments for each option,
  • the arguments against each option,
  • your recommended course of action
   and (this is the most important of all!)
  • your reasons for your recommendation.
   Now suppose you worked not at a university devoted to science, technology and engineering, but at a liberal arts college. Would your recommendation be different? Why or why not?

(2) In Chapter 9 you were asked to decide whether Clyde Tombaugh should receive the Nobel Prize posthumously for having discovered Pluto. Return to the answer you had given to this question. Do you wish to modify it in light of the discussion in this chapter? Why or why not?

12

A census of stars

In the previous chapter we studied one particular star: the nearest one, our Sun. We now transfer attention to all the others.

To the naked eye, stars look like nothing more than tiny points of light. That's what they look like through a telescope, too. Nothing in their appearance, even through our most powerful of telescopes, tells us much about them. Indeed, just as we saw in our study of the Sun, it is not easy to find out things about the stars. Our first task in this chapter will be to find techniques for answering questions about them – questions like "how far away is that star?" or "how bright is it?"

Our second task will be to use these techniques to conduct a census of stars. Just as pollsters do not interview each and every person in America, so it would be impossible to study each and every star in the sky. Accordingly, much of our time will be spent in developing ways to identify a representative sample of stars with which to conduct our census.

When we finally do so, a remarkable pattern will emerge, a pattern summarized in the so-called "Hertzsprung–Russell diagram." It will take us two full chapters to reach an understanding of this pattern.

Measuring the stars

In the previous chapter we began our study of the Sun with a list of all the things that we would like to know about it. These are also things we would like to know about the other stars.

(1) How far away are they?
(2) How big are they?
(3) How massive are they?
(4) How bright are they?
(5) How hot are they?
(6) Are they turning on their axes?
(7) What are they made of?

In the previous chapter we developed techniques to answer these questions. Some of these techniques will work for the distant stars as well. But others will not: to answer these questions we will have to develop new methods.

Let us begin by enumerating those techniques that we used for the Sun and *that can be used without modification* on the distant stars.

(4) *How bright are the distant stars?* If we knew a star's distance, we could find its luminosity just as we found that of the Sun: by measuring its apparent brightness and using the inverse square law.

(5) *How hot are the distant stars?* As we saw in the previous chapter, an object's temperature can be measured by using the Wien law: the bulk of the emission from a blackbody occurs at a wavelength $\lambda_{\text{most emission}}$ given by

$$\lambda_{\text{most emission}} = 0.003/T \text{ meters},$$

where T is the blackbody's temperature in kelvin. So if we measure the spectrum of a distant star we can find its temperature.

(7) *What are the stars made of?* As we know from our study of light, an object's composition can be determined from its spectrum.

Let us now enumerate those techniques that *cannot be used* on the distant stars.

(1) *How far away are the stars?* We answered this question for the Sun by exploiting the fact that the Earth orbits about it. Since we do not orbit the distant stars we will have to think of a new method to find their distances. This method is *parallax*: we will explore it below.

(2) *How big are the distant stars?* To the naked eye, bright stars appear large and dim stars appear small. So you might think that we could find out their sizes (once we know a star's distance) by measuring the angle subtended by a star and using the small-angle formula. Unfortunately this method will not work. The fact that bright stars appear large is an optical illusion. It has to do with the human perceptual system rather than the stars themselves. Indeed, the stars are so far away that, even through our biggest telescopes, they appear to be mere points of light. This means that *we have no way to measure their directly angular diameters.*

Let us calculate the angular diameters of stars in order to appreciate this. We will anticipate a bit, and take the known diameters and distances of the stars and use these to calculate the angle that they subtend. Our goal is to see if our telescopes are capable of resolving them. We will find that they are not.

← **Looking backward**
We studied the Wien law and spectroscopy in Chapter 4.

← **Looking backward**
We studied the resolving power of a telescope in Chapter 5.

The logic of the calculation

Step 1. We need to know stars' true diameters and distances. In doing our calculation, we will simply take their measured values.

Step 2. Solve the small-angle formula for the angle given the diameter and distance.

Step 3. Plug the star's diameter and distance into the resulting formula to find the star's angular diameter.

Step 4. Compare the result with the smallest angle our telescopes are capable of resolving.

As you can see from the detailed calculation, the answer we get is that even a nearby star subtends an angle of only 0.003 seconds of arc. This is far too small for our telescopes to resolve. Thus, even through our largest telescopes, stars appear just to be points of light.

⇐ Looking backward
We studied the Stefan–Boltzmann law in Chapter 4.

NOW YOU DO IT

What is the angular diameter of a star half the size of the Sun at a distance of 1.7×10^{14} kilometers?

Suppose a star with the same diameter was twice as far away. Would its angular diameter be twice as big, half as big or what?

Detailed calculation

Step 1. We need to know stars' true diameters and distances. In doing our calculation, we will simply take their measured values.

Although we have not yet seen this, it will turn out that the diameters of stars are similar to that of our Sun (1.4×10^9 meters). We will consider a nearby one, which should be the easiest to resolve: the distance to such a star is roughly 10^{14} kilometers.

Step 2. Solve the small-angle formula for the angle given the diameter and distance.

The small-angle formula is

$d = 2\pi D \, [\theta 360°]$.

Here d is the star's diameter, D is its distance and θ is the angle it subtends. Solving we find

$\theta = 360° \, [d/2\pi D]$.

Step 3. Plug the star's diameter and distance into the resulting formula to find the star's angular diameter.

We need to convert $D = 10^{14}$ kilometers into meters. Since a kilometer is a thousand meters, we find $D = 10^{17}$ meters.

Plugging in we then get

$\theta = 360° \, [1.4 \times 10^9 \text{ meters}/2\pi \, (10^{17} \text{ meters})] = 8 \times 10^{-7}$ degrees $= 0.003$ seconds of arc.

Step 4. Compare the result to the smallest angle our telescopes are capable of resolving.

In Chapter 5 we saw that the smallest angle a modern telescope is capable of resolving is very much larger than this. Thus, our telescopes are not capable of actually resolving stars: even through our largest telescopes, they appear merely to be points of light.

⇐ Looking backward
We studied the Doppler effect in Chapter 4.

How, then, can we measure the sizes of stars? A method that will work is to use the Stefan–Boltzmann law, which states that an object's luminosity is related to its size and temperature by

$$L = 0.57 \times 10^{-7} A T^4 \text{ watts.}$$

Here A is the star's surface area in square meters, given by $A = 4\pi R^2$ (R is the star's radius) and T the surface temperature in kelvin. So once we have measured a star's luminosity and temperature we can find its area and, from this, its size.

(3) *How massive are the stars?* We answered this question for the Sun by exploiting the fact that the Earth orbits about it, and using Newton's orbit formula. Since we do not orbit the distant stars we will have to think of a new method to find their masses. This method involves *binary stars*: we will explore it below.

(6) *Are the stars turning on their axes?* We answered this question for the Sun by directly observing the passage of sunspots across its surface. Since even our largest telescopes are incapable of resolving the distant stars, we cannot use this method.

A method that will work is to use the Doppler effect. As illustrated in Figure 12.1, some parts of a rotating star are approaching us and some receding from us. The wavelength of light from the approaching portions is shortened by the Doppler effect, while that of light from receding portions is lengthened. If the star is emitting a spectral line, in which radiation is concentrated at one wavelength, rotation will broaden the line so that it consists of a mixture of wavelengths (Figure 12.2). Furthermore, the faster the star is rotating the more the line is broadened. So by measuring the degree to which the line is broadened – its width – we can infer the star's rotational velocity.

We now turn to a method of measuring the masses of the stars.

Measuring the masses of the stars: binary stars

As we have seen, a new technique is needed to measure the distances to the stars. We now turn to this technique.

⇐ **Looking backward**
We studied Newton's law and the mass equation in Chapter 3.

this portion of the star is receding from us: wavelengths are lengthened by the Doppler effect

to the Earth

this portion of the star is approaching us: wavelengths are shortened by the Doppler effect

Figure 12.1 **A rotating star.**

⇐ **Looking backward**
We studied the concept of center of mass in Chapter 10.

Figure 12.2 **A spectral line is broadened by rotation.**

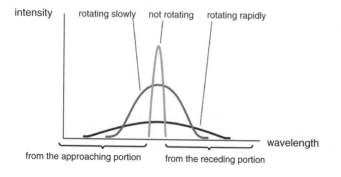

intensity
rotating slowly   not rotating   rotating rapidly

wavelength

from the approaching portion   from the receding portion

To the naked eye, stars appear to be individual objects, isolated from one another by the vast depths of interstellar space. Remarkably, this appearance is illusory. In fact, telescopic observations reveal that what to the naked eye appears to be a single star often consists of two stars lying very close to one another. These systems are known as binary stars. Just as we used Newton's laws to measure the mass of the Sun, so these laws can be applied to binary systems to measure the masses of the stars.

Contrary to what we might expect, binary stars turn out to be common: at least half, and perhaps as much at 80% of all stars are binary. Our own Sun, which is clearly not a member of a binary, is therefore not a good guide to the stellar population at large. The notion that most stars are single is an illusion. The illusion arises from the fact that the two stars making up a binary are very close to one another, so that to the naked eye they appear single.

Just as the planets do not merely lie close to the Sun, but actually orbit it, so too with the stars in a binary. As we discussed in our study of extrasolar planets, these orbits are about the center of mass of the two bodies. Orbits in a binary are elliptical and can have any eccentricity, ranging from circular to highly elongated.

The evidence for the binary nature of stars can come from a variety of observations.

- Sometimes the members of the binary pair are sufficiently widely separated that each can be seen through a telescope (Figure 12.3). Such a system is known as "visual binary." An example is the second star from the end of the handle of the Big Dipper: it is actually two stars, Mizar and Alcor.
- Sometimes we find a single star orbiting about "nothing." We presume that this "nothing" is in fact a second star too faint to be detected. Such a system is known as an "astrometric binary."
- The two stars might be so close together that, even through a telescope, their light is fused together into what appears to be a single star – but the spectrum shows spectral lines belonging to two different stars. Such a system is known as a "spectrum binary" or "spectroscopic binary" depending on whether the lines show evidence of the Doppler effect.
- The two stars might be so close together that even through a telescope their light is fused together into what appears to be a single star – but occasionally one star passes in front of the other: we detect a sudden drop in brightness as the more distant star is eclipsed by the nearer. Such a system is known as an "eclipsing binary."

Just as by observing the orbits of planets about the Sun we can use Newton's mass equation to measure the Sun's mass, so too can we measure the masses of stars by observing their orbits about one another. We now turn to this subject.

We have already introduced the concept of *center of mass* in Chapter 10, in which we discussed how a planet and star

Figure 12.3 **Albireo, a visual binary star.**

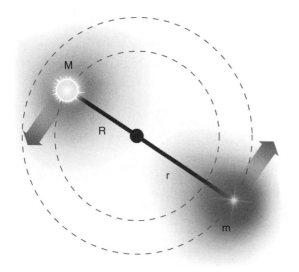

Figure 12.4 **Center of mass of two orbiting bodies lies on the line joining them. The bodies orbit about this center of mass. The higher the mass of the body, the closer it lies to the center of mass, and the smaller the orbit it moves in.**

orbit about their common center of mass. Let us remind ourselves of what we found there: the same concepts apply to the case of a binary system. Figure 12.4 illustrates two orbiting stars. Each star orbits about the center of mass, which lies on the imaginary line connecting them. For our purposes the most important fact is that the orbit depends on the mass of the star: the more massive the star, the smaller the orbit. Therefore, by measuring the diameters of their orbits we can learn something about the masses of the two stars.

As we saw in Chapter 10 the location of the center of mass can be found from the formula

$$m/M = R/r,$$

where m and M are the masses of the two stars, and r and R the radii of their orbits. So if we measure these radii, we can find the ratio of masses of the two stars.

But this is not enough to find the masses themselves. For example, suppose one star moves in an orbit twice as large as the other. This tells us that one star is half the mass of the other. But it does not tell us about the masses. One star could have a mass equal to that of the Sun, and the other half that – but the first could just as well have a mass ten times that of the Sun, and the second five times the Sun's.

Our equation for the center of mass equation has two unknowns – m and M. From mathematics we know that we need *two* equations to find two unknowns. The second equation, which comes from Newton's laws of motion, is

$$M + m = 4\pi^2 (R + r)^3 / GP^2,$$

where P is the period of orbit of each star. By combining this equation with the first we have enough information to solve for the masses of the two stars individually.

Detectives on the case

Parallax

In our ordinary experience the constellations are eternal. They never change their shapes. This means that – again, in our ordinary experience – the stars do not change their positions in the sky.

But this is one of the many instances in which high-accuracy measurements reveal entirely new phenomena. Let us point a very accurate telescope at some randomly

January 1

April 1

July 1

October 1

January 1

Figure 12.5 Observed motion of a star. Over the course of a year it appears to move sideways and back to its original position.

chosen region of the sky and take an image. Then let us wait a while and take a second image of the same region. If we compare the two, we will be able to see if any of the stars have changed their positions during the intervening period. And if we keep on doing this, we will be able to map out the motions of the stars.

Most of the time we see nothing at all: all the stars appear to be holding still. But occasionally we see something interesting. Figure 12.5 illustrates the phenomenon. As you can see, among all the stars in the image, one of them appears to be moving to and fro. It takes just a year to go through this motion. This is the phenomenon of parallax.

What can account for this behavior?

Let us try a first theory: *perhaps the star is orbiting about something we cannot see.* The unseen object about which our star orbits might be a black hole, or simply an ultra-dim star, one too faint for us to see. In either case, we know that all orbits take place in a plane: if we happen to be observing this plane edge-on, the motion will appear to be to and fro as observed.

Our first theory of parallax seems to make sense. From our study of binary stars we know that stars can orbit one another, and we know that in some cases one member of the pair can be too faint for us to observe. But now let us study other regions of the sky. As before, most of the time we see nothing remarkable. But eventually we stumble upon another star moving to and fro. And remarkably, this star turns out to require exactly the same amount of time to go through its cycle of motions: one year.

There is nothing in our theory of parallax that can account for this. If a star orbits about something, the time it takes to complete one orbit depends upon its distance from that "something," as required by Kepler's third law. This distance can be anything at all. There is nothing to force it to have any particular value. Perhaps we can accept the idea that, for one such star, the distance just happens to be exactly what is required to make the orbital period one year. But surely we cannot accept *two* such coincidences! And as we keep on searching the sky, we keep stumbling upon stars that exhibit parallax – and each turns out to do so over the same period of time.

Let us give up on our first theory, and seek a second. That one-year period of time is a clue. It is the time required for us to orbit the Sun. So let us try the theory that *parallax is caused not by the star's motion, but by our own.*

Can we produce for ourselves the same phenomenon by doing what the Earth does in its yearly orbit? Try moving your head about in a circular pattern so as to mimic the motion of the Earth. While moving your head in this way, fix your gaze upon some object. In this experiment, make sure that the circle your head traces out is fairly small – a few inches will do – and make sure that the object you are looking at is not too distant – a nearby bush, perhaps, or a nearby chair. Keep looking at it as you move. You will see that this object appears to move sideways, left and right against the more distant background. That object is doing exactly what the star is doing in

⇐ Looking backward
We studied Kepler's laws and orbits in Chapters 2 and 3.

Figure 12.5, and both are doing so for the same reason: objects appear to shift back and forth because we are observing them from different vantage points as we move about.

Testing the theory

Our theory of parallax makes a prediction: that *every* star must move to and fro with a period of one year. This is, of course, what your experiment of moving your head shows. Our observations, however, seem to show the opposite: only a very few stars show such apparent motion.

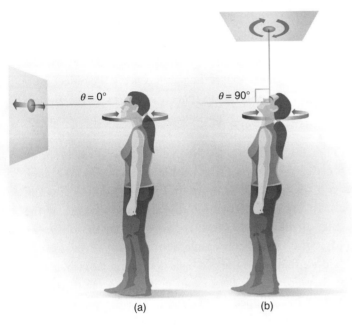

Why don't the others move too? They wouldn't if they were farther away, since the more distant an object the less it appears to move. (To demonstrate this, simply gaze at a more distant object as you move your head: you will see that its apparent motion is smaller.) Perhaps the "motionless" stars are so distant that their parallax motion is too small for us to detect.

Can we test this theory? To do so we need a means of measuring the distances to the stars. If we had such a method, we could use it to see if those stars that don't appear to move really are more distant than those that do. The difficulty is that it is exceedingly difficult to measure the distances to the stars. Indeed, as we will see in the next section, our best means of doing so is the very theory of parallax that we are trying to test!

Here is another way to evaluate our theory. Suppose we build a better telescope, capable of detecting smaller motions. If this better telescope finds that the "motionless" stars actually are moving to and fro, by smaller amounts but always with a period of one year, we would have found the evidence we seek.

Figure 12.6 **The apparent motion of parallax depends on the angle between the line of sight to the object and the plane in which your motion occurs. (a) If this angle is zero the apparent motion is along a line. (b) If it is 90° the motion is in a circle.**

When we do this, we find just what the theory predicts. The better the telescope we build, the more stars do we find that appear to move. Our theory has passed its first test.

A second test is suggested by the following. Go back to our experiment of moving your head in a circle. But now look at something not in front of you, but *above* you. Look straight up – at a spot on the ceiling, perhaps. As you move your head, you will see the spot moving. But it does not move in a line! Rather, it moves in a circle. This demonstrates that the apparent motion of parallax depends on the angle between the line of sight to the object and the plane in which your motion occurs (Figure 12.6).

⇐ **Looking backward**
The ecliptic was studied in Chapter 1.

The same must be true of the stars. The plane of the Earth's orbit is the ecliptic. So our theory predicts that all stars that lie in the ecliptic show parallax motion in a straight line. Similarly, all stars that lie perpendicular to the ecliptic ought to show parallax motion in a circle (Figure 12.7).

Here, too, when we perform the observations the theory passes the test.

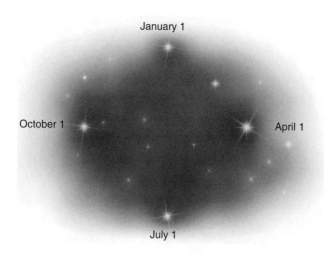

January 1

October 1   April 1

July 1

Figure 12.7 **Predicted motion of a star perpendicular to the ecliptic according to our theory.**

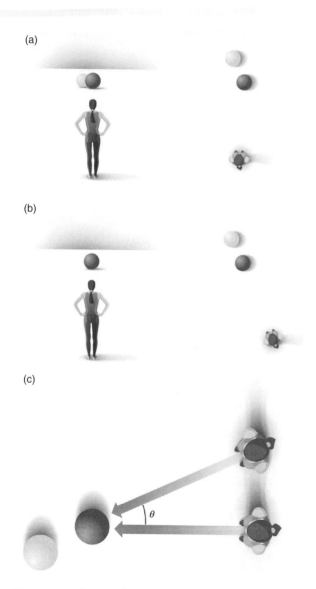

(a)

(b)

(c)

θ

Figure 12.8 **The meaning of parallax.**

NOW YOU DO IT
Consider stars intermediate between those in the ecliptic and those perpendicular to it. How does our theory predict that they would appear to move?

Measuring the distances to the stars using parallax

The phenomenon of parallax can be used to measure the distances to the stars. This is because parallax is large for nearby objects and small for faraway ones. If we turn this around we can say that if the parallax is big the object must be nearby. And this means that *we can measure the distance to a star by measuring how much it appears to move over the course of a year.*

Figure 12.8 illustrates the origin of parallax. It shows an observer looking at two objects: a nearby one such as a black ball, and a more distant one such as a white ball. In (a) and (b) we see the observer looking at the scene: at the left we see this from behind the observer, and at the right from above her.

Notice that in (a) the two balls do not line up, whereas in (b) they do. This is because in going from (a) to (b) the observer has stepped to her right. This "stepping to her right" is our analog of the swinging to and fro of your head, and of the circular motion of the Earth.

Finally, in (c) we illustrate the fact that in taking that step the observer has shifted her position by the angle θ. This is the angle by which the black ball appears to shift as the observer changes position. The same is true for stars: due to parallax, they change their positions in the sky by an angle θ as the Earth orbits the Sun. The angle θ is the crucial quantity that we need to think about: it will turn out that, if we can measure θ, we can infer the distance to the star.

The mathematics of parallax

In Figure 12.9 we illustrate the Earth orbiting the Sun, and the lines of sight to a nearby star. From the January position the line of sight to the star is indicated.

From the July position *two* lines of sight are indicated. The white line is the real one: it points to the star. The dark line, in contrast, is not the real one: it points to where the star used to be. That is, since it is parallel to the January line, this line indicates the direction in which the line

Earth in January

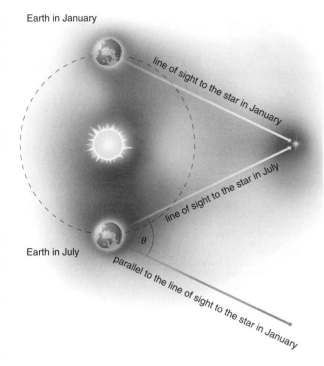

Figure 12.9 **The line of sight to a nearby star shifts by an angle θ as the Earth orbits the Sun.**

Earth in January

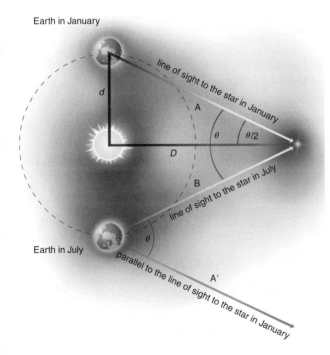

Figure 12.10 **The geometry of parallax. If we measure θ we can solve the indicated triangle to find D.**

of sight would have pointed, had the star not changed its apparent position between January and July. The angle between it and the true line of sight is the angle θ by which the star has changed its apparent position due to the Earth's motion. This is the angle we need to focus on.

In Figure 12.10 we illustrate the mathematics of parallax. In this figure we see that θ is also the angle between the two lines of sight A and B. This is because the parallel lines A and A' are cut by the line B, and geometry tells us that the interior angles made in such a cut are known to be equal. Notice also that we have constructed a right triangle in this figure. One side is d, the distance from the Earth to the Sun: this is the astronomical unit, which is known. Another side is D, the distance from the Sun to the star: this is the unknown we wish to find. Finally there is the angle $\theta/2$, which is known once we have measured the angle θ by which the star appears to shift position in the sky.

We can solve this triangle using the small-angle formula (while the angle θ in Figure 12.10 looks large, this is for purposes of illustrations only. In fact, measured values of θ are very small so that the formula is applicable):

$$D = [360\ d]/[2\pi(\theta/2)_{\text{in degrees}}].$$

Since we know d (the astronomical unit) and θ (the angle by which the star appears to shift its position over the course of a year) this formula tells us D, the distance to the star.

Intuitive mathematics. Notice that in this formula we are dividing by θ. So small values of θ are connected with large values of D. This conforms to our intuitive knowledge: faraway objects appear to shift their positions by only a little as we shift our heads from side to side.

New units: the parallax and the parsec

Astronomers employ new units to simplify this formula. The first is *the parallax p*, which is *the angle $\theta/2$ expressed in seconds of arc*. Since there are 3600 seconds of arc in a degree, our formula now reads

$$D = [(360)(3600)\ d]/[2\pi p],$$
$$D = [1\ 296\ 000\ d]/[2\pi p],$$
$$D = \left[\left(\frac{1\ 296\ 000}{2\pi}\right) \text{astronomical units}\right]\Big/ p.$$

In this formula, the distance D is in meters. But let us now define a new unit of distance, the *parsec* (from "parallax–second"), abbreviated pc:

1 parsec = 1 296 000/2 π astronomical units.

Since the astronomical unit is 1.496×10^{11} meters,

1 parsec = 3.09×10^{16} meters = 3.26 light years.

In terms of this new unit, our formula for the distance to a star whose parallax is p is as follows.

> The parallax formula : $D_{\text{in parsecs}} = 1/p$.

This formula is exceedingly simple to use. The distance to a star whose parallax is one second of arc is one parsec, the distance to a star whose parallax is $\frac{1}{10}$th of a second is ten parsecs, and so forth. Proxima Centauri, the closest star to us (other than the Sun!) has a parallax of 0.772 seconds, so its distance is 1/0.772 = 1.3 parsecs or 4.2 light years. The Hipparcos satellite, launched in 1989, was specifically built to measure parallax. It is capable of detecting parallaxes as small as 0.002 seconds of arc, which means it can measure the distances to stars out to 1/0.002 = 500 parsecs.

Currently such values of parallax are about the smallest we can detect. So other means must be used to measure the distances to stars more distant than this. We will discuss the Cepheid variable technique in Chapter 16 and the supernova technique in Chapter 18.

NOW YOU DO IT
Since Saturn is farther from the Sun than the Earth, its orbital motion carries it in a larger circle than that of the Earth. Suppose we lived on Saturn: would the parallax of a star be larger or smaller than the parallax of that same star as observed from the Earth? Explain your answer.
   (A) How many parsecs away is a star whose parallax is 0.4 seconds of arc?
(B) How many parsecs away is a star whose parallax is 0.04 seconds of arc?
(C) What is the parallax of a star 8 parsecs away? (D) What is the parallax of a star 80 parsecs away? (E) What is the parallax of a star 80 light years away?

THE NATURE OF SCIENCE

REPRESENTATIVE SAMPLES AND OBSERVATIONAL SELECTION

Representative samples

We have now put together a "tool chest": a set of tools for learning things about stars. We are ready to do a census of stars with the aid of these techniques. We wish to study the overall stellar population. How can we do it?

You might think that we should simply survey every star that exists. That would certainly tell us everything we need to know about them. But there are hundreds of billions of stars in our Galaxy, and hundreds of billions of galaxies known. There is no way we could survey each and every one of the stars in the Universe! Furthermore, it is important to emphasize that none of our tools is easy to use. It takes a lot of time and effort to measure the distance, luminosity and so forth of even a single star.

So we will need to find a *representative sample*: a group of stars that is representative of them all. Then, by doing a census of this limited group, we can learn things about stars in general. But how can we find this sample? Which stars should we study?

Two strategies for our astronomical census come to mind.

- We could study the closest stars – the ones in our corner of the cosmos. These are the stars of least distance.
- We could study the most obvious stars – the ones we immediately notice when we glance upwards at night. These are the stars of greatest apparent brightness.

Which would be the better census? The answer is not clear. So let us do both and compare them. They are presented in Table 12.1, which lists the 21 closest stars on the one hand, and the stars of greatest apparent brightness on the other.

As we look over these tables, our eyes probably glaze over. Who can pay attention to all these numbers? But let us look at the two lists intuitively, and try to see what they are telling us. There are a number of lessons to be learned from them.

STAR NAMES

The first lesson is that these are completely different lists! Hardly any of the stars that appear in one list also appear in the other. Among all these stars, only four appear in both lists.

Furthermore, notice what happens if we take a look at the names of the brightest stars. These names are beautiful and evocative, and they conjure up the romance and mystery of the nighttime sky: names like Aldebaran, Deneb and Procyon. On the other hand, the names of the closest stars are unlovely and mechanical: Ross 128, L 789–6 and so forth. These are merely catalog entries in a list.

Why the difference? Because the ancients who named the stars had no telescopes. They could see with their naked eyes the brightest stars – but they could not see the closest ones, and so did not get around to naming them. This is our first lesson: *most of the closest stars cannot be seen with the naked eye*. Their luminosity is so low that, even though they are very close, their apparent brightness is too small for us to see.

REPRESENTATIVE SAMPLES

The second lesson is that the stars in the "brightest" list have completely different properties from those in the "closest" list. Those in the "brightest" list are very luminous, but most of those in the "nearest" list are very dim. Every star in the

Table 12.1. **The 21 closest and brightest stars.**

The closest stars		
Name	Distance (parsecs)	Luminosity (solar units)
Our Sun	0	1
α Centauri A[a]	1.3	2.5
α Centauri B		0.4
α Centauri C		0.000 1
Barnard's star	1.8	0.000 5
Wolf 359	2.3	0.000 02
BD+36 21 47	2.5	0.01
L726–8 A[b]	2.7	0.000 1
L726–8 B		0.000 04
Sirius A[b]	2.9	23
Sirius B		0.004
Ross 154	2.9	0.000 6
Ross 248	3.1	0.000 1
L 789–6	3.2	0.000 1
e Eridani	3.3	0.4
Ross 128	3.3	0.000 3
61 Cygni A[b]	3.3	0.063
61 Cygni B		0.063
e Indi	3.4	0.16
Procyon A[b]	3.5	6.3
Procyon B		0.000 6

[a] This is a triple star system.
[b] These are double star systems.

The stars of greatest apparent brightness		
Name	Distance (parsecs)	Luminosity (solar units)
Our Sun	0	1
α Centauri A	1.3	2.5
Sirius A	2.6	23
Procyon A	3.5	6.3
Altair	5.2	11
Fomalhaut	7.7	17
Vega	7.8	50
Pollux	10	30
Arcturus	11	110
Capella	13	130
Aldebaran	20	150
Achernar	44	1 100
Spica	80	2 200
Acrux	92	4 000
Canopus	95	14 000
Mimosa	107	3 000
Betelgeuse	132	10 000
Hadar	153	12 000
Antares	184	11 000
Rigel	236	40 000
Deneb	920	250 000

"brightest" list has a luminosity greater than or equal to that of our Sun – but in the "closest" list only four out of the 21 do. This teaches us that *the stars we most readily notice at night are not representative*. They are abnormal, unusually brilliant stars – the ones that, even though they are quite far away, shine brightly in our sky.

We can understand this by means of an analogy. Let us imagine a census in which we study, not the brightness of a star, but the ability of a golfer. To make our analogy complete, imagine two censuses.

- A study of the abilities of the golfers who live closest to us.
- A study of the abilities of the most obvious golfers: those whose names are most prominent in the national consciousness.

The critical issue is that the golfers who live closest to us are most likely to be amateurs, people who play golf for fun. But those golfers whose names are most prominent in the national consciousness are most likely to be professionals: those who play in tournaments, win huge prizes and appear in the news most often. With this insight, we can readily understand our two lessons.

(1) Why do the two censuses yield completely different lists? In our analogy this is equivalent to asking why most of the amateur golfers never make it into the national consciousness. Because they are not good enough at the game for the news media to notice them. Most people in the "closest" list are not proficient enough to appear in the "most obvious" list, and most professional golfers do not happen to live nearby. Similarly, most stars in the "closest" list are not luminous enough to appear in the "brightest" list.

(2) Why are the golfers in the second list not representative of most golfers in the nation? Because *this is how they make it into that list* – by not being representative: by being unusually good at the game. And thus it is that the stars in the "closest" list have completely different properties from those in the "brightest" list.

We are now ready to answer our original question: which is the best survey to conduct? Clearly, the stars in the "brightest" list are not representative of stars in general: they are the unusual stars, those most readily visible by virtue of their unusually great luminosity. So to learn the most about stars in general, we should study not the most obvious ones, but those closest to us.

Observational selection

Both of the lists in Table 12.1 have been ordered in terms of increasing distance from us: they begin with the closest star (our Sun), and end with the most distant. But notice there is also another order present in one of these lists – the "brightest" list. This particular list is also nearly perfectly ordered in terms of luminosity! The nearest star in that list is also the least luminous (our Sun), and the farthest star is also the most luminous (Deneb). Furthermore, as we scan down the list from nearest to farthest stars, the luminosity nearly always increases. There are only four points at which this general trend reverses and the luminosity drops.

This is in contrast to the behavior of the "closest" list. In that list, there is no relationship between distance from us and luminosity. As we scan down that list, the luminosity changes erratically, showing no general trend.

We are finding that there is a correlation between the distance of a star and its luminosity. And we are finding that this correlation exists in one list – the list of brightest stars – but not in the other – the list of closest stars. Why?

You might think that this follows from the inverse square law, which tells us that the brightness of a star depends on distance. Consider an analogy in which stars are like light bulbs, and the bulbs are scattered about randomly. Doesn't the inverse square law predict a correlation between distance and brightness?

This line of argument has two errors.

- The inverse square law tells us that the closer a source of light is, the brighter it appears. But our correlation is just the opposite: the closer stars are the dimmer, not the brighter ones.
- This law refers to apparent brightness, not luminosity. A 100-watt light bulb that is brought closer to us appears brighter – but its luminosity is still 100 watts. The luminosity that appears in Table 12.1 measures the "wattage" of a star, not how bright it appears in the sky.

So there is something wrong with our analogy of stars as being like light bulbs.

The actual situation is that we have many different kinds of light bulbs, all of different wattage. A correct analogy would be that some of our sources of light are faint candles, others are normal light bulbs, and yet others are searchlights. And as we observe these light sources, we find a striking fact: those that happen to lie close to us turn out to be the candles, those at intermediate distances turn out to be light bulbs and the most distant ones turn out to be searchlights.

An amazing correlation! Why should the Universe be organized in this way?

In fact the Universe is not organized in this way! Our "discovery" that distant stars are unusually brilliant is actually false. It is false because our observations have been "contaminated" by a phenomenon known as observational selection.

Here is how observational selection has led us to our erroneous "discovery." Imagine in our analogy that all three types of light source – candles, light bulbs and searchlights – are in reality present everywhere, both near us and far from us. But imagine also that there are unequal numbers of these light sources:

(1) candles are most common,
(2) light bulbs are next most common,
(3) searchlights are few and far between.

And finally, imagine that they are spread over distances ranging from a mere few feet to many miles.

Consider the following.

(1′) When we look at the nearby sources, we see them all. But since candles are the most common type of source, mostly they are what we see. We therefore conclude that at nearby distances most sources are of low luminosity. Notice that this conclusion is correct.

(2′) When we look at intermediate distances, we are no longer able to see the candles. They are so far away that they are too faint to be seen. We do see the light bulbs and searchlights, however, and they are more luminous than the candles. So we conclude that at intermediate distances most sources are of intermediate luminosity. Notice that this conclusion is *not* correct.

(3′) When we look at the greatest distances, we are no longer able to see either the candles or the light bulbs: both are too faint to be seen. So we only see the searchlights. We therefore conclude that at the greater distances the sources are of the greatest luminosity. Notice that, here too, this conclusion is *not* correct.

It was this series of erroneous conclusions that led to our erroneous "discovery" of a relationship between luminosity and distance.

Why does this relationship not appear in the list of closest stars, so that this list is not contaminated by observational selection? Because even the dimmest stars are sufficiently bright for us to see them when they are close. Thus our survey of these nearby stars detects all of them. So the conclusions we draw from our study of the closest stars do not contain any errors: they reveal to us a representative sample of stars in general. This is yet another reason for choosing the "closest" survey to do our census.

Observational selection is always a problem in designing our studies. Unless care is taken to avoid it, it can render a study worthless. Here are two more examples of how it can undermine our conclusions.

First, an example from politics. In the 1948 presidential election between Harry Truman and Thomas Dewey, public opinion polls showed Dewey leading over Truman. Indeed, on the night of the election one newspaper went to press declaring Dewey the winner before the final count was in. The next morning Truman woke to find he had won the election, and in a famous photo was shown brandishing a newspaper with the headline "Dewey Defeats Truman" (Figure 12.11).

What had led the polls astray? One theory is that the polls had been conducted by telephone – and this was at a time when not everybody owned phones. Only the wealthier people could afford them and, statistically, the wealthier segments of the population preferred Dewey. Conversely, the less well

Figure 12.11 **Observational selection?**

off were more disposed to vote for Truman – but they were "invisible" to the pollsters, because they did not have phones. The method of observation selected people liable to vote for one of the candidates. A valid method would have used a technique that avoided this error.

Here is a second example. Until the mid twentieth century, astronomers thought of the Universe as being composed primarily of stars. Furthermore, this stellar cosmos was regarded to be static and unchanging, since stars were known to shine steadily for billions of years.

But in the late twentieth century astronomers were forced to revise their conception of the Universe. We now know the cosmos to be filled with vast interstellar clouds, black holes and other objects unknown to previous generations of astronomers. And we now know that the cosmos is from time to time shaken by violent, cataclysmic events: supernovae, quasar explosions and the like.

What had led the early astronomers astray? It was the fact that they used only optical telescopes. Telescopes that detect only visible light are not capable of detecting these other objects of the Universe. Interstellar clouds are too cool to emit visible light: they emit infrared radiation, and can only be detected by infrared telescopes. And the violent cataclysms that rock the cosmos mostly emit in the X-ray and gamma-ray regions of the spectrum. Here, too, until we conducted studies with the right kinds of instruments, we were unaware of these phenomena.

The lesson of observational selection is this: in conducting a study, we must be careful to avoid observational techniques that select for a certain result.

NOW YOU DO IT

(1) An analogy to stars in the sky might be a collection of automobile headlights. Imagine that you are standing beside a road at night and looking at the headlights of cars. You notice that some appear bright to you and some dim, and you notice that some cars are near you and some far down the road. You compile two lists: one of the closest cars, and one of the brightest headlights. Before taking a look at these lists, however, you do a little thinking.

According to the inverse square law, the apparent brightness of a source of light grows larger as the source is situated closer and closer to you. You therefore make a prediction: every car appearing on the list of nearby cars should also appear on the list of bright headlights. And, when you check, you find that they do.

Applying this reasoning to stars, you make a second prediction: that every star that appears on the list of nearby stars should also appear on the list of bright stars. But, when you check, you find that they don't!

Something is wrong with the reasoning behind your second prediction. What is your error?

(2) You are an officer in the US Census bureau, and you drive a top-of-the-line luxury automobile. There are two studies you would like to perform.

(A) In the first study, you wish to investigate the incomes of the United States population: how many people are there who make $40 000 per year, $80 000 per year, and so forth?

(B) In the second study, you wish to investigate the hair color of the United States population: how many people are there whose hair is brown, blond, etc.?

In each of these studies, in order to obtain a representative sample you can choose to survey one of two groups: (1) people whose initials are the same as yours, or (2) people who drive the same kind of car as you.

For each study, which is the better group to survey? Make sure you explain your reasoning.

- -

A census of stars

We now know how to conduct our census of stars: we should study the stars in our immediate vicinity. We now turn to a description of the results of these measurements. In this section we will concentrate on the most important general properties of stars, deferring to the next a detailed study.

The stars in general

The stars are very far away. In the previous chapter we explained that the Sun is very far away. But the distance to the Sun, large though it is by everyday standards, pales in comparison with that to the stars. A ray of light, traveling at 186 000 miles per second, requires 8 minutes to reach the Sun – but it requires 4 *years* to reach the second closest star, Proxima Centauri. Many stars easily visible to the naked eye are hundreds of light years distant. Deneb, a bright star in the constellation of Cygnus, lies a full 1600 light years from us. The light we see from it was emitted about the time of the fall of Rome.

If we were to build a scale model in which the Sun were 1 centimeter away, the distance to Proxima Centauri would be 2.6 kilometers (1.6 miles), and that to Deneb 1000 kilometers (621 miles).

The Sun is a typical star. Some stars are bigger than the Sun, and some smaller. Some are brighter and some dimmer, some more massive and some less massive. But as a general class, the stars are no different in kind from the Sun.

Different stars have different properties. They are not all alike. Stars range in brightness from tens of thousands of times brighter than the Sun to tens of thousands of times fainter. They range in diameters from a hundred times bigger than the Sun to a hundred times smaller. Their masses range from nearly $1/10$th of the mass of the Sun to 50 times its mass. Their surface temperatures range from 2000 kelvin to 30 000 kelvin.

The 40 closest stars

Table 12.2 lists the observed properties of a representative sample of stars: the 40 closest.

Table 12.2. **The 40 closest stars.**

Distance (parsecs)	Component (if multiple)	Luminosity (solar units)	Mass (solar units)	Radius (solar units)
0^a		1	1	1
1.3	A	2.5	1.1	1.2
	B	0.4	0.9	0.9
	C^b	0.000 1	0.1	?
1.8		0.000 5	?	?
2.3		0.000 02	?	?
2.5		0.01	0.35	?
2.7	A	0.000 1	0.11	?
	B	0.000 04	0.11	?
2.9	A^c	23	2.3	1.8
	B^d	0.004	1.0	0.02
2.9		0.000 6	?	?
3.1		0.000 1	?	?
3.2		0.000 1	?	?
3.3		0.4	0.9	?
3.3		0.000 3	?	?
3.3	A	0.063	0.63	?
	B	0.063	0.6	?
3.4		0.16	?	?
3.5	A	6.3	1.8	1.7
	B^d	0.000 6	0.6	0.01
3.5	A	0.004	0.4	?
	B	0.001 6	0.4	?
3.5	A	0.01	?	?
	B	0.000 6	?	?
	C	?	?	?
3.6		0.4	?	1.0
3.6		0.01	?	?
3.7	A	0.001 6	?	?
	B	?	?	?
3.7		0.000 02	?	?
3.8		0.000 3	?	?
3.8		0.025	?	?
3.9		0.004	?	?
4.0	A	0.001 6	0.27	0.51
	B	0.000 6	0.16	?
	C	?	0.01	?
4.0	A	0.000 6	0.14	?
	B	0.000 02	0.08	?
4.0		0.001 6	?	?

^{a} The first entry in this table is the closest star – our own Sun.
^{b} This is Proxima Centauri, the closest star other than our Sun.
^{c} This is Sirius A, the brightest star in the sky other than our Sun.
^{d} These are white-dwarf stars.

Let us study this table, trying to make intuitive sense of all the data it contains.

Observational selection

Notice first that *the table is incomplete.* Many entries in the table are question marks. These mark stars for which we have not been able to measure the quantity in question. Why should this be so?

One possibility is that astronomers have simply not gotten around to measuring these quantities. Perhaps, if they were only to work a little harder, they could do so. There may be some partial truth in this. But it cannot be the whole story, for there is a pattern to the question marks – a pattern that sheds light on the very nature of data-gathering. *The question marks preferentially appear for the fainter stars.* For the brighter stars the table is complete. For the dimmer ones it is incomplete. For example, notice that there are no question marks at all for those stars whose luminosity is greater than or equal to that of the Sun. Conversely, among all the faintest stars – those of luminosities less than 0.001 L_{Sun} – only one does *not* have a question mark.

We now understand the origin of the question marks. They are due to observational selection. Faint stars are hard to observe: that is why we have not "observed" them.

The mass function

Let us turn our attention to a second element of this table: it contains many stars of low mass, but only a few of high mass. Among the 40 stars it lists, only four have a mass greater than or equal to that of our Sun. All the rest are of lower mass. Figure 12.12 studies this in greater detail. As you can see, *low-mass stars are common: high-mass stars are rare.*

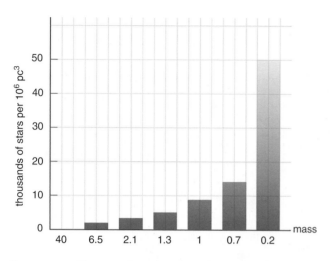

Figure 12.12 **The mass function plots the number of stars of a given mass. Low-mass stars are common, whereas high-mass star are rare.**

The mass–luminosity relation

A third interesting regularity in Table 12.2 is that *the more massive the star the brighter it tends to be.* The most massive stars in the table are also the most luminous ones. Conversely, the least massive stars also tend to be the least luminous. Notice, however, that this regularity is not universal. Two stars in the table do not follow this general trend. These are the white-dwarf stars, to which we will soon turn. All the other stars in Table 12.2 are main sequence stars (we will study these shortly), for which this correlation is valid.

Figure 12.13, in which we plot the luminosity of a main sequence star as a function of its mass, illustrates this in greater detail. The equation describing this relation is

$$(L/L_{Sun}) = (M/M_{Sun})^{3.5}.$$

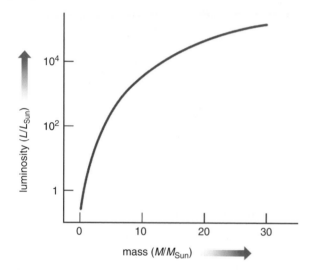

Figure 12.13 **The mass–luminosity relation for main sequence stars. The more massive a star the brighter it is.**

The mass–radius relation

Notice that in Table 12.2 the more massive the star the bigger it is. The most massive stars tend to be the biggest ones, the least massive the smallest. Again, the white-dwarf stars do not obey this general trend. Figure 12.14 illustrates the relation for main sequence stars.

· ·

NOW YOU DO IT

How much brighter than the Sun is a 2 solar mass star? How much dimmer is a ½ solar mass star?

  How much bigger than the Sun is a 2 solar mass star? How much smaller is a ½ solar mass star?

  What is the luminosity of a star twice the size of the Sun?

· ·

The Hertzsprung–Russell diagram

In 1911 the Danish astronomer Ejnar Hertzsprung plotted *the luminosities of stars versus their temperatures.* Two years later the American astronomer Henry Norris Russell did the same thing. Their plot has come to be known as the Hertzsprung–Russell, or H–R, diagram.

On an H–R diagram a particular star – the Sun, say – is represented by a single point, specified by the Sun's luminosity and temperature. A collection of stars, in turn, will be represented by a collection of points. We might have thought that these points would turn out to scatter more or less everywhere across the diagram. But, to their surprise, that is not what Hertzsprung and Russell found. The H–R diagram for stars in our vicinity is shown in Figure 12.15. (Notice that the temperature scale on the diagram is backward: hot stars are to the left and cool ones to the right. This

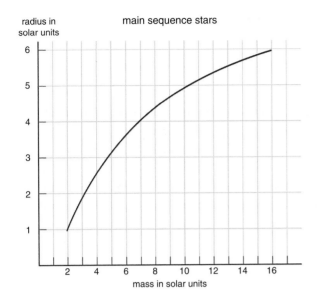

Figure 12.14 **The mass–radius relation for main sequence stars. The more massive a star the bigger it is.**

⟵ **Looking backward**
We studied the Stefan–Boltzmann law in Chapter 4.

Figure 12.15 **Hertzsprung–Russell diagram for stars in our vicinity. Such a diagram plots the luminosity of stars against their temperatures.**

luminosity

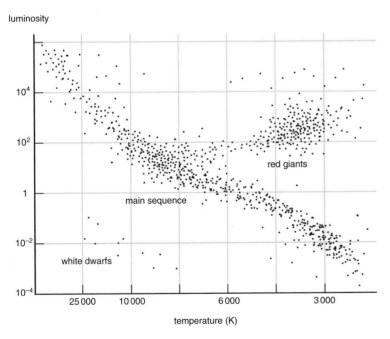

is due to the historical accident that what Hertzsprung and Russell actually plotted along the horizontal axis was a classification of a star's spectrum: only later was it realized that this classification was inversely related to temperature.) As you can see, stars form well-defined groups on the diagram.

The most prominent feature of the distribution is the striking band running from upper left to lower right. Stars that fall along this band are known as *main sequence* stars: our own Sun is such a star. Main sequence stars are by far the most common stellar type: roughly 90% of stars in our vicinity are main sequence. By observing main sequence stars in binary systems we can measure their masses. It turns out that there is a relation between a star's mass and its position along the main sequence. High-mass stars are found to occupy the upper-left portions of the main sequence: low-mass stars the lower-right portion.

Also visible in Figure 12.15 are a few stars in the upper-right and lower-left portions of the diagram. We can learn a little about them by reminding ourselves of the Stefan–Boltzmann law, which states that the luminosity of a star is given by

$$L = 0.57 \times 10^{-7} A T^4 \text{ watts.}$$

Here A is the star's surface area, given by $A = 4\pi R^2$ (R is the star's radius). Notice that this law states that the luminosity

- is large for high temperatures, but low for low temperatures,
- is large for large stars, but low for small stars.

Let us apply this to the stars in the upper-right part of the H–R diagram. They are very cool, but they are very bright. The only way a cool object can emit a lot of light is for it to have a large surface area. The stars in this region of the H–R diagram are therefore very large. Furthermore, because they are cool the light they emit is deep red in color. Hence their name: these are known as *red-giant* stars.

Conversely, the stars in the lower-left part of the H–R diagram are very hot (making them white in color) but very dim. If T is large, L can be small only if R is small: these are the *white-dwarf* stars.

....

NOW YOU DO IT
Where in the H–R diagram would we find stars that were very large and very hot? What would their color be?

Where on the H–R diagram would we find stars that were very small and very cool? What would their color be?

....

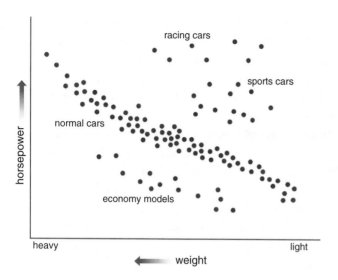

Figure 12.16 **An "H–W" diagram for cars. Horsepower is plotted against weight.**

⇒ **Looking forward**
We will study stellar structure in Chapter 14, and stellar evolution in Chapter 15.

The three categories of stars are indicated in Figure 12.15.

Comments

Clearly there is much to be understood here. The Hertzsprung–Russell diagram contains a number of systematic patterns that cry aloud for explanation. The astronomer Michael Seeds has invented an instructive analogy for this situation.[1] Suppose some extraterrestrial being came down to Earth, and decided to study the various objects it found on this strange new planet. Suppose it decided to study not stars but cars. Suppose it decided to measure the *horsepower* and the *weight* of cars, and then graphed them on an "H–W diagram". We can even imagine that the extraterrestrial, for some reason all its own, plotted the weight backward along the horizontal axis.

Figure 12.16 shows the result. The extraterrestrial would find that most cars turn out to lie on a band running from upper left to lower right – but a few lie in the upper right, and a few more in the lower left. Notice how similar this diagram looks to our H–R diagram!

How can we understand this? We understand the "main sequence for cars" by noting that the heavier the car, the more power it must have if it is to travel down the road at all. As for the "red-giant cars" – those in the upper-right part of the diagram – these are those few vehicles designed for racing: powerful but light. And finally the "white-dwarf cars" in the lower left are those underpowered heavy clunkers that we occasionally find lumbering down the highways.

This is how *we* understand Seeds' "H–W diagram." But would that mythical extraterrestrial figure this out all at once? Clearly, the creature would have a hard job ahead of it. And so do we. The task of understanding the full richness of the H–R diagram will prove to be a difficult one, and it will occupy us throughout two full chapters.

In these chapters we will develop a theory of stellar structure. We will see that stars evolve, and that as time passes they go through various stages in their evolution. We will find that

- main sequence stars are in the first stage of their evolution, burning hydrogen as their nuclear fuel,
- red giants are in the second stage of their evolution, burning helium as their nuclear fuel,
- white dwarfs are in a final stage, in which they have exhausted all their nuclear fuel.

---

[1] Michael A. Seeds, *Stars and Galaxies*, Wadsworth, 1999, p. 175.

Summary

Measuring the stars

- In many cases we can measure a property of the stars in the same way we measured it for the Sun.
  - Luminosity: if we know the star's distance, measure its apparent brightness (flux) and use the inverse square law.
  - Temperature: measure the spectrum and use the Wien law.
  - Composition: measure the spectrum.
- Measure the distances to the stars using parallax.
  - Observe a star over the course of a year.
  - Its apparent motion tells us its distance.
  - Parallax formula: $D = 1/p$ (D = distance in parsecs: p = apparent motion over a year in seconds of arc).
  - Note that p can be measured only for nearby stars.
- Measure the sizes of stars.
  - Even through a telescope they are too small to be resolved.
  - Can be determined from the Stefan–Boltzmann law once we have measured a star's luminosity and temperature.
- Measure the masses of stars.
  - Use binary stars.
  - Observe the orbits of both stars and determine their masses using a generalization of Newton's mass formula.
  - Measure the rotation of stars by means of the Doppler effect.

Binary stars

- At least half of all "stars" are actually two stars orbiting about each other.
- Four types of binary stars.
- Sometimes we can resolve them both.
- Sometimes we see one star "orbiting about nothing." We presume this "nothing" is a star too dim to be seen.
- Sometimes we see what appears to be a single star, but whose spectrum shows two sets of lines whose wavelength shifts (via the Doppler effect) reveal the presence of two stars.
- Sometimes we see brightness variations caused by one star eclipsing the other.
- The orbits of the stars tell us their masses.

A census of stars

- The stars are very far away: in a scale model in which the Sun were 1 centimeter away, the next closest star would be 2.6 kilometers away.
- The Sun is a typical star.
- Not all stars are alike. There is a wide range in their properties.

- Low-mass stars are common: high-mass stars are rare.
- Low-luminosity stars are common: high-luminosity stars are rare.
- The higher the mass of a star the bigger it is.
- Hertzsprung–Russell diagram.
  - Plots stars' luminosities against their temperatures.
  - Temperature scale is backward.
  - Most stars lie along the main sequence. As we will see, these generate their energy by fusing hydrogen into helium.
  - Red-giant stars are unusually large and cool. As we will see, these generate their energy by fusing helium into other elements.
  - White-dwarf stars are unusually small and hot. As we will see, these have run out of nuclear fuel.

DETECTIVES ON THE CASE

Parallax

Parallax is the observed shifting to and fro of a star over the course of the seasons.

- We first tried to account for this by postulating that the star is orbiting about another object too dim to be seen.
- But according to this idea the orbit could have any period at all. In reality, however, these stars are always observed to have periods of just one year.
- We then proposed that this arises because of the Earth's motion about the Sun: as we observe the star from different vantage points we see it at apparently different locations.

Testing the theory

- It predicts that those stars that do not appear to show parallax should do so, but to a smaller degree. This can be tested by building a better telescope.
- It predicts that stars in the ecliptic should appear to move along straight lines, while those perpendicular to the ecliptic should appear to move in circles.

The theory passes both tests.

THE NATURE OF SCIENCE

Representative samples and observational selection

- We cannot conduct a census of stars by studying all of them: that would take too much time.

- We need to conduct a census of a representative sample.
  - Often the things we most readily notice are not representative.
  - It requires much thought to select a sample that *is* representative.
- Observational selection.
  - Sometimes we find what appears to be a correlation but which in fact does not exist in nature.

- For instance, at first glance our census of stars appeared to show that the more distant a star is the brighter it is.
- In reality this is not so.
- The error arose because of observational selection: at great distances, only the brightest stars are visible.
- In choosing a representative sample, we must take care to avoid observational selection.

Problems

(1) We want to calculate the angular diameter of a star three times bigger than the Sun 10^{15} kilometers away. (A) Describe the logic of the calculation you will use. (B) Carry out the detailed calculation.

(2) Suppose a star suddenly moves twice as far away, but nothing else about the star changes. In what way will the following quantities change:
- its apparent brightness (flux),
- its absolute brightness (luminosity),
- its color,
- its mass,
- its parallax?

(3) Suppose we have a telescope capable of resolving 0.1 seconds of arc. Suppose we put that telescope on a space probe and send it off into space – and then turn it to point back toward us. We want to calculate how far the space probe can venture before the telescope is no longer capable of resolving the Earth. (A) Describe the logic of the calculation you will use. (B) Carry out the detailed calculation.

(4) In the following table we list various constellations. Identify these constellations on a celestial globe, and predict whether a star in that constellation would exhibit a parallax motion
- in a straight line,
- in a circle,
- in between a line and a circle.

Constellation	Your predicted parallax motion	Period of apparent motion
Virgo		
Ursa Minor		
Carina		
Fornax		
Sagittarius		
Cygnus		

In your table also identify which stars would have a period of apparent motion
- greater than a year,
- equal to a year,
- less than a year.

(5) Since Mercury is closer to the Sun than the Earth, its orbital motion carries it in a smaller circle than that of the Earth. Suppose we lived on Mercury: would the parallax of a star be larger or smaller than the parallax of that same star as observed from the Earth? Explain your answer.

(6) Suppose you were designing a space probe capable of measuring the distances to the stars using parallax. Based on your analysis of the previous problem, what sort of orbit would you choose for your space probe?

(7) Recall that it takes light a year to travel 1 light year, and 10 years to travel 10 light years. This means that, as we look at the stars, we are seeing them as they were some time ago. The bright star Rigel has a parallax of 0.0042 seconds of arc. Suppose an alien astronomer lived on Rigel and was observing the Earth. Suppose that astronomer's telescope was so amazingly accurate that the alien could read our calendars. What year would these calendars show it to be?

(8) Suppose there was a mirror in space 3 light years away, which just happened to be oriented so that we could see ourselves in it. (A) Suppose also that we had telescopes so amazingly accurate that we could read calendars in the reflection. What year would they indicate? (B) What is the parallax of that mirror?

(9) Suppose you could move the mirror of the previous problem farther and farther away from us. (A) How far would you have to move it till you could use it to see yourself being born? (B) What would be the parallax of the mirror in this case?

(10) In Table 12.1 identify the four stars that are *both* close to us and at great apparent brightness.

(11) In Table 12.1 identify the closest stars that are brighter than the Sun. What do you notice about their names?

(12) In Table 12.1, identify the four points in the list of brightest stars at which the luminosity decreases as we move from one star to the next in the list.

(13) In our treatment of observational selection, we discussed an analogy of a population containing large numbers of faint sources (candles), intermediate numbers of intermediate sources (light bulbs) and a very few bright sources (searchlights). Consider now a different population: one in which there were a large number of searchlights, an intermediate number of light bulbs and a few candles. Suppose that they were all scattered about, both near us and far from us.

In this situation, would we observe a correlation between the luminosity of a source and its distance from us? Why or why not?

(14) You are driving down a superhighway, and you are stuck in a slow-moving knot of cars. It extends for as far as you can see. You conclude that the traffic today is terrible, and that it moves slowly. But now you look across the divider strip at traffic moving the other way. You see that most of these cars are moving rapidly, and that only occasionally is there a tightly packed, slow-moving knot. So you reach a second conclusion: that the traffic moving in your direction is different from the traffic moving in the other direction.

In reality, both of these conclusions are erroneous: the traffic in both directions is the same, and it is not so terrible as all that. Your task in this problem is to explain how it was observational selection that had led you to these two errors.

(15) Use Figure 12.13 to find (A) the luminosity of a star of 0.2 solar masses, (B) the mass of a star of 3 solar luminosities.

(16) Use Figure 12.14 to find (A) the radius of a star of 10 solar masses, (B) the mass of a star of 3 solar radii.

(17) Make a rough sketch of the H–R diagram. On it indicate where we would find a star (A) the same temperature as the Sun, but far larger, and (B) the same luminosity as the Sun, but far hotter.

(18) Make a detailed sketch of the H–R diagram, and this time make sure you label the axes. On your diagram, indicate (A) a star the same temperature as the Sun, but with ten times its radius, and (B) a star four times the size of the Sun but with half its temperature.

13

The formation of stars and planets

Before turning to a study of the structure of the Sun and other stars, we discuss how they were formed. As we will see, the processes that form stars also form planets orbiting about them. So we are really discussing the origin of the Solar System as a whole.

Before the advent of modern science, such a study was really the province of myths and religion: ancient peoples made up stories about the creation of the world. It is important to emphasize that these stories had no basis in the kind of observation and deduction that forms the essential basis of scientific reasoning. Not so very long ago, however, the situation changed.

It changed for a variety of reasons. In the first place, the studies we have described of the structure of the Solar System showed that our Earth is but one among a whole panoply of moons, asteroids and planets – and that the configuration of this vast system gives us clues as to its origin. In the second place, the development of modern physics gives us the tools to reach an understanding of this origin. And in the third place, recently we have been able to actually observe other stars and planetary systems in the midst of their formation. These observations have added new and totally unexpected elements to our understanding.

Most importantly of all, though, is that our thinking about the origin of the world has totally changed. No longer does this origin lie in the realm of the magical or the supernatural. Now it is an empirical matter, one that we can study with the tools of science, and one that we are capable of understanding.

⇒ **Looking forward**
We will study stellar structure in Chapter 14.

When did the Solar System form?

We begin by asking when this formation process took place. How old are the Sun and planets? How long ago did they form? We can approach this question in a variety of ways.

Let us begin with the Sun. A little thought shows that it cannot be infinitely old. Because sunlight carries energy (this is the energy of solar power), were the Sun infinitely old it would have emitted an infinite amount of energy by now, which is impossible. Indeed, as we will see in our study of stellar structure, the nuclear fuel that powers the shining of the Sun can "only" last about ten billion years.

How far along this ten-billion-year span is the Sun? Detailed computer models of stellar structure allow us to answer this question. They predict that various

Figure 1.45 **An eclipse of the Sun.**

Figure 2.36 **Galileo's telescope alongside a modern equivalent.**

Figure 3.1 **Isaac Newton.**

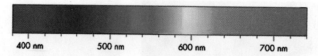

400 nm   500 nm   600 nm   700 nm

Figure 4.6 **The spectrum of visible light. One nm is 10^{-9} meters.**

Figure 5.1 **The Andromeda Galaxy.**

Figure 5.25 **Laser guide star. The Keck telescope in Hawaii projects a laser beam into the night sky. The galactic plane of the Milky Way is visible in the sky to the right of the image. The stars are trailed in this time exposure due to the rotation of the Earth.**

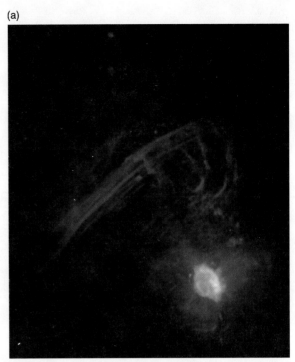

(a)

(b)

Figure 5.33 **Two views of the center of our Galaxy. (a) In radio waves. (b) In X-rays.**

Figure 5.29 **The Arecibo radio telescope.**

Figure 5.34 **The Hubble Space Telescope.**

Figure 5.35 **The James Webb Space Telescope. A full-scale model displayed at the American Astronomical Society Meeting in Seattle, January 2007.**

Figure 7.24 **The Earth photographed from the Apollo 17 mission as it was on its way to the Moon. Note how much H_2O is visible in this image.**

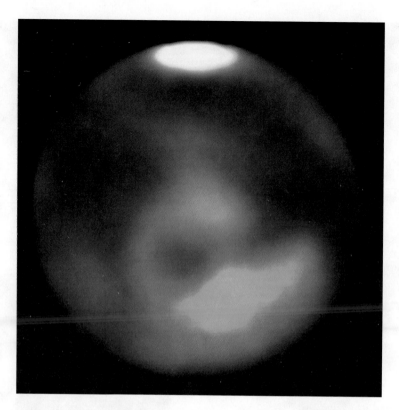

Figure 7.38 **Mars as photographed with a ground-based telescope. The image is blurred by atmospheric "seeing."**

Figure 7.40 **The surface of Mars as revealed by the Hubble Space Telescope.**

Figure 8.1 **Jupiter.**

Figure 8.2 **Jupiter's Great Red Spot is the size of two Earths.**

Figure 8.6 **Io.**

Figure 8.12 **Saturn.**

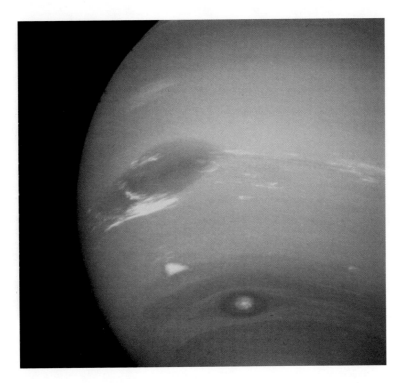

Figure 8.27 **Neptune.**

Figure III.1 **A universe of stars.**

Figure III.2 **The Pleiades.**

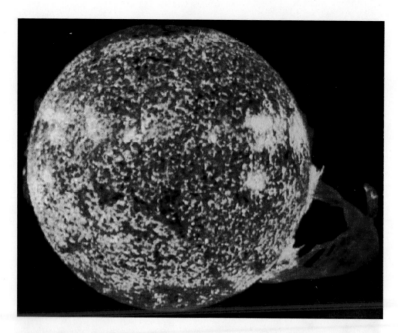

Figure 11.21 **Prominences.**

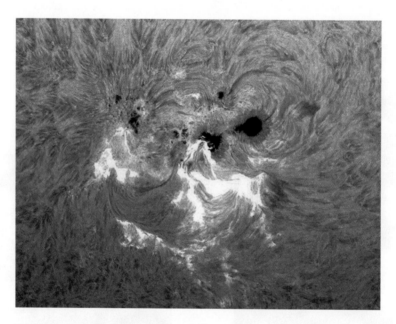

Figure 11.22 **A solar flare (white region) often takes place above a sunspot.**

Figure 11.23 **An aurora.**

Figure 13.1 **The Orion Nebula.**

(a)

(b)

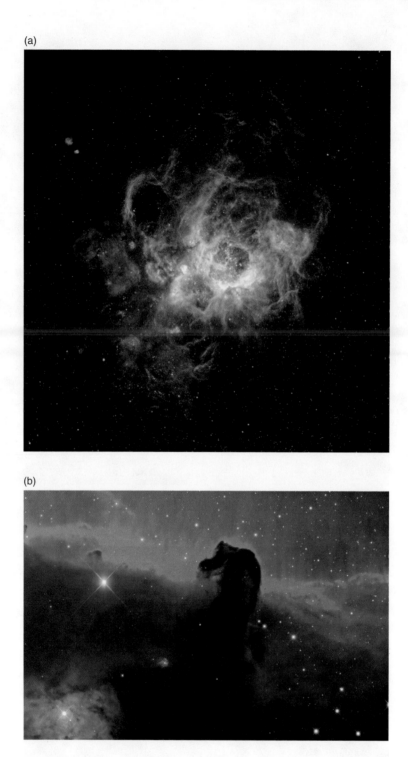

Figure 13.2 **Interstellar nebulae. (a) The Triangulum Nebula. (b) The Horsehead Nebula. (c) The Omega Nebula (overleaf).**

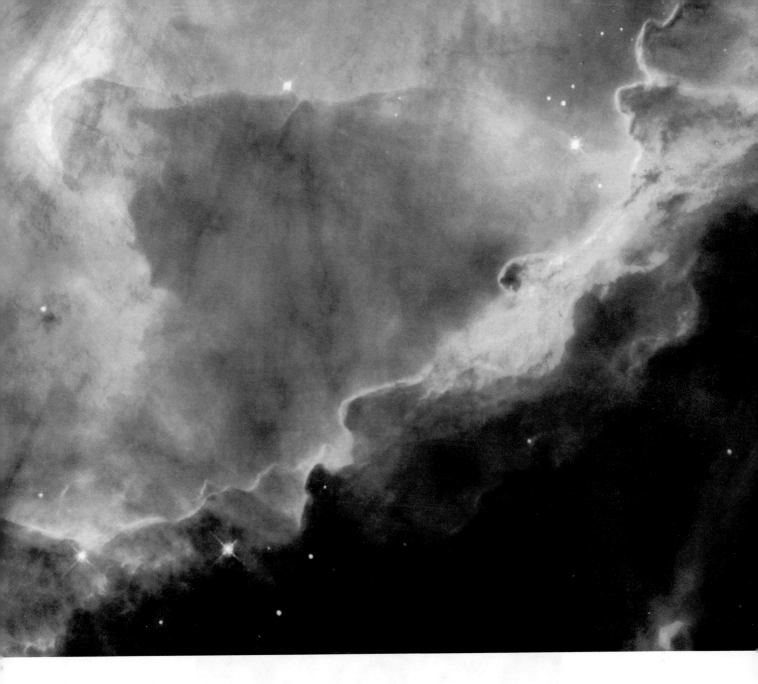

Figure 13.2 **(cont.)**

Figure 13.9 **The Eagle Nebula. Radiation from highly luminous new stars is evaporating the nebula from which they formed. The "evaporating gas globules" appear to contain protostars that might not survive this radiation.**
Behind it trails a "stalk" of material protected from the radiation by the globule's shadow.

Figure 13.24 **Solids begin to condense within the solar nebula.**
(Artist's conception by William K. Hartmann.)

(a)

(b)

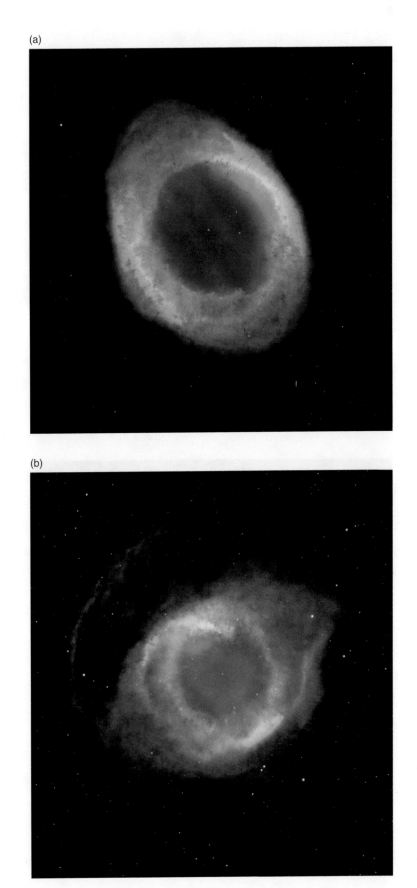

Figure 15.5 **Planetary nebulae. (a) The Ring Nebula. (b) The Helix Nebula. (c) The Eight Burst Nebula.**

(c)

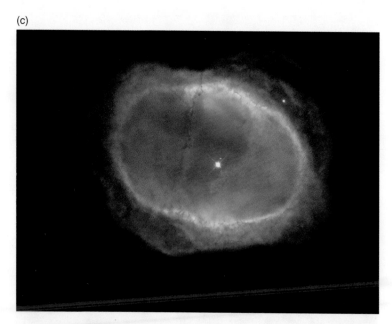

Figure 15.5 **(cont.)**

(a)

Figure 15.21 **Supernova remnants. (a) The Crab Nebula. (b) Remnant of Kepler's supernova. (c) Remnant of Brahe's supernova.**

(b)

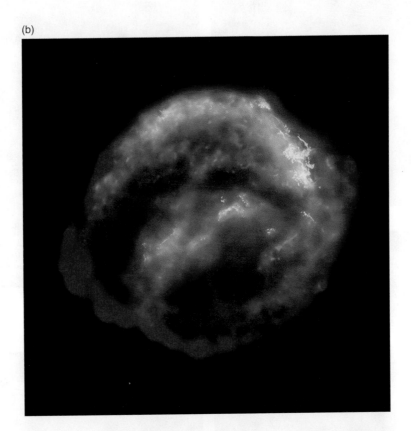

(c)

Figure 15.21 **(cont.)**

Figure IV.1 **Types of galaxies. (a) The spiral galaxy NGC 4414. (b) The elliptical galaxy ESO 325-G004 surrounded by smaller spiral galaxies. (c) The Clouds of Magellan are irregular galaxies. (d) The barred spiral NGC 1300.**

Figure IV.2 **The Hubble Ultra Deep Field.**

Figure 16.2 **Spiral nebulae: (a) the Whirlpool Galaxy. (b) NGC 6384. (c) Sombrero Galaxy. (d) NGC 2841.**

(c)

(d)

Figure 16.2 **(cont.)**

Figure 17.1 **The Andromeda Nebula.**

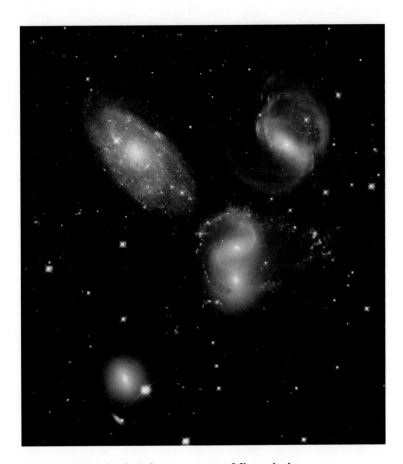

Figure 17.8 **Stephan's Quintet, a group of five galaxies.**

Figure 17.9 **The Antennae Galaxies.**

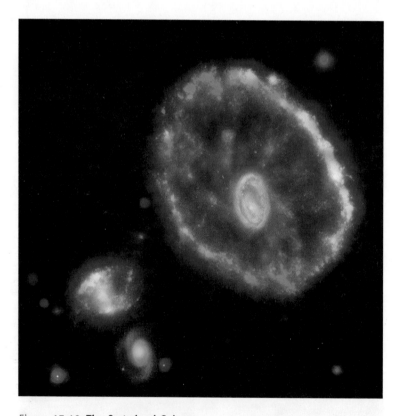

Figure 17.10 **The Cartwheel Galaxy.**

Figure 18.21 **The early Universe. Observations of the Cosmic Background Radiation by the WMAP satellite reveal minute (10^{-4} kelvin) fluctuations.**

Figure 18.25 **Simulation of structure in the Universe. (In this image, the scale bar is 43 megaparsecs long.)**

Figure 19.16 **The image contained in the message to the globular cluster M13.**

properties of the Sun slightly alter as it ages. These models predict that when it was formed the Sun's radius and luminosity were:

$$R_{\text{initial}} = 6.59 \times 10^8 \text{ meters} \qquad L_{\text{initial}} = 2.78 \times 10^{26} \text{ watts.}$$

But these are not their presently measured values! Rather, when we observe the Sun today we find

$$R_{\text{now}} = 6.96 \times 10^8 \text{ meters} \qquad L_{\text{now}} = 3.8 \times 10^{26} \text{ watts.}$$

The Sun has gotten slightly bigger and brighter since its formation. These computer simulations show that the Sun's radius and luminosity would have reached their present values after 4.5 billion years. This, then, is the age of the Sun: it is currently about halfway through its life span.

Let us now pass on to the ages of other bodies in the Solar System. Radioactive age dating allows us to measure the solidification age of a rock sample – the time since it last melted. Because the Earth is geologically active, material on its surface is continually being brought down into its deep interior and melted. Therefore many rocks, when age dated in this way, yield relatively young ages. The oldest rock sample from Earth that anyone has ever found has been dated to be slightly more than 4 billion years old. The Moon, on the other hand, is geologically inactive, so that rocks on its surface remain there, rather than being brought down into its interior. The time since the last melting of a rock on the Moon is therefore probably close to the Moon's actual age. Samples brought back by the Apollo missions have been radioactively age dated to be 4.5 billion years old. Similarly, the ages of meteorites can also be determined in this way. As we saw in Chapter 9 the oldest known meteorites are about 4.6 billion years old.

A difficulty with radioactive age dating is that it cannot be done remotely. In order to find the age of a sample, we must actually get our hands on it to measure its internal composition. This means we can determine the age of a body we have actually visited. Alternatively, a number of meteorites thought to originate on Mars have been age dated (see Chapter 19). While robotic spacecraft have landed on the surfaces of Mars, Venus and Titan, they have not carried onboard instruments capable of conducting such experiments. Nor have they returned samples to Earth for analysis. As things now stand, ages have been determined only for the Earth, Mars, the Moon and meteorites.

Nevertheless, from the limited data we currently have, a vitally important conclusion can be drawn. It must be significant that *all the above ages are very roughly equal*. The Sun, Earth, Moon and meteorites all appear to have more or less the same age. We therefore tentatively conclude that *the Solar System was formed as a unit*. In seeking to understand its origin, we should think in terms of all its various components having been formed simultaneously.

Ongoing star and planet formation

We can also think about other stars. As we will see in our study of stellar structure, the time during which nuclear reactions can keep a star shining depends on the star's mass: the more massive the star, the shorter its nuclear-burning lifetime. For example, a star 40 times the mass of the Sun can continue shining for only one million years.

A million years is long in human terms, but it is a mere 0.0001 the lifetime of the Sun. In astronomical terms, these high-mass stars survive for the merest blink of an

← **Looking backward**
We studied radioactive age dating in Chapter 9.

eye. As an analogy, 0.0001 of the lifespan of a person is a mere 3 days. The fact that such high-mass stars exist at all tells us that they were formed very recently on the age scale of the cosmos. We conclude that *star and planet formation is going on right now.* Just as new humans are continually being born, so too with stars and planets.

NOW YOU DO IT
How long is 0.0001 of the time since the American Revolution?

Where do stars and planets form?

Where in the Universe are the equivalents of our "birthing rooms"? To observe star and planet formation, we should look wherever we find these high-mass, short-lived stars. Because they shine so briefly, they cannot have wandered far from the places of their birth.

Let us analyze this in more detail. How far from the place of its birth can such a short-lived star travel during its lifetime?

The logic of the calculation

Step 1. Multiply the lifetime of such a star by the speed at which it moves.
Step 2. Express the answer in light years to get an intuitive feel for its significance.

As you can see from the detailed calculation, the answer we get is 33 light years – a fairly short distance in astronomical terms.

Detailed calculation

Step 1. Multiply the lifetime of such a star by the speed at which it moves.
As we have seen, the age of such a star is one million years. In one year there are

1 year = (365 days)(24 hours/day)(60 min/hour) (60 sec/min) = 3.15×10^7 seconds,

so that the age is

age = (10^6 years)(3.15×10^7 seconds/year) = 3.15×10^{13} seconds.

Stars are observed to move through space at about 10 kilometers per second, or 10^4 meters/second. Thus, during its million-year lifetime, a high-mass star travels:

distance = (velocity)(time)
        = (10^4 meters/second) (3.15×10^{13} seconds)
        = 3.15×10^{17} meters.

Step 2. Express the answer in light years to get an intuitive feel for its significance.
Since one light year is 9.46×10^{15} meters, this is:

distance = (3.15×10^{17} meters)/(9.46×10^{15} meters/ light year) = 33 light years.

NOW YOU DO IT
Suppose a star shines for 15 million years, and it travels at a speed of 5 kilometers per second. We want to calculate how many light years it can travel before it ceases shining.

Where, then, do we find these high-mass stars? Observations reveal that they are found in or near *interstellar clouds.* So these clouds are the "birthing rooms" of stars and planets.

Interstellar clouds

Perhaps the most prominent of all constellations is Orion the hunter. Hanging from Orion's belt is his sword, composed of what appear to the naked eye to be three stars. But if you were to look more closely at the sword's middle "star" – a small backyard telescope will do, or even a pair of binoculars – you would see that it is not a star at all. It is a nebula (Figure 13.1).

Figure 13.2 shows a gallery of interstellar nebulae (the word comes from the Latin for "clouds"). They are among the most beautiful of all astronomical

Figure 13.1 **The Orion Nebula.** 👁 **(Also see color plate section.)**

objects. They are vast clouds of gas and dust, many light years in extent. While these nebulae look substantial, in fact they are exceedingly rarified. An astronaut transported into the heart of one would still require a space suit. Indeed, within an interstellar cloud the density of material is far less than the best vacuum we have ever been able to produce here on Earth. But because they are large, they contain a good deal of matter: a typical cloud contains more matter than the Sun does. These beautiful structures are the birth-places of stars and planets.

Star formation: theory

We will begin by studying the process of star formation. In the next section we will move on to the formation of planets.

How can a large, diffuse interstellar cloud turn into a far smaller, far denser star? It must do so by contracting. Our first step in understanding the process of star formation is the recognition that the degree of this contraction is enormous. This is because an interstellar cloud is gigantic in comparison with the star it ultimately forms.

(a)

(b)

Figure 13.2 **Interstellar nebulae. (a) The Triangulum Nebula. (b) The Horsehead Nebula. (c) The Omega Nebula (overleaf).**
👁 **(Also see color plate section.)**

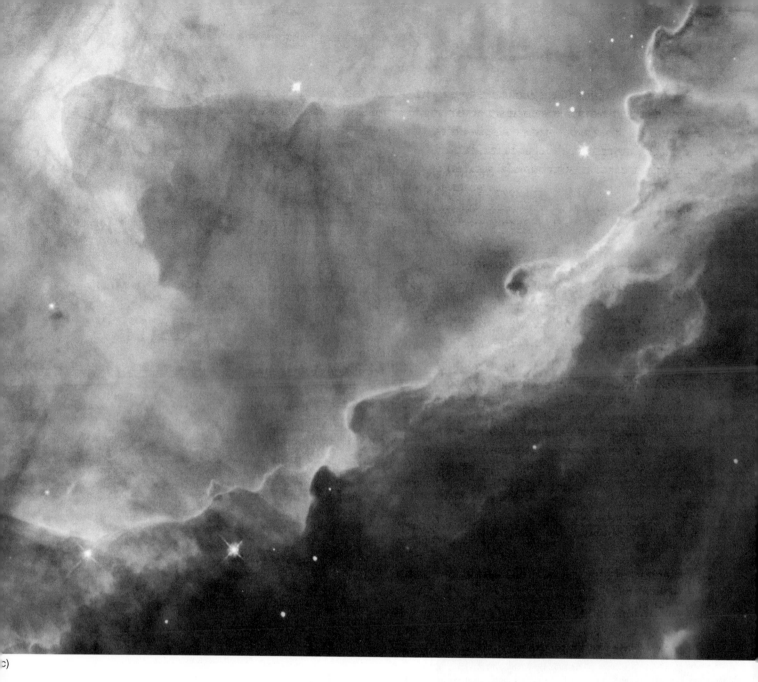

c)

Figure 13.2 **(cont.)**

Let us demonstrate this by calculating *the size of an interstellar cloud that contracted to form our Sun*.

The logic of the calculation

As we explained in Chapter 4, spectroscopy can tell us conditions in an astronomical object. In particular, we can use spectroscopy to measure the *density* of an interstellar cloud. Once we know this, we can calculate the size of a cloud that contains as much matter as our Sun. We then compare this with the size of our Sun.

Step 1. Since density is mass divided by volume, volume is mass divided by density. So we can find the volume of the cloud by dividing the mass of the Sun by the measured density of an interstellar cloud.

Step 2. Let us suppose that the cloud out of which the Sun formed was spherical. Then from its volume we can find its radius.

Step 3. Compare this radius with that of our Sun.

As you can see from the detailed calculation, the cloud out of which the Sun formed was more than 5 *million* times the size of the Sun. So this cloud must have contracted by a

Detailed calculation

Step 1. Since density is mass divided by volume, volume is mass divided by density. So we can find the volume of the cloud by dividing the mass of the Sun by the measured density of an interstellar cloud.

The measured density of an interstellar cloud is roughly 10^{-17} kilograms/meter^3, and the mass of the Sun is 2×10^{30} kilograms. We then find the volume V of the cloud out of which the Sun formed to be

V = mass of Sun/density of cloud
$= 2 \times 10^{30}$ kilograms/10^{-17} kilograms/meter^3
$= 2 \times 10^{47}$ meter^3.

Step 2. Let us suppose that the cloud out of which the Sun formed was spherical. Then from its volume we can find its radius.

For a sphere

$$V = \frac{4}{3}\pi R^3.$$

Solving this for the cloud's radius R we get

$$R = \sqrt[3]{\frac{3V}{4\pi}} = \sqrt[3]{\frac{(3)(2 \times 10^{47}\,\text{meters}^3)}{4\pi}}$$

$R = 3.6 \times 10^{15}$ meters.

Step 3. Compare this radius with that of our Sun.

Our Sun's radius is 6.96×10^8 meters. So we can find the ratio between the size of the cloud which formed the Sun to the size of Sun:

size of cloud/size of Sun = 3.6×10^{15} meters/6.96×10^8 meters = 5.2 million.

factor of more than 5 million in order to form the Sun. As an analogy, if we were to imagine the interstellar cloud as being a mile in diameter, the star it ultimately formed would be about the size of the period at the end of this sentence. Furthermore, if we were to do the same calculation for any other cloud which contracts to form a star, we would similarly find that the degree of contraction is enormous.

....

NOW YOU DO IT
Suppose the density of an interstellar cloud is just twice the density we have just considered. Calculate the size it must have if it will form a 2 solar mass star. (One way to do this is simply to re-do the above calculation. But see if you can think of an easier way!)

....

What could produce such enormous contractions? The answer is: gravitation.

In our daily lives, we experience gravitation as the force which attracts us to the Earth. Similarly, gravitation attracts the Earth to the Sun, and so bends its path into an orbit. But gravitation has another aspect as well. *It acts to compress things.* We can think of gravity as a pair of hands squeezing a tennis ball. It is this "pair of hands" which squeezes an interstellar cloud so much that it forms a star.

Recall that every object attracts every other object with a force of gravitation. Previously we have applied this law to objects like planets, satellites and so forth. But let us now focus our attention more closely, on the atoms in an interstellar cloud. After all, clouds are made of atoms – and even something so small as an atom attracts other atoms by the force of gravity.

Two atoms which lie close together attract with a stronger force than two other atoms which happen to lie farther apart. Furthermore, each force pulls in a different direction. The net result is a complicated network of forces. The situation is shown in Figure 13.3(a). Here we focus attention on one particular atom, and we sketch a few of all the other atoms which are attracting it. As we can see, the atom we are focusing on is being pulled in many directions at once. But we can also see that *the net result of all these competing forces is a tendency to be pulled inward towards the cloud's center* as shown in Figure 13.3(b). The same is true of any other atom we choose to think about. So the net result of gravitation is an inward force of compression.

As a result of this compressional force, the cloud begins to contract. Astronomers customarily refer to such a contracting cloud as a *protostar*. Let us follow this protostar in our minds as it contracts more and more.

It is a general principle that when a gas it compressed, it grows warmer. There are numerous examples of this in everyday life. If you have ever inflated a bicycle tire with a hand pump, for example, you may have noticed that the pump grew warm. As you pressed down on the pump, compressing the air it contains, you heated the air.

(a)

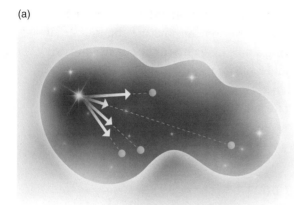

(b)

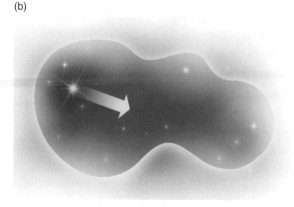

Figure 13.3 **Gravitational forces between atoms in a cloud. In (a) we choose a single atom and show the forces attracting it toward others. (The longer the arrow the stronger the force.) In (b) we show the sum of all these forces, an attraction toward the center of the cloud.**

Applying this principle to the contracting interstellar gas, we conclude that its contraction had the important effect of *heating the protostar*. Furthermore, since the degree of compression was so very great, the gas grew exceedingly hot.

Eventually it became so hot that it turned into a star. The dividing line between a protostar and a star is something of a matter of definition – both, after all, are nothing more than balls of gas. Astronomers define a star to be an object that generates its own energy via thermonuclear reactions. We will see in Chapter 14 that these reactions are triggered when a gas is heated to tens of million kelvin. Once our collapsing cloud attained this temperature, nuclear reactions commenced within it. These reactions generated yet more heat, and correspondingly a great pressure. This pressure was sufficient to resist the inward pull of gravity, and the collapse was halted. The protostar had turned into a star.

Astronomers have used computers to simulate the processes that occur during the collapse. The results of such computations are summarized in Figure 13.4. Let us study this figure carefully, for it contains a wealth of information. It shows the theoretically computed luminosities and surface temperatures of collapsing protostars. The various lines on this figure, called "evolutionary tracks," trace the computed histories of clouds of different masses. Along these tracks, the numbers give the times (in years) required to reach the indicated stages. The heavy line shows the main sequence, which we recall gives the location on this diagram of stars undergoing hydrogen-burning nuclear reactions: once a protostar in its evolution reaches this line it has become a star.

Several lessons can be learned from this figure.

Notice first that the evolutionary track of a protostar depends on its mass: what happens to a high-mass collapsing cloud is very different from what happens to a low-mass cloud. So the "birth history" of a star depends on its mass. Notice in particular that the higher the mass the faster the collapse. A high-mass protostar will become a star long before a low-mass one, even if they commenced collapsing at the same time. We will see soon that this has important consequences for the mechanism whereby star formation is initiated.

. . . .

NOW YOU DO IT
Figure 13.4 shows the evolutionary tracks of five different contracting protostars, of masses ranging from ½ to 15 solar masses. Suppose that at some time each of these commences contracting toward the main sequence. After one million years, which of them will have reached the main sequence to become stars? Which will still be contracting?

. . . .

Notice that there is a general pattern to the evolutionary tracks in Figure 13.4: an initial downward trend is followed, in most cases, by a more or less horizontal trend. What does this mean physically?

The initial downward trend in Figure 13.4 shows that, during the initial stages of its evolution, the temperature of a collapsing protostar remains roughly constant.

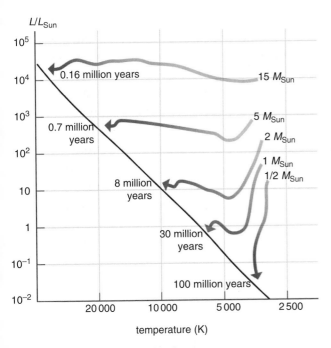

Figure 13.4 Computer models of star formation. The results of computer simulations of star formation are summarized on a Hertzsprung–Russell diagram. The numbers beside each dot give the time in years required to reach this stage of the collapse.

It is heated by the collapse, but only a little. Why should this be so? We know that every hot object emits blackbody radiation. Detailed calculations show that, during this initial phase, the protostar is so tenuous that it is transparent: the radiation it emits is free to escape the cloud. So the protostar has an efficient means of cooling itself, and its temperature does not increase very much during the collapse.

But ultimately the protostar contracts so much that it becomes very dense, and therefore no longer transparent. Because it is now opaque, its heat radiation is trapped within it. The protostar no longer has an efficient cooling mechanism, and it heats dramatically as it collapses yet further. This accounts for the second, nearly horizontal portion of the computed evolutionary tracks.

These general trends have an important effect on the *rate* of collapse. This collapse is governed by two forces: (1) gravity pulling inward, and (2) gas pressure pushing outward. The pressure resists the gravity. We need to analyze how both of these forces change as the protostar contracts.

(1) *Gravity.* Newton's formula for the tug of gravity is an inverse square law. The closer together are two particles, the stronger is the force between them. As illustrated in Figure 13.5, this tells us that the compressional force of gravity grows stronger as the protostar contracts.

(2) *Gas pressure.* We need to know two facts about gas pressure.

Gas pressure

The pressure exerted by a gas grows larger as

- the temperature of the gas grows greater,
- the volume of the gas grows smaller.

Let us now analyze the forces of gravitation and pressure within the protostar.

During the initial stages of the protostar's collapse, its volume is decreasing but its temperature is more or less unchanging. So the outward push of pressure is increasing – but not by very much. The inward tug of gravity, on the other hand, is increasing quite rapidly. Detailed calculations show that, *during this initial phase, gravity is stronger than pressure. Therefore the protostar collapses freely.* Indeed, this stage of the evolution is catastrophic: the protostar plunges inward upon itself like water pouring downwards in a waterfall.

NOW YOU DO IT
Consider a collapsing cloud. Since it is collapsing, its volume is decreasing. Which scenario would lead to a decrease of its pressure: (A) its temperature decreases as it shrinks, or (B) its temperature increases as it shrinks? Explain your answer.

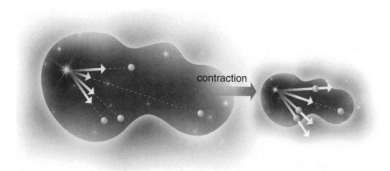

contraction

Figure 13.5 **The compressional force of gravity grows stronger as a protostar contracts because the atoms within it are closer together.**

Once the protostar has contracted enough to become opaque, however, the situation alters. Now its heat radiation is trapped within it, so that its temperature increases rapidly as it contracts further. So the pressure increases rapidly. *During this second stage of its evolution, the pressure within the protostar is sufficient to balance gravity,* and the cloud is no longer freely plunging inward. Rather *it slowly and smoothly contracts.* The final approach to the main sequence is a gradual creep.

We can see this behavior in the computer simulations. They show that the initial evolution requires a relatively short time, while the final approach to the main sequence is very slow.

Star formation: observations

The picture of star formation that we have described so far has been entirely theoretical. For many years it was not possible to perform observations which would test the validity of these theoretical ideas. There were three reasons for this lack of observational data.

- As you can see from Figure 13.4, the collapse of a protostar requires geological ages. We have not been around long enough to witness the complete process.

. . . .

NOW YOU DO IT
How long did the interstellar cloud that formed our Sun take to collapse?

. . . .

- Stars form in interstellar clouds, which are often opaque. If we think in terms of an analogy with human birth, we would have to imagine "birthing rooms" that have no windows into which we might peer.
- Figure 13.5 shows that the force of gravity acting to compress a dense protostar is stronger than the force compressing a more rarified one. This general principle also applies to a protostar that is denser in its center than its outer edges: the inner, dense regions contract more rapidly than the less dense outer layers. This tells us that it is the inner portion of the cloud that forms the protostar. Meanwhile, the more slowly contracting outer region forms a slowly contracting dense shroud, termed a "cocoon." This cocoon hides from our view the contracting protostar. Here, too, the "birthing room" has no windows.

In recent years this situation has improved, and we have begun making observations that yield insight into the formation of stars and planets. In large part this has been due to technological advances such as the development of infrared astronomy, which can penetrate into interstellar clouds, and of ultra-high-resolution telescopes, which give us a more detailed view. Note, however, a fundamental limitation to what we can achieve in this regard. Because the formation process is so slow, we will never

Figure 13.6 **Bok globules are seen silhouetted against an interstellar cloud with which they are associated. They may be clouds in the initial stages of collapse.**

be able actually to witness a star and its planets form. All we can hope for is a series of "snapshots" that yield insight into the process.

This said, we now turn to recent discoveries in this new and exciting field.

Bok globules

Named for the astronomer Bart Bok who studied them, Bok globules are the smallest and densest of the dark clouds (a type of interstellar cloud so dense that it blocks out background stars). A mere parsec or so in diameter, they contain 10 to 100 solar masses of material. Figure 13.6 shows them silhouetted against an interstellar cloud with which they are associated. Each has a density far greater than that of the surrounding cloud.

Do they constitute the first stage of the contraction process? Are they the "cocoons" that shroud the formation process? We are not sure. Notice that the globules are irregular in shape: gravity is not yet strong enough to contract them into a spherical shape.

T Tauri stars

These are a class of unusual star that lie just above the main sequence on the Hertzsprung–Russell diagram – precisely where we expect to find protostars that have just ignited their nuclear fuel. If this interpretation is correct, they must have only recently emerged from their cocoons.

There are several peculiarities to their spectra. On the one hand, they show strong infrared emission, which as we saw in Chapter 4 is characteristic of relatively low temperatures – far cooler than the stars themselves. What could this cool material be? Perhaps it is the remnant of the cocoon.

These spectra also show large Doppler shifts, indicating strong winds blowing away from the stars. These winds from T Tauri stars can be quite strong, blowing at up to 100 kilometers per second. It is probable that every star goes through such a T Tauri stage before settling down on the main sequence as a normal, garden-variety star. In our discussion of the formation of planets, we will see that planets are thought to have formed from a *protoplanetary disk* surrounding the newly formed star: these winds may be the agent that clears away the last remnants of the disk, after planets have formed. They also, of course, clear away the cocoons.

Bipolar jets

These are often associated with star formation. These are jets of matter, ejected in opposite directions from protostars (Figure 13.7). When emitted these jets are narrow: where they slam into the interstellar medium, far from the protostar, they generate turbulence.

Figure 13.7 Bipolar jets are often found associated with newly formed stars. This jet, observed in the infrared region of the spectrum, is associated with the young star DR 21.

A typical jet can be several light years long, and Doppler shift observations reveal the velocity of matter in one to be about 100 kilometers per second. We can use these observations to calculate the age of a jet.

The logic of the calculation

Concentrate attention on the outer tip of a jet: we need to calculate how long it took to get to its present location.

> Step 1. Express the jet's length and velocity in MKS units.
> Step 2. Divide the length by the velocity to find the time required for the outer tip to reach its present location.
> Step 3. Convert the answer to years.

As you can see from the detailed calculation, the "engine" powering the emission of a typical jet must be active for something like 9000 years.

Detailed calculation

Step 1. Express the jet's length and velocity in MKS units.
  If we consider a jet 3 light years long:

length of jet = (3 light years) (9.46×10^{15} meters/light year)
    = 2.84×10^{16} meters.

Suppose the velocity is 100 kilometers/second:

velocity = (100 kilometers/second) = 10^5 meters/second.

Step 2. Divide the length by the velocity to find the time required for the outer tip to reach its present location.

Time = distance/velocity
    = (2.84×10^{16} meters)/(10^5 meters/second)
    = 2.84×10^{11} seconds.

Step 3. Convert the answer to years.
  Since there are 3.15×10^7 seconds in a year,

time = (2.84×10^{11} seconds)/(3.15×10^7 seconds/year)
    = 9×10^3 years.

NOW YOU DO IT
Suppose a bipolar jet is 2 light years long, and material within it flows at 150 kilometers per second. Calculate for how long the "engine" powering the jet's emission has been active.

The origin of these jets is an unsolved mystery. In some way that we do not yet understand, a portion of the matter falling inwards to form a star is given a strong push and sent flying away at great velocities into space. We do understand, however, why the outflows form such narrow, highly collimated jets. As we will see, planets are thought to have formed from a disk of gas and dust surrounding the protostar. As illustrated in Figure 13.8, such a disk channels an outflow into two oppositely directed jets. These jets, therefore, can be taken as evidence for the existence of the disks out of which we believe planets form.

The formation of one star can prevent that of others

⇐ **Looking backward**
We conducted a census of stars in Chapter 12.

As we have seen from our census of stars, the most luminous stars emit an enormous quantity of light. The most brilliant can shine with as much as 10 000 times the brightness of our Sun. This intense radiation can have important effects on star formation.

Figure 13.8 **Jets arise when a wind from a protostar impinges upon a protoplanetary disk (artist's conception).**

Figure 13.9 **The Eagle Nebula. Radiation from highly luminous new stars is evaporating the nebula from which they formed. The "evaporating gas globules" appear to contain a protostar that might not survive this radiation. Behind it trails a "stalk" of material protected from the radiation by the globule's shadow.** 👁 **(Also see color plate section.)**

The beautiful Eagle Nebula (Figure 13.9) contains a number of recently formed stars. But the nebula is now being disrupted by its own progeny! Intense radiation from the most brilliant of these new stars is evaporating the nebula from which they formed.

Note in particular what appears to be several "eyes at the end of a stalk" protruding from the Eagle's pillars. As the nebula is stripped away by a nearby, highly luminous star, a small dense globule has been exposed. This globule is sufficiently opaque to shield the gas behind it from the star. Lying in the globule's shadow, this gas has survived to form the small thin "stalk."

Particularly intriguing is the "eye" at the tip of the stalk. This so-called "evaporating gas globule" appears to contain a protostar whose protective shroud has only recently been stripped away. Now exposed to intense radiation, it is evaporating. In some cases the evaporation is so rapid as to destroy the protostar before it has had time to condense to form a star in its own right.

What triggers star formation?

The formation of a star begins when an interstellar cloud begins contracting under the influence of its internal gravitation. But why did it begin contracting? That cloud had existed for countless eons in a presumably stable configuration. What disrupted this configuration to set it falling inward upon itself?

Recall from our discussion of Figure 13.5 that the gravitational force of compression within a cloud grows stronger if the cloud grows smaller. So if some outside agency were suddenly to compress an interstellar cloud, that cloud might begin contracting. Astronomers have identified three agencies that could do this.

- Just as the radiation from a recently formed, high-luminosity star can evaporate a nebula, it can also compress it.
- As we will discuss in Chapter 15, an ultra-luminous star has a very short lifetime, after which it explodes as a supernova. This immense explosion can also compress nearby interstellar clouds.

- As we will discuss in Chapter 17, the arms in spiral galaxies are regions of particularly high density. If an interstellar cloud enters such an arm, it will be compressed. Indeed, observations reveal that these spiral arms are, in fact, sites of star formation.

Notice that two of these three agencies involve other recently formed stars: the formation of one generation of stars can trigger that of others! *Star formation is contagious.* Paradoxically, newly formed stars can sometimes hinder the formation of others and, at other times, facilitate it.

Formation of the Solar System

⇐ **Looking backward**
We studied the Solar System in general in Chapter 6.

We have succeeded in building an understanding of how an interstellar cloud collapsed to form our Sun. We now turn to the rest of the Solar System – all those planets, moons and so forth, orbiting about the Sun in a vast, disk-shaped assemblage. How can we understand their origin? What have we left out of our story so far?

Let us begin by collecting together a list of things we want our theory to explain. This list is contained in Table 13.1.

Our theory of the formation of the Solar System must account for these regularities. Let us see how well we can do.

The first entry in our table tells us that *the formation of the planets seems to have been part and parcel of the formation of the Sun.* They were both components of the same process. So in developing our theory, we need to go back to our picture of the formation of the Sun and add something to it. The second entry in our table tells us that most of the material in the contracting cloud ended up as the Sun, with only a tiny amount ultimately ending up as planets. So what we add does not need to involve a lot of mass. In some loose sense the formation of the planets was a by-product of the formation of the Sun.

Let us attempt to build a theory of the formation of the Solar System. As before, there is no tried and true method. We will simply have to start somewhere and see what happens. Our first step will turn out to have a lot to do with the orbits of the planets.

Table 13.1. **Regularities in the Solar System.**

(1) The Sun and planets formed at about the same time.
(2) The Sun's mass is far higher than that of the planets.
(3) The Solar System is a disk: the orbits of all the planets lie in essentially the same plane, that of the ecliptic.
(4) The plane of the ecliptic coincides with that of the Sun's equator.
(5) The planets all orbit in the same direction.
(6) This direction is that of the Sun's rotation.
(7) The orbits of the planets are nearly circular, whereas those of meteors, asteroids and comets are quite elliptical.
(8) The orbits of most of the larger moons also lie in the ecliptic plane, and they orbit in the same direction as the planets.
(9) Many planets rotate in the same sense as does the Sun. There are, however, three exceptions to this general tendency: Venus, Uranus and Pluto.
(10) The surface of every object in the Solar System (other than the gaseous planets, and those with much erosion) is studded with impact craters.
(11) Comets are "dirty snowballs," composed of ices mixed with pebbles and dust.
(12) The asteroids lie at the junction between the inner and outer planets.
(13) There are two major groups of planets: the *inner planets*, which are composed of solid rock and have relatively low masses; and the *outer planets*, which are primarily composed of gas and have relatively high masses.

Detectives on the case

How can we understand the orbits of the planets?

Here's a first attempt at a theory: *as the protosun contracted, it left behind some little blobs, which ultimately turned into planets.*

Right now that theory is pretty vague. Perhaps we are imagining that little pieces of the interstellar cloud did not take part in its overall contraction. Or perhaps we are imagining that the blobs were somehow expelled by the contracting cloud. We will need to clarify all this. Furthermore, we are going to have to figure out how those blobs became planets. But there is no reason to do any of this if we decide that our theory has no merit. So let us go back to Table 13.1 and ask how our theory squares with it.

Figure 13.10 illustrates an interstellar cloud that has begun contracting, ultimately to form the Sun. Also illustrated in Figure 13.10 is a tiny blob that we are imagining will ultimately become a planet.

As the cloud continues contracting, it leaves behind more blobs (Figure 13.11).

What happens to these blobs? They can't just stay where they are. After all, they are gravitationally attracted to the contracting cloud. They are like some objects – pebbles, perhaps – that you hold in your hand and then let go. They will commence falling toward the source of the gravitational attraction. In our analogy the pebbles fall toward the Earth: in our theory the blobs fall toward the protosun. And as we know, falling is the same thing as orbiting. So *the left-behind blobs commence orbiting about the protosun.*

In Figure 13.12 we illustrate these orbits after the protosun has contracted to become the Sun. How do these orbits compare with those of the planets? Is our theory looking good?

It is looking very bad. The orbits illustrated in Figure 13.12 look nothing like the orbits of the planets. There are two problems.

(A) The orbits predicted by our theory are long, thin ellipses. This conflicts with item (7) in Table 13.1, which is that planets move in nearly circular orbits.

(B) They lie in many different planes. This conflicts with item (3) in Table 13.1, which is that the orbits of all the planets lie in essentially the same plane.

These problems are fatal. They show that our first theory is not going to work. Let us abandon it and turn elsewhere. But before we do so, we need to understand *why* our first theory made these erroneous predictions. Once we do, we can figure out how to find a better theory, one that avoids making them.

Let us remind ourselves of what we know about orbits. Figure 3.16 shows what happens if we launch a projectile with greater

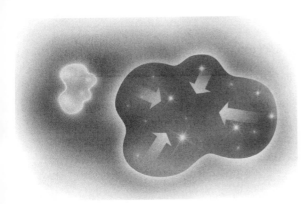

Figure 13.10 **A contracting interstellar cloud and a blob that we are theorizing ultimately becomes a planet.**

⇐ **Looking backward**
We studied orbits in Chapters 2 and 3.

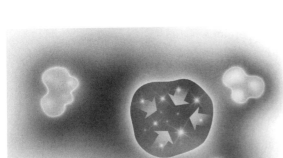

Figure 13.11 **The cloud, somewhat smaller, leaves behind more blobs. (Because the cloud is smaller its self-gravity is now stronger, and since gravity pulls things into a spherical shape, the cloud has become more nearly spherical.)**

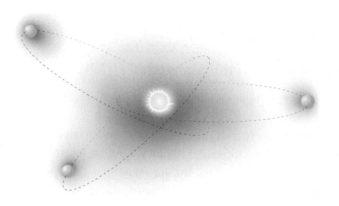

Figure 13.12 **Ultimate orbits of the left-behind blobs.**

and greater sideways velocity. With zero velocity it falls straight down. With small velocity it falls in a skinny ellipse – and this is what our blobs are doing. Finally, with a still greater velocity it moves in a circle – and this is what we want them to do.

So problem (A) arises because our blobs are not being launched sideways fast enough. We need to invent some other theory, one that makes them move (A) *with a greater sideways velocity*.

How about the theory's second difficulty (B)? What determines the plane in which an orbit lies? This is the plane that contains the orbiting object's velocity and the source of the gravitation – the protosun. Figure 13.13 illustrates an orbiting object, its direction of motion and the attracting body (the protosun). The plane in which the orbit takes place is the plane of the page.

So when we invent our new theory, we want all our blobs to be moving (B) *in the same plane*.

Let us scan over Table 13.1 again, listing the regularities we need this new theory to explain. Like all good detectives, we know it is filled with clues. Concentrate on (4).

(4) The plane of the ecliptic (i.e. the plane containing the planets' orbits) coincides with that of the Sun's equator.

The Sun is rotating on its axis: item (4) is telling us that its equatorial plane coincides with the plane of the Solar System. Is this an accident? Or is it telling us something?

Up to now we have not thought about rotation. But let us do so now. Interstellar clouds are observed to spin slowly. And as such a cloud begins to shrink, its rotation rate will increase – just like that of a figure skater, who spins faster as she draws her arms inward (Figure 13.14). The more the cloud collapses, the faster it must spin – and because, as we have calculated, the degree of compression is enormous, the cloud will be left spinning at an enormous rate.

This rotation has an important effect on the *shape* of the cloud. Notice the figure skater's long hair in Figure 13.14. In the initial panel, in which she is slowly rotating, it hangs straight down. But as she draws her arms in and spins more rapidly, it begins to flare out. By the final panel, in which she is rotating very rapidly, her hair is horizontal! A similar thing happens to our collapsing, rotating cloud. The more it collapses the more rapidly it spins, and the more it spins the more it bulges out at its equator. This behavior is illustrated in Figure 13.15. As you can see, by the final panel, the cloud has bulged so much that *it has become a disk*. Furthermore, at this point *the rotation is so rapid as to prevent further collapse*. Centrifugal effects keep the disk from shrinking any further. This rotating structure is known as a *protoplanetary disk*. An image of

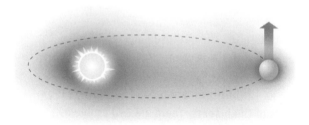

Figure 13.13 **The plane in which an orbit lies is the plane containing the direction of its motion and the attracting body.**

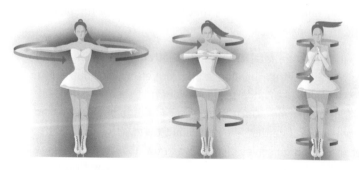

Figure 13.14 **A shrinking body increases its rotation rate.**

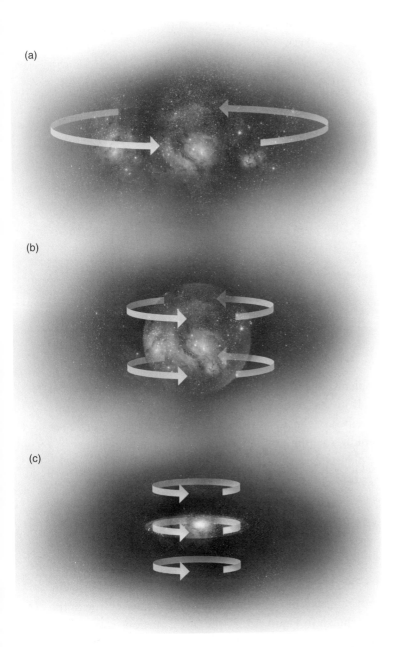

(a)

(b)

(c)

Figure 13.15 **Collapse of a rotating cloud. As the collapse proceeds, the cloud rotates more rapidly and becomes more flattened. Ultimately it becomes a disk whose rotation is so great as to prevent it from collapsing further.**

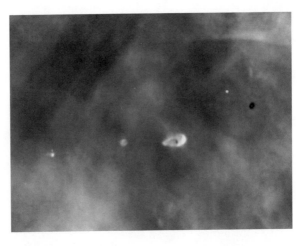

Figure 13.16 **Protoplanetary disks in the Orion Nebula.**

such disks in the star-forming Orion Nebula is shown in Figure 13.16.

In Figure 13.17 we have delineated three blobs within such a disk. Because the disk is rotating, we know that those blobs are moving. And now let us imagine that somehow they turn into planets. Figure 13.18 illustrates the orbit those planets will have.

Notice two things about the motion of these planets.

(A) The planets are moving sideways with a size-able velocity – because the disk was rotating rapidly.
(B) They are all moving in the same plane – the plane of the disk.

When we analyzed the reasons for our first theory's failure, we concluded that we needed a new theory in which the blobs were moving (A) faster, and (B) in the same plane. Our new theory does this for us. As needed, the orbits of the planets they form will be circular, and all in the same plane. Perhaps we have found a good theory!

A difficulty: why is the Sun rotating so slowly?

Our new theory is good, but it is not good enough. It solves the problem with the first theory – but at a price, for now we have a new problem. How can we account for *the Sun*?

The Sun today is rotating – but it is rotating slowly. It is rotating so slowly that it is essentially perfectly spherical. But the protoplanetary disk that our theory posits is rotating rapidly – so rapidly that it is a disk. That disk may well turn into a collection of planets. But how can it turn into the Sun? In our new theory *everything* partakes in this rapid rotation – and this means that everything will end up in circular orbits. There is no non-rotating portion left to form the Sun!

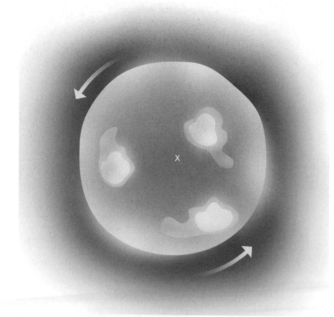

Figure 13.17 **Blobs in a rotating gaseous disk. We imagine that the blobs will turn into planets.**

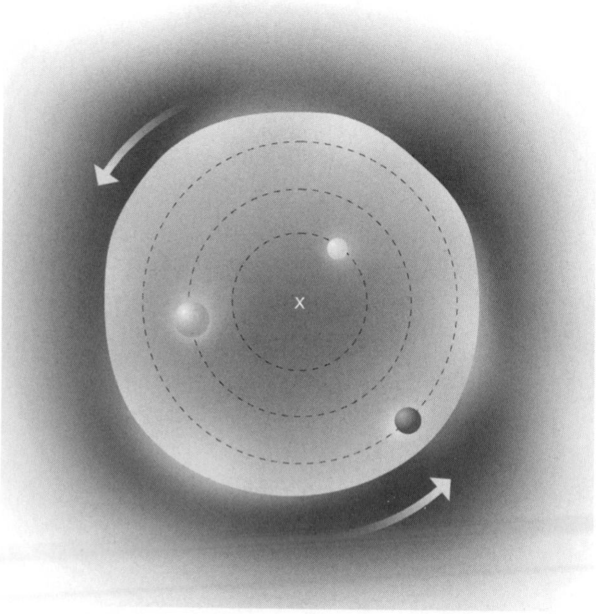

Figure 13.18 **Motion of planets formed from blobs in a rotating disk.**

How can we fix this problem? We need to think of some process that can *slow the rotation of part of the disk*. If we succeed, we can then posit that the slowly rotating portion evolves, as we described in the previous section, to become the Sun we see today. In contrast, the rapidly rotating portion evolves, in a way we will have to figure out, to become the planets.

Let us think about the nature of the rotation of the protoplanetary disk. Remember what we know of the orbits of the planets:

- Mercury orbits once every 0.24 years,
- Earth orbits once every year,
- Uranus orbits once every 84 years.

> ⇐ **Looking backward**
> We studied Newton's theory of orbits in Chapter 3.

The farther out a planet lies, the longer it takes to go around once. Indeed, as we saw from Newton's orbit formula, this is true of every orbit. And the same is true of the gas in our rotating disk. As shown in Figure 13.19, the inner portions rotate more rapidly than the outer ones.

It is the inner portions whose rotation we need to slow. How can we do this? A number of theories have been invented. In what follows we will consider one of them, known as "the magnetic braking theory." This theory proposes a connection between the various regions of the disk. The connection is a magnetic field (Figure 13.20).

> ⇐ **Looking backward**
> We studied plasmas and magnetic fields in Chapter 11.

We will describe this theory in terms of a simple model. The ideas that we develop for this simplified picture can be extended to cover the actual situation. The model deals not with a gas in which each portion rotates at a different rate, but merely with two disks: an inner one that rotates rapidly, and an outer one that rotates slowly. We will imagine that they are made of wood.

Recall what we learned about the behavior of a plasma (and our real disk is indeed a plasma) in a magnetic field.

> It is difficult for a plasma to move across magnetic field lines.

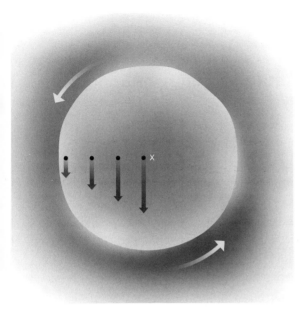

Figure 13.19 **The inner portions of the protoplanetary disk rotate more rapidly than the outer ones.**

Figure 13.20 **A magnetic field permeates the rotating disk.**

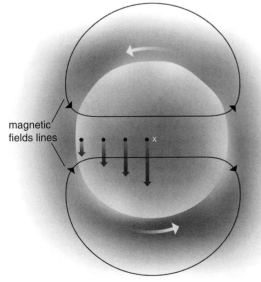

magnetic fields lines

We might think of the plasma and the lines of magnetic force as being "attached" to one another. In terms of our analogy, we can think of the field lines as rubber bands that have been nailed onto the wooden disks (Figure 13.21).

This figure illustrates the configuration at one moment. What does it look like later? After some time has passed, both disks will have rotated – but the inner one will have rotated more than the outer. And because the rubber bands are attached to the disks, this twists them (Figure 13.22).

Now let us recall a second thing we learned about the behavior of a plasma in a magnetic field.

> Magnetic lines of force in a plasma are elastic.

This is why we think of the field lines as being like rubber bands. Like all rubber bands, they resist being stretched! This means that the rubber bands exert a twisting force on the inner disk. And as you can see from Figure 13.23, *this force acts to slow the inner disk.*

We have succeeded in finding something that can slow the rotation of the inner portions of the protoplanetary disk. It is the magnetic field pervading the plasma.

NOW YOU DO IT
We have now described the magnetic braking theory of how the Sun arose out of a rotating disk. But as we mentioned, other such theories have been proposed. Here is an experiment that will give you insight into a completely different theory.

- *Take a pan of water.* **It is important that the pan be fairly large.**
- *Give the inner portion a stir* **with a spoon. When doing this, make sure that you do *not* stir the outer regions. Keep your spoon confined to the inner portions of the water.**
  - *Wait a while and watch what happens to the rotation of the water.*
  - **How do you** *explain what happens*?

  Now here is the most important part of all.

  - *What does this have to do with the origin of the Sun?*

As you can see, other processes have been thought of that can do the same. Which is the actual process? This is the subject of much current research, and we do not know the answer to this question. But whatever it is, that process resulted in a two-component structure: a slowly rotating portion (the protosun), which evolved to become the Sun as we have described in the previous section; and a rapidly rotating disk

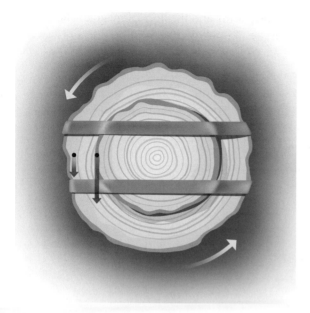

Figure 13.21 **Simplified model of a protoplanetary disk. We are replacing the actual gaseous disk plus magnetic field by two wooden disks connected by rubber bands.**

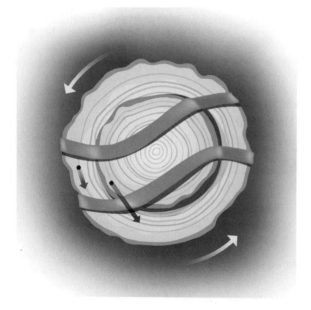

Figure 13.22 **The rubber bands are twisted by the differing rotation rates of the two disks.**

(the protoplanetary disk), which evolved to become the Solar System.

Planet formation: condensation and planetesimals

We now turn to the process whereby a rapidly rotating disk of gas and dust turned into the planets, comets and so forth that constitute our Solar System.

To understand this process we must add a further element to our picture: *the cooling of the protoplanetary disk*. As we have seen, compression heats a gas, so that the rotating cloud was heated as it contracted. But once it stopped contracting, stabilized by rotation, the trend was reversed. It ceased heating and began cooling as it radiated its energy away into the near absolute zero of space. So we must imagine the protoplanetary disk as a steadily rotating, steadily cooling gas.

Eventually it cooled so much that a new process commenced: tiny solid flakes and grains began to form within it.

This process, known as *condensation*, might sound unfamiliar, but it is actually quite ordinary. A blizzard is an example. Condensation occurs whenever a hot gas containing several different types of molecule cools. Within our atmosphere, for instance, molecules of water are dissolved; and when the atmosphere cools sufficiently they condense out to form snowflakes. In the same way, molecules of various substances were present within the protoplanetary disk. As it cooled, they condensed to form solids.

So the protoplanetary disk started filling up with tiny flakes and grains. Figure 13.24 is an artist's conception of what it might have looked like at this stage. Far off in the distance, the protosun has been heated by compression so much that it has begun to shine. In the outer reaches of the disk on the other hand, the temperature has dropped so much that condensation has begun. Perhaps the disk at this stage looked something like a snowstorm or a sandstorm. Tiny flakes and pebbles swirled about randomly.

From these tiny solids the Solar System formed. The initial stage of the formation process was a simple matter of adhesion. Snowflakes are sticky. They act like bits of "Velcro." Whenever two randomly swirling flakes happened to touch, they stuck together. When a third flake encountered the pair, it stuck to them. Bit by bit, the celestial equivalent of snowballs began to assemble within the disk. Whenever the slowly growing clump encountered a tiny solid grain, it captured that as well: the snowballs were somewhat dirty.

These dirty snowballs are known as *planetesimals*. They were the building blocks out of which the planets formed. Eventually the planetesimals grew so large that they began to exert a significant force of gravity on one another. Moving through the protoplanetary disk, their orbits bent by their mutual gravitational attraction, from time to time the planetesimals collided. If the collisions were at high speeds, the bodies were shattered by the impact. But if the collisions were at low speeds, the two amalgamated together to form a larger one. The more time passed, the larger the planetesimals became.

At this stage their internal gravity was also significant. Just as gravity compresses an interstellar cloud (Figure 13.3), so gravity compressed the larger planetesimals. And just as squeezing a snowball compacts it, so the planetesimal's loose, fluffy structures were crushed to form more compact solid bodies.

Ultimately the largest bodies, now nearly planets or satellites, exerted great gravitational forces. Objects were attracted into one another with great force, gouging immense craters when they collided. These final stages in the formation of the Solar System must have been spectacularly violent and cataclysmic, as massive bodies slammed together at high speeds.

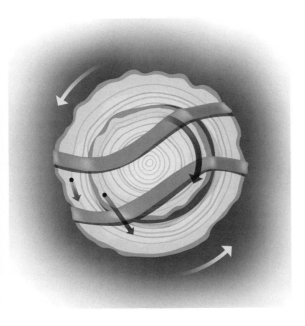

Figure 13.23 **The twisted rubber bands exert a force that slows the inner disk.**

Figure 13.24 **Solids begin to condense within the solar nebula. (Artist's conception by William K. Hartmann.)** 👁 **(Also see color plate section.)**

Inner versus outer planets

As we saw in Chapter 6, the Solar System is divided into inner and outer planets. The inner planets are solid, they are composed of minerals, and they have relatively low masses. The outer planets, in contrast, are gaseous, composed primarily of lighter elements, and have much higher masses. How can we understand these differences?

We must think more carefully about the condensation process within the protoplanetary disk. It depends on the disk's composition. Since the disk arose from an interstellar cloud, this composition was that of the interstellar medium. As we have discussed, hydrogen is the most abundant element in the cosmos, and helium the second most abundant. Indeed, these two elements taken together account for all but a few percent of the mass of the cosmos. Table 13.2 gives the abundances of the remaining elements.

In the condensation process, these elements combined to form solids. Let us think about what kinds of molecules were present within these solids, and how many there were of each variety.

Since hydrogen was the most abundant element, molecules formed from hydrogen must have been most abundant within the protoplanetary disk. But only one such molecule exists, H_2, and this is not a solid but a gas.

Since helium was the second most abundant element, you might think that reactions between hydrogen and helium would form the second most abundant molecules. But helium is a noble gas: it does not participate in chemical reactions! Rather we must think of reactions between hydrogen and the third

Table 13.2. **Composition of a protoplanetary disk.**

Hydrogen	Most abundant
Helium	
Intermediate elements such as carbon, nitrogen and oxygen	
Heavier elements such as iron, silicon and magnesium	Least abundant

most abundant group of elements (carbon, nitrogen and oxygen) to find the second most abundant compounds. Molecules formed in this way include water (H_2O), methane (CH_4) and ammonia (NH_3). Furthermore, unlike H_2 molecules these do form solids. *Compounds such as these were the primary constituents of the solids that formed within the protoplanetary disk.* Let us generically refer to these as "ices."

Similarly, compounds formed from the least abundant elements (such as iron, silicon and magnesium) were the least abundant within the planetesimals. Such compounds include quartz (SiO_2) and many other minerals. We will generically refer to solids formed from these as "dust grains."

Notice now a crucially important fact. The inner portions of the protoplanetary disk were strongly heated by the protosun – and ices cannot exist at high temperatures! Just as water ice melts and dry ice vaporizes at room temperature, just as comets vaporize when close to the Sun, so too were ices unable to form in the inner parts of the protoplanetary disk. Minerals, on the other hand, were quite capable of surviving such high temperatures. So *the dust grains that formed within the inner Solar System were composed only of the relatively low-abundance minerals.*

We now understand why the inner planets are composed of minerals – these were the only solids that existed in these inner regions of the Solar System. We also understand why the inner planets have relatively low mass – these minerals were relatively scarce within the protoplanetary disk.

The outskirts of the protoplanetary disk, on the other hand, were cold enough for water, methane and ammonia to freeze. So *the solids formed within the outer Solar System were composed of these far more abundant ices* (together with the minerals). This is why the outer planets contain these lighter compounds. And because these lighter compounds were so much more abundant than the minerals, the outer planets are more massive than the inner ones.

. . . .

NOW YOU DO IT
Suppose a protoplanetary disk were to form out of an interstellar cloud with a different composition, one lacking the heavier elements.

Hydrogen	Most abundant
Helium	
Intermediate elements such as carbon, nitrogen and oxygen	Least abundant

- **What are the most abundant molecules that would condense out of this disk?**
- **What are the most abundant solids that would form from these molecules?**

- What are the next most abundant solids that would form from these molecules?
- What sorts of planets would form within this disk? Would such a planetary system have the same distinction between inner and outer planets that ours does?

THE NATURE OF SCIENCE

THEORY AND OBSERVATION

There are two great facets to the enterprise of science. On the one hand, we make observations and conduct experiments. On the other, we build theories to explain what we have found. Our study of star and planet formation gives us a particularly good opportunity to study these two facets. As we have commented, our theoretical ideas were initially developed long before it was possible to make detailed observations, so they were created in a relative absence of observed facts.

Theory and observation interact with one another in many ways.

THEORIES MUST CONFORM WITH OBSERVATION

It is a hallmark of science that every theory must agree with observations. While complete agreement is often beyond our grasp, the more observed facts a theory is able to account for, the more confident we are entitled to feel of its general validity. Let us therefore test our theory of planet formation.

We begin by assembling a list of things we need our theory to explain. This list was given in Table 13.1. Our theory of the formation of the Solar System must account for these regularities. Let us see how well it does.

Our theory was built by using items (1) through (4), so it is not surprising that it conforms to them. But looking over Table 13.1, it is immediately clear that we have succeeded in accounting for many other items as well. For example, the fact of circular orbits all moving in the same direction (items (5) and (6)) easily follows from the fact that the protoplanetary disk rotated. Item (10), the prevalence of craters, is also easy. Indeed, it now acquires great significance: these craters are the scars left over from the final, violent stages of the formation of the Solar System.

Let us turn our attention to item (11) dealing with the composition of comets. The description given in (11) of a comet perfectly agrees with our theoretical picture of a planetesimal! So a comet must *be* a planetesimal. In the final stages of the formation of the Solar System, when the giant planets had been formed, numerous planetesimals must have passed near them without suffering actual collisions. During the encounters, the planets' strong gravitational pulls must have gripped the planetesimals and thrown them outward, far from the planets, there to orbit in what we know today as the Oort Cloud. Similarly, the Kuiper Belt contains those outlying planetesimals that did not suffer this fate. As an analogy, if we think of the Solar System as a building, the comets would be the remaining unused bricks left strewn about the building site. So our theory nicely accounts for item (11).

We now turn to item (12), the asteroids. It is probably significant that they lie near Jupiter, the largest planet and therefore the one with the strongest gravity. Many astronomers think that Jupiter's strong gravitational influence disrupted the formation of a planet close to it, leaving only the smaller bodies that today form the asteroid belt.

WHAT DO WE DO WHEN THEORY CONFLICTS WITH OBSERVATIONS?

So far our theory of planet formation is working fairly well. But it is defective in a vitally important respect. This theory nicely explains the distinction between inner and outer planets – but it entirely fails to account for the giant planets that have been discovered orbiting close to other stars.

As you recall, the theory predicts that gas giant planets must lie far from their star. In our Solar System they do – but in other planetary systems they do not! Indeed, as we discussed in Chapter 10, all of the extrasolar planets that have ever been discovered are very massive, and most lie abnormally close to their stars – far closer than we do to our Sun. This is entirely impossible according to our theory of the origin of the planets!

This theory was created at a time when only one system of planets was known – our own Solar System. This was long before the discovery of extra-solar planets. Only now, when we know of the existence of these other planetary systems, have we realized how poor is our theory of their formation. Indeed, "one theorist has admitted to me that he cannot think of a *single* prediction that he and his colleagues made about extrasolar planets that has been supported by observations," recently stated an astronomer involved in discovering extrasolar planets.[1]

What do scientists do when they realize that their theory has failed? One option is to "patch it up": to modify the theory, adding some new element in order to make it agree with the observations. This new element is *planetary migration*. According to this concept, the gas giants orbiting about other stars were indeed formed far from them, as predicted by theory – but they then migrated inwards toward their stars.

Various scenarios have been proposed to account for this migration. All of them lie at the forefront of current knowledge, and further research is needed to evaluate them. But we are not concerned here with their details. Our concern rather is that they might be viable – or they might not. One of these theories might turn out to be correct. But perhaps we will eventually decide that none of them is! Perhaps in the long run we will find that even our new concept of planetary migration does not quite account for all the data, and that a second new element needs to be added – and then a third.

When do we give up? How many modifications do we make before we decide that our ideas are *totally* wrong: that no amount of "patching up" will work, and

[1] Quoted in "Planetary harmony" by Robert Naeye, *Sky and Telescope*, January 2005, p. 45.

that we need to invent a completely new theory? When do we admit that we are facing a serious crisis of understanding?

The problem is that there is no clear answer to this question. We can never know for sure whether the next modification will finally be successful. Maybe, with one more new addition, we will suddenly realize that we have achieved a complete understanding – or maybe not. The central point we wish to emphasize is that usually there is no clear signpost telling us that we are barking up the wrong tree.

Furthermore, scientists are very reluctant to abandon a theory. This reluctance is entirely appropriate. It is all very well to say "we don't understand and we need new ideas," but this ignores the fact that there is a very great cost to giving up on our old ideas. The cost is that if we abandon our old theory, we are abandoning something that has much to recommend it. While our old theory doesn't explain everything, it does explain a lot. If we decide to throw out our theory, we are also throwing out all its many successes – and now we are faced with the task of inventing an entirely new set of ideas that duplicate these successes.

The ultimate goal of science is to find a theory that works perfectly, and that explains everything. But every scientist knows that we have not yet achieved this goal. There are things that we do not yet understand. There are unsolved mysteries. This does not mean that science has failed, however. It only means that we are not yet finished. It only means that we need to keep on working.

THEORIES GIVE WHAT WE OBSERVE MEANING

In the absence of any theoretical understanding, facts revealed by observation do not mean very much. We simply note them and move on. Only in the context of a theory do these facts assume significance.

As an example, consider the observed fact that most bodies in the Solar System are covered with craters. People knew for centuries that the Moon was heavily cratered whereas the Earth was not. But they did not take this very seriously, since the Moon differed from the Earth in so many other ways. The second body to be closely studied was Mars, and it came as a great surprise that this too turned out to be cratered. Only with the exploration of the Solar System by space probes did people realize that craters are the rule. It is the Earth that is the exception, its craters having been erased by erosion and geological activity.

But so what? What is this fact telling us? It is our theory of planet formation that makes this fact meaningful. We now understand that these craters are leftover scars from the cataclysmic creation of the Solar System. They are mute testimony of that violent, catastrophic epoch billions of years ago, when large bodies slammed into one another with terrible force.

As a second example, consider comets. Observation tells us that they are unlike other bodies in the Solar System: "dirty snowballs" rather than terrestrial planets or gas giants, usually orbiting far from the Sun in ultra-cold darkness, and only rarely venturing inward in highly elliptical orbits. Here, too, our theoretical understanding of planet formation gives these facts new meaning. We now understand that the comets are the planetesimals, the ancient building blocks out of which the Solar System formed.

Crime novels are filled with episodes in which the detective suddenly stiffens with excitement, as he notices some apparently trivial fact. Because he has already formulated a theory of the crime this fact, so meaningless to everyone else, is for him an important clue. It is the same with science.

OBSERVATIONS ALWAYS SURPRISE US

We have already commented (Chapter 2) that scientists are very lucky in having observations against which to test their theories. Other fields, such as philosophy, are not so fortunate. No matter how well developed our theories may be, once we start making observations, we find surprises – things we failed to predict, which guide us in developing better theories.

As an example, prior to the space program nobody realized how fascinating the moons of other planets would turn out to be. Even through our greatest Earth-bound telescopes they were merely points of light, with few distinguishing features. People tended to think of them as being like our Moon, with perhaps a few differences here and there.

Now we know how wrong we were. Io is the nearest thing to Hell we have ever found: violently active, festooned with volcanoes, a surface of molten sulfur. Ganymede is covered with a network of parallel grooves, intersecting and overlaying each other in complex patterns. Europa has no dry land, but is entirely covered with a frozen ocean. Titan, larger than Mercury, has an atmosphere: its surface may be covered with an organic sludge containing molecules that on Earth were precursors to life.

Similarly, no theory had prepared us for the existence and importance of winds from newly formed stars. They are something of a mystery. What generates such massive outpourings of matter? What accelerates them to such high velocities – several hundred thousand miles per hour – and returns a good fraction of the mass of an entire star back to the interstellar medium from which it came? We are not sure.

Theory in the absence of observation is barren. The more we observe, the more we learn how rich the Universe is. It is richer than our theories of it.

Summary

When did the Solar System form?
- Computer models of the Sun tell us it is about 4.5 billion years old.
- Radioactive age dating tells us that certain other bodies in the Solar System have similar ages.
- We tentatively conclude that the Solar System formed as a unit.

Where do stars and planets form?
- A high-mass star cannot wander far from the place of its formation during its nuclear-burning lifetime.

- These stars are found in or near interstellar clouds.

Star formation: theory
- A star is formed when an interstellar cloud contracts under the influence of its gravity.
- The degree of this contraction is very great.
- As the cloud contracts, it heats.
- When the temperature becomes sufficiently great, nuclear reactions commence and the protostar has become a star.

- Computer models of the collapse of the protostar have been developed.

Star formation: observations

- "Bok globules" may be interstellar clouds in the early stages of collapse.
- "T Tauri stars" may be recently formed stars.
- "Bipolar jets" are observed emanating from recently formed stars.
- In some circumstances, the formation of one star can prevent that of others.
- In other circumstances, the formation of one star can trigger that of others.

The formation of the Solar System

- Our theory of the formation of the Solar System must account for all the regularities we find in it today.
- These regularities point to a theory in which the contracting cloud formed a rotating disk, out of which the Sun and planets formed.
- The Sun formed by decelerating the rotation of part of this disk.
- The planets formed by condensation within the disk to form planetesimals.
- Inner versus outer planets:
  - the most abundant molecules formed by condensation were ices such as water, methane and ammonia;
  - but these could not have existed in the inner portions of the protoplanetary disk: it was too hot;
  - the condensation also formed dust grains, but these were far less abundant;
  - only these could have existed in the inner regions of the disk.

DETECTIVES ON THE CASE

How can we understand the orbits of the planets?

- We first proposed that the planets formed from "blobs" left behind by the contracting cloud.
- We then realized that the orbits of these blobs would be utterly unlike those of the planets.
- We realized that this problem arose because their motion was too slow, and not all in the same direction.
- We then realized that a rotating interstellar cloud would rotate faster as it contracted, and ultimately form a disk.
- Planets formed from such a disk would have the correct orbits.
- Several theories, including the theory of "magnetic braking," have been developed to account for the existence of the Sun in this picture.

THE NATURE OF SCIENCE

Theory and observation

- Theories must conform with observations:
  - our theory of the formation of the Solar System conforms with some of the observations;
  - but it fails to conform with the observation of massive extrasolar planets close to their stars;
  - planetary migration has been proposed to account for this, but we do not know if this theory is correct;
  - it is never easy to know when to abandon a theory.
- Theories give what we observe meaning:
  - example: the ubiquity of craters is accounted for by the violent final stages of planet formation;
  - example: comets are leftover planetesimals.
- Observations always surprise us:
  - theory in the absence of observation is barren;
  - the Universe is richer than our theories of it.

Problems

(1) We want to find how many light years the Solar System has traveled since it was formed. (Take its speed to be 10 kilometers per second.) (A) Describe the logic of the calculation you will use. (B) Perform the detailed calculation.

(2) Suppose an interstellar cloud is 10 light years in radius, and a star forms in its center. The star is moving at 10 kilometers per second. We want to calculate how long it takes the star to leave the cloud. (A) Describe the logic of the calculation you will use. (B) Perform the detailed calculation. (C) If the star's lifetime is a million years, will it leave the cloud before it ceases shining?

(3) Suppose the density of an interstellar cloud is 10^{-18} kilograms per meter^3. We want to calculate the size it must have if it will form a 2 solar mass star.

(A) Describe the logic of the calculation you will use. (B) Perform the detailed calculation. (C) If the cloud were half that size, what mass star would it form? (See if you can answer this question *without* performing the detailed calculation of (B)!)

(4) Suppose an interstellar cloud is half a light year in radius, and its density is 10^{-17} kilograms/meter^3. We want to calculate the mass of the star it will form. (A) Describe the logic of the calculation you will use. (B) Perform the detailed calculation.

(5) Figure 13.4 shows the evolutionary tracks of five different contracting protostars, of masses ranging from ½ to 15 solar masses. Suppose that at some time each of these commences contracting towards the main sequence. Which will have reached the main sequence after a billion years?

(6) Figure 13.4 shows the evolutionary track for a 1 solar mass contracting protostar. How much brighter than the Sun was such a protostar when it began contracting? How much cooler?

(7) Figure 13.4 shows the evolutionary track of a 15 solar mass contracting protostar. As you can see, for most of the contraction the track is horizontal on the H–R diagram. This means that the protostar's luminosity remains constant, but that it grows steadily hotter. But we know from the theory of blackbody radiation that an object radiates more light as it grows hotter. How does the protostar manage to keep on radiating the same amount of light as it heats?

(8) In our discussion of Bok globules, we mentioned that a typical globule might be 1 parsec in diameter and contain 10 solar masses of material. We want to find its density. (A) Describe the logic of the calculation you will use. (B) Perform the detailed calculation. Now suppose that such a globule is in fact the first stage in the contraction of a cloud that will ultimately form a group of stars, and that its density before it began collapsing had been that of an interstellar cloud, which is about 10^{-17} kilograms per meter^3. We want to know its size back then. (C) Describe the logic of the calculation you will use. (D) Perform the detailed calculation.

(9) Suppose the "engine" powering a bipolar jet is active for 11 000 years, and it shoots out material at 75 kilometers per second. We want to calculate the length (in light years) of the jet it produces. (A) Describe the logic of the calculation you will use. (B) Perform the detailed calculation.

(10) Suppose a protoplanetary disk formed with the following composition.

Hydrogen	Least abundant
Helium	
Intermediate elements such as carbon, nitrogen and oxygen	
Heavier elements such as iron, silicon and magnesium	Most abundant

- What are the most abundant molecules that would condense out of this disk?
- What are the most abundant solids that would form from these molecules?
- What are the next most abundant solids that would form from these molecules?
- What are the least abundant molecules that would condense out of such a disk?
- What sorts of planets would form within this disk? Would such a planetary system have the same distinction between inner and outer planets that ours does?

YOU MUST DECIDE

You are a planetary scientist. You are deeply worried by the complete failure of our old theory of the origin of the planets to account for the presence of gas giant planets lying close to other stars. You are also worried by the fact that our current theory does not account for all of our own Solar System's properties.

It is late at night, but you cannot sleep. In order to get your thoughts in order, you are writing a letter to your parents.

Your task is to write this letter. In it, do the following.

- Identify those general features of the Solar System listed in Table 13.1 that we have not adequately discussed.
- See if you can find an explanation for these features based on our current ideas of the origin of the planets.
- If you are not successful in explaining them all, what are you going to do?
  - Do you propose to keep on accepting this theory, and admit that our understanding is incomplete and more research is needed?
  - Or do you propose to completely abandon our theory of the origin of planets, and search for a new one? If you adopt this strategy, how are you going to account for those items in Table 13.1 that our current theory explains so well?

14

Stellar structure

In this chapter we explore the basic principles governing the structure of stars.

⇐ **Looking backward**
We studied the Sun in
Chapter 11.

We have already realized that the Sun – and like it, every star – is nothing more than a ball of superheated gas. You might think that such a ball would be pretty uninteresting, but you would be wrong. Indeed, stars are quite remarkable things.

Perhaps the most remarkable thing about them is the enormous amount of energy they emit. In a tiny fraction of a second, for instance, the Sun emits more energy than the human race has used in all the centuries since the industrial revolution. For many years scientists were at a loss to account for all this energy. Only recently have we realized that stars are powered by nuclear reactions occurring in their cores. So in a very real sense, what we term "solar energy" is in fact nuclear energy. The only difference is that, in this case, the nuclear reactor is floating off in space, 93 million miles away.

With this insight in hand, scientists have constructed computer models of stars. These models balance the enormous inward pull of gravity by the equally enormous outward push of gas pressure. Similarly, the generation of energy from nuclear reactions is balanced by the emission of energy via starlight. These models have revealed that main sequence stars are in what we might call the first phase of their evolution, in which they derive their energy from nuclear reactions involving hydrogen. Our own Sun is such a star. In the next chapter we will consider what happens once the supply of hydrogen runs out.

Detectives on the case

What powers the shining of the stars?

Perhaps the most surprising thing about the stars is the truly gigantic quantity of energy they emit. Let us consider our own Sun: it is a representative star. Even at its great distance, its emission is sufficient to ward off the bitter cold – more than 450 °F (232 °C) below zero – of interplanetary space. It is this energy source that heats us to sweltering midsummer temperatures, gives sunburns and through photosynthesis supplies the energy needs of every plant on Earth. All this from a distance of nearly 100 million miles (about 161 million kilometers)!

⟵ **Looking backward**
In Chapter 11 we measured the luminosity of the Sun.

The emission of energy from the Sun – its luminosity

$$L_{\text{Sun}} = 3.8 \times 10^{26} \text{ watts}$$

is so enormous that it is difficult to comprehend intuitively. We will begin with a quick calculation in order to attempt to appreciate this immense outpouring of energy.

We have often throughout this book used the watt as the unit of luminosity. Recall that *the watt is also a unit of rate of emission of energy.* Light carries energy: the brighter the light the more energy it carries. If we have two light bulbs, one of which has twice the luminosity as the other, the brighter bulb is emitting twice as much light energy each second as the dimmer. In the MKS system that we use in this book *the unit of energy is the joule.* A 1-watt light bulb then emits 1 joule each second: this is the power it emits. Similarly, a 40-watt bulb emits 40 joules each second, and so on.

MKS system: light power

Unit of energy: the joule.
One watt = one joule per second.

The Sun, therefore, is emitting energy at a rate

$$L_{\text{Sun}} = 3.8 \times 10^{26} \text{ joules/second.}$$

Recall that in Chapter 11 we noted that this rate is 38 000 000 000 000 times greater than the total power requirements of our entire planet-wide industrial civilization. To make this yet more vivid, let us ask *how long would it take the Sun to emit enough energy to power our entire worldwide industrial civilization for a century?* We will find that it takes a very short time indeed.

Detailed calculation

Step 1. We need to find how rapidly energy is being used worldwide by our industrial civilization.

Numerous attempts have been made to evaluate this. A very rough estimate is 10^{13} joules per second.

Step 2. Multiply the rate of use of energy by one century to find the total energy our civilization uses in a century. (This assumes that the rate of use of energy is constant.)

Let us call the total energy used by industrial civilization in a century $E_{\text{one century}}$. We calculate

$$
\begin{aligned}
E_{\text{one century}} &= (10^{13} \text{ joules/second}) \text{ (one century)}\\
&= (10^{13} \text{ joules/second}) \text{ (100 years)(365 days/}\\
&\quad \text{year)(24 hours/day)(60 minutes/hour)}\\
&\quad \text{(60 seconds/minute)}\\
&= 3.15 \times 10^{22} \text{ joules.}
\end{aligned}
$$

Step 3. Divide this total energy by the rate the Sun is emitting energy to find the time required for the Sun to emit this much energy.

$$
\begin{aligned}
\text{Time} &= E_{\text{one century}}/L_{\text{Sun}} = 3.15 \times 10^{22} \text{ joules}/3.8 \times 10^{26}\\
&\quad \text{joules/second}\\
&= 0.000\,08 \text{ seconds.}
\end{aligned}
$$

The logic of the calculation

Step 1. We need to find how rapidly energy is being used worldwide by our industrial civilization.

Once we know this, we do the following.

Step 2. Multiply our rate of use of energy by one century to find the total energy our civilization uses in a century. (This assumes that the rate of use of energy is constant.)

Step 3. Divide this total energy by the rate the Sun is emitting energy to find the time required for the Sun to emit this much energy.

As you can see from the detailed calculation, it takes a mere 0.000 08 seconds for the Sun to emit enough energy to power the entire worldwide industrial civilization for a century.

NOW YOU DO IT

How long would it take for the Sun to emit enough energy to power the entire worldwide industrial civilization for a month?

(A) How many joules has the Sun emitted since you were born? (B) For how long would this power our industrial civilization (assuming our rate of use of energy does not change in the future)?

Where could such a prodigious output possibly come from? We can think of the Sun as a furnace, burning fuel and so producing heat and light. What sort of fuel could power so intense an emission? Because we know from Chapter 11 that the Sun is composed primarily of hydrogen, we must think of ways in which this element can be made to yield so much energy.

Could stars shine by chemical reactions?

Let us first suppose that the fuel might be something with which we are familiar in our daily lives. Let us imagine that the Sun is a furnace in which hydrogen undergoes chemical reactions to produce energy. Is this possible? We can easily show that it is not. The reason is that hydrogen, when it undergoes chemical reactions, does not release enough energy to power so great an energy output.

A gas consisting solely of hydrogen can undergo only one chemical reaction: the formation of the hydrogen molecule from two hydrogen atoms:

$$2H \rightarrow H_2.$$

In this reaction, a kilogram of hydrogen releases

$$E_{chemical} = 220 \text{ million joules}$$

of energy. Let us use this fact to calculate *how long the Sun could continue shining* if it were using this kind of fuel. We will find that it could not have continued shining up to the present day.

The logic of the calculation

We know the quantity of energy released when a certain mass (one kilogram) of fuel reacts. To find the *total* quantity of energy the Sun could ever release, we need to multiply this by the mass of all the fuel the Sun contains. Of course we do not know this. But we do know that the mass of the fuel could certainly be no more than the mass of the entire Sun, which is 2×10^{30} kilograms. Thus, we can do the following.

  Step 1. Multiply (a) the energy released when 1 kilogram of hydrogen reacts by (b) the mass of the Sun. This will give us the maximum total amount of energy the Sun could possibly emit.

We also know the *rate* at which the Sun is emitting energy – this is its luminosity. To find how long the Sun could continue shining at this rate we do the following.

  Step 2. Divide this maximum total amount of energy the Sun could possibly emit by the rate it is emitting energy. This will give us the greatest amount of time it could possibly continue shining.

Detailed calculation

Step 1. Multiply (a) the energy released when 1 kilogram of hydrogen reacts by (b) the mass of the Sun.

We will call the result E_{max}: it is the maximum total amount of energy the Sun could possibly emit throughout its entire history.

E_{max} = (220 million joules/kilogram) (1.989 × 10^{30} kilograms) = 4.38 × 10^{38} joules.

Step 2. Divide E_{max} by the Sun's luminosity L_{Sun}.

This will give us "max lifetime," the greatest amount of time the Sun could possibly continue shining:

max lifetime = E_{max}/L_{Sun} = 4.38 × 10^{38} joules/3.8 × 10^{26} joules/second

= 1.1 × 10^{12} seconds = 35 thousand years.

⇒ **Looking forward**
In Chapter 19 we will study the origin of life.

NOW YOU DO IT

Suppose the Sun were made of wood, and it derived its energy from the burning of that wood. One kilogram of wood released 14 million joules when burned. Calculate how long it could continue shining in this case.

As you can see the result of the detailed calculation is 35 thousand years. But we know that the Sun has been shining for far longer than that. Evidence from geology and paleontology tells us that the Earth has been warmed by the Sun for billions of years. Clearly there is something missing in our attempt to understand the shining of the Sun.

Searching for a better theory

Let us look over our calculation and identify the point at which it might be altered in order to produce so much longer a period of time.

If the Sun has been shining for more than 35 thousand years, it has emitted more energy than allowed for in our calculation. The only way this could be is that the fuel it is consuming releases *more energy per kilogram* than we assumed. So to understand the Sun's great luminosity over such enormous periods of time, we need to invoke some way in which hydrogen can be made to release more energy.

How much more energy? As we will see in Chapter 19, our planet has been warm enough to support life for roughly 3.5 billion years. This means the Sun has been shining for at least this long. So we can set up a proportion:

$$\frac{\text{Energy per kilogram of the actual reaction}}{\text{Energy per kilogram of our calculation}} = \frac{\text{3.5 billion years}}{\text{35 000 years}}$$
$$= 100\,000.$$

In fact there are plenty of more powerful fuels than hydrogen. But none is remotely close to being 100 000 times more powerful. None is sufficiently powerful to have enabled the Sun to continue shining for geologic ages.

To resolve this paradox we need to think of a wholly new form of energy production, one whose output enormously exceeds that of all these fuels. Only nuclear energy can supply so great an amount of energy. In the following section we will see that nuclear reactions release 2.9 million times more energy than chemical ones. So we are led to a new theory: *stars derive their energy from nuclear reactions*.

Testing the theory

In the following section we will develop our understanding of nuclear energy. But for now let us ask how we might test this new theory.

The first thing might be simply to look: train our telescopes on the Sun and search for evidence that nuclear reactions are taking place. But this will not work, since these reactions are taking place deep within the Sun. When looking at the Sun we can only see its outer layers, but the Sun's deep interior, where its energy is generated, is hidden from our view.

We are forced to rely on a variety of indirect tests.

(1) *Modeling*. We can produce detailed computer models of the structure of a star, based on our theory of nuclear energy generation. These models are described below. The fact that their predictions are in agreement with observation is evidence for the theory's correctness.

(2) *Solar seismology*. Just as seismic waves propagate through the Earth, so do pressure waves propagate through the Sun. And just as the study of seismic waves tells us much about the Earth's deep interior, so too can we probe the Sun's depths by studying its vibrations. As we discuss below, detailed studies of these waves confirm the theory of nuclear energy generation.

(3) *Solar neutrinos*. Neutrinos are elementary particles with near-zero mass that interact exceedingly rarely with ordinary particles. They are emitted by the same nuclear reactions that our theory says power the shining of the Sun. Studies of these neutrinos reveal that they are being emitted at the predicted rate, thus confirming the theory.

Nuclear energy generation in stars

We now turn to a detailed study of how nuclear reactions power the shining of the stars.

Energy production via nuclear reactions differs fundamentally from energy production via chemical reactions. The energy we get from such fuels as hydrogen, gasoline and burning wood is released via *chemical* reactions. In such a reaction, atoms join together to form molecules. But in a *nuclear* reaction, the nuclei within these atoms merge into one another, or split apart, to form new nuclei.

Atomic and nuclear structure

⇐ **Looking backward**
We studied the structure of the atom in Chapter 4.

Recall that an atom consists of a nucleus surrounded by electrons. The mass of the nucleus is very much greater than that of the electrons, so that the nucleus contains most of the atom's mass. On the other hand, the nucleus is very much smaller than the region through which the electrons move, so that most of the volume of the atom is empty space.

A nucleus, in turn, consists of neutrons and protons, subatomic particles of nearly equal mass. In another regard, however, these two particles differ radically from one another, for the proton possesses a positive electric charge whereas the neutron has no charge. The electrons, finally, are negatively charged. The atom as a whole is electrically neutral, so that the number of electrons equals the number of protons.

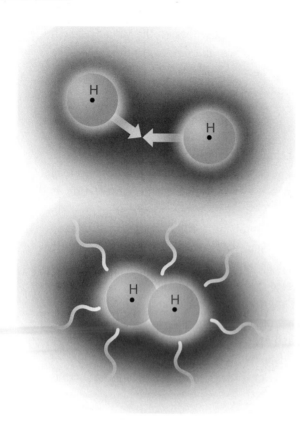

Figure 14.1 **In a chemical reaction atoms stick together. We illustrate the reaction of two hydrogen atoms to form a hydrogen molecule.**

The number of electrons, protons and neutrons within an atom depends on the chemical element. Hydrogen consists of one electron, one proton and no neutrons. The next element up the periodic table, helium, consists of two electrons, two protons and two neutrons.

Chemical and nuclear reactions

In Figure 14.1 we show schematically the chemical reaction we have been discussing:

$$2H \rightarrow H_2$$

in which two hydrogen atoms react to form a hydrogen molecule. In the initial state the atoms are far apart. But then the atoms approach one another and "stick together" to form the molecule. In the process, energy is released – chemical energy.

Notice that, after this reaction has occurred, the two hydrogen atoms still exist – they are simply bound together in close proximity. But in a *nuclear* reaction this is no longer the case. In a nuclear reaction it is not the atoms that approach one another to "stick," but the nuclei; and when they have stuck together the nucleus they form is that of an entirely new element. In Figure 14.2 we show schematically the actual reaction that powers the shining of the Sun:

$$4H \rightarrow He$$

in which four hydrogen nuclei react to form helium. Here, after the reaction has occurred, the hydrogen has entirely disappeared.

This figure illustrates the process of *fusion*, the joining together of nuclei. Conversely, Figure 14.3 diagrams nuclear *fission*, in which a heavy nucleus splits apart into fragments. On Earth, the fission of heavy elements such as uranium is what powers atom bombs and nuclear reactors, and the fusion of light elements is what powers the hydrogen bomb. Because heavy elements are almost entirely nonexistent in the stars, fission plays little role in astronomy. Fusion, on the other hand, is all-important: it is what powers the shining of most stars.

Energy release in nuclear reactions

A remarkable feature of this reaction arises if we consider the masses of the hydrogen and helium nuclei. The mass of the nucleus of hydrogen – the proton – is 1.673×10^{-27} kilograms. Since four of them join together to form helium, we would expect the mass of the helium nucleus to be four times as great, or 6.693×10^{-27} kilograms. But it is not! The mass of the helium nucleus is somewhat less, 6.645×10^{-27} kilograms.

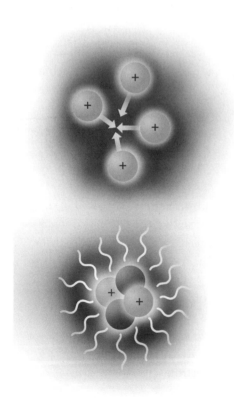

Figure 14.2 **In a fusion nuclear reaction nuclei stick together. We illustrate the reaction of four hydrogen nuclei to form a helium nucleus.**

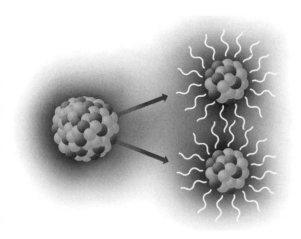

Figure 14.3 **In a fission nuclear reaction heavy nuclei split apart.**

What has happened? We are witnessing the conversion of mass into energy, as described by Einstein's famous formula

$$E = mc^2.$$

Not all of the mass of the hydrogen nuclei remains in the helium nucleus: a small fraction has vanished, and in its place energy has appeared. This is the energy released by nuclear fusion.

We can calculate the quantity of energy as follows:

initial mass = 4 × (mass of hydrogen nucleus)
$$= 6.693 \times 10^{-27} \text{ kg},$$
final mass = mass of helium nucleus
$$= 6.645 \times 10^{-27} \text{ kg},$$
mass lost = 0.048×10^{-27} kg.

If we multiply the mass lost by c^2 we get the energy released in a single reaction 4H → He:

$$(0.048 \times 10^{-27} \text{ kg})(3 \times 10^8 \text{ m/s})^2 = 4.3 \times 10^{-12} \text{ joules}.$$

Notice that the mass lost in a reaction, 0.048×10^{-27} kilogram, is 0.007 times the mass of four hydrogen nuclei. This means that, in such a nuclear reaction, 0.7% of the reacting mass is converted into energy.

This is not a lot of energy. On the other hand, so far we have been considering only a single reaction. In the previous section we considered the energy released when an entire kilogram of hydrogen underwent a chemical reaction. In the same spirit, let us now find the energy released when a kilogram of hydrogen undergoes a *nuclear* reaction.

The logic of the calculation

In order to do this, we need to know how many hydrogen nuclei are contained in a kilogram of hydrogen.

Step 1. The number of hydrogen nuclei contained in a kilogram of hydrogen is 1 kilogram divided by the mass (in kilograms) of a hydrogen atom.

Now we find the number of reactions that can take place if we have this many nuclei.

Step 2. Since it takes 4 hydrogen nuclei to produce one reaction, the number of reactions that can take place within this kilogram is ¼ the number of hydrogen nuclei.

And finally we find the energy released by these reactions.

Step 3. The total energy released when a kilogram of hydrogen undergoes fusion is the number of reactions times the energy released in one reaction.

Detailed calculation

Step 1. Divide 1 kilogram by the mass (in kilograms) of a hydrogen atom to find N, the number of hydrogen atoms in a kilogram.

$$N = 1 \text{ kg}/1.673 \times 10^{-27} \text{ kg} = 5.98 \times 10^{26} \text{ atoms}.$$

Step 2. Divide N by 4 to find $N_{reactions}$, the number of reactions that can take place within a kilogram of hydrogen.

$$N_{reactions} = N/4 = 5.98 \times 10^{26}/4 = 1.49 \times 10^{26} \text{ reactions}.$$

Step 3. Multiply $N_{reactions}$ by the energy released in one reaction to find $E_{nuclear}$, the energy released when a kilogram of hydrogen undergoes a nuclear reaction.

$$E_{nuclear} = (1.49 \times 10^{26} \text{ reactions}) (4.3 \times 10^{-12} \text{ joules/reaction}) = 6.43 \times 10^{14} \text{ joules}.$$

Step 4. To get an intuitive feeling for the significance of our result, compare it to the energy released in a chemical reaction. Let us find the ratio $E_{nuclear}/E_{chemical}$. We calculate

$$E_{nuclear}/E_{chemical} = 6.43 \times 10^{14} \text{ joules}/220 \text{ million joules} = 2.9 \times 10^6.$$

We see that nuclear reactions release 2.9 *million* times more energy than chemical ones.

Step 4. To get an intuitive feeling for the significance of our result, compare it to the energy released in a chemical reaction.

As you can see from the detailed calculation, the energy released by nuclear reactions is very great. In the previous section we calculated that if the Sun produced its energy via chemical reactions, it could continue shining at its present rate for only 35 thousand years. We can now do the same calculation for the case of nuclear reactions. The straightforward way to do this is to repeat the above calculation, but with $E_{chemical}$ replaced by $E_{nuclear}$. But there's an easier way. Since nuclear reactions release 2.9 million times more energy then chemical ones, we can immediately conclude that the maximum possible lifetime of the Sun is 2.9 million times our previous result, or (2.9 million) (35 thousand years) = 10^{11} years = 100 billion years.

We emphasize that this is the maximum possible lifetime – the lifetime that would obtain were all the hydrogen in the Sun available as fuel. In the next section we will explain why only 10% of the Sun's hydrogen is actually available. So the actual time the Sun can continue shining at its present rate is 10% of 10^{11} years, or 10 billion years. But this is still a very long time – far greater than the age of the Earth. We conclude that nuclear reactions are so enormously powerful that they are easily capable of fueling the shining of the Sun over geologic ages.

Main sequence lifetimes

Let us turn our attention away from the Sun, and to stars of other masses. We want to repeat the same calculation for them in order to find how long they could continue shining as main sequence stars. We will reach an important insight: *the more massive a star, the more rapidly does it use up its nuclear fuel.*

Let us recall how we calculated the nuclear-burning lifetime of the Sun. We did this by dividing the maximum total amount of energy it could possibly emit (E_{max}) by the rate at which it is emitting energy (L_{Sun}). Suppose we do the same for other stars. The logic of the calculation will be just the same.

Let us first re-do the calculation for a star just twice as massive as the Sun. Each of the factors in our calculation will be different.

• *Energy.* The more massive the star the more fuel it contains. A star twice as massive as the Sun has twice as much fuel. So E_{max} will be just twice what it was for the Sun.
• *Luminosity.* The mass–luminosity relation (Chapter 12) tells us that the more massive a star the greater its luminosity.

$$L/L_{Sun} = (M/M_{Sun})^{3.5}.$$

If we insert $M = 2M_{Sun}$ into this relation we obtain

$$L/L_{Sun} = (2)^{3.5} = 11.$$

(We can also get this result simply by reading the luminosity off of Figure 12.13.) Notice that our 2 solar mass star can emit twice the energy as the Sun – but it is doing so 11 times faster. So it will use up its fuel more rapidly. Specifically, the lifetime of such a star, as compared to that of the Sun, is

$$\text{lifetime of a 2 solar mass star/lifetime of Sun} = (E_{max}/E_{max\,sun})/(L/L_{Sun})$$
$$= 2/11 = 0.18.$$

Similarly, if we were to find the lifetime of a star just half as massive as the Sun, we find that the lifetime is greater:

$$\text{lifetime of a ½ solar mass star/lifetime of Sun} = 5.6.$$

Notice that the less-massive star shines for longer than the Sun, and the more-massive star shines for a shorter period of time.

· ·

NOW YOU DO IT

Find the main sequence lifetimes of (1) a star 3 times as massive as the Sun, and (2) a star ⅓ as massive as the Sun.

· ·

The role of temperature in thermonuclear reactions

Let us now consider a paradox. We have seen that hydrogen can be used as a fuel of enormous power. But hydrogen is common on the Earth – after all, water is H_2O, and the oceans cover much of the Earth's surface. So our planet is replete with the very fuel used by both thermonuclear weapons and the stars. Why doesn't all this fuel detonate in a hydrogen bomb of terrible proportions?

The answer is that on Earth the fuel is cold, whereas in weapons and in the stars it is hot. Indeed, the temperature of the interior of the Sun, where the fusion reactions are occurring, is 15 million kelvin (about 27 million °F). Just as the heat of a match is required to set paper burning, this enormous temperature is required to initiate nuclear fusion reactions.

Let us see why. Figure 14.4 illustrates two hydrogen nuclei approaching one another. If they touch, nuclear fusion reactions will occur. But there is a force acting to prevent their touching! Recall that hydrogen nuclei are protons, and protons possess a positive electrical charge – and like charges repel one another.

This force acts to prevent the protons from touching – which is to say that it acts to prevent fusion reactions (Figure 14.4a). How then do two protons approach one another sufficiently closely to undergo nuclear fusion? By moving very rapidly. It is like throwing a ball upward: the ball's momentum allows it to rise even though it is being dragged downwards by gravity. Similarly, a proton's

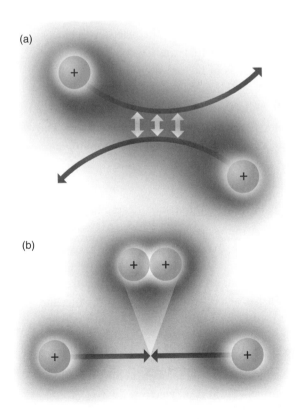

(a)

(b)

Figure 14.4 **Electrical repulsion inhibits nuclear fusion. Nuclei are repelled by electrostatic forces. (a) At low temperatures they are moving so slowly that they never touch. But at higher temperatures (b) the particles are moving sufficiently rapidly to touch and undergo fusion.**

momentum allows it to approach another proton, even though it is being pushed away by electrical repulsion. Furthermore, the hotter a gas, the more rapidly do the particles within it move. Only at high temperatures do they move sufficiently rapidly to touch (Figure 14.4b). Thus it is that fusion reactions are *thermo*nuclear reactions.

This circumstance has an important bearing on our calculation of the nuclear-burning lifetime of the Sun. This calculation assumed that *all* the hydrogen in the Sun was available to it as fuel. But only the hydrogen in the core is sufficiently hot to participate in thermonuclear reactions. It turns out that only about 10% of the Sun's hydrogen is this hot. Thus the nuclear-burning lifetime of the Sun is not 10^{11} but 10^{10} (ten billion) years.

Finally, since the Sun is 4.5 billion years old, we see that it will continue shining for another 5.5 billion years. At present the Sun is roughly halfway through its lifetime.

The proton–proton chain

Let us return to the nuclear reaction that powers the shining of the Sun:

$$4H \rightarrow He.$$

A little thought shows that there must be more going on in the reaction than this. The nucleus of hydrogen is simply the proton, so that the left-hand side of the reaction consists of four protons. But the nucleus of helium does not: it consists of two protons and two neutrons.

In fact the above is merely a schematic indication of the actual nuclear reactions that occur in the core of the Sun. Let us pass on to a more detailed description. The actual reaction occurs in steps. The sequence of steps is known as the *proton–proton chain.*

The *first step* in the chain is the most complex. It is:

$$\text{proton} + \text{proton} \rightarrow \text{deuteron} + \text{positron} + \text{neutrino} + \text{energy}.$$

Here two hydrogen nuclei have fused to produce three different products (plus energy!). Let us discuss each of these products in turn.

- The *deuteron* is the nucleus of heavy hydrogen, an isotope (alternative form) of the element hydrogen. The deuteron consists of a proton plus a neutron. Since the mass of the neutron equals that of the proton, we see that the deuteron has twice the mass of ordinary hydrogen. Note also that, since the deuteron contains a neutron, we are starting to see where the neutrons in helium come from.
- The *positron* is the antiparticle of the electron: it has the same mass of the electron, but the opposite charge. When a positron encounters an electron the two *annihilate* in the reaction

$$\text{positron} + \text{electron} \longrightarrow \text{energy}.$$

We are witnessing here the complete conversion of mass into energy, as predicted by Einstein's famous $E = mc^2$. Note that the Sun is filled with electrons, so that in fact this reaction happens all the time within it: its released energy is part of the energy supplied by the conversion of hydrogen into helium.

- The *neutrino* is an elementary particle possessing an exceedingly small mass and no electric charge. It interacts only weakly with matter, and flies out through the Sun and into space. As we described above, the detection of these neutrinos provides a test of our theory of stellar structure.

The *second step* in the proton–proton chain is far less complex. It is:

$$\text{deuteron} + \text{proton} \rightarrow \text{helium-3} + \text{energy}.$$

Just as the deuteron is an isotope of hydrogen, so helium-3 is an isotope of helium. It consists of two protons and one neutron. Since the deuteron consists of one proton and one neutron, we see that this reaction is simply a matter of the proton "sticking" to the deuteron to form helium-3: in the process, energy is released.

The *third step* in the proton–proton chain is merely that the above two reactions cycle through again, so producing a second helium-3. And then the *final step* is

$$\text{helium-3} + \text{helium-3} \rightarrow \text{helium-4} + \text{proton} + \text{proton} + \text{energy}.$$

We can now see how the schematic reaction $4\text{H} \rightarrow \text{He}$ summarizes this chain. The first step in the proton–proton chain combines two protons, and the second step adds a third. The final step repeats the first two, so that six protons are involved. But in this final step, two protons are released. Thus, in the final analysis, only four protons have been consumed to make one helium nucleus.

The structure of a star

Every star is a ball of superheated gas. It exists in a state of balance – a double balance, in fact. On the one hand is a balance of *forces*, in which gas pressure balances gravitation. On the other hand is a balance of *energy rates*, in which the rate of energy production by nuclear reactions balances the rate of energy emission by starlight. We now consider these two balances.

Balancing forces within a star

We have seen that the Sun's great temperature is what allows it to utilize its nuclear fuel, and so continue shining for billions of years. The same is true of every other star. In this section we explore a further issue: it is this great temperature that allows a star to *exist*. If by some magical means we could suddenly cool a star, it would not just cease reacting its nuclear fuel. It would also catastrophically collapse inward.

To understand why, we turn our attention to the structure of a star. This structure is determined by two forces, (1) gravity and (2) gas pressure: gravitation compresses the star inward, while pressure balances gravity by expanding it outward. And the star's pressure depends on its temperature: the greater the temperature, the greater the pressure. Only a high temperature generates sufficient pressure to counteract the huge pull of gravity.

Let us consider each of these forces in turn.

(1) Gravitation in a star

As we discussed in Chapter 13, gravitation acts to compress things. The situation is illustrated in Figure 14.5(a). Here we focus attention on one particular atom in a

(a)

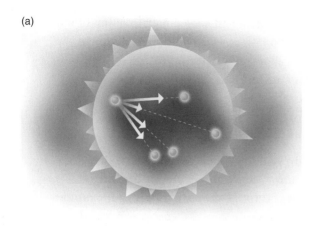

(b)

Figure 14.5 **Gravitational forces between atoms in a star. In (a) we choose a single atom and show the forces attracting it toward others. (The longer the arrow the stronger the force.) In (b) we show the sum of all these forces, an attraction toward the center of the star.**

star, and we sketch a few of all the other atoms that are attracting it. As we can see, the atom we are focusing on is being pulled in many directions at once. But we can also see that *the net result of all these competing forces is a tendency to be pulled inward toward the star's center* as shown in Figure 14.5(b). The same is true of any other atom we choose to think about. Thus, the net result of gravitation is an inward force of compression.

This is the force acting to compress the star. In the process of star formation that we studied in the previous chapter, nothing opposes it and the interstellar cloud collapses. Similarly, if nothing opposed it in a star, the star would collapse inward upon itself. But something does oppose it in a star: gas pressure. Let us now turn to this, the second of the two forces that determine the structure of a star.

(2) Gas pressure in a star

We know that the stars must be gaseous because they are so hot: hot enough to vaporize everything within them. Furthermore, we know that a gas exerts a pressure. It is this pressure that resists gravitation within a star, and prevents it from collapsing inwards.

We will need to know two facts about gas pressure.

Gas pressure

The pressure exerted by a gas grows larger as

- the temperature of the gas grows greater,
- the volume of the gas grows smaller.

The critical issue here is the first: the hotter the gas, the greater is the pressure it exerts. Only at high temperatures is this pressure sufficiently great to counteract gravity. Thus it is that high temperatures allow a star to remain in equilibrium between these two competing forces.

· ·

NOW YOU DO IT
Suppose something were to make a star hotter. Would the star grow larger, keep the same size or grow smaller? Why?

Suppose something were to make gravitation within a star weaker. Would the star grow larger, keep the same size or grow smaller? Why?

· ·

Balancing energy rates within a star

We have so far discussed one kind of balance within a star: that of pressure against gravity. There is a second kind of balance operating in the stars as well: a balance between (1) *the rate of generation of energy by nuclear reactions* and (2) *the rate of emission of energy by starlight.* If nuclear reactions generated less energy than was emitted, a star would cool off as it lost the energy radiated away by starlight. Similarly, if they generated more energy than was emitted, it would heat up as it gained the energy generated by nuclear reactions. Since in reality the stars are growing neither cooler nor hotter, we conclude that these two rates must balance.

Let us now consider each of these rates in turn. Both of them depend on the star's temperature.

(1) As for the first, that of *energy generation by nuclear reactions*, we know that there is no energy generation unless the temperature is very high: it turns out that the rate grows higher as the temperature increases, once the reactions have begun.

(2) As for the second of these rates, that of *emission of energy by starlight*, it also depends on temperature through the Stefan–Boltzmann law, which states that an object's luminosity is related to its temperature by

$$L = 0.57 \times 10^{-7} A T^4 \text{ watts}$$

(here A is the star's surface area). Notice that the luminosity too grows higher as the temperature is increased.

In order to balance these rates, we need to adjust the star's temperature until they are equal. We emphasize, however, that the first rate depends on the temperature of the *core* of the star, for that is where the nuclear reactions take place. Similarly, the second of these rates depends on the temperature of the *surface*, for that is what emits the starlight. Therefore, we need to consider the ways in which a star's surface is heated by its core – that is, the ways in which heat flows outward from the core to the surface of a star.

Heat flow within a star

This heat flow takes place in three ways.

- *Conduction.* Just as the high temperature of scalding coffee conducts outward through a coffee cup, making it hot to the touch, so too does heat from the core of a star conduct outward to its surface. To calculate the rate of this flow, we need to understand the insulating properties of the material making up the star.
- *Convection.* Just as in a fireplace, in which rising columns of air heated by a fire carry its heat up the chimney, so too do currents of superheated gas flow outward from the core to the surface of a star. (As we have seen in Chapter 11, the granulation observed on the Sun's surface represents the tops of these rising plumes of gas.) To calculate the rate of this flow, we need to understand the velocity of these convecting currents.
- *Radiation.* The interior of a star is filled with brilliant light. Just as the light emitted by a lamp carries energy away from it, so too within a star does the light within the core carry energy outward toward its surface. But unlike the emission from a lamp, within a star light does not flow unimpeded. Rather, it is continually being absorbed and then re-emitted by atoms as it propagates outward. To calculate the rate of this flow, we need to understand the degree to which atoms in the star absorb and then re-emit light.

Putting it all together: computer models of main sequence stars

The task of combining all the elements of stellar structure that we have been discussing is clearly complex. It is, in fact, so complex that it can only be done with a computer. In this section we discuss the results of these computations.

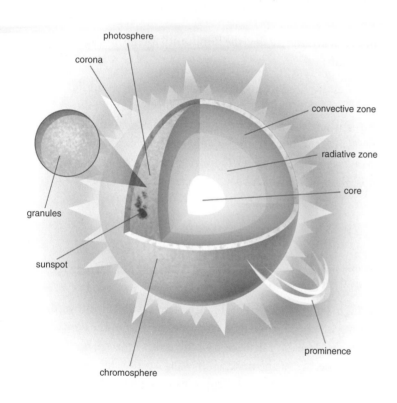

Figure 14.6 **A computer model of our Sun.**

In Figure 14.6 we present the results of extensive computer simulations of the structure of one particular star: our Sun. At the center of the Sun lies its core, in which nuclear reactions take place. This core is at a temperature of 15 million kelvin. At such a high temperature, of course, the gas exerts an enormous pressure. Nevertheless, so gigantic is the weight of the overlying layers that it is compressed to a density of some 150 000 kilograms per cubic meter – 20 times that of iron! As we proceed outward the temperature and density drop. Eventually the temperature is too low to support nuclear reactions, and we pass beyond the core into the *radiative zone*, where heat is transported outwards by radiation. Roughly halfway outward to the Sun's surface the density has dropped to 1000 kilograms per cubic meter, the density of water: the temperature here is still several million kelvin. Somewhat beyond this point lies the base of the *convection zone*, within which heat is transported outwards by rising plumes of gas. Finally, at the Sun's surface, the density has become exceedingly low: 2×10^{-4} kilograms per cubic meter, roughly one ten-thousandth the density of air: the temperature here is 5800 kelvin.

Confirming the calculations: solar seismology

It is important to note that what we have been describing is a *theory*. As always with a scientific theory, it is essential to test it – and the more different ways we can do so, the better. We will describe three such tests.

(1) Recall that on the Hertzsprung–Russell diagram the main sequence consists of stars that are undergoing hydrogen fusion in their cores. The Sun is such a star. Its theoretically predicted location on the H–R diagram can be found by computing the surface temperature and luminosity of the model Sun. Conversely, the Sun's actual location on the H–R diagram can be found by measuring its temperature and luminosity. The theoretically predicted location agrees well with observation.

(2) The same computer program that produced a model of the Sun can be modified to produce models of any other main sequence star. The only difference lies in the star's mass: as we have seen, stars on the upper-left portion of the main sequence are of high mass, and stars on the lower-right portion are of low mass. Detailed computations have been performed. Here, too, theoretical results are in good accord with the observed locations of stars along the main sequence.

. . . .

NOW YOU DO IT

Such a computer calculation makes another prediction: it predicts the mass of a star whose location on the H–R diagram is known. What observations can you think of that could test this prediction?

. . . .

(3) A far more detailed check on the validity of these computer calculations has been made, involving a technique known as solar seismology (or helioseismology).

Just as seismic waves propagate through the Earth, so do pressure waves propagate through the Sun. And just as the study of seismic waves tells us much about the Earth's deep interior, so too can we probe the Sun's depths by studying its vibrations. By analogy with the Earth, this study is termed "solar seismology."

⇐ **Looking backward**
We studied seismic waves in Chapter 7.

The pressure waves within the Sun cause its surface to vibrate up and down. These oscillations can be detected by using the Doppler effect. Observations reveal the motion of the Sun's surface to be complex. It vibrates with many different periods of oscillation. Periods ranging from 2.5 to 11 minutes have been detected, each with a different amplitude. Each period corresponds to a different mode of oscillation of the Sun's deep interior. Figure 14.7 illustrates one such mode.

The motion induced by these waves is quite slow in astronomical terms: roughly 20 centimeters per second (less than half a mile an hour). As a consequence, the up-and-down motion of the surface is no more than a kilometer or so. It is a very difficult task to detect this motion.

Nevertheless, because of the subject's importance a variety of efforts are under way to conduct these observations. One, named GONG for "Global Oscillation Network Group," is a network of small telescopes strategically arrayed across the world (Figure 14.8). Located in Australia, Hawaii, California, Chile, India and an island off Africa, they are placed so that

Figure 14.7 **Theoretical model of the vibration of the Sun. This computer-generated image shows one of many different ways in which the Sun can oscillate. Dark regions move outwards, light regions inwards. Note that the deep interior oscillates, as well as the surface.**

at least one of them is in daylight, and hence able to conduct observations, throughout the 24 hours of the day. Another is the SOHO spacecraft, which we described in Chapter 11.

· ·

NOW YOU DO IT
(A) Explain why it would be advantageous to place telescopes at *both* the north and south poles in order to study these solar oscillations – and explain why it makes little sense to place only one such telescope at one of the poles. (B) Since this seems to be such a good idea, why do you think it hasn't been done already?

· ·

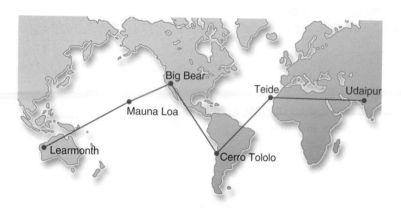

Figure 14.8 **The GONG network.**

How can these observations be used to probe the Sun's deep interior? After all, it is hidden from our view by hundreds of thousands of kilometers of opaque gas. As an analogy, imagine that you are standing on a beach and watching the ocean waves roll in. Close observation reveals many different sorts of waves. Some are large, and have a wavelength of many feet. Others are tiny, closely spaced ripples. Some come from one direction, and some from another. Each such wave is produced by a different mechanism: the tiny ripples are produced by that faint breeze even now blowing against your cheek, while the large waves are the faint remnants of a faraway storm. Remarkably, it is actually possible to deduce the existence of the breeze, and of that faraway storm, solely from the pattern of surf. And in a similar way, it is possible to deduce conditions deep within the Sun from the pattern of vibrations of its surface.

By such means, we have pinpointed the precise point at which convection – the upward-welling plumes of superheated gas that transport heat from the interior to the Sun's surface – ceases (30% of the way from the surface to the center). We have been able to measure the abundance of helium in the Sun's deep interior (it equals the abundance at its surface). These quantities are also predicted by the detailed computer models of the Sun: the measured values are in accord with these predictions.

Brown dwarfs

Recall the question we asked in Chapter 8: "What is the difference between a planet and a star?" We had found that the difference is somewhat a matter of arbitrary definition. Specifically, we had noted the following points.

(1) One might think that planets are small and stars are big. But in Figure 8.5 Jupiter's size lies midway between that of the Earth and the Sun.
(2) One might think that planets are solid and stars are not. But Jupiter is almost entirely a gas surrounding a liquid.

(3) One might think that stars are composed of different chemical elements from planets. But Jupiter's composition is almost identical to that of the Sun.

(4) One might think that stars emit light and planets do not. But Jupiter emits its own (infrared) light.

All this implies that there is a gray area lying between what we normally think of as planets and stars. The so-called "brown dwarfs" occupy this gray area.

One way to think of these objects is as "failed stars." Astronomers have agreed to define a star as an object generating energy via the thermonuclear fusion of hydrogen (or heavier elements). Our best estimate is that any object whose mass is 80 times that of Jupiter (0.08 times that of the Sun) qualifies to be a star by this definition. Similarly, we define a planet to be an object of a mass so low as not to fuse deuterium (this reaction occurs at lower masses). By this definition, any object of mass less than 13 times that of Jupiter is a planet. Brown dwarfs, therefore, are objects whose mass lies between 13 and 80 times that of Jupiter.

Such objects would generate only trace amounts of energy, making them very cool. Furthermore, they would be very small. Via the Stefan–Boltzmann and Wien laws they would, therefore, have very low luminosity and emit in the infrared region of the spectrum. Finally, it is difficult to distinguish one from a normal star of very low luminosity.

All this makes such objects exceedingly hard to discover. For years astronomers postulated them but were unable to produce unambiguous evidence for their existence. Indeed, a number of claims were made that ultimately proved to be mistaken. But in 1995 came the first clear detection of a brown dwarf – an object known as Gliese 229 B. Gliese 229 B orbits Gliese 229, a normal star that has 58% the mass of the Sun. But Gliese 229 B, the brown dwarf, has only 20–50 Jupiter masses. Its surface temperature is a mere 950 kelvin.

By now hundreds of such brown dwarfs are known.

> ⇐ **Looking backward**
> We studied the Stefan–Boltzmann and Wien laws in Chapter 4.

Summary

DETECTIVES ON THE CASE

What powers the shining of the stars?
- The watt measures the rate of emission of energy: the unit of energy is the joule so 1 watt is 1 joule emitted per second.
- The Sun emits energy at an enormous rate: for instance, in just 0.000 08 seconds it emits enough energy to power our worldwide industrial civilization for a century.
- If the Sun generated this energy via chemical reactions, it could only continue shining for a relatively short time.
- To understand how the Sun can continue shining for geological ages, we need a source of energy hundreds of thousands of times more efficient than chemical reactions.

- This source is nuclear reactions.
- Three tests have been performed of this theory: it passes all the tests.

Nuclear energy generation in stars
- An atom consists of a nucleus surrounded by electrons.
- In a chemical reaction, atoms stick together to form molecules, or molecules split apart, releasing energy as they do so.
- In a nuclear reaction, nuclei stick together or split apart, releasing energy as they do so.
- When nuclei join together, they lose a little mass: this is the source of the released energy (via $E = mc^2$).

- The main sequence in the Hertzsprung–Russell diagram consists of stars fusing hydrogen to form helium.
- The Sun will use up its nuclear fuel in about 5 billion years.
- Along the main sequence, the more massive the star the more rapidly does it use up its nuclear fuel.
- Hydrogen must be heated to millions of degrees to initiate such nuclear reactions.
- In the Sun the reactions proceed via the "proton–proton chain."

The structure of a star

- A star is a ball of gas.
- In a star the inward pull of gravity is balanced by gas pressure.
- In a star the emission of energy via starlight is balanced by the generation of energy via nuclear reactions.
- This energy flows from where it is generated (the core of the star) to where it is emitted (the surface) by conduction, convection and radiation.

- A computer model of the Sun.
  - At center: temperature is 15 million kelvin and density is 20 times that of iron.
  - Halfway out to the surface: temperature is several million kelvin and density is about that of water.
  - At surface: temperature is 5800 kelvin and density is one ten-thousandth the density of air.
  - Inner regions: heat transported by radiation.
  - Outer regions: heat transported by convection.
- Solar seismology.
  - Sound waves (seismic waves) are continually propagating through the Sun.
  - GONG ("Global Oscillation Network Group") is a network of telescopes arrayed across the world continuously observing these waves.
  - These observations give us information about the Sun's deep interior.

Brown dwarfs

These occupy the gray area between planets and stars.

Problems

(1) (A) Suppose you have two light bulbs: one emits three times as much light energy in a minute as the other. How much greater is its luminosity? (B) Suppose you have two light bulbs: one emits four times as many joules per second as the other. How much greater is its luminosity? (C) Suppose you have two light bulbs. Each is connected to its own battery, both of which have stored the same amount of energy. One light bulb is twice as bright as the other. The brighter one stops shining after an hour; for how long will the dimmer one keep shining?

(2) We want to calculate how long it would take the Sun to emit enough energy to power the entire worldwide industrial civilization for a week. (A) Describe the logic of the calculation you will use. (B) Carry out the detailed calculation.

(3) Suppose we knew that the Sun had been shining for 100 000 years, and we wanted to identify the type of fuel that would allow it to do so. In particular, we would need to calculate the amount of energy released by consuming one kilogram of this fuel. (A) Describe the logic of the calculation you would use. (B) Carry out the detailed calculation. (C) Suppose alternatively the Sun were composed of a fuel that emitted three times as much energy per kilogram. For how long could it continue shining?

(4) The table below gives the amount of energy released when various fuels are burned. Choose one fuel, and imagine that the Sun is composed of this substance. We want to calculate the greatest possible length of time it could continue shining were this the case. (A) Describe the logic of the calculation you will use. (B) Carry out the detailed calculation.

Fuel	Energy released by burning one kilogram (joules)
crude oil	43 million
coal	29 million

(5) We will see in the next chapter that certain stars known as red giants derive their energy not from nuclear reactions involving hydrogen, but by reactions involving helium. Specifically, red-giant stars are powered by the reaction

$$3 \text{ helium} \rightarrow \text{carbon.}$$

Using the data in the table below, we want to calculate (1) the mass lost in one such reaction, (2) the energy released in one such reaction, (3) the energy released when a kilogram of helium fuses to become carbon, (4) the ratio between this and the amount of energy released when a kilogram of hydrogen fuses to become helium, and finally (5) the length of time a 10 solar luminosity star of 1 solar mass could continue shining if it were powered by this reaction.

Nucleus	Mass (kilograms)
Helium	6.645×10^{-27}
Carbon	1.9922×10^{-26}

For each, (A) describe the logic of the calculation you would use, and (B) carry out the detailed calculation.

(6) We want to find the mass of a star that continues shining for $\frac{1}{10}$th the lifetime of our Sun. (A) Describe the logic of the calculation you will use. (B) Carry out the detailed calculation.

Stellar evolution and death

We closed Chapter 12 on the "census of stars" with a question: what is the significance of the three categories of stars: main sequence, red giant and white dwarf? In Chapter 14 we reached an understanding of main sequence stars: they are powered by thermonuclear reactions that transform hydrogen into helium. In this chapter we move on to red giants and white dwarfs.

We can think of the hydrogen in a main sequence star as fuel that powers its shining. In the previous chapter we explained that a star has lots of hydrogen, so that it can continue shining for a long time. But no matter how long this can go on, eventually the star will run out of this fuel. You might think that, once this happens, the star will simply go out. But it turns out that helium, the residue of hydrogen reactions, is itself a fuel. In order to use this new fuel, the star must readjust its structure to become a red giant. Indeed, the subsequent evolution of a star is governed by a whole series of other such readjustments.

But this process cannot go on forever. Eventually, no more fuel will be left to the star. Again, you might think that, once this happens, the star will simply "die." But it does not die. Instead, the star is utterly reborn as a new and exciting "corpse" – a white dwarf, a neutron star, a black hole or a cataclysmic supernova.

After the main sequence: running out of hydrogen

Recall that every star floats in the near absolute zero of interstellar space. We might think of it as being analogous to a furnace, burning fuel to heat a house in midwinter. When such a furnace runs out of heating oil, the house quickly loses its heat to the bitter cold outside. Is that what happens to a star? Once it has used up its hydrogen, does its "furnace" go out, allowing the star to cool rapidly?

Remarkably, it does not. There is a difference between a star and a furnace, one that concerns the *exhaust* from the burning of fuel:

- in a furnace, this exhaust is expelled up a chimney, and it wafts away;
- in a star, the "exhaust" is helium, which remains in the star;
- helium itself is a fuel, which can undergo thermonuclear reactions as well.

So, once a star has run out of hydrogen, it has not run out of fuel. Indeed, precisely the opposite: it still has plenty of fuel remaining to power its shining.

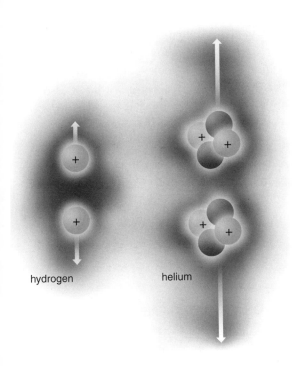

hydrogen   helium

Figure 15.1 **The repulsion between two helium nuclei is greater than that between two protons. So higher temperatures are required to fuse helium than hydrogen.**

But this new fuel cannot be used until the star has grown hotter. Why? We showed in the previous chapter that fusion reactions that transform hydrogen into helium require very high temperatures. We will now show that even higher temperatures are required to fuse the resulting helium into something else.

As we discussed, the force of electrical repulsion acts to prevent nuclear reactions. This force is countered by the effects of temperature: the hotter a gas, the more rapidly do the particles within it move, and the more they are able to overcome the effects of this repulsion. Only at high temperatures do they move sufficiently rapidly to undergo nuclear fusion. So it is that fusion reactions are *thermo*nuclear reactions.

Let us now apply these ideas to the fusion of *helium* nuclei. Figure 15.1 illustrates two of them near each other. Because the helium nucleus has two protons it has twice the charge of the proton – and since the charge is greater, the electrical force of repulsion is greater. So the temperature required to overcome this greater repulsion is greater. *Higher temperatures are required to initiate helium fusion than hydrogen fusion reactions.*

. . . .

NOW YOU DO IT
The nucleus of an *isotope* of an element has the same number of protons, but a different number of neutrons. Thus helium-4, the normal helium nucleus, consists of two neutrons and two protons, but the isotope helium-3 has two protons and only one neutron. Is the electrical repulsion acting to inhibit fusion reactions between two helium-3 nuclei greater, less or the same as the repulsion inhibiting the fusion of two helium-4 nuclei? Explain your answer.

Fission is the process whereby a very heavy nucleus, with lots of protons and neutrons, splits into several smaller nuclei (which are known as "fission products"). Explain why, once this happens, the fission products fly away from one another at great speed.

. . . .

Thus we see that the termination of hydrogen fusion reactions in a star does not mean that it has run out of fuel. What it means is that the star has *temporarily* run out of fuel. But the star must grow hotter in order to ignite its supply of this new source of energy.

A star has a way of reaching such higher temperatures. Return to our analogy between a star in the cold of space and a house in the dead of winter. Just as the house rapidly cools once it has run out of fuel, so we might expect our star to cool once it has run out of its hydrogen. But in fact it does just the opposite! We will now show that *once it has exhausted its hydrogen, the inner region of a star grows smaller and hotter.*

Let us seek to understand this strange fact. A second difference between a star and a house furnace is that, unlike a furnace:

• a star is compressed by gravity,
• it resists this compression by internal pressure,
• this pressure exists because the star is hot.

Recall what we know of gas pressure from Chapter 14: the pressure in a gas grows larger as the temperature increases. Conversely, the pressure grows less as the temperature decreases. So once the star runs out of fuel it cools – not greatly, only by a tiny amount – and as soon as this happens, the pressure within it also decreases – again, not greatly, but by a tiny amount. Once this has happened *the pressure is no longer enough to balance gravity.* As a consequence the star begins to gradually shrink.

We now add a further element. A gas that shrinks is, of course, being compressed – and *compressing a gas heats it.* So the exhaustion of hydrogen within a star makes its core shrink and heat. It will continue to do so until its temperature is sufficiently great to ignite the burning of helium. This helium burning is the second phase in the history of a star.

Red giant stars

There is yet another element we need to add to our picture of this second phase of stellar evolution. It is illustrated in Figure 15.2. Recall that the temperature of a star is greatest in its center, and grows progressively lower as we move outwards. Once the star has shrunk sufficiently to ignite helium fusion in its core, the temperature in the intermediate regions has also increased. Indeed, it has increased enough to promote *hydrogen* fusion reactions. So the star develops a layered structure, with helium reactions proceeding in its core, and hydrogen reactions proceeding in a shell surrounding the core. This new source of heat, generated at intermediate depths, heats the very surface of the star, causing it to expand. So we see that the evolution of the star once it has run out of hydrogen is somewhat complex: its inner regions shrink, while its outer regions expand.

Just as compressing a gas heats it, so too does expanding a gas cool it. And from our study of blackbody radiation, we know that light emitted by a low-temperature blackbody is redder than that from a high-temperature one. Thus, as seen from outside, our star has grown *bigger* and *redder.* It has become a *red-giant star.*

The nuclear fusion reaction that powers red-giant stars is

$$3\,^4\text{He} \rightarrow\, ^{12}\text{C}.$$

Thus the element carbon, so essential to life, is being created within the cores of these stars. Figure 15.3 shows the results of a computer model of the evolution of a 1 solar mass star that has exhausted its hydrogen fuel.

During this evolution, the star expands to enormous sizes, reaching more than 50 times its original radius. This means that the Sun, in the red-giant phase of its evolution, will be nearly as large as the orbit of the planet Mercury. For comparison, if we represent the initial main sequence star by a volleyball, it would expand to more than 10 yards in diameter as it evolves into a red giant. At the same time its surface temperature drops from 5780 to 3500 kelvin, and its color becomes a deep red. Finally, it grows very much brighter, reaching some 400 solar luminosities – indeed the red-giant stars are among the brightest known stars in the sky. The star has reached the

Figure 15.2 **Nuclear reactions in a red-giant star. The temperature of the core is sufficient to promote helium fusion, while the temperature in a surrounding somewhat cooler layer is sufficient to promote hydrogen fusion.**

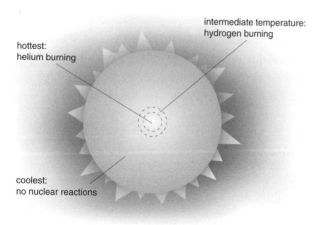

intermediate temperature:
hydrogen burning

hottest:
helium burning

coolest:
no nuclear reactions

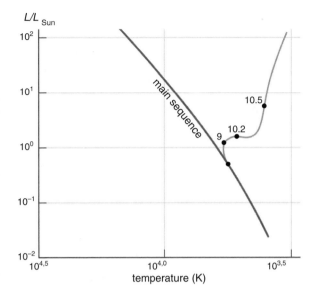

Figure 15.3 **Evolution from the main sequence to the red-giant stage of a 1 solar mass star. The numbers beside the dots give the time, in billions of years, to reach the indicated stage.**

Figure 15.4 **The end of the world. Once the Sun reaches the end of its evolution on the main sequence 5.5 billion years from now, it will evolve into a red-giant star, growing more luminous as it does so. The Earth, therefore, will grow hotter.**

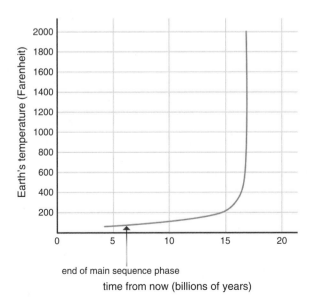

second stage in its evolution, in which it occupies the red-giant portion of the Hertzsprung–Russell diagram.

The end of the world

As the Sun grows more luminous, the Earth will grow hotter. Eventually it will grow too hot to sustain life. So as the Sun evolves toward its red-giant stage, all life will come to an end on the planet Earth.

You might think that since the Sun's surface is growing cooler the Earth would too. But the Earth's temperature does not really depend on the temperature of the Sun. It depends on the amount of energy the Sun generates – on its luminosity. Recall our analogy of the Sun as being like a furnace, generating heat: the more heat it generates, the more it warms the planets around it. Calculations reveal that when the Sun reaches its full luminosity of $400L_{Sun}$ as a red giant, the Earth will reach a temperature of 1300 kelvin, or 1880 degrees Fahrenheit.

This heating process is recorded in Figure 15.4. As you can see, it will take a long time. The Sun's expansion from the main sequence to the red-giant stage will be gradual, requiring billions of years. Initially, as the Sun slowly expands and brightens, it is only the Earth's equatorial regions that will grow too hot to sustain life. The north and south poles, in contrast, will become pleasantly warm. Far in the future we can imagine that most life will be confined to the poles, with an uninhabitable belt around the equator separating the Earth into two isolated ecosystems. Sunlight in those far-distant ages will be a deep, angry red; the Sun an immense bloated ball, approaching 50 times its present size in the sky.

Eventually, as the ages roll by and the Sun grows yet more luminous, the Earth will become monstrously hot, an utterly hostile environment. All liquid water will evaporate away as oceans, lakes and rivers boil. Eventually our home planet will become so hot that lead will melt.

This is the fate in store for Earth. There is no escaping it, and no way to ward it off. Since the Sun is presently only halfway through its ten-billion-year main sequence lifetime, we might take comfort in the thought that this fate lies many billions of years in the future. Nevertheless, it is no exaggeration to describe this fate as the end of the world.

After the red-giant stage

Just as the main sequence stage terminates when the star runs out of hydrogen, so does the red-giant stage end when the helium in the star's core is used up. The "exhaust" from the fusion of helium is carbon; and just as the "exhaust" from hydrogen burning is itself a fuel, so too with carbon. Once again, the star's core shrinks and heats till this third source of energy ignites.

Subsequent stages in the evolution of the star grow more and more complex, as more and more elements undergo fusion in more and more layers within the star. Simulating on a computer these late stages of stellar evolution is difficult, and we are currently quite unsure of many details. In what follows we concentrate on a few specific phenomena that are known to occur in these stages.

Cepheid variable stars

Most stars shine steadily. A few are known, however, whose luminosity fluctuates in time. They grow brighter and dimmer in a regular pattern. Observations reveal that these fluctuations in brightness are accompanied by a fluctuation in size: these stars are regularly expanding and contracting, changing their temperatures as they do. Since the Stefan–Boltzmann law states that the luminosity of a star depends on its radius and temperature, the fluctuations in size and temperature are accompanied by variations in brightness.

Polaris, the pole star, is such a star, of a particular type known as a *Cepheid variable* (named in honor of the first such star to be discovered, which lay in the constellation of Cepheus). It varies in brightness by 10% over a period of 4 days. In general, a Cepheid variable can have a period of oscillation lying between 3 and 50 days, and a luminosity lying between 1000 and 10 000 solar luminosities. Cepheid variables are high-mass stars (3 to 18 solar masses) undergoing helium fusion in their cores.

What accounts for their oscillations? Let us start with an analogy. Suppose I am an excitable sort of person, and I realize that I have been spending too much of my income. I decide to cut back on frills. But being so excitable, I do so excessively. I deny myself everything. I begin by cutting out movies and dinners in restaurants, but soon I decide to skip lunch and stop buying pretty much everything. I have a car but I stop using it to save money on gas, and I no longer download even the cheapest of apps to my smartphone. After several weeks of this extreme regime my financial situation has started to improve. But by this time I have become so perpetually miserable that I revolt. I go on a buying binge. Pretty soon my debt has crept up again. One day, I realize what has happened... and I start saving all over again. And so the cycle continues, my finances oscillating up and down.

My error is that I am *excessive*. I save too much, and then I spend too much. Had I behaved more moderately, I would never have gotten stuck in this perpetual oscillation. Similar factors are at play in stellar structure: normal stars are like most people, whereas variable stars are like people in this analogy.

In this analogy my spending habits have been playing the role of the *pressure within the star*. This pressure responds to variations in the star's size: when the star contracts the pressure builds up, resisting the contraction; and when the star expands the pressure drops, allowing it to contract. In normal stars this variation in pressure is moderate: in variable stars it is excessive.

Let us consider what happens to a star that, for some reason, contracts. As we recall from our discussion of stellar structure, every star exists in an equilibrium between two forces: gravity pulls it in and gas pressure pushes it out. When the star has the right size, these forces just balance. Imagine now that we disturb this balance, by making the star contract slightly. What happens?

⇐ **Looking backward**
We studied the principles lying behind stellar structure in Chapter 14.

As we saw in Chapter 14, the gas pressure depends on two factors,

- the volume of the star: the smaller the volume, the more the pressure;
- the temperature of the star: the hotter the star, the more the pressure.

If we start with a star at the right size and then shrink it, its volume decreases. Furthermore, since compressing a gas heats it, the star's temperature increases. For both of these reasons, *when a star shrinks its pressure increases.* So in its contracted state, the pressure is more than sufficient to balance gravity, and the star expands.

A normal star behaves like most people. The increase in pressure is moderate. But in a variable star an additional factor is at work, one that makes the increase in pressure excessive. This factor has to do with the manner in which heat flows outwards from the interior of a star. Cepheid variables have the peculiar property of growing *less able to transmit heat outwards in their compressed state.* When compressed, their internal heat is dammed up. Thus, when compressed, the star is even hotter than it would have been in normal circumstances. As a consequence, its pressure is even greater than it would have been in normal circumstances, so driving the perpetual cycle of oscillation.

Cepheid variable stars possess a property that makes them particularly important as distance indicators. We will return to this topic in Chapter 16.

> ⇒ **Looking forward**
> We will study Cepheid stars in Chapter 16.

Detectives on the case

What are planetary nebulae?

Figure 15.5 illustrates a variety of so-called planetary nebulae. (We note that in reality these nebulae have nothing to do with planets. Their names arise from a historical accident. In the early days of astronomy, when telescopes were small, these nebulae appeared as small faint disks: people sometimes confused them with planets. We now know they are quite different.) What are these strange objects?

Let us take as our first idea that *these nebulae are rings*. As an analogy we might think of them as being like donuts. That is certainly what they look like in Figure 15.5. Does this theory make sense?

Figure 15.6 illustrates a collection of "cosmic donuts" scattered through space.

Notice in this figure that the "donuts" are oriented randomly relative to the Earth. Some just happen to be facing us. But others are tilted relative to our line of sight. Indeed, some we are seeing edge-on.

Each of these configurations will lead to a different appearance. A face-on donut will look circular from our vantage point. A tilted configuration will appear to be an elongated oval. And one we see edge-on will appear as a line. So the prediction of this theory is that planetary nebulae should have this distribution of shapes.

This prediction does not accord with observations. This is not what planetary nebulae look like. *None* of the planetaries in Figure 15.5 appear as a highly elongated oval or a line. Rather, all are more or less circular.

What is wrong with our first theory? Is there anything we can do to reconcile it with observations? Figure 15.7 shows one possibility. In it, the "donuts" have been oriented very carefully. Now they are all facing us. As illustrated in the figure, if this

(a)

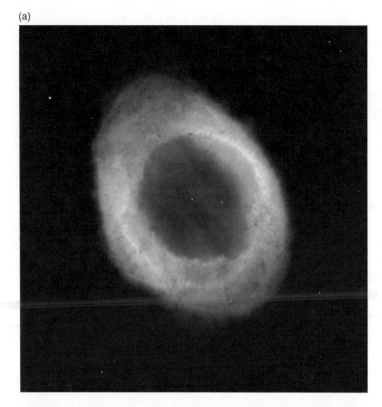

(b)

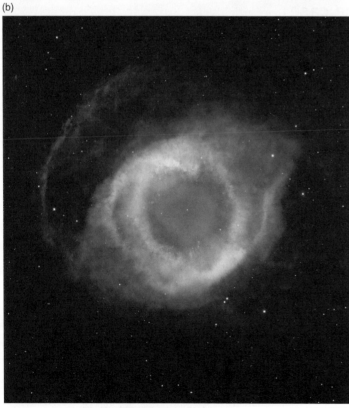

Figure 15.5 **Planetary nebulae. (a) The Ring Nebula. (b) The Helix Nebula. (c) The Eight Burst Nebula.** 👁 **(Also see color plate section.)**

were the configuration every planetary nebula would be circular as seen from our vantage point.

This second theory is certainly possible. But is it likely? The theory posits that we lie in a preferred position relative to the distribution of donuts. They seem to "know where we are" and are turned to face us. Is this likely?

It is not likely. After all, just what agency could have so carefully oriented each one of these "donuts" to face us? There is no such agency. There is no reason why every ring should just happen to be oriented so carefully relative to us. This principle, that we do not occupy a special place in the Universe, is so significant that it has been given a name: the *Copernican Principle*, in honor of Copernicus, who first removed the Earth from its position as the center of the cosmos.

Our conclusion is that our second theory makes no sense. The Copernican Principle tells us that planetary nebulae cannot be donut shaped. What, then, is their shape? If we do not accept the notion that we lie in some special position, we must think in terms of a model in which we are observing the nebulae in all orientations, as opposed to the special orientations posited in Figure 15.7. We must think of *a shape that looks like a donut when seen from every vantage point*.

As a third theory think of a sphere. Any sphere will do: a basketball, perhaps, or an orange. Put the sphere on a table and walk about it. No matter where you stand, the sphere looks the same. So a sphere has one of the properties we need: it presents the same appearance from all vantage points.

· ·

NOW YOU DO IT
Consider some object which is a long thin tube – a straw, perhaps. Does such an object appear to be a long thin tube from every vantage point? From most vantage points? Or only from a particular vantage point?

· ·

But that appearance is wrong. A sphere does not appear to be a ring: it appears to be a disk. How can we modify our third theory?

We need to keep some of the elements of that theory, and add new ones that patch up its defects. In particular, we need to keep the concept of

(c)

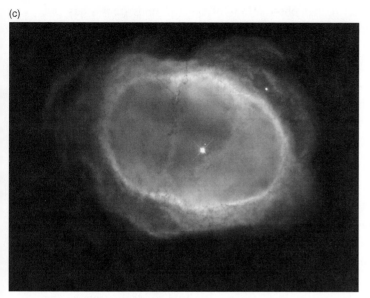

Figure 15.5 **(cont.)**

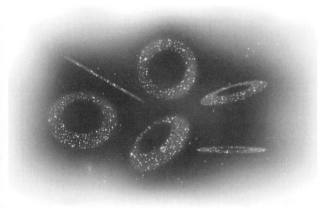

Figure 15.6 **If planetary nebulae were donut shaped we would expect to see them in all orientations.**

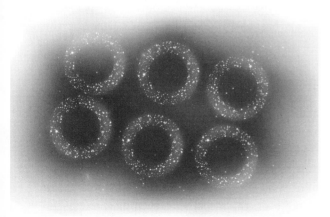

Figure 15.7 **Only if the "donuts" were specially oriented to face us would we observe all of them to be circular.**

sphericity. But now suppose that a planetary nebula is *a partially transparent spherical shell*. The fact that it is spherical means it will present the same appearance no matter what our vantage point. And the fact that it is a shell means that some lines of sight will pass through only a little bit of it, whereas other lines of sight will pass through lots of it (Figure 15.8). If the shell is partially transparent, the inner regions will appear almost empty. The outer regions, on the other hand, will appear to be filled in.

NOW YOU DO IT
Make your own version of this diagram, in which you indicate (A) the line of sight that passes through the most material, and (B) the line of sight that passes through the least. (C) Now consider a line of sight passing through the very center of the shell: does it pass through the most material, the least material or something in between?

This is just what planetary nebulae look like. Our fourth theory seems to be making sense. How can we test it? Notice that according to it our line of sight through these inner regions does not pass through zero material, but rather only a little. So our theory makes a prediction: that the inner regions of the "donut" should not be completely empty. And this prediction is borne out: notice in Figure 15.5 that the inner regions of each "donut" are partially filled.

What could these spherical shells be? We can glean some clues from more careful observations.

- Notice that in Figure 15.5 every planetary nebula seems to have a number of stars within it. But this appearance is illusory. In fact many of the stars in this figure lie in front of the nebula, and many behind it. But there is always one star that actually lies at the very center of the nebula! (We can demonstrate this by measuring the distances to the nebula and the stars: only if they are the same does the star lie within the nebula.) We conclude that *a planetary nebula is a shell with a star lying at its center*.

- We learned in Chapter 4 that observations of spectral lines can be used to determine the composition and state of a material. Observations of these lines from planetary nebulae reveal that the nebulae are composed of gas.

⇐ **Looking backward**
We studied spectral lines in Chapter 4.

- We also learned in Chapter 4 that observations of spectral lines from a gas can reveal its density. Combining this with measurements of the nebula's volume, we can infer the masses of planetary nebulae. *Measured masses of planetary nebulae amount to a significant fraction of the mass of a star.*

- As we also learned in Chapter 4, Doppler shift observations of spectral lines from a gas can reveal how it is moving. Such observations of planetary nebulae reveal that *planetary nebulae are expanding* at velocities of several tens of kilometers per second.

These observations lead us to the idea that a planetary nebula is a shell of gas expanding away from a star. If we calculate the energy required to give this much matter such a high velocity, we find the required energy to be very big. So our theory is that *a planetary nebula is the debris from an explosion that has happened on a star.* It is an expanding cloud, containing a good fraction of the star's mass, resulting from a stellar explosion. Such an explosion is called a "nova."

(This name, from the Latin for "new," is actually a misnomer. If the star had been previously so faint as to have escaped notice, it would suddenly seem to "be born" as it brightened during the explosion. But in reality the star had been there all along.)

Testing the theory

A perfect test of our theory would be actually to witness the nova explosion of a star; and to come back years later and make sure that a planetary nebula had appeared. Unfortunately this is impossible. There are two reasons.

(1) Only a very few stars undergo nova explosions. Furthermore, these explosions happen very rarely. So we would have to look at a great number of stars for huge periods of time before we were lucky enough actually to witness such an explosion.

(2) Once an explosion has occurred, huge periods of time are required for the cloud of debris to expand to become a planetary nebula.

We are forced, therefore, to use indirect means to test our theory. The best means is to construct a detailed computer model of the explosion, and to compare it with observed properties of planetary nebulae. This has been done: the results bear out the theory.

The causes of nova explosions

What could cause such an outburst? After a star has used up all its helium fuel in its core, the helium in a spherical shell surrounding the core continues to undergo nuclear fusion. Just as in a Cepheid variable star, such a configuration can respond excessively to a slight compression. In this case, however, the response is even more excessive: instead of leading to a mild expansion of the star, it can lead to a violent outburst that entirely blows its outer layers into space.

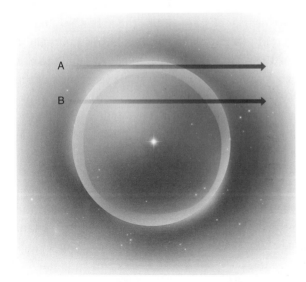

Figure 15.8 **Our final theory of a planetary nebula is that it is a partially transparent spherical shell. It appears to be a "donut" because the lines of sight through the outer regions (A) pass through more material than those through the inner ones (B).**

A

B

Let us remind ourselves of the source of a variable star's excessive behavior. If such a star is compressed slightly:

(1) its volume decreases and temperature increases;
(2) the temperature increase is excessive, owing to a decreased ability to transmit heat outwards;
(3) the resulting pressure increase pushes the star back outwards.

In our current situation the two factors (1) and (3) still operate. But factor (2) does not. In its place is a *new factor (2): the increased temperature leads to an increase in the rate of energy generation by the nuclear reactions.* And since these nuclear reactions themselves liberate heat, we see that the initial slight warming quickly leads to a far larger heating. Detailed calculations show that the nuclear reaction rate depends very sensitively on the temperature: even a slight increase leads to a huge increase in the energy generation rate.

This is an inherently unstable situation. A star in this state is prone to wild bursts of energy generation. Each burst leads to a sudden strong heating of the outer layers of the star, which are violently expelled away into space to form a planetary nebula.

....

NOW YOU DO IT
Suppose the rate of generation of energy by nuclear reactions did not grow larger as the temperature increased. Would such a star be prone to explosions? Explain your answer carefully.

....

Four ways a star can die

We have been surveying what might be termed a star's "middle age." We now turn to what we might term the star's "death": the end of its nuclear-burning evolution. What happens once the star entirely runs out of nuclear fuel?

White dwarf stars

Sirius B
Not far from the constellation of Orion lies the brightest star in the sky, Sirius the Dog Star. In 1862 the American telescope builder Alvin Clark found a faint point of light close beside it. This object is known as Sirius B – or sometimes more affectionately as "the pup." It is very unusual.

Clark noticed two striking facts about Sirius B:

(1) it is white in color,
(2) it is very faint.

Taken in isolation, neither of these is very strange. But in combination they are strange indeed.

Begin with fact (1). As we know from our study of blackbody radiation, the whiter a blackbody the hotter it must be. So observation (1) tells us that Sirius B is very hot. But this is in conflict with fact (2), for we also know that the

⇐ **Looking backward**
We studied blackbody
radiation in Chapter 4.

hotter the blackbody the brighter it should be. The luminosity of a blackbody is given by

$$L = 0.57 \times 10^{-7} A T^4 \text{ watts},$$

where A is its surface area. Since T is big, how can L be small?

The only possibility is that Sirius B is small – for according to this formula, the smaller is the surface area the smaller is L. Let us calculate how small the star must be. Detailed observations have succeeded in measuring the temperature and luminosity of Sirius B:

$$T = 24\,790 \text{ kelvin}$$
$$L = L_{\text{Sun}}/41 = 9.22 \times 10^{24} \text{ watts}.$$

We can use these to calculate its radius. Since L and T are measured, we can find the area and then the radius of Sirius B.

The logic of the calculation

We want to find the size of our blackbody: our formula contains its surface area, which we can use to find this size. The steps are as follows.

Step 1. Solve our formula for the area A. It depends on L and T.

Step 2. Plug in the known values of L and T and find A.

Step 3. We know a formula for the area of a sphere in terms of its radius. Solve this formula for the radius in terms of A.

Step 4. Plug in A and calculate the radius. Is it small?

As you can see from the detailed calculation, Sirius B is very small. Indeed, it is smaller than the Earth!

> **NOW YOU DO IT**
> Suppose we observe a white-dwarf star with twice the temperature and half the luminosity of Sirius B. We want to find out its radius. One way to do this is simply to repeat the above calculation. But there is an easier way. See if you can find it.

Could it be, then, that Sirius B is a planet? We can show that this is not possible. Observations over very long periods of time reveal that Sirius A and B form a binary system. Figure 15.9 shows their orbits. As you can see, they are orbiting about a common center of mass. By measuring the diameters of the two orbits we can measure the two stars' masses.

Measurements reveal that the diameter of the orbit of Sirius A is 42% that of Sirius B. Our formula for center-of-mass motion

$$m/M = R/r = 0.42$$

then tells us that the mass of Sirius B is 42% that of Sirius A. Since the measured mass of Sirius A is

Detailed calculation

Step 1. Solve our formula for the area A.
We obtain $A = L/[0.57 \times 10^{-7}\, T^4]$ meters^2.

Step 2. Plug in the known values of L and T and find A.
We calculate

$$A = 9.22 \times 10^{24}/[(0.57 \times 10^{-7})\,(24\,790)^4] \text{ meters}^2$$
$$A = 4.28 \times 10^{14} \text{ meters}^2.$$

Step 3. We know a formula for the area of a sphere in terms of its radius R. Solve this formula for R in terms of A.
This formula is $A = 4\pi R^2$.

Solving it we find $R = \sqrt{[A/4\pi]}$.

Step 4. Plug in A and calculate R.
We calculate

$$R = \sqrt{[4.28 \times 10^{14}/4\pi]}$$
$$R = 5.84 \times 10^6 \text{ meters}.$$

This is indeed small – it is 92% the radius of the Earth!

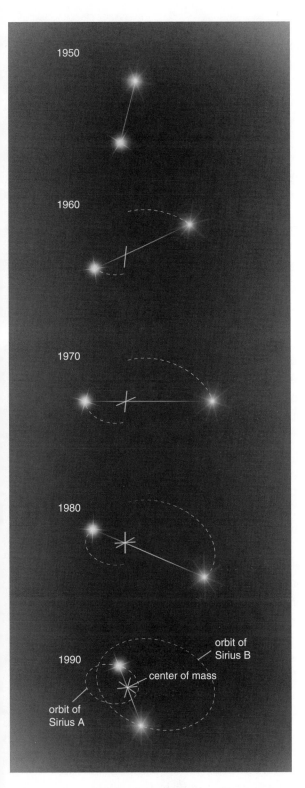

Figure 15.9 **Mutual orbits of Sirius A and B.**

$2.4M_{Sun}$, we conclude that the mass of Sirius B is (0.42) (2.4 M_{Sun}), which works out to be just the mass of our Sun.

Now we understand why Sirius B is so unusual: it is *a star the size of a planet.*

NOW YOU DO IT
Think back and review the train of logic that we used to reach this conclusion. Describe the ways in which it is similar to the train of logic that we used in Chapter 10, in our discussion of planets beyond the Solar System.

Sirius B is a white-dwarf star. Such stars occupy the lower-left portion of the Hertzsprung–Russell diagram. They are extraordinarily compressed. It is difficult to imagine just how compressed they are. Let us try to get an intuitive sense of this by computing the density of Sirius B. We will find that, even though it is a gas (because it is so hot), it is overwhelmingly denser than lead.

The logic of the calculation

Density is mass divided by volume. We have measured both of these quantities, so we know how to calculate the density.

We will want our density in MKS units, and so far we have mass and volume in other units. Furthermore, once we get our answer it will be in MKS units: to get an intuitive feeling for the answer it would be good to convert it to something with which we are more familiar, namely pounds. So the only problem is that we need to convert units.

The detailed calculation shows that a chunk of a white dwarf one meter on a side – and a meter is very similar to a yard – would weigh 5.2 billion pounds if brought back to Earth.

NOW YOU DO IT
Suppose a star had twice the density of Sirius B, and you had a chunk of it half a meter on a side. We want to know how much that chunk of it would weigh on Earth. One way to do this is simply to repeat the above calculation. But there is an easier way. See if you can find it.

The end of stellar evolution: running out of fuel

Where do white-dwarf stars fit into our scheme of stellar evolution? They are instances of what happens to a star when it runs out of fuel. Recall that, once a star has used up its supply of nuclear energy, it must necessarily shrink. This is why white-dwarf stars are so highly compressed.

Detailed calculation

$M_{\text{Sirius B}} = M_{\text{Sun}} = 1.989 \times 10^{30}$ kilograms,

$R_{\text{Sirius B}} = 5.84 \times 10^{6}$ meters.

The formula for density is $D_{\text{Sirius B}} = M_{\text{Sirius B}}/V_{\text{Sirius B}}$.

And $V_{\text{Sirius B}} = (4/3)\,\pi\,(R_{\text{Sirius B}})^{3} = (4/3)\,\pi\,(5.84 \times 10^{6}$ meters$)^{3} = 8.34 \times 10^{20}$ meters^{3}.

So we calculate the density:

$D_{\text{Sirius B}} = (1.989 \times 10^{30}$ kilograms$)/(8.34 \times 10^{20}$ meters$^{3})$
$= 2.38 \times 10^{9}$ kilograms/meter^{3}.

For comparison, the density of lead is 1.1×10^{4} kilograms/meter^{3} – a white dwarf is more than 100 000 times denser than this!

At this density, a single cubic meter would have

mass = (density)(volume)
    = $(2.38 \times 10^{9}$ kilograms/meter$^{3})(1$ meter$^{3})$
    = 2.38×10^{9} kilograms.

Since a kilogram is 2.2 pounds, we see that one cubic meter of a white dwarf would weigh:

weight = (mass in kilograms)(2.2 pounds/kilogram)
    = $(2.38 \times 10^{9}$ kilograms$)(2.2$ pounds/kilogram$)$
    = 5.2 billion pounds.

As we have explained running out of fuel is not the same thing as running out of hydrogen. The evolution of a star is driven by successive stages of nuclear fusion. In the first stage the star converts hydrogen, the lightest element, to helium, the second lightest. This is the main sequence phase of stellar evolution. In the next red-giant phase, helium is converted to carbon. Yet later phases involve the fusion of carbon to yet heavier elements.

How long can this go on? Does the evolution of a star *ever* come to an end? A little thought shows that it must. After all, a star cannot continue shining forever! If it did, it would have emitted an infinite amount of energy, which is impossible.

As we have mentioned the "exhaust" of the fusion of one form of thermonuclear fuel is itself a fuel. But since this cannot go on forever, there must be some end-point to this process – and this means that there must be some form of exhaust that is not itself a fuel. What is this critical form of exhaust?

Let us think more carefully about nuclear energy, and how it is used on Earth. Recall that there, are two forms: fusion and fission. Fusion involves the joining together of two nuclei; fission the splitting apart of one. A hydrogen bomb is an example of the use of fusion: notice that it uses a light element (hydrogen) as its explosive material. Conversely, an atom bomb uses the fission of a heavy element (uranium). Both of these uses are for weapons: in contrast, all of our non-military power generation employs nuclear fission alone.

Notice that there is a pattern here: nuclear energy is released if two *light* elements *fuse together*, or if a *heavy* element *splits apart*. Conversely, processes that operate in the opposite direction do not release energy. Indeed, just the opposite: energy is not released but absorbed to fission a light nucleus, or to fuse two heavy nuclei.

Figure 15.10 illustrates the processes of fission and fusion. It shows a line along which the elements of the periodic table are arranged in order of increasing mass: lightest at the left, heaviest at the right. Fusion moves from left to right; fission from right to left. As we can see, energy is released if we move from the edges of the diagram toward the middle. Conversely, energy is absorbed if we move from the middle toward either edge. Clearly, there must be a critical point on the diagram, somewhere near its middle, from which no energy can be extracted. The element occupying this critical point turns out to be iron. If we begin with iron and move toward the left (fission), energy is absorbed. Similarly, if we move from iron toward the right (fusion), energy is also absorbed. *Iron is not a fuel.*

Figure 15.10 **Energy is released by the fusion of light elements and the fission of heavy elements. Neither fission nor fusion of iron releases energy.**

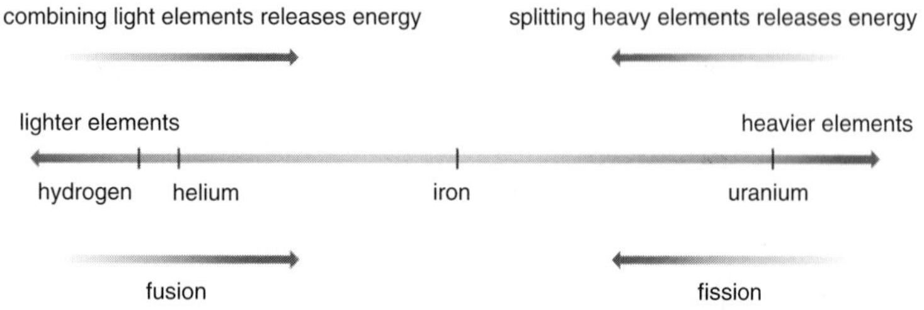

combining light elements releases energy   splitting heavy elements releases energy

lighter elements   heavier elements

hydrogen   helium   iron   uranium

fusion   fission

Iron is the end-point of stellar evolution. Once the various stages of stellar evolution have converted its core to pure iron, a star has no more sources of nuclear energy. From this point on, it must contract. We discussed this phenomenon above, where we saw that it drove the transition from main sequence to red giant. But now there is no new source of thermonuclear energy that can turn on and halt the contraction. The star will continue contracting until it has become a white dwarf.

White dwarf stars are supported by so-called *degeneracy pressure*, a purely quantum-mechanical pressure that arises in certain gases compressed to ultra-high densities. This new form of pressure has the important property of not growing weaker as the gas cools. Since under these circumstances the white dwarf's pressure does not decrease as its temperature drops, the star will not shrink further as it cools.

Billions of years in the future, after our Sun has passed through its red giant and subsequent stages of evolution, it will ultimately use up its supply of nuclear fuel. Then it will commence its final, slow contraction to the white-dwarf stage. As it does so it will grow hotter – because a gas is heated by compression – and fainter – because it is shrinking. On the Hertzsprung–Russell diagram, therefore, it moves downwards and to the left. As for the Earth, as its source of warmth fades the Earth will grow steadily colder, ultimately reaching a temperature of many hundreds of degrees below zero Fahrenheit.

NOW YOU DO IT

There seems to be a paradox here: the Sun grows hotter but the Earth grows colder. We encountered the same apparent paradox earlier in this chapter. Do you remember where this was?

As seen from the Earth, the Sun will have become a mere point, casting a faint light on a frozen world. Thereafter, the Sun will slowly cool, growing yet fainter and redder as the billions of years pass. Ultimately it will so dim as to be entirely invisible as seen from the Earth. The Earth, having endured the searing temperatures of the Sun's red-giant phase, will cool down to nearly absolute zero.

Pulsars and neutron stars

Degeneracy pressure is capable of supporting a white-dwarf star only so long as it is not too massive. Detailed calculations show that white dwarfs can be supported by this new form of pressure only so long as their masses are less than 1.4 times the mass of the Sun, the so-called *Chandrasekhar limit*, named after the astronomer who discovered it. So our own Sun will ultimately become a white dwarf. But what happens to a more-massive star when it runs out of fuel?

Stars more massive than 1.4 solar masses cannot be supported by degeneracy pressure against the inward tug of gravitation. Such stars must keep on contracting to densities even higher than those of white dwarfs.

As they do so, the very structure of matter is transformed. In ordinary circumstances matter is composed of atoms – all matter: matter in our bodies, in the ocean, at the very center of the Earth. This is true in the astronomical context as well: meteors, other planets and dark interstellar clouds are all composed of atoms.

← **Looking backward**
We discussed the structure of an atom in Chapter 4.

But an atom is mostly empty space: as we calculated, if we were to imagine a scale model in which an atomic nucleus were an inch (2.54 centimeters) in diameter, the atom as a whole would be 1 mile (1.6 kilometers) across.

But in the process of contraction, a star of greater than 1.4 solar masses crushes its atoms together until they lose their identity. The matter within them becomes a uniform mixture of electrons and atomic nuclei. Upon yet further contraction, the electrons combine with the protons within the nuclei to make neutrons. The final result is that the star consists almost entirely of neutrons: they are called *neutron stars*. Furthermore, the neutrons within a neutron star are packed so closely together that they actually touch. A neutron star is the only example in the Universe of matter that is *not* mostly empty space.

A brief calculation will show the truly extraordinary density of this matter. Let us calculate the size of a neutron star.

Intuitive mathematics

We can do this by thinking in terms of *density*. If the neutrons within a neutron star are actually touching, the density of the neutron star will equal that of a neutron. We can then use this density to calculate the size the star must have, if it is to have the mass of our neutron star.

Let us carry out this process to find the size of a neutron star of 1.4 solar masses.

The logic of the calculation

The steps of our calculation are as follows.

Step 1. Look up the mass and radius of a neutron.
Step 2. Calculate the neutron's volume.
Step 3. Divide the neutron's mass by its volume to find its density.
Step 4. Imagine a star with this density: find the volume it must have in order to have a mass of 1.4 solar masses.
Step 4. What must be the star's radius for it to have this volume?

As you can see from the detailed calculation, a neutron star is extraordinarily small: a mere few miles in radius. It is also extraordinarily dense: a chunk the size of a sugar cube would weigh 100 million tons.

Detailed calculation

Step 1. Look up the mass and radius of a neutron.
The neutron's mass is $M_{neutron} = 1.67 \times 10^{-27}$ kilograms, its radius is about $R_{neutron} = 10^{-15}$ meters.

Step 2. Calculate the neutron's volume.
The neutron's volume is $V_{neutron} = (4/3) \pi (R_{neutron})^3$

$$V_{neutron} = (4/3) \pi (10^{-15} \text{ meters})^3 = 4.19 \times 10^{-45} \text{ meters}^3.$$

Step 3. Divide the neutron's mass by its volume to find its density.
The density is $D_{neutron} = M_{neutron}/V_{neutron} = 1.67 \times 10^{-27}$ kilograms/4.19×10^{-45} meters^3

$$D_{neutron} = 4 \times 10^{17} \text{ kilograms/meters}^3.$$

Step 4. What volume must a star have to possess a mass of 1.4 solar masses, if it has this density?
We use the density formula in reverse:

$$V_{star} = M_{star}/D_{star}$$

and we use

- for *M*: 1.4 solar masses, or (1.4) $(1.989 \times 10^{30}$ kilograms$) = 2.78 \times 10^{30}$ kilograms;
- the same density for the star that we found for the neutron: 4×10^{17} kilograms/meters^3.

We find

$$V_{star} = 2.78 \times 10^{30} \text{ kilograms}/4 \times 10^{17} \text{ kilograms/meters}^3$$
$$= 6.97 \times 10^{12} \text{ meters}^3.$$

Step 5. What must be the star's radius for it to have this volume?
We use the volume formula in reverse:

$$V_{star} = (4/3) \pi (R_{star})^3.$$

So that

$$R_{star} = \sqrt[3]{V_{star}/(4/3) \pi}$$
$$= \sqrt[3]{6.97 \times 10^{12} \text{meters}^3/(4/3) \pi} = 1.18 \times 10^4 \text{ meters},$$

which is about 7 miles. More detailed calculations, using more realistic calculations of a neutron star's density, get an only slightly different result.

NOW YOU DO IT
Suppose that neutrons actually had twice the above mass, but the same radius. How would all the above calculations be altered?

The discovery of pulsars

In 1967 Jocelyn Bell, a graduate student working with a group of astronomers in England, discovered a new source of radio signals in space. The source was unusual in that, rather than emitting steadily, its emissions came in brief bursts. Following this discovery, other such pulsing sources of radio signals were soon discovered; they came to be known as pulsars.

Three features of pulsar emission captured the attention of astronomers.

- The bursts were exceedingly brief, each lasting a small fraction of a second.
- They recurred once every few seconds.
- They recurred with extreme regularity. In this regard, pulsars acted like very accurate clocks.

Detectives on the case

What are the pulsars?

Throughout the "Detectives on the case" sections of this book, we have invoked the analogy that scientists are like detectives: they use clues to solve mysteries. We have discussed how they invent theories to account for new discoveries, and we watched the process in action by seeing how they go about answering scientific questions. We want to do the same here. Our question is now "What are the pulsars?" and we will follow the actual path scientists took in answering it.

From the very outset, scientists felt that a crucial question was the nature of the pulsar "clock." What was the timing mechanism that accounted for the extreme regularity of their pulses? Three different ideas were proposed.

Figure 15.11 **Orbital model of pulsar emission.**

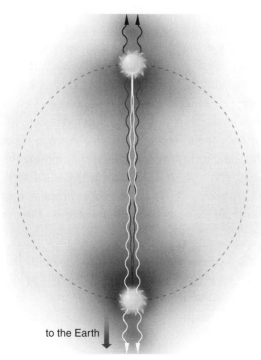

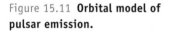
to the Earth

(1) The clock might be provided by the *orbital motion of two stars about each other*. This proposal was that a pulsar might be a binary star system in which radio emission from one star was somehow guided by the other. As illustrated in Figure 15.11, this would produce a beacon that swept about: each time it pointed at the Earth we would observe a pulse.

(2) The clock might be provided by the *vibration of a star*. This proposal was that a pulsar might be something like the Cepheid variable stars we studied above. Somehow a burst of radio signals might be emitted each time the star vibrated. This model is illustrated in Figure 15.12.

(3) The clock might be provided by the *rotation of a star*. This proposal was that a pulsar was a star that for some reason emitted a narrow beam of radio signals. As illustrated in Figure 15.13, this would produce a beacon that swept about: each time it pointed at the Earth we would observe a pulse.

Scientists also realized that, no matter what the timing mechanism, the stars in question could not be ordinary

to the Earth

Figure 15.12 **The vibration model of pulsar emission.**

to the Earth

Figure 15.13 **The rotation model of pulsar emission.**

ones. Only highly compact objects such as white dwarfs or neutron stars were possible. Let us see why.

(1) Begin by thinking about the first proposal. It must be able to account for the fact that pulsars emit bursts once every few seconds. This means that the two stars posited by the model must orbit about one another every few seconds. Compare this to the orbit we are most familiar with: the Earth's orbit about the Sun. *Our* orbit requires a full year to be completed!

How can we go from an orbit lasting a year to one lasting a few seconds? We know from our study of the Solar System that the closer a planet is to the Sun, the shorter the orbital period. There are two reasons:
- the circle they move in is smaller,
- they move faster.

The same is true with binary stars: the closer they are to one another the shorter the period. So if we are to accept this first proposal we must think of a pair of stars so close together that they zoom about each other every few seconds. Is this possible?

The smallest the period could ever be comes when the two stars are nearly touching as they orbit. And in this case, the more compact the stars the shorter the period will be (Figure 15.14). Because they are more compact, the stars can approach one another even more closely – and therefore
- the circle they move in is even smaller,
- they move even faster,

making the period even shorter. We conclude that the denser the star, the shorter the orbital period. Only white dwarfs and neutron stars turn out to be dense enough to account for the extreme rapidity of pulsar emission.

Furthermore, we reach a similar conclusion within the context of the other two proposals.

(2) Just as a swinging pendulum has a natural period of oscillation, so too with a star. A normal star such as the Sun, if set into vibration, would oscillate roughly once an hour. A denser star, on the other hand, would vibrate more rapidly.

(3) The more rapidly an object rotates, the more it bulges out at its equator: if it rotates too rapidly, it flies apart. A normal star such as the Sun cannot rotate more rapidly than once an hour without flying apart. A denser star, on the other hand, can rotate more rapidly.

In every case we reach the same conclusion: *the stars responsible for pulsar emission cannot be of the normal sort. They must be exceedingly small: they can only be white dwarfs or neutron stars.*

Thus, up to a year after the discovery of pulsars, scientists had six different theories as to what they might be.

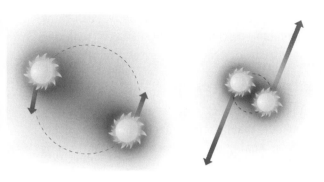

Figure 15.14 **The more compact a star the shorter its orbital period if it orbits nearly skimming the surface of its companion.**

orbiting white dwarfs
orbiting neutron stars
vibrating white dwarfs
vibrating neutron stars
rotating white dwarfs
rotating neutron stars

Among these six theories, only one could be correct. But which one? Two new discoveries answered the question:

(i) A new pulsar was found, the so-called "Crab Pulsar," which pulsed an amazing 30 times a second.

(ii) Pulsars were found to be slowing down. Their rate of pulsation was steadily decreasing.

Let us begin with the first discovery: the ultra-rapid Crab Pulsar. As we have seen, the extreme rapidity of pulsar emissions can only be accounted for by positing a very dense star: the more rapid the pulsation, the greater the required density. White dwarfs are capable of accounting for a pulsation once every second, but they are utterly incapable of dealing with a pulsar emitting 30 bursts a second. Astronomers could therefore rule out three of their six theories.

~~orbiting white dwarfs~~
orbiting neutron stars
~~vibrating white dwarfs~~
vibrating neutron stars
~~rotating white dwarfs~~
rotating neutron stars

Pass on to the second of these discoveries: that pulsars are slowing down. If we think of a pulsar as being a sort of cosmic clock, we can think in terms of an analogy with ordinary clocks. What might cause a clock to slow down? Some sort of friction operating within it. Let us think about friction operating in the context of each of the proposed timekeeping mechanisms.

If we posit a timekeeping mechanism based on *orbital motion*, we might imagine some gas surrounding the orbiting stars, exerting a frictional force upon them as they move. This is what happens to an artificial satellite orbiting the Earth if it gets too close: it enters our upper atmosphere, which exerts a drag upon it. In response, the satellite spirals inwards toward the Earth. But as we have seen, the closer a satellite is to the Earth, the more rapidly it orbits. Thus the net result of friction operating in this model would be to *increase* the rate of pulsation. This is exactly the opposite of what is observed!

Thus, scientists were able to eliminate one of their three remaining theories.

~~orbiting white dwarfs~~
~~orbiting neutron stars~~
~~vibrating white dwarfs~~
vibrating neutron stars
~~rotating white dwarfs~~
rotating neutron stars

Similarly, if we posit a timekeeping mechanism based on *oscillation*, we can think in terms of an analogy with a pendulum clock. We have all seen swinging pendulums, and we have all seen how friction steadily decreases the amplitude of their motion. But if we observe such a pendulum carefully, we will see that the rate of oscillation does not change as this occurs. Even though the amplitude of the swing is decreasing, the pendulum continues to require the same amount of time to execute each swing. We conclude that friction operating in this model would have *no effect* on pulsar periods. Again, this is the opposite of what is observed.

At this point scientists had eliminated one of their two remaining theories.

~~orbiting white dwarfs~~
~~orbiting neutron stars~~
~~vibrating white dwarfs~~
~~vibrating neutron stars~~
~~rotating white dwarfs~~
rotating neutron stars

Only one theory is left. Does it account for the slowing-down of pulsars? It does! If we posit a timekeeping mechanism based on *rotation*, we can think in terms of an analogy with a spinning top, and we have all seen how friction steadily slows a top's rate of spin. So the net result of friction operating in this model is to *decrease* the rate of pulsation – and this is in agreement with what is observed.

Scientists had found a model of pulsar emission: *pulsars are rotating neutron stars*.

~~orbiting white dwarfs~~
~~orbiting neutron stars~~
~~vibrating white dwarfs~~
~~vibrating neutron stars~~
~~rotating white dwarfs~~
rotating neutron stars   ✓

Comments

In our previous discussions of how scientists invent theories to account for new discoveries, we made several general comments about the process. The very same comments apply to our present case of answering the question "What are the pulsars?"

- Creating our theory was a matter of trial and error. There was a lot of groping around involved. We tried six different ideas, and we finally found one that worked.
- Often we found ourselves thinking in terms of analogies. We thought of a pulsar as being like a pendulum clock or a spinning top.
- We tested each theory quantitatively. Only a detailed quantitative analysis was capable of telling us that white dwarfs are not capable of emitting bursts 30 times a second.
- We never finish asking questions. Now we need to ask: *why* does a spinning neutron star generate a beam of radio signals? Even today, decades after the discovery of pulsars, we do not have a detailed answer to this question.

Indirect evidence and confirmation

The evidence that pulsars are neutron stars is highly indirect: all we know so far is that they couldn't be anything else. What assurance do we have that this theory is the right one? After all, so far in our discussion nothing has told us that "Yes, pulsars are indeed rotating neutron stars."

As we discussed in "The nature of science" in Chapter 10, indirect evidence is often all we have in science. But it would be wonderful if we could find direct confirmation of this theory.

Two lines of such confirming evidence have been found.

(1) Pulsars have been found in binary systems, in which they orbit with other stars about their common center of mass. This has allowed us to measure the pulsars' masses. The masses turn out to be comparable to those of normal stars, which is what we expect for neutron stars.

(2) Blackbody radiation has been detected coming from pulsars. Measuring the intensity and spectrum of this radiation allows us to measure the radius of the emitting body. The measured radius turns out to equal that predicted above for neutron stars.

NOW YOU DO IT
Explain how it is that measuring the intensity and spectrum of the blackbody radiation from a pulsar allows us to measure the radius of the emitting body.

Black holes

Just as white-dwarf stars can exist only if their mass is less than 1.4 solar masses, so a neutron star can only exist if its mass is less than a few solar masses. A more-massive star, once it runs out of nuclear fuel, has no way to resist the inward tug of gravity. It will contract – and in this case, there will *never* be enough pressure to halt the collapse. It will not be able to become a white dwarf or a neutron star. Rather, the star will contract so greatly that it turns into a black hole.

Newton's inverse square law of gravitation tells us that the gravitational attraction between objects grows stronger as they come closer together. So as the star contracts and its various parts approach one another, gravitation about the star becomes stronger. Ultimately, the gravitation is so intense that Einstein's theory of general relativity comes into play. In particular, if a star becomes smaller than its *Schwarzschild radius* R_{Sch}, given by

$$R_{Sch} = 2GM/c^2,$$

the star turns into a black hole (Figure 15.15).

We emphasize that the Schwarzschild radius of a star is not its actual radius. The speed limit along a highway does not tell us how fast cars are actually driving: rather it separates legal from illegal driving. In the same way, the Schwarzschild radius of a star is a marker, a certain critical size that separates two cases:

• a star larger than its Schwarzschild radius behaves normally,
• a star smaller than its Schwarzschild radius is a black hole.

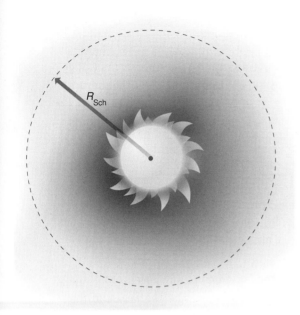

Figure 15.15 **A black hole forms when a star is smaller than its Schwarzschild radius.**

Under normal circumstances black holes do not form. A star must be compressed until it is extraordinarily small in order to turn into a black hole. To see this let us calculate the Schwarzschild radius of a normal star – our Sun – and compare it with the Sun's actual radius. We will see that the Sun is far larger than its Schwarzschild radius, so that it behaves normally.

The logic of the calculation

Step 1. Look up the Sun's mass.

Step 2. Multiply this by $2G/c^2$. The result will be the Sun's Schwarzschild radius in meters.

Step 3. To get an intuitive feel for the result, convert it to kilometers.

Step 4. Compare this with the actual radius of the Sun.

As you can see from the detailed calculation, the Schwarzschild radius of the Sun is nearly 3 kilometers. This is one 236-thousandth the actual radius of the Sun! So the Sun is nowhere near becoming a black hole. It would have to be compressed by 236 000 in order to turn into a hole.

Detailed calculation

Step 1. Look up the Sun's mass.
The Sun's mass is 1.989×10^{30} kilograms.

Step 2. Multiply this by $2G/c^2$. The result will be the Sun's Schwarzschild radius in meters.
We get
$$R_{Sch} = (2)(6.67 \times 10^{-11})(1.989 \times 10^{30})/(3.00 \times 10^8)^2$$
meters
$$= 2950 \text{ meters.}$$

Step 3. To get an intuitive feel for the result, convert it to kilometers.
2950 meters is 2.95 kilometers, i.e. about 3 kilometers.

Step 4. Compare this with the actual radius of the Sun.
We calculate the ratio
$$R_{Sch}/R_{Sun} = 2950 \text{ meters}/6.96 \times 10^8 \text{ meters} = 1/236\,000.$$

NOW YOU DO IT
What is the Schwarzschild radius of the Earth? What is the ratio between this and the Earth's actual radius?

No one can think of a process that could achieve this compression – in the case of the Sun, and of stars like it. But as for a star of more than a few solar masses, it is easy to understand how this could happen. Once such a star runs out of nuclear fuel, its own gravitation will pull it together into such a highly compressed state. So such a star will turn into a black hole.

Once this happens, general relativity predicts a variety of extraordinary properties of black holes.

(1) Light is trapped by a black hole.
(2) Matter is trapped by a black hole.
(3) A singularity forms at the center of a hole, in which matter is crushed to zero size and infinite density.
(4) Time slows down in the vicinity of a hole.

We will study each in turn.

(1) Light is trapped by a black hole

According to Newtonian ideas, light travels in straight lines. According to Einstein's relativity, however, light rays are bent by gravitation. Weak gravitation bends the rays only slightly, but strong gravitation bends the rays a lot. Figure 15.16 illustrates

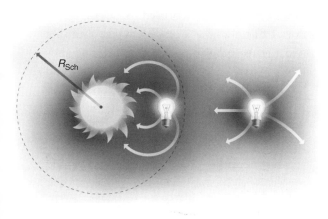

Figure 15.16 **Light rays are bent in the vicinity of a black hole. Rays emitted from outside the black hole are bent slightly. But rays emitted from within the hole are trapped.**

how a black hole traps light. If, say, a light bulb is outside the hole, its rays are attracted somewhat. But if it lies within the hole, *all* its rays are attracted by gravitation: none of them escape.

Suppose you were hovering in space beside a star which was smaller than its Schwarzschild radius. You would not be able to see the star. Even if it were a mere few miles from you, it would be invisible. You would simply see an area of blackness. Furthermore, if you were to take some object – a brick, say – and toss it toward that star, the brick would vanish from your view as it approached the hole's edge. *Everything inside a black hole is invisible.*

(2) Matter is trapped by a black hole

According to Newtonian ideas, gravitation grows stronger as you approach a body. This is also true according to relativity, but with this difference: *gravitation within the Schwarzschild radius is infinitely strong.* Inside a black hole, no force, no matter how great, is sufficient to oppose gravitation. So if, intrigued by the area of invisibility before you, you decided to venture into the hole to see what lay inside, you would be forever trapped within. Indeed, you would helplessly fall onto the star below and be killed.

In the previous section we imagined tossing a brick into a black hole. Suppose rather that you tossed a bomb into the hole, timed to go off in a few seconds. Once that bomb passed inside the hole it would vanish from your view – and the force of its explosion would also be trapped. Indeed, you would have not the slightest evidence that the explosion had ever occurred.

A black hole acts like a "cosmic vacuum cleaner," sucking up all matter that comes too close. Once it has fallen inside, such matter cannot emit energy. But as it plummets inwards, prior to reaching the Schwarzschild radius, the matter is highly compressed and it definitely does emit. In Chapter 17 we will study quasars and radio galaxies, bizarre objects that might be powered by such a process of energy emission.

(3) The singularity and the mystery of the ultimate nature of matter

In our study of stellar structure we explained that every star exists in a perpetual state of balance: gravitation acts to compress it, and gas pressure acts to resist the compression. This is also true of the star whose gravitation produces the remarkable effects we have been discussing. But now there is a difference. Because inside the Schwarzschild radius gravitation is infinite, *no* pressure is capable of withstanding it. So the star that forms the black hole cannot possibly exist in a state of balance. It must necessarily collapse inwards upon itself. No matter how much it has collapsed, it cannot cease contracting. This process goes on until the star has contracted all the way to zero size and infinite density (Figure 15.17). Furthermore, because the force of gravity is so huge, the star does not contract gradually: it catastrophically plummets inwards at the speed of light. The ultimate state it reaches is known as a "singularity."

This presents us with a major crisis in our understanding. In the collapse to the singularity, the star is crushed to greater and greater densities. At first the atoms out

⇐ **Looking backward**
We studied the balance of forces in a star in Chapter 14.

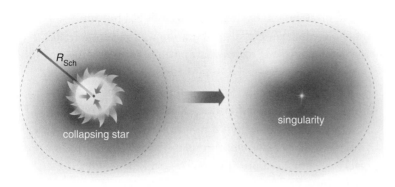

Figure 15.17 **A star smaller than its Schwarzschild radius collapses to zero size and infinite density. This state is known as "a singularity."**

of which it is made are forced to overlap. What does this do to the atoms? It is like crushing two eggs together: their shells crack and they merge. So in the collapse the atoms merge and all that is left is their constituent parts – electrons and nuclei. Now the star is composed of these particles. It collapses further. Within a fraction of a second it shrinks so much that the nuclei within it are forced together and merge. Now the star is composed solely of their constituent parts – neutrons and protons. The star has attained the state of a neutron star. But it does not stay that way! It keeps on collapsing. Very quickly the neutrons and protons are forced to overlap. Physicists believe that these are composed of yet smaller particles known as quarks: if so, the neutrons and protons dissolve into them. Enormous pressure develops within the star – but the pressure is not enormous enough, and even here the collapse has not terminated. The quarks are rammed together.

Are quarks themselves made of yet smaller particles? If so, the star is soon composed of them. Alternatively, if quarks are truly fundamental, and are not made of anything smaller, what happens when they are crushed together? What happens when every single particle of an entire star is forced to occupy exactly the same point? What happens when matter is crushed so greatly that it no longer occupies any space at all?

Nobody knows the answers to these questions. Within a fraction of a second any star smaller than its Schwarzschild radius collapses to an unknown state. And within a fraction of a second any matter that falls onto the star is crushed to this state as well. Our understanding of the nature of matter is utterly insufficient to comprehend what occurs.

(4) Time slows down in the vicinity of a hole

Figure 15.18 illustrates a person holding a clock at the end of a rope, and slowly lowering it toward a black hole. He finds that the closer the clock is to the hole, the more slowly it ticks. Eventually, if he lowers it till it hovers just above the Schwarzschild radius, the clock seems to cease ticking altogether.

You might think this is occurring because the intense force of gravitation is distorting the clock mechanism. Could the inner workings of a clock be affected by gravity? Unfortunately this theory will not work. It is not merely clocks that undergo this strange transformation. If you were to suspend an egg in a pan of boiling water close to a black hole, you would find that it takes half an hour to become soft boiled. If you were to suspend a person close to a hole, you would find her living centuries before reaching old age.

It is not this process or that process that is slowed by gravitation. It is time itself. The closer something is to the edge of a black hole, the more slowly does time there pass.

This influences the behavior of matter falling into a black hole. Suppose a hole encounters a star. The star is

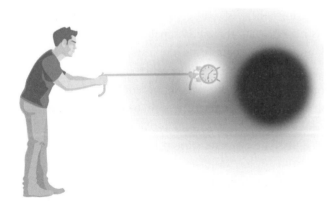

Figure 15.18 **Time slows down close to a black hole. The closer a clock is to the edge of a hole, the more slowly it ticks.**

torn to bits by the hole, and its gases commence plummeting inwards toward it. But as they approach the hole, the gases appear to slow. Indeed, so great is the slowing of time that, as seen from far away, ultimately they would hover just above its edge. Their motion takes place *in time*, and this time is passing more and more slowly as they fall closer and closer to the Schwarzschild radius.

It passes more and more slowly relative to time far away. But if you yourself were to jump into the hole, your time would stretch as you fell inwards. In terms of your time, events would proceed at a normal pace. Very rapidly you would fall through the Schwarzschild radius – a process that takes forever as seen from the outside.

Supernovae

Supernova 1987a

On the night of February 24, 1987, the astronomer Ian Shelton was observing a nearby small galaxy, the Large Magellanic Cloud. He noticed what appeared to be a new star in the galaxy's outskirts (Figure 15.19).

Astronomers have known for centuries of such "new stars" appearing in the skies. We now know that they arise when an explosion takes place on a star. If it had been initially too dim to be noticed, a new star would seem suddenly to appear, so leading to the name for these outbursts: *nova*, from the Latin for "new." We stress, however, that the name is a misnomer: novae mark not the creation of a new star, but an explosion on an already existing one.

Such explosions do not represent a particularly violent event in the history of a star. After its nova explosion a star settles down to its initial state, and can even undergo a number of subsequent outbursts. Once in a very long while, however, a far more violent explosion occurs. These, termed *supernovae*, are so gigantic that they destroy the star entirely.

Shelton's supernova is named Supernova 1987a, the "a" indicating that it was the first to be discovered in that year. Cataclysms such as these are very rare. In all of recorded history, there have been only three visible to the naked eye. The first exploded in 1054 and was noted by Chinese astronomers: it was so brilliant that it was visible (in daylight!) for 23 days, only gradually dimming to obscurity. The remnant of this explosion, the so-called Crab Nebula, contains the Crab Pulsar, which as we have seen was so important to astronomers seeking to determine the nature of pulsars. The second was observed by Brahe in 1572 and the third by Kepler in 1604. A total of four nearby explosions in nearly a thousand years does not add up to very many! And because they are so rare they usually take place far away, so that astronomers are forced to study them in distant galaxies. As a consequence, we know very little about them.

Shelton had found the first nearby supernova since the scientific revolution. He immediately recognized that he had made an extraordinary discovery. Within a few hours, astronomers

Figure 15.19 **Supernova 1987a. The image on the right shows a portion of the Large Magellanic Cloud prior to the outburst: the image on the left during the outburst.**

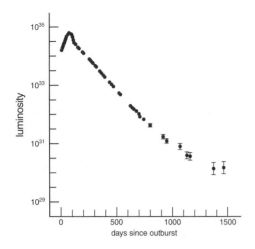

Figure 15.20 **Brightening and fading of Supernova 1987a.**

across the globe had been notified. Every available telescope, on the ground and in space, was pressed into emergency service. Over the succeeding months, as the explosion first brightened and then gradually dimmed, astronomers had a golden opportunity to study an extraordinary cataclysm.

A supernova can be nearly as bright as the entire galaxy in which it is situated. But a galaxy contains billions of stars! Detailed observations have shown that, at maximum brightness, a supernova's luminosity can be up to 10^{10} times that of the Sun. Furthermore, as shown in Figure 15.20, this supernova maintained these extraordinary luminosities for months. If we add up the total amount of light energy emitted by a typical supernova over the duration of its outburst, we get several times 10^{43} joules. It takes our Sun nearly a billion years to emit this much energy.

. . . .

NOW YOU DO IT
Suppose a supernova emits 10 billion solar luminosities for 3 weeks. We want to calculate (1) the total number of joules it emits and (2) how long it would take the Sun to emit this much energy. For each, (A) describe the logic of the calculation you will use, and (B) carry out the detailed calculation.

. . . .

But in fact a supernova emits even more energy than this.

- Great amounts of matter – a good fraction of the mass of the entire star – are exploded away at velocities of several tens of thousands of kilometers per second, or 20 to 40 *million* miles per hour. The energy of motion of so much matter moving at such great speeds can be up to 10^{44} joules.
- Neutrinos, elementary particles emitted from the deep interior of the exploding star, were detected from Supernova 1987a. The total energy carried by these neutrinos was about 10^{46} joules. Other supernovae are thought to emit neutrinos of similar energies.

As we have mentioned, the debris from a supernova expands outwards until it forms a *supernova remnant*. Figure 15.21 illustrates a variety of such remnants. They can be many light years across. Perhaps the most famous such remnant, the Crab Nebula, contains the ultra-rapid Crab Pulsar.

So violent are supernova explosions that they can pose a severe danger to life. If our own Sun were to explode in this way, our entire Earth would be badly damaged, and possibly torn to shreds. Every living being on the planet would be wiped out instantaneously. Even if the explosion were more distant – on a nearby star, say – it would still pose a danger. The burst of gamma rays and high-energy particles from the outburst would cause severe damage to Earth's ozone layer, and severe mutations to all life forms. Furthermore, recall that a supernova remnant can be several light years across. If a star within several light years were to explode, the expanding cloud of debris would eventually engulf our Solar System. Because these remnants contain great numbers of high-energy particles, our Earth would again be exposed

(a)

(b)

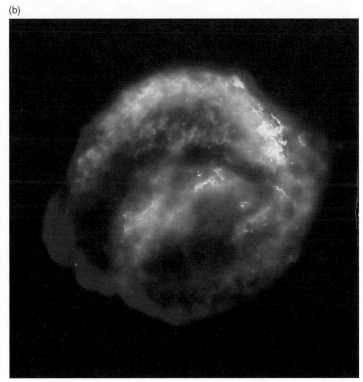

Figure 15.21 **Supernova remnants. (a) The Crab Nebula. (b) Remnant of Kepler's supernova. (c) Remnant of Brahe's supernova.** 👁 **(Also see color plate section.)**

to ionizing radiation. We emphasize, however, that because supernovae are so very rare, there is very little chance of these dangers ever coming to pass.

Supernovae are found to be of two types, based primarily on their spectra. Recall from Chapter 4 that spectroscopy can reveal evidence of which chemical elements are present in a gas. Type I supernovae do not show strong lines indicating the presence of hydrogen, whereas Type II supernovae do. The two types also exhibit differences in their curves of brightness versus time. And finally, Type I supernovae are found to occur in all types of galaxies, and everywhere throughout a galaxy, whereas those of Type II are found only in the arms of spiral galaxies. (We will study galaxies in detail in Chapter 17.)

This last difference is an important clue. In Chapter 13 we explained that high-mass stars do not have time to wander very far from their places of formation (because they only shine for a relatively brief amount of time), and that such stars are only found in or near interstellar clouds. Furthermore, interstellar clouds are only found in the spiral arms of galaxies. We therefore can understand the fact that Type II supernovae occur in spiral arms if we postulate that *Type II supernovae occur in high-mass stars*. Similarly, the fact that Type I supernovae are observed to occur everywhere indicates that they cannot be of high mass: *Type I supernovae occur in low- and intermediate-mass stars*.

What causes supernovae?

How can we understand these enormous explosions? We are very uncertain. Two problems stand in the way of advancing our understanding. On the theoretical side, detailed computer simulations of supernovae have been carried out – but because these explosions are so complex, modern computers do not have enough power to realistically simulate them. Our theoretical models are therefore necessarily incomplete. And on the observational side, because supernovae are so very rare, we have very little observational evidence about them. This is particularly troubling in that science always requires observational evidence. Recall in this regard our discussion in "The nature

(c)

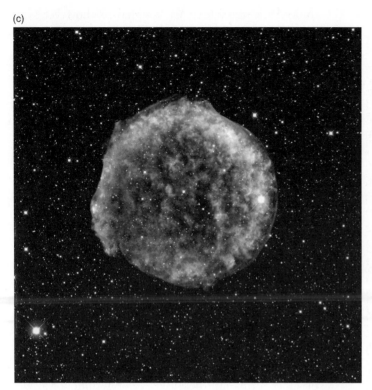

Figure 15.21 **(cont.)**

of science. Theory versus observation" (in Chapter 13). There we emphasized that theory in the absence of observation is barren. Observations always surprise us, and they always help us discard bad theories. Given the lack of extensive observational evidence, astronomers are acutely aware that their theoretical ideas will undoubtedly be modified in the course of future research.

With this proviso, we now discuss two of the many theoretical models that have been proposed to account for supernovae.

A model of a Type I supernova

The steady shining of a normal star is powered by nuclear reactions proceeding *gradually and steadily* within it. One model of supernovae postulates that *the explosion is caused when all the nuclear fuel within a star reacts at once.* So much energy is released when this happens that the star is blown to shreds.

More specifically, it is a white-dwarf star whose fuel is supposed suddenly to react. The event that triggers this cataclysm is the collapse of the star: as it falls inwards, it heats so much that all the fuel within it suddenly and catastrophically reacts.

The scenario that leads to this collapse begins by imagining a white-dwarf star in orbit about a normal star. Such examples of white dwarfs in binary systems are well known: an example is Sirius B, which, as we have seen, is in a mutual orbit with Sirius A. What happens to such a system when the normal star exhausts the hydrogen in its core? As we have seen, it expands to become a red giant. Under normal circumstances, the result would merely be a white dwarf orbiting in a binary system with a red giant, not a particularly unusual system. But what happens if the red giant and white dwarf lie very close to one another?

In this case, there is more to the story. Before the normal star expanded to become a red giant, it had possessed a tidal bulge, raised by gravity from the white dwarf. While we are all familiar with tides on the Earth's oceans, we also noted that they arise in many astronomical situations. In our discussion of the origin of Saturn's rings, for instance, we invoked the theory that these rings arose when a moon wandered too close to the planet and was pulled apart by tides. We will now show that, once the normal star expands to its red-giant stage, it will similarly be pulled apart by tides raised upon it by the white dwarf.

How big is the tide raised on the normal star? There are two forces acting upon it:

(1) its internal gravity pulls it together into a spherical shape;
(2) the tidal force from the white dwarf pulls it apart into a bulged shape.

Consider first a case in which the two stars are far apart. Then force (2), the tidal force, is small, and the normal star will be essentially spherical. But if the two stars

⇐ **Looking backward**
We studied tides in Chapters 7 and 8.

Figure 15.22 **Material falls onto a white dwarf from its companion red giant, due to the tidal force from the white dwarf on the red giant. Because the two stars are in orbit about one another, the material does not fall directly onto the white dwarf, but spirals down onto it from an accretion disk.**

are close together, the tidal force is big and the normal star will have a moderate bulge. That is the situation envisaged in our model of a supernova.

What now happens when the normal star, with its moderate tidal bulge, exhausts the hydrogen in its core and expands to become a red giant? Now force (1) grows *smaller* – since the magnitude of this gravitational compression force obeys Newton's inverse square law, and if the radius of the star grows larger, the gravitational force grows smaller. Furthermore, force (2) grows *larger* – since the edge of the red giant closest to the white dwarf has now gotten even closer. The net result is that, as the star expands to its red-giant stage, its tidal bulge grows even larger. Eventually, the bulge grows so large that it stretches out all the way to the white dwarf!

Once this happens, matter from the outer regions of the red giant starts being pulled by gravity up off its surface and down onto the white dwarf (Figure 15.22). The consequence is that the white dwarf starts steadily gaining in mass. Eventually its mass grows larger than the Chandrasekhar limit of 1.4 solar masses. And when this happens, the white dwarf has become too massive to be supported by degeneracy pressure – and it collapses catastrophically inwards.

This is the first half of the scenario that leads to the supernova explosion. To understand the second half, we ask: what is the fate of the collapsing white dwarf?

As we have emphasized throughout this chapter, stars are composed of thermonuclear fuel: all stars, the collapsing white dwarf in particular. (As we have seen, iron is not a thermonuclear fuel – but even though a white dwarf is an end-point of stellar evolution, it still contains a good deal of nuclear fuel: it is only the central core that is composed of iron.) What happens to the fuel of which it is composed once it begins collapsing? Two principles are at work.

- When a gas is compressed it heats: in a sudden collapse the heating is sudden.
- The hotter a gas, the more rapidly do nuclear reactions within it proceed.

So the moment the white dwarf starts to collapse, the nuclear reactions within it start to generate heat.

Under normal circumstances this increasing temperature would lead to an increase in the white dwarf's pressure, which would oppose the collapse. But as we mentioned above, in a degenerate star the pressure does not depend on the temperature! So as the thermonuclear reactions generate more and more heat, no additional pressure is generated that might oppose the collapse. So the star collapses further and further, and it grows hotter and hotter, and the nuclear reactions within it proceed more and more rapidly. Very quickly, all the material of the collapsing white dwarf undergoes nuclear reactions.

The catastrophic collapse has triggered a sudden burning of *all* the nuclear fuel in the star. This burning liberates so much energy that it blows the star to shreds in a supernova explosion.

Confirmation of the theory

We can easily see how this model accounts for the observed facts of Type I supernovae.

(1) As we saw, Type I supernovae arise from low- to intermediate-mass stars. It is indeed the case that white-dwarf stars are of this mass.

(2) We also understand why Type I supernovae show no hydrogen lines in their spectra. Because white dwarfs are end-points of stellar evolution, they have exhausted all their hydrogen.

These points amount to a partial confirmation of the theory. But they are by no means definitive. As we noted above, more detailed observations are needed.

A model of a Type II supernova

The second model we will explore posits that *a Type II supernova marks the cataclysmic formation of a neutron star or a black hole.*

If a boulder falls off a cliff, pulled downwards by gravity, it releases energy when it lands – the "crash." Similarly, when a star collapses, pulled inwards by gravity, it releases a very great deal of energy. Neutron stars and black holes are formed in such a catastrophic collapse of an ordinary star: the energy released in the collapse is thought to power Type II supernovae.

As we have seen, Type II supernovae arise on high-mass stars. Such stars are too massive to end their lives as white dwarves. What happens to such a star once it has exhausted all its nuclear fuel?

Computer simulations have tracked the evolution of such a star. The result of one such simulation is shown in Table 15.1. It shows the various stages in the burning of the various nuclear fuels available to a star. The first, the fusion of hydrogen to form helium, is the main sequence phase that we studied in the previous chapter. The second, the fusion of helium to form carbon, is the red-giant phase we explored above. Notice that this second stage is marked by higher temperatures and densities than the first, and that it lasts for a shorter amount of time – only about a tenth as long.

This pattern continues throughout all the succeeding stages of nuclear burning, in which yet heavier elements are formed. At each stage, the core of the star has shrunk to yet higher densities and temperatures; and each stage proceeds more rapidly than the previous one. Toward the end the star is evolving at a furious pace: neon burning lasts only seven years! And the final step, the fusion of silicon into iron, takes place in less than a week.

This final step marks the beginning of the end for the star. As we have seen, no nuclear energy can be obtained from iron. Once the core of the star is composed of

Table 15.1. **Evolution of a 15M_{Sun} star.**

Fuel	Central temperature (millions of kelvin)	Central density (kg/m^3)	Duration	
Hydrogen	34	5.9×10^3	12 million years	Main sequence stage
Helium	160	1.3×10^6	1.3 million years	Red-giant stage
Carbon	620	1.7×10^8	6300 years	
Neon	1300	1.6×10^{10}	7 years	Further stages
Oxygen	1900	9.7×10^{10}	1.7 years	
Silicon	3100	2.3×10^{11}	6 days	Iron forms: beginning of the end
Collapse	8300	6.0×10^{12}	0.3 seconds	Neutron star birth

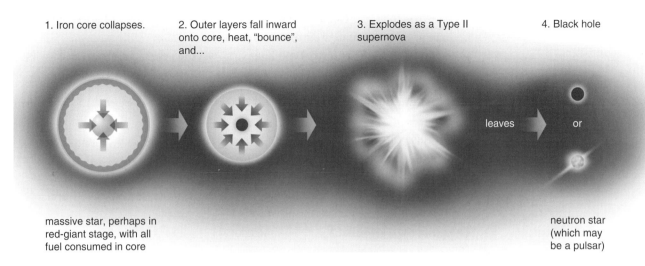

1. Iron core collapses. 2. Outer layers fall inward onto core, heat, "bounce", and... 3. Explodes as a Type II supernova 4. Black hole

leaves   or

massive star, perhaps in red-giant stage, with all fuel consumed in core

neutron star (which may be a pulsar)

Figure 15.23 **Computer model of a Type II supernova.**

this element, it has no further means of generating heat. Detailed computer calculations show that this core is so dense that it is degenerate. Thus it can be supported by degeneracy pressure – but only for a while. Because the star as a whole is so massive, its core has a mass approaching the limiting mass that can be supported by degeneracy pressure, 1.4 solar masses. The more time passes, the denser the star's central regions have become, and the more massive the degenerate core has grown. The star is edging closer and closer to catastrophe.

Eventually the core mass exceeds 1.4 solar masses, and it can no longer be supported by degeneracy pressure. It collapses inwards. The evolution of the star, which began with its leisurely 12-million-year residence on the main sequence, culminates in a collapse lasting no more than a fraction of a second. In this climactic instant, the star's inner regions are rushing inwards upon themselves at speeds approaching that of light.

Figure 15.23 illustrates the results of a detailed simulation of the formation of a neutron star. When these regions reach neutron-star densities, they develop a great pressure that reverses the collapse. Enormous numbers of neutrinos are generated, which propagate outwards through the infalling material. The core of the star "bounces." Infalling matter rains down upon the newly formed neutron star, and rebounds violently. The outwards-moving material explodes to form a supernova.

Confirmation of the theory

How well does this model account for the observed properties of Type II supernovae?

(1) Type II supernovae are observed to be associated with high-mass stars. This model specifically deals with the evolution of such stars.

(2) Type II supernovae are observed to show strong hydrogen lines in their spectra. This is understandable, because the outer regions of a highly evolved star are thought to be composed of hydrogen: it is only the inner regions that have produced heavy elements through nuclear reactions.

(3) The model posits a connection between supernovae and neutron stars. This connection has been verified observationally: several supernova remnants are

known – the Crab Nebula is the most famous – that contain pulsars. And as we have seen, pulsars are neutron stars.

(4) The model predicts that a Type II supernova emits great numbers of neutrinos. These particles were in fact detected when Supernova 1987a exploded.

These amount to a partial confirmation of the theory. But as before, more detailed observations are sorely needed.

As of this writing, the cloud of debris from Supernova 1987a is opaque. However, it will soon have expanded sufficiently to allow us to peer inside. It will be very interesting to see if a newly born neutron star lies at the heart of the debris. Very soon, we will have an opportunity to check the validity of our theoretical ideas about these immense explosions.

Summary

After the main sequence: running out of hydrogen

- Once a star has fused all its hydrogen into helium it has not run out of fuel, for helium is a fuel.
- But this new fuel cannot be used until the star has grown hotter.
  - This is because the nuclei that undergo nuclear reactions repel one another.
  - It takes high temperatures to overcome the repulsion between hydrogen nuclei, and even higher temperatures to overcome that between helium nuclei.
- Once the hydrogen fuel is used up, the inner regions of the star contract and heat till the helium commences reacting.
- As this happens the outer regions expand until the star becomes a red giant.
  - Such a star converts helium into carbon.
  - Its surface is cooler than that of the main sequence star it used to be.
  - It is larger and more luminous than the main sequence star it used to be.
  - When the Sun does this, the Earth will grow too hot to support life.

After the red-giant stage

- Cepheid variable stars.
  - These stars expand and contract in a regular cycle.
  - As they do so, their luminosities vary.
  - This occurs because they are less able to transmit heat outwards when compressed.
  - The heat is therefore dammed up inside and the pressure grows excessively, thus forcing the star to expand excessively.

- Nova explosions.
  - Form planetary nebulae.
  - Arise because within these stars a slight increase in temperature causes a great increase in the rate of nuclear reactions.
  - Such a situation is unstable, and the reactions are liable to "run away," leading to an explosion.

Four ways a star can die

- White dwarf stars.
  - Sirius B is a faint, white star in a binary orbit with Sirius A.
  - It is a white-dwarf star: a star the size of a planet.
  - In spite of being gaseous, it is exceedingly dense.
  - Once a star has used its nuclear fuel to make iron, it has no more nuclear energy available to it.
  - White-dwarf stars are one of the things that can happen to a star when it runs out of nuclear fuel in this way.
  - A white dwarf cannot have more than 1.4 solar masses (the "Chandrasekhar limit").
- Neutron stars and pulsars.
  - A star more massive than the Chandrasekhar limit will contract until it becomes a neutron star.
  - Such a star is far more dense than a white dwarf.
  - A neutron star the mass of the Sun has a radius of a mere 7 miles.
  - It is composed almost entirely of neutrons: atoms have been crushed out of existence.
  - A neutron star cannot have too great a mass: a yet more massive star collapses to become a black hole.
- Black holes.
  - A star's "Schwarzschild radius" is $R_{Sch} = 2Gm/c^2$.

- If the star's actual radius is less than R_{Sch} it becomes a black hole.
- Light is trapped by a black hole.
- Matter is trapped by a black hole.
- A singularity forms at the center of a hole, in which matter is crushed to zero size and infinite density.
- Time slows down in the vicinity of a hole.
- Supernovae.
  - Supernovae are explosions so great as to entirely destroy a star.
  - They are very rare: the supernova of 1987 was the first nearby supernova since the scientific revolution.
  - Their remnants are expanding clouds of gas.
  - Theoretical model of a Type I supernova:
    - they are thought to occur when a white dwarf orbits a red giant,
    - material from the red giant is transferred to the white dwarf,
    - once its mass exceeds the Chandrasekhar limit, the white dwarf collapses,
    - when it does so, all its nuclear fuel reacts at once.
  - Theoretical model of a Type II supernova:
    - succeeding stages in the nuclear evolution of a star take shorter and shorter amounts of time,
    - the final stage is the formation of iron,
    - after this the star has no means of staving off collapse,
    - the collapse to form a neutron star is cataclysmic,
    - the collapse releases enough energy to blow away the outer regions of the star.

DETECTIVES ON THE CASE

What are planetary nebulae?
- So-called planetary nebulae have nothing to do with planets.
- They appear to be donut shaped.
- But they cannot be donut shaped, for if so many would appear elliptical in shape.

- Their true shape is one that looks like a donut when seen from every vantage point: this shape is a hollow shell.
- We advanced the theory that a planetary nebula is a shell of gas exploded away from a star.
- Such an explosion is called a "nova."

DETECTIVES ON THE CASE

What are the pulsars?
- Pulsars are sources of radio emission that emit very brief pulses once every few seconds with great regularity.
- Three ideas were proposed to account for the "pulsar clock":
  - it could involve the orbital motion of two bodies about each other,
  - it could involve the vibration of a star,
  - it could involve the rotation of a star.
- Each of these ideas required a highly compact body: no normal star could produce a pulse once every few seconds by such means.
- The only stars sufficiently compact were white dwarfs or neutron stars.
- The discovery of the Crab Pulsar resolved the mystery.
  - It was found to pulse 30 times per second:
    - white dwarfs, no matter what the "clock mechanism," cannot pulse so rapidly.
  - It was found to be slowing down:
    - "clock mechanisms" based on orbital motion or vibration speed up as they lose energy.
  - The only theory left: pulsars are rotating neutron stars.
- This idea has been confirmed by two independent tests.
  - Pulsars have been found in binary systems. This allows us to measure their masses, which turn out to be what we expect.
  - Blackbody radiation has been detected from pulsars: this allows us to measure their radii, which turn out to be what we expect.

Problems

(1) The carbon nucleus of C^{12} contains six protons. Consider the following fusion reactions:
   (A) reactions between He^4 and C^{12},
   (B) reactions between C^{12} and C^{12},
   (C) reactions between He^4 and He^4.
   (i) Order these reactions in terms of increasing electrical repulsion acting to inhibit them.
   Explain your reasoning. (ii) Now order them in terms of increasing temperature required to initiate them. Again, explain your reasoning.

(2) If compressing a gas heats it, then expanding a gas cools it. See if you can design a refrigerator based on this logic.

(3) Suppose the outer layers of a star shrink, thus compressing them. (A) Will the surface grow hotter or cooler? (B) How will the star's color change?

(4) When the Cepheid variable star Polaris is biggest, is it hottest or coolest? Is it bluest or reddest?

(5) Once the Sun expands to become a red-giant star, the Earth will grow too hot to sustain life, as shown in Figure 15.4. (A) How far in the future will the oceans boil? (B) How far in the future will it be hot enough to melt lead? (The melting temperature of lead is 621 °F.) (C) What will the Earth's temperature be 17 billion years from now?

(6) Suppose a star had the property that it was *more* able to transmit heat outwards if it were to shrink slightly. Do you think such a star would oscillate like a Cepheid variable? If not, what *would* such a star do?

(7) A star has a surface temperature of 10 000 kelvin and a luminosity of $L_{Sun}/40$. We want to find out how big it is. (A) Describe the logic of the calculation you will use. (B) Carry out the detailed calculation.

(8) Suppose a star had a mass half that of the Sun, and a radius of 3×10^6 meters. We want to find (i) its density, and (ii) how much a cubic meter of this stuff would weigh. For each, (A) describe the logic of the calculation you will use, and (B) carry out the detailed calculation.

(9) We want to find the radius of a neutron star of 1 solar mass. (A) Describe the logic of the calculation you will use. (B) Carry out the detailed calculation. (C) Consider a second star with the same mass but a radius three times bigger: would its density be bigger than, smaller than or the same as that of the neutron star?

(10) We want to (i) find the Schwarzschild radius of our entire Milky Way Galaxy (mass 1.57×10^{42} kilograms), (ii) express this in light years to get an intuitive feel for it and (iii) compare it with the Galaxy's actual radius (about 40 000 light years). (A) Describe the logic of the calculation you will use. (B) Carry out the detailed calculation.

(11) Suppose a supernova emits 1 billion solar luminosities for 6 weeks. We want to calculate (i) the total number of joules it emits and (ii) how long it would take the Sun to emit this much energy. One way to do this is to repeat the calculation you were asked to do in our discussion of supernovae. But there is an easier way: see if you can find it.

· ·

WHAT DO YOU THINK?

(1) Project your mind billions of years into the future, when the Sun is expanding into its red-giant phase, and the Earth has grown horribly hot. In particular imagine the epoch during which the Earth's equator is too hot to allow for life, but the poles are pleasantly warm.

Imagine that the north pole is inhabited, and that these inhabitants are not particularly technologically advanced. Perhaps they wonder if the south pole might also be inhabited by some kind of "aliens." Suppose they decide to mount an expedition to the south pole to find out. Write a science fiction story in which you imagine such an expedition.

(2) Suppose you observe a planetary nebula, and you would like to find out how long ago the explosion took place that created it. What observations would you need to conduct, and what calculations would you need to do, in order to accomplish this?

(3) Some people think of the nova explosion of a star as being like the detonation of a hydrogen bomb. Why do you think they make this analogy?

· ·

You must decide

· ·

(A) You are a minister preparing a sermon. You are struck by the prediction that, when the Sun leaves the main sequence, all life on the Earth will come to an end. A passage by the philosopher Bertrand Russell has caught your attention:

… all the labors of the ages, all the devotion, all the inspiration, all the noonday brightness of human genius, are destined to extinction in the vast death of the solar system… How in such an alien and inhuman world, can so powerless a creature as

Man preserve his aspirations untarnished? [from *A Free Man's Worship*, by Bertrand Russell]

You have resolved to address this issue in your sermon. In particular, two problems plague you:
- How can God allow such a terrible thing?
- How does anything that we mere humans do matter?

What will you say in your sermon? *In responding to this question, make sure that you pay attention to our discussion in "Science and religion are separate" in Chapter 2.*

(B) We all know the difference between "matter" and "not matter." Thus, the chair you are sitting on is made of matter, as is the air you breathe. Similarly, happiness and death certainly exist, but they are not *things* in the sense that chairs and air are things. In particular, they are not made of matter.
- Write an essay in which you define matter.

Suppose now you drop some matter into a black hole. It immediately (in terms of its own time) reaches the singularity that forms within the hole, and it is crushed to zero size. But it is still matter!
- In the light of this fact, are you satisfied with your previous definition of matter? If not, how would you revise it?

Part IV

Introducing galaxies and the Universe

So far we have surveyed what might be called our corner of the cosmos: a region of space extending outward several thousand light years. We now extend our vision much farther – to the very edge of the known Universe. In doing so we will find a new and previously unsuspected structure: galaxies. Everything we have so far studied – the Earth, the Sun and all the Solar System; the stars visible to the naked eye and the more distant stars that telescopes reveal; interstellar clouds – all these are part of an enormous structure known as the Milky Way Galaxy. Lying beyond our Galaxy lie other galaxies, billions and billions of them, stretching out to the farthest bounds of the Universe.

It is difficult to comprehend the immensity of the distances we are about to encounter. A ray of light, which can cross the Atlantic Ocean in 0.02 seconds, would require a hundred thousand years to cross our Galaxy. Even the nearby galaxies lie millions of light years from us. Light from a distant galaxy began its journey toward us long before the Earth was formed.

It is also difficult to comprehend the sheer scale of numbers in the astronomical Universe. Our own Milky Way Galaxy contains 200–400 billion stars. Roughly five hundred billion other galaxies exist in the observable Universe. It is worthwhile to multiply these two numbers to find the total number of stars – each perhaps with planets.

> Number of stars in the observable Universe =
> 100 000 000 000 000 000 000 000 to 200 000 000 000 000 000 000 000.

We emphasize that this is merely the number of stars in the *observable* Universe – the cosmos accessible to our telescopes. As we discuss below, the entire Universe may extend outwards infinitely far.

There are three types of galaxies: spirals, ellipticals and irregulars (Figure IV.1). Sometimes a spiral contains a bar: our home Milky Way Galaxy is such a barred spiral.

Galaxies can be the seat of violent activity. Immense explosions have been witnessed in their cores, leading to great outpourings of energy. Quasars, which appear to be giant black holes consuming stars, lie at the hearts of galaxies; it may be that every galaxy once was a quasar. Indeed, a giant black hole has been found lying at the center of our home Galaxy.

Figure IV.1 **Types of galaxies. (a) The spiral galaxy NGC 4414. (b) The elliptical galaxy ESO 325-G004 surrounded by smaller spiral galaxies. (c) The Clouds of Magellan are irregular galaxies. (d) The barred spiral NGC 1300.** 👁 **(Also see color plate section.)**

Figure IV.2 shows one of the most famous astronomical images ever made, the Hubble Ultra Deep Field. To record this spectacular image the Hubble Space Telescope was trained on a tiny ($\frac{1}{10}$th the diameter of the full Moon) region of the sky in the constellation of Fornax for one million seconds, an extraordinary commitment of observing time on the world's most famous telescope.

The Hubble Ultra Deep Field is one of the most distant looks we have ever taken into the depths of space. Nearly every object seen in this image is a galaxy, most so distant that they had never been seen before. They swarm like snowflakes in a blizzard, receding seemingly without limit into the distance. Indeed, no matter how deeply we have probed into the cosmos, we have never found an end to it.

Is there a limit to the Universe? Or does the swarm of galaxies glimpsed in the Hubble Ultra Deep Field extend outward infinitely? No one knows.

Figure IV.2 **The Hubble Ultra Deep Field.**   ◉ **(Also see color plate section.)**

Edwin Hubble's discovery of the expansion of the Universe is one of the great triumphs of twentieth-century science. (Remarkably this expansion had been predicted by Einstein's relativity theory, but before Hubble's discovery: so unnerved was Einstein by what he took to be a failure of his theory that he modified relativity to avoid the "error." Later he retracted the modification.) Galaxies are flying away from each other: the more distant a galaxy the more rapidly does it move. This expansion began in the Big Bang, a state of infinite density and temperature. It may eventually reverse, and far in the future galaxies may commence approaching one another, ultimately to come together in a Big Crunch: more likely the expansion will continue forever. We do not know if the Big Bang marked the creation of the Universe.

Recent discoveries have revealed that everything we can see in the Universe – all the planets, moons, stars, nebulae and galaxies – amounts to a small fraction of the total. Far more plentiful is the so-called "dark matter" – and far more plentiful than *that* is the so-called "dark energy", which is causing the expansion to accelerate. We have only the faintest idea of what they comprise.

No evidence has ever been found for life elsewhere in the cosmos. But this does not mean life does not exist, or has not existed in the past. Searches are currently under way for evidence of life elsewhere in the Solar System, and for signals from extraterrestrial civilizations.

The Milky Way Galaxy

So far in this book we have surveyed what might be called our corner of the Universe: a region of space extending outward several thousand light years. An important result of that survey was that there was no overall pattern to the distribution of stars in space. Stars were found to scatter essentially randomly about us.

In this chapter we extend our survey much farther – out to roughly one hundred thousand light years' distance. In doing so we will reach a dramatically different conclusion. In a survey over such gigantic distances, a pattern emerges. We will find that everything we have so far studied – the Earth, the Sun and all the Solar System; the stars visible to the naked eye and the far more distant stars that telescopes reveal – all these are part of an enormous structure: the Milky Way Galaxy.

The discovery of our Galaxy is one of the triumphs of twentieth-century astronomy. But why was the Galaxy so difficult to discover? After all, we live within it! One answer is that much of the Galaxy is hidden from our view by interstellar clouds. But another reason is simply that it is so big. The Milky Way Galaxy is too big to see. Until recently we had been something like a race of intelligent ants, crawling across the face of a mountain but completely unaware of its existence.

Constituents of the Galaxy

Before recounting the story of this discovery it is worthwhile to pause a moment, and review the state of our knowledge at the time of the discovery of the Milky Way Galaxy. Of what did the astronomical Universe, as understood at the time, consist?

So far in this book we have encountered:

- planets, moons, etc.;
- stars;
- clusters of stars – two types are known: so-called "open clusters" such as the Pleiades (Figure III.2) and "globular clusters" (Figure III.3);
- interstellar clouds (e.g. Figure 13.1);
- remnants of stellar explosions: planetary nebulae (Figure 15.5) and supernova remnants (Figure 15.21).

There were also two more important constituents of the Universe that we have not yet mentioned.

Figure 16.1 **The Milky Way as seen through a telescope.**

- The Milky Way. To the naked eye this appears to be a faint band of light arcing across the sky. A telescope reveals it to be composed of huge numbers of distant stars (Figure 16.1).

- Diffuse structures known at the time as spiral nebulae (Figure 16.2). Through the telescopes of the day these bore some resemblance to interstellar clouds. They appear to glow with a uniform, pearly light and, and at the time we are considering, no individual stars could be discerned within them. Unlike the interstellar clouds they have very regular shapes: they are circular thin disks, something like frisbees. They are shot through with winding spiral arms. We will study them in detail in the next chapter.

NOW YOU DO IT
How do we know that spiral nebulae are shaped as circular, thin disks like frisbees, as opposed to being more elongated in shape such as footballs? (In answering this question, include a discussion of the issues raised in Chapter 15 on the shapes of planetary nebulae.)

At this stage of our analysis it is difficult to know what to make of the fact that individual stars could not be discerned within spiral nebulae. This might mean that they are gaseous. Alternatively, it might mean that they are composed of stars, but are so far away that our telescopes are not powerful enough to resolve the stars individually. We will return to this later in this chapter.

These are the constituents of the astronomical Universe. How are they distributed through space?

The Herschels' "star gauging"

An important first step toward the discovery of the Milky Way Galaxy was taken in 1785 by the English astronomers William and Caroline Herschel. Their method was to count the numbers of stars visible in various directions.

The logic behind such a method is as follows. Suppose that you do not have access to a map, and that for some reason you have managed to forget the name of the state in which you live, and your home town's location within that state. But suppose that you do have data, culled from a recent US Census, that tell you *how many people live in various directions from you*. As illustrated in Figure 16.3, simply by counting people in various directions, you can figure out the shape of the state in which you live, and your location within that state.

(a)

(b)

Figure 16.2 **Spiral nebulae: (a) the Whirlpool Galaxy. (b) NGC 6384. (c) Sombrero Galaxy. (d) NGC 2841.** 👁 **(Also see color plate section.)**

NOW YOU DO IT
Suppose you live in an imaginary state whose shape is rectangular. This state has four counties: A, B, C and D.

Suppose you assume that people are uniformly spread around the state: i.e. there are the same number of people per square mile everywhere. You want to find out where you live. To determine this, you count how many people live in various directions. You find:

750 000 people live to your north
250 000 people live to your south
830 000 people live to your east
170 000 people live to your west.

In what county do you live? Explain your reasoning.

This is not an infallible method. It relies on an assumption: the assumption that the density of people is the same everywhere – i.e. that there are the same number of people per square mile in one region as in another. But this may not be true, and if it isn't the Herschels' method will yield false results.

To see this, let us apply their method to deduce *our location within the state of Massachusetts*. It turns out that people are distributed very non-uniformly across this particular state: Massachusetts is heavily populated in its eastern regions, and very rural toward the west. In Figure 16.4 we show a map of the state of Massachusetts, showing the various counties. Within each county, its population (in millions) is indicated. Suppose that you lived somewhere in Worcester county. As you can see from the map, this county is located at the geographical center of the state – but it is nowhere near the *population* center of the state. Indeed, most Massachusetts residents live to the east of Worcester county. But if we

(c)

(d)

Figure 16.2 **(cont.)**

did not know this, and if we assumed that within Massachusetts there were the same number of people per square mile everywhere, the Herschels' method would have us conclude that we lived in the far western portion of the state.

NOW YOU DO IT

Suppose you live in Worcester county, Massachusetts, and you wish to know your location. Unfortunately you have no access to any maps. You do, however, have data from a recent US Census. You decide to find your location by a variation of the Herschels' method: you count the number of people living in various directions.

Unfortunately this method is going to lead to a false result, because the eastern parts of Massachusetts are far more densely populated than the western parts. But suppose you don't realize this, and you assume that the population density is the same everywhere. By consulting Figure 16.4, find the total number of people who live in various directions from Worcester county. By using your results, where would you conclude you lived were you to use this method and assumed the population density to be the same everywhere?

(a)

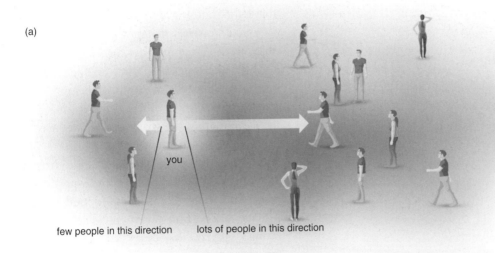

few people in this direction   lots of people in this direction

(b)

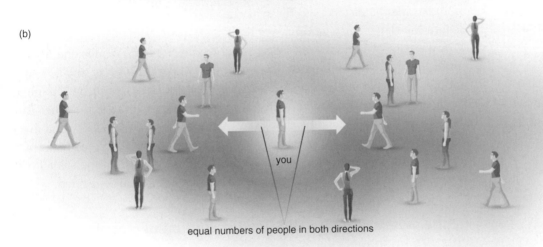

equal numbers of people in both directions

(c)

fewer people in this direction than in this   the same number of people in both directions

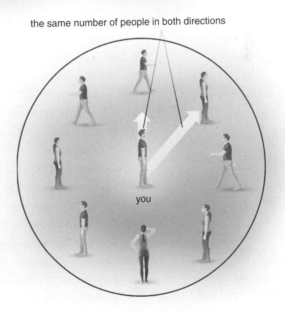

Figure 16.3 **By counting people in various directions you could determine the shape of the state in which you live, and your location within the state. In (a) you are off to one side of the state, so that you count different numbers of people in different directions. In (b) you are at the state's center, and you count equal numbers in all directions. In (c) you can tell whether your state is square or circular in shape.**

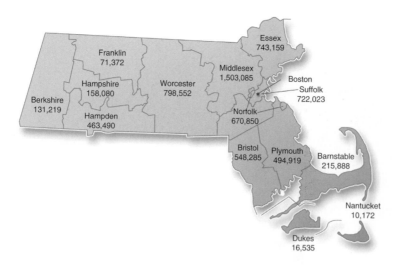

Figure 16.4 **Massachusetts counties. The figures give the populations (2005) in millions.**

The Herschels knew all this perfectly well. But there was nothing they could do about it, for they had no way to test the assumption that stars were distributed uniformly. Recognizing full well the potential pitfalls of their method, they went ahead and counted the number of visible stars in 683 different directions. Figure 16.5 is a map of the structure they deduced.

As you can see, their structure had the shape of a thick wheel, far shorter in one direction than in the perpendicular directions. It came to be known as "the grindstone model" of the Universe. The Sun lay roughly at the grindstone's center. Its cross section was quite irregular. Because the Herschels had no means of measuring the distances to the stars they were counting, they were not able to deduce the size of their grindstone.

In the light of current knowledge, we now know that the grindstone model is entirely wrong. The Sun is not at the center of the Milky Way Galaxy, and its shape is not that illustrated in Figure 16.5. Three errors led the Herschels astray. On the one hand, stars are not distributed uniformly through the Galaxy. On the other hand, their survey did not extend far enough to see the outlines of the Galaxy. And finally, dark interstellar clouds (of which they were entirely unaware) obscured many of the regions in which they counted stars. It is these clouds that are responsible for the irregularities in the grindstone structure they thought they had discovered.

Cepheid variable stars and the distribution of globular clusters in space

The true shape of our home, the Milky Way Galaxy, was not discovered until 1917, when the American astronomer Harlow Shapley mapped *the distribution of globular clusters in space*. He did this by exploiting a new method of measuring distance, one involving Cepheid variable stars.

Globular clusters lie about us in all directions. But there are more in some directions than in others. Applying the Herschels' logic, one might deduce that we are located off to one side of their distribution. Aware of the method's pitfalls, Shapley decided to test this conclusion more carefully. He did this by not just counting the numbers of clusters in various directions, but by actually measuring their distances from us. Once these distances were known, he could assemble a map of their real distribution in space.

Our best means of measuring distance is parallax. Unfortunately the method will work only for relatively nearby stars: our most sensitive measurements are currently capable of measuring parallax only out to about 500 parsecs. But Shapley's clusters were more distant than that. He needed a new method.

The one he used involved the fact that the more distant a source of light, the fainter it appears to be. So by measuring the source's apparent brightness one can deduce its distance. But this will only work if one knows the *intrinsic* brightness of the source. After all, if at nighttime you see a faint glimmer of light, you might decide you are observing a distant searchlight – but it could

⇐ **Looking backward**
We studied parallax in Chapter 12.

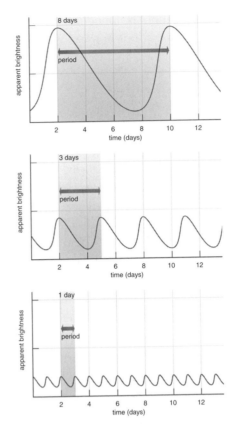

Figure 16.6 **Cepheid variable stars grow brighter and dimmer in a periodic pattern. As illustrated, the brighter the star the longer the period of variation.**

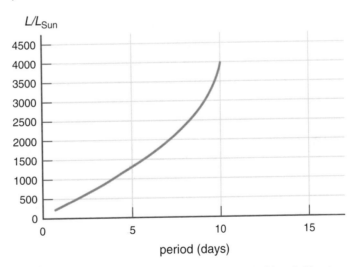

Figure 16.7 **Period–luminosity relation for Cepheid variable stars.**

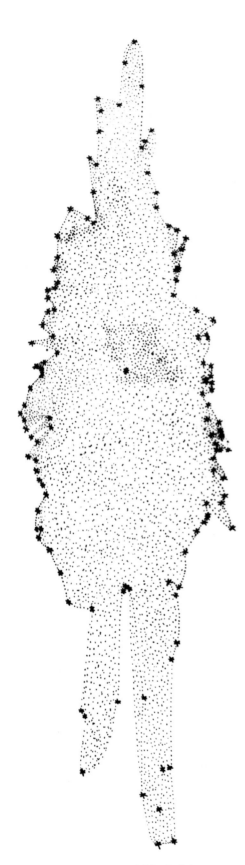

Figure 16.5 **The Herschels' "grindstone model" of the Universe.**

just as well be a nearby candle. Mathematically we say that the inverse square law

$$f = L/4\pi R^2$$

can be solved to yield the distance R in terms of luminosity L and flux f:

$$R = \sqrt{[L/4\pi f]},$$

⇐ **Looking backward**
We studied the inverse
square law in Chapter 4.

⇐ **Looking backward**
We studied Cepheid
variable stars in Chapter 15.

but R cannot be found until we know *both* f and L. It is easy to measure the apparent brightness f. But how can we know the light source's luminosity L?

It turns out that we can do this for Cepheid variable stars. As we know, these are stars, in advanced stages of evolution, that periodically vary in brightness. Luckily, this variation gives us what we might call a "signpost" sticking out of the star upon which is written its luminosity.

In the early years of the twentieth century, the American astronomer Henrietta Leavitt was studying Cepheid variables in certain stellar groups known as the Magellanic Clouds. During this study, she discovered a remarkable property of their variation. This property is illustrated in Figure 16.6, in which we see this waxing and waning of three different Cepheids.

It is immediately clear from this figure that some Cepheids are brighter than others. It is also clear that some vary in brightness more rapidly than others. As illustrated in Figure 16.6, there is a pattern: *the slower the variation, the brighter the Cepheid.*

Shapley realized that this relationship could provide the "signpost" sticking out of the star that he needed. Suppose that we find a Cepheid variable whose period of variation is just a week. Suppose furthermore that we can determine the luminosity of this star. We will then know the luminosity of *every* one-week Cepheid. If we now find one in a distant globular cluster, we will have our "signpost": we will know its luminosity and we can use it to deduce its distance by measuring its apparent brightness. In this way we will have measured the distance to the cluster.

The critical step here is finding some means of measuring the luminosity of that first one-week Cepheid. The only way to do this is to find one so close to us that we can find its distance using parallax: once this is known, we can find its luminosity from its apparent brightness. Shapley undertook a long-term program of searching for such nearby Cepheid variables, measuring their parallaxes, and determining their luminosities. Once he had done this, he was able to assemble the relation between the period and luminosity illustrated in Figure 16.7.

. . . .

NOW YOU DO IT

Suppose you are observing a globular cluster, and you find within it a Cepheid variable star whose period of variation is one week. What is its luminosity? Give your answer both in solar luminosities, and in watts.

  Now suppose that you measure the apparent brightness (the flux) of this Cepheid and find it to be 10^{-8} watts/meter^2. You want to use this to find the distance to the cluster. (A) Describe the logic of the calculation you will perform. (B) Do the detailed calculation: give your answer both in meters and in light years.

. . . .

Shapley's next step was to search for Cepheid variable stars in the globular clusters (and a second category with similar properties, the so-called RR Lyrae stars). Luckily, these are among the most luminous of all stars, so that they could be discerned out to great distances. Nevertheless, the farther globulars were so distant that he was unable to identify any individual stars within them. For these clusters, Shapley was forced to rely on other, less-certain methods of estimating their distances. He was further plagued by the fact, unknown to him initially, that there

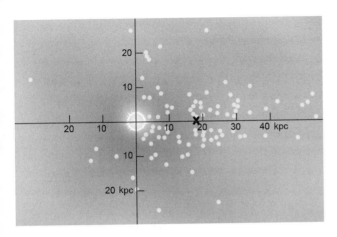

Figure 16.8 **Shapley's map of the distribution of globular clusters in space. (The distance scale is now known to be in error, due to an error in the method Shapley used in measuring the distances to the clusters. The modern value for the distances is about half the indicated values.)**

were in reality two classes of Cepheid variable stars, with different period–luminosity relations. Nevertheless, Shapley was able ultimately to produce a map showing the distribution of globular clusters in space.

This map is shown in Figure 16.8. It shows that the clusters occupy a roughly spherical region of space. We are located near the edge of this sphere. The center of the distribution of globulars, indicated by the cross, lies within the Milky Way, and it is an immense 26 000 light years from the Sun.

Shapley's next step was a great imaginative leap. In making this leap, he paid attention to two suggestive facts.

- The center of the globular cluster distribution lies in the Milky Way.
- Telescopes reveal the Milky Way to be composed of great numbers of stars.

Shapley proposed that we live in a gigantic disk-shaped structure composed of stars. He named this structure the Milky Way Galaxy. He proposed that the Milky Way is the disk seen edge-on. He proposed that the disk is imbedded in a spherical distribution of globular clusters. The center of the distribution of clusters coincides with the center of the disk: it is the center of our Galaxy.

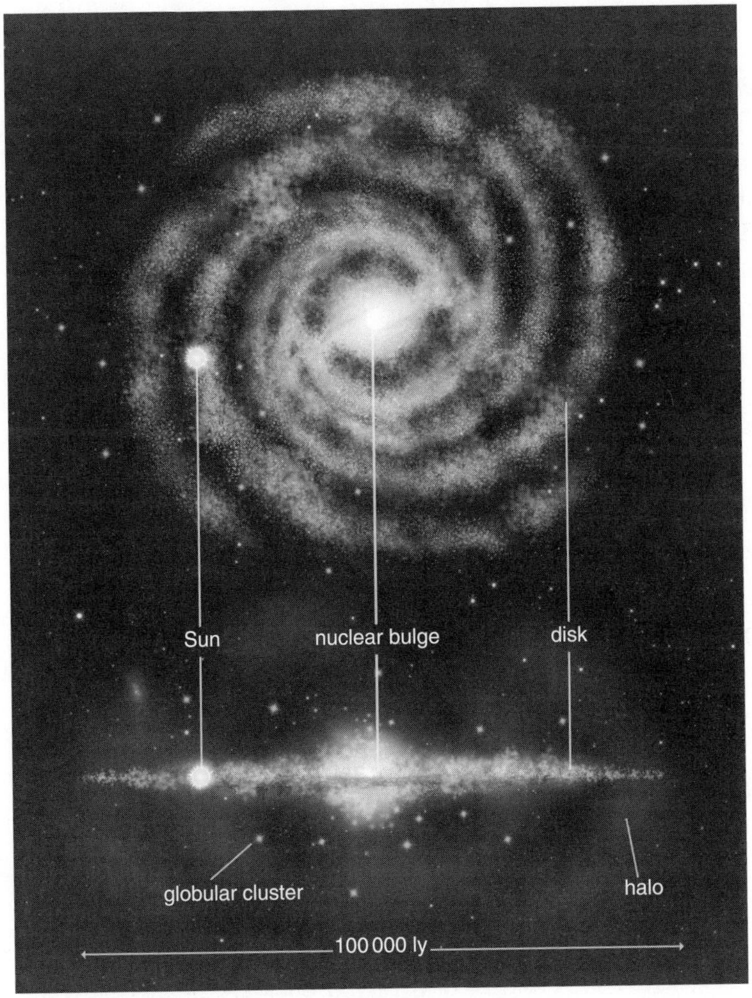

Architecture of the Milky Way Galaxy

A diagram of this immense structure is shown in Figure 16.9. It has two components: the *disk* and the *halo*.

- The disk is where we live: it is composed of stars, and it is shot through with a bar and winding spiral arms. Also to be found in this disk are many of the other constituents we listed above: the open clusters, the interstellar clouds and the remnants of stellar explosions. Every star that we can see with the naked eye lies in the disk.
- The second component is the halo. It is composed of globular clusters – and, as we will shortly see, some individual stars as well. But it contains very few interstellar clouds and no spiral arms.

The Milky Way Galaxy is about 100 000 light years across. A ray of light, which moves so fast that it can travel from Miami to Seattle in 0.02

Sun   nuclear bulge   disk

globular cluster   halo

—————— 100 000 ly ——————

Figure 16.9 **Architecture of the Milky Way Galaxy.**

seconds, requires 26 000 years to reach us from its center. When we observe its distant edge we are seeing light emitted during the age of the Neanderthals. This is our great home, our address in the Universe.

How was it that Shapley discovered this structure while the Herschels had not? The answer lies in the instruments available to them. In the Herschels' day telescopes were not sufficiently powerful to make out Cepheid variable stars. Nor were they sufficiently accurate to measure the parallaxes of nearby Cepheids. And finally, the Herschels lacked instruments with which to measure accurately apparent brightness.

Rotation of the Galaxy

Our orbit through the Galaxy

If we observe the Doppler shift of a globular cluster, we can determine the relative velocity between it and us. When these observations are performed for many clusters, it is found that they all have different relative velocities. Some of these clusters are moving rapidly relative to us, while some are moving slowly. Some are moving toward us, others away from us.

If, however, we look at the results of the measurements as a group, a striking regularity is found. This regularity is illustrated in Figure 16.10. As shown in that figure, in certain directions *there are more redshifts than blueshifts*, while in just the opposite direction *there are more blueshifts than redshifts*. Furthermore, if we survey at 90° to these special directions, equal numbers of red- and blueshifts are obtained. We might say that there seems to be a "special axis" in space. Along this axis the relative velocities exhibit a striking regularity, while perpendicular to it they exhibit no such regularity.

How can we interpret these results? On the one hand, the fact that every cluster has a different Doppler shift than every other tells us that they are moving, each with its own velocity. The systematic regularity, on the other hand, tells us that something else is going on. In addition to these individual motions, there is a general drift.

There are two possible interpretations of this drift.

- We are moving (from right to left in Figure 16.10).
- It is the globular clusters that are moving (from left to right in Figure 16.10). They are acting like a swarm of insects blown sideways by the wind: within the swarm each insect is flying about randomly – but the swarm as a whole is drifting sideways.

How to decide between these two interpretations? A clue is provided by the fact that *the direction to the galactic center turns out to be at right angles to the direction of drift* (Figure 16.11).

It is difficult to imagine a reason why the entire population of globular clusters should be moving crosswise to this direction. But it is easy to imagine a reason why we should be: because we are moving in a circle about the galactic center! More precisely, *we are orbiting through the Galaxy* (Figure 16.12).

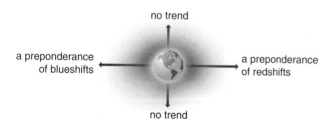

no trend

a preponderance of blueshifts ⟵ ⟶ a preponderance of redshifts

no trend

Figure 16.10 **Doppler shifts of globular clusters show a systematic trend.**

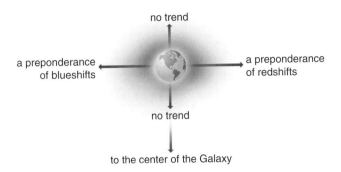

Figure 16.11 Direction to the center of the Galaxy is perpendicular to the axis of drift.

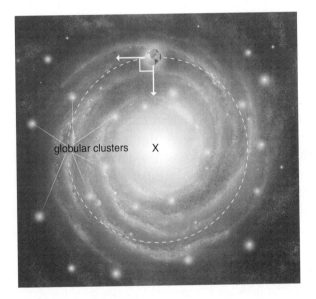

Figure 16.12 We are orbiting about the center of the distribution of globular clusters.

NOW YOU DO IT
In both of the interpretations we are considering, when we look perpendicular to the "special axis" we should see a sideways velocity of the globular clusters. Why don't Doppler observations reveal this motion?

As the Sun orbits, it carries along with it the Earth and all the other planets of the Solar System. The Sun's velocity about the galactic center is measured to be 220 kilometers per second – an enormous velocity, about half a million miles per hour. Nevertheless, so vast are the dimensions of this orbit that we require more than 200 million years to complete one swing about the Galaxy. In the entire history of the Sun, some 4.5 billion years, we have only completed 20 orbits: we can say that the Sun is 20 "galactic years" old.

NOW YOU DO IT
(A) Describe the logic of the calculations that led us to these conclusions. (B) Carry out detailed calculations to verify them.

As for the globular clusters, they too are orbiting about the galaxy – but not in circular paths. They are in highly eccentric orbits, essentially toward and away from the galactic center. It is their velocities in these in-and-out orbits that account for the various velocities relative to us of each individual cluster as measured by the Doppler effect.

Orbits in the Galaxy

The picture of the Galaxy we have drawn so far is that of a rotating disk embedded in a non-rotating sphere containing the globular clusters. Think of a spinning frisbee inside a stationary basketball. A little thought, though, shows that this is not a good analogy. After all, the disk of the Galaxy is not solid. It does not really *rotate*. Rather, every star within it *orbits*.

A better analogy to motions within the galactic disk might be to the orbits of planets in the Solar System. Let us compare these two analogies – frisbee and Solar System – to understand the difference between them.

Consider first a rotating frisbee. Each point on that frisbee moves in a circle. Notice that each of these points takes the same amount of time as all the others to swing about in its circle. But the outer points move in bigger arcs! So they must be moving faster. We see that rotational velocities within the frisbee *increase* with distance from its center.

NOW YOU DO IT

Consider two points on a rotating frisbee: one on its outermost edge, and the other halfway in. Suppose the frisbee is 6 inches in radius (Figure 16.13).

Suppose the outer point on the frisbee rotates once per second. (A) How long does it take the inner point to rotate about once? (B) We want to calculate the velocity of each point. Describe the logic of the calculation you will perform to find this. (C) Carry out the detailed calculation.

Notice that you are finding that the outer point moves faster than the inner one!

⇐ **Looking backward**
We analyzed orbits in Chapter 3.

This motion is illustrated in Figure 16.14.

The situation within the Solar System is just the opposite: velocities within the Solar System *decrease* with distance from its center. As we have seen, the inner planets orbit rapidly while the outer ones orbit slowly. These motions are illustrated in Figure 16.15.

NOW YOU DO IT

We want to calculate the velocities of planets in their orbits about the Sun, in order to verify the trend illustrated in Figure 16.15. Appendix III gives the data you will need. (A) Describe the logic of the calculation you will perform. (B) Select any two planets and carry out the detailed calculation. Notice that you are finding that the outer planet moves more slowly than the inner one!

NOW YOU DO IT

Use the Web to look up the orbital velocity of several Earth-orbiting artificial satellites. Do they obey the rule that the outer body moves more slowly than the inner one?

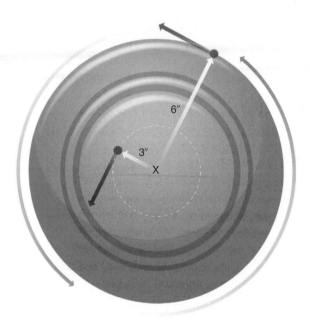

Figure 16.13 **A frisbee.**

velocity

distance from center

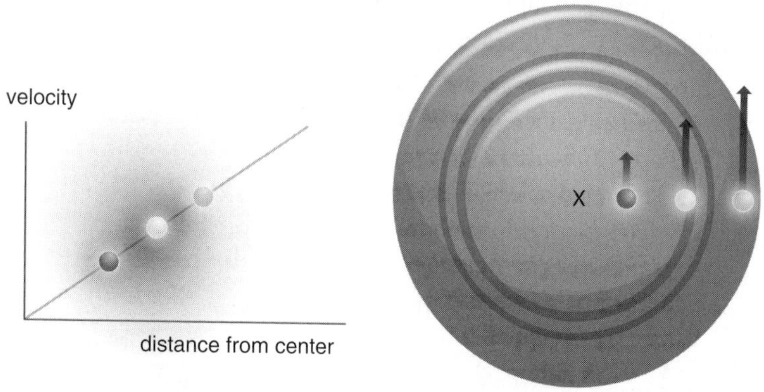

Figure 16.14 **The rotation of a frisbee. The outer regions move faster than the inner regions.**

Which of these analogies – frisbee or Solar System – more closely matches the actual situation? Figure 16.16 shows the observed orbital velocities within the Galaxy. Remarkably, motions in the Galaxy bear no resemblance to *either* of them!

It is not so surprising that orbits in the Galaxy differ from the rotation of a solid body. After all, the Galaxy is not solid. But why do they differ from orbits in the Solar System? What is the difference between the Galaxy and the Solar System?

The difference between the two cases has to do with the nature of the gravitational forces within them – and it is these forces that determine the paths of orbiting bodies. Figure 16.17

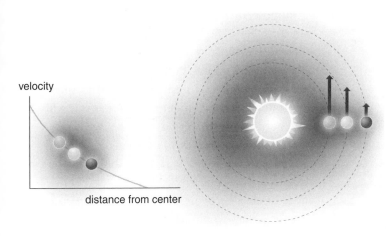

velocity

distance from center

Figure 16.15 **The "rotation" of the Solar System bears little resemblance to the rotation of a frisbee. The outer portions orbit more slowly than the inner ones. (The velocities are drawn to scale, but the distances within the Solar System have been altered for clarity.)**

illustrates the gravitational forces acting on (a) a planet in the Solar System, and (b) a star in the Galaxy. As we can see from (a), the planet is acted on by *only one force* – that from the Sun. But in (b) the star is acted on by *many forces* – those from all the other stars in the Galaxy. The gravitational forces within the Galaxy are more complex than those within the Solar System. So it is not surprising that the orbits are so different: they take place in radically different environments.

Spiral arms

The spiral arms winding through the disks of galaxies are among their most striking features. What causes them? The first thing to note is that *they cannot be permanent structures*. This is because of the peculiar nature of the orbits of stars within the Galaxy.

In Figure 16.18 we illustrate a galaxy and one of its spiral arms. Initially (Figure 16.18a), the arm is loosely wrapped. What will happen to the arm as time passes? Were it to be a permanent structure composed of stars, we could answer this question by thinking about how these stars move.

In Figure 16.18 we have identified two stars for consideration, the outer one just twice as far out as the inner. Recall from Figure 16.16 that all stars within our Galaxy orbit with the same velocity. Since the outer star travels in a circle whose circumference is twice that of the inner, it takes twice as long as the inner to complete one revolution.

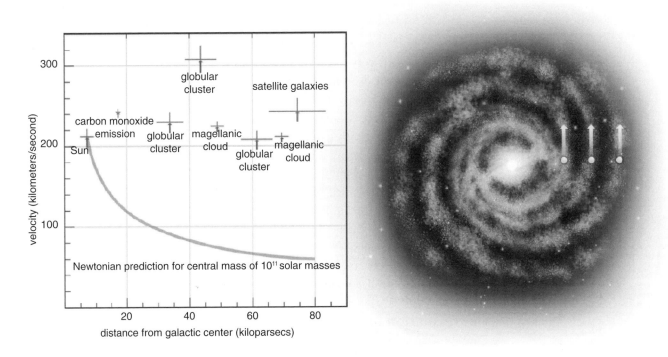

Figure 16.16 **The observed "rotation" of our Galaxy differs from both that of a frisbee and that of the Solar System.**

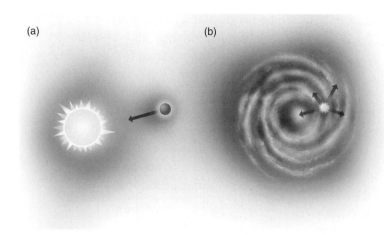

Figure 16.17 **Gravitational forces are simpler in the Solar System (a) than in the Galaxy (b).**

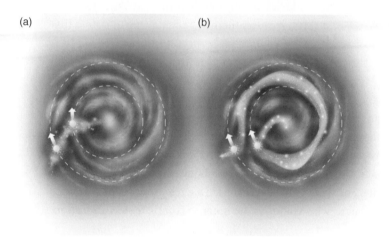

Figure 16.18 **Spiral arms would be wrapped by galactic rotation were they permanent structures composed of stars.**

Suppose, then, we wait until the outer star has had enough time to orbit *once*. In this time interval, the inner star will have orbited *twice*. In Figure 16.18b we illustrate what the galaxy will look like at this point in time. The spiral arm leading from the outer to the inner star will have been wrapped around once.

The more time passes, the more tightly wound does the arm become. Each time the outer star completes an orbit, the arm wraps once more about itself. Were the spiral arms permanent structures, they would have wrapped many times about themselves by now. But the arms we actually observe are nowhere near so tightly wrapped. We conclude that they cannot be thought of as structures at all.

Density waves

What, then, are the spiral arms? Modern theory regards them as *density waves propagating through the galactic disk.* Because they are waves in the pattern of stars, rather than structures composed of stars, they escape being wound into tight spirals by the galactic rotation.

In many ways these density waves are similar to sound waves. Like sound waves, they produce a compression in the medium through which they travel. A sound wave produces a compression in air, whereas a density wave produces a compression in the distribution of stars. Also like sound waves, density waves travel with a certain velocity. They turn out to travel more slowly than the orbiting stars. The wave rotates about the Galaxy, maintaining shape as it does so. Individual stars move about the Galaxy as well – but more rapidly than the wave. As they catch up to it, they are bunched more closely together; once passing through, they return to their original distribution.

A density wave is in many ways analogous to a traffic jam. Imagine a road and a slowly moving truck. Behind the truck is a dense jam of cars, which travels down the road at the speed of the truck, maintaining its pattern as it does so. But the jam does not always consist of the same cars! Rather, each individual car travels more rapidly than the truck, and so catches up to it from behind and becomes caught in the jam. But it does not stay caught in the jam forever. Rather, each car moves slowly forward, eventually passing the truck and returning to its original velocity (Figure 16.19a). Stars behave in a similar manner in the density wave theory (Figure 16.19b).

Star formation in spiral arms

We can get an important insight into the nature of spiral arms by studying the masses of stars found within them. These masses turn out to exhibit a striking regularity. Low- and intermediate-mass stars are spread uniformly through the entire galactic disk. We find them both within spiral arms, and outside the arms. But *high-mass stars are concentrated within spiral arms.*

In our discussion of star formation we encountered a similar situation. There we noted that high-mass stars are always found in or near interstellar clouds. We interpreted this regularity as showing that these clouds are where stars form.

> **⇐ Looking backward**
> We studied star formation in Chapter 13.

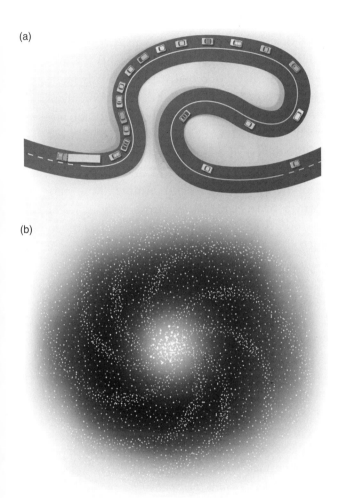

(a)

(b)

Figure 16.19 **Density waves (b) are analogous to traffic jams (a). Cars travel more rapidly than the truck until they reach the jam, then move at the same speed. Once the cars overtake the truck they return to their original velocity.**

> **⇐ Looking backward**
> We studied spectroscopy in Chapter 4, and the composition of the Sun in Chapter 11.

NOW YOU DO IT
What was the logic we used in our discussion that led us to this conclusion?

We can use the same logic here, and conclude that *star formation takes place only in spiral arms.*

In Chapter 13 we noted that the sudden compression of an interstellar cloud can trigger star formation. We now see that the passage of a density wave provides this compression. Just as the density wave squeezes the distribution of stars, so too does it compress interstellar clouds.

The short lifetime of high-mass stars is intimately related to the very fact that we can see the spiral arms. These stars survive so briefly because they are consuming nuclear fuel at an enormous rate. Consequently, they are very bright (recall the mass–luminosity law of Chapter 12). It is this high luminosity that enables us to see the spiral arms. The high-mass stars turn out to be so brilliant as to outshine all the others: when we see a spiral arm, we are primarily seeing them.

High- and low-velocity stars

Just as Doppler shift observations of globular clusters can reveal their motion relative to us, so too with the stars. We can measure the velocity of each.

When this is done it is found that, like the globular clusters, each star is found to have its own velocity relative to us. The average velocity of these stars is relatively small: about 10 kilometers per second. Thus their name: these are called the *low-velocity stars.* There is no systematic trend to the directions of their velocities: roughly half are moving toward us, and half are moving away from us.

However, once in a while a star is found that does not obey this general rule. These so-called *high-velocity stars* travel far more rapidly: not 10 kilometers per second but up to 150 kilometers per second. Furthermore, there is a systematic

(a)

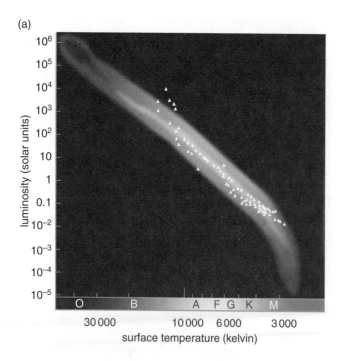

(b)

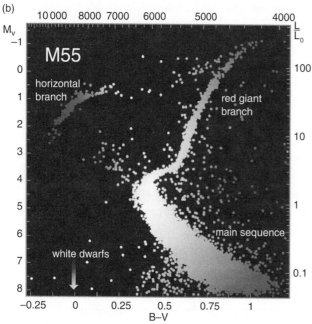

Figure 16.20 **H–R diagrams of the Pleiades (a) and the globular cluster M55 (b).**

trend to their velocities – indeed, the same trend as found for the globular clusters (Figure 16.10).

The high- and low-velocity stars form two distinct groups, distinguished by their orbits. And as we will now see, the two groups differ in another way as well.

Stellar populations

Recall that stellar spectra provide us with a means of determining which chemical elements are present in stars. Recall too that stars are mostly made of hydrogen and helium. All the other elements of the periodic table add up to no more than one percent or so of a star. Nevertheless, an analysis of the abundances of these heavier elements reveals a striking regularity. Some stars turn out to have far less of them than others. Furthermore, those stars in which these elements are particularly deficient turn out to differ systematically from stars in which they are more prevalent.

The chemical elements with which we are concerned are all those heavier than helium: we will lump them together and refer to them as "the heavy elements." By analyzing the Sun's spectrum, we can deduce that these heavy elements make up less than 2% of its mass. On the other hand, certain other stars possess far fewer heavy elements. We call the Sun a *population I star*, and those with far fewer heavy elements *population II stars*.

The Sun is not the only population I star. If we survey the Galaxy, we find many others. Indeed, *most low-velocity stars turn out to be population I. The high-velocity stars, on the other hand, turn out to be population II.*

This regularity persists if we study not individual stars, but clusters of stars. We will concentrate on two: the Pleiades, and the globular cluster M55. The Pleiades are an open cluster that resides in the galactic disk. The globular cluster, on the other hand, resides in the halo. Spectral analysis reveals that the stars making up the Pleiades all have relatively high heavy-element abundances, so that they are population I. In contrast, our globular cluster stars possess relatively few heavy elements, and so are population II.

But a new twist emerges if we compile Hertzsprung–Russell diagrams of these clusters (Figure 16.20). A glance at these diagrams shows that the Pleiades contains only main sequence stars, whereas the globular cluster also contains red giants and

Table 16.1. **Systematics of stellar populations.**

	Population I	Population II
Heavy-element abundance	Relatively high	Relatively low
Velocity	Low	High
Examples	The Sun, open clusters such as the Pleiades	Halo stars, globular clusters
Age	Young	Old

⇐ **Looking backward**
We studied the
Hertzsprung–Russell
diagram in Chapters 12, 14
and 15.

horizontal branch stars. As we know, the main sequence is the first stage of stellar evolution, whereas the giant and horizontal branch stages come later. So our globular cluster must be older than the Pleiades. This turns out to be true in general: *population II stars are older than population I.*

We can summarize all this in Table 16.1.

Detectives on the case

High- and low-velocity stars, and stellar populations

We have stumbled upon a striking regularity. How can we understand it? Why are low-velocity stars population I, and high-velocity stars population II?

Stellar motions in the Galaxy

Sometimes it is helpful to think in terms of analogies. Think of an analogy in which stars are like cars. We are in one of these cars – it is our own Sun together with its planets. Should we think of our car as moving? Or is it holding still?

Until recently we would have said that the Sun is holding still. But with the discovery of our Milky Way Galaxy we have realized that this is not so. In reality the Sun is orbiting through the Galaxy. Indeed, we have seen that it is moving quite rapidly: at 220 kilometers per second.

So when we speak of "low-velocity cars" we mean cars whose velocity is low relative to us. Since we are moving, they must be too. They are not really "low velocity" at all – actually they are moving very fast. But they are moving just like us: all at more or less the same speed, and more or less in the same direction.

Now add a new element to our analogy. Let us imagine that up ahead is a cross street. Cars are driving down it: they are going just as fast as us. What is their velocity relative to us? Because these cars are moving crossways to us, this velocity will be quite large.

So our analogy has given us a new way of thinking about these two classes of stars. It has helped us realize that *all the stars are moving, but they are moving in different directions.* What we term the low-velocity stars are moving along the same path as us. What we term the high-velocity stars are moving crossways to our path.

These paths are *orbits* – orbits of stars through the Galaxy. In Figure 16.21 we indicate these orbits. This figure shows our own Sun, accompanied by its planets, orbiting the Galaxy. It also shows a low-velocity star: its orbit differs very slightly from ours, giving rise to the slow motion of this star relative to our own.

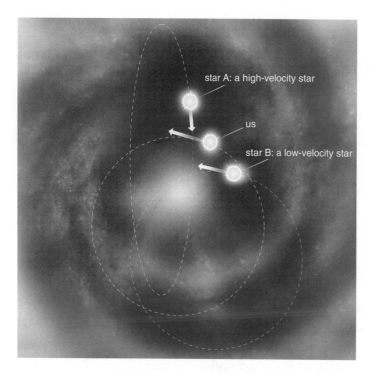

Figure 16.21 **High- and low-velocity stars. Because star B is moving in nearly the same direction as us, and with nearly the same velocity, its speed relative to us is low. Star A, on the other hand, is moving crossways relative to our velocity, so it is moving quite rapidly relative to us.**

The high-velocity stars, in contrast, are traveling in very different orbits. Unlike stars such as the Sun, whose motion is primarily circular, these stars are primarily moving toward and away from the galactic center. Furthermore, while the orbits of the low-velocity stars all lie in the plane of the Galaxy's disk, those of the high-velocity stars entirely fill the Galaxy's spherical halo. The high-velocity stars are visitors from faraway portions of the Galaxy, passing by us as they plunge through the Galaxy's disk in their orbits.

We began this section with a question: why is it that low-velocity stars are population I, and high-velocity stars are population II? We have not answered that question, but we have been able to re-phrase it. Now the question is: *why are stars in the halo population II, while stars in the disk are population I?* We can now replace Table 16.1 with Table 16.2.

Nuclear evolution of the Galaxy

Let us move on to the problem of stellar populations. Think of the Pleiades (population I) and the globular cluster M55 (population II). Why are they different? Where do the heavy elements in the Pleiades come from?

⇐ Looking backward
We studied the origin of the elements in Chapters 14 and 15.

Think back over what we know about the creation of elements in stars. A star like our Sun takes hydrogen and turns it into helium. Can we postulate that the heavy elements in the Pleiades came from this process? And could it be that the absence of heavy elements in the globular cluster indicates that it has not yet gotten started in this process?

A little thought shows that this cannot be so. Figure 16.21 shows us that the Pleiades have a main sequence, and the globular cluster has red giants. But a Hertzsprung–Russell diagram tells us the evolutionary state of a star: the main sequence comes first, and the red giants come second. This means that the Pleiades must be young and the globular cluster must be old. The Pleiades are just starting out

Table 16.2. **Systematics of stellar populations, revised.**

	Population I	Population II
Heavy-element abundance	Relatively high	Relatively low
Velocity	Low	High
Examples	The Sun, open clusters such as the Pleiades	Halo stars, globular clusters
Location	Disk of the Galaxy	Halo of the Galaxy
Age	Young	Old

on their evolution, and the globular cluster has been evolving for a long time. This is just the opposite of what we have posited! Our theory of nuclear energy in stars predicts that the more time passes, the more heavy elements accumulate within a star – but we are finding that it is the *younger* stars that have more heavy elements!

If the Pleiades are young we can imagine going back in time to a moment before they were formed. At that time the globular cluster was young: it had only a main sequence. But unlike the main sequence of the Pleiades, this one had no heavy elements. Now let time pass, until the Pleiades have appeared. What could have happened that gave heavy elements to these new main sequence stars?

Could it have been the nuclear evolution of the stars within the Pleiades? Did they create the heavy elements we see within them? No! The Pleiades are main sequence stars, which means they are turning hydrogen into helium, and they are not making anything beyond helium in the periodic table. But when we speak of heavy elements we are talking about precisely these "trans-helium" elements.

So if the heavy elements in the Pleiades don't come from the Pleiades' own nuclear reactions, they must come from somewhere else. Where else is there to come from?

Could they be raining down onto the stars from interstellar space? Let's take a look at our own Sun: it is a main sequence star, just like those in the Pleiades. Is interstellar matter flowing into the Sun? As we know from Chapter 11 the situation is just the opposite: the Sun is *expelling* matter, via the solar wind.

⇐ **Looking backward**
We studied the solar wind in Chapter 11.

The only other possibility is that the heavy elements in the Pleiades came from the interstellar cloud from which the Pleiades formed. We are postulating that they were present in that cloud before the Pleiades existed. Let us think of our mystery in a new way. Let us think of it not as a mystery, but as a clue. Our clue is telling us that *interstellar clouds are growing more and more filled with heavy elements as time passes*.

Why would this be true?

We know that the shining of the stars is accompanied by the creation of heavy elements. So as time passes, there are more and more heavy elements in the Universe. But these heavy elements reside in the depths of stars! How do they get out of the stars, and into the interstellar clouds?

⇐ **Looking backward**
We studied novae and supernovae in Chapter 15.

Go back to our Sun and its solar wind. That wind is one candidate: it is transporting matter from the Sun out into interstellar space. And other stars, stars that do create heavy elements by nuclear reactions, have winds as well. Furthermore, when we discussed stellar evolution we encountered the phenomena of novae and supernovae. Previously we had thought of these as explosions, but now let us think of them in a new light: they are further types of transport mechanisms, whereby material from the depths of a star is returned to the interstellar medium.

Evolution of the Galaxy

So as time passes, interstellar space becomes seeded with heavy elements. We are thinking of the heavy elements as a kind of gradually accumulating "pollution" within an interstellar cloud. That is why stars formed later in galactic history (such as those in the Pleiades) have more heavy elements than stars formed early (such as those in our globular cluster). Furthermore, we can think of the abundances of the heavy elements as a kind of clock: the more heavy elements a star contains, the more recently it must have been formed.

Let us return to our question: why are stars in the halo population II, while stars in the disk are population I (Table 16.2)? Nothing in the theory we have developed so far tells us that star formation must begin in a spherical pattern, only later moving on to take place in a disk. Once again, we can take this to be a clue. This surprising regularity must be telling us something about the evolution of the Galaxy as a whole. It must be telling us that at first the Galaxy was spherical, and that it later became disk-shaped.

What could have happened to turn the sphere into a disk? Recall that the Galaxy's halo is larger than its disk. If the halo turned into the disk, it must have done so by shrinking. And had the halo been slowly rotating, this shrinkage would have made it rotate faster – just as a figure skater spins faster when she pulls in her arms.

The final step is to recognize that rotation can alter the shape of a body. The more rapidly an object spins, the more it bulges at its equator. As you can see from Figure 8.12, a rapidly rotating planet like Saturn bulges visibly at its equator. Similarly, in our analysis of the origin of the Solar System, we saw that a slowly rotating interstellar cloud would become disk-shaped as it shrank to form the protoplanetary disk (Figure 13.15). The galactic disk can be thought of as a far larger version of this phenomenon.

Let us now put these insights together, and develop a scenario depicting the entire evolution of the Galaxy.

Early in its history, the Galaxy must have been a slowly rotating, spherical cloud of gas. In this cloud, an initial generation of stars formed. They fell inward toward the cloud's center, zoomed past it and swung out again in elongated, highly elliptical orbits to form a spherical distribution. We can think of the Galaxy's halo as a kind of fossil, in which this initial shape is preserved. Also within this cloud, globular clusters formed: these too moved in highly elliptical orbits about the cloud's center. The fact that these early stars are so deficient in heavy elements tells us that the gas from which they formed must have been composed solely of hydrogen and helium. These are the population II, high-velocity stars.

Slowly, the cloud shrank. As it did so, its rotation rate gradually increased, and it commenced bulging at its equator. The more it shrank the more bulged did it become, until eventually it had become a rapidly rotating disk. While this was going on, new stars were continually forming. These partook of the disk's rapid rotation, and so moved in circular, rather than elliptical, orbits. Also while this was going on, stars were busy creating heavy elements and returning them to the interstellar medium via stellar winds and explosions such as novae and supernovae. So the newer stars, formed within the disk, were enriched with these elements. These are the population I, low-velocity stars. Our Sun is one of them.

One final point. As we will see in the next chapter, galaxies can encounter one another and merge. There is evidence that some fraction of the Galaxy actually consists of stars that originally resided in another galaxy, which have been amalgamated into our own. The insight we have gained here is only an initial one, and it is clear that the actual situation is a good deal richer than this.

> **⇐ Looking backward**
> We studied the formation of our Solar System in Chapter 13.

> **⇒ Looking forward**
> We will study galactic mergers in Chapter 17.

NOW YOU DO IT

Suppose the Galaxy had initially been slowly rotating and spherical, but that all its stars formed at the same time – in particular, a very long time ago, before the Galaxy had shrunk. Suppose further that since this initial burst of star formation, no more stars had formed. Describe the stellar populations we would have found, and present a new version of Table 16.1.

THE NATURE OF SCIENCE

THE NATURE OF SCIENTIFIC THEORIES

The summary in Table 16.2 of the observed properties of stellar populations is a strange list indeed. It appears to be no more than a confused jumble of unrelated facts. Learning these facts might seem just as difficult, and just as meaningless, as memorizing a list of random numbers.

But we have already explained[1] that a scientific theory brings coherence to such lists of facts. A theory gives us something that we can understand, something that sets the facts of the list in a new light. The situation is a little like that in a murder mystery. At the beginning of the mystery the various clues add up to nothing more than a series of disconnected facts – but once the detective has solved the mystery, the clues all make sense, and they have come together to form a comprehensible story. In the same way, a scientific theory is a story we tell ourselves that gives observed facts meaning. This is one of the most important functions of a theory: it replaces *facts* with *understanding*. In this case, the understanding is profound indeed: an insight into the multi-billion year history of our Milky Way Galaxy.

Dark matter

At the beginning of this chapter we listed the constituents of the astronomical Universe: planets, stars, nebulae and so forth. We will now show that this list is woefully incomplete, and that there is something else in the Universe, something that emits no light, and that has entirely eluded detection till recently. Astronomers term this mysterious stuff "dark matter." Remarkably, there is far more of this dark matter than everything else: if we were to add up all the planets, stars, nebulae and so forth in the Universe, we would get a total far less than the quantity of dark matter. Most unsettling of all is the fact that, in spite of years of intense effort, we still have no idea whatever what the dark matter might be.

We can demonstrate the existence of dark matter in our own Milky Way Galaxy by measuring

(1) the mass of all the visible matter in the Galaxy,
(2) the total mass of the Galaxy.

We will find the total mass to be far greater than the visible mass.

(1) The mass of all the visible matter in the Galaxy

As we have seen, in our own Solar System the mass of the Sun is far greater than the total mass of everything else: if we were to add up all the planets, asteroids, comets and so forth we would get a total far less than the mass of the Sun. This is universally true: stars constitute almost all of the visible mass of the Galaxy. So in order to find the mass of all the visible matter in the Galaxy, we need to find the mass of all the stars in the Galaxy.

---

[1] In "The nature of science: The understanding that science brings," in Chapter 11, "Our Sun."

Chapter 16

Detailed calculation

Step 1. Measure the luminosity of the Galaxy.
The measured value of the Galaxy's luminosity is

$$L_{galaxy} = 3.8 \times 10^{37} \text{ watts.}$$

Step 2. Divide L_{galaxy} by the luminosity of the Sun.
The Sun's luminosity is $L_{Sun} = 3.8 \times 10^{26}$ watts, so we find that the number of stars in the Galaxy is roughly

$$N_{stars} = L_{galaxy}/L_{Sun} = (3.8 \times 10^{37} \text{ watts})/(3.8 \times 10^{26} \text{ watts})$$
$$= 10^{11}.$$

Step 3. Find the mass of all these stars by multiplying their number by the mass of a typical star.

$$M_{stars} = N_{stars} M_{Sun}.$$

The mass of the Sun is $M_{Sun} = 1.989 \times 10^{30}$ kilograms, but since our value for the number of stars is only an estimate, there is no need to use such a precise value for its mass. Let us simply say that

$$M_{Sun} = 2 \times 10^{30} \text{ kilograms.}$$

We then find our final estimate for the mass of all the stars in the Galaxy:

$$M_{stars} = N_{stars} M_{Sun} = (10^{11}) (2 \times 10^{30} \text{ kilograms})$$
$$= 2 \times 10^{41} \text{ kilograms.}$$

The most straightforward way to find this would be to count all the stars in the Galaxy, and then multiply their number by the mass of a typical star. But this method is not practicable, since our telescopes are simply not capable of resolving each and every star. We can, however, find the number of stars in another way: by measuring how much starlight the Galaxy emits, and calculating how many stars it would take to produce this much light.

The logic of the calculation

Step 1. Measure the total amount of light emitted by the Galaxy – i.e. the Galaxy's luminosity.

Step 2. Divide the Galaxy's luminosity by that of a typical star. This will give us an estimate of the number of stars in the Galaxy.

Step 3. Multiply the number of stars by the mass of a typical star.

(Since our own Sun is typical, we can use its luminosity and mass in the calculation.)

As you can see from the detailed calculation, this method shows that there about 10^{11} stars in the Milky Way Galaxy, and that their combined mass is about $M_{stars} = 2 \times 10^{41}$ kilograms.

. . . .

NOW YOU DO IT

(A) Suppose the measured value of the Galaxy's luminosity were twice as great. How would our estimate for the mass of all the stars in the Galaxy be changed? (B) Suppose that the luminosity of a typical star were not that of our Sun, but rather twice as great. How would our estimate for the mass of all the stars in the Galaxy be changed?

. . . .

(2)   The total mass of the Galaxy

← **Looking backward**
We studied Newton's laws in Chapter 3.

This can be found from Newton's laws. Recall that Newton's mass formula tells us the mass of a body in terms of the motion of something orbiting about it. If the orbiting object moves with velocity V in a circle of radius r, then the mass is given by

$$M = rV^2/G.$$

So by measuring the velocity and orbital radius of a star moving through the Galaxy, we can find the Galaxy's mass. And since the gravitational force of the Galaxy arises from *everything* within it, not just the stars, the mass we get will be the total mass.

We must, however, deal with the fact that a star orbiting in the Galaxy is moving through a continuous distribution of matter. As we have seen (see Figure 16.17),

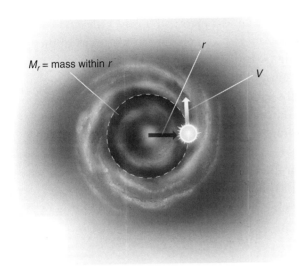

M_r = mass within r

Figure 16.22 The mass within a circle of radius r in a continuous distribution of matter can be found from the usual orbit formula applied to a star orbiting around the perimeter of that circle.

this complicates the analysis. In particular, we cannot use the usual orbit formula. Rather, in such a situation the correct formula is

$$M_r = rV^2/G,$$

where M_r means *the total mass lying within the orbit of the star* (Figure 16.22).

If we apply this formula to an orbiting star lying somewhere deep within the Galaxy, we will succeed in measuring only the total mass lying within the star's obit – i.e. within the Galaxy's inner portions. To find the total mass of the entire Galaxy, we need to study the orbits of its very outermost members. As shown in Figure 16.16, objects have been found orbiting a full 70 kiloparsecs from the galactic center – far out in the halo, beyond the outer edge of the disk. Their orbital velocity is about 220 kilometers per second. If we use these values in our formula, we will be able to find the total mass of the entire Galaxy.

The logic of the calculation

Step 1. Convert 220 kilometers per second and 70 kiloparsecs to MKS units.
Step 2. Use the orbit formula.

As you can see from the detailed calculation, the total mass of the entire Galaxy is 1.57×10^{42} kilograms.

NOW YOU DO IT
(A) Suppose the orbital velocity of stars in the Galaxy were bigger. Would our estimate for the total mass of the Galaxy be bigger or smaller? (B) Suppose the orbital velocity were just twice as great. By how much would our estimate for the total mass of the Galaxy be changed?

Detailed calculation

Step 1. Convert 220 kilometers per second and 70 kiloparsecs into MKS units.

r = 70 kiloparsecs = (70 000 pc)(3.086×10^{16} meters/pc)
= 2.16×10^{21} meters.

V = 220 kilometers/second = (220 kilometers/second) (1000 meters/kilometer)

= 2.2×10^5 meters/second.

Step 2. Use the orbit formula.

$M_r = rV^2/G,$

where $G = 6.67 \times 10^{-11}$ meters³/second² kilogram.

We calculate

$M_r = M_{\text{total}} = (2.16 \times 10^{21}$ meters) $(2.2 \times 10^5$ meters/second)²$/(6.67 \times 10^{-11}$ meters³/second² kilogram)

= 1.57×10^{42} kilograms.

Dark matter in the Galaxy

Let us look closely at our two results:

$$M_{\text{stars}} = 2 \times 10^{41} \text{ kilograms},$$

$$M_{\text{total}} = 1.57 \times 10^{42} \text{ kilograms}.$$

The total mass is far larger than the mass in stars! Indeed, their ratio mass in stars/total mass = (2×10^{41} kilograms)/(1.57×10^{42} kilograms) $\approx$ 1/8 shows that visible matter in the Galaxy accounts for only about ⅛th of its mass. The remaining ⅞th of its mass is dark: *there is seven times as much dark matter as visible matter in our Galaxy.* Truly this is an astonishing result. All the planets, meteors and comets, all the stars, gas clouds and nebulae – all this adds up to a mere ⅛th of the Galaxy.

What do we conclude from this discrepancy? It can only mean that most of the mass of the Galaxy is contained in objects that emit much less light than the Sun. Stars make up only part of the story: most of the matter in the Galaxy is dark.

Furthermore, ours is not the only galaxy that has been found to contain enormous quantities of dark matter. The masses and luminosities of other galaxies have been measured, and similar discrepancies discovered. Depending on the galaxy, something like ten times more dark matter than visible matter exists. Nor is this all. We can also measure the quantity of matter filling the spaces *between* galaxies (by observing them as they orbit about each other). Here too, unseen matter is found. Most shocking of all is its quantity: hundreds of times more of it than visible matter.

But what could this dark matter be?

What is the dark matter?

As we noted above, in spite of years of intense effort we still have no idea what the dark matter might be. Indeed, the mystery of the dark matter is one of the central unsolved problems facing astronomers today. So our attempts to answer the question will not lead to a clear-cut answer. Nevertheless, let us see what we can do.

One thing we know about the dark matter is that it is dark – it does not emit any light. Let us begin our investigations by listing a few of the astronomical objects we have encountered in this book that do not emit their own light:

(A) planets,
(B) interstellar dust,
(C) black holes.

Could the dark matter consist of any of these? In answering this question, we will use the fact that there is seven times as much dark matter as visible matter in the Galaxy.

(A)   Planets

Perhaps this is a tenable hypothesis. After all, we know that many stars have planets. Could it be that every star is accompanied by great numbers of planets, most of which have not yet been discovered?

How many planets would there have to be in order for them to be the dark matter? Let us answer this question in the context of our own Solar System: this should give us a good idea of the general situation. How many planets do we have to postulate orbiting the Sun for the Solar System's total mass to be eight times that of the Sun?

The logic of the calculation

Step 1. If the total mass of the Solar System is eight times the Sun's, then the mass of the proposed planets must add up to seven times that of the Sun.

Step 2. Find their number by dividing their total mass by the mass of a single planet.

Detailed calculation

Step 1. Find the total mass of the proposed planets.

This mass is seven times that of the Sun. Since the Sun's mass is 1.989×10^{30} kilograms, the mass of all the planets taken together is

$$M_{planets} = M_{Sun} \times 7 = (1.989 \times 10^{30} \text{ kilograms})(7)$$
$$= 1.4 \times 10^{31} \text{ kilograms.}$$

Step 2. Find the number of planets by dividing their total mass by that of a single planet.

We have to decide what sort of planet we are thinking of. Let us choose the Earth as a representative example. Since the mass of the Earth is $M_{Earth} = 5.974 \times 10^{24}$ kilograms, we find the number of planets $N_{planets}$ to be $N_{planets} = M_{planets}/M_{Earth} = 1.4 \times 10^{31}$ kilograms$/5.974 \times 10^{24}$ kilograms $= 2.3$ million.

As you can see from the detailed calculation, this proposal forces us to hypothesize that our own Sun – and every other star in the sky as well – possesses a full 2.3 *million* planets orbiting about it. This is so far-fetched as to be impossible.

. . . .

NOW YOU DO IT

Suppose you chose a less-massive planet, such as Mercury, as a representative example. How would your estimate for the number of planets accompanying each star be changed?

. . . .

(B)　Interstellar dust

This, too, might be a tenable hypothesis. Dark interstellar grains, also termed interstellar dust, emit no light, and they are very difficult to detect. Surely there is much dust of which we are entirely unaware.

As with planets, one way to evaluate this possibility is to calculate how much dust we would have to postulate if it were to be the dark matter. We could then calculate the degree to which it would obscure our view of the distant stars. The interstellar dust would act like atmospheric dust on a hazy day, which obscures our view of the distant hills. The details of this calculation are too technical to repeat here, but the result is unambiguous. Were the dark matter to be composed of interstellar dust, we would be able to see no more than a few thousand light years before our view of the Universe was blocked by this "interstellar haze." Since in reality we can see much farther than that, we conclude that this proposal is impossible.

(C)　Black holes

This, too, might be tenable. Black holes by their very nature emit no light, and they are exceedingly small in astronomical terms. Even if many were to exist in the Galaxy, most of them would have escaped detection till now.

To evaluate this possibility, we again need to calculate how many such holes we must postulate. We can then think about the process that created these holes. Since every time a black hole is formed, a supernova explosion results, we can exploit the fact that supernovae are observed to be very rare, and ask whether the required number of supernovae would be too great.

As we have seen, the observed number of supernovae in the Galaxy is a few per century. Let us calculate how many supernovae per century the black hole proposal requires. If our answer turns out to be more than what is observed, we will have to discard the black hole hypothesis.

⇐ **Looking backward**
We studied black holes and supernovae in Chapter 15.

The logic of the calculation

Step 1. Find the number N_{holes} of black holes in the Galaxy, by dividing the mass of all the dark matter by the mass of a single hole.

Step 2. This number equals the number of supernovae $N_{supernovae}$ that have ever occurred in the Galaxy.

Detailed calculation

Step 1. Find the number N_{holes} of black holes in the Galaxy, by dividing the mass of all the dark matter by that of a single hole.

We have already shown that the mass of the dark matter in the Galaxy is ⅞th the Galaxy's total mass, which we have measured to be $M_{total} = 1.57 \times 10^{42}$ kilograms. Thus

$$M_{dark\ matter} = (7/8)(1.57 \times 10^{42} \text{ kilograms})$$
$$= 1.37 \times 10^{42} \text{ kilograms.}$$

To find the number N_{holes} we need to know the mass of a single one. Let us choose a 5 solar mass hole as a representative example. Then the mass of a single hole is

$$M_{hole} = 5\ M_{Sun} = (5)(1.989 \times 10^{30} \text{ kilograms})$$
$$= 9.9 \times 10^{30} \text{ kilograms.}$$

And the number of black holes in the Galaxy is

$$N_{holes} = M_{dark\ matter}/M_{hole} = 1.37 \times 10^{42} \text{ kilograms}/9.9 \times 10^{30} \text{ kilograms} = 1.4 \times 10^{11}.$$

Step 2. This equals the number $N_{supernovae}$ that have ever occurred in the Galaxy.

$$N_{supernovae} = N_{holes} = 1.4 \times 10^{11} \text{ supernovae.}$$

Step 3. Find the rate of supernovae by dividing their number by the period of time in which they have been occurring.

This period of time is the age of the Galaxy, which is about 10 billion years. Thus the rate is given by

$$\text{supernova rate} = N_{supernovae}/10^{10} \text{ years} = 1.4 \times 10^{11}$$
$$\text{supernovae}/10^{10} \text{ years} = 14 \text{ supernovae per year.}$$

Step 4. Compare this with the observed rate.

Let us find the ratio between our predicted rate and the observed rate, which we will take to be about 2 per century:

$$\text{predicted rate/observed rate} = 14 \text{ per year}/2 \text{ per century}$$
$$= 14 \times 100 \text{ per century}/2 \text{ per century}$$
$$= 700.$$

Step 3. Find the rate of supernovae by dividing their number $N_{supernovae}$ by the period of time in which they have been occurring.

Step 4. Compare this with the observed rate.

As you can see from the detailed calculation, the black hole hypothesis requires that the Galaxy would have 700 times more supernova explosions per century than is actually observed. While a few such explosions are probably hidden from our view by dark interstellar clouds, there is no possibility that we would have missed so many of them.

NOW YOU DO IT

(A) Suppose our Galaxy were actually twice as old as we think. (B) Suppose there were actually twice as many supernovae per century in our Galaxy as we think. In either case, would the black hole hypothesis be tenable? (C) If we were to adopt *both* assumptions, would the black hole hypothesis be tenable? (One way to answer these questions is to re-do the detailed calculations. But there is an easier way! See if you can find it.)

Comments on these arguments

We have repeatedly used the analogy that a scientist is like a detective. When faced with a mystery, a good detective starts by assembling a list of possible suspects. She then winnows down the list, closing in on the culprit by finding reasons to declare some of these suspects innocent. We have been doing the same. In our attempt to understand what the dark matter might be, we began by assembling a list of three candidates, and we then looked for reasons to rule one or the other of them out. Unfortunately, we have managed to rule *all* of them out.

But have our arguments been ironclad? In fact they have not! Each of our arguments has relied on certain assumptions – assumptions that may be false. There are objections that can be raised to each of our three arguments.

(1) In evaluating the planets hypothesis, we tacitly assumed the planets to be orbiting stars. But what if they are not orbiting stars? What if they are wandering freely through interstellar space?

(2) In evaluating the dust hypothesis, we tacitly assumed that the proposed dust particles have the same size as those in known interstellar clouds. But the obscuring power of dust depends on the size of the particles: were we to propose larger particles, there would be fewer of them, and it turns out they would be less effective in blocking our view of the distant Universe.

(3) In evaluating the black holes hypothesis, we tacitly assumed the holes to have been forming at a steady rate throughout time. But we could alternatively assume that they were all formed early in the history of the Galaxy. In this case, all the supernovae would have occurred long ago, when we were not around to witness them.

What are we to make of these objections? Let us evaluate each of them in turn.

⇐ **Looking backward**
We studied star formation and planet formation in Chapter 13.

(1) The idea that planets might exist wandering freely through interstellar space entirely ignores the theory we developed of the formation of stars and planets. In that theory, the formation of planets is regarded as being part of the process whereby stars formed. In particular, planets are thought to have formed from the rotating disk of gas and dust surrounding the star, so that they were formed orbiting about their stars. They could not, therefore, be wandering freely about.

(2) The idea that great amounts of large dust particles exist in space entirely ignores the fact that the bulk of the matter in the cosmos has been found to be hydrogen and helium, and that heavy elements make up no more than 1% of the Universe. Hydrogen and helium, however, are gaseous, so that dust particles can only be composed of just those rare heavy elements. Therefore dust particles could not possibly account for more than 1% of the mass of the Galaxy.

(3) The idea that great numbers of black holes formed early in the history of the Galaxy fits in nicely with current theories of the evolution of galaxies, which posit an initial burst of star formation early in a galaxy's history.

For these reasons, most scientists believe that the dark matter could not possibly consist of planets or dust. On the other hand, it could very well consist of black holes. But this does not mean that the dark matter *is* black holes: all we know is that we have found no reason to discard this hypothesis.

We now turn to two projects that have attempted directly to investigate the nature of the dark matter.

MACHOs

MACHO stands for "MAssive Compact Halo Object." The MACHO search was designed to test the hypothesis that the dark matter consists of great numbers of unseen objects filling the Galaxy's halo. These objects are hypothesized to be massive (roughly the mass of a star) and compact (no larger than a normal star, and maybe far smaller). But they are hypothesized to emit much less light than a normal star.

What could these objects be? One possibility is that they are very low mass stars. Recall from Chapter 12 that the luminosity of a low-mass star is far less than that of the Sun, so that such stars would be "dark" in the sense that they are too faint to be seen were they to reside in the galactic halo. But this is not the only possibility. Another is that the MACHOs might be stellar "corpses" – stars that have exhausted their nuclear fuel. As we know, there are three sorts of corpse: white-dwarf stars, neutron stars and black holes. White dwarfs and neutron stars emit very little light, and black holes emit none at all. So these stellar corpses are also "dark."

⇐ **Looking backward**
We studied white dwarfs, neutron stars and black holes in Chapter 15.

The strategy employed by the MACHO search involves a prediction of Einstein's theory of relativity that light is influenced by gravity. Any massive object therefore bends the light rays passing by it. We can think of a MACHO as a kind of lens – a

(a)

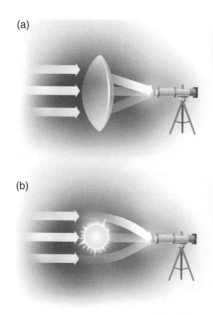

(b)

Figure 16.23 **A lens gathers light over a large area and focuses it to a point, thus making the light brighter. This is true whether the lens is made of glass (a) or gravity from a star (b).**

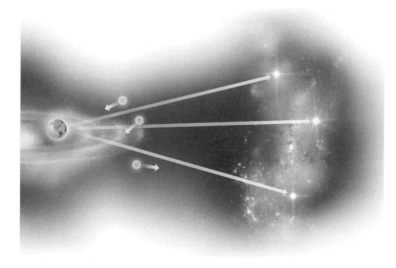

Figure 16.25 **As MACHOs wander about, occasionally one crosses our line of sight to a star in a distant galaxy. When this happens, the star briefly appears to brighten.**

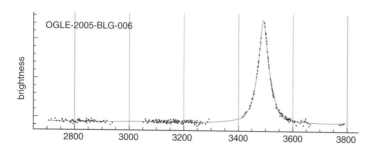

Figure 16.26 **A brief increase in brightness of a distant star, interpreted to be caused by passage of a MACHO in front of it.**

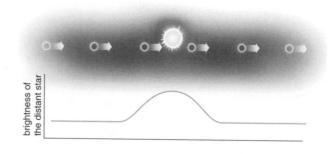

Figure 16.24 **A gravitational lens brightens the light from a background star as it passes in front of it.**

gravitational lens. As illustrated in Figure 16.23, the lens focuses passing rays of light from a second star lying beyond the MACHO.

The consequence of this focusing is to make the more distant star appear brighter. We do not see the MACHO – it is too far away and too dim for us to see. But we can detect it through its effects: it increases the brightness of any star behind it. As the MACHO wanders about – it is orbiting through the galactic halo – the focusing grows stronger and then weaker as the lens passes in front of a distant star (Figure 16.24). If we see a distant star brighten and then dim, we conclude that this was caused by a gravitational lens – a MACHO – passing in front of it. So we detect a MACHO, not by looking at it, but by looking at stars behind it.

Figure 16.25 illustrates MACHOs orbiting through the halo of our Galaxy, and the lines of sight to stars in a distant galaxy. As you can see, in this figure most MACHOs do not fall on any line of sight, and they produce no noticeable brightening of background stars. But one MACHO in its wandering is just about to pass in front of a distant star, which will cause it to brighten.

In Figure 16.26 we illustrate data of a lensing event obtained by the MACHO researchers. The graph shows a brief increase in the apparent brightness of a star in a nearby galaxy. Such an event cannot be caused by some variation within the star itself. The reason is that variable stars are known to vary cyclically, whereas an event like this happens only once. Rather, each such event is evidence for the passage of a gravitational lens – a MACHO – in front of the distant star.

THE NATURE OF SCIENCE

SCIENTISTS NEED LOTS OF DATA

The MACHO researchers have unquestionably discovered dark matter – after all, MACHOs are indeed dark. But there is another question: have they solved the mystery of dark matter? Does the dark matter in our Galaxy consist of *nothing but* MACHOs? To answer this question we need to know how many MACHOs there are. If there are a lot of them, then we have found out what the dark matter is: it is MACHOs. But if there are not very many of them, then the mystery remains, for part of the dark matter in our Galaxy must consist of something else – something whose nature we still do not understand.

So how many MACHOs are there? We do not know the answer to this question. The reason is that great amounts of data would be required to answer it – more than we have. Furthermore, it is very hard to get these additional data.

It is a general feature of science that scientists always require great amounts of data. We first explored this issue when we discussed a test of astrology and realized that, in order to conduct this test, we would require large numbers of subjects. We now return to the same issue, but in the context of the MACHO project. Let us see why great amounts of data are required to find out how many MACHOs there are.

In the present context, this means that we must observe a great number of stars. Figure 16.27 illustrates three MACHOs and a few stars. The MACHOs are, of course, invisible to us: all we see are the stars. But one of those MACHOs happens to lie in front of a star, and is causing it to brighten. This allows us to detect that particular MACHO (we have indicated this by the exclamation mark). But the other MACHOs do *not* happen to lie in front of stars, so we have no idea they are there. So in this configuration, we have only discovered one of the three MACHOs.

What if there were more data – i.e. more stars? Figure 16.28 illustrates what would happen in this case: the same three MACHOs, but lots of stars. Notice that now *every* MACHO has a star behind it. So in this configuration we have discovered all the MACHOs.

Return to our discussion of the test of astrology. There we considered the possibility that astrology was false; and we realized that even so an astrologer might get lucky and, purely by random guessing, just happen to make correct predictions. But we noted that this was very unlikely if the astrologer was asked to make a lot of predictions. Conversely, if the astrologer was only asked to make a few predictions, it would not be so rare for them all to just happen to turn out right. This was why great amounts of data were required to evaluate the truth of astrology.

We find the same logic now as we think about MACHOs. Go back to Figure 16.27 illustrating the few-stars situation.

⇐ **Looking backward**
We conducted a test of astrology in Chapter 2.

Figure 16.27 **Small amounts of data. If we observe only a few background stars, we will probably detect only a small fraction of the foreground MACHOs. (The MACHO we do detect is indicated by the exclamation mark.)**

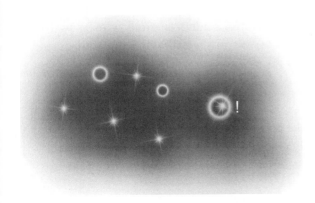

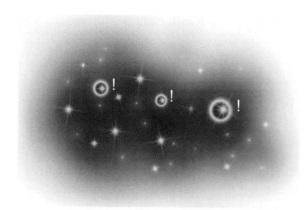

Figure 16.28 **Great amounts of data. If we observe many background stars, we will very likely detect most of the foreground MACHOs. (Those MACHOs we do detect are indicated by the exclamation marks.)**

It is always possible that, purely by luck, every MACHO might just happen to have a star behind it, and so be detected. This is not so unlikely. Similarly, in the many-stars case (Figure 16.28), it is also possible that every MACHO might fail to have a star behind it, and so not be detected – but this is *very* unlikely. This means that if we have few data, almost certainly the true number of MACHOs will be very different from the number we detect; but if we have lots of data, almost certainly the true number of MACHOS will equal the number we detect.

> Lots of data give us great confidence in our conclusions. Small amounts of data leave us with very little confidence in our conclusions.

In practice, the MACHO researchers did not have very much data. Thus, while they had proved that MACHOs exist, they could not reliably estimate how many of them there are. How could they obtain more data? They would need to observe more stars. But the team had observed 8.5 million stars for two full years, and ended up finding a mere eight lensing events. To monitor the brightness of far more stars required a far better telescope.

In subsequent years, two separate developments occurred.

Notice that the gravitational bending of light tells us nothing about the nature of the object producing the lens beyond its mass. The objects responsible for the lensing events found by the MACHO researchers could be black holes, which emit no light at all. Or they could be more ordinary stars of very low luminosity, such as white dwarfs or neutron stars. But in this case highly sensitive surveys can be mounted, searching for their light. Recently such a search was performed – and it did not find the light predicted if the objects were low-luminosity main sequence stars.

Notice that this finding does not demonstrate that the dark matter cannot be composed of MACHOs. It demonstrates that, if it is, the MACHOs must be objects such as black holes that emit far less light than an ordinary star.

The second development involved more sensitive searches for lensing events – one by the original MACHO collaboration, and another by a separate group. These searches failed to find the same amount of lensing. Here, too, we see the consequences of performing more detailed observations: often, quite different results are obtained.

As we will explain in the following chapter, scientists often wrestle with conflicting evidence and incomplete data. We are witnessing here the same issue.

> ⇒ **Looking forward**
> We will discuss the problem of incomplete and conflicting evidence in Chapter 17.

. . . .

NOW YOU DO IT

You are given two coins. One of them is fair: that is, it lands heads up or heads down with equal probability. The other coin, however, is very slightly unfair: it lands heads up slightly more often than heads down. You want to

figure out which is which, and you will do this by flipping them both and recording the number of heads you get.

Explain why you must flip the coins a great number of times in order to make this decision with confidence.

WIMPs

WIMP stands for "Weakly Interacting Massive Particle." As we have discussed in Chapter 4, ordinary matter is composed of three fundamental particles: the neutron, the proton and the electron. Many others can be produced by collisions in a particle accelerator, or in the decays of these particles. Certain theories of elementary particles, however, posit the existence of entirely new, as yet undiscovered, particles. They would be extremely massive compared to the known particles, but weakly interacting so that they would have escaped detection so far. We emphasize that these theories are highly speculative, and we do not know whether they are correct. In particular, the particles predicted by these theories have never been observed.

But perhaps they have been observed! Perhaps they are the dark matter. If so, we would not need to search for the dark matter between stars, or far off in the galactic halo. Rather it would fill all space. In particular, it would regularly bombard the Earth. Indeed, these particles would be passing through our very bodies at this very moment. Because WIMPS interact so weakly, they would produce not the slightest of effects as they flew through us, and we would be entirely unaware of their existence. But they would also be flying though physics laboratories. If we could build an exceedingly sensitive detector, perhaps we could record their traces.

This is the strategy of the WIMP projects that have been mounted. Many groups of researchers are engaged in this effort. As of this writing, a few groups have claimed to have detected WIMPs, but most have not.

If WIMPs do exist, and if they are indeed the dark matter, we will have reached a remarkable conclusion. We will have realized that the Universe consists of two worlds. On the one hand there is the world of ordinary matter. Everything in our experience – rugs and newspapers, the oceans and our bodies and the Moon and the Sun – belongs to this world. But all this adds up to no more than a fraction of the totality. If the dark matter is composed of WIMPs, the bulk of the Universe would consist of a second world, a world made of strange, elusive particles of whose nature we know next to nothing. This is a shadow world, one of which we are entirely unaware. It coexists with ours, occupying the same space, but passing through us all like ghosts.

Summary

The discovery of the Milky Way Galaxy
- At first glance there is no pattern to the distribution of stars about us.

- In 1785 William and Caroline Herschel attempted to find a pattern by counting the number of stars visible in various directions.

- This will delineate the shape of any structure in which we are located – but only if the density of stars is the same everywhere.
- The Herschels had no way to tell if this was so.
- Their results became known as the "grindstone" model of the Universe.
- We now know this model is false.

Cepheid variables and the distribution of globular clusters in space

- The parallax method of measuring distances only works for nearby stars.
- Cepheid variable stars provide another means:
  - the period of variation of the Cepheid's brightness tells us its luminosity,
  - the inverse square law then tells us its distance.
- Harlow Shapley used this to determine the pattern in which globular clusters are arranged.
- Architecture of our Galaxy:
  - globular clusters lie in a sphere,
  - Shapley proposed that we live in a disk imbedded in this sphere,
  - the Milky Way is this disk seen edge-on,
  - it is about 100 000 light years across.

Rotation of our Galaxy

- Doppler shifts of globular clusters reveal that the Solar System is orbiting about the Galaxy's center.
- 200 million years are required to complete one orbit.
- The orbital velocity of a star in the Galaxy:
  - depends on the radius of its orbit,
  - the dependence is different from those of planets,
  - this is because the gravitational force on the star is different from that on a planet.

Spiral arms

- Cannot be permanent structures: if they were they would be tightly wrapped by now.
- In fact they are density waves propagating through the galactic disk.
- High-mass stars are found only in spiral arms.
- These arms are the sites of star formation.

High- and low-velocity stars, and stellar populations

- We find two groups of stars:
  - low-velocity stars: velocity relative to us about 10 kilometers per second,
  - high-velocity stars: about 150 kilometers per second.
- Stellar populations:
  - population I: heavy elements make up about 1% of the star's mass,
  - population II: far fewer heavy elements.

- Low-velocity stars are found to be population I, high-velocity stars are population II.
- Hertzsprung–Russell diagrams reveal that population II clusters are older than population I clusters.

Dark matter

- We find the mass of all the visible matter in our Galaxy by measuring its luminosity and dividing by the luminosity of a star.
- We find the total mass of the Galaxy by observing the orbit of an object in its outskirts.
- We find the total mass of the Galaxy to be far more than that of the visible matter.
- This is also true of other galaxies, and of the matter lying between galaxies.
- What is the dark matter?
  - We still do not know what the dark matter might be.
  - We considered and rejected the possibility that it could be planets or interstellar material.
  - It might conceivably be black holes, if they were formed more often in the past than now.
  - The MACHO project found gravitational lensing events caused by MAssive Compact Halo Objects.
  - One group has reported the detection of WIMPS, Weakly Interacting Massive Particles. Other groups have not been able to confirm this finding.

DETECTIVES ON THE CASE

High- and low-velocity stars, and stellar populations

- Low-velocity stars move just as rapidly as high-velocity stars, but in the same direction as that of our motion.
- This tells us that low-velocity stars reside in the galactic disk, while high-velocity stars reside in the halo.
- The mystery: why are disk stars population I and halo stars population II?
- A population I star cluster contains heavy elements, but these cannot have been created within the stars themselves.
- Rather they must have been present in the cloud from which these stars formed.
- Such clouds must have been "seeded" with heavy elements produced in an earlier generation of stars.
- Evolution of our Galaxy:
  - initially it was a slowly rotating, nearly spherical cloud,
  - stars formed in this cloud now make up the Galaxy's halo: they are population II,
  - the cloud shrank, increasing its rotation rate to form a disk,
  - stars formed in this stage now make up the Galaxy's disk: they are population I.

THE NATURE OF SCIENCE

The nature of scientific theories

Scientific theories replace lists of meaningless facts with understanding.

THE NATURE OF SCIENCE

Scientists need lots of data

- The MACHOs are indeed dark matter – but do they constitute *all* the dark matter?

- We do not reliably know how many MACHOs there are.
- This is because:
  - lots of data are required for us to have confidence in our conclusions,
  - with only a few data we can have little confidence in our conclusions.
- In order to gather enough data to find out if the dark matter consists entirely of MACHOs, we would require a dramatic improvement in technology.

Problems

(1) Suppose you live in an imaginary state whose shape is square. This state has four counties: A, B, C and D (Figure 16.29).

Suppose you assume that people are uniformly spread around the state: i.e. there are the same number of people per square mile everywhere. You want to find out where you live. To determine this, you count how many people live in various directions. You find:

750 000 people live somewhere to your north
250 000 people live somewhere to your south
830 000 people live somewhere to your east
100 000 people live somewhere to your west.

Looking at these figures, you realize that you have made a mistake in your counting. How did you know this?

(2) Consult Figure 16.4 showing the distribution of people in Massachusetts. Where in the state would you actually be living, if you counted the same number of people to the north of you as to the south, and the same number to the east of you as to the west?

(3) In this problem, too, we suppose that you live in a state whose shape is rectangular, and we assume that people are uniformly spread around the state: i.e. there are the same number of people per square mile everywhere. Suppose that you want to find out the shape of your state: i.e. is it nearly perfectly square, or an elongated rectangle? To determine this, you count how many people live in various directions. You find:

750 000 people live directly north of you.
750 000 people live directly south of you.
500 000 people live directly east of you.
500 000 people live directly west of you.

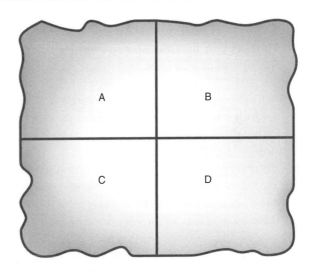

Figure 16.29 **An imaginary state.**

Draw a rough map showing the proportions of your home state. Place a dot on it indicating your home town. In both cases, explain your reasoning.

(4) Suppose you are observing a globular cluster, and you find within it a Cepheid variable star whose period of variation is two weeks. What is its luminosity? Give your answer both in solar luminosities, and in watts.

(5) Consider the Cepheid variable star in the globular cluster discussed in the previous problem. You measure the apparent brightness (the flux) of the Cepheid and find it to be 3×10^{-8} watts/meter^2. You want to use this to find the distance to the cluster. (A) Describe the logic of the calculation you will perform. (B) Do the detailed calculation: give your answer both in meters and in light years.

(6) You have already considered two points on a frisbee 6 inches in radius: one point on its perimeter, and the

other halfway in. You found the rotational velocities of these two points, and you verified that the outer one was moving faster than the inner one. (A) Suppose the frisbee was 12 inches (30.48 centimeters) in radius: in what way would your results be changed? (B) Suppose you considered not a point halfway in from the edge of a 6-inch frisbee, but one ⁹⁄₁₀th of the way in. In what way would your results be changed? (Note: both of these questions can be answered by re-doing your previous calculation – but there is an easier way! See if you can find it.)

(7) Consider two stars within our Galaxy: one 20 000 light years from its center, and the other 10 000 light years from its center. Both orbit with velocity 220 kilometers per second. We want to calculate the time required for each to orbit around once. (A) Describe the logic of the calculation you will perform. (B) Carry out the detailed calculation. Next, suppose the Galaxy is 13 billion years old. We want to calculate how many times each star has orbited. (C) Describe the logic of the calculation you will perform. (D) Carry out the detailed calculation for each star. (E) If a spiral arm passes through these stars, by how many times will it have wrapped about by now?

(8) Consider a track meet on an oval racetrack. These racetracks are laid out such that an outer runner starts from a point ahead of an inner one (Figure 16.30). Why are racetracks laid out in this manner?

(9) Use the Web to figure out the orbital velocities of the moons of Jupiter. (If you can find a site that actually tells you these velocities, fine. If not, you can certainly find one that tells you their orbital radii and periods, and you can simply calculate the velocities.) Do these moons also obey the law that an outer body moves more slowly than an inner one?

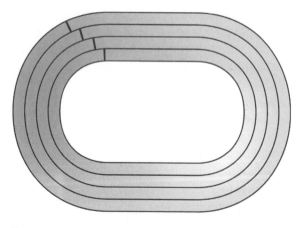

Figure 16.30 **A race track.**

(10) In our discussion of the two stellar populations described in Table 16.1, we postulated a history for the evolution of the Galaxy. In this problem we postulate some alternative histories.
   • Suppose that no stars had formed until the Galaxy had shrunk a good deal.
   • Suppose that the initially slowly rotating, nearly spherical Galaxy had not shrunk, and that stars had formed at a steady rate, some early in its history, and some later.
   • Suppose that the Galaxy had begun spherical and then shrunk, and that stars had formed at a steady rate throughout its history – but that our Galaxy was so young that no star has had the time to progress beyond the main sequence stage of its evolution.
   Your task is to describe the stellar populations we would have found, and to present new versions of Table 16.1, for each of these alternative histories.

(11) Suppose the luminosity of a hypothetical galaxy is 2×10^{37} watts. We want to calculate the mass of all the *visible* matter in this galaxy. (A) Describe the logic of the calculation you will perform. (B) Carry out the detailed calculation.

(12) Suppose that stars 50 000 light years from the center of the galaxy of the previous problem orbited at 300 kilometers per second. We want to calculate the mass of *all* the matter in this galaxy. (A) Describe the logic of the calculation you will perform. (B) Carry out the detailed calculation.

(13) Combine your results from the two previous problems to find the ratio between the total mass and the visible mass in that galaxy.

(14) Suppose the dark matter in our Milky Way Galaxy were made up of planets the mass of Jupiter. We want to calculate how many such planets must accompany each star in order for this to be true. (A) Describe the logic of the calculation you will perform. (B) Carry out the detailed calculation. (C) Suppose alternatively the dark matter were made up of planets half the mass of Jupiter. How many such planets must accompany each star in this case? See if you can do this last part without going through the detailed calculation all over again!

(15) Suppose the dark matter in our Milky Way Galaxy were made of black holes the mass of the Sun. We want to calculate how many supernovae must have occurred within our Galaxy per century in order for this to be true. (A) Describe the logic of the calculation you will perform. (B) Carry out the detailed

calculation. (C) Compare this with the observed rate of supernovae in our Galaxy. (D) Suppose alternately that the holes were more massive than the Sun: would the rate of supernovae be greater, less or the same? (E) Based on your results for (C) and (D), would we need to postulate that the holes are more massive, less massive or as massive as the Sun in order to account for the dark matter? Explain your reasoning.

(16) The MACHO researchers found a mere eight lensing events, in which a distant star brightened as a gravitational lens passed in front of it, in two years of observing. We want to calculate the number of years they would have to observe in order to find 100 brightenings. (A) Describe the logic of the calculation you will perform. (B) Carry out the detailed calculation.

WHAT DO YOU THINK?

In our discussion of the data presented in Figure 16.11, we considered two possible interpretations. Both interpretations make a prediction: that when we look perpendicular to the "special axis" we should see a sideways velocity of the globular clusters. You have already explained why Doppler observations cannot reveal this motion. But, as we repeatedly have emphasized, it is always essential to test our theories. Can you think of some *other* kinds of observations we could use to reveal this motion?

You must decide

You are a program officer at the National Science Foundation committed to furthering research on the nature of the dark matter. Three proposals have landed on your desk.

(1) The first is a proposal to build a telescope specifically designed to find interstellar planets. The authors of this proposal are entirely aware of the objections to this possibility we outlined above, but they emphasize that these objections are not ironclad. In particular, the proposers point out that our theory of the formation of planets concerns events that happened billions of years ago, and that this theory has received very few observational tests. The theory may well be wrong, the proposers assert.

(2) The second is a proposal to search for great numbers of large dust grains in space. The authors of this proposal point out that the "fact" that only 1% of the Universe consists of heavy elements may well be false, because it is based on observations only of objects that we can see – namely, observations of stars and known interstellar clouds. They propose that objects we cannot see, namely the dark matter, may well be different.

(3) The third is a proposal to build a telescope capable of doing a search for MACHOs far more rapidly than has been done so far. In this way, the authors claim, they will be able to gather vast amounts of data and so determine whether the dark matter consists of MACHOs.

Unfortunately, each of these proposals requests a great amount of money, and you have funds at your disposal to support only one of them. Your task is to write a memo explaining which proposal you have decided to support, and – more important – the reasons for your decision.

17

Galaxies

In the previous chapter we discussed our home in the Universe: the Milky Way Galaxy. But ours is not the only galaxy. Here we discuss all the others. Some are like our own: others are completely different.

The number of galaxies defies the imagination. Immense, faintly glowing, slowly rotating, they drift about, influenced by their mutual gravitational attraction. Occasionally they actually encounter one another. You might think of them as resembling snowflakes in a blizzard.

Our own Galaxy is disk-shaped, with a central bar and winding spiral arms. Other galaxies lack the central bar; yet others are more nearly elliptical and also lack the spiral arms. Yet still others are completely irregular in shape. Some have lots of interstellar gas; some hardly any.

And some are the seat of violent, hugely powerful explosions. These objects emit vast quantities of energy as optical light, radio waves and other forms of radiation. These emissions probably arise from matter falling into giant black holes lying at their cores.

Galaxies lie all about us: indeed, some are so close they can be seen with the naked eye. But for years nobody knew what they were. We begin with the story of how astronomers came to discover their true nature.

"Spiral nebulae"

We mentioned in the previous chapter the structures known at the time as spiral nebulae (Figure 16.2): disk-shaped objects shot through with winding spiral arms. Today we know that they are galaxies, just like our own. But no one knew this at first. The path to the discovery of the nature of these nebulae is long and complex, and it was filled with many twists and turns. There were clues that pointed one way, and other clues that pointed the opposite way. Because it beautifully illustrates the nature of scientific discovery, we will follow the path of this discovery in detail. Nothing was simple about the process. (Although the term "spiral nebula" is no longer used, we will use it in keeping with the debate that actually occurred at the time.)

Why was it so hard to realize that these nebulae are just like our own Milky Way Galaxy? There were two difficulties.

(1) Astronomers could not see the structure of our own Milky Way Galaxy, so they could not see that it looks just like the spiral nebulae. Much of our Galaxy is hidden from view by interstellar clouds. Nowadays we have been able to trace out its spiral arms by using, for instance, radio telescopes, which can "see" right through these clouds. But these methods were not available back then. Indeed, as we recounted in the previous chapter, Shapley's discovery of the Milky Way Galaxy was highly indirect: he never actually saw its structure.

It was a little like being locked inside a room in a house and looking out the window. You can see other houses nearby. But you can't see any part of your own home beyond your room, and you have no way of knowing whether you live in a house just like them!

> ⇐ **Looking backward**
> We discussed the origin of the Sun and Solar System in Chapter 13.

(2) Astronomers had an alternative idea of what the spiral nebulae might be: stars and planetary systems in the process of forming. Recall that the planets are thought to have formed from a disk of gas and dust rotating about a condensing protosun. This theory was well known in those days. Many astronomers thought that the central bulges that can be seen in spiral nebulae were protostars, and the knot-like structures that can be seen in their outer parts were protoplanets.

So there were two competing theories of the nature of the spiral nebulae.

- Perhaps they were planetary systems in the process of forming. This theory came to be known as the "Nebular Hypothesis."
- Perhaps they were galaxies, just like our own. This theory came to be known as the "Island Universe Hypothesis."

In deciding between these rival theories, much of the effort involved trying *to measure the distances to the spiral nebulae.* A result that placed them within our own Milky Way Galaxy would be evidence in favor of the Nebular Hypothesis: a result that placed them far outside our Galaxy would point toward the Island Universe Hypothesis.

Detectives on the case

What are the spiral nebulae?

In what follows, we will focus on the Andromeda Nebula (Figure 17.1), the largest nearby galaxy. Most of the important research that led to a resolution of the problem was done on it.

First clue: a possible nova in Andromeda. In 1885 a "new star" was discovered in the Andromeda Nebula. What could that object be?

> ⇐ **Looking backward**
> We studied novae in Chapter 15.

Astronomers knew in those days of novae, stars that suddenly increased in brightness because they had undergone explosions. People quickly realized that, if so, the "new star" would provide a means of measuring the distance to Andromeda. The intrinsic brightness – the luminosity L – of novae was known. By measuring the apparent brightness – the flux f – they could then find the distance R by solving the inverse square law

$$f = L/4\pi R^2$$

for R:

$$R = \sqrt{[L/4\pi f]}.$$

⇐ **Looking backward**
We studied the inverse square law in Chapter 4.

Novae were known to have relatively small luminosities. Inserting representative values of L for a nova, and the measured value of f of the "new star," it was found that the Andromeda Nebula was relatively close. So this piece of evidence was in favor of the Nebular Hypothesis.

Intuitive mathematics

We can intuitively understand why the relatively low value for the luminosity of a nova leads to a relatively small value for its distance from us. Let us study how the answer we get for R depends on L. Our formula for the distance to a source of light tells us that R is proportional to the square root of L. Suppose we have reason to think that L is small. Then the square root of L is also small, and R will be a short distance. Alternatively, if we have reason to think that L is big, then R will be big.

This accords with our intuitive understanding. Suppose we see a faint light at night. If we assume it to be a candle we will decide it is nearby. But if we assume it to be a searchlight, we will decide it is very far away.

NOW YOU DO IT
Suppose you see a faint light at night. You consider two possibilities: maybe it is a 1-watt candle, or maybe it is a 100-watt light bulb. Suppose that its apparent brightness leads you to conclude that, if it were a candle, it would be 20 feet away. How far away would it be if it were a light bulb?

Figure 17.1 **The Andromeda Nebula.** 👁 (Also see color plate section.)

⇐ **Looking backward**
We studied absorption lines in Chapter 4.

Second clue: Andromeda might contain stars. In 1897, however, a new piece of evidence was discovered – one that pointed toward the opposite conclusion! It was a spectrum of the Andromeda Nebula. This spectrum showed absorption lines. As we saw in our study of spectroscopy, such lines are found in the spectra of stars. So this spectrum told astronomers that Andromeda contained at least some stars. This was evidence pointing toward the Island Universe Hypothesis.

Third clue: maybe the first clue was wrong. In 1910 a further piece of evidence pointing toward the same hypothesis was discovered. A new and very large telescope (60 inches – big for those days) was used to study the Andromeda Nebula. Several more "new stars" were found within it – but these were very much fainter than the "new star" of 1885. What if *these* were novae, and the 1885 object something else?

Because these new objects were so very faint, they implied that the earlier measurement of Andromeda's distance was far too small. The newly calculated distance worked out to an enormous value – far larger than the dimension of our own Milky Way Galaxy. So this was evidence pointing to the Island Universe Hypothesis.

Intuitive mathematics

We can intuitively understand why a lower value for the measured flux of a nova in Andromeda leads to a larger value for its distance from us. If we realize that our earlier measurement of the flux was too large, and now assign a smaller value to f, we will get a larger distance R. This is because our formula for the distance R tells us that to find R we must divide something by the square root of f. If f is small, we are dividing by a small number and we will get a large number.

This accords with our intuitive understanding. Imagine not one but two lights off in the distance at night: a relatively bright "1885 light" and a far dimmer "1910 light." If we suppose the "1885 light" is a candle we will judge it to be fairly close. But if it turns out that it is the far fainter "1910 light" that is the candle, we will judge it to be much farther away.

. . . .

NOW YOU DO IT
Suppose that you have seen a faint glimmer of light and mistakenly think it to be a 100-watt light bulb one mile away. Suppose, however, that you suddenly realize that you had been wrong, and actually a different glimmer, whose apparent brightness (flux) were 25 times less, was the light bulb. How far away would you conclude it to be?

. . . .

An error.   The debate between the Nebular and Island Universe hypotheses was muddled by an unfortunate mistake. Notice that one of the methods for deciding the debate turned on measuring the spiral nebulae's distances *and comparing them with the size of our Milky Way Galaxy*. But what is this size?

It was in 1917 that Shapley proposed his structure for the Milky Way Galaxy. As we have recounted, he found its size by using Cepheid variable stars to measure the distances to the globular clusters. Unfortunately, in doing so he made two errors.

⇐ **Looking backward**
We found the size of the Milky Way Galaxy by using Cepheids in Chapter 16.

(A) He assumed that the period–luminosity relation for Cepheid variables in globular clusters was the same as that for the nearby variables that he had used to calibrate his method. Unfortunately, it was not. This led him to overestimate the luminosities of Cepheids in his globular clusters.

(B) He did not appreciate the ubiquitous nature of interstellar gas and dust, which act like an interstellar "haze" and dim the light of distant stars (a phenomenon termed "interstellar extinction").

What effect did these errors have on Shapley's measurements of the distances to his globular clusters? Both led him to *overestimate* their distances, and thus the size of our Milky Way Galaxy. They led him to inflate our Galaxy in his mind, until it had expanded to encompass the spiral nebulae. Thus Shapley argued in favor of the Nebular Hypothesis.

Intuitive mathematics

It is easy to see why both Shapley's errors led him astray in this direction.

(A) Shapley thought that the Cepheid variable stars he was observing were brighter than in fact they were. What effect did this error have? Our formula for the distance tells us to take the square root of the luminosity. If his value for L was too big, his result would be too big.

This accords with our intuitive understanding. Return to our faint light in the distance at night. If we mistakenly assume it to be a searchlight we will decide it is very far away. In reality, however, it is a candle, so that it is actually very close.

(B) Shapley thought that the apparent brightness of his Cepheids was exceedingly small. In fact this apparent brightness was not really so low: it was interstellar extinction that made it appear that way. In reality they were not really so faint, and f was actually bigger than Shapley thought. Our formula for the distance tells us that to find R we must divide something by the square root of f. If f is too small, we will get too large an answer.

This too accords with our intuitive understanding. Suppose we are not even aware that, as we gaze at our faint light at night, there is a fog that is dimming it. Because it is faint we will decide it is very distant. But in fact it is not dim because it is far away: it is dim because of the fog.

....

NOW YOU DO IT
Suppose you are looking at a 100-watt light bulb at night. Suppose further that the night is foggy, and that the fog allows only 1% of the light from the bulb to reach you – but you are not aware of this fact. If you estimate from the apparent brightness that the light bulb is a mile away, how far away is it in reality?

....

Another error.   Throughout the 1920s, evidence in favor of the Nebular Hypothesis came from a series of observations of the sideways motions of bright knots in the outer regions of spiral nebulae. Adrian Van Maanen, an expert in such work, claimed he had detected such motion, and he attributed it to the rotation of these disks (recall that both protoplanetary disks and our own Milky Way Galaxy rotate).

Van Maanen's measurements can be used to calculate the velocity of this rotation. As illustrated in Figure 17.2, the more distant the nebula the greater is the velocity we will calculate. Van Maanen's measurements implied that, were Andromeda a distant Island Universe, it would be rotating at enormous velocities. If, however, we accept the Nebular Hypothesis and take it to be closer, the velocities are smaller and more reasonable.

In the light of present-day knowledge, we now know that Van Maanen's measurements were entirely erroneous. Even now, with far larger telescopes and far more accurate methods of measurement, we have still not been able directly to observe the rotation of a spiral nebula. Nobody knows what led Van Maanen astray. He was an exceedingly careful worker, and an expert in this type of work. But he was working at the very limit of technology, and in such circumstances mistakes are hard to avoid. Edwin Hubble, a contemporary of Van Maanen's, put it this way.

angle the knot
has moved in
a few years

Figure 17.2 **The more distant the spiral nebula, the faster is its rotation, implied by Van Maanen's measurements. He found that bright knots had moved by a certain angle over a few years. As illustrated, the more distant we assume the spiral nebula to be, the greater is the velocity implied by this measurement. At distances required by the Island Universe Hypothesis, the implied velocity is too great.**

Thus the exploration of space ends on a note of uncertainty. And necessarily so... With increasing distance, our knowledge fades, and fades rapidly. Eventually, we reach the dim boundary – the utmost limits of our telescopes. There, we measure shadows, and we search among ghostly errors of measurement for landmarks that are scarcely more substantial.

[*The Realm of the Nebulae* by Edwin Hubble, Silliman Memorial Lectures, Yale University Press (1982).]

Hubble solves the problem. In 1924 Hubble made a discovery that led to the final resolution of the problem. In a masterpiece of understatement, he wrote in a letter to Shapley that[1] "You will be interested to hear that I have found a Cepheid variable in the Andromeda Nebula." Since Cepheids provided an unambiguous means of measuring distance, this discovery pointed the way to the correct answer. Five years later he published an immense paper in which he reported on an analysis of 40 Cepheid variable stars in Andromeda. His result for its distance placed it far beyond the outer limits of our own Milky Way.

Hubble had answered the question. Spiral nebulae were Island Universes – galaxies in their own right, similar to our own.

THE NATURE OF SCIENCE

THE PROCESS OF DISCOVERY IN SCIENCE

People often believe that science is cut-and-dried, and that to every question there is one and only one right answer. In actual practice, however, the process of discovery is never so simple. Astronomers in their debate on the nature of the spiral nebulae confined themselves to purely technical bones of contention. But as they did so, they were plagued by many subtleties. These subtleties profoundly influenced their reasoning – and the reasoning of every scientist faced with an unsolved problem.

How much weight should we give evidence? A good scientist knows that every line of argument is potentially suspect. After all, as they debated, astronomers were discussing two pieces of "evidence" that we now know are false – Shapley's excessive size of the Milky Way Galaxy, and Van Maanen's erroneous "discovery" of the rotation of Andromeda. Furthermore, the three other clues we have recounted above are also suspect. Let us return to these clues and evaluate them critically. How much weight should we give to each?

---

[1] Quoted from *Coming of Age in the Milky Way* by Timothy Ferris, p. 172.

(1) Our first clue involved the discovery of a "new star" in Andromeda. Historic-ally, astronomers interpreted this as a nova and so deduced that Andromeda was nearby. But they did not have conclusive evidence that it was a nova. Suppose it was something else?

(2) Our second clue involved evidence that Andromeda exhibited absorption lines. Historically, astronomers interpreted them as coming from stars. But they did not know with certainty that they came from stars. Suppose they came from something else?

(3) Finally, our third clue involved the discovery of a second class of "new star" in Andromeda. They were then faced with the question: which were novae? Was it possible that neither were?

This suspect nature of evidence makes all scientific discovery difficult. Scien-tists often find themselves in the position of having to make judgment calls.

Occam's Razor: how many new hypotheses must we make? The so-called "Occam's Razor" often figures in scientific debates. It is the general principle that one should always give precedence to simplicity – that, between two competing theories, we ought to choose the one that requires us to make *the smallest number of new postulates*. Galileo, for instance, employed it in his argumentation.

Notice, however, that the Island Universe Hypothesis forces us to accept *two* extraordinary new postulates.

(A) That the scale of the Universe is incomparably greater than anything people had believed before. If they are indeed galaxies just like our own, the mil-lions upon millions of spiral nebulae receding into the depths of space enormously expand our conception of the size of the cosmos.

(B) That explosions take place of a magnitude overwhelmingly greater than any-thing that had been witnessed before. If Andromeda really is so very distant, then the 1885 "new star" that had appeared within it must have been tens of thousands of times more violent than any nova. We now know that such explo-sions do in fact exist – they are the supernovae. But nobody knew this in 1920.

Had scientists used Occam's Razor they would have rejected the Island Universe Hypothesis – a hypothesis that we now know to be correct.

Psychological factors. Quite aside from such a general philosophical principle, it can be personally terrifying to take so big a mental leap as the Island Universe required. Indeed, scholars who have studied the history of this debate feel that people on both sides were finding it difficult to accept the immense scale of the cosmos.

Certain scholars believe that there was also a psychological element to Shapley's preference for the Nebular Hypothesis. They feel that he was proud of the Milky Way Galaxy he had discovered, and impressed by its great size. He referred to it as the "enormous, all-comprehending" Galaxy. In his mind it constituted the entire Universe. Such a psychological factor is present in all of us. Imagine, for instance, the pride you would feel had you been the person who discovered the world's highest mountain – as opposed to the lesser pride had you discovered the second highest.

The role of evidence in science. Nevertheless, I personally do not believe that such psychological factors weigh very heavily in the process of scientific discovery. While they unquestionably influence individual scientists, I believe that they do not influence the final conclusion. This is because in science *evidence can be found* – evidence that is clear and compelling, and no longer subject to ambiguity. Such evidence overcomes all the psychological factors, and it brings a debate to a conclusion. In the present case, this evidence was provided by the Cepheid variable stars in Andromeda.

Notice, however, that this sort of evidence is not easy to find. In our present case, Hubble was able to find his Cepheids only by using what was at the time the world's biggest telescope. Smaller telescopes, on the other hand, were incapable of spotting Cepheids at such enormous distances. Thus it is that astronomers are constantly building bigger and bigger telescopes, and ever more sensitive detectors. Without such new – and expensive – instruments, scientific progress is impossible.

Types of galaxies

Having established the nature of galaxies, we turn to a study of them. It happens that spiral galaxies are only one among a number of types. These types were illustrated in Figure IV.1. Let us study them.

The galaxy illustrated in (a) is termed a *spiral galaxy*. It has spiral arms. The one shown in (d) does too, except that the spiral arms do not lead all the way into the nucleus, but rather to a bar running through the galaxy. Such galaxies are known as *barred spirals*. (Our own Milky Way Galaxy is a barred spiral.) We note that the bars in such galaxies are composed solely of stars: there is no "rigid rod" running through them! On the other hand, (b), illustrating an *elliptical galaxy*, shows an object quite unlike our own Milky Way Galaxy. It is in many ways far simpler in shape: ellipticals possess no spiral arms and no central nuclear bulge; and rather than consisting of a disk imbedded in a halo, they consist solely of an elliptical assemblage of stars. Finally, (c) illustrates an *irregular galaxy*, which has no well-defined shape at all.

Most of the light from galaxies comes from stars. But gas and dust can also be seen in these images. The dust, for instance, can be seen silhouetted against the uniform glow of the stars. Notice in this figure that the different galaxy types contain different amounts of gas and dust. Ellipticals, in particular, contain very little. We will return to this point soon.

Galaxy classification

Let us study each of these four types in greater detail. Hubble invented a diagram in which the classifications of spirals, barred spirals and ellipticals can be summarized. This diagram, known for obvious reasons as *Hubble's "tuning fork diagram,"* is shown in Figure 17.3.

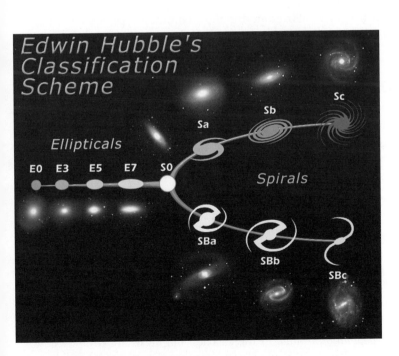

Figure 17.3 **Hubble's "tuning fork diagram" of galaxy classification.**

Table 17.1. **Spiral arm classification.**

Spiral type	Spiral arms	Nucleus
a	Smooth, ill defined, tightly wound	Large
b	↓	↓
c	Clumpy, well defined, loosely wound	Small

Table 17.2. **Gas content of galaxies (by mass).**

Ellipticals	Very small
Spirals	4%–16%
Irregulars	50%–90%

There are clear differences among the galaxies shown in this figure, differences that allow us to classify them into types "a," "b" and "c."

These classifications relate to the *shapes of the spiral arms* and *the relative size of the nucleus* (Table 17.1). Sa and SBa galaxies possess arms that are smooth, poorly defined and tightly wound, whereas in Sc and SBc galaxies the arms are far more clumpy, well defined and loosely wound. Furthermore, in Sa and SBa galaxies the central nucleus is quite large, whereas in type "c" galaxies it is smaller.

In Figure 17.3 we also illustrate elliptical galaxies. Here the differences among the types relates to the *degree of sphericity of the galaxy*: type E0 ellipticals are perfectly spherical, E3's more elongated, and so forth.

Finally, in Figure 17.4 we illustrate irregular galaxies. Here no classification scheme is possible: each is different from all the others.

When he proposed it, Hubble thought his diagram represented an actual evolution of galaxies, in which when formed a galaxy was of type E0 and then evolved towards the right in this diagram into either a spiral or a barred spiral. We now know this is not the case: we will return to this shortly.

A census of galaxies

We now turn to more detailed study of the properties of these four classes of galaxy.

How many of what type?
Surveys have shown that irregular galaxies are the most common type, spirals the next most common and ellipticals the least common (Figure 17.5).

Sizes
Figure 17.6 illustrates the diameters of galaxies. As you can see, our own Milky Way Galaxy is somewhat on the large side for spirals. Irregulars are typically somewhat smaller than spirals. Elliptical galaxies have the greatest range of all: the very largest galaxies are elliptical, and no type is smaller than the smallest.

Gas content
We have already noted that elliptical galaxies contain far less gas and dust than spirals. Table 17.2 shows the gas content of galaxies. As you can see, irregular galaxies contain the greatest amount of gas and dust, and ellipticals the least. Spirals are intermediate.

Colors and stellar populations
There is a systematic trend in the colors of galaxies as we pass along Hubble's "tuning fork diagram." As summarized in Table 17.3, elliptical galaxies are composed of the reddest stars, and irregulars of the bluest. Spirals, again, are intermediate. These colors are related to the *stellar population content* of the galaxies. Ellipticals are composed almost exclusively of old, population II stars, spirals of both population I and II and irregulars almost exclusively of population I.

(b)

Figure 17.4 **Irregular galaxies. (a) The galaxy NGC 1569.**
(b) The Magellanic clouds, two irregular galaxies close to our own.
(c) Barnard's galaxy.

Star formation in galaxies

Let us focus attention on the regularity documented in Table 17.3: the systematic variation in color as we pass from elliptical to irregular galaxies. We have by now put together a good deal of understanding about stars. Let us apply what we have learned to see what this regularity is telling us.

Let us begin by thinking about blue stars. We know three things about them.

(1) Blue stars are hot. (This is because the light from a star is blackbody radiation and, as we saw in Chapter 4, the hotter a blackbody the bluer it is.)

(2) Hot stars have high mass. (This is because hot stars lie along the left-hand portion of the main sequence in the Hertzsprung–Russell diagram, which is occupied by high-mass stars (Chapter 12).)

(3) High-mass stars have short lifetimes. (This is because (Chapter 14) they are very luminous, and they consume their nuclear fuel at a great rate.)

Our conclusion is that *blue stars have short lifetimes* – far shorter, in fact, than the age of a galaxy. Conversely, *red stars have long lifetimes* – longer than the age of a

(c)

Figure 17.4 **(cont.)**

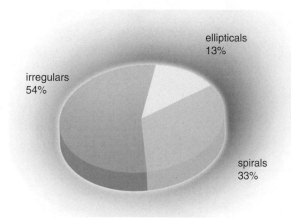

Figure 17.5 **The population of galaxies in our vicinity (out to 9.1 Mpc from us).**

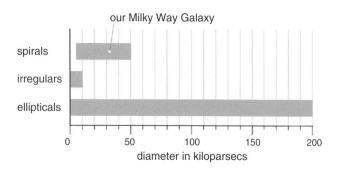

Figure 17.6 **Distribution of galaxy diameters.**

Table 17.3. **Colors and populations.**

	Color	Stellar populations
Ellipticals	Reddest	Old population II
Spirals	↓	Both population I and II
Irregulars	Bluest	Young population I

galaxy. We now understand what the data in Table 17.3 are telling us: all the stars in ellipticals are very old, whereas in spirals there are both young and old stars. *Star formation is still going on in irregular and spiral galaxies; in ellipticals it ceased a long time ago.*

Since stars form from interstellar clouds (Chapter 13), we see that this ties in with the great amount of gas and dust seen in irregular and spiral galaxies, and the absence of gas and dust in ellipticals. For some reason, elliptical galaxies used up all their star-forming material in a single burst of star formation early in their history. Conversely, spiral and irregular galaxies have been using up their star-forming material more slowly, so that star formation is still going on.

We can use this argument to understand why Hubble was wrong in thinking that galaxies evolved from left to right along his "tuning fork diagram." It is indeed the case that elliptical galaxies formed their stars a long time ago. But it is not the case that spiral galaxies have formed all their stars only recently. Notice that spirals contain *both* red and blue stars, both population I and II stars. This tells us that only some of their stars were formed recently: others were formed long ago.

Galaxy formation

How did galaxies form, and what accounts for the differences between the various types? We do not have good answers to these questions. We have detailed theories of the origin of planets and stars – but our theory of the formation of galaxies is nowhere near as complete. We are only beginning to understand how these immense objects were created.

Two theories have been proposed. In one, the so-called *top-down* theory, galaxies are thought to have formed from the collapse of huge clouds of gas – in somewhat the same way that we believe stars form from the collapse of interstellar clouds. In the other, the so-called *bottom-up* theory, galaxies are thought to have formed by the amalgamation of smaller bodies – in somewhat the same way that we believe planets formed by the amalgamation of planetesimals. Each theory has its merits and difficulties. Perhaps both are partially right: perhaps neither are. We will consider each in turn.

The top-down theory of galaxy formation

Historically, this was the first theory to have been proposed. We have, in fact, already described it – in the previous chapter, when we discussed the origin of our Milky Way Galaxy. Here is a quotation from that section.

> Early in its history, the Galaxy must have been a slowly rotating, spherical cloud of gas. In this cloud, an initial generation of stars formed. They fell inward toward the cloud's center, zoomed past it, and swung out again in elongated, highly elliptical orbits to form a spherical distribution. We can think of the Galaxy's halo as a kind of fossil, in which this initial shape is preserved. Also within this cloud, globular clusters formed: these too moved in highly elliptical orbits about the cloud's center. The fact that these early stars are so deficient in heavy elements tells us that the gas from which they formed must have been composed solely of hydrogen and helium. (In Chapter 18 we will be able to explain this fact.)
>
> Slowly, the cloud shrank. As it did so, its rotation rate gradually increased, and it commenced bulging at its equator. The more it shrank the more bulged did it become, until eventually it had become a rapidly rotating disk. While this was going on, new stars were continually forming. These partook of the disk's rapid rotation, and so moved in circular, rather than elliptical, orbits. Also while this was going on, stars were busy creating heavy elements and returning them to the interstellar medium via stellar winds and explosions such as novae and supernovae. So the newer stars, formed within the disk, were enriched with these elements. These are the population I stars: our Sun is one of them.

The top-down theory postulates that this was how *every* spiral galaxy formed. How does it account for elliptical galaxies? Notice that the above description postulates several stages of star formation: an initial burst when the cloud was large and spherical, and a later stage when the cloud had shrunk, and was rapidly rotating and disk shaped. Notice, however, what happens if we postulate an alternate scenario for star formation. Imagine a different type of galaxy, in which *all* the gas in the cloud turned into stars early in its history, before the cloud had time to collapse. In this case

- all the stars would be in a spherical pattern,
- there would be no disk,
- all the stars would now be old,
- there would be no gas and dust left at present.

Notice that this is just a description of elliptical galaxies! We conclude that in the top-down theory *the difference between elliptical and spiral galaxies is that in ellipticals star formation happened all at once a long time ago, and it used up all the gas and dust. In spirals, on the other hand, star formation was a continuous process that is still going on.*

Problems with the top-down theory

The top-down theory suffers from a number of problems.

(1) It must postulate that the cloud out of which a spiral galaxy formed was rotating. This is because the disk of a spiral rotates, and it was formed by the collapse of this cloud. Were the cloud not rotating, neither would be the disk. (Think of a figure skater who has drawn in her arms. If she had initially been rotating slowly, she would now be rotating rapidly. But if she had initially been motionless, she would still be motionless.)

As we said, our Galaxy's halo is a kind of fossil that preserves the cloud's initial state. The problem is that our halo is *not* rotating. Just as many stars within it orbit in the same sense as the disk as in the opposite sense. The top-down theory predicts that most should orbit in the same sense.

(2) Within the theory, the disk of our Galaxy formed all at once. But there appear to be *two* disks within our Galaxy, and they have different ages. One is less than 12 billion years old, while a second, far thicker disk, is more than 14 billion years old. That 2 billion year spread in ages is difficult to account for on the basis of the standard top-down theory.

(3) The top-down theory predicts that all the globular clusters within our Galaxy's halo should have formed at the same time. But they did not all form at the same time: their ages have a definite range. This is difficult to account for on the basis of the standard top-down theory.

The bottom-up theory of galaxy formation

In the bottom-up theory, galaxies are thought to have been created by the amalgamation of smaller bodies – in somewhat the same way that we believe planets were created by the amalgamation of planetesimals.

As we will see in Chapter 18, the expansion of the Universe is thought to have begun in a Big Bang, an expanding cloud of gas. Although initially the gas was very hot, the more the cloud expanded the cooler it became. Let us think about the force of gravity within this cloud. What were the effects of this force? Because gravity is a force of attraction, it acted to slow the cloud's expansion. The more time passed, the more slowly did the Universe expand.

Let us now add another crucial element to this picture: the Universe was not perfectly uniform. Rather, certain parts were slightly denser than others. Consider two regions within the expanding cloud of debris from the Big Bang. Suppose that they had the same size but different densities. The one with the higher density contained more matter – and since the force of gravity is greater for large masses than small ones, the force acting to slow its expansion was greater. So the parts with high density expanded more slowly than those with low density

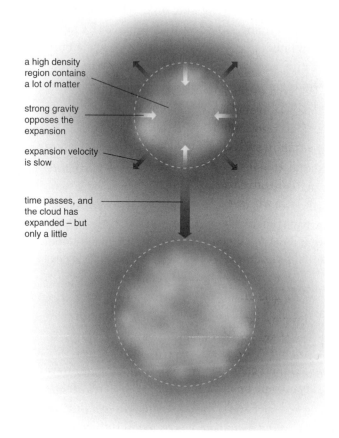

Figure 17.7 **Expansion of the early Universe. High-density regions expanded slowly, low-density ones more rapidly. Eventually the high-density regions ceased expanding and collapsed, while the low-density ones continued expanding.**

a high density region contains a lot of matter

strong gravity opposes the expansion

expansion velocity is slow

time passes, and the cloud has expanded – but only a little

(Figure 17.7). Ultimately, the high-density regions simply stopped expanding and started to contract. Meanwhile, those with low density continued to expand.

The net result was that, within the cloud expanding from the Big Bang, small dense knots separated out. *These knots were the "building blocks" out of which galaxies are thought to have formed.* Theory suggests that each contained between 1 and 100 million solar masses of material (although stars had not yet begun to form within them). Gravity from each attracted the others. They coalesced together to form the galaxies.

· ·

NOW YOU DO IT

We want to calculate how many "building blocks" are required to form a galaxy, if each "block" had a mass of 1 million solar masses. Suppose the galaxy we are considering has a mass equal to that of our Milky Way Galaxy (about 800 billion solar masses). (A) Describe the logic of the calculation you will perform. (B) Carry out the detailed calculation. (C) If it formed over a time span of 5 billion years, how many years must have elapsed between mergers of these building blocks during its "construction phase"?

· ·

It is important to emphasize that in this bottom-up picture, *the process of galaxy building is still going on.* It began shortly after the Big Bang, and initially it was very rapid. As time passed, it proceeded more slowly – but it has never stopped. Even today, we can witness its final stages, as galaxies merge together. We now turn to this topic.

Figure 17.8 **Stephan's Quintet, a group of five galaxies.** 👁 **(Also see color plate section.)**

Galaxy mergers

Figure 17.8 is an image of Stephan's Quintet, a collection of five galaxies. The object at the upper left portion of this image is a foreground galaxy that by chance happens to lie in front of the others. But the remaining four galaxies are physically close to one another. As for these four, none of them is normal. Each has long curving filaments of stars, and distorted shapes. Two of them actually seem to overlap. These are *colliding galaxies.*

What happens when galaxies collide? It is not like the crashing together of two cars. Rather it is more like what happens when a flock of birds flies into a second flock. They merely interpenetrate. Furthermore, during this interpenetration the birds do not collide with each other either, since within each flock there is a lot of empty space between the birds. So in the encounter the flocks simply pass through one another. Similarly, in an encounter the galaxies pass through one another, and the stars within them do not collide. So it is better to refer to the event as an encounter between galaxies, rather than a collision.

Figure 17.9 **The Antennae Galaxies.** 👁 (Also see color plate section.)

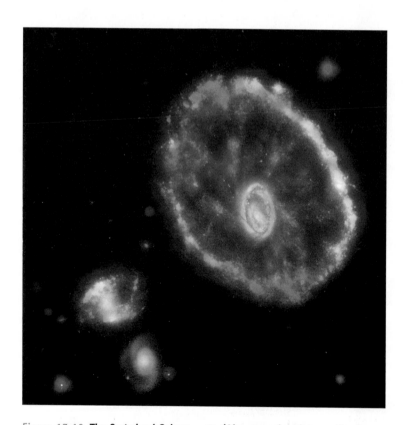

Figure 17.10 **The Cartwheel Galaxy.** 👁 (Also see color plate section.)

But unlike flocks of birds, galaxies do not emerge from the encounter unchanged. Because of the gravitational interactions between them, the encounter drastically alters their structure. The galaxies in Stephan's Quintet are examples of what can happen when this occurs. Figures 17.9 and 17.10 show more examples. As you can see, encounters between galaxies have strange and interesting consequences. *The bottom-up theory of galaxy formation postulates that galaxies formed and grow as a result of these interactions.*

We now need to study each of these issues more carefully. We will first show that (A) encounters between galaxies are common, and in an encounter direct collisions between stars do not occur. And we will then study (B) what happens when galaxies encounter one another.

(A)   The rate of encounters between galaxies and stars

Our goal in this section is to understand why it is that *encounters between galaxies are common,* but that in an event *actual collisions between stars do not occur.*

Figure 17.11 illustrates a collection of objects moving about randomly. How often do they encounter one another? As shown in this figure, the answer depends on two factors.

(1) It depends on *the distances between the objects*. If they are close together, encounters will be common, whereas if they are far apart encounters will be rare.

(2) It also depends on *their sizes*. We can think of each object as a "target," and the others as "bullets flying towards the target." If the "target" is big, the "bullets" are more likely to strike it than if it is small.

This teaches us that we need to think about the ratio κ:

$$\kappa = (\text{distance between objects})/(\text{size of objects}).$$

Notice that there are two ways to make κ big: to make the distance between objects large, or to make the size of each object small. But both of these lead to a very small rate of collisions! Similarly, there are two ways to make κ small: to make the distance between objects small, or to make the size of each object large – and both of these lead to a very large rate of collisions. So we see that *if κ is large, collisions will be rare: if it is small, collisions will be common.*

Let us begin with galaxies, and ask how often we expect them to encounter one another. We will find out by calculating κ.

The logic of the calculation

Step 1. Find the diameter of galaxies and the distance between them.

Step 2. Divide the distance by the diameter to find their ratio κ.

Step 3. Is κ big or little? If it is big, encounters will be rare: if it is little, encounters will be common.

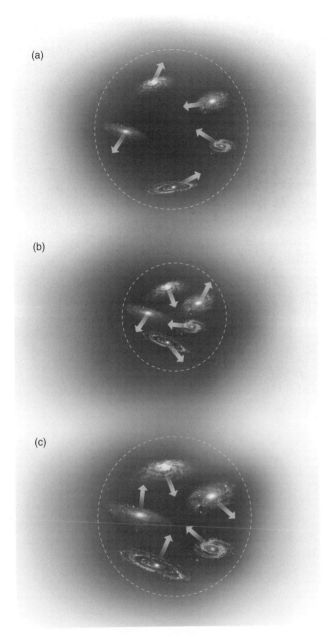

Figure 17.11 **Encounters between randomly moving objects. In configurations (b) and (c) encounters happen more frequently than in (a). This teaches us that the rate of these encounters depends on two factors. If we compare (a) with (b) we see that the rate depends on the distance between objects; if we compare (a) with (c) we see that it depends on their size.**

As you can see from the detailed calculation, κ is large but not exceedingly so. If we wait long enough – and in the astronomical context we have billions of years to wait – encounters will occur.

Now we pass on to our second question: what happens once two galaxies encounter one another? Will the stars within them actually collide?

Detailed calculation

Step 1. Find the diameter of galaxies and the distance between them.

Our Galaxy is average, and our neighborhood is typical. So let us use in our calculations the diameter of our Galaxy and a typical distance between it and a nearby large neighbor, which is about a megaparsec.

The diameter of the Milky Way Galaxy is D_{Galaxy} = 100 000 light years and the distance between galaxies is 1 MPC, or

$D_{\text{neighbor galaxy}}$ = 1 MPC = (3.26 light years/parsec) × (1 million parsecs/MPC)
= 3.26 × 10^6 light years.

Step 2. Divide the distance by the diameter to find their ratio κ.

κ = 3.26 × 10^6 light years/100 000 light years = 33.

Step 3. Is κ big or little?

κ is big but not exceedingly big: encounters will not be particularly rare.

Detailed calculation

Step 1. Find the diameter of a star and the distance between stars.

Our Sun is an average star, and our neighborhood is typical. So you might think that we should use in our calculations the diameter of our Sun and the distance between it and its nearest neighbor, Alpha Centauri (4 light years). But notice that, when two galaxies pass through each other, the stars from both are jammed into the same region of space. This means that the distance between stars will be smaller than that between the Sun and Alpha Centauri. We will take a distance between stars to be 3 light years rather than 4 light years.

The diameter of the Sun is

D_{Sun} = 2 R_{Sun} = (2) (6.96 × 10^8 meters) = 1.39 × 10^9 meters and the distance to the nearest star will be

$D_{\text{other star}}$ = (3 light years) (9.46 × 10^15 meters/light year)
= 2.84 × 10^16 meters.

Step 2. Divide the distance by the diameter to find their ratio κ.

κ = 2.84 × 10^16 meters/1.39 × 10^9 meters = 20 million.

Step 3. Is κ big or little?

κ is extremely big: collisions will be very rare.

The logic of the calculation

We need to repeat the same calculation, but now thinking about the stars within each galaxy – and here we really are thinking about actual collisions, rather than encounters!

NOW YOU DO IT
Our analysis is valid for *any* collection of randomly moving objects. So let us apply it to (A) snowflakes in a blizzard, and (B) grains of sand in a sandstorm. If each flake is ¼ inch across, and they are 6 inches apart, what is κ? If each sand grain is 1 millimeter across and they are ½ centimeter apart, what is κ? In which case are encounters more common?

As you can see from the detailed calculation, κ is very large: 20 million. This tells us that, when galaxies encounter one another, *collisions between the stars within them essentially never happen.*

(B)   Encounters between galaxies

We now turn to a study of what happens when galaxies encounter one another. We have seen some of the many strange and wonderful things that can result. How can we understand them?

It is important to realize that we cannot simply wait and watch what happens. Previously we likened an encounter between two galaxies as being like what happens when two flocks of birds pass through one another. But this is not really a very good analogy, for as we will now see this encounter takes place over geologic ages. As shown in the detailed calculation, an astronomer would have to live for hundreds of millions of years in order to observe what happens. So we are forced to rely on computer simulations as opposed to direct observation to build our understanding.

The logic of the calculation

Divide the diameter of a typical galaxy by their relative speed. This will tell us how long it takes one galaxy to pass through another.

Detailed calculation

Divide the diameter of a typical galaxy by their relative speed. This will tell us how long it takes one galaxy to pass through another.

As they wander through space, galaxies move at relative speeds of about 100 kilometers/second. So we calculate:

collision time = D/V.

As we know, a typical galaxy's diameter is 100 000 light years, which is

D = (100 000 light years)(9.46 × 10^{15} meters/light year)
   = 9.46 × 10^{20} meters.

The velocity is

V = (100 kilometers/second)(1000 meters/kilometer)
   = 10^5 meters/second.

We then find the collision time to be

collision time = D/V = 9.46 × 10^{20} meters/10^5 meters/second = 9.46 × 10^{15} seconds = 300 million years.

⇐ **Looking backward**
We studied tides in Chapter 7.

NOW YOU DO IT
Suppose an encounter between two objects, with a relative velocity of 50 kilometers per second, takes place over 50 000 years. How big are they?

Following are some of the principles that come into play in these simulations.

Tides

Tides arise when two astronomical objects lie close to one another. As we recall, under the action of tidal forces each body develops two bulges. We also saw (in our study of Saturn's rings) that if a body approaches another too closely, the tides can be so great as to rip it to shreds.

This occurs even if the bodies in question are not planets or moons, but galaxies. Figure 17.12 is a series of stills from a computer simulation of spiral galaxies encountering one another.

As the galaxies approach, stars within each galaxy begin to orbit about the other. As you can see, in this process each galaxy raises a tide on the other. Rather than being manifested as bulges, however, these take the form of long curving tails. The differences between these results and our experience with tides in moons and planets arise because a galaxy is not a solid body, but rather a collection of orbiting stars. Notice how similar this simulation is to the Antennae Galaxies (Figure 17.9): we now understand that the "antennae" are tidal bulges.

Galactic cannibalism

Under certain circumstances, a large galaxy can "eat" a smaller one, in a process known as "galactic cannibalism." Imagine a small galaxy on a path that will carry it right through a larger one. In terms of our analogy discussed above, we might liken this to a small flock of birds passing through a big flock. That analogy, however, is not entirely accurate. It turns out that, even if there is no collision between stars in the encounter, there is a decelerating force that acts to slow the intruding galaxy. We might think of this force as a kind of "friction" (it is termed "dynamical friction"). Ultimately this "friction" brings the smaller galaxy to rest, and it is absorbed into the larger.

Figure 17.13 illustrates the origin of dynamical friction. The figure shows an intruding small galaxy moving through the larger one. As it moves, the intruder gravitationally attracts stars within the larger galaxy toward it. Responding to this force, these stars follow curving paths about the intruder. As indicated, these paths intersect one another in a region behind the moving small galaxy. We might say that the paths are "focused" into this region by the small galaxy's gravity. The net result

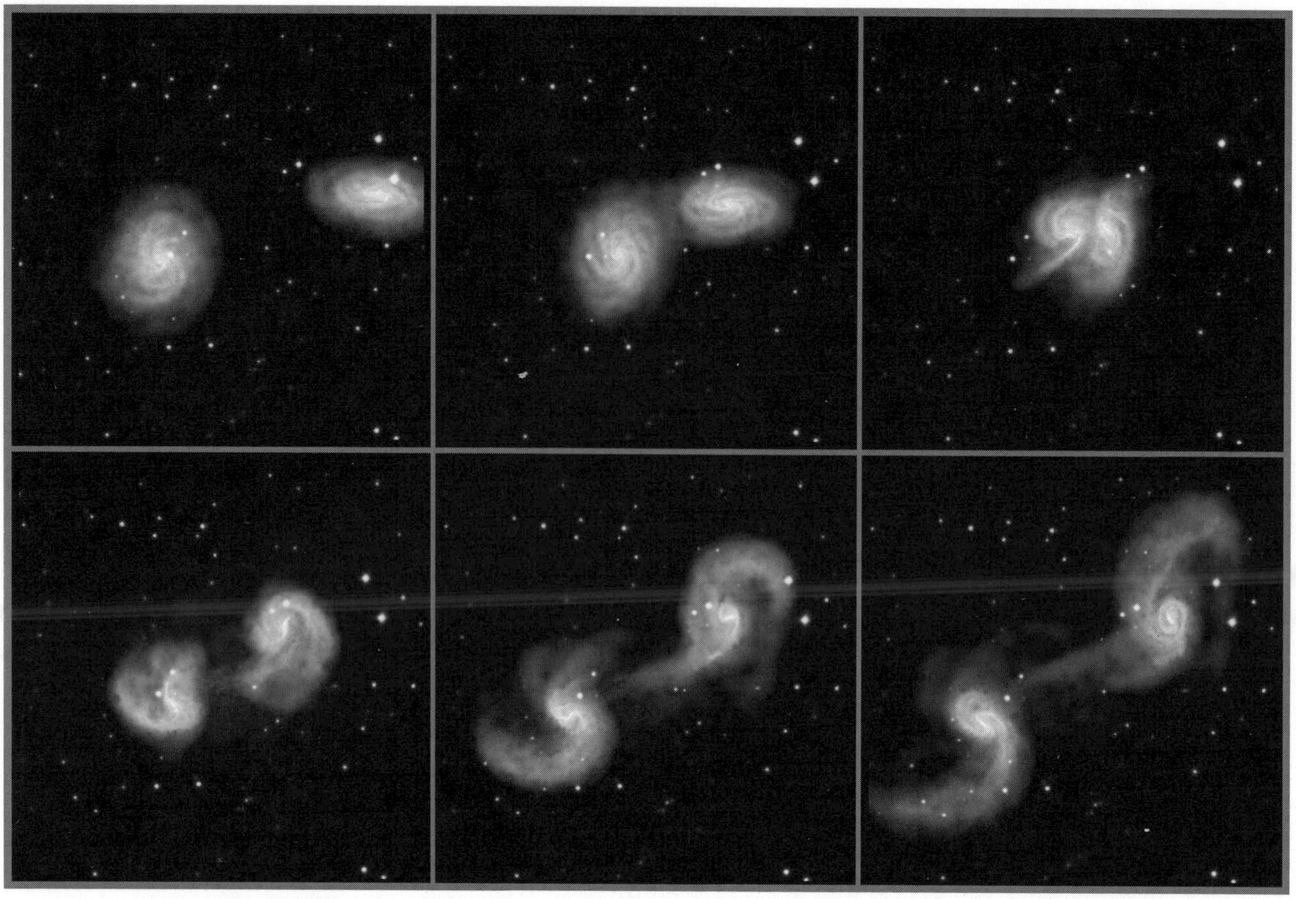

Figure 17.12 **Tides in galaxies. A computer simulation of two galaxies encountering one another. Each raises a tide on the other, leading to long curving tails.**

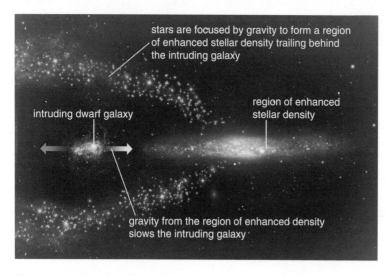

stars are focused by gravity to form a region
of enhanced stellar density trailing behind
the intruding galaxy

region of enhanced
stellar density

intruding dwarf galaxy

gravity from the region of enhanced density
slows the intruding galaxy

Figure 17.13 **Dynamical friction.**

is that stars from the larger galaxy are concentrated into a "wake" following behind the smaller one. This "wake" gravitationally attracts the intruder toward it – which is to say, *the wake slows the motion of the intruder*. This decelerating force is our dynamical friction.

The process of galactic cannibalism is particularly efficient if the small galaxy remains in the vicinity of the larger one. This would occur if it were orbiting about the larger, rather than simply passing through it. In this case, even though the "friction" might be weak, it has billions of years in which to act. The result is a steady slowing of the motion of the orbiting galaxy, which smoothly spirals inwards until it is brought to rest within the larger one and is absorbed into it. This process appears to be occurring right now in our own Milky Way Galaxy. It has been calculated that dynamical friction is slowing the motion of the Clouds of Magellan, small galaxies in our vicinity, and that they will spiral into our own in some 10 billion years.

Star formation

Let us turn now to the 'Cartwheel Galaxy' illustrated in Figure 17.10. This appears to be a galaxy through which a smaller galaxy has passed. We might liken the Cartwheel to a "target" through which a "bullet" has been shot. Presumably the "bullet" is one of the two small galaxies on the left of the figure, although there is some argument as to which is the culprit.

Observations with the Hubble Space Telescope have shed light on the striking ring surrounding the Cartwheel. Hubble has found within the ring (1) huge holes in the interstellar medium blown by exploding supernovae, together with (2) large numbers of extremely blue stars. As we saw in Chapter 15, supernovae arise from the deaths of short-lived, high-mass stars. And as we have seen above, blue stars are extremely short-lived and massive. Both (1) and (2) are telling us that the stars in the ring are young – which means that there was a recent burst of star formation within the ring.

Within the Cartwheel Galaxy, dynamical friction appears not to have been strong enough to have brought the intruding smaller galaxy to rest. Rather it passed right through and emerged on the other side. But the encounter deposited a great amount of energy in the "target" galaxy, resulting in an expanding wave (200 000 miles per hour!) in the gas sur-

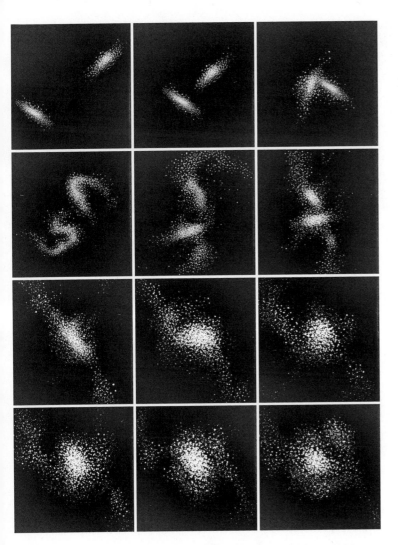

Figure 17.14 **Computer simulation of the merger of two spiral galaxies to yield an elliptical.**

rounding it. This wave of compression has triggered a burst of new star formation in the ring. (The faint spokes joining the inner portions of the Cartwheel to the ring appear to be the first traces of spiral structure beginning to re-form.) The Hubble Telescope has also identified regions of new star formation in the Antennae Galaxies (Figure 17.9). Indeed, there is a whole class known as "starburst galaxies," in which a recent encounter has triggered an intense burst of new star formation.

Formation of elliptical galaxies

In some cases a merger between two spiral galaxies can produce an elliptical. The many gravitational interactions among stars destroy the orderly pattern of disk and spiral arms, leaving a roughly spherical distribution. Figure 17.14 shows a computer simulation of just such a process.

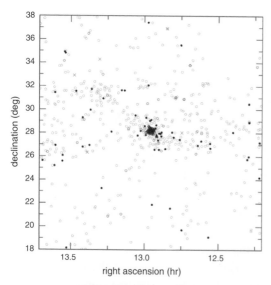

Figure 17.15 **Elliptical galaxies prefer high-density regions. In this map of the distribution of galaxies in the Coma cluster of galaxies, each filled circle represents an elliptical galaxy and each open circle a spiral. Ellipticals are clearly more prevalent in the inner, denser regions where collisions are more common.**

Figure 17.17 **An incoming dwarf galaxy warps the disk of a spiral.**

Figure 17.16 **A warped spiral galaxy.**

Evidence for the bottom-up theory of galaxy formation

Two lines of evidence have been found that support this theory of the formation of galaxies.

(1) *Elliptical galaxies are preferentially found in regions where collisions are expected.* Figure 17.15 is a map of a cluster of galaxies. Each circle represents a galaxy: a filled circle an elliptical galaxy, and an open circle a spiral. As you can see, ellipticals are more prevalent in the inner regions of the cluster, and spirals in the outer regions. But the cluster's inner regions are just where galaxies are more closely packed together, and where collisions should be more common. Similarly, in the cluster's outer regions galaxies lie far from one another, and collisions should be rarer. Figure 17.15 is therefore just what we would expect if ellipticals resulted from collisions between spirals.

(2) *The theory nicely explains warped disks.* Often the disks of spiral galaxies are found to be warped. Figure 17.16 shows an example. Surveys have shown that about half of all spirals possess such warped disks.

Alternatively, the top-down theory gives a possible explanation of why this should be so. Figure 17.17 illustrates an encounter in which a dwarf galaxy passes through a spiral. As illustrated, as the dwarf passes above the disk it exerts a gravitational force twisting its near edge upwards. After the dwarf has passed below the disk, it exerts a force twisting its far edge downwards. The fact that the theory can so easily explain such disks is evidence in its favor.

THE NATURE OF SCIENCE

HOW MUCH WEIGHT SHOULD WE GIVE EVIDENCE?

In our discussion of the process of discovery in science, we emphasized that every line of evidence is potentially suspect. Much "evidence" that scientists think they have discovered turns out in the long run to be simply false – it fails to hold up to subsequent tests, or is shown to be wrong in the light of new, previously unavailable, information. This suspect nature of evidence makes all scientific discovery difficult. Scientists often find themselves in the position of having to make judgment calls.

We wish to return to this issue here, and keep it in mind as we discuss the above two lines of evidence. Do they really argue in favor of the bottom-up theory of galaxy formation? Both of these two *observations* are ironclad. But the *interpretations* of these observations are not. In particular, alternate theories have been proposed, which attempt to explain these observations without recourse to the bottom-up theory.

(1) The fact that elliptical galaxies are preferentially found in regions where collisions are expected does not have to mean that they were formed from collisions. An alternate theory has been developed in which galaxies formed via the top-down process – but the high densities characteristic of the centers of clusters favored the formation of ellipticals.

(2) As illustrated in Figure 17.18 if a spiral galaxy has an elliptical halo of dark matter, and if the halo is for some reason tilted relative to its disk, gravity from the halo will warp the disk.

So we see that neither of these arguments in favor of the theory is ironclad. No scientist is happy with such a situation. The goal of science is to find airtight arguments: reasoning we can rely on. Until we have done so, our work is not complete.

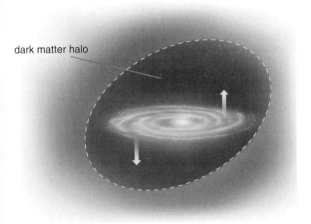

dark matter halo

Figure 17.18 **Alternate theory of warped disks. A tilted elliptical halo can warp a disk via its gravitational pull.**

Violent activity in galaxies

Our Milky Way Galaxy is quiescent – but certain other galaxies are found to be the seats of violent activity, and they emit vast amounts of energy. We now turn to this topic.

Radio galaxies

The radio source Cygnus A

Among the very first sources of extraterrestrial radio emissions ever to be detected were Cas A, so named because it is the brightest radio source in the constellation of Cassiopeia, and Cygnus A, the brightest radio source in Cygnus. Cas A is a nearby supernova remnant. Cygnus A, on the other hand, is a distant galaxy. Indeed, the galaxy is so far away that it is difficult to obtain a good image of it: Figure 17.19 is representative of the best that was possible for years. (The image is a negative, so white appears black and vice versa.)

It is remarkable, therefore, that such an enormously distant galaxy emits enough energy to make it the second brightest radio source in the sky. Indeed, it is nearly as bright as the brightest: the galaxy's flux is fully 70% of that of the far closer supernova remnant. But Cygnus A lies 750 million parsecs from

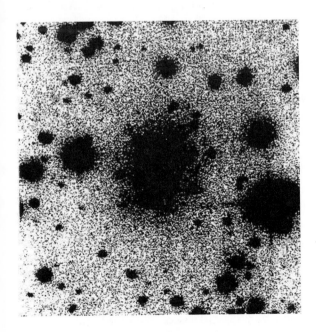

Figure 17.19 **Optical image of Cygnus A.**

us, while Cas A is a mere 3 thousand parsecs distant. The ratio of these two distances is very large. Indeed, if we imagine an analogy in which the closer source lay one meter away – about at arm's length – the other would have to lie 250 000 meters, or fully 150 miles, off in the distance. And yet they are nearly the same brightness! We are forced to conclude that Cygnus A must be a very intense source of radio emissions indeed:

$$\frac{\text{distance to Cygnus A}}{\text{distance to Cas A}} = \frac{750 \text{ million parsecs}}{3 \text{ thousand parsecs}} = 250\,000.$$

The radio galaxy grows even more remarkable if we study it at radio wavelengths. Figure 17.20 is an image of the radio emission from Cygnus A. As we see from this image, the emissions do not come from the galaxy itself. Rather they come from two lobes lying symmetrically about it. These lobes are far larger than the galaxy, and they are connected to it by two oppositely directed, ultra-thin jets. No optical emission is observed from these radio-emitting lobes: they appear to contain no stars at all.

Detailed observations demonstrate that the emissions from these lobes come from ultra-high-speed electrons moving through a magnetic field. Some object within the central galaxy must have ejected the field and electrons (together with an equal number of protons that emit little radiation). Presumably the jets linking the galaxy to the lobes are the "pipelines" through which the field and particles were ejected. Observations also indicate that this ejection process did not take place all at once, but was spread out over time: indeed, it may be going on right now.

Many radio galaxies like Cygnus A have been discovered. All exhibit similar characteristics, in which an elliptical galaxy appears to have ejected two blobs of magnetic field plus particles in opposite directions. Often a jet of radio emission can be detected connecting the blobs to the galaxy.

The energy requirements of radio galaxies

The radio-emitting lobes of radio galaxies do not contain planets, stars, dust or anything else that can be detected by optical telescopes. Nevertheless, they do contain matter in the form of rapidly moving particles and fields. And because moving particles carry kinetic energy, and a magnetic field carries magnetic energy, these radio-emitting lobes contain energy. Indeed, measurements reveal that they contain a very great deal of energy indeed.

Let us concentrate on the specific example of Cygnus A. The total energy content of its two radio-emitting lobes is measured to be roughly 10^{53} joules. To gain an appreciation of this, let us ask how long our own Milky Way Galaxy would take to emit so much energy. Since the luminosity of our Galaxy is $L_{\text{Milky Way}} = 3.8 \times 10^{37}$ joules per second, we see that the time required is 10^{53} joules$/3.8 \times 10^{37}$ joules per second $= 2.6 \times 10^{15}$ seconds, or 84 million years. This is a very great length of time indeed.

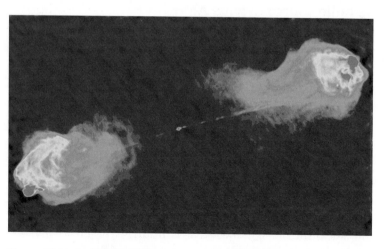

Figure 17.20 **Radio Map of Cygnus A. The optical galaxy is the small dot from which the jets emanate.**

Figure 17.21 **3C 273 lies somewhere within this image, but the early radio surveys were incapable of precisely pinpointing its location. We now know that the radio source is the tiny dot indicated by the arrow.**

Figure 17.22 **Location of the Moon at the moment it occulted 3C 273. The radio source must lie somewhere along the Moon's edge.**

What object within Cygnus A could emit high-speed particles and fields carrying such a huge amount of energy? Clearly, the object must be very remarkable. We will return to this question shortly.

Quasars

The radio source 3C 273

Early radio astronomers compiled extensive catalogs of astronomical sources of radio emission. One such catalog was the 3C catalog of radio sources, so named because it was the third to be produced by astronomers at Cambridge University in England. 3C 273 is the 273rd radio source listed in that catalog.

As we have explained in Chapter 5, it is exceedingly difficult to build a radio telescope with good resolving power. As a consequence, the early radio telescopes were not capable of precisely identifying the objects responsible for the emissions they had detected. Rather, they were only capable of saying that their radio emitting object lay within some general area. Figure 17.21 shows the region of the sky in which 3C 273 lay. As you can see, it contains a great number of stars: nobody knew which one of them was emitting the radio signals.

Eventually the precise location of the source of these radio emissions was pinned down by means of an ingenious technique. It so happened that the Moon, in its orbit about the Earth, occasionally passed through the region illustrated in Figure 17.21. As it did so, radio astronomers trained their telescopes on 3C 273, carefully monitoring its emissions. They found that, at a certain instant, these emissions abruptly ceased.

The object, of course, had not really ceased emitting its radio signals. Rather, its emissions had been cut off by the Moon. Just as the Moon when it passes in front of the Sun causes a solar eclipse, so had the Moon now eclipsed the radio source 3C 273. Such events are termed "occultations."

Because astronomers knew the orbit of the Moon very accurately, they knew precisely where it was at the instant it occulted 3C 273. Figure 17.22 shows the location of the Moon's edge at that moment. Astronomers now knew that the source of radio emissions lay somewhere along the curving path defined by this edge.

As you can see, there are still many objects that might be 3C 273. How could astronomers narrow down the range of possibilities? They did this by waiting till the *next* occultation of the radio source. Because the orbit of the Moon is constantly changing, it was in a different position at the moment of this second occultation than it had been at the first. So the second curving path indicating the possible locations of 3C 273 was different from the first. As shown in Figure 17.23, there is only one object lying on the edge of *both* paths. This object is the radio source 3C 273.

As you can see, 3C 273 looks like a star. Some of the other radio sources in the 3C catalog also looked like stars. Because of their appearance, astronomers called them

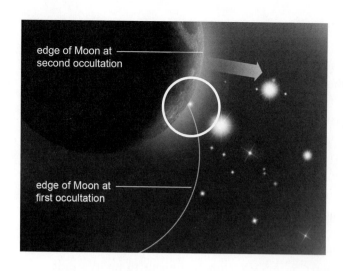

edge of Moon at second occultation

edge of Moon at first occultation

Figure 17.23 **Pinpointing the position of 3C 273. It must lie at the intersection of the two curving paths.**

Quasi-Stellar Radio Sources, or QSRS for short. If you pronounce this acronym, you will hear yourself saying "Quasars."

What are quasars? We will answer this question by continuing to focus on one of them: 3C 273. We will begin by asking how far away it is.

Where are the quasars?

Because 3C 273 looks like a star, astronomers thought at first that it was a star located within our Galaxy. Eventually they realized that they were drastically mistaken, and that it lay far outside the limits of our Galaxy.

In 1963 the American astronomer Maarten Schmidt studied the spectrum of 3C 273 (Figure 17.24). Initially he found the spectrum difficult to interpret, because the spectral lines were unfamiliar. Eventually Schmidt realized that the lines were unfamiliar because they ordinarily had very short wavelengths. Indeed, the wavelengths were so short that they lay in the ultraviolet region of the spectrum and were absorbed by our atmosphere: for this reason they had never before been observed in astronomical objects. But in 3C 273 these lines were redshifted by an enormous amount, into the visible region of the spectrum. Schmidt found that the lines were shifted by

$$\Delta\lambda/\lambda = 0.16.$$

If we use the Doppler effect formula

$$\Delta\lambda/\lambda = v/c$$

we see that the velocity v of 3C 273 is 0.16 of the speed of light. The object, which looks like a star, was flying away from us at 16% the speed of light.

This enormous speed tells us something about the distance to 3C 273. Let us ask if the quasar could lie somewhere within our own Milky Way Galaxy. We can easily see that it probably does not. The reason is that the enormous velocity we have found for 3C 273 is far greater than the escape velocity from our Galaxy. If our Galaxy contained the quasar, its gravity could not bind the quasar to it. Rather 3C 273 would fly out of the Galaxy and into space.

There is a more likely possibility. As we will see in the next chapter, the entire Universe is expanding. According to Hubble's law of the expansion of the Universe,

Figure 17.24 **Spectrum of 3C 273. Note that the three emission lines are greatly shifted toward longer wavelengths.**

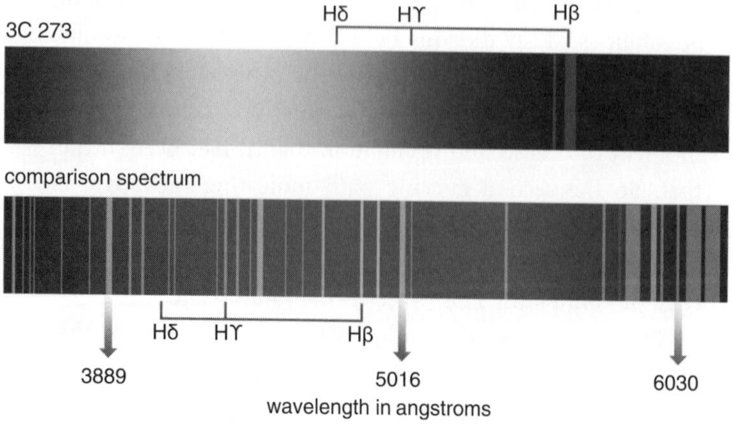

the more distant an object the faster it is moving away from us. So we can understand the quasar's great velocity by postulating that it is very distant. By using Hubble's law, we can calculate how far away 3C 273 must lie in order for it to have such a velocity. The answer is that 3C 273 is 2.4 billion light years away.

At the time Schmidt did this calculation, this was greater than the distance to any known galaxy. He had discovered the farthest object in the Universe. Furthermore, by now we know of many quasars even

3C 273

Hδ   HΥ   Hβ

comparison spectrum

Hδ   HΥ   Hβ

3889   5016   6030

wavelength in angstroms

farther away than this. As a class, the quasars are the most distant objects we have ever found.

How bright are the quasars?

Return to Figure 17.19, showing the quasar 3C 273. As the name "quasar" (QSRS) implies, it looks like a star. In particular, it is as bright as the true stars in that image. But we now know that it is far more distant than they are. This means that it must be very bright.

Let us compare the distance of 3C 273 to that of the stars in Figure 17.19. Typically, a star in our vicinity of the Milky Way Galaxy lies several thousand light years from us. How does 3C 273's distance of 2.4 billion light years compare to this? Concentrating on a representative star, say one 2400 light years distant, we calculate

$$\frac{\text{distance to 3C 273}}{\text{distance to the star}} = \frac{2.4 \text{ billion light years}}{2400 \text{ light years}} = 1 \text{ million}.$$

There is only one way something a million times more distant than a star can appear to be as bright as the star: by being enormously more luminous. Indeed, as we will now see, the quasars are the most brilliant objects in the Universe.

Let us use the inverse square law (Chapter 4) to calculate the luminosity of 3C 273.

The logic of the calculation

Step 1. Solve the inverse square law

$$f = L/4\pi R^2$$

for the luminosity L as a function of the apparent brightness f and distance R.

Step 2. Plug the measured values of f and R into the result, and find the luminosity.

Step 3. To get an intuitive feel for the answer, compare the answer for the luminosity to that of our entire Milky Way Galaxy.

Detailed calculation

Step 1. *Solve the inverse square law for the luminosity.*
We have

$f = L/4\pi R^2$.

Solving for L we get

$L = 4\pi R^2 f$.

Step 2. *Plug the measured values of f and R into this and calculate the luminosity.*
The measured apparent brightness of 3C 273 – its flux f – is

$f = 1.5 \times 10^{-12}$ watts/meter^2

and its distance is

$R = 2.4$ billion light years $= (2.4$ billion light years$) \times$ $(9.46 \times 10^{15}$ meters/light year$)$
$= 2.3 \times 10^{25}$ meters.

So we calculate

$L = 4\pi (2.3 \times 10^{25}$ meters$)^2 (1.5 \times 10^{-12}$ watts/meter$^2)$
$= 9.7 \times 10^{39}$ watts.

Step 3. *To get an intuitive feel for the answer, compare the answer for the luminosity of that of our entire Milky Way Galaxy.*
The luminosity of our Milky Way Galaxy is

$L_{\text{Milky Way}} = 3.8 \times 10^{37}$ watts.

So we calculate the ratio

$L_{3C\ 273}/L_{\text{Milky Way}} = 9.7 \times 10^{39}$ watts$/3.8 \times 10^{37}$ watts $= 256$.
The quasar is 256 times as bright as our Galaxy.

NOW YOU DO IT
Suppose the quasar were actually at half the above distance, and its apparent brightness were 10 times greater. We want to calculate the ratio between its luminosity and that of our Galaxy. One way to do this is to re-do the above calculation. But there is an easier way! See if you can find it.

As you can see from the detailed calculation, the quasar 3C 273 is 256 times as luminous as our entire Milky Way Galaxy. We get similar results for the other quasars as well: they are the most brilliant known objects in the Universe.

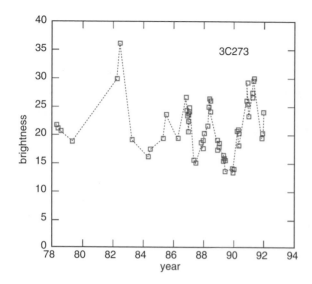

Figure 17.25 **3C 273 varies in brightness over short periods of time.**

How big are the quasars?

If 3C 273 is brighter than a galaxy, we might think that it is bigger than a galaxy. Remarkably, this turns out not to be the case. Indeed, the astonishing thing about quasars is that, even though they generate extraordinary amounts of energy, they are comparatively speaking very small.

We can get an idea of the size of a quasar by studying the *variations in its brightness*. Figure 17.25 shows the measured apparent brightness of 3C 273 over the course of several years. As you can see, it does not shine steadily. It flares and fades.

What does the variability of 3C 273 have to do with its size? It turns out that *if an object is observed to brighten over a certain period of time, its radius cannot be greater than the distance a light ray travels in this time.* Since 3C 273 is observed to brighten over a short time, its size must be comparatively small.

Let us see why this is so. Suppose we turn on a light bulb. The bulb turns on all at once. But we do not *see* it turn on all at once. Light reaching us from the near edge of the bulb has a shorter distance to travel than light reaching us from its top. This means that we will receive the light from the near edge before we receive the light from the top.

As illustrated in Figure 17.26, the extra distance the light from the top has to travel is R, the radius of the bulb. If c is the velocity of light, it takes an extra time R/c for the light from the top to travel this extra distance. This tells us that if an object of radius R brightens instantaneously, we perceive it to brighten over a time R/c. We can also turn this logic around: if an object is seen to brighten in a time t, its radius cannot be greater than $R = ct$. We recognize that ct is the distance that light travels in a time t.

In everyday life this principle has no practical significance. Let us apply it, for instance, to the process of turning on a light bulb.

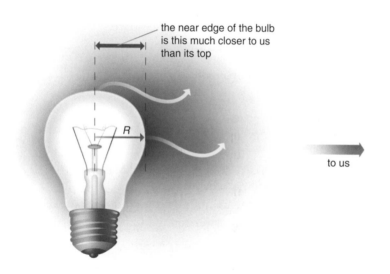

the near edge of the bulb is this much closer to us than its top

R

to us

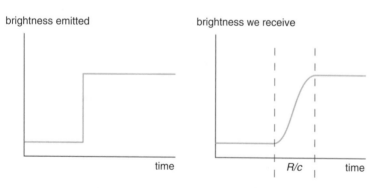

brightness emitted

brightness we receive

time

R/c   time

Figure 17.26 **Turning on a light bulb. Because different parts are at different distances from us, we do not see it brighten all at once.**

The logic of the calculation

Step 1. Measure the radius of a light bulb.

Step 2. Divide this radius by the speed of light to find the time over which it appears to brighten when we turn it on.

As you can see from the detailed calculation, the bulb appears to brighten over an extraordinarily short period of time: a mere 8.5×10^{-11} seconds. This is instantaneous for all practical purposes. But if we repeat this calculation for something larger we get a longer time. If, for instance, the Sun were suddenly to brighten, we would observe it gradually to brighten over a period of several seconds.

Now let us apply the reverse logic to 3C 273. As we can see from Figure 17.25, it brightens and dims over a period of several months. What does this tell us about its size?

Detailed calculation

Step 1. Measure the radius of a light bulb.

A light bulb is about an inch in radius.

Step 2. Divide this radius by the speed of light to find the time over which it appears to brighten when we turn it on.

The speed of light is 3×10^8 meters/second. So we calculate

time = (1 inch) (0.0254 meters/inch)/(3×10^8 meters/second) = 8.5×10^{-11} seconds.

The logic of the calculation

Step 1. Measure the time interval over which 3C 273 brightens.

Step 2. Multiply c times this time to find its greatest possible size.

Step 3. Compare this size to that of a galaxy to get an intuitive idea of the significance of our result.

As you can see from the detailed calculation, 3C 273 is extraordinarily small: about 1/400 000th the size of a galaxy. And in spite of this small size, it generates hundreds of times as much energy as a galaxy!

Detailed calculation

Step 1. Measure the time interval over which 3C 273 brightens.

As we can see from Figure 17.25, in 1986 the quasar brightened in about 3 months.

Step 2. Multiply c times this time to find the quasar's greatest possible size.

We calculate

maximum size = (3.00×10^8 meters/second) (about 3 months) (about 30 days/month) (24 hours/day) (3600 seconds/hour) = about 2.3×10^{15} meters.

Step 3. Compare this size to that of a galaxy to get an intuitive idea of the significance of our result.

The diameter of our Galaxy is 100 000 light years. We calculate

diameter of our Galaxy = (100 000 light years) (9.46×10^{15} meters/light year) = 9.46×10^{20} meters.

Let us calculate the ratio of the size of our Galaxy to that of 3C 273:

ratio = 9.46×10^{20} meters/about 2.3×10^{15} meters = about 400 000.

3C 273 is about 1/400 000th the size of a normal galaxy.

NOW YOU DO IT

Suppose I live one mile from the Wal-Mart parking lot, and suppose there are two people, one at the near edge of the parking lot and the other at the far edge. At a certain moment, both start walking toward me at 4 miles per hour (6.44 kilometers per hour). One person reaches me at noon, and the other 3 minutes later. We want to calculate how big the parking lot is. (A) Describe the logic of the calculation you will perform. (B) Carry out the detailed calculation. (C) Why is this exercise located here in our study of quasars?

This is the central mystery of quasars: *how can they generate such huge amounts of energy in such small regions of space?*

Quasars as the nuclei of galaxies

Because quasars are so far away it is difficult to see them clearly. Nevertheless, by utilizing our most powerful telescopes, it is just barely possible to study them in some detail. These studies have revealed that

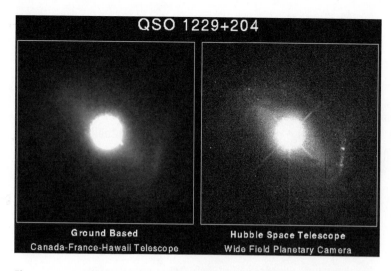

Figure 17.27 **The faint "fuzz" around a quasar appears to be a galaxy. Such images indicate that quasars reside at the nuclei of galaxies. This quasar, QSO 1229+204, lies about 6 billion light years away.**

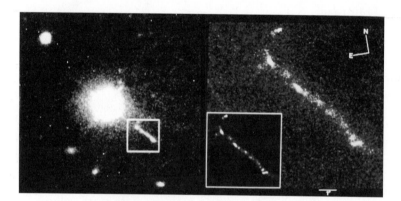

Figure 17.28 **Optical image of the quasar 3C 273.**

quasars are not really individual points of light. Rather, they are points of light surrounded by a faint "fuzz." We believe that this "fuzz" is starlight, and that *quasars reside in the nuclei of galaxies.*

Figure 17.27 shows one such quasar, QSO 1229+204. This image has been greatly overexposed, so that the light from the quasar, which in reality is point like, here appears to be spread out over a large region. But this overexposure has allowed us to capture the far fainter light from the galaxy in which the quasar resides. Clearly visible in this image is a faint ring, presumably arising from the stars in an arm of the galaxy.

Figure 17.28 is an image of 3C 273. Notice the jet of material that has been apparently ejected from this quasar.

Such studies lead to the view that quasars are phenomena that are part and parcel of the evolution of galaxies. Similarly, we have seen that radio emission is often observed from objects that are clearly galaxies. Apparently something can happen within the nucleus of a galaxy that can lead to violent outbursts and the emission of great amounts of energy. What could this "something" be?

Detectives on the case

What powers radio galaxies and quasars?

By "central engine" of a radio galaxy or quasar we mean the object, whatever it might be, that releases such great amounts of energy from such a small region of space. What could possibly power such an "engine?"

First theory: nuclear energy?

We have already studied one very powerful source of energy: the nuclear reactions that power the shining of the stars. Let us explore the possibility that *the energy output of radio galaxies and quasars is supplied by a great number of stars packed into a small amount of space.* We are imagining that within the central region of a radio

⇐ **Looking backward**
We studied nuclear energy
in Chapters 14 and 15.

galaxy or quasar lies a highly compacted group of stars. Could such a group emit as much energy as we observe coming from radio galaxies and quasars?

How many stars must we postulate in our group? Let us concentrate on the quasar 3C 273. As we saw, it emits 256 times as much energy as our entire Galaxy. So we would have to postulate that its group contains 256 times as many stars as our Galaxy. We also know something about the quasar's size: it can be no bigger than 2.3×10^{15} meters in radius. So we know how many stars we must pack into this small a region of space. Is this possible?

We will now show that it is not. There are two separate reasons: (1) the required mass of such a group of stars is greater than the mass of the entire galaxy it is postulated to inhabit, and (2) such a group of stars cannot exist; it would collapse to form a giant black hole and emit no light at all.

(1) We can work out the mass of all the stars we are postulating, and compare this to the mass of an entire galaxy. If the mass postulated by our model exceeds that of a galaxy, we know that our model cannot be correct.

But since our model is postulating that the central region of 3C 273 contains 256 times as many stars as our Galaxy, their mass must add up to 256 times the mass of our Galaxy. This is impossible: it is like postulating that the weight of all the gas in your car is greater than the weight of your car (including the gas).

⇐ **Looking backward**
We measured the mass of
our Galaxy in Chapter 16.

Our own Milky Way Galaxy is representative: we showed in the previous chapter that its total mass is 1.57×10^{42} kilograms. But the mass of our postulated group of stars is far greater than this! In a previous chapter we saw that there are about 10^{11} stars in our Galaxy, so we need to postulate that our quasar consists of $256 \times 10^{11} = 2.56 \times 10^{13}$ stars. The mass of all these stars is $2.56 \times 10^{13}\ M_{\text{one star}}$ and, since our Sun is typical, we can set $M_{\text{one star}} = M_{\text{Sun}} = 1.989 \times 10^{30}$ kilograms. So we find the mass of all the stars in our model quasar to be

$$M_{\text{our model quasar}} = (2.56 \times 10^{13})(1.989 \times 10^{30}\ \text{kilograms}) = 5.09 \times 10^{43}\ \text{kilograms}.$$

But this is greater than the measured mass of our Galaxy – and of most other galaxies as well.

. . . .

NOW YOU DO IT
Suppose we propose a theory in which a quasar contains 55 times the number of stars as our Galaxy: what would be its mass?

. . . .

(2) The second reason why our postulated model is untenable is that, if we imagine packing this many stars into such a small region of space, we would not end up with a brilliantly shining group of stars. We would end up with a black hole – which emits no light at all!

⇐ **Looking backward**
We studied black holes in
Chapter 15.

As we saw in our studies of black holes, if an object is compressed so greatly that it becomes smaller than its Schwarzschild radius it turns into a black hole. This is also true if we compress not a single object, but a collection of objects such as our group of stars. As we saw in Chapter 15, the Schwarzschild radius of an object of mass M is

$$R_{\text{Schwarzchild}} = 2GM/c^2.$$

So the Schwarzschild radius of our postulated assemblage of stars would be

$$(2)(6.67 \times 10^{-11})(5.09 \times 10^{43})/(3.00 \times 10^8)^2 = 7.54 \times 10^{16} \text{ meters.}$$

But the radius of our quasar is smaller than this! We conclude that, if we tried to pack enough stars into such a small region of space as we know a quasar occupies, they would turn into a giant black hole and emit no energy at all.

. . . .

NOW YOU DO IT
Calculate the Schwarzschild radius of our collection of 55 times the number of stars as exist in our Galaxy. Is your result smaller than the radius of 3C 273?

. . . .

Our first theory cannot be correct. We must seek another one.

Searching for a new theory

Because it is hard to pack a lot of fuel into a small volume, the "fuel" that powers the "engine" must be highly efficient – that is, it must liberate a great amount of energy from a given mass of fuel. There are only two such highly efficient sources of energy known: *nuclear energy* and *gravitational potential energy*. Let us therefore study the efficiency of these two sorts of energy.

We have already studied nuclear energy: in our study of the Sun we showed that if a mass m of fuel undergoes nuclear reactions in which hydrogen is converted to helium, the energy liberated is 0.7% of mc^2. We now need to do a similar calculation for gravitational potential energy. We will find that it is even more efficient – i.e. that it releases even more than this.

In this scenario we imagine that our "central engine" is some large object onto which mass is continually falling. The situation is analogous to what happens at the base of a waterfall. The plummeting water, accelerated by the force of gravity, lands with high velocity and liberates a good deal of energy. The source of this released energy is the gravitational potential energy of the falling water. Similarly, our postulated mass "m" is also falling – not downwards but inwards: sucked toward some large object. As it is crushed together, great amounts of its gravitational potential energy are released. How much – and in particular, what fraction of mc^2?

The formula for the gravitational potential energy released when a mass m falls onto a central mass M whose radius is R is

$$PE = GMm/R.$$

We need this to be as big as possible. Clearly, the bigger is M and the smaller is R, the greater is the energy PE released by the infalling mass. This means that we need our central object to be *very massive* and *very small*. So let us imagine that *the central object is a very massive black hole*.

It is important to emphasize the difference between this picture and the model we discussed above, in which a large group of stars is so compressed that it becomes a black hole. Now we are not imagining that the emissions come from a group of stars: rather we are imagining that its emissions come from matter falling onto the hole. Nor are we imagining that hole once consisted of 256 times more stars than an entire galaxy. In fact it will turn out that the hole is far less massive than a galaxy.

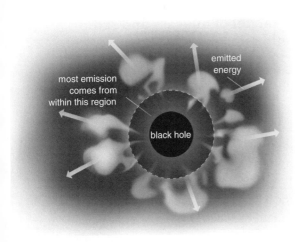

most emission comes from within this region

black hole

emitted energy

Figure 17.29 The central engine of a quasar (schematic): matter falling onto a giant black hole. The bulk of the emission comes from the dotted region.

Figure 17.29 illustrates this model of the central engine. We imagine a giant black hole residing at the center of a galaxy. Matter is continually falling into the hole, sucked inwards by its gigantic gravitational pull. Presumably this matter consists of stars that wandered too close, were captured into orbits about the hole and were torn to shreds. The "shreds" are now falling inward toward the hole. (That this figure is schematic only: in Figure 17.34 below we will depict a more realistic model of the central engine.) As the infalling matter falls, it collides with other infalling matter, is greatly heated and emits energy. As we have seen in our study of black holes, no emission is possible once the matter has fallen inside the hole. But we can receive emission from matter before it reaches its edge.

The emission we are receiving from a quasar or a radio galaxy must therefore come from matter that has fallen to somewhere not far beyond the edge of a giant black hole. As a specific example, let us suppose that it comes from five times the hole's Schwarzschild radius. Let us calculate the energy released in this model. We can do this by setting R equal to $5R_{Schwarzschild}$ in our formula for the gravitational potential energy released when a mass m falls into the hole (when we are thinking about black holes, we must use Einstein's theory of relativity to be fully accurate, but the formula is a good deal more complicated than this, and the final answer turns out to be not very different):

$$PE = GMm/R$$
$$= GMm/5R_{Schwarzschild.}$$

Using our formula for the Schwarzschild radius of a black hole of mass M,

$$R_{Schwarzschild} = 2GM/c^2,$$

and inserting this into our formula for PE, we get

$$PE = \frac{GMm}{5}\frac{c^2}{2\,GM}$$

$$PE = 0.1\,mc^2.$$

This shows that if a mass m falls onto a giant black hole, about 10% of its mc^2 is released.

Let us now compare this with our previous result.

- If a mass m undergoes nuclear fusion from hydrogen into helium, 0.7% of mc^2 is released.
- If a mass m falls onto a black hole, 10% of mc^2 is released.

Clearly, gravitational potential energy is *more efficient* than nuclear energy. For this reason, most astronomers believe that *the central engine powering radio galaxies and quasars is a giant black hole consuming large amounts of infalling matter*. More detailed calculations indicate that the central black hole of a quasar such as 3C 273 has a mass perhaps 900 million times that of the Sun.

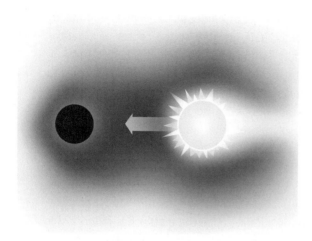

Figure 17.30 **An unlikely scenario: an approaching star is aimed directly at a black hole.**

The black hole model of quasars and radio galaxies: accretion disks

In Figure 17.29 we sketched our model of the "central engine" of quasars and radio galaxies. That figure, however, was drastically oversimplified, for it depicts matter falling directly into the hole. Far more likely is that the matter will fall indirectly into the hole, via a so-called *accretion disk*.

Let us think more carefully about how a black hole swallows matter – a star, for example. In Figure 17.30 we illustrate a star approaching a black hole. Note that in this figure the star is traveling directly toward the hole. This, however, is an unlikely scenario. It is analogous to an arrow flying directly toward the bull's eye of a target. Since in reality there is no "archer" to aim the approaching star so carefully, the stars are actually moving randomly. So it is far more likely that the star will miss the hole by some margin. Rather than hitting it, the star will *go into orbit* about the black hole (Figure 17.31).

We now add a second crucial element to our scenario: once the star is close to the black hole, it is pulled apart by the hole. More precisely, it is pulled apart by *tidal forces* from the hole. We have already studied tides in our study of the Earth, where we showed that, if one celestial body approaches another, tidal forces will raise bulges on its near and far parts. So a star, as it nears the black hole, will be stretched into the configuration illustrated in Figure 17.32.

Eventually the star is entirely pulled apart by gravity from the hole, and it forms a so-called *accretion disk* about it (Figure 17.33).

Figure 17.34 illustrates our full model of the central engine of quasars and radio galaxies.

The infalling star has been stretched into an accretion disk surrounding the black hole. From the inner regions of this disk, matter rains down upon the hole. Close to

> ← **Looking backward**
> We studied tides in Chapter 7.

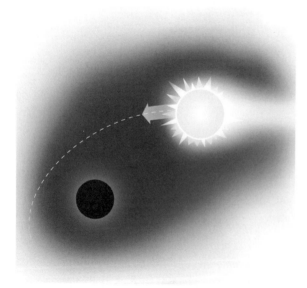

Figure 17.31 **A more likely scenario: the approaching star is not directly aimed at the black hole, but rather goes into orbit about it.**

Figure 17.32 **Tides distort the star as it approaches the black hole.**

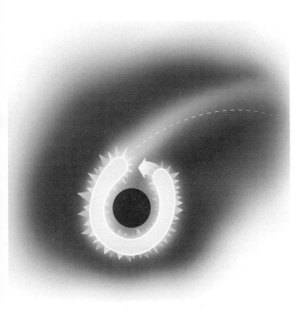

Figure 17.33 **An accretion disk is formed as material from the star is wrapped about the black hole.**

Figure 17.34 **The central engine of quasars and radio galaxies.**

Figure 17.35 **Jets escaping from an accretion disk.**

the hole, enormous quantities of energy are released by the infalling matter. Particles are accelerated to velocities close to that of light, and they stream outward. But most of these particles ram into the accretion disk, and are blocked by it. Only at the top and bottom openings of the disk do the particles escape. They stream away from the black hole as two jets (Figure 17.35). Eventually, each of these jets rams into the intergalactic medium at great distances from the hole. *When they do, radio emission is generated.*

In this way, we can understand the double lobes of emission we observe in radio galaxies. But how can we understand that in some cases only *one* of the two jets is seen? This arises from the fact that rapidly moving particles radiate more strongly in the direction in which they travel than in the perpendicular direction. So if the jets happen to point crosswise to us we will see both; but if they happen to point towards and away from us, the one pointing away will appear far fainter than the other and so be missed.

Testing the theory

We have now described the current theory of the enormous emission of energy observed from radio galaxies and quasars. But what reason do we have to think that the theory is correct? What evidence do we have that these objects actually do contain giant black holes? We will conduct two tests.

(1) *Rate of consumption of matter.* We will calculate how rapidly such a black hole must consume matter in order to account for the great emission that we observe. Suppose the answer we get tells us that the hole must swallow stars at an unreasonably large rate. This would lead us to believe that our theory cannot be true.

(2) *Searches for black holes.* We will search the nuclei of radio galaxies and quasars to see if we can spot black holes within them.

(1)   Rate of consumption of matter

Let us continue focusing on the quasar 3C 273. We want to find the quantity of matter that falls into its hole in a certain period of time – say, one year. In order to do this we need to (A) find the total amount of energy 3C 273 emits in one year, and then (B) adjust the quantity of matter that falls into it in one year to yield this energy.

The logic of the calculation

Step 1. Measure the rate of energy emission from 3C 273 – i.e. the quantity of energy emitted per second. This is its luminosity.

Step 2. Multiply this by the number of seconds in a year to find the total quantity of energy emitted in a year.

Step 3. Set this equal to 10% of mc^2, where m is the quantity of matter that falls into the hole in one year.

Step 4. Solve the resulting equation for m.

Step 5. To get an intuitive feeling for the result, compare it to the mass of a typical star.

NOW YOU DO IT

Suppose a quasar emits 10^{40} watts. How many stars must it swallow in a month to account for this?

As you can see from the detailed calculation, 17 solar masses must fall into the hole each year in order to generate enough energy to power the shining of 3C 273. Presumably this means that every year the hole "eats" 17 stars. Is this too much? Seventeen is certainly not a very large number, and observations reveal that there are huge numbers of stars in the central regions of all galaxies. There seems to be nothing in our result that would lead us definitively to abandon our theory.

(2)   Searches for black holes in galaxies

We are unable to conduct this search *directly*. We cannot train our telescopes on a radio galaxy or quasar and see black holes within them. The difficulty is that radio galaxies and quasars are very far away, so that it is very hard to study them. Furthermore black holes, even giant ones, are very small – and they are, of course, black. We have, therefore, no *direct* evidence

Detailed calculation

Step 1. Measure the rate of energy emission from 3C 273. We have already done this. We found

$$L_{3C\ 273} = 9.7 \times 10^{39} \text{ watts} = 9.7 \times 10^{39} \text{ joules/second.}$$

(Since one watt corresponds to one joule of energy emitted per second.)

Step 2. Multiply this by the number of seconds in a year to find the total quantity of energy E emitted in a year. We calculate

$$E = (9.7 \times 10^{39} \text{ joules/second}) (365 \text{ days/year}) (24 \text{ hours/day}) (3600 \text{ seconds/hour})$$
$$= 3.1 \times 10^{47} \text{ joules.}$$

Step 3. Set this equal to 10% of mc^2, where m is the quantity of matter that falls into the hole in one year. We calculate

$$3.1 \times 10^{47} \text{ joules} = (0.1) (m \text{ kilograms}) (3.0 \times 10^8 \text{ meters/second})^2 = 9.0 \times 10^{15} (m) \text{ joules,}$$

where m is in kilograms.

Step 4. Solve the resulting equation for m. We calculate

$$m = (3.1 \times 10^{47} \text{ joules}/9.0 \times 10^{15} \text{ joules}) \text{ kilograms}$$
$$= 3.4 \times 10^{31} \text{ kilograms.}$$

This is the quantity of matter falling into the hole in one year.

Step 5. To get an intuitive feeling for the result, compare it to the mass of a typical star. Since the Sun is a typical star, we divide our result for m by M_{Sun} (1.989×10^{30} kilograms) to find the infalling mass in solar units:

$$3.4 \times 10^{31} \text{ kilograms}/1.989 \times 10^{30} \text{ kilograms} = 17 \text{ solar masses.}$$

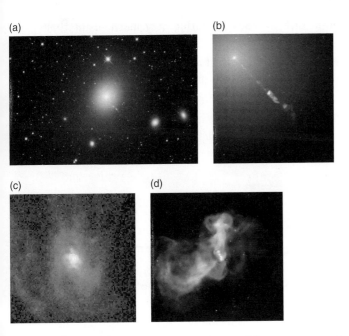

(a)   (b)

(c)   (d)

Figure 17.36 **The galaxy M87. This elliptical galaxy is a source of radio emission. A long exposure (a) reveals the galaxy's outskirts. The shorter exposure shown in (b) reveals its inner regions, and in particular a jet. In (c) we see an even closer view of the inner regions reveals an elongated gaseous structure, and finally in (d) a view of the radio emission from the galaxy.**

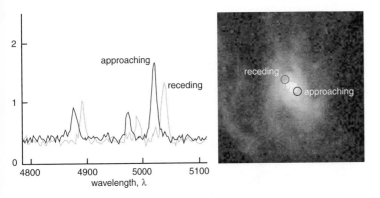

Figure 17.37 **A spectrum of the central regions of M87 shows enormous Doppler shifts, indicating that the elongated cloud is rotating about some central mass.**

whatsoever that radio galaxies or quasars contain giant black holes.

But we do have *indirect* evidence of giant black holes in certain nearby galaxies. One of them, known as M87, is a source of radio emission. The other is neither a quasar nor a radio galaxy, but it is one we know very well: it is our own Milky Way Galaxy. We will consider each in turn.

The galaxy M87 is one of the closest radio galaxies, and as such has received a good deal of attention. Figure 17.36 shows a series of views of this galaxy. As shown in Figure 17.36(a), it appears at first glance to be a normal elliptical galaxy. But this figure was intentionally overexposed to reveal the galaxy's faint outskirts. Figure 17.36(b), in turn, is a shorter exposure revealing its brighter inner regions. Clearly visible in this image is its brilliant central core, from which extends a jet, reminiscent of the jet of the quasar 3C 273. This jet is composed of knots moving outwards at very high speeds. In Figure 17.36(c) we see an extreme close-up of the galaxy's very inner regions. This image shows an elongated cloud of gas: the long axis of this cloud is perpendicular to the jet. Radio emission has been observed coming from the galaxy, as shown in Figure 17.36(d). As you can see, this galaxy is similar to the radio galaxies: like them, it has two jets extending outwards from the central galaxy, although in this case the two lobes are replaced by an extended cloud of radio emission.

Clearly, M87 shares many features with both radio galaxies and the quasar 3C 273. Furthermore, in 1994, observations made using the Hubble Space Telescope gave evidence that a giant black hole resides at the very center of this galaxy.

This evidence comes from a spectrum that was obtained of the emissions from the central gas cloud. The spectrum, shown in Figure 17.37, shows Doppler shifts. Light from one side of the gas cloud is blueshifted, while that from the other side is redshifted. This means that one side of the cloud is moving toward us, and the other side away from us. Clearly the cloud is rotating on its axis – an axis that coincides with the jets.

This cloud is an accretion disk seen edge-on. It is a swarm of material orbiting about some central object – an object that cannot be seen in Figure 17.36, but that must lie at the very heart of the cloud. The velocity of orbit can be measured by using the Doppler shift formula: it turns out to be 600 kilometers per second. This velocity is enormous as compared with those of most astronomical objects. Our own Earth, for example, orbits the Sun a mere $\frac{1}{20}$th as rapidly, while stars in our vicinity travel even more slowly. Clearly, the accretion disk in M87 is rotating very much faster than these. What could cause such great velocities?

As we saw in our study of Newton's laws, the velocity at which bodies orbit about some central object depends on the mass of that central object. (This is true whether the orbiting bodies are planets, moons or gas in an accretion disk.) The more massive the central object, the more rapidly things orbit about it. Conversely, a high velocity of orbit tells us that the central object must have a high mass. Newton's formula for the mass M of the central body is

$$M = RV^2/G.$$

> **⇐ Looking backward**
> In Chapter 3 we saw how measuring orbits about a body can tell us the mass of that body.

In this formula R is the distance of the orbiting material from the central mass (the radius of the accretion disk) and V its orbital velocity. Using measured values of R and V we find that the mass of the unseen object that lies at the heart of M87's accretion disk is two billion times the mass of the Sun.

Could the "central object" simply be two billion stars? It could not, for such a great number of stars would emit a great deal of light – far more light than is actually observed coming from the central regions of M87. Whatever it is, the central object about which the accretion disk orbits cannot be emitting very much light at all. The only possibility that people can think of is that it is a 2 billion solar mass black hole.

Our Milky Way Galaxy. Remarkably, there is similar evidence that a giant black hole lies at the center of our very own Milky Way Galaxy. In 1995 a group of astronomers used the world's largest telescope, the Keck Observatory in Hawaii, to observe its very center. Unlike M87, they found no cloud of gas there – no accretion disk. All they found were stars.

But these stars were moving in a very suggestive way. As we have noted, stars in a galaxy orbit about its center. The astronomers studied these orbits, by measuring the positions of the stars in 1995 to very great accuracy – and by then observing them again during the next two years. They found that the stars had moved sideways during that two-year interval.

By measuring the amount by which the stars had moved, it was possible to deduce their orbital velocities. These velocities turned out to be very large, larger even than that of M87's accretion disk. One star was discovered to be moving at an astonishing 1400 kilometers per second. Newton's mass formula, when applied to these stars, showed the central object about which they are orbiting to have a mass of 4 million times that of the Sun. As in the case of M87, this central object cannot possibly be a collection of 4 million stars, for this many stars would emit far more light than is observed coming from the nucleus of our Galaxy. It can only be a black hole.

If our own home Galaxy has a giant black hole at its center, why is it not a radio galaxy or a quasar? Presumably this is because the hole in our Galaxy is not consuming any stars. Because we find no accretion disk about it, we know that there is nothing falling into the central hole that would emit great quantities of energy.

Is the black hole theory correct?

The fact that so unprepossessing a galaxy as our own contains a giant black hole at its center raises the possibility that all the other galaxies do too. Perhaps a giant black hole lies at the center of *every* galaxy. If this is correct, there is only a difference of degree between normal galaxies and those that exhibit violent activity. Galaxies whose central hole happens not to be swallowing stars are normal: those whose hole is undergoing accretion are quasars or radio galaxies. It is also possible that every galaxy, including our own, once was a quasar or radio galaxy – and may be so again in the future.

It is important to stress however, that no one has ever observed a black hole in a quasar. There are only three things that we know for sure.

- The energy emitted by radio galaxies and quasars is enormous, and it is difficult to think of a process capable of providing so much energy: accretion onto black holes releases more energy than any other known process.
- Certain galaxies are known to contain large concentrations of matter at their centers, concentrations that emit little or no radiation. There is nothing other than black holes that these might be.
- Great amounts of indirect evidence have been found pointing to the correctness of this theory. As we have repeatedly emphasized, indirect evidence can be very powerful – but it is, indeed, indirect.

These points are powerful – but it is very likely that we will learn more as our technological capabilities increase. If someone were to think of an even more powerful source of energy, our ideas might very well change. Surprises are probably in store.

Summary

DETECTIVES ON THE CASE

What are the spiral nebulae?
- Many spiral nebulae were known. There were two theories of what they might be.
  - Perhaps they were planetary systems in the process of forming.
  - Perhaps they were galaxies, just like our own.
- First clue: a "new star" was discovered in the Andromeda spiral galaxy.
  - Perhaps this was a nova like those in our Galaxy.
  - If so, its apparent brightness implied that the nebula was nearby, i.e. not a galaxy in its own right.
- Second clue: the nebula might contain stars. If so, it was a galaxy.
- Third clue: perhaps the first clue was wrong.
  - A more powerful telescope revealed fainter "new stars." If these were the novae, then the first "new star" must have been something else.
  - If so, the nebula was a galaxy.
- Error (1): an error in the Cepheid variable method of measuring distances led scientists to overestimate the size of our own Galaxy.

- Error (2): it was thought that small knots in the Andromeda spiral nebula were moving sideways. If so, the nebula would have to be fairly close.
  - Both errors led scientists toward the idea that the nebula was not a galaxy.
- Edwin Hubble ultimately solved the problem by using the Cepheid variable method to measure the distance to the Andromeda Nebula. He found it to be a galaxy.

THE NATURE OF SCIENCE

The process of discovery in science
- People often believe that science is cut-and-dried, and that to every question there is one and only one right answer. In actual practice, however, the process of discovery is never so simple.
- How much weight should we give evidence? What we call "evidence" can sometimes be simply wrong. This makes all scientific discovery difficult. Scientists often find themselves in the position of having to make judgment calls.

- "Occam's Razor": in choosing between two competing theories, we ought to choose the one that requires us to make the smallest number of new postulates. However, the final solution to the mystery of the spiral nebulae forced us to accept not one but two extraordinary new assumptions.
- Psychological factors can play a role in the process of discovery.
- But this role is not absolute, since in the long run evidence can be found that overcomes all psychological factors.

Types of galaxies

- Four types of galaxies:
  - spiral,
  - elliptical,
  - irregular,
  - barred spiral.
- Classification into subtypes Sa through Sc, E0 through E9, Sba through SBd and Irregulars.
- Most galaxies are irregular, fewer are spiral, fewest are elliptical.
- Gas content: ellipticals very low, spirals intermediate and irregulars have the most gas.
- Stellar populations: ellipticals old population II, spirals both population I and II, irregulars young population I.
- Star formation is still going on in irregular and spiral galaxies; in ellipticals it ceased a long time ago.

Galaxy formation

- The top-down theory.
  - Postulates that galaxies formed by the collapse of a slowly rotating cloud of gas.
  - Postulates that the difference between elliptical and spiral galaxies is that in ellipticals star formation happened all at once a long time ago: in spirals it is a continuous process that is still going on.
  - Problems with the theory.
    - The Galaxy's halo is not rotating.
    - Our Galaxy appears to contain two disks.
    - Globular clusters appear not all to have the same age.
- The bottom-up theory.
  - Postulates that galaxies form by the amalgamation of smaller objects.
  - These were perhaps formed in the early stages of the expansion of the Universe.
  - Galaxy mergers.
    - Galaxies can "collide" with one another.
    - When they do, stars do not collide.
    - But gravity from each galaxy distorts the other through tides and galactic cannibalism.
  - Encounters between galaxies can trigger star formation.
  - Mergers between spirals can produce ellipticals.

Violent activity in galaxies

- Radio source Cygnus A.
  - Very luminous: even though it is very distant it is very bright.
  - Emission comes from two lobes on either side of galaxy.
  - Emission produced by high-speed electrons in a magnetic field.
  - These electrons possess enormous energy.
- Quasars.
  - "Quasi-Stellar Radio Source."
  - Doppler shift tells us they are very distant.
  - Rapid variability tells us they are very small.
  - Apparent brightness plus great distance tells us they are very luminous.
  - They reside in the nuclei of galaxies.

DETECTIVES ON THE CASE

What powers radio galaxies and quasars?

- We explored a first theory: that they derive their power from a great number of stars packed into a small volume of space.
- This turned out to be impossible:
  - the required mass of such a group of stars is greater than the mass of the entire galaxy it is postulated to inhabit,
  - such a group cannot exist: it would collapse to form a giant black hole and emit no light at all.
- Gravitational potential energy:
  - is the energy released when something falls,
  - can be more efficient than nuclear energy.
- We then explored the theory that the central engine powering radio galaxies and quasars is a giant black hole consuming large amounts of infalling matter.
  - Matter falling into a black hole goes into an accretion disk.
  - Matter in the disk spirals into the hole.
  - As it does so (and before it reaches the edge of the hole) it emits energy.
  - Testing the theory.
    - If such a hole "eats" 17 stars per year, enough energy can be produced to power a typical quasar.
    - Certain galaxies are known to contain giant black holes in their centers: examples are M87 and our own Milky Way Galaxy.
    - However, we have no direct evidence at all that radio galaxies and quasars contain giant black holes.

Problems

(1) Suppose you see a faint light off in the distance at night. There are two possibilities: it might be (i) a 100-watt light bulb, or it might be (ii) a 10 000-watt searchlight.

(A) Suppose hypothesis (i) leads you to conclude that the light is one mile (1.6 kilometers) away. How far away would it be if you adopted hypothesis (ii)?

(B) Suppose hypothesis (ii) leads you to conclude that the light is one mile (1.6 kilometers) away. How far away would it be if you adopted hypothesis (i)?

(2) Suppose that the apparent brightness and period of a Cepheid variable star leads you to conclude that it is 15 000 light years away. But suppose that interstellar matter is actually absorbing half of the light the star is emitting. How far away is it in reality?

(3) Figure 17.38 shows a slightly more detailed version of Figure 17.2, representing the distance (X or Y) moved by a bright knot in the Andromeda Nebula in a few years.

We want to know how much bigger the distance Y is than X. One way to do this is to build a scale model, and measure these two distances.

Draw on a piece of paper the above figure: let the shorter base represent the nebula's 100 000 light year distance, and the longer base represent a 1 million light year distance. Now use a ruler actually to measure X and Y. What is the *ratio* between Y and X? (Warning: Figure 17.38 is not to scale, but your drawing must be!)

(4) In Problem 10 of Chapter 16 you described the stellar populations that would be found assuming a number of alternative scenarios for the history of a galaxy. Do any of your answers correspond to known galaxy types?

(5) What are the differences between an elliptical galaxy and (A) the halo of our Milky Way Galaxy, and (B) the nucleus of our Milky Way Galaxy?

(6) Suppose galaxies had a diameter of 100 000 light years, and they were 0.8 Mpc apart. We want to calculate the ratio κ. (A) describe the logic of the calculation you will perform. (B) Carry out the detailed calculation.

(7) Suppose galaxies had a diameter of 100 000 light years, and their relative speeds were 75 kilometers per second. We want to calculate how long it would take for one to pass through another. (A) Describe the logic of the calculation you will perform. (B) Carry out the detailed calculation.

(8) Suppose a radio-emitting body named "Swan A" were 200 million parsecs away, and another named "Pigeon A" were 4 thousand parsecs away. (A) Calculate the ratio of their distances. (B) Imagine a scale model in which Swan A were half a yard (0.46 meters) away: how far would it be to Pigeon A?

(9) Suppose the radio-emitting lobe of a radio galaxy contained 3×10^{52} joules of energy. We want to calculate how long it would take for our Milky Way Galaxy to emit this much energy. (A) Describe the logic of the calculation you will perform. (B) Carry out the detailed calculation.

(10) Suppose we were to discover a quasar whose distance was 3 billion light years, and whose flux were 6×10^{-12} watts/meter^2. We want to calculate how much brighter it is than our Galaxy – i.e. the ratio between its luminosity and that of our Galaxy. (A) Describe the logic of the calculation you will perform. (B) Carry out the detailed calculation. (C) Suppose the quasar were actually twice as far away. Now find how much brighter it is than our Galaxy. See if you can do this without actually carrying out the detailed calculation all over again!

(11) Suppose a parking lot is ¼ mile across, and there are two bicyclists, one at the lot's near edge and the other at its far edge. At noon, they both start bicycling toward my home at 10 miles per hour.

• Suppose I live 10 miles from the lot's near edge. We want to calculate when each bicyclist arrives at my home, and we want to calculate the

Figure 17.38 **A more detailed version of Figure 17.2 (not to scale).**

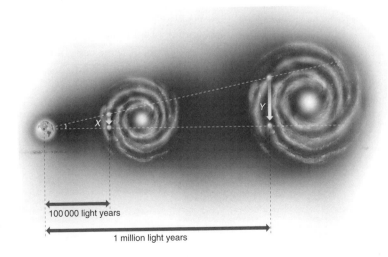

100 000 light years

1 million light years

difference between these two times. (A) Describe the logic of the calculation you will use. (B) Carry out the detailed calculation.

- Suppose I live 30 miles from the lot's near edge. We want to calculate when each bicyclist arrives at my home, and we want to calculate the *difference* between these two times. (A) Describe the logic of the calculation you will use. (B) Carry out the detailed calculation.

- Notice that the times the bicyclists arrive are different in these two cases – but the *differences* in these times are not. Why is this?

(12) We wrote that if the Sun were to brighten instantaneously, we would perceive it to brighten gradually over a period of several seconds. Your task is to verify this. (A) Describe the logic of the calculation you will perform. (B) Carry out the detailed calculation.

(13) Suppose a quasar has a luminosity of 6×10^{38} watts. We want to know how many stars it swallows per year. (A) Describe the logic of the calculation you will perform. (B) Carry out the detailed calculation. (C) Suppose it suddenly starts swallowing six times as many stars per year. What is its new luminosity?

(14) We saw that the radio-emitting lobes of Cygnus A contain 10^{53} joules of energy. We want to calculate how many stars must fall on a black hole to release this much energy. (A) Describe the logic of the calculation you will perform. (B) Carry out the detailed calculation.

WHAT DO YOU THINK?

We have described three problems facing the top-down theory of galaxy formation. Do you regard these difficulties as being fatal? Or can you think of some modifications of the theory that would solve these problems?

You must decide

(1) Recall that it was Edwin Hubble's discovery of Cepheid variable stars in the Andromeda Nebula that demonstrated that spiral nebulae were actually distant galaxies. In this problem let us imagine what might have happened had Andromeda not happened to contain any Cepheid variables.

Imagine that you were an astronomer living at that time, and you were trying to find some other evidence that would decide between the Nebular Hypothesis and the Island Universe Hypothesis. You have been granted a strictly limited amount of observing time on the world's largest telescope, with no possibility of asking for more time.

- What research program would you carry out with your precious observing time in order to resolve the debate?
- Why do you feel this program has the best chance of success?

(2) You are a program officer of the National Science Foundation. Previously you had been in charge of supporting research on the formation of stars, but now you have been assigned to a new office dedicated to supporting research on the formation of galaxies. From your previous experience, you are aware of how much observational evidence has been gathered on the subject of star formation – and, in your current position, you are painfully aware of how little observational evidence we have to help us with our theories of galaxy formation.

As a consequence, you recently put out a Request for Proposals, in which you announced funding for *observational* projects (i.e. actual observations with actual telescopes, as opposed to such things as computer simulations or new theories) designed to further our understanding of how galaxies formed.

Now a pile of proposals sits on your desk. You have enough money at your disposal to fund only one of them. Your task now is to write a memo to your superior, explaining

- which proposal you decided to fund.

And, equally important,

- your reasons for choosing this proposal over all the others.

18

Cosmology

So far in this book we have been concerned with individual things – with moons and planets, with stars and nebulae and galaxies. Now we move on to study, not Things, but Everything. What is the nature of the Universe as a whole?

Our answers to this question have continually evolved. When we were babies "the world" consisted of little more than our homes and families. As we grew older, our worlds expanded to include other families, school and friends, our home town. This kind of expansion is also true historically. Primitive peoples regarded their immediate vicinity to be "the world." Early societies, such as the ancient Greeks, drew larger maps, and as we see in Figure 18.1 their maps grew ever more comprehensive. With the scientific revolution (Figure 18.2) the view expanded immensely. But even here the view was strictly limited: in Figure 18.2 the entire Universe beyond the orbit of Saturn is represented merely as a single sphere of "fixed stars."

This book has followed the same path. We began with a study of the relatively nearby – the Solar System – and progressively moved to more and more distant objects. Perhaps we are ready for yet another giant expansion. It is the greatest of them all – out into what may well be infinite.

Olbers' Paradox: why is it dark at night?

Perhaps the most profound question that we can ask about the Universe is whether it is infinite. Does the cosmos extend endlessly far into the depths? Or does it have an edge?

There is a fascinating paradox related to this question. If the Universe is infinite it would contain an infinite number of stars, and taken together they would emit an infinite amount of light. *So why don't we receive an infinite amount of light*? Why is it dark at night?

This question has been discussed by many thinkers over the centuries. Nowadays we know it as Olbers' Paradox, after the German physician and astronomer Heinrich Olbers who wrote of it in 1823. This argument seems to prove that the cosmos cannot be infinite. But as we will see, the paradox contains an error, so that we cannot in fact draw this conclusion.

Let us suppose that the Universe is indeed infinite, and let us calculate the net amount of light we would receive in that case. Our goal is to calculate the

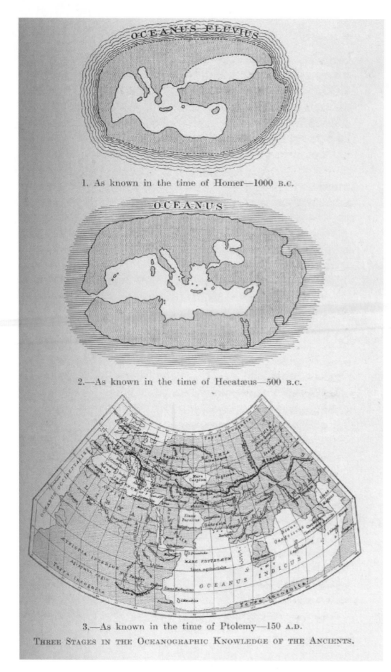

1.—As known in the time of Homer—1000 B.C.

2.—As known in the time of Hecatæus—500 B.C.

3.—As known in the time of Ptolemy—150 A.D.

THREE STAGES IN THE OCEANOGRAPHIC KNOWLEDGE OF THE ANCIENTS.

Figure 18.1 **Early maps reveal an ever-expanding view of what constitutes "the world." Three stages are shown in the oceanographic knowledge of the Ancient Greeks.**

Figure 18.2 **Image of the heliocentric model from Copernicus'** *De revolutionibus orbium coelestium.*

total apparent brightness – the total flux – we would observe from all this infinite number of stars. As we will see, the answer turns out to be that we should receive an infinite flux. Of course in reality we don't: this is Olbers' Paradox.

To find the total flux we receive, we must add up all the light from all the stars. This is going to be hard since there are an infinite number of them. Here's a trick. Many of these stars lie the same distance from us – namely, those that lie in the thin shell indicated in Figure 18.3. Suppose that all stars have the same luminosity (it turns out that this does not affect the final answer). Then we receive the same flux from each of them.

So to calculate the net flux we receive from all the stars in this particular shell, we need to

(1) find the flux from one star in the shell,
(2) find the number of stars in the shell,
(3) multiply (1) and (2) together.

Finally, to find the total flux from the whole cosmos, we add up all the shells.

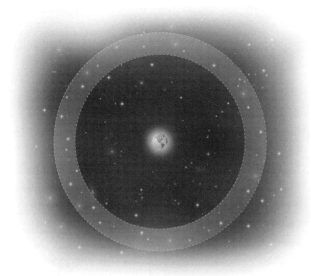

Figure 18.3 **All the stars in the indicated shell lie the same distance from us so that we receive the same flux from each.**

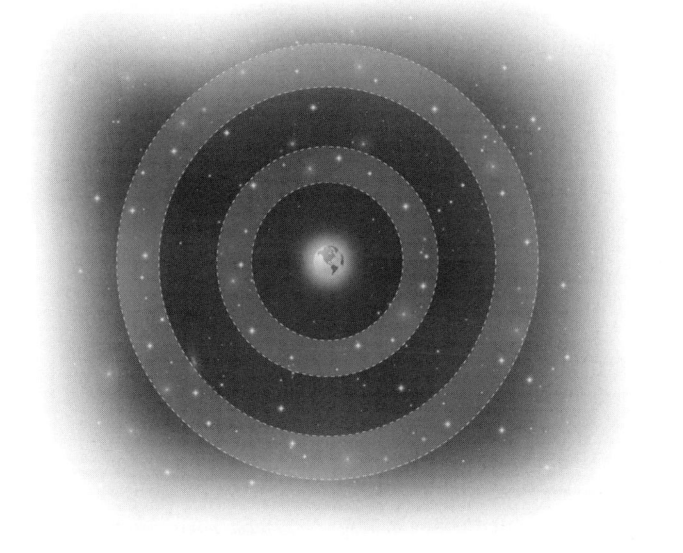

Figure 18.4 **There are more stars in the bigger shell than in the smaller one.**

Step 1. Find the flux from one star in the shell

This is given by the inverse square law

$$f = L/4\pi R^2,$$

where L is the average luminosity of a star, and R its distance from us – i.e. the radius of the shell.

Step 2. Find the number of stars in the shell

To get a feeling for this, consider two such shells, one bigger than the other, as in Figure 18.4.

If you count the stars in this figure, you will find that there are more in the bigger shell than in the smaller one. Why is this so? Because there is more space available for them in the bigger shell! This tells us that to find the number of stars in a shell, we need to know how much space it contains – and this is the shell's volume. The volume of a shell of radius R is $4\pi R^2$ times its thickness. We conclude that *the number of stars in a shell is proportional to $4\pi R^2$.*

Step 3. Multiply (1) and (2) together

Notice that (1) is inversely proportional to $4\pi R^2$ while (2) is *proportional* to $4\pi R^2$. If we multiply these together, the factors of $4\pi R^2$ cancel out. We conclude that *the total flux we get is the same for a big shell as for a small one.* We get less light from each star in the bigger shell, but there are more of them, making the total just the same. In the sidebar we consider a detailed example to make this point clearer.

To find the flux from the entire Universe, we need to add the fluxes from all the shells. But suddenly we realize that this is very easy. We merely multiply the flux from one shell times the number of shells. But the number of shells in an infinite cosmos is infinite! We conclude that *we would receive an infinite amount of light from all the stars in an infinite Universe.*

Detailed calculation

Let us make our calculation of the flux from a pair of shells more vivid by considering a specific example. Suppose

- the flux we receive from a star in the smaller shell is 10^{-9} watts/meter^2,
- the smaller shell contains 100 stars,
- the larger shell is twice as big as the smaller shell.

Then the flux we receive from the smaller shell is $(100) \times (10^{-9} \text{ watts/meter}^2) = 10^{-7}$ watts/meter^2.

Consider now the larger shell, which is twice as far away. The flux from each of its stars is $1/2^2 = 1/4$ that from stars in the smaller shell, i.e. the flux is 2.5×10^{-10} watts/meter^2. The larger shell contains $2^2 = 4$ times as much space as the smaller one, so it contains not 100 but 400 stars. Then the flux we receive from the larger shell is $(400) (2.5 \times 10^{-10} \text{ watts/meter}^2) = 10^{-7}$ watts/meter^2. As you can see, these two fluxes are the same. Finally, if we multiply the flux of 10^{-7} watts/meter^2 by infinity, we get infinity as the flux from the entire Universe.

Of course we know that in fact the total amount of light we receive at night is not infinite at all: indeed it is very small. This is Olbers' Paradox.

NOW YOU DO IT
Consider two spherical shells. Suppose that the flux we receive from a star in the smaller shell is 1.3×10^{-11} watts/meter^2, and suppose that it contains 480 stars. Suppose that the larger shell is 10 times bigger than the smaller one. We want to calculate the net flux we receive from all the stars in each shell. (A) Describe the logic of the calculation you will perform. (B) Carry out the detailed calculation. (C) Is the flux the same for both of them? (D) Suppose the Universe is not infinite, but that it contains only 700 shells of stars – and beyond this set of shells, there is simply nothing at all. We want to calculate the net flux we receive from all the stars in the Universe. (E) Describe the logic of the calculation you will perform. (F) Carry out the detailed calculation.

What is the error in Olbers' Paradox?

There must be something wrong with Olbers' line of reasoning. After all, nighttime is dark! Where does his error lie?

One thought might be that the light from a very distant star is very faint – too faint to have any effect. After all, the very distant stars are too distant to be seen! Therefore we should ignore them, and we should concentrate our attention only on the closer stars. Since there are only a limited number of them, the total light we receive from them is not infinite. As for all the rest, their light is too dim to matter.

But this line of thinking is false. The amount of light we get from a star is governed by the inverse square law, and according to this law no matter how far away the star, we still get a certain amount of light from it. It does not matter that the light from a single star is too faint for us to detect. When we add it to all the equally faint glimmers from all the other, equally distant stars, we still get a significant quantity.

A second possibility might be that something could be blocking the light. Dark interstellar clouds, for instance, block our view of much of our own Milky Way Galaxy (Chapter 16). Perhaps many more such clouds lie at great distances from us, shielding us from the blaze of light from the distant Universe.

Unfortunately this argument too will not work. To see why, notice that any light falling on these hypothetical clouds would heat them up. As an analogy, hold your hand up close to an incandescent light bulb. Your hand is blocking some of its light – and you can feel your hand warming as a consequence. How much would our hypothetical clouds be heated? Since they are blocking an infinite amount of

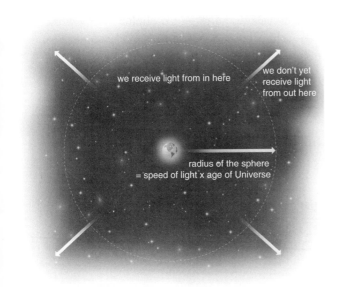

Figure 18.5 If all stars turned on at the same time, we would only receive light from stars within the indicated sphere.

light, they would be infinitely heated! And we know that a hot body emits light. Indeed, by the Stefan–Boltzmann law, an infinitely hot cloud would emit an infinite amount of light.

We now turn to yet a third objection to Olbers' argument – and this is an objection that will turn out to be correct. The total amount of light from all those portions of the Universe that lie a finite distance from us adds up to a finite number. It is only when we consider the infinitely distant stars that we get an infinite amount of light. But *the light from an infinitely distant star takes an infinite amount of time to get here.* The Universe is simply not old enough for the light level to have built up to an infinite value.

Let us return to our picture of an infinite Universe, and imagine that every star within it turned on at some moment. Just before that moment the sky would have been dark. It would also have been dark just *after* that moment, since none of the light would yet have reached us. The Sun, for instance, is so far away that 8 minutes are required for light from it to reach us: therefore we would not see the Sun until 8 minutes after it turned on. Similarly, since Alpha Centauri is 4 light years distant, we would not see it for 4 years. The more time passed, the more light we would receive as more and more stars came into view.

Figure 18.5 illustrates those stars from which we would receive light. They lie in a sphere whose radius equals the distance the light had traveled since they turned on. This radius is the speed of light multiplied by the time the light had been traveling, which is the age of the Universe.

Figure 18.5 illustrates the situation at a particular moment of time. Notice that, at that moment, we receive light from only a finite number of stars, so that the total amount of light we receive is finite. As time passes the sphere grows bigger, and we receive more light. *But the sphere is never infinitely big, and the light is never infinite.*

Furthermore, as we saw in Chapter 15, *every star eventually consumes its nuclear fuel and goes out.* Taken together, these two facts resolve Olbers' Paradox.

The expansion of the Universe

Hubble's great discovery

The discovery of the expansion of the Universe is one of the great triumphs of twentieth-century science. In 1912, the astronomer Vesto Slipher began a program of measuring the Doppler shifts of galaxies. This work went slowly, for the galaxies were so distant that their light was very faint, and it was exceedingly difficult to take the requisite spectra. By 1923 he had measured the Doppler shifts of 41 galaxies. Almost all showed a redshift, indicating motion away from us. Their velocities were amazingly great, amounting to thousands of kilometers per second.

Spurred on by this discovery, Edwin Hubble began a program of measuring not just the velocities, but also the distances to the galaxies by using Cepheid variable

⇐ **Looking backward**
We learned how to use
Cepheid variable stars to
measure distances in
Chapter 16.

stars within them. Over the next several decades he extended Slipher's early work, pushing to ever more distant reaches of the Universe. By 1955 he and his colleagues had measured redshifts and distances of more than 800 galaxies.

But even by 1929 Hubble had discovered that there was more to the expansion of the Universe than the mere fact of recession. There was a striking regularity to this motion: *the more distant the galaxy, the faster it was moving.* If one galaxy was moving at such-and-such a speed, a second, twice as distant, turned out to be moving twice as fast.

Let us work through an example, in order to understand Hubble's method.

Hubble chose a certain galaxy and searched for a Cepheid variable star within it. Once he found one, he measured the period of its variation, and once he had done this he used the period–luminosity relation for Cepheid variable stars (Chapter 16) to find its luminosity. Next he measured the apparent brightness (the flux) of that star. Finally, he measured the wavelength $\lambda_{observed}$ of one of its spectral lines.

The logic of the calculation

Step 1. Solve the inverse square law, using the star's luminosity and flux, to find its distance R.

Step 2. Plug in the measured values of L and f to find R. Since the star lies within the galaxy, this is also the galaxy's distance.

Step 3. Find the difference $\Delta\lambda = \lambda_{observed} - \lambda$ between the observed wavelength of the spectral line and its true value.

Step 4. Solve the Doppler effect formula $\Delta\lambda/\lambda = v/c$ for the galaxy's velocity v.

Step 5. Plug in the measured values of $\Delta\lambda$ and λ and find v.

The megaparsec

In cosmology we commonly reckon distances in megaparsecs, abbreviated Mpc. One megaparsec is

- a million parsecs,
- or 3.26 million light years,
- or 3.08×10^{19} kilometers.

The megaparsec is a natural unit of distance in cosmology, since galaxies lie roughly one Mpc apart from one another.

Step 6. For convenience, express the distance R in megaparsecs, and the velocity v in kilometers per second. Then find the ratio between the galaxy's velocity v and distance R. Let us name the ratio H, for Hubble.

Let us suppose that the Cepheid variable star that Hubble was working with had

- a period of one day,
- a flux of 2.52×10^{-18} watts/meter^2,
- a spectral line with wavelength that ought to be $\lambda = 6562$ angstroms and was observed to have a wavelength $\lambda_{observed} = 6565.06$ angstroms.

Detailed calculation

Step 1. Solve the inverse square law, using the star's luminosity and flux, to find its distance R.

The inverse square law is

$$f = L/4\pi R^2.$$

We solve this to find the distance R:

$$R = \sqrt{L/4\pi f}.$$

Step 2. Plug in the measured values of L and f to find R. Since the star lies within the galaxy, this is also the galaxy's distance.

Consulting the period–luminosity relation (Chapter 16), we find that a star with a period of 1 day has a luminosity of 316 L_{Sun}. Since $L_{Sun} = 3.8 \times 10^{26}$ watts, we find the star's luminosity to be $L = 1.20 \times 10^{29}$ watts. Plugging in this and $f = 2.52 \times 10^{-18}$ watts/meter^2 we find

$$R = \sqrt{(1.20 \times 10^{29}\text{watts}/4\pi 2.52 \times 10^{-18}\text{watts/meter}^2)}$$
$$= 6.16 \times 10^{22} \text{ meters.}$$

Step 3. Find the difference $\Delta\lambda$ between the observed wavelength of the spectral line and its true value.

Since the measured $\lambda_{observed} = 6565.06$ angstroms and its true value is $\Delta\lambda = 6562$ angstroms, we find

$$\Delta\lambda = \lambda_{observed} - \lambda = (6565.06 - 6562) \text{ angstroms}$$
$$= 3.06 \text{ angstroms.}$$

Step 4. Solve the Doppler effect formula $\Delta\lambda/\lambda = v/c$ for the galaxy's velocity v.

We find

$$v = c\,[\Delta\lambda/\lambda].$$

Step 5. Plug in the measured values of $\Delta\lambda$ and λ and find v.

We find

$$v = c\,[\Delta\lambda/\lambda] = (3.00 \times 10^8 \text{ meters/second})(3.06 \text{ angstroms/} 6562 \text{ angstroms})$$
$$v = 1.40 \times 10^5 \text{ meters/second.}$$

Step 6. For convenience, express the distance R in megaparsecs, and the velocity v in kilometers per second. Then find the ratio between the galaxy's velocity v and distance R.

Since 1 megaparsec is 3.08×10^{22} meters, 6.16×10^{22} meters is

$$6.16 \times 10^{22} \text{ meters}/3.08 \times 10^{22} \text{ meters per Mpc} = 2 \text{ Mpc.}$$

Since 1 kilometer is 10^3 meters, 1.40×10^5 meters/second is 140 kilometers/second.

We find the ratio to be

$$H = v/R = (140 \text{ kilometers/second})/2 \text{ Mpc}$$
$$= 70 \text{ kilometers/(second megaparsec).}$$

As you can see from the detailed calculation, Hubble found this galaxy to be 2 megaparsecs away, and moving at 140 kilometers per second away from us. The ratio between the galaxy's velocity and distance H worked out to be 70 kilometers/(second megaparsec).

NOW YOU DO IT
Here are some other Cepheid variable stars in other galaxies. Choose either of them, and work through the same process with it. In particular, pay attention to the value you find for the ratio between the galaxy's velocity and distance: do you get the same value that we did above?

Galaxy	Wavelength (angstroms)	Cepheid period (days)	Cepheid flux (watts/meter^2)
A	6567.8	3.2	2.21×10^{-18}
B	6578.8	8.0	6.58×10^{-19}

Hubble's law

If you worked through the above problem, you were probably amazed at the result you got. Different galaxies are at different distances, and they are moving away from us at different velocities. But no matter which galaxy you worked on, you found that the ratio H between a galaxy's velocity and distance was always the same!

Let us explore this mathematically. We are finding that

$$H = V/R$$

is always the same. We can solve this for the velocity:

$$V = HR,$$

where R is the distance to a galaxy and V its velocity. We see that the velocity of recession of a galaxy is proportional to its distance. The constant of proportionality H is known as *the Hubble constant*. This relation is known as *Hubble's law*. Figure 18.6 shows Hubble's data. As you can see, a graph of the velocities of galaxies versus their distance is a straight line.

Let us get some practice with Hubble's law.

Intuitive mathematics

Notice in Hubble's law that to find the velocity of a galaxy we multiply H by its distance R. For a nearby galaxy R is relatively small and we get a small result for V:

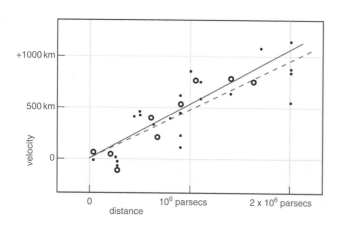

Figure 18.6 **Expansion of the Universe. This is Hubble's original graph showing that the velocity of a galaxy is proportional to its distance. (Notice that Hubble had accidentally mislabeled the vertical axis as kilometers rather than kilometers per second! Furthermore, bedeviled by the same problem that had misled Shapley (Chapter 16) he had underestimated the galaxies' distances.)**

conversely, for a distant galaxy R is big and we get a big value for V. So Hubble's law predicts that the more distant galaxies are moving away from us faster than the nearby ones.

The logic of the calculation

To use Hubble's law we need to know the numerical value of the constant H. This value depends on the system of units we use. In cosmology we commonly reckon distances in megaparsecs. Adopting this unit of distance, and measuring velocity in kilometers per second, we express the value of H in kilometers per second per megaparsec. The numerical value of H is exceedingly difficult to determine; in what follows we will take the value

$$H = 70 \text{ kilometers/(second megaparsec)}.$$

The units of H are somewhat confusing. In the sidebar we get some practice with Hubble's law in order to understand them.

NOW YOU DO IT
(1) **According to Hubble's law, what is the velocity of recession of a galaxy 25 megaparsecs away?**
(2) **Suppose you observe a galaxy whose velocity of recession is 550 kilometers per second. We want to use Hubble's law to find how far away it is from us. (A) Describe the logic of the calculation you will perform. (B) Carry out the detailed calculation.**

Detailed calculation

Consider a galaxy 10 megaparsecs away. Since we want its distance in Mpc, we set $R = 10$. Hubble's law then tells us that

$V = HR$
= [70 kilometers /(second megaparsec)] ×
  (10 megaparsecs)
= 700 kilometers per second.

Now let us think about a more distant galaxy: say, one twice as far away. For this galaxy $R = 20$, and Hubble's law tells us

$V = HR$
= [70 kilometers /(second megaparsec)] ×
  (20 megaparsecs)
= 1400 kilometers per second.

We see that the more distant galaxy is moving faster.

We are not the center of the Universe

Hubble's law makes it appear that we are at the center of the Universe. After all, according to it, every galaxy is moving away *from us*. But this appearance is illusory. It is certainly true that galaxies are moving away from us – but they are also moving away from each other! To see this, turn to Figure 18.7. Each dotted line in this figure shows the distance between a pair of galaxies. Notice that these dotted lines are growing longer as time passes, indicating that these galaxies are drawing away from one another.

Imagine that we were to get in touch with an alien astronomer living on a distant galaxy. Let us suppose this astronomer's name is Elbbuh. We can imagine sending Elbbuh a radio message asking whether galaxies were moving away from her. Elbbuh would reply that they are. If Elbbuh were to measure the distances of various galaxies from her, and their velocities relative to her, the alien might very well come to the conclusion that her galaxy, rather than ours, was the center of the

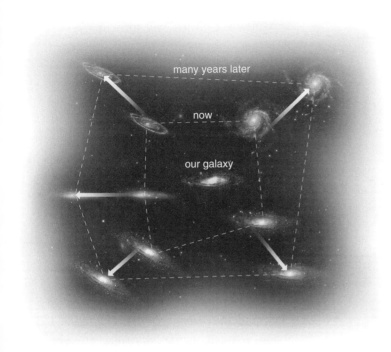

Figure 18.7 **Galaxies are expanding away from each other, as well as from us.**

Universe. Indeed, if there were astronomers living on *every* galaxy, each of them would consider herself the center of the expansion.

Relativity and the expanding Universe

Perhaps the most astonishing thing about the expansion of the Universe is that Albert Einstein had predicted it – and then retracted his prediction.

Einstein's prediction was based on his general theory of relativity. Shortly after announcing it in 1915, he realized that this theory predicted the Universe to be expanding. But at the time, Slipher and Hubble's discoveries lay far in the future. In 1915 everybody thought that the Universe was static. Einstein accordingly modified his theory, adding what he termed the "cosmological constant." He made this addition solely for the purpose of avoiding the prediction of an expanding Universe; years later, when Hubble discovered the expansion, Einstein retracted his cosmological constant, declaring it to be the greatest blunder of his life.

Remarkably, the modern discovery of the acceleration of the Universe has renewed interest in the cosmological constant. Before discussing this, however, we describe Einstein's original cosmological models.

Decelerating expansion: three models of the expansion of the Universe

According to standard relativity theory (without the cosmological constant), the expansion of the Universe is decelerating. The more time passes, the more slowly do galaxies move apart from one another. It is easy to see why this is so. Every galaxy exerts a force of gravity on all the others. This force opposes their outward motion, thus slowing them down. So in the future the cosmos will be expanding more slowly than it is now. Conversely, in the past it had been expanding more rapidly.

We can distinguish three cases.

Case I If the gravitational attraction between galaxies is very weak, their outward motion will be slowed only a little. Indeed, far in the future, when galaxies are very widely separated, their mutual gravitational attraction will be negligible, and they will coast outwards steadily.

Case II If the attraction between galaxies is moderately weak, the Universe will continue expanding forever, albeit at a steadily decreasing rate.

Case III If the attraction between galaxies is sufficiently strong, eventually their outward motion will cease – and then they will start moving together! The

Table 18.1. **Gravitation and the expansion of the Universe.**

Case I	Weak gravitational attraction	Eternal expansion at a rate that diminishes initially, but then becomes constant
Case II	Intermediate gravitational attraction	Eternal expansion at a steadily diminishing rate
Case III	Strong gravitational attraction	Expansion followed by contraction

Table 18.2. **Density and the expansion of the Universe.**

Case I	Weak gravitational attraction	Low density	Eternal expansion at a rate that diminishes initially, but then becomes constant
Case II	Intermediate gravitational attraction	Intermediate density	Eternal expansion at a steadily diminishing rate
Case III	Strong gravitational attraction	High density	Expansion followed by contraction

Table 18.3. **Density, critical density and the expansion of the Universe.**

Case I	Weak gravitational attraction	Low density	D_{actual} less than $D_{critical}$	Eternal expansion at a rate that diminishes initially, but then becomes constant
Case II	Intermediate gravitational attraction	Intermediate density	D_{actual} equals $D_{critical}$	Eternal expansion at a steadily diminishing rate
Case III	Strong gravitational attraction	High density	D_{actual} greater than $D_{critical}$	Expansion followed by contraction

scale of the Universe

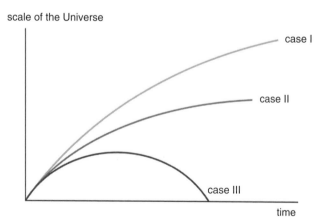

Figure 18.8 **Expansion of the Universe. Three possible cases.**

expanding Universe will turn into a contracting Universe. (As we will shortly see, the recent discovery of the acceleration of the Universe shows that the Universe cannot be of this type.)

These three cases are summarized in Table 18.1 and Figure 18.8.

The critical density

According to relativity, it is the *density* of the cosmos that determines the force of gravity acting to slow its expansion. The greater the density the stronger the gravitational force. This, too, is easy to understand: the greater the density the more tightly are galaxies packed together, and the stronger is the gravitational attraction between them. Table 18.2 summarizes the three cases.

We see that to ascertain the future of the Universe we need to measure how tightly packed together galaxies are, i.e. the overall density of the cosmos. Let us call the result of this measurement D_{actual}. According to Table 18.2, if D_{actual} is very low, the Universe is case I. If it has a certain intermediate value, the Universe is case II. And if D_{actual} turns out to be greater than this intermediate value, the Universe is case III.

This critical intermediate value is referred to as the *critical density*. We emphasize that this critical density is *not* the actual density D_{actual}. Rather, it is merely a reference value. Just as the speed limit on a highway does not tell us how fast cars are actually traveling, but merely separates legal from illegal driving, so too does the critical density merely separate eternal expansion from ultimate collapse. According to relativity theory, to determine the ultimate fate of the Universe we must measure D_{actual} and then compare it with $D_{critical}$ as shown in Table 18.3.

What is the result of this comparison? How does the actual density compare with its critical value? Unfortunately, we do not know for certain. The reasons for our present state of uncertainty are instructive, and they shed light on the nature of science. We now turn to a discussion of them.

THE NATURE OF SCIENCE

UNCERTAINTY IN SCIENCE

We emphasized in "The nature of science. Certainty and uncertainty in science" (in Chapters 3 and 9) that, contrary to popular opinion, uncertainty often plagues scientific research. Many people believe that, while most things in daily life are ambiguous, science allows us to reach conclusions that are absolutely certain. We explained, however, that while science does indeed bring certainty, it does so only in the long run. In the short run, scientists working at the cutting edge of knowledge are incessantly plagued by ambiguity.

Einstein's relativity theory tells us exactly what we must do in order to determine the future of the Universe: we have to measure its density. Sadly, we are unable to do this with any confidence. The reason is that it is very hard to measure D_{actual}. In this section we will explore why this is so.

Density is mass divided by volume. So to measure density we need to measure both mass and volume. Specifically, to measure the density of the Universe, we must (1) delineate a certain region of space, (2) add up all the mass the region contains and (3) divide this mass by the region's volume to obtain the density. It is the first two steps in this process that are hard. Let us see why.

To be specific, suppose we wish to measure the density of that portion of the Universe lying within 100 Mpc of us.

In order to *delineate a certain region of space* we need to identify all the galaxies lying within 100 Mpc of us. To do this we must observe each candidate galaxy and measure its distance: those within 100 Mpc we keep in our tally and those farther away we do not. But throughout this book we have repeatedly emphasized that the measurement of distance is often difficult in astronomy. As we have seen, the technique for measuring the distances of galaxies involves the so-called Cepheid variable stars. Can this method be used to delineate our 100 Mpc region?

In order to do so, we need to be able to observe Cepheid variable stars in galaxies 100 Mpc away. But since these distances are so great, the light from the Cepheid will be very faint and we will need a very sensitive telescope. Just as we cannot see exceedingly faint lights with our naked eyes, so too with telescopes. Speaking in more technical language, we say that the most sensitive telescopes we possess are only capable of detecting fluxes of light brighter than a certain limit. Let us call this limit f_{limit}. If the flux of light from a distant Cepheid variable star is greater then f_{limit} our best telescopes can detect it – but if it less than f_{limit} they cannot. Numerically f_{limit} is about 10^{-18} watts per meter^2.

Are our telescopes capable, then, of spotting Cepheid variable stars 100 Mpc away? To find out we need to calculate the actual flux f_{actual} we would get from such a star at this distance, and compare it to f_{limit}. To calculate this actual flux we can use the inverse square law:

$$f_{actual} = L/4\pi R^2.$$

Intuitive mathematics

Let us examine our procedure, to see what it says about detecting distant stars.

Detailed calculation

Step 1. Look up the luminosity L of a Cepheid variable star.
We are lucky in that Cepheid variables are among the brightest stars known, reaching luminosities of 10 000 solar luminosities. So we calculate

$L = 10\,000\ L_{Sun}$
$L_{Sun} = 3.8 \times 10^{26}$ watts.

So that $L = (10\,000)(3.8 \times 10^{26}$ watts$) = 3.8 \times 10^{30}$ watts.

Step 2. Take a value of 100 Mpc for its distance R.

Since 1 Mpc $= 3.08 \times 10^{22}$ meters, we calculate
$R = (100)(3.08 \times 10^{22}$ meters$) = 3.08 \times 10^{24}$ meters.

Step 3. Use the inverse square law to calculate the flux f_{actual} we receive from such a star.

$f_{actual} = L/4\pi R^2$
$\qquad = 3.8 \times 10^{30}$ watts$/4\pi\ (3.08 \times 10^{24}$ meters$)^2$
$\qquad = 3.18 \times 10^{-20}$ watts/meter^2.

Step 4. Is this bigger than f_{limit}? If so, we can detect the star.
The flux we calculated is less than f_{limit}, which is 10^{-18} watts/meter^2. So we cannot detect Cepheid variable stars at such great distances.

(1) Notice first that the law tells us to multiply by the star's luminosity L. So bright stars (big L) produce bigger fluxes than dim ones (small L). Since we need our f_{actual} to be bigger than our limiting value f_{limit}, this means it is more easy to see bright stars than dim ones. This conforms to our intuitive expectations.

(2) Notice also that the law tells us to divide by the star's distance squared. So near stars (small R) produce bigger fluxes than far ones (big R). Since we need our f_{actual} to be bigger than our limiting value f_{limit}, this means it is more easy to see near stars than far ones. This also conforms to our intuitive expectations.

The logic of the calculation

Step 1. Look up the luminosity L of a Cepheid variable star.

Step 2. Take a value of 100 Mpc for its distance R.

Step 3. Use the inverse square law to calculate the flux we receive from such a star, f_{actual}.

Step 4. Is this bigger than f_{limit}? If so, we can detect the star.

....

NOW YOU DO IT
We want to calculate the flux from a Cepheid variable star whose luminosity is 8000 times that of the Sun if it is 50 Mpc away. **(A)** Describe the logic of the calculation you will perform. **(B)** Carry out the detailed calculation. Is this star bright enough for our telescopes to detect it? **(C)** If this star is exceedingly far from us we would not be able to observe it. Conversely, if it is very close to us we would. What is the dividing line between these two cases: that is, at what distance is the flux from such a star just barely enough for our telescopes to detect? **(D)** Describe the logic of the calculation you will perform. **(E)** Carry out the detailed calculation.

....

As we can see from the detailed calculation, we cannot detect Cepheid variable stars 100 Mpc away. This means that we cannot reliably measure the distances of these faraway galaxies using this technique. There are other techniques, but they too suffer from drawbacks. Thus it is hard to accomplish the first step in our process of measuring the density of the Universe.

Let us now pass on to the *second step*, and *add up all the mass the region contains*. To do this, we need to measure all the mass lying within 100 Mpc of us. As we saw in Chapter 3, Newton's law of gravitation shows us how to do this. It gives us a formula (the mass formula) for the mass of a body about which other bodies orbit. In Chapter 16 we used this to measure the masses of galaxies.

The difficulty with this procedure is that this method only allows us to measure the mass contained *in* galaxies – but it tells us nothing about any matter lying *between* them. Furthermore, the discovery of dark matter

(Chapter 16) tells us that much of the mass of the Universe emits no light and cannot be detected by ordinary means. This means that we have no means of reliably measuring the *total* amount of matter contained in our 100 Mpc region of space. All we can measure is that portion of it lying within galaxies.

For these two reasons, the task of measuring the density of the Universe is exceedingly difficult. In spite of the best efforts of astronomers for decades, we have still not succeeded in doing it reliably. In order to do so, we need to develop alternate methods of measuring both distances and masses. The methods of measuring distance need to work for faraway galaxies, and the methods of measuring masses need to be able to find matter lying outside of galaxies. None of this is impossible. People are working right now on both projects. Ultimately they will succeed. But until they do we are not sure of the future of the Universe.

The theory of the Inflationary Universe provides a different means of attacking this problem. As we will see below, it predicts that the Universe should be case II. However, this theory, while widely accepted, relies on an extrapolation of our knowledge to ultra-high energies, and it has not been well tested experimentally.

The geometry of the Universe

Within the general theory of relativity, space can possess a geometry different from the familiar Euclidian that we all learned in school. Each of the three cases we have been considering turns out to have a different geometry.

The case II possibility is simplest: it has our old familiar Euclidian geometry. Because this geometry describes figures drawn on a plane, it is known as flat space. The case I possibility, on the other hand, has the geometry of a saddle (Figure 18.9). It is known as open space because the saddle, like the plane, extends outward without limit.

Finally, the case III possibility has the geometry of a sphere (Figure 18.10).

These geometries are listed in Table 18.4.

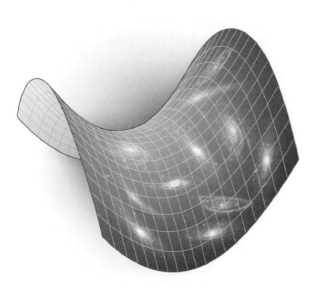

Figure 18.9 **Saddle geometry obtains if the density of the Universe is less than critical.**

Figure 18.10 **Spherical geometry obtains if the density of the Universe is greater than critical.**

Table 18.4. **Density, critical density, expansion and geometry of the Universe.**

Case I	Weak gravitational attraction	Low density	D_{actual} less than $D_{critical}$	Eternal expansion at a rate that diminishes initially, but then becomes constant	Saddle geometry
Case II	Intermediate gravitational attraction	Intermediate density	D_{actual} equals $D_{critical}$	Eternal expansion at a steadily diminishing rate	Euclidian geometry
Case III	Strong gravitational attraction	High density	D_{actual} greater than $D_{critical}$	Expansion followed by contraction	Spherical geometry

Is the Universe infinite?

The paradox of the edge

Is there an end to the Universe, or is it infinitely big? The first thing to be said is that we have never found an end to the Universe. The farther into space our telescopes have peered, the more galaxies we have found.

Furthermore, the moment we even ask whether there might be an end to the cosmos, we are plunged into confusion. For no matter what we imagine the answer to be, paradoxes arise.

Suppose we imagine that the Universe is finite. This means that it has some sort of edge, beyond which there lies nothing. Let us journey in our imaginations out to this edge. Floating there at the limits of the cosmos, we would find ourselves gazing outward into a terrifying void, containing not a single star, not the tiniest grain of sand, and indeed not even the faintest glimmer of light. But what is to keep us from moving outward, and penetrating beyond the very "edge of the Universe" we had postulated? We might imagine some sort of wall marking the outer edge of the cosmos – but what is there to prevent an atom from detaching itself from this wall and wandering off into space? Indeed, recall that the Stefan–Boltzmann law guarantees that this wall must be emitting light, so that it must be steadily broadcasting energy out beyond the "edge" – and energy and matter are connected via $E = mc^2$.

For all these reasons it seems impossible to imagine that the Universe might be finite. Let us then consider the only other possibility, and suppose that it has no outer limit.

This possibility does not involve us in logical paradoxes. But it is certainly dizzying. Faced with this infinity, our mightiest telescopes are capable of studying only an infinitesimal fraction of the totality. If the Universe is infinite, then it must contain an infinite number of planets that look just like ours. Many of these planets would not contain life – but many would. Indeed, in an infinite Universe there must be an infinite number of planets that contain creatures exactly like us: creatures that wonder about the nature of the cosmos just as we do, and that find themselves just as baffled as we.

Spherical geometry: finite but unbounded

Remarkably, the spherical geometry of case III shows us a way out of these paradoxes. Within this geometry the Universe is finite – and at the same time, it has no edge.

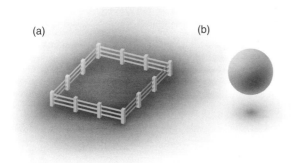

(a)   (b)

Figure 18.11 **A finite area on a plane (a) requires a fence. But on a sphere (b) it does not.**

Figure 18.12 **In Euclidian geometry a straight line can be extended indefinitely.**

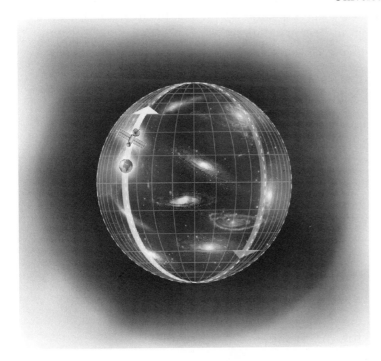

Figure 18.13 **In spherical geometry a straight line cannot be extended indefinitely. Rather, it eventually returns to its starting point.**

Although we did not make it explicit, our Paradox of the Edge relies on the principles of Euclidian geometry. To see this, let us consider how a finite amount of space is delineated – first in this geometry, and then in a different one. Figure 18.11 depicts a limited amount of space in both Euclidian and spherical geometry.

To be specific, imagine that our "limited amount of space" is a single acre. In Euclidian geometry (Figure 18.11a), which concerns figures drawn on a plane, we delineate this acre by fencing it off. But spherical geometry (Figure 18.11b) concerns figures drawn on the surface of a sphere. And suppose this sphere happens to have a surface area of one acre. Then we do not need to fence our acre off!

The spherical geometry has the remarkable property of *possessing a finite amount of space without having an edge.* Figure 18.11(a) requires a fence to delineate a finite space – but that fence is nowhere to be seen in the spherical geometry of Figure 18.11(b). In Euclidian geometry we needed a fence to delineate our single acre because there exists plenty of space in addition to that acre. But in spherical geometry there simply isn't any more space than the acre. This is why no fence is needed.

The fence in Figure 18.11(a) is analogous to the "edge of the Universe" in our Paradox of the Edge. We now realize that our need for an edge arose because we had been using Euclidian concepts. But in spherical geometry this need is gone: the Universe is finite without having any edge at all.

We can understand this further by imagining a spacecraft designed to explore the geometry of the Universe. Let us call this craft a Cosmic Geometry Explorer – a CGE. Imagine that we were to launch our CGE into space, and send it off in a perfectly straight line leading away from the Earth. This spacecraft would follow the path shown in Figure 18.12: a straight line extending outward endlessly. This is because it is a principle of Euclidian geometry that a straight line can be extended indefinitely. But in spherical geometry that principle is absent: a straight line *cannot* be extended indefinitely. Rather it returns to its starting place (Figure 18.13). Our CGE spacecraft, which ages ago left on its voyage away from the Earth, would eventually return from the opposite direction! It would have traveled a finite distance, and in this travel it would have crossed all the space that exists.

Table 18.5. **Density, critical density, expansion, geometry and the amount of space in the Universe.**

Case I	Weak gravitational attraction	Low density	D_{actual} less than $D_{critical}$	Eternal expansion at a rate that diminishes initially, but then becomes constant	Saddle geometry	Infinite amount of space
Case II	Intermediate gravitational attraction	Intermediate density	D_{actual} equals $D_{critical}$	Eternal expansion at a steadily diminishing rate	Euclidian geometry	Infinite amount of space
Case III	Strong gravitational attraction	High density	D_{actual} greater than $D_{critical}$	Expansion followed by contraction	Spherical geometry	Finite amount of space

One might suppose that somehow the CGE achieved this feat by traveling in a circle. Its path certainly *looks* like a circle in Figure 18.13. But this figure is misleading. The printed page is Euclidian, and figures drawn on it cannot fully represent non-Euclidian geometries. Figure 18.13 is merely a model, meant to suggest certain features of a non-Euclidian geometry that properly can only be approached mathematically. In reality, the CGE had been traveling in a perfectly straight line.

We summarize our new results in Table 18.5. Notice that the task of measuring the density of the cosmos and comparing it to $D_{critical}$ now assumes added importance. Once we achieve this, we will find out whether the Universe is finite or infinite in extent.

The accelerating Universe: "dark energy" and the cosmological constant

As we have noted, according to Einstein's theory without the cosmological constant the expansion of the Universe should be decelerating. Because galaxies attract one another gravitationally, the more time passes, the more slowly should they be moving apart. But in 1998 two teams of astronomers announced that the expansion of the Universe is in fact accelerating! This acceleration shows that something capable of resisting gravity must exist in the Universe. This new "something" might be Einstein's cosmological constant: more generally it is termed *dark energy*.

To show that the Universe is accelerating we must measure the expansion rate and show that it is more than it had been in the past. To measure the expansion rate in the past, in turn, requires us to survey the Universe at great distances. For example, if we observe the cosmos one billion light years away we are receiving light emitted one billion years ago, and we are observing the expansion of the cosmos as it was one billion years in the past.

But to survey the cosmos at such great distances we cannot use Cepheid variables. As we have seen, these stars cannot be seen more than a few hundred million light years away. We need a new means of measuring distances, one that will allow us to observe yet more remote portions of the cosmos.

Supernovae: a new means of measuring distance

This new means is provided by supernovae. As we saw in Chapter 15, these exploding stars are exceedingly luminous, thus allowing us to observe them very far away.

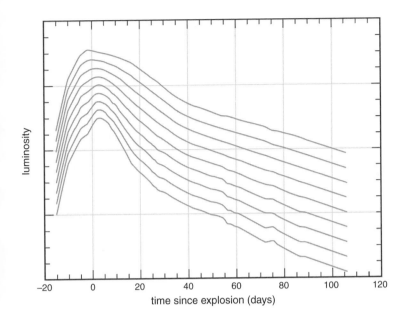

Figure 18.14 **Brightness versus time for Type Ia supernovae. The brighter the supernova, the longer it takes for its brightness to decline. So by measuring the rate of decline, we can infer the brightness.**

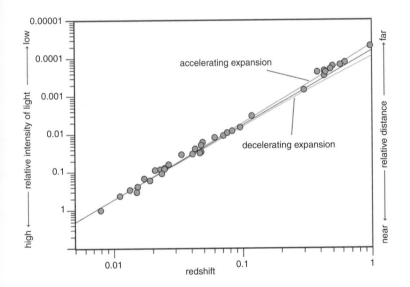

Figure 18.15 **Acceleration of the Universe. The brightness of a galaxy is plotted against its redshift. The two lines show theoretical predictions for a decelerating and an accelerating Universe.**

Furthermore, certain supernovae (those of Type Ia) possess a property that allows us to measure their intrinsic luminosity L. Once this is done, we can measure their apparent brightness f and use the inverse square law to infer their distance.

The means of measuring the intrinsic brightness of a supernova is illustrated in Figure 18.14. As shown in this figure, the brighter the supernova, the longer it takes for its brightness to decline. So by measuring this rate of decline, we can infer the brightness. And if we combine this with a measurement of redshift, we can obtain the results given in Figure 18.15.

The figure shows that the supernovae of a given brightness show smaller redshifts than expected. The interpretation is that the Universe has expanded less than expected during the time in which their light has been traveling toward us. And this implies that currently the expansion of the Universe is accelerating.

"Dark energy" and the cosmic repulsion

An accelerating Universe can only be understood by postulating some form of repulsion that acts to overcome gravity, a repulsion that has been named "dark energy." An important feature of this universal repulsive force is that it can remain constant as the Universe expands. But gravity is an inverse square force, and as galaxies recede from one another their mutual gravitational attraction diminishes. So early in cosmic history, when galaxies lay close together, gravity was more important than the repulsion. Late in cosmic history on the other hand, when galaxies are widely separated, gravity has become weaker than the repulsion.

We conclude that the effects of the dark energy were insignificant in the past, but have grown ever more important since then. Far in the cosmic future it is gravity that will be insignificant, and the fate of the Universe will be dictated by the repulsion. The Universe will expand forever – indeed, it will accelerate forever.

Giving this universal repulsion the name "dark energy" does nothing to explain it. It is an entirely new phenomenon, previously unknown to physics. What is dark

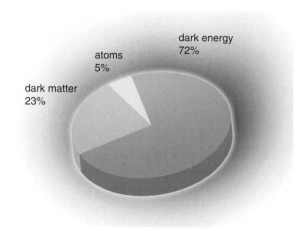

Figure 18.16 **Contents of the Universe.**

energy? The task of answering this question lies at the very forefront of physics today, involving modern, highly speculative extensions of our knowledge.

If future work confirms that the cosmic acceleration and its interpretation in terms of dark energy is correct, a mystery we discussed in Chapter 16 has grown even more perplexing. In that chapter we described the evidence for the existence of great quantities of dark matter, of whose nature we are ignorant, pervading the cosmos. As we noted in that chapter, observations reveal there to be more dark matter than ordinary matter. But the dark energy is even more abundant than the dark matter! It is sobering to realize that all the ordinary substance of the Universe – the matter out of which our bodies are made, the rocks in the Earth and all the other planets, the gases in the Sun and every other star, all the interstellar clouds and galaxies that observation reveals – all this adds up to a small portion of the Universe (Figure 18.16). And as for the truly significant portion, we have only the faintest conception of what it might be.

The infinite future of the Universe

Let us try to imagine what the cosmos will be like in the far distant future.

To the best of our knowledge, the cosmos will continue forever. Many millions of years from now, biological evolution will have transformed most species beyond recognition – including, presumably, our own. But suppose that creatures capable of observing the Universe still exist. What will they see?

Millions of years from now, the random motion of stars will have distorted the constellations into unfamiliar shapes. Hundreds of millions of years from now, continental drift will have done the same to the map of the Earth. But even these extraordinary spans of time are short compared to the time scales over which the Universe itself evolves.

Five billion years in the future, the Sun will exhaust the hydrogen in its core and commence turning into a red-giant star. By this time, the same will have happened to many other stars, which will have reached the various end-points of stellar evolution we described in Chapter 15. As we saw in that chapter, when the Sun becomes a red giant temperatures on the Earth will first grow exceedingly hot, but afterwards will drop as the Sun contracts to a dim white dwarf. To survive, our hypothetical astronomers will be forced to migrate to a more hospitable planet. Assuming they do so, as they continue to observe the heavens they will notice fewer and fewer stars in the sky, as more and more stars go out. The nighttime sky will be emptier than it is today.

The same will be happening in the other galaxies, which will grow steadily fainter. Furthermore, they will become yet more distant. So our hypothetical astronomers will find it more and more difficult even to notice the faint, distant galaxies.

The more time passes the fewer and fewer stars will remain shining, and the fewer and fewer galaxies will be observable. Ultimately, every star in the Universe will go out, and the cosmos will become utterly dark. At this point every planet will be deprived of its source of life-giving warmth, and life in the Universe will come to an end. But even though beings capable of observing the Universe will no longer exist, and even though the distant galaxies will be emitting no light whatsoever, these galaxies will still exist, and they will continue flying apart forever.

The past history of the Universe: the Big Bang

Now let us turn our attention in the other direction: starting from the present and moving backward into the past.

Since galaxies are expanding away from one another, they used to be closer together. The farther back in time we project our minds, the closer they had been, which means that the more compressed the Universe had been. Indeed, in all three of the cases we have been discussing, the scale of the Universe was *zero* at some point in the past. This state of infinite compression is the Big Bang. Figure 18.17 sketches the history of the cosmos, beginning shortly after the Big Bang and continuing to the present day.

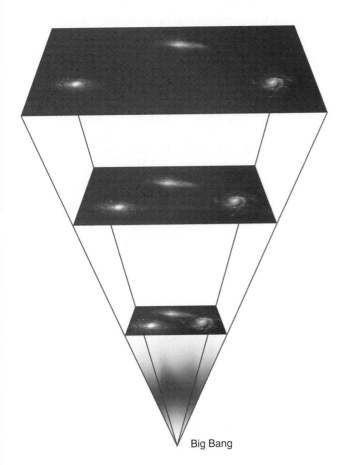

Big Bang

Figure 18.17 **History of the cosmos.**

A voyage backward in time: the early Universe

Referring to this figure, let us approach the Big Bang in stages, projecting our minds ever farther backward in time. Between 1 and 2 million years ago, the earliest hominids walked the Earth. Several hundred million years ago, drifting continents were joined together into a single super-continent (Chapter 7). 4.5 billion years ago, the Solar System formed (Chapter 13). These are immense spans of time indeed – but they are only the beginning of our survey of the history of the cosmos.

The farther out in space we look, the farther back into the past we are seeing. Return to the Hubble Deep Field (Figure IV.2 of "Introducing galaxies and the Universe," preceding Chapter 16). This is one of the farthest looks we have ever taken into the depths of space. The light recorded in this image was emitted long before the Earth was formed. In the faint smudges shown in this image we may well be witnessing galaxies in the earliest stages of their evolution.

Beyond the Hubble Deep Field our telescopes are not yet sufficiently powerful to penetrate. But we can continue casting our scientific theories back to yet earlier epochs in the history of the Universe. Several hundred million years after the Big Bang, it is thought that galaxies formed. Presumably the first stars formed at about the same time in the dense central regions of these newly formed galaxies.

Prior to the appearance of these first stars was an epoch of unrelieved, stygian darkness. This epoch is sometimes known as the "Dark Ages": we are only beginning to be able to study it observationally. We know very little about the conditions that obtained, but we do know that the cosmos was filled with a cold, roughly uniform gas. The density of this gas was very low – lower, for instance, than that of air.

While cold, this gas was not at absolute zero – and the farther back in time we project our thoughts, the warmer it must have been. Fifteen million years after the Big Bang, the Universe was at room temperature. Earlier, 1.5 million years after the Big Bang, the temperature was 1200 kelvin, and the Universe was filled with a faint, deep red glow. This light was the blackbody radiation (Chapter 4) emitted by the hot gas filling the cosmos. At even earlier epochs, the radiation was yet brighter, and the primal Universe was flooded with light.

A critical transition stage was reached about 380 000 years after the Big Bang. Prior to this point, the Universe was opaque. The great heat of the cosmos was sufficient to ionize atoms, stripping electrons away from their nuclei; the many freely roaming electrons were highly effective absorbers of light. A time traveler, magically transported to this epoch, would have been able to see only short distances through the brilliantly glowing gases. In later epochs, on the other hand, the Universe was sufficiently cool for atoms to exist; since atoms are not so efficient at absorbing light, the Universe was transparent in these later epochs. This critical transition between opaque and transparent was reached when the temperature was 3000 kelvin and the Universe was filled with a brilliant, yellow light. It is known as the epoch of decoupling, since it was then that the blackbody radiation became effectively disconnected from the matter. It will turn out to be particularly significant when we turn to a discussion of the cosmic background radiation.

Recall that light carries energy. The farther back in time we project our thoughts, the greater was the energy carried by the blackbody radiation. During the first 100 000 years of the history of the cosmos, this energy exceeded that carried by the matter. The primal Universe, therefore, did not just contain light: it essentially *was* light. The matter, in contrast – all the matter eventually destined to condense into the stars, planets and indeed our own bodies – was a dilute, relatively inconsequential mixture suspended in that blaze of pure radiance.

At exceedingly early epochs, the great heat of the cosmos was sufficient not just to strip electrons away from nuclei, but to shred the very nuclei themselves into their constituent particles. At these early times, the cosmos consisted merely of protons, neutrons and electrons flying rapidly through the blackbody radiation. Several minutes after the Big Bang, when the temperature was 1 billion degrees, these particles underwent nuclear reactions in which helium and a few other of the lightest elements were formed.

The hotter a gas is, the more rapidly do its constituents move. During the first few minutes everything was moving at nearly the speed of light. To study these primal epochs researchers turn to elementary-particle accelerators, in which particles are slammed into one another at high energy. Recent years have seen the emergence of a remarkable union of high-energy physics and cosmology, in which advances in each field yield fruit in the other.

Perhaps the most remarkable of these fruits is the concept of cosmic inflation. This is a brief period in which the early Universe is thought to have expanded at a far greater rate than normal. This ultra-rapid expansion is thought to have begun a remarkable 10^{-36} seconds after the Big Bang, at which time the temperature was an astonishing 10^{28} kelvin and the density an equally astonishing 10^{78} times that of water. The ultra-rapid inflation is thought to have ceased 10^{-34} seconds after the Big Bang, after which the cosmos resumed its normal, far slower expansion. During the period of inflation, the cosmos expanded enormously: we are not sure by how much, but estimates range from factors of 10^{21} to 10^{43}. For comparison, a single proton inflated by 10^{43} would be larger that the presently observed Universe!

An important prediction of the theory of inflation is that the geometry of the Universe will be Euclidian. We can understand why this is so by returning to our discussion of the geometry of the cosmos. In case III this rapid inflation would have expanded the sphere representing the geometry to enormous sizes: just as the surface of the Earth *looks* flat to our limited viewpoint, so too this geometry becomes essentially flat (case II) even over distances our telescopes are capable of surveying. The same thing happens in case I: in this case, it is the saddle that is expanded enormously.

Figure 18.18 summarizes our understanding of the history of the Universe.

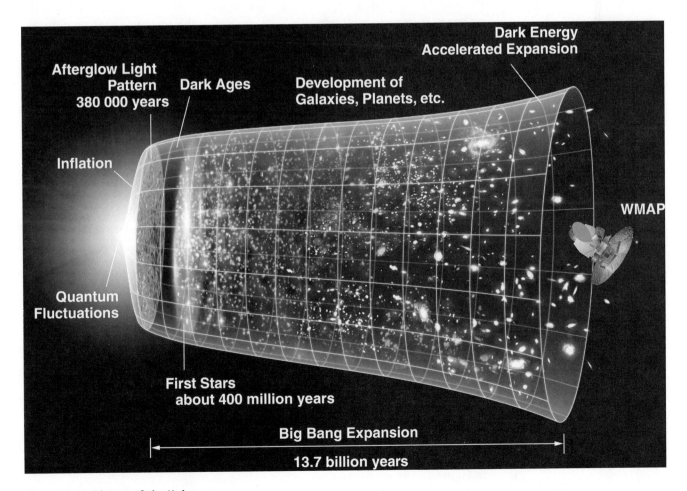

Figure 18.18 **History of the Universe.**

The moment of the Big Bang

Our voyage backward in time, which began at the present, has now carried us to within a fraction of a second of the Big Bang. Further advances in high-energy physics will allow us to push our understanding yet further, to still earlier times and even higher temperatures. But at present our understanding of these very stages is speculative and incomplete.

Preceding everything was the Big Bang, a state of literally infinite density and temperature. What do we know about it?

With this question, we are brought face to face with a profound mystery. One of our most fundamental conceptions of matter is that it occupies space. Every object has a certain definite size, a size that may be small but is never zero. At the instant of the Big Bang, however, every particle in the Universe was forced into a region of space of literally no size at all. We do not have the slightest idea how to understand this extraordinary state. We do not understand what the matter could have been like at that climactic instant, and we do not understand how it could have evolved from that into its present prosaic form.

As we discussed in Chapter 15, the same sort of state forms at the center of a black hole. While in Chapter 15 we had no idea how to understand this, we were able to comfort ourselves with the notion that the mysterious singularity was hidden from our view within the Schwarzschild radius. To some degree, therefore, the mystery was of academic interest only. In the case of the Big Bang we do not have that comfort. Indeed, the situation is just the opposite: all the matter that exists, including the matter in your very own body, was once crushed to this incomprehensible state.

It may be, however, that the situation is not quite so dire as all that. This possibility arises because Einstein's relativity theory, on which the concept of the Big Bang is based, does not take into account quantum mechanics (Chapter 4). Nobody knows how to unite these two theories, but many scientists believe that when this is accomplished we will learn that the degree of compression in the Big Bang, while enormous, was not literally infinite. The effects of quantum mechanics are likely to be significant only up to 10^{-43} seconds after the Big Bang. At this extraordinarily early time the density was 10^{94} times that of water, and the temperature was 10^{32} kelvin. During this era of "quantum cosmology," our very ideas of space and time are altered: according to the uncertainty principle even the concepts of "before" and "after," and "near" and "far" become inappropriate. Further progress in understanding the Big Bang will not be possible until we learn how to live in this strange new world.

Was the Big Bang creation?

Was the Universe created in the Big Bang? Or did it exist prior to that moment? We do not know, and we have no way of finding out until we learn both how to unite general relativity with quantum mechanics, and how to understand the state of matter compressed to enormous, and possibly infinite, densities. For now, our ideas are purely speculative.

One possibility is that the Universe existed prior to the Big Bang, but in a state of contraction rather than expansion. Perhaps the galaxies moved inward over cosmic ages, the Universe growing steadily denser and hotter until it crushed inward – but then "bounced" outward into its present expansion. Such a possibility is illustrated in Figure 18.19.

Yet another possibility is that nothing existed prior to the Big Bang – that the Universe was created at that moment. It is important to appreciate what is meant by

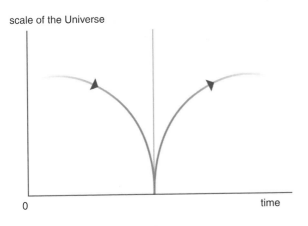

scale of the Universe

0   time

Figure 18.19 **A bouncing Universe.**

this suggestion. We are speculating that prior to the Big Bang nothing whatsoever existed: not a single grain of sand, not a single elementary particle, not the faintest glimmer of light.

Thinking about creation brings us face to face with profound mysteries. Did the laws of physics exist before creation? What would it mean to say that two bodies attract each other through the force of gravity when there were no bodies? What would it mean to say that an object of mass m has an energy mc^2 when there was no such thing as mass, and no light to have a velocity? Indeed, did space and time themselves exist before creation? And if time did not exist, what sense does it make to speak of "before" the moment of creation?

We mentioned that recent years have seen the emergence of a remarkable union of high-energy physics and cosmology, in which advances in each field yield insights into the other. Recent, highly speculative theories of high-energy physics have suggested that the very laws of physics are not inviolate, but that they might themselves have been created at the instant of the Big Bang.

Yet other theories speculate that the collapse of a black hole might trigger the formation of an entire new Universe – a "child Universe" – which expands outward into its own space, completely unrelated to that in which the black hole collapsed. According to this notion one cosmos can give rise to many others. Researchers speak of multiple Universes branching outward from one another, all existing simultaneously but entirely disconnected from one another. Perhaps ours is not the first Universe, but is itself the child of some unknown and unknowable parent.

It is important to emphasize that all of these ideas are highly speculative. Facing the mystery of creation, we have not progressed all that far beyond the ancient Hindus who wrote the following in the *Rig Veda*.

> But after all, who knows, and who can say
> whence it all came, and how creation happened?
> The gods themselves are later than creation,
> so who knows truly whence it has arisen?
> Whence all creation had its origin,
> he, whether he fashioned it or whether he did not,
> he, who surveys it all from highest heaven
> he knows – or maybe even he does not know

The Cosmic Background Radiation

Filling the Universe is a faint, nearly uniform glow of electromagnetic radiation known as the Cosmic Background Radiation (CBR). It is the tenuous remnant of the blackbody emission that flooded the early Universe: the blaze of the Big Bang.

The spectrum of this CBR, as measured by radio telescopes, is blackbody. As we noted in Chapter 4, blackbody radiation can be emitted only by dense objects. The fact that the CBR's spectrum is blackbody therefore demonstrates that the Universe once was in a highly compressed state. The CBR therefore provides decisive confirmation of the basic correctness of our picture of the past history of the Universe as outlined above. Surely it is remarkable that our theories, developed when the cosmos was many billions of years old, have turned out to yield a valid understanding of the Universe when it was half a million years old!

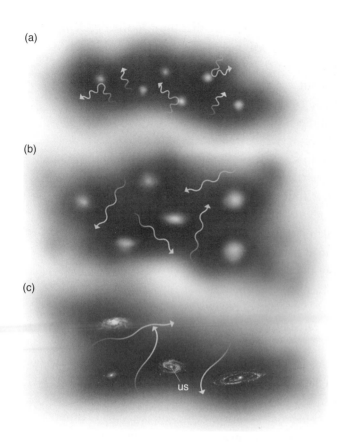

(a)

(b)

(c)

us

Figure 18.20 **Blackbody radiation in the Universe. (a) Early in its history, the radiation scattered frequently against particles: the Universe was opaque. (b) After the epoch of decoupling the blackbody radiation did not scatter against particles: the Universe was transparent. Notice that the wavelength of this radiation has been slightly increased, and its density slightly decreased, by the expansion. (c) The Cosmic Background Radiation that we observe today is the dilute, long-wavelength residue of the radiation depicted in (b).**

Let us try to picture the blackbody radiation filling the cosmos. In Figure 18.20(a) we illustrate the early Universe. In this figure the dots represent particles – the electrons, neutrons and protons making up the early Universe – and the wiggly lines represent light rays. These rays were traveling in all directions through the uniform distribution of particles. This figure is a "snapshot," taken prior to the epoch of decoupling, when the electromagnetic radiation propagated only a short distance before interacting with the dense, hot gas. Thus each ray shown in the figure travels only a short distance before scattering off a particle.

Conversely, radiation emitted after the epoch of decoupling traveled unimpeded through space. Figure 18.20(b) represents this later state, when the cosmos had become transparent. Here each light ray travels in a straight line, and the light rays do not scatter off the particles. Notice in (b) that both the particles and the light rays are more widely separated than those in (a). This is because the Universe has expanded somewhat during the time separating the two "snapshots." Notice also that the wavelength of the light in (b) is longer than in (a). This is because the expansion of the Universe stretches wavelengths. The high-frequency, intense light of the early Universe has become a little lower in frequency, and a little less intense.

Finally, Figure 18.20(c) shows a "snapshot" of the Universe right now. At this epoch the subatomic particles depicted in (a) and (b) have cooled and condensed into the galaxies we observe today. The rays of electromagnetic radiation are far more sparse, and their wavelengths have been stretched by the cosmic expansion. Calculations show that they have been stretched all the way from visible to radio wavelengths. This dilute, stretched radiation is the CBR.

To observe this radiation, we therefore need a radio telescope. In what direction must we point our telescope? Notice in Figure 18.20 that the radiation is traveling in every direction. We conclude that the Cosmic Background Radiation is falling upon the Earth from every direction. To observe emission from a particular astronomical object, we need to point our telescope right at it – but to observe the early Universe, we can point our telescope anywhere at all.

There is another way to think about the Cosmic Background Radiation. Recall that the Big Bang occurred about 13.7 billion years ago. Suppose we could build a super-sensitive telescope, capable of seeing 13.7 billion light years. Since light travels 13.7 billion light years in 13.7 billion years, this telescope would be capable of detecting the light emitted when the Universe was young. Since by Hubble's law, distant objects are highly redshifted, this radiation would not be visible light, but at far longer wavelengths. So surrounding us in every direction is a ghostly radio-frequency image of the epoch of decoupling, half a million years after the Big Bang.

Figure 18.21 **The early Universe. Observations of the Cosmic Background Radiation by the WMAP satellite reveal minute (10^{-4} kelvin) fluctuations.** 👁 **(Also see color plate section.)**

Figure 18.21 shows this image, as observed by the WMAP satellite (the Wilkinson Microwave Anisotropy Probe satellite). It is presented in the form of an equal-area projection map of the sky, in which the galactic plane forms the map's equator, and the directions perpendicular to this form the map's poles. In this extraordinary image we catch a glimpse of the cosmos in its infancy. The most striking feature of this image is that *the early Universe was not precisely uniform*. The brighter regions of this image are very slightly hotter (10^{-4} kelvin) than the darker. These hotter regions are the most ancient structures we have ever discovered. Their detailed structure has been shown to be in accord with the predictions of the theory of cosmic inflation.

Large-scale structure of the Universe

Throughout this book – and throughout the history of astronomy – the scale of our vision has grown steadily larger. We began by mapping the structure of our Solar System and then moved on to mapping the distribution of stars in our vicinity. We moved from our Galaxy to other galaxies. We now expand our vision yet more. If we were to draw the largest-scale map of all, a map showing how galaxies are distributed throughout the entire cosmos, what would it show?

As we have repeatedly seen in this book, the problem of mapping the astronomical Universe is one of measuring distances – and measuring distances is hard. The parallax technique works only for nearby stars, and the Cepheid variable technique works only for nearby galaxies. If we wish to measure distances out to the very limits of our greatest telescopes, we are going to need yet another method.

Redshift survey

One technique, involving supernovae, was discussed earlier in this chapter. Unfortunately, this technique is fraught with difficulties. It requires great amounts of observing time, which is inconsistent with the needs of a survey involving great numbers of galaxies. Accordingly, recent work on mapping the cosmos has taken a different approach. Researchers measure distances indirectly. They do this by measuring a distant galaxy's velocity (by measuring its redshift) and then by *assuming* Hubble's law to be valid for this galaxy. If we know a galaxy's velocity, and if we assume that velocity is proportional to distance, then we can find the distance). Using such a technique, one can draw a map of the distribution of large numbers of galaxies in space. Such a technique is called a *redshift survey*.

THE NATURE OF SCIENCE

THE DESIGN OF OBSERVATIONS

As we have discussed several times already, scientists don't just go out and measure everything in sight. It is hard to gather data, and it is important to be careful in choosing *which* data to gather. Let us study this again with regard to designing our redshift survey.

Suppose that, using a large telescope with modern equipment, we can measure the redshift of a galaxy in 10 minutes. And suppose that we want to spend no more than a year in drawing our map of the distribution of galaxies in the Universe. How big a map can we draw – i.e. how many redshifts can we measure in a year? As we will see, the answer is that we will be able to map only a disappointingly tiny fraction of the cosmos.

The logic of the calculation

Step 1. Calculate how many 10-minute intervals there are in a night.

Step 2. Multiply this by the number of nights in a year to find the number of redshifts we can measure in a year.

As you can see from the detailed calculation, 26 280 redshifts could then be measured in a year. What sort of map will this allow us to draw? Another way to put this question is to ask: what fraction is this of the total number of galaxies that we can see? Let us call this fraction "f."

There are about 100 billion galaxies in that portion of the Universe accessible to our telescopes. Therefore

f = number of galaxies whose distance we have measured/total number of galaxies accessible to our telescopes

f = 26 280 galaxies/100 billion galaxies

f = 1/3.8 million.

Our survey, which took an entire year, is only capable of mapping one 3.8 millionth of the Universe!

To put this in perspective, let us consider an analogy in which some alien, who has never visited our planet before, were to be given a map showing one 3.8 millionth of the Earth. How big an area would this map cover?

⇐ Looking backward
We considered the design of observations in Chapters 10 and 12, when we discussed "The nature of science."

Detailed calculation

Step 1. Calculate how many 10-minute intervals there are in a night.

We need to know how many hours there are in a night. This depends on how long a night is. In winter the nights are long, and in summer they are short. Let us take the *average* length of a night, which is 12 hours. Then

number of 10-minute intervals = (12 hours) (60 minutes/hour)/(10 minutes) = 72.

Step 2. Multiply this by the number of nights in a year to find the number of redshifts we can measure in a year.

There are 365 nights in a year. Thus:

number of redshifts we can measure in a year = 72 × 365 = 26 280.

(Of course this assumes that no nights are cloudy!)

Detailed calculation

Step 1. Find the surface area of the Earth.

The surface area of a sphere of radius R is $A = 4\pi R^2$. The Earth's radius is $R = 6378$ kilometers $= 6.378 \times 10^6$ meters. So $A = 4\pi (6.378 \times 10^6 \text{ meters})^2 = 5.11 \times 10^{14} \text{ meters}^2$.

Step 2. Divide this area by 3.8 million. This is the area shown on the map we gave the alien.

$$A_{\text{map}} = 5.11 \times 10^{14} \text{ meters}^2/3.8 \times 10^6 = 1.34 \times 10^8 \text{ meters}^2.$$

Step 3. If this map is square, we can find the length of a side of the map by taking the square root of the area (since the area of a square is the length of its side squared).

The length D of a side of the map is

$$D = \sqrt{A_{\text{map}}} = \sqrt{1.34} \times 10^8 \text{ meters}^2 = 1.16 \times 10^4$$
meters $= 11.6$ kilometers $= 7.2$ miles.

The logic of the calculation

Step 1. Find the surface area of the Earth.

Step 2. Divide this area by 3.8 million. This is the area shown on the map we gave the alien.

Step 3. If this map is square, we can find the length of a side of the map by taking the square root of the area (since the area of a square is the length of its side squared).

As you can see from the detailed calculation, the map would be tiny, covering a mere 7.2 miles (11.6 kilometers). This map would utterly fail to give our alien a representative look at our world. If it happened to be of a portion of our globe covered by ocean, it would fail to show land; if it happened to be of desert, it would fail to show jungle; and if it happened to be of a city, it would fail to show wilderness.

NOW YOU DO IT

Suppose it took you a mere 7 seconds to measure a redshift, and you wanted to complete all your observations in one month. (A) How many redshifts could you measure, and (B) what fraction of the total number of galaxies in the Universe is this? (C) Now consider an analogy, in which we represent the cosmos by the United States. Look up its surface area on the Web: how big an area have we mapped in our analogy?

If we want our map of the cosmos to do better than this, we are going to have to come up with a better strategy. The people who design redshift surveys adopt a different technique. Suppose we do not give our alien a square map, but a map that is in the form of a thin strip wrapping entirely around the Earth. As you can see from Figure 18.22(a), this map would pass from ocean to land, from desert to mountain, and from city to wilderness. Our alien would still have information covering only the same tiny fraction of our world – but it would be a fraction containing a lot more information than contained in the square map!

(a)

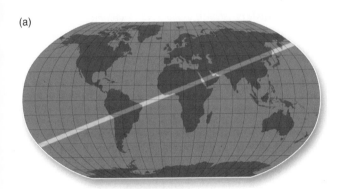

(b)

we are here ➡

Figure 18.22 **Strategy of a redshift survey is to map (a) a representative thin strip of the Earth and, in practice, (b) a wedge of the sky.**

Results of the redshift survey: a map of the cosmos

In terms of mapping the sky, this means that it is smart to measure redshifts of galaxies lying in a wedge, as illustrated in Figure 18.22(b). Recently several of these surveys have been undertaken. Figure 18.23 shows the results of one such survey.

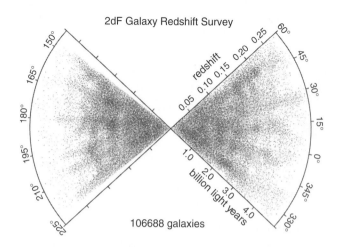

Figure 18.23 A slice of the Universe. Redshifts have been measured for over 100 000 galaxies lying along a thin ribbon arcing across the sky. This map of their distribution in space was obtained by converting the results to distance through Hubble's law. We are located at the point of the "pie wedge."

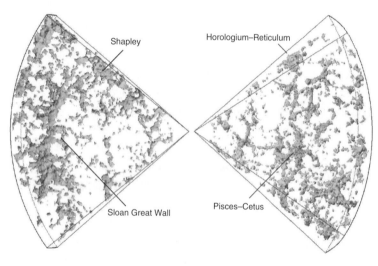

Figure 18.24 The largest known structure, the "Sloan Great Wall," can be seen in this three-dimensional reconstruction of the distribution of galaxies in the Universe.

It is difficult to conceive of the scale of this map. It is the largest map that has ever been drawn. In it the entire Earth is far too small to see – it would be smaller than a single atom on this scale! Our entire Milky Way Galaxy, 100 thousand light years across, has been shrunk to a single dot. The farthest galaxies shown on this extraordinary figure are more than 4 billion light years distant. The ancient light by which we are seeing them was emitted when the very Earth had only recently been formed.

In Figure 18.23, galaxies appear to be relatively sparse at great distances, but this is an example of observational selection (Chapter 12): faraway galaxies are hard to observe, so that we spot only the most prominent ones. But the filamentary, spongy structure revealed on this map is real. This amazing pattern is known as "the cosmic web." Galaxies congregate along gigantic winding filaments or sheets, each extending for hundreds of millions of light years. The locations at which these filaments meet are dense clusters of galaxies. Also revealed on this map are holes in the distribution: regions apparently free of galaxies. These immense voids can be up to 400 million light years in diameter. Apparently, however, they are not truly empty. Recent observations have revealed the presence of great quantities of hydrogen gas within them. Is this gas that for some reason failed to condense into galaxies? Or does it reside in the halos of dwarf galaxies too faint to be observed? Only time will tell.

Figure 18.24 presents a three-dimensional reconstruction of the inner portions of Figure 18.23. In it we can see the so-called "Sloan Great Wall," currently the largest known structure in the Universe. It is nearly 1.4 billion light years long, and it is about a billion light years away.

The origin of structure in the Universe: "lambda cold dark matter" simulations

We are only beginning to understand how the gigantic structures that constitute the cosmic web arose. The "mottles" of the cosmic background radiation (Figure 18.21) must represent the seeds from which they formed. A critical uncertainty in this work is the dark matter, which must have played an important role in their formation. This is for two reasons. On the one hand, there is more dark than visible matter, so that the structures we observe must be mere tracers of other, unseen, structures of dark matter. On the other hand, since the dark matter does not interact well with radiation, it was able to start condensing earlier than the visible matter in the history of the Universe.

31.25 Mpc/h

Figure 18.25 **Simulation of structure in the Universe. (In this image, the scale bar is 43 megaparsecs long.)** 👁 **(Also see color plate section.)**

Detailed computations of these processes are currently under way. Early work postulated a so-called "hot dark matter" model, the term "hot" referring to the assumption that the dark matter particles that constituted the bulk of the cosmos were moving at speeds close to that of light. Computer simulations within this picture showed that the earliest structures to form were the largest, which then fragmented into the smaller ones we see today. The results of these simulations, however, were ultimately found to conflict with the observations. Accordingly, most contemporary research focuses on the so-called "lambda cold dark matter" model, in which the dark matter moves slowly compared to the speed of light ("lambda" refers to the inclusion of the cosmological constant in the calculations). Within this picture, the earliest structures that formed were the smallest, and they have been amalgamating into larger ones ever since. According to this picture the geometry of the Universe is flat or very slightly open, it will expand forever and it is accelerating.

The results of one such simulation are shown in Figure 18.25. It was produced by a group using the largest supercomputer in Europe, at the German Astrophysical Virtual Observatory. In a computation lasting over a month, the group simulated the history of the Universe in a cube over 2 billion light years on a side, holding 20 million galaxies. We emphasize that this image shows not visible structures, but the distribution of the dark matter itself: the galaxies we see in Figure 18.23 are mere tracers of this distribution.

Summary

Olbers' Paradox: why is it dark at night?
- We can calculate the total amount of light we would receive from all the stars in the Universe if the Universe is infinite.
- The result we get is infinity!
- What is the error in our reasoning?
  - We might think that the light from a very distant star is too small to make a difference – but this ignores the inverse square law.
  - We might think that intervening material blocks the light – but this ignores the fact that this material would then be heated and emit its own light.

  - The correct resolution: light from an infinitely distant star takes an infinite time to get here.
  - Since stars shine for only a finite amount of time, the total amount of light we get is finite even if the Universe is infinite.

The expansion of the Universe
- Edwin Hubble measured the distances and velocities of galaxies.
- He found they are all moving away from us.
- Hubble's law:
  - the farther away the galaxy the greater is its velocity,
  - $V = HD$ (where H is "Hubble's constant").

- We are not the center of the Universe: galaxies are expanding away from each other as well as from us.

Relativity and the expanding Universe

- Einstein's general theory of relativity had predicted the expansion before it was discovered.
- Believing the Universe to be static, Einstein modified his theory to avoid the prediction (the "cosmological constant").
- Once Hubble discovered the expansion, Einstein retracted the cosmological constant.
- Relativity without the cosmological constant predicts three possibilities.

The future of the Universe

- The Universe will continue forever.
  - Continental drift will remake the map of the Earth.
  - Evolution will remake species.
  - Ultimately the Sun will go out and all life on Earth will cease.
  - All stars will go out, and the Universe will be utterly dark.
  - It will continue expanding endlessly.

The past history of the Universe: the Big Bang

- In the past, galaxies were closer together.
- Several million years ago early hominids walked the Earth.
- About 4.5 billion years ago the Solar System formed.

Case I	Weak gravitational attraction	Low density	D_{actual} less than $D_{critical}$	Eternal expansion at a rate that diminishes initially, but then becomes constant	Saddle geometry	Infinite amount of space
Case II	Intermediate gravitational attraction	Intermediate density	D_{actual} equals $D_{critical}$	Eternal expansion at a steadily diminishing rate	Euclidian geometry	Infinite amount of space
Case III	Strong gravitational attraction	High density	D_{actual} greater than $D_{critical}$	Expansion followed by contraction	Spherical geometry	Finite amount of space

- Is the Universe infinite?
  - The Paradox of the Edge: it seems logically impossible to think that the Universe is finite.
  - Non-Euclidian (spherical) geometry is finite but has no edge, so it resolves this problem.
  - In such a Universe you could travel in a straight line and get back to where you started.

The accelerating Universe: dark energy

- Certain supernovae (Type Ia) have "signposts" that tell us their intrinsic luminosity.
  - This "signpost" is the rate of decrease of its brightness: the slower the decrease, the brighter the supernova.
  - Once we know a supernova's luminosity, we can measure its distance merely by measuring its apparent brightness (flux) – which is easy.
- This method of measuring distance allows us to study the expansion of the Universe far away – i.e. far back in time.
- We find that the rate of expansion is accelerating.
- "Dark energy" is the term given to whatever powers this acceleration.
- We have only the faintest idea of what it might be.

- Several hundred million years after the Big Bang the first galaxies are thought to have formed.
- The early Universe had a temperature: the earlier we project our minds, the higher this temperature had been.
- Some 500 000 years after the Big Bang the temperature was 3000 kelvin and the Universe was filled with a brilliant, yellow light.
- At this point the Universe made a transition from being opaque to transparent.
- The very early Universe was super-hot and expanding at a great rate.
- The very early Universe may have undergone an ultra-rapid "cosmic inflation."
- The moment of the Big Bang:
  - "the singularity,"
  - infinite temperature and density,
  - all matter in the Universe was crushed to zero size,
  - we do not have any idea how to understand this,
- A union of quantum mechanics and relativity may tell us how to solve this problem.
- We do not know if the Big Bang was creation, or whether the Universe existed prior to it.

The Cosmic Background Radiation
- The early Universe was hot.
- All hot bodies emit blackbody radiation.
- This radiation is still here:
  - its intensity has been reduced by the cosmic expansion,
  - its wavelength has been increased by the cosmic expansion,
  - it is now at radio wavelengths.
- Observation of this "Cosmic Background Radiation" gives us a "snapshot" of the early Universe.
- It was not precisely uniform, but contained hotter and colder regions. These regions are the most ancient structures we have ever witnessed.

Large-scale structure of the Universe
- It is very hard directly to measure the distance to a galaxy.
- If, however, we assume that Hubble's law is valid, all we need to do is measure a galaxy's redshift (which is easy).
- Redshift surveys study galaxies lying along a slice.
- They reveal structures resembling foam: galaxies lie on curving sheets and filaments surrounding voids.
- "Lambda cold dark matter" computer simulations seek to understand this so-called cosmic web.

The nature of science. Uncertainty in science
Relativity theory tells us what we have to do to determine which of its three cases is the true one. But we are not able to do this at present.

- We need to measure the actual density of the Universe and compare it to the critical density.
- To measure density we need to measure mass divided by volume.
- To measure volume we need to measure the distances to galaxies, and this is hard because we cannot even detect Cepheid variable stars at these distances.
- Other methods of measuring distance exist, but they are not yet perfected.
- To measure the mass contained in a volume we need a reliable means of measuring the mass not lying in galaxies, and of accounting for all the dark matter.

THE NATURE OF SCIENCE

The design of observations
- Scientists don't just go out and measure everything in sight. It is hard to gather data, and it is important to be careful in choosing which data to gather.
- Even with a good telescope and modern equipment, we could measure only the redshifts of one 3.8 millionth of the Universe in a year!
- If we merely observed the nearest portion of the cosmos in that year, our results could not possibly be representative of the whole Universe.
- If, however, we observed galaxies lying along a thin slice through the cosmos, we would have a far more representative view.

Problems

(1) Suppose there are two clusters of stars, one cluster ten times farther away than the other. All the stars have the same luminosity. (A) How much brighter do stars in the nearby cluster appear than those in the farther one – i.e. what is the ratio of their fluxes? (B) Suppose the clusters have the same number of stars: how much brighter is the nearby cluster than the farther one – i.e. what is the ratio of their fluxes? (C) Suppose alternatively that the two clusters appear equally bright to us. How many more stars does the more distant one have than the nearer one – i.e. what is the ratio of their numbers? (D) Suppose all the stars do not have the same luminosity, and suppose that the two clusters have the same number of stars. If the clusters appear equally bright to us, how much brighter are stars in the more distant cluster than those in the nearer one – i.e. what is the ratio of their luminosities?

(2) It is exceedingly difficult to measure the Hubble constant, H. Suppose we have made a mistake, and actually its value is 75 kilometers per second per megaparsec. (i) According to this law, what is the velocity of recession of a galaxy 25 megaparsecs away? (ii) Suppose you observe a galaxy whose velocity of recession is 550 kilometers per second. We want to use this new Hubble constant to find how far away it is from us. For both questions, (A) describe the logic of the calculation you will perform and (B) carry out the detailed calculation.

(3) Suppose we build a finer telescope, one capable of detecting fluxes even fainter than we can today. Suppose, in particular, that f_{limit} for our new telescope is 100 times fainter than our current f_{limit}. Would this telescope be capable of detecting a $10\,000\ L_{Sun}$ Cepheid variable star 100 Mpc away?

(4) Suppose our model of a spherical geometry has a surface area of one acre, and we want to delineate a region of half an acre. How would we do this?

(5) Suppose our model of a spherical geometry has a surface area of one acre, and we want to delineate a region of two acres. How would we do this?

(6) Explain why a creature living on an asteroid might know that its world was round, whereas a creature living on a planet might think that its world was flat.

(7) In discussing Figure 18.20 we described how blackbody radiation evolves as the Universe expands. Give a similar description of how a light ray, generated by nuclear reactions at the core of a dense, hot star, makes its way out through less dense, less hot outer layers to the star's surface, and then flies off into space.

(8) Suppose you wish to conduct a telephone survey of the residents of New York City: each phone call will take you one minute.
- If you phone only between the hours of 6 p.m. – 11 p.m. on weekdays, how long will it take you to survey every resident? (The population of New York is 8.1 million.)
- Alternatively, if you want to finish your survey in one year, how many people can you contact? What percent of the total population of New York is this?

For each question (A) describe the logic of the calculation you will perform, and then (B) carry out the detailed calculation.

(9) In our discussion of redshift surveys, we calculated the number of redshifts that can be measured in a year, by assuming that on average a night is 12 hours long. One way to measure more redshifts is to use different telescopes at different times of year.
(A) Which is the better strategy?
- Use a telescope in Arizona in July, and one in Chile in December.
- Use a telescope in Arizona in December, and one in Chile in July.

Make sure that you explain your answer.
(B) Suppose that your strategy gives you 14 hours of nighttime 365 days per year. We want to calculate how many redshifts you can observe.
- Describe the logic of the calculation you will use.
- Carry out the detailed calculation.

(C) What fraction does this represent of the total number of galaxies accessible to our telescopes?
(D) Suppose we give our hypothetical alien a map covering this fraction of the Earth's surface. We want to calculate the length of a side of this map, if the map is square.
- Describe the logic of the calculation you will use.
- Carry out the detailed calculation.

(10) Suppose the redshift survey had shown *more* galaxies at great distances from us than nearby. Would this have been an example of observational selection? Explain your answer.

You must decide

. . . .

(1) In Chapter 15 you wrote on the following issue.
  You are a minister preparing a sermon. You are struck by the prediction that, when the Sun leaves the main sequence, all life on the Earth will come to an end. A passage by the philosopher Bertrand Russell has caught your attention:
    . . . all the labors of the ages, all the devotion, all the inspiration, all the noonday brightness of human genius, are destined to extinction in the vast death of the solar system. . . How in such an alien and inhuman world, can so powerless a creature as Man preserve his aspirations untarnished?
  [from *A Free Man's Worship*, by Bertrand Russell]

You have resolved to address this issue in your sermon. In particular, two problems plague you.
- How can God allow such a terrible thing?
- How does anything that we mere humans do matter?
  What will you say in your sermon?

You should now return to this question. Your sermon had been responding to the fact that our Sun will eventually go out, and all life *on Earth* will come to an end. However, our study of cosmology has now taught us that eventually all life in *the entire Universe* will come to an end.

- In cases I and II the Universe will continue forever, but eventually every star will go out.
- In case III the Universe will eventually collapse into a "Big Crunch."

Does this lead you to change your plans for your sermon? In what ways? In responding to this question, make sure that you pay attention to our discussion in "Science and religion are separate" in Chapter 2.

(2) Write a science fiction story in which you imagine that you are an astronomer living billions of years in the future. Depending on whether the Universe is of case I or II on the one hand, or case III on the other, you will discover very different things about the cosmos. Try to imagine what your discovery will mean to people in human, emotional terms. What will be your view of life, and the view of life of "the person on the street?"

19

Life in the Universe

Are we alone, or does life exist elsewhere in the Universe? We have never found convincing evidence of life on other worlds. But there are reasons to think that extraterrestrial life is possible. On the one hand, the Earth does not appear to be unique in any way: what happened here can very well also happen somewhere else. And on the other hand, there are an awful lot of these "somewhere elses" in the Universe: there are many stars in our Galaxy, and many galaxies in the cosmos.

In our discussion we will deal solely with life as we know it, similar to the form we find on Earth: life that evolves by mutation and natural selection, life based on carbon chemistry and employing DNA as the carrier of genetic information. Other kinds are at least conceivable. There have been speculations about a form of life based on silicon. Science fiction writers have imagined far stranger possibilities. But we will not consider these, for the simple reason that we don't know anything about them. The only life about which we know anything at all is the kind that exists on Earth: concerning other varieties, we can only speculate.

Searches have been conducted for life on Mars, and searches are under way right now for signals from extraterrestrial civilizations. These searches have found nothing persuasive so far, and nobody thinks their task will be easy. But it is no exaggeration to say that the discovery of life elsewhere in the Universe would be one of the greatest scientific triumphs of all time.

The origin of Life

We have essentially no direct evidence concerning the origin of life. No scientist has ever managed to create it in the laboratory. Nor has anything ever been found in the fossil record documenting life's origin. But we do have two important pieces of evidence.

(1) In geological terms, life arose rapidly here on the Earth. Although we cannot be sure, it may be that this means that, when an environment is suitable, it is not so hard for life to arise. If this interpretation is correct, the chances are good that it arose on other planets as well.

(2) Important precursors of life have been created in the laboratory, as demonstrated by the Stanley Miller experiment.

We will consider each of these in turn.

(1) A clue: life on Earth arose rapidly

The evidence supporting this contention is as follows.

(A) Life could not possibly have existed prior to about 4.1 billion years ago.
(B) Life did exist by 3.5 billion years ago, and possibly by 3.9 billion years ago.

This means that life on Earth arose within a time span of between 200 million and 600 million years. While this is long in human terms, it is quite rapid in geological or astronomical terms.

We now turn to the evidence for each of these assertions.

(A) Life could not possibly have existed prior to about 4.1 billion years ago

> ⇐ **Looking backward**
> We studied the formation of planets in Chapter 13.

As we saw in our study of the origin of the Solar System, the formation of planets was a violent, catastrophic process. In the final stages gigantic bodies slammed into one another with great force. Here on the Earth, almost all the craters produced by these impacts have been erased by geological processes, but on the Moon they are still there for us to see: mute testimony to the violence of the final stages of the formation of the Solar System (Figure 7.2). This process of bombardment appears to have come to an end about 4.1 billion years ago.

To get an idea of the magnitude of these catastrophic impacts, consider the Moon's south pole/Aitken basin (Figure 19.1). It is about 2200 kilometers across, and it was formed by the impact of a body about 200 kilometers across. This impact was clearly a violent catastrophe – and if one such body had slammed into the Moon, we can be sure that many had slammed into the Earth.

- The Earth is bigger than the Moon: it presents a bigger "target" to the randomly orbiting bodies filling the early Solar System.
- The Earth's gravity is stronger than the Moon's: it would have more strongly attracted these bodies toward it.

Detailed calculations show that, for every one such body hitting the Moon, about 20 would have hit the Earth.

We will now show that, during this period of incessant bombardment, life could not possibly have existed. The reason is that these giant impacts would have heated the Earth to temperatures so great as to sterilize the entire planet.

To show this we need to do two things.

(i) Calculate the amount of energy released when such a body hits the Earth.
(ii) Calculate the consequences of the release of this much energy.

Figure 19.1 **The Moon's south pole/Aitken basin is the dark area at the bottom of this image of the Moon's far side. It was produced by the impact of a body 200 kilometers in diameter.**

Detailed calculation

Step 1. Calculate the mass of a body 200 kilometers across. Do this by multiplying the volume of such a body by its density.

Let us suppose that the body was more or less spherical. The formula for the volume of a sphere is

volume = $(4/3)\pi R^2$

where R is its radius (half its diameter). Taking $R = 100$ kilometers = 10^5 meters,

we calculate

volume = $(4/3)(3.14)(10^5)^3 = 4.19 \times 10^{15}$ cubic meters.

For the body's density, recall that most bodies in the Solar System (asteroids, comets, etc.) are solid. Many solids have densities of roughly twice that of water, or

density = 2000 kilograms/cubic meter.

We then find

M = (volume)(density) = $(4.19 \times 10^{15}$ cubic meters)(2000 kilograms/cubic meter)
= 8.38×10^{18} kilograms.

Step 2. Calculate the velocity with which it hit the Earth.

In Figure 9.23 we showed that the relative velocity between the Earth and another body orbiting the Sun depends on the directions they are traveling: in general the relative velocities lie between 12 and 72 kilometers per second. Since we have no idea in what direction the impacting body had been moving, let us take an average of 40 kilometers per second = 4×10^4 meters per second. (In fact the body, when it hit the Earth, would have been moving faster since gravity had been pulling it toward the Earth: it turns out that this does not alter the velocity very much.)

Step 3. Plug these into the formula for kinetic energy.

We calculate

kinetic energy of impacting body = $(1/2) MV^2 = (1/2) \times (8.38 \times 10^{18})(4 \times 10^4)^2 = 6.70 \times 10^{27}$ joules.

We now turn to each of these.

(i) To calculate the amount of energy released when such a body hits the Earth, we use the fact that a body of mass M moving with velocity V has an energy of motion, its so-called kinetic energy, given by

kinetic energy = $(1/2) MV^2$.

The logic of the calculation

Step 1. Calculate the mass of a body 200 kilometers across. Do this by multiplying the volume of such a body by its density.

Step 2. Calculate the velocity with which it hit the Earth.

Step 3. Plug these into the formula for kinetic energy.

As we see from the detailed calculation, the result is
kinetic energy = 6.7×10^{27} joules.

. . . .

NOW YOU DO IT
(1) **Suppose the asteroid had a density twice as great. How much kinetic energy would it release?**
(2) **Suppose the asteroid had the same density as (1), but twice the radius. How much kinetic energy would it release?**
(3) **Suppose the asteroid had the same density and radius as (2), but twice the velocity. How much kinetic energy would it release?**

[*Note: one way to answer these questions would be to re-do the calculations. But there is an easier way! See if you can find it.*]

Suppose an asteroid has the shape of a perfect cube, and it is 100 kilometers on a side. (A cubical asteroid is impossible, but we are simply playing with ideas here!) Calculate its kinetic energy if it is traveling at 30 kilometers per second, and has a density of 2000 kilograms per cubic meter.

. . . .

(ii) We now calculate the consequences of the release of this much energy.

As we know, most of the Earth's surface is covered by oceans. So the most likely scenario is that the body landed in an ocean. While there is evidence that oceans existed at the time of the impact, we have no idea of their extent. But let us imagine that they were as extensive as they are today, and ask what this amount of energy deposited into the oceans would have done.

The answer is that much of the kinetic energy released by the impact would have been transformed into heat – lots of heat: so much that *the oceans would have entirely boiled away*. The atmosphere would have been supersaturated with scalding vapor. This vapor would have acted as a greenhouse gas and heated the planet still further. Detailed calculations show that the Earth would have been heated to temperatures of thousands of degrees, and it would have remained that hot for thousands of years! This would have entirely killed any life that might have been present. And we emphasize that this must have happened not once, but many times during the early history of the Earth.

Let us demonstrate that the energy liberated by a single impacting body would have been more than enough to boil away all the oceans of the Earth.

Detailed calculation

Step 1. Look up the amount of energy required to boil away one kilogram of water.

It requires $E_{1\,kg} = 2.26 \times 10^6$ joules of energy to boil one kilogram of water.

Step 2. Look up the number of kilograms in the oceans.

The oceans today have a volume of 1190 million cubic kilometers, or

$V = 1.19 \times 10^9$ cubic kilometers.

We need to find the volume in cubic meters. Since there are 1000 meters in a kilometer, there are $(1000)^3 = 10^9$ cubic meters in a cubic kilometer. So the volume of the oceans is

$V = (1.19 \times 10^9$ cubic kilometers) $(10^9$ cubic meters/cubic kilometer) $= 1.19 \times 10^{18}$ cubic meters.

To find the mass of the oceans, we multiply their volume by the density of water, which is 1000 kilograms per cubic meter:

$M = (1.19 \times 10^{18}$ cubic meters) $(1000$ kilograms/cubic meter) $= 1.19 \times 10^{21}$ kilograms.

Step 3. Multiply these together: the result is the amount of energy required to boil away all the oceans of the Earth.

$E_{oceans} = (2.26 \times 10^6$ joules/kilogram) $(1.19 \times 10^{21}$ kilograms) $= 2.69 \times 10^{27}$ joules.

Step 4. Compare this to the energy deposited in the oceans by the impacting body.

Let us find the ratio

kinetic energy of impacting body/$E_{oceans} = 6.70 \times 10^{27}$ joules/2.69×10^{27} joules $= 2.5$.

The energy released by the impact was more than twice what is needed to boil away all the oceans of the world today.

The logic of the calculation

Step 1. Look up the amount of energy required to boil away one kilogram of water.

Step 2. Look up the number of kilograms in the oceans.

Step 3. Multiply these together: the result is the amount of energy required to boil away all the oceans of the Earth.

Step 4. Compare this to the energy deposited in the oceans by the impacting body.

As you can see from the detailed calculation, the energy deposited by the impacting body is far more than the energy required to boil away all the oceans.

NOW YOU DO IT

Would the energy released by the cubical asteroid you analyzed above be enough to boil away all the oceans of the world?

Look up on the Web the quantity of water in the Atlantic Ocean. How big an asteroid would be required to boil it all away? Assume the asteroid is travelling at 40 kilometers per second, is spherical and has a density of 2000 kilograms per cubic meter.

We have shown that the life that now exists on the Earth could not possibly have survived such a bombardment by giant meteors. We now turn to the second step in our argument.

Figure 19.2 **Stromatolites. Photographed in Western Australia.**

(B) Life did exist by 3.5 billion years ago, and possibly by 3.9 billion years ago

To repeat: since the Earth was uninhabitable prior to 4.1 billion years ago, this means that life arose within a time span of between 200 million and 600 million years. While this is long in human terms, it is quite rapid in geological or astronomical terms. Although we cannot be sure, it may be that this means that, when an environment is suitable for life, it is not so hard for life to arise. If this interpretation is correct, the chances are good that life arose on other planets as well.

As we have mentioned, the geological record of the first billion years of the history of the Earth is essentially non-existent. We now turn to some of the earliest records that we do have. Remarkably, they show evidence of life. There are three lines of evidence pointing to the existence of life in these ancient times.

(i) *Stromatolites* (Figure 19.2) are structures formed by colonies of bacteria. They are quite rare, but can be found in a few locations, notably along the coastline of Australia. The colonies consist of different types of bacteria that thrive just below the ocean surface. As sediments are deposited on top of the colony, the bacteria migrate upwards to maintain their preferred depth. The result is a series of layers that can form a structure as much as a foot across. Fossils have been found that appear identical in form to these structures. The oldest such fossils are between 3.3 and 3.5 billion years old.

(ii) *Fossil cells* have been found in rocks 3.5 billion years old (Figure 19.3). These cells are similar in appearance and structure to bacteria living today: they are of similar size and shape, and they have cell walls and contain organic carbon. There is some evidence that they contain fossilized remnants of DNA. The contemporary cells these fossils most closely resemble obtain their energy from photosynthesis, which leads us tentatively to conclude that photosynthetic organisms existed in these ancient times.

(iii) *Isotopes of carbon.* Carbon, widely used by all organisms, comes in two isotopes: carbon-12, whose nucleus contains six neutrons and six protons, and carbon-13, which contains an extra neutron. All contemporary organisms have a biochemistry that favors carbon-12, which causes them to have slightly more of this isotope than normal. We have found a very few rocks

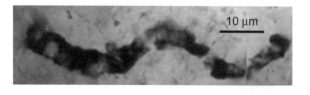

10 μm

Figure 19.3 **Fossil cells in rocks 3.5 billion years old.**

that are 3.85 billion years old – and they have a similar excess of this isotope. These rocks have been subjected to such intense pressure and temperature as to have destroyed whatever fossils they might have contained, but these geological processes would have not altered their composition in terms of isotopes. This excess may have arisen as a result of some non-biological process, so this line of evidence is suggestive but not conclusive.

It is important to note that both stromatolites and bacteria are advanced forms of life. In spite of their small size, they are not primitive organisms. They pass on their hereditary information via DNA, and they employ the same complex biochemical reactions as life today. So the conclusion is that *advanced life arose on Earth within a geologically short period of time.*

(2) The Stanley Miller experiment

We now turn to our second piece of evidence: important precursors to life have been created in the laboratory. In 1953, an experiment (Figure 19.4) was performed by the chemist Stanley Miller, who was at the time a graduate student working in the laboratory of Harold Urey. Urey had developed a theory that the Earth's primitive atmosphere contained large amounts of methane, ammonia and hydrogen, and that the first step in the origin of life might have been chemical reactions among these molecules. Miller filled a glass flask with them and then sterilized it. Within the flask a heater boiled water, producing vapor, and an electrical discharge simulated the effects of lightning. The experiment was run for several weeks, and the resulting mixture analyzed. Miller and Urey found that a variety of organic molecules had been formed, as well as five different amino acids (building blocks of proteins). Subsequent experiments have shown that similar results can be obtained by exposing such a mixture to ultraviolet radiation, or the kind of sudden heating that would have been produced by the impact of a medium-sized meteor. There are also experiments that show such compounds can be produced deep in the oceans, near so-called "hydrothermal vents," where superheated plumes of water rise.

More recent work has cast doubt on Urey's original hypothesis concerning the gases that made up the Earth's atmosphere at the time at which life originated. The mixture he proposed is known by chemists as reducing, but nowadays, the weight of the evidence points to an atmosphere that is its opposite, oxidizing.

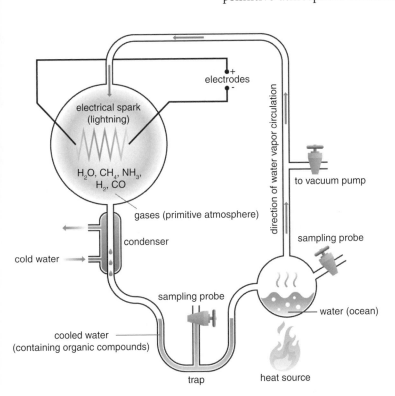

Figure 19.4 **The Stanley Miller experiment. A sterilized glass flask contained what was thought at the time to be the main constituents of Earth's primitive atmosphere. An electrical discharge simulated the effects of lightning. A variety of amino acids and organic molecules were produced.**

Experiments analogous to Miller's have been performed on such mixtures: they also produce amino acids and organic compounds, although the yield is lower.

There is an immense gap between the compounds produced by these experiments and the full biochemical machinery of life. In spite of their name, organic compounds are not alive. Rather, they are merely molecules containing covalent carbon–hydrogen bonds. Their name arose as the result of a historical mistake: originally it had been thought that they can only be formed by living organisms, but nowadays we know that this is entirely untrue. Indeed, such molecules have been found in space. One meteorite that landed in Australia, for instance, turned out to contain more than 70 different amino acids: no one would contend that this meteorite harbored life.

A major difficulty is that we have essentially no geological record at all of the epoch in which life formed. And as we emphasized in "The nature of science. Theory and observation" (Chapter 13), observations are vital to the progress of science. No matter how well developed our theories may be, once we start making observations, we find surprises – things that our theories did not predict, or ways in which they are entirely wrong. Theory in the absence of observation is barren, and the more we observe, the more we learn. But as far as the origin of life is concerned, we have no observations.

Nevertheless, based on the Miller experiment and others like it, a tentative theory has been advanced. It is thought that the amino acids and organic compounds, however produced, were washed by rain into shallow pools, where they concentrated into a "primordial soup" out of which life arose.

Timeline of the evolution of life on Earth

Let us follow the history of life on Earth, starting with the planet's formation 4.5 billion years ago and proceeding on to today. We will think in terms of an analogy, in which this entire time span is compressed into a single year. Imagine that the Earth was formed at midnight on December 31, and that by now a year has passed and it is midnight on the next December 31. As this eventful year proceeds, we will watch the history of life from its earliest, unknown beginnings to its present state.

In terms of this analogy, the giant impacts that rendered our planet uninhabitable took place during the month of January, and gradually petered out in early February. By the middle of February we find an excess of carbon-12 in rocks, possibly indicating the existence of life. By mid March we know that life had arisen, as evidenced by fossil stromatolites and cells.

Starting in the middle of June, the geological record shows evidence of a major increase in the amount of oxygen in the atmosphere. This was probably caused by photosynthesis, in which case we conclude that plant life was becoming common. Sometime in the fall, multicellular life appears, and by mid November life with hard parts – bones and shells. By Thanksgiving life has expanded out of the oceans onto dry land: early mammals are found by early December. The extinction of the dinosaurs took place during Christmastime. Early primates put in an appearance about this time.

By this point in our analogy nearly a full year has passed, and we need a finer view. Hominids begin to walk upright in the late afternoon of December 31. Our species, *Homo sapiens*, appears at 11.20 p.m. on New Year's Eve. The earliest cave paintings put in an appearance at 11.56 p.m. The entire edifice of science is developed during the scant four seconds prior to midnight: the industrial revolution takes place two seconds before midnight.

To this list we add one last crucial milestone: the ability of life on Earth actively to search for life elsewhere in the Universe. As we will see, it was not until the mid twentieth century that we developed the technological capacity to do this. In terms of our analogy this did not happen until half a second before midnight.

Life elsewhere in the Solar System

A caution: extremophiles

Recently organisms have been discovered that survive in environments that we would have considered utterly lethal to life. Such organisms are called "extremophiles."

There are organisms that can live in environments of extremely high acidity, and others that are exceedingly basic. Life forms have been found that can survive temperatures in excess of boiling, and others that flourish within permafrost. Multicellular organisms have been discovered clustered about deep sea hydrothermal vents, where the pressure is hundreds of atmospheres, temperatures exceed 100 °C and sunlight is entirely absent. Arsenic-eating organisms have been discovered in a California lake.

All this teaches us that, as we consider the search for life elsewhere in the Solar System, we must be careful in our thinking to avoid assuming that all living creatures must be similar to those with which we are familiar – people, insects, trees and so forth. Life is amazingly resilient and adaptable.

With this caution in mind, let us ask: other than the Earth, where could life exist in the Solar System? We must search for an environment with a temperature not too different from those found here. And we must search for an environment with plenty of liquid water. Let us see why.

The importance of liquid water

Liquid water is essential to life as we know it on Earth. Our own bodies are 60% water. Take away the bones and the percentage is even higher. Blood is remarkably similar in chemical composition to seawater. Nor is blood the only place where water is found in our bodies: it is present in every cell. The same is true of all other organisms – birds, insects, plants: all contain great amounts of liquid H_2O.

Why is water so important to life? The most important reason is that water is the "conveyer belt" that transports various substances through an organism. Oxygen, inhaled into our lungs, is transported to cells. Carbon dioxide, a waste product of metabolism, is carried from the cells back to the lungs and exhaled. Nutrients in food are broken down in the stomach and carried to cells. Similarly, nutrients in soil are taken up by the roots of a plant and transported upwards through its trunk and branches, and nutrients produced by photosynthesis in leaves are transported throughout the plant.

All this transport takes place by *dissolving* these substances in water, and then moving the water about the organism. In this regard water is unique. It dissolves a wider variety of substances, and it dissolves them more readily, than any other commonly occurring liquid. Furthermore, among other commonly occurring

Table 19.1. **Freezing and boiling temperatures (in °C).**

	Freezing temperature	Boiling temperature	Temperature range over which the substance remains liquid
Water	0	100	100
Ethane (C_2H_6)	−183	−89	94
Ammonia (NH_3)	−78	−33	45
Methane (CH_4)	−182	−164	18

liquids, water remains liquid over an unusually wide range of temperatures (Table 19.1).

And finally, the actual temperatures at which water is liquid are particularly appropriate to the requirements of life. Ethane, for instance, remains liquid over a similarly wide range of temperatures, but it does so only when very cold – and the lower the temperature, the more slowly do the biochemical reactions within an organism proceed. An organism that utilized ethane in place of water could survive exceedingly low temperatures, but it would "live more slowly": billions of years may not be enough time for evolution to proceed very far for such creatures. Conversely, something that remained liquid at far higher temperatures than water would also be inappropriate, but for a different reason: many organic molecules split apart at high temperatures.

The search for life on Mars

⇐ **Looking backward**
We studied Mars in
Chapter 7.

This makes it all the more important that there is abundant evidence that great amounts of water existed on Mars in the past, and that at least some remains today. Furthermore, conditions on Mars are the most similar to conditions on Earth. For this reason Mars, among all the planets of the Solar System, is the best place – or perhaps we should say, the least bad place – to search for signs of life. We are lucky that Mars is also one of the closest planets to us, making it less difficult to get there.

There are two questions: does Mars harbor life now, and did Mars harbor life in the past?

Life now

At present Mars is an extraordinarily harsh environment for life. (1) It is very cold, (2) it has no liquid water, (3) there are only trace quantities of oxygen in its atmosphere and (4) it has no ozone layer to protect organisms from solar ultraviolet radiation.

To place these conditions in perspective, it is helpful to think of the one place on Earth that even remotely approaches conditions on Mars: Antarctica. Studies have shown that many Antarctic plains are utterly sterile, and contain no life whatsoever. This in spite of the fact that Antarctica has (3) plenty of oxygen in its atmosphere and (4) an ozone layer protecting it (until recently!).

But life is amazingly adaptable. Biologists have found life in Antarctica – not on its plains, but *inside* rocks, living in tiny crevasses deep within their interiors.

Figure 19.5 **A Viking Mars lander. The astronomer Carl Sagan is standing in front of a model.**

Elsewhere, life forms have been discovered miles beneath the ocean's surface, where no sunlight ever penetrates, clustered around superheated plumes rising from undersea vents. Many organisms are known that breathe CO_2, the primary constituent of the Martian atmosphere – plants of course, as well as certain bacteria. Indeed we believe that, when life originated, the Earth's atmosphere was utterly lethal to the kind of life forms that exist today, including ourselves. The Earth's early atmosphere contained no oxygen, for example, and no ozone layer.

Perhaps, therefore, life has been able to adapt to the even harsher conditions found on Mars. In 1976 two *Viking missions* landed on the planet, and conducted experiments designed to search for microorganisms in the Martian soil (Figure 19.5).

Each lander was equipped with a mechanical arm, which reached out and dug up a scoop of dirt. It then deposited the soil into one of three experimental chambers.

(A) One experiment searched for evidence of organisms "inhaling" Martian air. For five days the soil was exposed to a specially designed artificial atmosphere, consisting of carbon dioxide in which the normal carbon was replaced with the radioactive isotope carbon-14. The experiment then flushed out the radioactive air, and a device searched for radioactivity *in the soil*. Had it been found, the conclusion would have been that something in the soil was taking up atmospheric CO_2.

(B) A second experiment searched for evidence of organisms "eating" a nutrient. The soil sample was wetted with an artificial liquid food that contained radioactive carbon. After a while, the experiment searched for radioactivity *in the chamber atmosphere*. Had it been found, the conclusion would have been that something in the soil was taking up the nutrient and releasing a gaseous product. Many organisms here on Earth release various gases upon metabolism.

(C) In a third and somewhat similar experiment, liquid nutrient was fed to a soil sample and the released gases were chemically analyzed. While this was going on, a lamp illuminated the soil: were any microscopic plant life to be present, photosynthesis would have resulted in the release of oxygen.

All three experiments yielded positive results. Does this mean that there is life on Mars?

As we discussed in "The nature of science. The importance of skepticism" (Chapter 2), skepticism is one of the most cherished scientific virtues. Every scientist is skeptical, and every scientific theory is provisional and subject to doubt. In that discussion, we used the analogy of shaking a ladder before climbing it. The point is not to knock the ladder down – the point is to make sure that nothing *else* can knock it down. We want our scientific knowledge to be as certain as we can make it, and we do this by testing it in every possible way. This is doubly true when the purported "discovery" is as important as the detection of extraterrestrial life.

Let us return, therefore, to each of the three Viking biology experiments, and "shake" them. Only if they provide *irrefutable* evidence for life on Mars will we accept them.

(A) This experiment showed that something in the soil was taking up atmospheric CO_2. But was this "something" alive? One way to find out would be to sterilize the soil and run the experiment again. The experiment did this by heating the soil to a temperature sufficient to kill any microbe within it. It was found that atmospheric CO_2 was still being taken up.

(B) This experiment showed that something in the soil was taking up the nutrient and releasing the products into the chamber. But did this "exhalation" persist? After all, we would expect organisms to keep on eating. The experiment added yet more nutrient, but there was no further release of radioactivity. The release turned out to be a one-time event.

(C) This experiment found evidence of the release of oxygen into the atmosphere, which would be expected were photosynthesis going on. But could some other process, not involving photosynthesis, be responsible for the release? An experiment was performed in which photosynthesis could not have been operating: the oxygen kept on being released.

NOW YOU DO IT
What do you think this experiment was?

All three "shakes" have shown the evidence for life on Mars to be weak. We note that the soil samples gathered by the Viking experiments came from a mere few inches below the Martian surface, and they came from only two locations. Conceivably life does exist on Mars, but farther below its surface, or at other locations. But at present the inescapable conclusion is that we have no firm evidence for the presence of life on Mars.

Life in the past: the meteorite from Mars

Even though Mars is hostile to life now, we have seen that there is abundant evidence for an earlier warm wet period in its history. Could life have existed then, only to have been destroyed in a global catastrophe in which the planet lost its atmosphere and cooled? In 1996 a group of researchers claimed to have found evidence for fossilized life forms in a rock from Mars.

This rock was not brought home by a sample return mission – there has never been a sample return mission to Mars. Rather, the rock got here on its own. Some enormous explosion must have blasted the rock entirely away from Mars and out into space. Perhaps the explosion was a volcanic eruption, or perhaps an impact by an asteroid or comet. At any event, the rock wandered for geological ages through the Solar System in its own orbit about the Sun. Ultimately, and purely by chance, this orbit brought it close to the Earth. Pulled by our gravity, it fell to the ground as a meteorite.

Meteorites are usually hard to find. The difficulty is that to the untrained eye they look like terrestrial rocks – you may have seen one yourself, and not even known it! If you want to go hunting for meteorites, a particularly good place to do so is Antarctica. If you find a stone on some of its windswept plateaus, it is not likely to

(a)

(b)

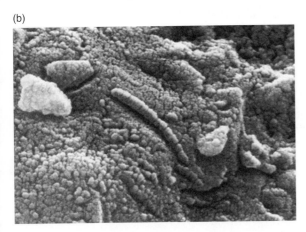

Figure 19.6 **The meteorite from Mars. (a) The Allen Hills meteorite. (b) An electron micrograph showing what might be fossilized primitive life forms.**

be from this planet, for in Antarctica the Earthly rocks lie buried beneath many thousands of feet of solid ice. Rather, that stone is likely to have fallen there from outer space. Furthermore, the motion of glaciers in certain regions tends to funnel them together.

One of the meteorites, discovered there in 1984, is known as ALH 84001 (Figure 19.6a). Initially nobody knew where it came from, but eventually the composition of gases trapped in tiny bubbles within ALH 84001 was measured. It was realized that this composition matched that of Martian air, as measured by the Viking missions.

Analysis shows that ALH 84001 is very old: it was formed about 4.5 billion years ago. It was blasted away from Mars about 16 million years ago, and it wandered through interplanetary space for most of the intervening time, falling to Earth some 13 thousand years ago. The 1996 claim that it contains fossils of primitive life forms is based on four bits of evidence.

(A) Microscopic globules of carbonate are found lying along fractures within the meteorite. These could have been formed had liquid water slowly oozed through the rock in ages past.

(B) Organic molecules called polycyclic aromatic hydrocarbons (PAHs) are found in the vicinity of these globules.

(C) Other organic molecules are also found, including certain amino acids.

(D) Strange elongated structures, ranging from 20 to 200 nanometers in size, are seen in electron micrographs (Figure 19.6b). They are concentrated in the carbonate globules, and they vaguely resemble terrestrial microorganisms. (A nanometer is 10^{-9} meters.)

None of these bits of evidence is overwhelmingly convincing. (A): we already knew that liquid water existed on Mars. (B) and (C): in spite of their name, organic molecules are not exclusively the domain of living organisms. They can be formed by mechanisms not involving life. (D): many biologists have disputed the claim that the striking structures seen in electron micrographs of ALH 84001 can only be fossilized life forms. They might be something else.

As we noted in our discussion of the Viking biology experiments, if we are to accept the claim that Mars once contained life, unshakable evidence must be presented. At present the general consensus is that the case for fossil life in ALH 84001 does not meet this test.

Jupiter, Saturn and their satellites

We know next to nothing about possible life elsewhere in the Solar System. The following are some interesting locations.

- Several of Jupiter's and Saturn's moons are known to contain great quantities of water. Furthermore, beneath their frozen surfaces this water is warm enough to melt: geysers of "molten ice" – i.e. liquid water – have been found erupting from them. Conceivably, life might exist in underground oceans within them. Although the technical difficulties are enormous, NASA has considered the possibility of sending a space probe to the surface of Jupiter's moon Europa, and of attempting to drill through the ice floes to reach the ocean beneath.

- The visible surface of Jupiter is very cold, but the planet is warmer deep within its interior. At a depth of 60 kilometers, it is room temperature. Jupiter's atmosphere is composed of compounds lethal to us, and indeed to every organism that exists today. But it is not necessarily lethal to all forms of life. Jupiter does not have any solid surface for living things to live on, but perhaps it is conceivable that some form of life floats permanently within this narrow warm zone among its clouds.

- Titan, one of Saturn's moons, is particularly interesting because the chemistry of its atmosphere and surface involves organic compounds. As we have mentioned, these compounds are not produced only by living organisms. Nevertheless, Titan's organic chemistry may be similar to that of the primitive Earth. Studies of Titan may shed light on our world's early evolution, and the chemical processes that preceded the origin of life here.

Life beyond the Solar System: requirements of life

Let us now move beyond the Solar System, and consider the possibility of searching for life on planets orbiting the distant stars. Does *every* star provide a habitat conducive to life? Or are some stars unsuitable? To answer this question, let us compile a list of requirements an environment must possess in order for life to arise and evolve. We already know two: the temperature must be right, and there must be liquid water. But there are other requirements as well: as we will see, not all stars provide good habitats for life.

Life requires a source of energy. For us this source is the Sun. It warms the Earth sufficiently for liquid water to exist. Furthermore, it is the ultimate source of the energy utilized by all living organisms to power their biochemical processes. On Earth this energy – solar power – is captured by plants through photosynthesis: animals eat the plants to gain the energy for themselves. This means that in searching for life we must look for environments near stars – presumably planets or satellites.

This source of energy must be available for very long periods of time. Life cannot flourish if the star runs out of fuel too quickly. For instance, as we have seen in Chapter 14, high-mass stars shine for only very brief periods of time: far too short to allow life to arise and evolution to proceed. The same is true of the red-giant phase of the evolution of stars of all masses.

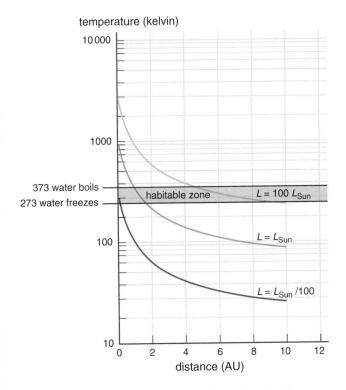

Figure 19.7 **Predicted temperatures of planets. The closer a planet lies to its star, the hotter it is expected to be. Furthermore, the brighter the star, the hotter its planets are expected to be. The habitable zone is the range of distances over which the temperature allows water to be liquid.**

The planet must orbit at the right distance from its star in order for it to have the right temperature to allow life. Too close and the planet is too hot: too far and it is too cold. Thus, in our Solar System, Mercury is too close to the Sun for life to exist, and Neptune too far.

This leads to the concept of a *habitable zone*: the zone within which a planet must orbit if it is to be capable of supporting life. This zone depends on the luminosity of the star. The brighter the star the more energy it emits, and the farther from it lies the habitable zone. So we stay far from a roaring bonfire, but move up close to a faint pile of embers.

It is possible to predict how the temperature of a planet depends on its distance from a star. This prediction is shown in Figure 19.7. On it we have indicated the two benchmarks delimiting the habitable zone: the temperature at which water boils, and the temperature at which it freezes. Within these two benchmarks water can exist as a liquid, making life at least possible.

Notice from this figure that the habitable zone is thin for dim stars, and thick for bright ones (Problem 4). So far as we know, whether a planet lies within a habitable zone is a matter of chance. If so, the wider the zone the greater the chances that a planet orbits within it. We conclude that low-luminosity stars are less likely to possess life than high-luminosity ones. If we combine this with our earlier conclusion that high-luminosity stars are too short-lived to harbor life, we see that intermediate-luminosity stars are the best bet for harboring life. Perhaps, therefore, it is not surprising that our own Sun is an intermediate-luminosity star.

We must, however, add a major warning. The predictions shown in Figure 19.7 are sometimes wrong. A case in point is Venus. As shown in Problem 5, Venus is a lot hotter than our prediction. This is because the prediction did not take into account the massive greenhouse effect which has heated that planet so greatly (Chapter 7). But this greenhouse effect could not have been foreseen had we not already known a great deal about Venus. The conclusion is that we should not be too confident of our knowledge about planets, concerning which we know hardly anything, orbiting distant stars.

Life around binary stars?

Can life exist on a planet orbiting a binary star? Consider the orbit illustrated in Figure 19.8, in which a planet moves in a "figure-8 pattern" about both members of the binary pair. As you can see, such a planet doesn't keep the same distance from its sources of warmth: sometimes it is close to one star,

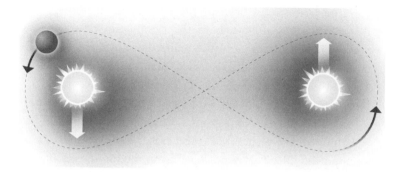

Figure 19.8 **An orbit about a binary star exposes the planet to great swings in temperature, possibly preventing life.**

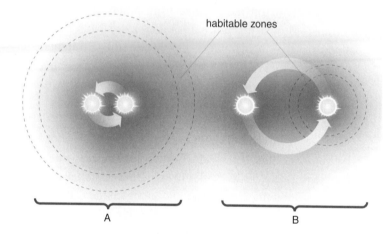

Figure 19.9 **Planetary orbits with steady temperatures around binary stars are possible in the two configurations. In (A) the planet orbits far outside the two binaries' orbits, in (B) it orbits about only one of them.**

sometimes to the other, and sometimes it is far from both of them. Furthermore, at some points in the planet's orbit, there are two suns in the sky at the same time: at other points it is never nighttime (Problem 6). This means that the temperature on such a planet would undergo extreme swings from very hot to very cold – far greater than the variation between summer and winter on Earth. At some points in its orbit the planet might be hotter than the Sahara in midsummer: at others colder than Antarctica in midwinter. Could such an environment support life? Probably not.

Notice, however, that there are other possibilities. Suppose the binary stars are very luminous, so that the habitable zone lies far from them. In particular, suppose the habitable zone lies farther away than the distance between the two stars. Then (Figure 19.9a) a planet in the habitable zone would be orbiting about *both* stars, and it would keep more or less the same distance from each. Such a planet would not suffer huge swings in its temperature.

Alternatively (Figure 19.9b), if the stars are very dim, the habitable zone lies very close to each. Suppose furthermore that the stars lie much farther away from each other than the radius of this habitable zone. Then a planet would orbit about *only one* star, and it too will keep roughly the same temperature throughout its orbit.

These issues are explored further in Problem 13.

Searching for life around other stars: interstellar space travel?

How shall we search for life on planets orbiting the distant stars? We are currently searching for life on Mars by sending space probes there: is such a strategy possible for life elsewhere in the Universe? We will see that, while not physically impossible, sending a space probe to a distant star is so hard that there is no means of doing this in the foreseeable future.

Let us begin with a simple calculation. As we saw in Chapter 8, the Cassini mission to Saturn, one of the most ambitious space probes ever launched, took seven years to reach its destination. At this rate, how long would Cassini have taken to reach the nearest star, Alpha Centauri?

The logic of the calculation

As we can see from Figure 3.21, Cassini did not travel in a straight line from Earth to Saturn. Nevertheless, to get a crude idea of the answer, we can use a simple proportion:

$$\frac{\text{time to Alpha Centauri}}{\text{time to Saturn}} = \frac{\text{distance to Alpha Centauri}}{\text{distance to Saturn}}$$

So the steps in the calculation are as follows.

Step 1. Solve this equation for the time required to reach Alpha Centauri.
Step 2. Plug in the distances to Saturn and Alpha Centauri.

As you can see from the detailed calculation, it would take more than two hundred thousand years to reach the closest star at this rate. While this is not an exact answer to our question, the general point is incontrovertible: if we are to take seriously the idea of interstellar travel, this is not the way to do it.

Detailed calculation

Step 1. Solve the equation for the time required to reach Alpha Centauri.
We get:

time to reach Alpha Centauri = $(D_{\text{Alpha Centauri}}/D_{\text{Saturn}})$ (7 years).

Step 2. Plug in the distances to Saturn and Alpha Centauri.
Saturn lies 9.5 AU from the Sun (Appendix III), and the Earth lies 1 AU from the Sun. So the straight-line distance from Earth to Saturn is 8.5 AU. Similarly, Alpha Centauri lies 4 light years away. We need to convert these both to MKS units in order to find their ratio. We calculate

D_{Saturn} = (8.5 AU) $(1.496 \times 10^{11}$ meters/AU) = 1.27×10^{12} meters.
$D_{\text{Alpha Centauri}}$ = (4 light years) $(9.46 \times 10^{15}$ meters/light year)
$= 3.78 \times 10^{16}$ meters.

Plugging in we find:

time to reach Alpha Centauri = 209 000 years.

NOW YOU DO IT
Suppose we wished to send off a spacecraft capable of reaching Alpha Centauri in 10 years. We want to calculate its velocity. (A) Describe the logic of the calculation you will use. (B) Perform the detailed calculation. Express your answer in miles per hour to get an intuitive feel for it.

We will need to find a way to travel faster. The problem is that it would require enormous amounts of rocket fuel to accelerate a spacecraft to the huge velocities required to reach the stars in short amounts of time. Indeed, if we wished to accelerate Cassini to a velocity so great that it would reach Alpha Centauri in a thousand years, the quantity of fuel required would be overwhelmingly greater than the entire fossil fuel reserves of the world!

Methods involving nuclear power, ion propulsion and the like have been studied. These are conceivable, but they require technologies far beyond our present capabilities. There is no currently available technology capable of sending a space probe to even the closest star.

Intelligent life in the Universe

CETI: Communicating with ExtraTerrestrial Intelligence

There is another way to search for life in the Universe: to search for *intelligent* life. Suppose that intelligent creatures exist on a planet orbiting a distant star: are we technologically capable of communicating with them? We will now show that this is eminently feasible: while we cannot travel to a distant star, we are entirely capable of

(a)   (b)

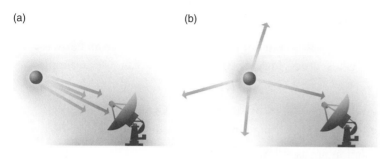

Figure 19.10 A beamed signal is more intense than an unbeamed one. Because beaming a signal (a) concentrates its rays into a narrow cone, more of them strike the detector than if the signal is unbeamed (b).

entering into a "conversation" with any extra-terrestrial civilizations that exist in our vicinity. *Communicating with an alien civilization is technologically within our grasp right now.*

We will begin by examining a specific example: that of using a radio telescope to communicate with a hypothetical alien civilization orbiting a nearby star. We will find that if such a civilization possesses telescopes as sensitive as ours, they would be capable of receiving signals we sent them. Conversely, if they are capable of sending signals as powerful as those we can send, we would be capable of receiving these messages.

It is not particularly difficult for current radio telescopes to *broadcast* a radio signal with a luminosity of one million watts. Similarly, it is not particularly difficult for us to *detect* a radio signal with a flux of 10^{-21} watts per meter^2. Let us imagine that our hypothetical alien civilization possesses these same technological capabilities. Suppose we sent out a message with a luminosity of one million watts: could they detect it? To answer this question, we need to calculate the flux produced by our signal at the distance of the star. If this flux is greater than 10^{-21} watts per meter^2, the answer is Yes. They would be able to hear us, and we would be able to enter into an exchange of signals with them.

To find the flux the alien astronomers would receive from our broadcast, we use the inverse square law. But there is an improvement we need to discuss. Flashlights are capable of casting strong beams even though their light bulbs are really very weak. They do this by *focusing* their emissions. Rather than sending the light out in all directions, they beam the light into a narrow cone. As illustrated in Figure 19.10, this has the effect of making the flux stronger. In (a) all of the emitter's rays reach the telescope; in (b) most miss it. Flashlights do this by mounting the light bulb in front of a reflector. Radio telescopes do it the same way.

The formula giving the flux for beamed radiation is more complicated than the simple inverse square, so we will not give it. But we can intuitively understand three aspects of this formula.

- The greater the luminosity of our emitter, the greater the flux the alien receives.
- The narrower the cone into which we beam our emissions, the greater the flux the alien receives.
- The closer it lies to us, the greater the flux the alien receives.

So we want to send a *strong* signal into a *narrow beam* aimed at a *nearby star* in order to make the flux big enough to be received by any hypothetical aliens that might exist.

Figure 19.11 shows the flux versus distance for a variety of beams. We present in this graph calculations for a relatively wide beam (one minute of arc) and a narrower one

flux (watts/meter squared)

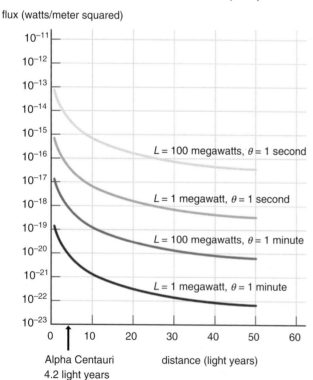

L = 100 megawatts, θ = 1 second

L = 1 megawatt, θ = 1 second

L = 100 megawatts, θ = 1 minute

L = 1 megawatt, θ = 1 minute

Alpha Centauri
4.2 light years

distance (light years)

Figure 19.11 Beamed radiation. The apparent brightness (flux) of various beams.

(one second of arc), and a not particularly powerful emission (one million watts) and a more powerful one (100 million watts). We emphasize that none of these choices are beyond current technology.

Let us use these results to analyze the feasibility of sending a message to the closest star, Alpha Centauri. Its distance (4.2 light years) is indicated in Figure 19.11. Suppose we sent out a one-million-watt signal with a one minute of arc beam. Figure 19.11 tells us that the flux at the distance of Alpha Centauri is 10^{-20} watts per meter^2. This is much greater than the minimum detectable flux: the aliens could easily detect our signal. And more powerful broadcasts, or narrower beams, make it even easier for the aliens.

NOW YOU DO IT

(A) Suppose an alien civilization on a planet orbiting Alpha Centauri were less technologically advanced than us, and was only able to detect fluxes of 10^{-18} watts per meter^2. And suppose we sent them a message with a power of one million watts in a beam of one minute of arc. Would they be able to detect it? (B) How about a 200-million-watt message in a 1-minute beam?

We reach the same conclusion if we analyze communication using visible light signals. High-powered lasers can emit exceedingly narrow beams of such intensity as to be easily visible at the distances of the nearby stars. So no matter what kind of radiation we use, *communicating with an alien civilization is technologically within our grasp right now.*

SETI: Searching for ExtraTerrestrial Intelligence

The Drake equation

The hard part is not talking to them: it is finding them in the first place. The fact that communication with an extraterrestrial civilization is technically feasible does not mean that it is going to be easy to locate one. Indeed, our best estimate is that this will be an exceedingly difficult task. This is because, to the best of our knowledge, the chances are very small that any given star possesses a communicating civilization at the current time. Therefore we are going to have to study a lot of them before we find one with which we can exchange messages. Since a good deal of time and effort is required to investigate even a single star, the task of studying the required number will be very difficult.

Suppose we point our telescope at a particular star. Does it harbor a communicating civilization? In order for this to be the case, all the conditions listed in Table 19.2 must obtain. The problem is that this is a long list of conditions – and the probability of their all being met is very low.

To find the probability that a given star harbors a communicating civilization, we use the rule of multiplication of probabilities: *the probability that a list of things happens is the product of the probabilities that each of them happens.* This tells us that to find our probability we need to multiply together the probabilities of all the items in Table 19.2's list.

Let us call the result $P_{\text{communicating civilization}}$. We get:

$$P_{\text{communicating civilization}} = P_{\text{planets}} P_{\text{suitable}} P_{\text{life}} P_{\text{intelligence}} P_{\text{technology}} P_{\text{timing}}.$$

Since each probability in this product is low, the net probability $P_{\text{communicating civilization}}$ will be *very* low. This is the central problem facing SETI.

Table 19.2. **Conditions for a communicating civilization.**

Condition	Rationale	Probability
The star must possess planets	Life as we know it cannot exist in interstellar space, on a star, etc. Only planets (and maybe satellites) are suitable abodes for life	$P_{planets}$
At least one of these planets must be capable of supporting life	Not all planets or satellites provide suitable abodes for life	$P_{suitable}$
Life must have arisen on this planet		P_{life}
Intelligence must have arisen on this planet		$P_{intelligence}$
This species must have developed the technological capability of communicating across interstellar distances	Humanity arose as a species about a million years ago – but we developed the technology required for interstellar communication only in the past few decades. The aliens could be deeper philosophers than Plato and smarter than Newton – but if they don't have telescopes as powerful as ours we will never be able to communicate with them	$P_{technology}$
This capability must have arisen at the same cosmic epoch that it did on Earth	A civilization that arose millions of years ago, but that by now has vanished, is not one with which we can communicate. Conversely, the civilization may not yet have arisen	P_{timing}

Table 19.3. **Coins.**

Denomination	Face
Nickel	**Heads**
Nickel	**Tails**
Dime*	**Heads***
Dime	**Tails**
Quarter	**Heads**
Quarter	**Tails**

*This is the combination we seek.

The rule of multiplication of probabilities

Before we use this rule, let us get used to it. Suppose you have a box containing a nickel, a dime and a quarter: each can be heads up or heads down. Suppose you close your eyes, reach into the box, and choose a coin at random. What is the probability that this coin will be a heads-up dime? First we will analyze this problem directly. Then we will use the rule of multiplication of probabilities, to make sure that it gives the same answer. This will give us some confidence that our rule makes sense.

First let us analyze the problem directly. If we choose a coin at random it can be either the nickel, the dime or the quarter. Similarly, it can be heads up or heads down. Table 19.3 lists all the possibilities.

As you can see, there are six entries in the table, and only one of them is the combination we seek. So the probability of finding this combination is 1/6.

Let us now analyze this using our rule for the multiplication of probabilities, to make sure it gives the same answer. We need to know two individual probabilities. (1) Our randomly chosen coin has three possibilities for its denomination. So the probability of its denomination being what we are looking for is $P_{denomination} = 1/3$. Similarly, (2) it has two possibilities for its face. So the probability of its face being what we are looking for is $P_{face} = 1/2$. The rule of multiplication of probabilities now tells us that the probability of choosing a heads-up dime is the product of these individual probabilities:

$$P_{\text{heads-up dime}} = P_{denomination}\, P_{face}$$
$$= (1/3)\,(1/2) = 1/6.$$

As you can see, the rule gives the same result as our direct calculation.

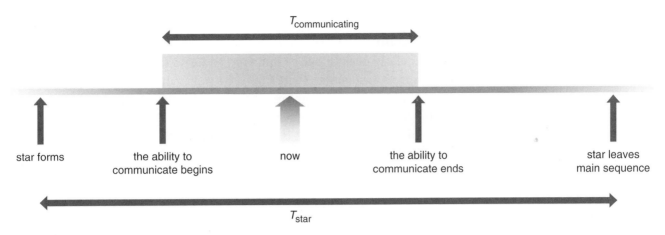

Figure 19.12 Time lines. A star provides a suitable environment for life for a time T_{star}. A civilization capable of interstellar communication exists for a time $T_{communicating}$ (shaded area). The light arrow indicating the present time can fall anywhere within this figure: if it falls within the shaded area, we can enter into a dialog with the alien civilization.

. . . .

NOW YOU DO IT
Suppose you have a pair of dice, one red and the other green. Choose one of these dice at random. We want to calculate the probability that you have chosen a red 4. Your task is to find this probability in two different ways, and to check that you get the same answer. (1) Draw up a table showing all the possibilities for the randomly chosen die, and identify the one that corresponds to a red 4. What fraction is this of the total number of possibilities? (2) Find (a) the probability that a randomly chosen die is red, and (b) the probability that a randomly chosen die is 4. Then multiply these together.

. . . .

The "timing" probability

Before we use this rule, we need to find a formula for P_{timing}, the probability that the alien civilization is capable of communicating with us right now. Figure 19.12 shows a time line illustrating the development of the ability to communicate. The line begins with the formation of the star and its planetary system, and it ends when the star leaves the main sequence, wiping out any life that might exist on its planets. The shaded area on this figure illustrates the period of time during which a communicating civilization exists. The question is, what is the probability that the present moment falls within this shaded area?

Clearly "now" falls somewhere along the longer line illustrating the main sequence phase of the star's evolution (we ensure this by only searching for life about main sequence stars). But we have no idea where along this line "now" falls: it merely lies at some random point along this line. So the probability P_{timing} that it falls within the shaded area is

$$P_{timing} = \text{the fraction of the line occupied by the shaded area}$$
$$= T_{communicating}/T_{star}.$$

If, for instance, the shaded area occupies ½ the line, then the chances are 1/2 that a randomly chosen "now" falls within it. Similarly, if the shaded area only occupies 1% of the line, the chances are only 1% that the communicating civilization exists at present.

Because our own Sun is typical, we can take T_{star} to be its main sequence lifetime: 10^{10} years.

. . . .

NOW YOU DO IT
Mary regularly goes to bed at midnight and gets up at 8 in the morning. But she is an insomniac. Every night she wakes up for half an hour. But the times at which she awakens are random: sometimes she awakens after only a few minutes of sleep, at other times just before she finally climbs out of bed, and at other times she is awake in the middle of the night. (A) What is the probability that she is awake at 3.15 a.m.? (B) What does this have to do with communicating with an alien civilization?

. . . .

We are finally ready to formulate the Drake equation. It gives *the total number of communicating civilizations in our Galaxy*. Since there are around 10^{11} stars in our Milky Way Galaxy, and the chances that any one of them has a communicating civilization is $P_{\text{communicating civilization}}$, this number $N_{\text{civilizations}}$ is:

$$N_{\text{civilizations}} = 10^{11} P_{\text{communicating civilization}}$$
$$= 10^{11} P_{\text{planet}} P_{\text{suitable}} P_{\text{life}} P_{\text{intelligence}} P_{\text{technology}} P_{\text{timing}}.$$

This is the Drake equation.[1]

We can also find *how many stars we must search in order to have a good chance of finding a communicating civilization*. In a coin toss the probability of getting heads is 1/2, and on average you must toss the coin 2 times to get heads. Similarly, in tossing a die the probability of getting some particular result is 1/6, and on average you must toss the die 6 times to get that result. Similarly, if the probability that a randomly chosen star possesses a communicating civilization is $P_{\text{communicating civilization}}$, we will have to search

$$N_{\text{search}} = 1/P_{\text{communicating civilization}}$$

in order to find one.

Guessing probabilities

We are now ready to evaluate $P_{\text{communicating civilization}}$. To do this we need to assign values to each of the separate probabilities occurring within the Drake equation. Since we know very little about most of the conditions listed in Table 19.2, we will not be able to do this with confidence. All we can do is guess intelligently. In what follows we will assign definite values to these probabilities, but there is little reason to take these values seriously. In the problems you will be invited to examine the consequences of making different choices.

We now consider each of Table 19.2's conditions in turn.

The star must possess planets.   As we discussed in Chapter 10, many planets have been discovered orbiting other stars. We do not yet know what fraction of all stars possess planets, but this fraction is probably not low. We will guess $P_{\text{planets}} = 1/10$.

At least one of these planets must be capable of supporting life.   As we saw in our discussion of habitable zones, life can only exist within a limited range of

---

[1] You may be more familiar with a different version of this equation. If we use our formula for P_{timing} we obtain $N_{\text{civilizations}} = (10^{11}/T_{\text{star}}) P_{\text{planets}} P_{\text{suitable}} P_{\text{life}} P_{\text{intelligence}} P_{\text{technology}} T_{\text{communicating}}$, which is the usual form of the Drake equation. The factor in parentheses is the rate at which stars have been formed in the Galaxy. Finally, Drake also multiplied his formula by the number of habitable planets orbiting a star.

temperatures. This means that the planet cannot be too close to or too far from its star. It also means that phenomena such as Venus' runaway greenhouse effect must not occur. Within our Solar System the Earth has the right temperature, and Mars might once have. It is also conceivable that life exists within the interiors of planets such as Jupiter, or on certain moons. Perhaps it is reasonable to guess that at least one planet is going to have conditions conducive to life, so that $P_{suitable} = 1$.

Life must have arisen on this planet. As we saw in our discussion of the origin of life, we have no fossil evidence at all concerning the origin of life on Earth. On the other hand, we do know that life arose fairly rapidly once the appropriate conditions had been reached. Lacking further knowledge, let us guess that $P_{life} = 1/5$.

Intelligence must have arisen on this planet. In contrast to the origin of life, the origin of intelligence appears to have been a very unlikely event. As we saw in our time line of life on Earth, on Earth intelligence arose only very recently – with the appearance of our species, *Homo sapiens*. Recall that in terms of our time line of life on Earth, in which the entire history of our planet was compressed into a single year, this happened at 11.20 p.m. on New Year's Eve. The interval of time between then and now – 40 minutes in terms of our analogy – is a tiny fraction of the full span of time during which life has existed. This presumably means that the evolution of intelligence was very unlikely.

A second line of evidence rests on the estimate that the total number of species that have ever existed on the Earth is roughly one billion – but out of this great number, only one has developed the means of communicating across interstellar distances: ourselves. This, too, is reason to believe that the emergence of intelligence is not at all guaranteed in the course of evolution.

While it is very dangerous to rely on a single case on which to base our estimate, both these arguments lead us to take the probability of intelligence arising to be very small. We will take it to be one in a million. Thus $P_{intelligence} = 10^{-6}$.

This species must have developed the technological capability of communicating across interstellar distances. On Earth this happened only very recently – long after the appearance of intelligence. Is it possible that a race of intelligent beings on another world would *never* develop the required technology? Perhaps it is. In what follows we will guess that $P_{technology} = 1/2$.

This capability must have arisen at the same cosmic epoch that it did on Earth. This probability is the lifetime of a communicating civilization divided by the lifetime of a star, which we will take to be 10^{10} years.

We are now ready to collect all our guesses and estimate the probability that a given star has a communicating civilization.

$$P_{communicating\ civilization} = P_{planets}\ P_{suitable}\ P_{life}\ P_{intelligence}\ P_{technology}\ P_{timing}$$

$$= (1/10)\ (1)\ (1/5)\ (10^{-6})\ (1/2)\ (T_{communicating}/10^{10}\ \text{years})$$

$$= 10^{-8}\ (T_{communicating}/10^{10}\ \text{years})$$

To evaluate this we need to estimate $T_{communicating}$, the length of time a typical civilization possesses the means of interstellar communication. Of course we have no way to know this. But let us try a few guesses.

As we have noted, on Earth this ability developed only recently – within the past half century at most. Suppose we are pessimistic, and imagine that our technologically advanced civilization will come to an end very soon. Perhaps it will be wiped out in a global thermonuclear war, or a massive environmental catastrophe. Specifically, suppose that $T_{communicating}$ is merely a century. Then $T_{communicating}/10^{10}$ years $= 100$ years$/10^{10}$ years $= 10^{-8}$ and

$$P_{communicating\ civilization} = (10^{-8})(10^{-8}) = 10^{-16}.$$

This is an extraordinarily low probability. It tells us that there is only one chance in 10 000 000 000 000 000 that a given star possesses a civilization with which we can exchange messages. In order to find this civilization we would then have to search 10 000 000 000 000 000 stars. But this is greater than the number of stars in our Milky Way Galaxy! The nearest communicating civilization then lies in some other galaxy. But there is no possibility of our being able to communicate across such gigantic distances.

We conclude that, if a typical communicating civilization only lasts for a period of time measured in centuries, we will never find one.

Now let us adopt a more optimistic approach, and imagine that a typical communicating civilization survives for enormous periods of time – say, one billion years. In this view humanity has only just entered its full maturity, a maturity destined to stretch onwards into the unimaginably distant future. In this case $T_{communicating}/10^{10}$ years $= 10^{9}$ years$/10^{10}$ years $= 1/10$ and

$$P_{communicating\ civilization} = (10^{-8})(1/10) = 10^{-9}.$$

This is still a low probability. There is only one chance in a billion that a given randomly chosen star possesses a civilization with which we can exchange messages. To find the number of communicating civilizations in our Milky Way Galaxy we use our formula

$$N_{civilizations} = 10^{11} P_{communicating\ civilization} = 10^{11} 10^{-9} = 100.$$

So we are not alone: there are 100 other communicating civilizations at this very moment in our Galaxy.

But it is going to be hard to find them. To find even one we need to examine

$$N_{search} = 1/P_{communicating\ civilization} = 10^{9} \text{ stars} = \text{one billion stars.}$$

How long does it take to examine a billion stars? Suppose we need to point our radio telescope at one for an hour in order to carefully "listen" for any signals it may be sending us. Then it will take a billion hours to study them all. But a billion hours is 114 000 years!

We emphasize that this is based on a very optimistic assumption – that a typical advanced civilization survives for a billion years. Nevertheless, we are reaching a very pessimistic conclusion: even in such a case, it will take huge amounts of time and effort to find one.

SETI projects

Scientists involved in searches for extraterrestrial civilizations have a saying: "the chances of finding an alien civilization are small – but if we don't search, the chances are zero." Clearly, the strategy is to find a way to minimize the time required to check any given star for intelligent signals, so that many stars can be surveyed in a reasonable length of time. We will mention only a few of the searches that have been performed. Not surprisingly, none of them has found evidence for intelligent life in the Universe.

Project Ozma was the first search for signals from an intelligent civilization. Conducted by the astronomer Frank Drake in 1960, it was named after the imaginary kingdom of the classic "Wizard of Oz." The project searched for radio signals from two nearby stars over several months.

This is a very small number of stars. But Project Ozma was significant not for the details of what it did, but for the fact that it happened at all. It was humanity's first serious attempt to search for alien civilizations elsewhere in the cosmos – a punctuation mark in the history of our species.

Project Phoenix. In 1993, the US Congress cut funding for an ambitious NASA SETI program. Project Phoenix, conducted by a private organization known as the SETI Institute, was designed to achieve the goals of NASA's canceled program. Entirely funded by non-governmental sources, it operated between 1995 and 2004, employing many of the world's largest radio telescopes. It searched a carefully selected list of about 840 stars within 240 light years. It was so sensitive that it would have detected a transmitter with a power similar to our military airport radars. At the time, Phoenix was the most ambitious search for extraterrestrial intelligence ever undertaken.

In contrast, the *Optical SETI* project works at visual wavelengths. Funded by another private organization, The Planetary Society, it employs a 72-inch telescope: this is the world's largest optical telescope fully dedicated to SETI, and indeed it is the largest optical telescope in the eastern USA. As opposed to the "targeted search" technique of Project Phoenix, this project surveys the entire sky, taking 200 clear nights to do so: once a survey is complete, the telescope simply begins all over again. The telescope's operation is automated, so that astronomers rarely need actually to visit it. The heart of the data-gathering system is a set of photomultiplier tubes, similar to the charge-coupled devices discussed in Chapter 5, but far more sensitive. Each is divided into 1024 pixels, each of which works as an independent camera. The electronics are so sophisticated that it can detect light pulses as short as a billionth of a second.

The Allen Telescope Array (Figure 19.13), named for Microsoft Corporation co-founder Paul Allen, who supplied much of its funding, is planned to be an array of 350 small radio telescopes joined together to form a single giant interferometer (Chapter 5). Each individual radio telescope is 6 meters in diameter, which is quite small. However, they will be connected via high-performance electronics. Only as the result of recent technological advances has the cost of this electronics dropped sufficiently to make such a concept of small dishes possible. This is known as "replacing steel with silicon."

When completed, the Allen Telescope Array will be one of the most powerful telescopes in the world. It will be able to examine many target stars simultaneously for signs of intelligent signals, and at the same time to do traditional radio astronomy. As of 2007, 42 dishes had been completed, and the array had begun performing observations. However, as is always the case with large projects, funding is always a problem, and in 2011 the telescope was placed in hibernation due to budget problems. At the time of this writing short-term funding has been found.

Figure 19.13 **The Allen Telescope Array (artist's rendering).**

CETI: interstellar messages

If we ever do make contact with an extraterrestrial civilization, the messages we exchange cannot be like the ordinary ones we send each other every day. The aliens will not understand English. How can we communicate without sharing a common language? The task of CETI is to *design a message that itself tells the recipient how to decipher it*. We will describe three such messages.

A mathematics lesson

Here is one way to do this.

Suppose we were to broadcast into space a series of flashes of light, some of which are short and the others long. Using an "S" to denote a short flash and "L" to denote a long one, here is the series.

```
LSLLSLLLSLLLLSLLLLLSLLLLLLSLLLLLLLSLLLLLLLLSLLLLLLLLLSSSSSSS
LLLLLLLLLSSLLLLLLLLSSLLLLLLLSSLLLLLLSSLLLLLSSLLLLSSLLLSSLLSS
LSSSSSSS
LSLLSSSSSSSSLSLLLLLSSSSSSSSLLSLLLSSSSSSSSLLSLLLLLSSSSSSSSLLLSLLL
LLSSSSSSSSLLLLSLLLLLSSSSSSSSLLLSSLSSSSSSSSLLLLLSSLSSSSSSSSLLLSSLLS
SSSSSSLLLLLSSLLSSSSSSSLLLLLLSSLLLSSSSSSSSLLLLLSSLLLLSSSSSSSSLSSSL
SSSSSSSLLSSSSLLSSSSSSSSLLLSSSLLLSSSSSSSSLLLLSSSLLLLSSSSSSSSLLLLLSSS
LLLLLSSSSSSSLSSSSLSSSLLSSSSSSSSLSSSSLSSSSLSSSLLLSSSSSSSSLSSSSLSSS
SLSSSSLSSSLLLLLSSSSSSSSLSSSSLSSSSLSSSSLSSSSLSSSLLLLLSSSSSSSSLSSSSL
LSSSLLLSSSSSSSSLLSSSSLLSSSSLLLLSSSSSSSSLLSSSSLLLLLLSSSLLLLLLLLSSS
SSSSLLLLLSSSSSLSSSLLLLSSSSSSSSLLLLLSSSSSLLSSSLLLSSSSSSSSLLLLLLSSS
SSLLLSSSLLSSSSSSSLLLLLSSSSSLLLLLSSSLSSSSSSSLLLSSSSLLLLLLSSSLLLLL
SSSSLLLSSSSSSSLLLSSSSLLLLSSSLLLLSSSSLLLSSSSSSSLSSSSSSLLSSSSSSSLL
LSSSSSSLLLLSSSSSSLLLLLLSSSSSSSSLLLLLSSSSSSLLLLSSSSSSSLLLSSSSSSSLLS
SSSSSLSSSSSSSSLLLSSSSLLLLLLSSSSSSSLLLLSSSSLLLLLLSSSSSSSSLLLSSSSLLLL
LSSSSSSSLLSSSSLLLLLLSSSSSSS
```

What could such a strange pattern mean? Let us start with just the first line.

```
LSLLSLLLSLLLLSLLLLLSLLLLLLSLLLLLLLSLLLLLLLLSLLLLLLLLLSSSSSSS
```

There is an obvious regularity to this series: the message is counting from one to nine. First there is a single long pulse, then two, then three and so forth. So let us take a guess that one long flash represents the number one, two represents two, and so on. Then the first step in decoding the message would be as follows.

```
1 S 2 S 3  S  4  S  5    S  6    S  7      S    8      S    9        SSSSSSS
↕  ↕  ↕   ↕     ↕        ↕          ↕              ↕                    ↕
LSLLSLLLSLLLLSLLLLLSLLLLLLSLLLLLLLSLLLLLLLLSLLLLLLLLLSSSSSSS
```

Now turn attention to the short flashes. What might they represent? So far the message does not tell us. There are many possibilities.

(1) An S might be something like the spaces we insert between words in a sentence. In this case the line might simply be the digits 1 through 9 in order.

(2) An S might mean "is less than." In this case the line would mean something different: it would read "1 is less than 2, which is less than 3" and so on.

(3) An S might mean "is the number just before." In this case the line would mean "1 is the number just before 2, which is the number just before 3," and so on.

How can we resolve this ambiguity, and decide what a short pulse of light means? Let us scan through the message to see other places where a single S appears. We find them starting at the beginning of the message's fourth line.

... LSLLSSSSSSSLSLLLLLSSSSSSSSLLSLLLSSSSSSSSLLSLLLLLSSSSSSSSLLLSLLLLL SSSSSSSLLLLSLLLLLSSSSSSS...

Which we know means the following.

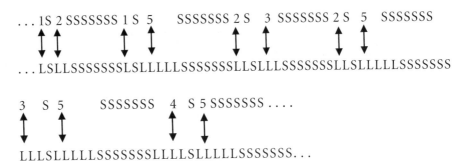

This is enough to eliminate possibilities (1) and (3), and tell us that a short flash represents "is less than." So this portion of the message reads as follows.

1 is less than 2 SSSSSSS 1 is less than 5 SSSSSSS 2 is less than 3 SSSSSSS
2 is less than 5 SSSSSSS 3 is less than 5 SSSSSSS 4 is less than 5 SSSSSSS

And finally, it is now clear that seven short pulses functions like a period or a comma: it separates different mathematical statements. So the full translation of this part of the message is as follows.

1 is less than 2. 1 is less than 5. 2 is less than 3. 2 is less than 5. 3 is less than 5. 4 is less than 5.

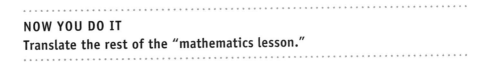

NOW YOU DO IT
Translate the rest of the "mathematics lesson."

The Pioneer interstellar plaques

In 1972 and 1973 two space probes, Pioneer 10 and Pioneer 11, were launched towards Jupiter. Two years later they swept by it and then continued outwards into interstellar space. Even now they are still traveling onwards endlessly, far beyond the orbit of the farthest planet.

As we have discussed, traveling to even a nearby star in a reasonable period of time is technologically impossible. Furthermore, the Pioneer probes are not aimed at any star in particular. So there is essentially no possibility that they will ever enter another planetary system and be discovered. Nevertheless, it seemed appropriate that these spacecraft, the first man-made objects ever to leave the Solar System, carry some indication of where they came from, and of the strange race that built them. Accordingly, each of these spacecraft carries a plaque (Figure 19.14).

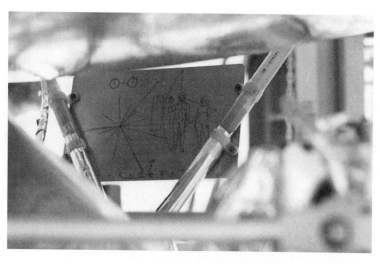

Figure 19.14 **The Pioneer interstellar plaques. We see a plaque as it was mounted on the spacecraft and the message engraved upon it.**

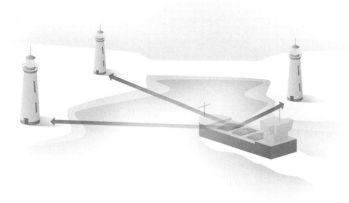

Figure 19.15 **A ship's location can be specified by giving the directions and distances to various lighthouses.**

On each plaque are two people, illustrating that there are two types of human being. They stand in front of a schematic drawing of the spacecraft, which gives an indication of our size. Along the bottom runs a representation of the Sun and planets of the Solar System (this was when Pluto was still classified as a planet), and a curving line indicating that the spacecraft came from the third planet and swept past the fifth. The spacecraft's home planet is also represented by a second drawing of the probe, oriented to point its antenna toward the Earth. The tiny figures alongside each planet give their distances from the Sun in binary notation.

The striking star-shaped pattern in the diagram is a description of our location within the Milky Way Galaxy. The strategy is similar to that illustrated in Figure 19.15. Mariners know that they can find their position at sea if they can determine the directions and distances to some lighthouses. Similarly, the 14 lines comprising the star-shaped pattern in the Pioneer plaques give the directions and distances to 14 pulsars (Chapter 15). Of course, Figure 19.15's strategy will not work unless the mariner knows *which* lighthouses she is observing. At sea, this information is provided by arranging for each lighthouse to flash in its own, unique pattern. Similarly, each pulsar is known to flash at its own, unique rate. On the plaques this rate is indicated in binary alongside the line to each pulsar. And finally, since every pulsar is steadily slowing its rate of pulsation, the Pioneer plaques tell not only *where* we lived, but *when*: it is only at this particular moment of cosmic history that these pulsars emit at these rates.

An important issue is that we measure pulsation rates in seconds. But these are our unit of time, and the aliens have no way of knowing how long our seconds are. The pattern on the upper left-hand portion of the plaques is an attempt to deal with this. It is meant to represent the internal readjustment in an atom which produces one of the most important units of time known to physics: the period of vibration of the electromagnetic wave associated with the most important spectral line of the most common element in the Universe, hydrogen. The pulsar periods in Figure 19.15 are all expressed not in seconds, but in terms of this unit of time.

Figure 19.16 **The image contained in the message to the globular cluster M13.** 👁 **(Also see color plate section.)**

A message to M13

In 1974 the Arecibo radio telescope was used to transmit an interstellar message to the globular star cluster M13. It consisted of a series of 1679 short pulses of radio signals. Some of these pulses were at one frequency, and the rest were at a slightly different one.

Why 1679 pulses? Why not 1678 or 1680? People of a mathematical bent will recognize that this number has a special property. Many numbers are the product of other numbers, and most are the product of *many different pairs* of factors. For example, 1000 is the product of

- 2 times 500
- 4 times 250
- 5 times 200

and so on. But 1679 is the product of only *one* pair of factors, 23 times 73.

The senders of this message hoped that any civilization receiving it would notice this fact, and try arranging the various pulses in a grid 23 cells across and 73 cells down. This grid is shown in Figure 19.16, where we have colored a cell black for a pulse at one frequency, and various colors for a pulse at the other frequency. As you can see, the cells form a pattern: in Figure 19.16 various elements of the pattern have been color coded to make them easier to see.

A great deal of information is contained in this pattern (you will need to refer to the version of the figure in the color plate section when working through this section).

(1) At the bottom of every group in the pattern is a "marker cell," which is either black or colored. Black means that the group is to be taken as a picture – as, for example, the stick figure representing the person in the middle of the message. In contrast, colored means that the group is to be taken as a binary number: a black cell indicating a 0 and a colored cell a 1.

With this we are ready to interpret the entire message. Scanning from the top to the bottom:

(2) Arrayed along the top of the message in white is the sequence 001, 010, 011 and so on. These are the numbers 1 through 10 in binary notation.

(3) Next in purple come the numbers 1, 6, 7, 8 and 15 in binary. These are the atomic numbers of the chemical elements most important in our biochemistry: hydrogen, carbon, nitrogen, oxygen and phosphorous.

(4) Next in green come formulas for the most important molecules in our biochemistry: the four bases of DNA, and the phosphate group and the sugar deoxyribose.

(5) Next in blue comes a crude representation of the double helix of DNA: running along its central line in white is an enormous binary number giving the number of nucleotides in our DNA.

(6) Next in red comes a crude image of a human being: in white to its left is the population of the Earth in binary, and to its right the height of a person. Just as the Pioneer plaques required a unit of time, so here a unit of distance is needed: this unit is the wavelength of the radio signal transmitting the message.

(7) Next in yellow is an image of the Solar System: the third planet is displaced upward toward the person. (Pluto was still considered a planet when the message was sent!)

(8) Next in purple is an image of the telescope used to send the message.

(9) Finally, running along the bottom of the pattern in blue and white is a number representing the diameter of this telescope, again in units of the radiation used to transmit the message.

Comments on interstellar communication

Our first example of an interstellar message, the "mathematics lesson," is easy to decipher. The last example, in contrast, is quite hard. Indeed, the author of the message to M13 happened to show it to a number of his colleagues: none of them managed to fully translate it! This points to an important general issue: if we ever do detect a signal from an alien civilization, this signal will be hard to understand fully. Some portions of it may prove easy for us to translate. Other portions, on the other hand, may elude our comprehension forever.

Furthermore, even our message to M13 contains only a tiny bit of all the information about humanity that we might want to transmit. Imagine the difficulty of informing an alien about all the myriad facets of our world! The same will be true of that alien's civilization. A full message, containing such great amounts of information, is likely to be enormously long and complex, and exceedingly difficult to translate.

Notice that any intelligent civilization is likely to lie very far from us. If, for example, there are only a few civilizations within our Milky Way Galaxy, then even the closest lies tens of thousands of light years away. This means that a message takes tens of thousands of years to travel between us and them! So communication with alien civilizations is nothing like a *conversation*. Rather, it is more like a series of long monologues.

Finally, recall that any civilization may survive only for a limited amount of time. If, for example, a civilization lasts for only a thousand years, and we receive a message from one lying ten thousand light years from us, there is no reason for us to respond: by the time we receive the signal, the civilization that has transmitted it has come to an end. We will be receiving a message from an extinct civilization.

Summary

The origin of life

- We have no direct evidence at all for how life began.
- But we do know that life on Earth arose rapidly (in geologic terms).
  - Life could not have existed prior to 4.1 billion years ago.
    - Early in its history Earth was continually undergoing giant impacts.
    - Any one of these would have released so much energy as to have sterilized the entire planet.
  - Life did exist 3.5 billion years go, and possibly 3.85 billion years ago.
    - Fossil stromatolites and cells have been found in rocks 3.5 billion years old.
    - An excess of an isotope of carbon favored by living organisms has been found in rocks 3.85 billion years old.
- The Stanley Miller experiment synthesized amino acids and organic molecules by natural means.

Time line of the evolution of life on Earth

- An analogy: the Earth was formed at midnight on December 31, and by now a year has passed and it is midnight on the next December 31.
- In this analogy the Earth became habitable during February.
- Mid March: life existed.
- Mid June: evidence for photosynthesis (plants).
- Fall: multicellular life.
- Early December: mammals.
- Late afternoon December 31: Hominids.
- Scientific revolution: 4 seconds before midnight.

Life elsewhere in the Solar System

- Liquid water is essential to life as we know it.
- Life on Mars at present.
  - Conditions on Mars are very harsh, and probably inimical to life.
  - Viking missions found no convincing evidence for life on Mars.
- Life on Mars in the past.
  - We have much evidence that Mars in the past was warmer, had a thicker atmosphere, and possessed liquid water.
  - It has been claimed that a meteor from Mars contains fossilized life forms.
- Many of the moons of Jupiter and Saturn possess liquid water in their deep interiors.
- The chemistry of Titan (moon of Saturn) may resemble that of the primitive Earth.

Life beyond the Solar System

- Requirements for life.
  - A source of energy, i.e. a star.
  - The source must last a long time – i.e. not high-mass stars or red-giant stars.
  - Life can only exist within a narrow range of temperatures, so the planet must lie in the "habitable zone" around its star.
  - Life requires a stable temperature, so many orbits around binary stars are not conducive to life.
- Searching for life around other stars.
  - Interstellar space travel is technologically impossible at present, and will remain so for a long time.
  - Communication with extraterrestrial civilizations is technologically possible right now.
  - The "Drake equation" tells us the number of communicating civilizations in the Galaxy.
  - The chances are very small that any given star possesses a communicating civilization now.
  - Therefore the task of finding one is exceedingly difficult.
  - A few searches have been conducted and/or are under way.
- Communicating with extraterrestrial intelligence.
  - We must design messages that can be decoded by beings that know nothing about us.
  - Several such messages have been sent.
  - If we do make contact with an extraterrestrial civilization, the "conversation" will be slow and complex.

Problems

(1) We want to find out whether the impact of a body 100 kilometers across would have liberated enough energy to vaporize all the Earth's oceans, if it had the same velocity and density as we assumed in the text.
(A) Describe the logic of the calculation you will use.
(B) Perform the detailed calculation.

(2) Suppose an approaching body had twice the velocity of that in Problem 1, but the same mass. (A) How much more energy would the collision have released?
(B) How much more water would it have boiled away?
(C) What are your answers to these questions if the body had the same velocity as in Problem 1, but twice the mass?

(3) In the text we described an analogy in which the 4.5-billion-year history of the Earth was likened to a single year. Let us now use a different analogy, in which we imagine that you take a 1-mile (1.5 kilometers) walk: the beginning of the walk corresponds to the formation of the Earth, and the end of the walk to "now." In terms of this analogy, we want to calculate how far would you would have walked corresponding to when:
- the bombardment preventing the origin of life ceases (4.1 billion years ago),
- photosynthesis is taking place (2.5 billion years ago),
- the scientific revolution occurred (about the year 1600).
(A) Describe the logic of the calculation you will use.
(B) Perform the detailed calculation.

(4) In Figure 19.7 identify the closest and farthest distance from each of three stars at which water can be liquid.

Now subtract these distances to find the width (in astronomical units) of the habitable zones about each of these stars. Notice that *the brighter the star, the wider the zone.* So bright stars give more "room" for life than dim ones.

(5) In Figure 19.7 read off the *predicted* temperature of Venus as it orbits our Sun. What is the *actual* temperature of Venus? How good is our prediction?

(6) Identify the points in the orbit depicted in Figure 19.8 at which there are two suns in a planet's sky at the same time. Identify the points at which it is never night. Identify the points at which one sun is directly overhead as the other one is setting.

(7) You have already considered a spacecraft capable of reaching Alpha Centauri in ten years. How long would it take before it exited the Solar System – i.e. before it reached the orbit of Neptune? (A) Describe the logic of the calculation you will use. (B) Perform the detailed calculation.

(8) Suppose four people live in a house: Alice, Bob, Candice and Dan. Suppose each one of them spends exactly half the time being happy, and the other time being gloomy. One day you put in a telephone call to the house. (A) What is the probability that you reach Candice in a good mood? (B) What is the probability that you reach a woman in a bad mood? (You should analyze each of these questions in two different ways: first by listing all the possibilities, and then by using the rule of multiplication of probabilities. Do they get the same answer?)

(9) You have already thought about Mary, who regularly goes to bed at midnight and gets up at 8 in the morning. But recall that she is an insomniac. Every night she wakes up for half an hour. But the times at which she awakens are random: sometimes she awakens after only a few minutes of sleep, at other times just before she finally climbs out of bed, and at other times she is awake in the middle of the night. Here are some more questions about her. (A) Suppose you yourself awaken at random times during the night. What is the probability that, the moment you awaken, she is awake? (B) How many nights pass before (a) Mary is awake at 3.15 a.m., and (b) you find her awake when you awaken? For each question, carefully explain your reasoning.

(10) In our study of the Drake equation we guessed that the evolution of intelligence is a rare event, and we took $P_{intelligence}$ to be one in a million. Suppose that we take a more optimistic view, and suppose that $P_{intelligence} = 1/100$. If all the other guesses for the individual probabilities are the same, find:

- the probability that a given randomly chosen star has a communicating civilization with which we can enter into a dialog;
- the number of stars we must search before we have a reasonable chance of finding a communicating civilization;
- the amount of time this will take, assuming that a star must be studied for 30 minutes in order to tell if it is sending us a message;
- the number of communicating civilizations that exist right now in our Galaxy.

For each of these, study the two cases $T_{communicating}$ = a century, and a billion years.

(13) You are in charge of deciding how the Allen Telescope Array is to be used to search for signals from an extraterrestrial civilization. You have decided that it should be pointed at selected "target stars." Below is a list of stars. For each one, your task is to rate it:

(i) definitely include this star in the list of target stars to be studied,

(ii) definitely do not include this star in the list,

(iii) it is unclear whether or not we should include this star in the list.

In every case, describe your reasons for your rating.

Star	Rating	Reason
A very distant Sun-like star		
A very near Sun-like star		
A very near red-giant star		
An intermediate-distance star about to become a supernova		
An intermediate-distance very-high-mass main sequence star		
A very near very-low-mass main sequence star		
A nearby variable star that changes its brightness by a factor of ten over a one-year period		
A binary star system consisting of two stars 1 astronomical unit apart orbiting about each other: each of the stars is one solar mass		
A binary star system consisting of two stars 1 astronomical unit apart orbiting about each other: each of the stars is $1/10$th of a solar mass		

(14) (A) Is there anything in the "mathematics lesson" we presented in our discussion of CETI to indicate that the creatures that sent it have ten fingers? (B) Is there anything in it to indicate that we have two arms?

(15) How would you design an interstellar message indicating that (A) the Earth has only one moon? That (B) there are 365 days in a year? (C) How would you guard against misunderstandings, in which the aliens might conclude that the Earth has 365 moons and rotates only once per year?

WHAT DO YOU THINK?

(1) We discussed in this chapter what little we know concerning the origin of life. But we did not discuss the following theory.

> Life arose on Earth because God made it happen.

We have often emphasized that, in order for a theory to be a *scientific* theory, it must be capable of being tested. In the light of this –

- is the above theory a scientific theory;
- if it isn't, what is it?

(2) Re-read our discussion of the Viking mission and its search for life on Mars. There are two possible interpretations of its results:

- we have evidence that there is no life on Mars,
- we don't know whether there is life on Mars or not.

Which is your interpretation of Viking's results – and why?

You must decide

(1) You are a program officer at NASA charged with supporting the search for life in the Universe. You are contemplating funding a mission to Mars to continue the search for life there. Haunting your thoughts, however, is the memory of the Viking missions. You are acutely aware that these missions returned ambiguous results. Their biology experiments found results that might have been caused by living organisms – but they might also have been caused by processes not involving life. Thus, these experiments failed unambiguously to demonstrate the existence of life on Mars. Critics have argued that they didn't prove anything at all!

Furthermore, the Viking missions were tremendously expensive. You need to decide whether to proceed with an equally expensive follow-up mission, also designed to search for signs of life on Mars. You are acutely aware that, no matter what you decide, you will be severely criticized by *someone*.

Your task is to write a memorandum analyzing whether to proceed with such a mission. In your memorandum you should give a detailed account of

- the arguments for proceeding with the mission,
- the arguments for not proceeding,
- the decision that you reached,

and (this is the most important of all!)

- your reasons for your decision.

(2) You are an astronomer working at a radio telescope. You are studying a run-of-the-mill star, one of hundreds of billions with nothing in particular to attract anyone's attention to it. Remarkably, however, you realize that an interstellar message is coming from it. As you decipher the message, you realize that the transmitting civilization is 100 million years more advanced than ours. Indeed, the civilization is so advanced that it has solved every problem facing our own. It has

- eradicated all diseases,
- learned how to stave off old age indefinitely, so that each individual lives forever,
- solved the problem of global warming,
- eradicated war,
- eradicated poverty,
- solved every one of the scientific problems we have discussed in this book,

and so forth. Indeed, these problems have been so completely solved that these beings do not even *remember* that they ever were problems!

That night you cannot sleep. You have been suddenly seized with worry. Perhaps, you think, this is not a great discovery but a devastating blow to humanity. Perhaps people will feel that all their aspirations, all their efforts, add up to nothing. Perhaps they will feel that there is nothing more to live for and to strive for. Perhaps, you think, you should keep your discovery secret.

What do you decide to do: do you announce your discovery or do you keep it a secret? What is the justification for your decision?

In the Preface, I gave my "short list" of the truly essential facts about astronomy and the nature of science. Now that you have spent many months studying astronomy, this list, which probably did not mean very much to you back then, will have assumed much more meaning. Here it is again.

Three BIG FACTS about the Universe

The Universe is very big

It is probably impossible to appreciate the immensity of the astronomical Universe. If we represent the entire Earth by a dot a mere ¹⁄₂₅th of an inch (1 millimeter) across, the Sun would by 40 feet (12.2 meters) away, and the nearest star a full 1840 miles (2961 kilometers) distant. Our Milky Way Galaxy would be an astonishing 46 million miles (about 74 million kilometers) in diameter. Beyond this lies the void of intergalactic space and untold billions of other galaxies. We have never found an end to these oceanic immensities. Indeed, the Universe might be infinite in extent.

The Universe is very old

It is also probably impossible to appreciate the immensity of the age of the cosmos. Our Earth is more than four billion years old: that is thousands of times longer than the span of time our human race has been in existence. If we shrink the lifetime of a person to a single minute, the Big Bang occurred nearly four centuries ago.

We are not the center of the Universe

Nothing about the Earth is unique. Our home planet lies in the outskirts of our Galaxy. We revolve about the Sun, which orbits about the Galaxy, which itself moves through space. Immense numbers of other planets revolve around their home stars.

You must decide

What is your personal reaction to the view of humanity's place in the cosmos provided by this list?

Three BIG FACTS about the nature of science

The Universe is knowable

It is actually possible to find out something about the cosmos.

We do this by making observations and formulating theories to explain them

These observations require ever-more sensitive telescopes and ever-more sophisticated techniques. The theories often involve concepts unfamiliar to us in daily life.

These theories are tested

Once we have formulated a theory, we do not simply believe in it. Rather, we test it, and the tests are repeated over and over again. The more tests the better: the more different kinds of tests the better. Only those theories that withstand this process are accepted. There is a great deal of evidence in their favor. Nevertheless, we are always learning new things.

You must decide

Has this second list led you to change your personal reaction to the view of humanity's place in the cosmos provided by astronomy? If so, why? If not, why not?

In conclusion

As you began your study of astronomy you took a few minutes to write yourself a letter in which you discussed (1) why you had decided to study astronomy and (2) what you hoped to get out of this study.

You kept your letter in a safe place. Now, at the conclusion of your study, please take it out and read it. Have you gotten what you wanted? Have you changed your ideas about what you want from a study of astronomy?

The small-angle formula

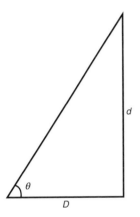

Figure AI.1 **A right triangle.**

Figure AI.1 shows a right triangle of sides D and d and angle θ. We want to find a relationship between these three quantities.

If you are familiar with trigonometry you will recall that this relationship is

$$\tan(\theta) = d/D.$$

But if you are not familiar with trigonometry – or even if you are! – you will be happy to learn that there is a simpler formula that works if the angle θ is small. In astronomy it turns out that the angles we work with are normally very small indeed, so we will be able to use this simpler formula just about all of the time.

The formula is known as the small-angle formula.

The small-angle formula

$$\theta = \frac{d}{2\pi D} \, 360 \text{ degrees}.$$

Let us try to understand where this formula comes from. Notice that it contains a term $2\pi D$, and it contains 360 degrees. Both of these make us think of a circle – specifically, a circle of radius D, since $2\pi D$ is the circumference of a circle of radius D. Let us use this as a clue and rewrite the small-angle formula to read

$$\frac{\theta}{360 \text{ degrees}} = \frac{d}{2\pi D}.$$

Does this make sense? It would if d were the arc length subtended by the angle θ, since then the formula would say

$$\frac{\text{angle subtended by an arc length}}{360 \text{ degrees}} = \frac{\text{arc length}}{\text{circumference of the circle}}.$$

And this is certainly true.

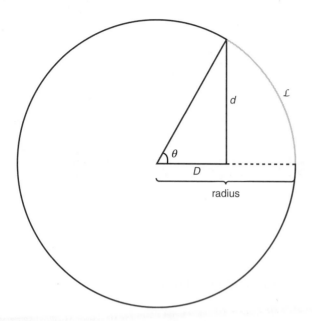

Figure AI.2 Our triangle imbedded in a circle. Notice that the angle θ is large.

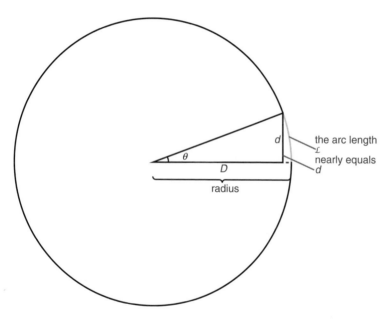

Figure AI.3 Our triangle imbedded in a circle. Notice that the angle θ is small.

This leads us to try imbedding our triangle in a circle (Figure AI.2).

When we do so, we realize that there are two problems.

- The arc length subtended by θ is not d: it is $\mathcal{L}$.
- The radius of the circle is not D: it is greater than D, the excess indicated by the dotted line.

So we don't seem to be getting anywhere.

But notice that the angle θ in Figure AI.2 is large. The small-angle formula, however, is supposed to work only when θ is small. In Figure AI.3 we work with a small value of θ. And as you can see, both of these problems are resolved in Figure AI.3.

- The arc length $\mathcal{L}$ is nearly d.
- The radius of the circle is nearly D.

Furthermore, the smaller the angle θ becomes the closer D comes to the radius, and the closer $\mathcal{L}$ comes to d. So we see that the small-angle formula makes sense: it is nothing more than a simple statement of proportionality.

Exponential notation

In astronomy we often deal with very large or very small numbers.

Distance to the Sun = 149 600 000 000 kilometers.
Mass of a proton = 0.000 000 000 000 000 000 000 000 001 67 kilograms.

All those zeros make these numbers hard to work with. Exponential notation makes it easier.

Writing numbers

Powers of ten

Recall that a number squared is that number multiplied by itself. Let's apply this to the number 10:

$$10^2 = 10 \times 10 = 100.$$

Notice that the result (100) has two zeros in it – and "2" is the exponent on the left side of this equation

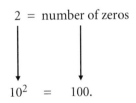

$$2 = \text{number of zeros}$$

$$10^2 = 100.$$

It also works for any other power:

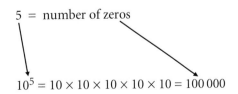

$$5 = \text{number of zeros}$$

$$10^5 = 10 \times 10 \times 10 \times 10 \times 10 = 100\,000$$

So we have a general rule: 10^n is a "1" followed by n zeros. (10 is simply 10^1: 1 is 10^0.)

We can also define what we mean by a negative exponent, 10^{-n}:

$$10^{-n} \text{ is defined to be } \frac{1}{10^n}.$$

Let's work this out for $n = 2$:

$$10^{-2} \text{ is defined to be } \frac{1}{10^2} = \frac{1}{100} = 0.01.$$

Notice that the exponent is minus 2, and that there is $2 - 1 = 1$ zero to the right of the decimal point here

exponent = minus 2   number of zeros to the right of the decimal point = $2 - 1 = 1$

$10^{-2} = 0.01.$

This, too, applies to any other power:

exponent = minus 5   number of zeros to the right of the decimal point = $5 - 1 = 4$

$10^{-5} = 1 / (10 \times 10 \times 10 \times 10 \times 10) = 1 / 100\,000 = 0.000\,01$

So we have a general rule: 10^{-n} is a decimal point followed by $n - 1$ zeros and then a "1". (0.1 is 10^{-1}: 1 is 10^{-0}, which is the same thing as 10^{+0}.)

Writing numbers

Let's go back to the distance to the Sun: $149\,600\,000\,000$ kilometers. We can write this as

$$149\,600\,000\,000 \text{ kilograms} = 1.496 \times 100\,000\,000\,000 \text{ kilograms}$$
$$= 1.496 \times 10^{11} \text{ kilograms,}$$

which is a much more compact way to write it – we've gotten rid of all those zeros! Similarly, the mass of the proton is

$$0.000\,000\,000\,000\,000\,000\,000\,001\,67 \text{ kilograms}$$
$$= 1.67 \times 0.000\,000\,000\,000\,000\,000\,000\,000\,001 \text{ kilograms}$$
$$= 1.67 \times 10^{-27} \text{ kilograms,}$$

which is also much better – again, we've gotten rid of the zeros.

Exponential arithmetic

Multiplying

Consider:

$$100 \times 100 = 10\,000.$$

Count the zeros: there are four on the left (two for each "100") and four on the right. Now let's write this in exponential notation. We get

$$10^2 \times 10^2 = 10^4$$

and notice that two and two is four:

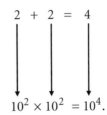

$$2 + 2 = 4$$
$$10^2 \times 10^2 = 10^4.$$

This is an example of a general rule: *to multiply two numbers written in exponential notation, add the exponents.*

Let's try this again, just to make sure. Let us find $10 \times 100 \times 10\,000$ first in the normal way:

$$10 \times 100 \times 10\,000 = 10\,000\,000$$

and then using our rule:

$$1 + 2 + 4 = 7$$
$$10^1 + 10^2 + 10^4 = 10^7.$$

It works!

Now we know how to multiply more complicated numbers. Let's try $16\,700\,000 \times 23\,500\,000\,000$.

First write these in exponential notation. We get

$$(1.67 \times 10^7) \times (2.35 \times 10^{10}).$$

Now group these as follows

$$(1.67 \times 2.35)(10^7 \times 10^{10}).$$

We multiply the first group of numbers in the ordinary way

$$1.67 \times 2.35 = 3.9245$$

and we multiply the second group using our rule

$$10^7 \times 10^{10} = 10^{7+10\ =\ 17}$$

and we combine the two groups to get our result

$$16\,700\,000 \times 23\,500\,000\,000 = 3.9245 \times 10^{17}.$$

Dividing

Consider:

$100/1000 = 1/10 = 0.1$

writing this in exponential notation

$10^2/10^3 = 10^{-1}$

In dividing numbers written in exponential notation, subtract the exponents:

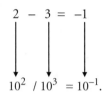

$$2 - 3 = -1$$

$$10^2 / 10^3 = 10^{-1}.$$

Let's use this to divide two numbers, say

$16\,700\,000/23\,500\,000\,000 = (1.67 \times 10^7)/(2.35 \times 10^{10}).$

Using the same strategy as we used for multiplication:

$= (1.67/2.35)(10^7/10^{10})$

$= (1.67/2.35 = 0.0711)/(10^{\,7-10\,=-3}).$

As a final refinement let's re-write the first group:

$0.0711 = 7.11 \times 10^{-2}$

and combine it with the second group:

$(0.0711)(10^{-3}) = (7.11 \times 10^{-2})(10^{-3}) = 7.11 \times 10^{-2-3\,=\,-5}$

we get our final answer

$16\,700\,000/23\,500\,000\,000 = 7.11 \times 10^{-5}.$

The Solar System

Table AIII.1. **Basic data for the planetary system.**

Name	Mean distance from Sun (km)	Mean distance from Sun (AU)	Mass, M_P (kg)	Mass, M_P ($M_E = 1.0$)	Orbital period (days or years)	Orbital eccentricity	Orbital inclination (deg)	Equatorial diameter (km)	Equatorial rotation period	Number of satellites (Dec. 2010 data)
Mercury	57 900 000	0.387	3.3010×10^{23}	0.055 8	87.97 d	0.21	7.0	4878	58.6 d	0
Venus	108 200 000	0.723	4.8673×10^{24}	0.815	224.7 d	0.006 8	3.4	12 104	243.2 d	0
Earth	149 598 000	1.000	5.9722×10^{24}	1.000	365.25 d	0.017	0.0	12 756	23h 56m 4s	1
Mars	227 940 000	1.524	6.4169×10^{23}	0.107 4	687.0 d	0.093	1.8	6794	24h 37m 23s	2
Jupiter	778 340 000	5.203	1.8981×10^{27}	317.9	11.86 y	0.048	1.3	143 884	9h 50m 30s	63
Saturn	1 427 000 000	9.539	5.6832×10^{26}	95.15	29.5 y	0.056	2.5	120 536	10h 14m	61
Uranus	2 869 600 000	19.18	8.6881×10^{25}	14.54	84.0 y	0.047	0.77	51 118	17h 14m	27
Neptune	4 496 700 000	30.06	1.0241×10^{26}	17.23	164.8 y	0.008 6	1.77	50 538	16h 6m	13
(Pluto)	5 913 520 000	39.44	1.309×10^{22}	0.002 2	247.92 y	0.25	17.1	2390	6.387 d	3

Sources
Moore, P. & Rees, R. (2011). *Patrick Moore's Data Book of Astronomy*. Cambridge: Cambridge University Press.
Lang, K. R. (2011). *The Cambridge Guide to the Solar System*, 2nd edn. Cambridge: Cambridge University Press.
de Pater, I. & Lissauer, J. J. (2010). *Planetary Sciences*, 2nd edn. Cambridge: Cambridge University Press.
Pluto, B. W. (2010). *Pluto: Sentinel of the Outer Solar System*. Cambridge: Cambridge University Press.

Table AIII.2. **Basic data for the satellites of Mars.**

Name	Mean distance from host planet (km)	Mass (10^{20} kg)	Orbital period (d)	Orbital eccentricity	Orbital inclination (deg)	Radius (km)	Equatorial rotation period*
Phobos	9 376	1.08×10^{-4}	0.318 910	0.015 1	1.075	$13.1 \times 11.1 \times 9.3$	S
Deimos	23 458	1.80×10^{-5}	1.262 441	0.000 2	1.788	$7.8 \times 6.0 \times 5.1$	S

* "S" indicates synchronous rotation: the rotation period equals the orbital period.

Table AIII.3. **Basic data for the satellites of Jupiter.**

Name	Mean distance from host planet (km)	Mass (10^{20} kg)	Orbital period (d)	Orbital eccentricity	Orbital inclination (deg)	Radius (km)	Equatorial rotation period*
Io	421 800	893.3	1.769 138	0.0041	0.036	1 821.3	S
Europa	671 100	479.7	3.551 810	0.0094	0.466	1 565	S
Ganymede	1 070 400	1 482	7.154 553	0.0013	0.177	2 634	S
Callisto	1 882 700	1 076	16.689 018	0.0074	0.192	2 403	S

* "S", synchronous rotation.

Table AIII.4. **Basic data for the satellites of Saturn.**

Name	Mean distance from host planet (km)	Mass (10^{20} kg)	Orbital period (d)	Orbital eccentricity	Orbital inclination (deg)	Radius (km)	Equatorial rotation period*
Mimas	185 539	0.38	0.942 4218	0.019 6	1.574	109 × 196 × 191	S
Enceladus	238 042	0.65	1.370 218	0.000 0	0.003	249.1	S
Tethys	294 672	6.27	1.887 802	0.000 1	1.091	533	S
Dione	377 415	11.0	2.736 915	0.002 2	0.028	561.7	S
Rhea	527 068	23.1	4.517 500	0.000 2	0.333	764	S
Titan	1 221 865	1 345.7	15.945 421	0.028 8	0.306	2 575	~S
Hyperion	1 500 933	0.054	21.276 609	0.023 2	0.615	180 × 133 × 103	C
Iapetus	3 560 854	18.1	79.330 183	0.029 3	8.298	736	S

* "S", synchronous rotation, "C", chaotic rotation.

Table AIII.5. **Basic data for the satellites of Uranus.**

Name	Mean distance from host planet (km)	Mass (10^{20} kg)	Orbital period (d)	Orbital eccentricity	Orbital inclination (deg)	Radius (km)	Equatorial rotation period*
Ariel	190 900	13.53	2.520	0.001 2	0.041	581.1(0.9) × 577.9(0.6) × 577.7(1.0)	S
Umbriel	266 000	11.72	4.144	0.003 9	0.128	584.7	S
Titania	436 300	35.27	8.706	0.001 1	0.079	788.9	S
Oberon	583 500	30.14	13.463	0.001 4	0.068	761.4	S
Miranda	129 900	0.659	1.413	0.001 3	4.338	240(0.6) × 234.2(0.9) × 232.9(1.2)	S

* "S", synchronous rotation.

Table AIII.6. **Basic data for the satellite of Neptune.**

Name	Mean distance from host planet (km)	Mass (10^{20} kg)	Orbital period (d)	Orbital eccentricity	Orbital inclination (deg)	Radius (km)	Equatorial rotation period*
Triton	354 759	214.7	5.876 854	0.0000	156.865	1 352.6	S

* "S", synchronous rotation.

Table AIII.7. **Basic data for the satellites of Pluto.**

Name	Mean distance from host planet (km)	Mass (10^{20} kg)	Orbital period (d)	Orbital eccentricity	Orbital inclination (deg)	Radius (km)	Equatorial rotation period
Charon	17 536	—	—	0.002 2	0.001	—	—
Nix	48 708	—	—	0.003 0	0.195	—	—
Hydra	64 749	—	—	0.005 1	0.212	—	—

Sources

Jet Propulsion Laboratory, California Institute of Technology http://ssd.jpl.nasa.gov/?phys_data

de Pater, I. & Lissauer, J. J. (2010). *Planetary Sciences*, 2nd edn. Cambridge: Cambridge University Press.

APPENDIX IV

The closest and brightest stars

Table AIV.1. **The 21 closest and brightest stars.**

The closest stars		
Name	Distance (parsecs)	Luminosity (solar units)
Our Sun	0	1
α Centauri A[a]	1.3	2.5
α Centauri B		0.4
α Centauri C		0.000 1
Barnard's star	1.8	0.000 5
Wolf 359	2.3	0.000 02
BD+36 21 47	2.5	0.01
L726–8 A[b]	2.7	0.000 1
L726–8 B		0.000 04
Siruis A[b]	2.9	23
Sirius B		0.004
Ross 154	2.9	0.000 6
Ross 248	3.1	0.000 1
L 789–6	3.2	0.000 1
e Eridani	3.3	0.4
Ross 128	3.3	0.000 3
61 Cygni A[b]	3.3	0.063
61 Cygni B		0.063
e Indi	3.4	0.16
Procyon A[b]	3.5	6.3
Procyon B		0.000 6

Source: Impey, C. & Hartmann, W.K. (2000). *The Universe Revealed,* Appendix B-2, Instructor's edition. Brooks-Cole.

[a] This is a triple star system.

[b] These are double star systems.

Table AIV.2. **The stars of greatest apparent brightness.**

Name	Distance (parsecs)	Luminosity (solar units)
Our Sun	0	1
α Centauri A	1.3	2.5
Sirius A	2.6	23
Procyon A	3.5	6.3
Altair	5.2	11
Fomalhaut	7.7	17
Vega	7.8	50
Pollux	10	30
Arcturus	11	110
Capella	13	130
Aldebaran	20	150
Achernar	44	1 100
Spica	80	2 200
Acrux	92	4 000
Canopus	95	14 000
Mimosa	107	3 000
Betelgeuse	132	10 000
Hadar	153	12 000
Antares	184	11 000
Rigel	236	40 000
Deneb	920	250 000

Physical and astronomical constants

Table AV.1. **Astronomical constants.**

Symbol	Value in SI units	Quantity
AU	1.496×10^{11} m	Astronomical unit of distance
ly	9.4605×10^{15} m	Light year
pc	3.086×10^{16} m	Parsec
$M_\odot$	1.989×10^{30} kg	Solar mass
$R_\odot$	6.96×10^{8} m	Solar radius
$L_\odot$	3.827×10^{26} J s^{-1}	Solar luminosity
$\mathcal{F}_\odot$	1.37×10^{3} J m^{-2} s^{-1}	Solar constant
$M_\oplus$	5.976×10^{24} kg	Earth's mass
$R_\oplus$	6.378×10^{6} m	Earth's equatorial radius

Table AV.2. **Physical constants.**

Symbol	Value in SI units	Quantity
c	2.997925×10^{8} m s^{-1}	Speed of light
G	6.674×10^{-11} m^3 kg^{-1} s^{-2}	Gravitational constant
h	6.626069×10^{-34} J s	Planck's constant
k	1.380650×10^{-23} J deg^{-1}	Boltzmann's constant
m_e	9.109382×10^{-31} kg	Mass of an electron
m_p	1.672622×10^{-27} kg	Mass of a proton
m_n	$1.67492716(13) \times 10^{-27}$ kg	Mass of a neutron
q	6.626176×10^{-19} C	Electron charge

Table AV.3. **Mathematical constants.**

Symbol	Value	Quantity
π	3.141592653589	Pi
rad	57.2957795 degrees	Number of degrees in one radian

Table AV.4. **Mathematical formulas.**

Circumference of a circle	$c = 2\pi r, c = \pi d$
Area of a circle	$A = \pi r^2$
Surface area of a sphere	$A = 4\pi r^2$
Volume of a sphere	$V = 4/3\, \pi r^3$

Source
de Pater, I. & Lissauer, J. J. (2010). *Planetary Sciences*, 2nd edn.
Cambridge: Cambridge University Press.

Conversion factors

Table AVI.1. **Temperature conversions.**

From degrees Celsius[a]	$T_K = T_C + 273.15$	T_K temperature in kelvin
		T_C temperature in degrees Celsius
From degrees Fahrenheit	$T_K = \frac{T_F - 32}{1.8} + 273.15$	T_F temperature in degrees Fahrenheit

[a] The term "centigrade" is not used in SI, to avoid confusion with "10^{-2} of a degree."

Table AVI.2. **Other conversions.**

1 meter = 39.37 inches
12 inches = 1 foot
5280 feet = 1 mile
1 kilometer = 1000 meters
1 kilogram = 2.205 pounds
1 newton = 0.2248 pounds
1 kilometer/second = 2.237 miles/hour
1 watt = 1 joule/second
1 inch = 2.54 centimeters = 0.083 feet
1 foot = 0.3048 meter = 0.33 yards
1 yard = 0.9144 meters
1 mile = 5280 feet = 1.6093 kilometers

Constellation maps

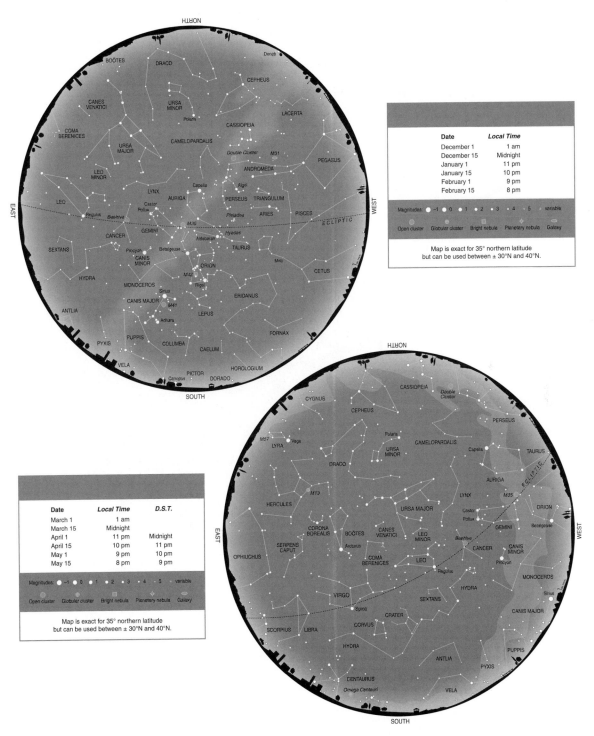

Figure AVII.1

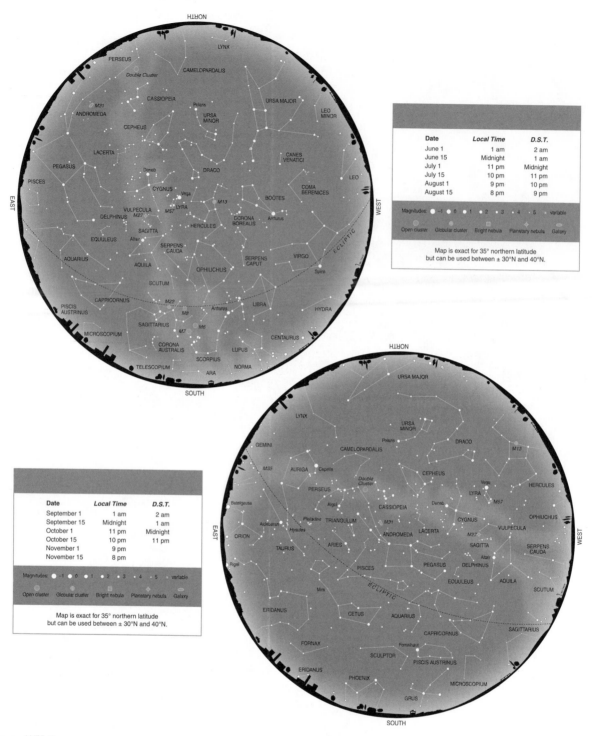

Figure AVII.2

GLOSSARY

absolute zero	The lowest possible temperature, where almost all molecular motions stops. This marks the start of the Kelvin temperature scale and is equivalent to $-273.5\ ^\circ$C on the Celsius scale and $-459.67\ ^\circ$F on the Fahrenheit scale.
absorption lines	These appear as dark lines across a continuous spectrum. They are produced when photons of light are absorbed at specific wavelengths (or frequencies) that correspond to changes in energy levels of atoms or molecules. Observing absorption lines can help us learn about the chemical compositions of the object and the medium that the light passes through.
accelerating Universe	The Universe is expanding and so the distance between faraway objects is increasing. Observational evidence for this was found by Edwin Hubble in 1929. In the late twentieth century, it was found that the rate of this expansion is itself increasing. Scientists have suggested dark energy as the cause but there are ongoing debates as to its nature.
accretion disk	An accretion disk is formed when material such as gases and dust orbit around a more-massive central object, spiraling inwards due to gravitational forces. They are found in many astronomical situations, such as when new solar systems form around young stars or matter orbits a black hole.
active optics	This is a computer process that uses feedback to keep a telescope mirror in the optimum shape. Before active optics, large mirrors (above 5 or 6 meters in diameter) would deform or sag under their own weight.
Adams, John Couch (1819–1892)	British astronomer whose most famous contribution was predicting the existence of Neptune based solely on discrepancies in the orbit of Uranus. At about the same time, French astronomer Urbain Le Verrier made similar calculations, leading to his observation of Neptune in 1846.
adaptive optics	A system in some modern ground-based telescopes where a computer-controlled sensor array connected to a deformable mirror is used to compensate in real time for atmospheric disturbances or the quality of "seeing."
albedo	A measurement of the reflecting qualities of a surface, such as a planet. It is defined as the amount of reflected light divided by the amount of incoming (incident) light to hit the surface. If 100% of the light is reflected, the object has an albedo of 1.0. Ocean ice has an albedo of around 0.6, whereas the Moon's albedo is measured at 0.12, about the same as asphalt.
Allen Telescope Array (ATA)	A radio interferometer under construction in northern California. When completed it should consist of about 350 separate 6-meter dishes, working together to act as a powerful single 900-meter dish. A large part of the initial funding came from Microsoft co-founder Paul Allen.
Almagest	An early treatise on astronomy and cosmology by Ptolemy in the second century AD. It is important as a comprehensive source on ancient mathematical astronomy. It puts forward a

geocentric model of the Solar System that was to last unchallenged for 1200 years and it contains a star catalog listing over 1000 bright stars.

Alpha Centauri	The brightest star in the constellation Centaurus and noted for being our closest neighbor. Alpha Centauri and Proxima Centauri are the nearest stars to the Sun at 4.37 and 4.24 light years away, respectively. Alpha Centauri is now known to be a double star system but the companion star cannot be determined without a telescope.
Andromeda Galaxy	A spiral galaxy that forms the largest member of the Local Group of galaxies, to which the Milky Way belongs. Given the Messier number M31, it is visible to the naked eye (under dark skies) as a fuzzy blob in the constellation Andromeda.
angular diameter	The apparent size of an object, measured as an angle between the viewing point and two opposite edges of the object. For example, the angular diameter of the Moon from Earth's surface is about half a degree.
aphelion	The point on Earth's (or another object's) elliptical orbit where it is at its farthest point from the Sun. The opposite is perihelion.
Apollo program	The NASA spaceflight program in the 1960s and early 1970s. Its goal was to send humans to the Moon (and bring them back safely). This was accomplished on July 20, 1969, by two of the crew of Apollo 11. There were a further five successful Moon landings. To date it is the only spaceflight program to have taken humans beyond a low-Earth orbit.
archeoastronomy	A branch of astronomy concerned with how different civilizations and cultures understood or used night-sky phenomena. Many prehistoric and ancient sites might have had some astronomical significance, for example, in marking the passage of time and the changing of the seasons.
Arecibo Observatory	A large radio telescope in Puerto Rico, boasting the largest single-aperture dish at 305 meters (1000 feet) diameter.
Aristarchus of Samos (310 BC–c.230 BC)	A Greek astronomer and mathematician who put forward the first recorded heliocentric model of the Solar System. He also came up with a geometric argument for determining the relative distances between the Earth, Moon and Sun but lacked the tools needed to make the measurements with sufficient accuracy.
Aristotle (384 BC–322 BC)	Greek philosopher and student of Plato who laid the foundations for much of Western philosophy and learning.
asteroid belt	The location in the Solar System at about 2.0 and 3.3 AU between Mars' and Jupiter's orbits, where the main population of asteroids are found.
asteroids	Small bodies (tens of meters to kilometers across) in the Solar System orbiting the Sun, which are thought to be the smashed-up remnants of planetesimals. There are millions of asteroids in the Solar System. Asteroids, unlike comets, show no active features as comets do and are generally believed to be made of rock, ices and metals.
astrology	The belief that celestial objects and their motions can affect or influence human personalities and events. In ancient cultures, astrological study was linked to observational astronomy as a means of keeping a calendar. Astrology and astronomy parted ways at the end of the seventeenth century. Today it is not taken seriously as a science.
astronomical unit (AU)	A measure of distance equaling the mean Sun–Earth distance, around 1.49×10^{11} m. Today it is defined technically by other more precise means. It is a convenient unit for measuring orbits within the Solar System.

atmosphere	The outer layer of gases or vapors surrounding a body of sufficient size, held in place by gravity. In stars, the stellar atmosphere is the outermost layer of a star beyond its photosphere.
aurora	A natural phenomenon of shimmering green- or red-colored lights seen in the atmosphere at high latitudes (near the Earth's poles). They are caused by charged particles from the solar wind entering the Earth's magnetic fields and emitting photons when they interact with molecules in the upper atmosphere. They are most common when the Sun is active.
barred spiral galaxy	A galaxy type defined by its shape, in which the spiral arms are connected by a "bar" of stars running through the galaxy's center. Our Galaxy, the Milky Way, is believed to be of this type.
Big Bang	The most commonly accepted cosmological model for the early Universe, in which it began as an incredibly hot and dense state and expanded rapidly outwards and continues to expand today. Evidence for the Big Bang comes from the expansion of the Universe, the large-scale distribution of galaxies, detailed measurements of the Cosmic Background Radiation and the abundance of chemical elements.
binary stars	A star system made up of two stars orbiting around their common center of mass. Many, or even most, stars are believed to be binaries or multiple-star systems.
bipolar outflow	In an accretion disk, as material is pulled inward toward the more-massive object by gravity, energy and excess angular momentum are released as jets of hot gaseous material flowing from the poles, i.e. the disk's rotational axis. These may be observed as dumbbell-shaped planetary nebulae.
blackbody (or black body) radiation	This describes the characteristic bell-shaped energy curve that a theoretical object emits when it absorbs all the radiation to strike it and re-emits it at all wavelengths. Plotting these wavelength and intensity curves can tell us a characteristic temperature for the object. Many objects, such as stars, behave like blackbodies.
black hole	An object in space with a gravitational field so strong that even light cannot escape. They are predicted by Einstein's theory of general relativity. Although we cannot directly observe black holes there is abundant indirect evidence that they exist; for example, through observing the gravitational interactions of stars in the nearby vicinity. Some massive stars will end their lives as black holes following a supernova explosion. Black holes can grow in "size" through absorbing more material. Many galaxies, including our own, are thought to have massive black holes at their centers.
blueshift	According to the Doppler effect, something is described as blueshifted if its relative motion is toward the observer. Light and other forms of radiation appear to have a higher frequency (or shorter wavelength) than they do in reality. The opposite of this is redshift.
Brahe, Tycho (1546–1601)	A Danish nobleman and observational astronomer who made the most accurate (pre-telescope) astronomical measurements of his time from his observatory Uraniborg. He also made important observations of the supernova of 1572, which appeared as a bright new star in the constellation Cassiopeia. Johannes Kepler used Brahe's measurements to derive his laws of planetary motion.
butterfly diagram	When sunspot locations on the Sun are plotted against time over a period of several decades or more, the resulting graph shows a regular symmetrical pattern resembling butterflies, which repeats every 11 years. It shows that sunspots appear at the same latitude above the equator and is evidence for an 11-year cycle of solar activity.
Cassegrain telescope	An optical set-up in a reflecting telescope in which light is collected in a concave primary mirror and focused in a secondary convex mirror located within the telescope tube. Many

amateur astronomy telescopes favor this design as it is relatively compact for its power compared to other types of telescope.

Cassini–Huygens mission	An international space mission launched in 1997 to study Saturn and its moons. The Huygens probe landed on Saturn's largest moon Titan in 2005, making it the first man-made object to land on any planet's moon beyond our own. The Cassini orbiter continues to study Saturn, its rings and moons and its magnetic field.
CCD	The common acronym for a charge-coupled device, a type of electronic chip that converts light to electrical signals. CCD sensors are found in many astronomical imaging devices, effectively replacing photographic film. More commonly, they are used in digital cameras, video recorders and cell-phone cameras.
celestial equator	A circular reference line on the celestial sphere formed by projecting the plane of the Earth's equator outward in all directions. The celestial equator has a declination of zero.
celestial poles	The points on the celestial sphere where the Earth's axis of rotation meets it, i.e. directly above Earth's north and south poles.
celestial sphere	An imaginary sphere with Earth at its center which projects Earth's latitude and longitude coordinate system onto the sky. Objects in the sky are given the coordinates of declination, Dec (analogous to Earth's latitude, with negative declinations in the southern celestial hemisphere) and right ascension, RA (similar to longitude), but measured in angular units of hours, minutes and seconds, where 24 hours equals a full circle of 360°.
Cepheid variable star	A particular class of variable star with a known relationship between their luminosity and their period of pulsation. By measuring the period, you can find out the star's true luminosity and, by comparing this with the apparent luminosity, calculate the distance to the star. Cepheid variables played an important role in working out the extragalactic distance scale and in deriving Hubble's law.
Chandrasekhar limit	The maximum mass for a stable white-dwarf star, approximately 1.4 solar masses. Above the Chandrasekhar limit, the pressure forces that resist the star's own gravity are not sufficient to keep it from collapsing into a neutron star or a black hole. It is named after the Indian-born astrophysicist and Nobel laureate Subrahmanyan Chandrasekhar (1910–1995), who first predicted it in 1930, before he had even undertaken any graduate studies.
circumpolar star	A star that does not set (i.e. pass below the observer's horizon) from a given latitude. A star will be circumpolar if its declination plus the observer's latitude is greater than 90° for observers in the northern hemisphere, or if the declination plus the observer's latitude is less than −90° for the southern hemisphere. If you are standing at the north or south poles, all stars will be circumpolar; at the equator, none will be circumpolar.
Cocoon Nebula	An emission nebula in the constellation Cygnus with the official designation IC 5146. Clouds of ionized gas are heated by a nearby star or stars, causing reactions that can emit light. The color of the light depends on the chemical composition of the gas.
comet	A small icy body orbiting the Sun, which when nearing the Sun shows a blurry tail of dust and gas streaming away from it in the opposite direction to the Sun. Comets are made up of a mixture of rock, dust and a variety of ices. Some comets are on periodic orbits, such as Halley's comet, which is visible to us every 75 or 76 years. Its next perihelion will be in 2061.
constellation	Historically this referred to any grouping of prominent stars that formed a pattern, with different cultures giving them their own names and stories behind them. Today, the celestial sphere is divided into 88 official constellations, whose boundaries are determined by the International Astronomical Union.

convection	The movement of heat in fluids (liquids and gases) by the movement of the fluid itself. Convection can be seen in astronomy, for example, in the solar "granules" visible on the Sun's surface, which are giant convection cells made up of hot plasma.
Copernicus, Nicolaus (1473–1543)	A Renaissance scholar whose book, translated as *On the Revolutions of the Celestial Spheres*, laid the framework for a Sun-centered Solar System that replaced the old Ptolemaic model. It played a key role in the scientific revolution. Copernicus himself died at around the time the book was published.
corona	A tenuous plasma atmosphere streaming away from the Sun. Heated to over a million kelvin, the corona extends millions of kilometers into space. It is best seen during a total solar eclipse, when the Sun's body is blocked by the Moon, or with a coronagraph, a telescope adapted with a mask to block out direct sunlight. Distinct coronal features, such as loops, prominences and flares have been observed in association with sunspot activity, indicating that they are related to the Sun's magnetic field.
Cosmic Background Radiation	The residual background temperature of the Universe, filling space in every direction at a chilly 2.725 K. It is thought to be left-over electromagnetic radiation and so forms a key part of the evidence for the Big Bang cosmological model. It is sometimes called the Cosmic Microwave Background, as it is most concentrated in the microwave part of the spectrum.
cosmology	The study of the origin, structure and evolution of the Universe. Modern cosmology spans from the Big Bang 13.7 billion years ago, through the present day and on to the eventual fate of the Universe.
crater	The round or oval concave scar left by the impact of a small Solar System body on a larger one, such as is left when an asteroid or comet hits a terrestrial planet. Many planets and moons are covered in craters, bearing testimony to the violence of the early Solar System. The Earth also has many impact craters but they are less obvious due to weathering processes and, over longer periods of time, they disappear due to tectonic processes in which the surface rock is recycled.
critical density	In physical cosmology, the critical density refers to the amount of mass in the Universe needed in order for it to stop expanding after a finite amount of time. If there is less mass than this, the Universe will keep on expanding forever; if there is more mass, the Universe will one day begin to collapse in on itself.
dark energy	The name given to a constituent part of the Universe's composition, along with ordinary matter and dark matter. Evidence from the WMAP satellite suggests the Universe is 73% dark energy, 23% dark matter and 4% ordinary matter. Several theories for the nature of dark matter are being investigated: it could be a property of empty space or a wholly new type of matter. Or it could be that our current theories of how gravity works are wrong for larger scales.
dark matter	A type of matter that cannot be directly observed like stars and planets, nor does it interact with light in the usual way. Nevertheless it exerts a gravitational influence on matter at the galactic scale. Among other theories, two are that dark matter might be ordinary matter hidden in brown dwarfs or "massive compact halo objects" around galaxies (nicknamed MACHOs), or else they could be a new form of weakly interacting massive particle (termed a WIMP).
Doppler effect	The change in frequency observed when an observer is moving relative to a wave source, i.e. toward it or away from it. The frequency is increased when an object is moving toward the observer and decreased when moving away from it. In astronomical terms, visible light is blueshifted, i.e. moves toward the blue end of the spectrum when an object is moving toward and redshifted when an object is moving away. The same principle holds true for

wavelengths beyond the visible spectrum. The phenomenon is named after Austrian physicist Christian Doppler (1803–1853).

Drake equation	An attempt within SETI studies to estimate the probability of finding extraterrestrial civilizations by multiplying the likelihoods for different factors that come into play. The equation is named after Frank Drake (1930–), a pioneer of SETI efforts.
Earth	The name of our home planet, the third planet from the Sun.
eccentricity (e)	In mathematics, eccentricity is a measure of how circular or how flattened an ellipse is. A circle can be thought of as an ellipse with an eccentricity of zero ($e = 0$).
eclipse	An eclipse occurs when one astronomical object passes in front of another relative to the observer. In a solar eclipse, the Moon passes in front of the Sun, which casts the Moon's shadow on the Earth's surface. In a lunar eclipse, the Earth passes between the Moon and the Sun, so that the Moon moves through Earth's shadow.
ecliptic	The plane on the celestial sphere containing both the Earth and the Sun. Simply put, it is an extension of the path the Sun takes across the sky. The planets' motions loosely follow the ecliptic as seen from Earth, as this is approximately the plane of the Solar System.
Einstein, Albert (1879–1955)	German-born theoretical physicist who developed several monumental theories in modern physics, including the photoelectric effect and the theories of special and general relativity. His theory of general relativity describes gravity as the warping of spacetime, and plays a key role in astrophysics and cosmology.
electromagnetic radiation	A form of energy that can travel through space, composed of an electric field and a magnetic field. The most familiar form is visible light, but the electromagnetic spectrum extends to invisible wavelengths: shorter wavelengths correspond to higher energies: ultraviolet, X-rays and gamma rays; at longer wavelengths are infrared, microwaves and radio waves. In modern astronomy each of these plays a role.
elliptical galaxy	A type of galaxy characterized by it smooth oval shape without the notable distinguishing arms of spiral and barred spiral galaxies. They are made up of mainly older stars, with little evidence of star formation.
emission line	In spectroscopy, an emission line is produced when photons of light are emitted at specific wavelengths (or frequencies) corresponding to changes in energy levels of atoms or molecules. Observing emission lines can help us learn about the chemical compositions of the emitter in question and its environment, as the energy levels are characteristic of specific atoms or molecules.
epicycle	In the system of the Solar System proposed by Ptolemy, some corrections were required in order to explain some of the observations of planets: the retrograde motion (apparent changes in direction compared to the background stars) and changes in the apparent sizes due to distance. Epicycles, small circles on larger circles, were used in calculations of the planets' positions.
equator	The line running around the Earth at 0° latitude. A plane passing through the Earth's center that is perpendicular to the Earth's axis of spin meets the surface at the equator.
equinox	This marks the position on Earth's orbit around the Sun when the Earth's axis of tilt exactly matches the plane of the equator, so the season is neither summer nor winter, and the length of night and day are approximately equal. This happens twice per year: the Vernal (Spring) equinox usually occurs around March 20, while the Autumnal equinox is around September 21.
ESA	The European Space Agency, an international collaboration of 18 European countries (Canada is an associate member).

escape velocity	The minimum speed required by an object to escape from Earth's (or another astronomical body's) gravity. To escape the bounds of Earth, the escape velocity is 11.2 kilometers per second or about 40 000 kilometers per hour.
expanding Universe	The Universe at the largest scales is expanding, i.e. the distance between distant objects is increasing. The theoretical framework for this model was developed independently in the 1920s by Alexander Friedmann (1888–1925) and Georges Lemaître (1884–1966) as a solution to Einstein's general relativistic field equations. Observational verification was later provided by Edwin Hubble.
extrasolar planet or exoplanet	A planet orbiting around a star other than the Sun, in another solar system. The first definitive discovery of an exoplanet around an ordinary star was made in 1995. Since then, over 500 exoplanets have been confirmed, with many more candidate stars being investigated. The most common detection method is the "radial velocity" method, which utilizes the Doppler shift caused by a planet's gravitational interactions with its host star. The easiest planets to detect with this method are large and close-in, as these have the fastest orbits and induce the biggest wobble.
galaxy	A gravitationally bound system of stars. Dwarf galaxies may contain only a few million stars, whereas giant galaxies may contain a hundred trillion stars. Our host galaxy is called the Milky Way Galaxy.
Galileo Galilei (1564–1642)	Italian physicist and astronomer who played a key role in the scientific revolution, through his use of experiments and observation combined with mathematics. His astronomical contributions include developing the telescope for astronomical use, discovering Jupiter's four largest moons, observing the phases of Venus and making detailed observations of sunspots. He championed Copernicus' model for a Sun-centered cosmos, which led to trouble with the religious authorities of the time.
gamma ray	The highest energy form of electromagnetic radiation. Gamma rays have the highest frequency and the shortest wavelength. As such they are usually associated with high-energy phenomena, such as pulsar beams, accretion onto a black hole or supernovae explosions.
gas giant planet	The outer planets: Jupiter, Saturn, Uranus and Neptune are collectively known as the gas giants. They are many times the size of Earth, and composed of a mixture of gases (hydrogen and helium) for Jupiter and Saturn and ices (water, ammonia and methane) for Uranus and Neptune.
globular cluster	A tightly gravitationally bound association of stars, often spherical in shape, with a high density of stars at their cores. They are usually found orbiting a host galaxy. Smaller and looser groups of stars are called "open clusters."
granulation	In solar physics, this describes the grainy appearance of the surface of the Sun's photosphere, caused by convection cells about 1000 miles wide. Hot plasma rises at the center of each cell, cools and then falls at the darker edges.
gravitation or gravity	One of the four fundamental forces in the Universe, gravitation (or gravity) acts as a force of attraction between all objects that have mass. Galileo demonstrated that the gravitational acceleration on Earth acts upon all objects equally regardless of their mass. Newton formulated the inverse square law for gravity, showing that the gravitational force between two objects is proportional to the product of their masses and inversely proportional to the square of the distance between them. Our current theory of gravitation stems from Einstein's theory of general relativity where gravitational effects are the result of the bending of space (spacetime) by matter.
greenhouse effect	This describes the heating of a planet where the radiation it receives is absorbed by gases (such as water vapor, carbon dioxide, methane and ozone) in the atmosphere and re-radiated

at lower wavelengths that cannot escape the atmosphere, causing it to heat up. The planet Venus has a thick carbon dioxide atmosphere; the greenhouse effect is thought to be responsible for its high surface temperature of over 450 °C (850 °F).

Halley's comet or Comet Halley	A short-period comet whose journey through the inner Solar System repeats every 75 to 76 years. It is named for English astronomer Edmond Halley (1656–1742) who first predicted its return.
helioseismology	The science of studying waves transmitted through the Sun, through careful analysis of its surface. It is analogous to seismology here on Earth and yields information about the Sun's internal structure.
Herschel, Caroline (1750–1848)	The sister and assistant of Sir William Herschel. She was a noted observer in her own right and made several comet discoveries. She was the first woman to do so.
Herschel, Sir William (1738–1822)	German-born English astronomer who built the largest telescopes of his day. He undertook the first systematic sky surveys to sweep the skies for interesting "nebulous" objects, assisted by his sister Caroline Herschel; during one of these surveys he discovered the planet Uranus.
Hertzsprung–Russell (H–R) diagram	A plot of stars' magnitudes (or luminosities) against their color (which correlates with their temperature). The resulting graph reveals different populations of stars, and provides evidence of how stars evolve during their lifetimes. It is named after the astronomers who first developed it, Ejnar Hertzsprung (1873–1967) and Henry Norris Russell (1877–1957).
high-velocity star	Stars in our neighborhood fall into two groups: high- and low-velocity stars. High-velocity stars have speeds relative to us of up to 150 kilometers per second. They are of population II.
Hipparchus (c. 190 BC–120 BC)	Greek astronomer and astrologer, who applied geometry and trigonometry to study the motions of, and compute distances to, the Sun and the Moon.
Hubble, Edwin Powell (1889–1953)	American astronomer who made several contributions to astronomy, not least confirming that galaxies lie far beyond the Milky Way through the use of Cepheid variables. His system of classifying galaxies by their shape is still in use today. Hubble also discovered that the redshift of distant galaxies is proportional to the distance to them. This is called Hubble's law, an observation that confirmed the theory of an expanding Universe.
Hubble Space Telescope (HST)	One of NASA's Great Observatories, HST is a 2.4-meter telescope operating in low-Earth orbit. The benefit of putting a telescope in space is that there is no atmosphere, allowing it to take amazingly clear and detailed images. Named after astronomer Edwin Hubble, it has made unparalleled contributions to science since its launch in 1990 as well as demonstrating the visual beauty of the cosmos as never before. HST's planned successor is the James Webb Space Telescope.
Hubble's law	This maps the relationship between a distant galaxy's velocity away from us (related to its redshift) and its distance from us, due to the expansion of the Universe.
infrared radiation	The part of the electromagnetic spectrum lying at longer wavelengths than red visible light, commonly manifesting itself as heat energy. In astronomy, infrared detectors are used to see through dusty regions of space and to detect the most distant galaxies.
interferometer	A set of astronomical instruments designed to take advantage of the properties of electromagnetic waves, generally consisting of several small telescopes (or commonly radio dishes, for radio interferometry), which can act together to simulate a much larger telescope (or radio dish). Examples of radio interferometers being built are the Allen Telescope Array and the Very Large Array.
interstellar dust	The dust component of the interstellar medium, estimated to be less than 1% (the rest is gas).

interstellar medium (ISM)	The low-density material that occupies the space between stars (most of space!), comprising atomic and molecular gas (mainly hydrogen and helium), ionized gas and dust.
ion	An atom or molecule that has electric charge, due to gaining or losing electrons. Ions in gaseous form are called plasmas and are affected by magnetic fields.
ionization	The process through which an ion is formed.
irregular galaxy	A galaxy that does not have a distinguishable spiral or elliptical shape. Irregular galaxies are believed to be spiral or elliptical galaxies that have been gravitationally disrupted through collisions or near-collisions with neighboring galaxies. About a quarter of all galaxies are classed as irregulars.
James Webb Space Telescope (JWST)	The planned successor to the Hubble Space Telescope. JWST, an international collaboration, will have a 6.5-meter mirror and observe in visible light and infrared. Its main science goals are to study the birth of stars and planets, galaxy evolution and the very first stars and galaxies. It is named after NASA administrator James Webb (1906–1992) who played an important role in the Apollo program.
Jansky, Karl (1905–1950)	American physicist and engineer who first detected radio waves from the Milky Way around 1933. Unfortunately, radio astronomy would not take off until the second half of the twentieth century.
Jupiter	The fifth planet from the Sun and the largest planet in the Solar System at over 11 times the diameter of the Earth and about 1300 times Earth's volume. Jupiter is a gas giant, comprised mainly of hydrogen and helium. Jupiter's largest moons are visible through amateur telescopes, as are its surface features such as its stripy cloud layers and its "Great Red Spot," a massive storm that has lasted centuries and is bigger than the Earth.
Keck Observatory	More formally called the W. M. Keck Observatory, it consists of two 10-meter telescopes atop Mauna Kea's summit in Hawaii. Working together, they form an interferometer, equivalent to a single telescope with an 85-meter mirror.
Kepler, Johannes (1571–1630)	German mathematician and astronomer who formulated the laws of planetary motion. He came to the conclusion that planets orbit in elliptical orbits.
Kepler's laws	These are Kepler's descriptions of planetary motion around the Sun. The first states that planets move in elliptical orbits, with the Sun at one of the foci (points within an ellipse equivalent to the center of a circle). The second shows mathematically that planets move fastest when nearer the Sun (at perihelion) and more slowly when farthest away (at aphelion). The third law shows the relationship between the orbital period (i.e. the time a planet takes to complete an orbit) and the average distance of the planet's orbit.
Kirkwood gaps	Gaps in a plot of the distribution of main-belt asteroids (which lie between Mars and Jupiter) at certain orbits, due to orbital resonances with the planet Jupiter, which, over time, eject the asteroids from these orbits.
Kuiper Belt or Edgeworth–Kuiper Belt	A region beyond Neptune in the outer reaches of the Solar System dominated by small icy bodies, remnants of the Solar System's formation. Pluto is considered to be the largest Kuiper Belt Object, although several are known to be nearly as large as Pluto. The name comes from the scientists who first hypothesized the nature of this region: Dutch–American astronomer Gerard Kuiper (1905–1973) and Irishman Kenneth Edgeworth (1880–1972).
Lagrangian points	Specific points in an orbital gravitational system (e.g. Sun–Earth or Earth–Moon) where there is no net gravitational pull, so a smaller object will remain in the same position relative to the larger bodies. There are five such stable points. Several current space probes and observatories are placed at the Lagrangian points. They are named for the French–Italian

mathematician Joseph Louis Lagrange (1736–1813) who developed a new system of classical mechanics.

le Verrier, Urbain Jean Joseph (1811–1877)
French astronomer who successfully predicted the existence and location of the planet Neptune based on tiny discrepancies in Uranus' orbit. Similar calculations were done independently by British astronomer John Couch Adams.

low-velocity star
Stars in our neighborhood fall into two groups: high- and low-velocity stars. Low-velocity stars have speeds relative to us of about 10 kilometers per second. They are of population I.

light year
A unit of distance equal to the distance that light can travel (in a vacuum) in a year. The speed of light is approximately 3×10^8 meters per second; there are $365.25 \times 24 \times 60 \times 60$ seconds in a year, yielding a distance of 9.4607×10^{12} meters, or 63 240 AU.

mare (plural maria)
Smooth dark surface features on the Moon, named from the Latin for "sea." They are made of up solidified lava flows from early in the Moon's history.

Mars
The fourth planet from the Sun, lying between Earth and Jupiter. It is about half the diameter of Earth, and has a thin atmosphere and an average surface temperature of around −60 °C. Observationally it has a red hue due to the iron oxide content of the Martian soil. Evidence from recent orbital, lander and rover missions suggests that Mars once had a wet past. Traces of liquid and ice water have been found on its surface.

Mauna Kea observatories
At the summit of the Mauna Kea volcano on the Big Island of Hawaii are sited about a dozen of the world's best ground-based telescopes, operated by many countries and collaborations. The site offers dry, clear atmosphere above the clouds, ideal conditions for observing. The siting has not been without controversy, however, due to local cultural concerns.

Maunder minimum
A period from around the years 1645 to 1715 when the observed number of sunspots was very low (tens per year instead of more than 100). It is named after solar astronomer Edward Maunder (1851–1928).

Mercury
The closest planet to the Sun. It has been revealed to be very dense with a heavily cratered surface and has the highest range of surface temperatures, around −180 °C to over 400 °C. The second spacecraft to study Mercury reached its destination in 2011 so we are still learning a lot about it.

meteor
The streak of light created when a small meteoroid (dust-grain sized to a few meters across) is heated upon entering the Earth's atmosphere. The everyday name for the phenomenon is a "shooting star." At particular times of year regular meteor showers occur, where the meteor rate can be many times higher than usual.

meteor shower
On several nights per year, meteors are seen to radiate from a particular part of the sky at a rate higher than usual. Meteor showers are named after the stars and constellations in which they appear. They occur when Earth's orbit intersects the path of a comet, which leaves a trail of meteoroids behind it.

meteorite
The remnants of a large meteoroid or asteroid that survives its fall to Earth. The composition depends upon the nature of the original object; the commonest types are stony, iron–nickel or a mixture of both.

meteoroid
Small solid bodies (smaller than asteroids) in the Solar System that orbit the Sun in interplanetary space. When they are found on the Earth's surface after an impact or a fall, they are called meteorites.

Milky Way
The thin pale band dense with stars visible in the night sky (from dark locations). It is our view-from-within of the disk-shaped galaxy we inhabit, the Milky Way Galaxy.

Milky Way Galaxy	The name of our home Galaxy. The Milky Way is a barred spiral galaxy 100 000 light years wide, containing around 300 billion stars. The Galaxy's center lies in the direction of the constellation Sagittarius.
minute of arc or arcminute	Angular measurement equal to one sixtieth of a degree. Minutes of arc may be further subdivided into sixty seconds of arc, or arcseconds.
Moon	As a proper noun this refers to the Moon, Earth's only satellite. The Moon is about one quarter the diameter of Earth and orbits at a distance of approximately 385 000 km ($\pm$20 000 km). More generally, any of the natural satellites orbiting a planet or larger body in the Solar System may be called a moon.
nadir	The point vertically beneath an object (usually defined by the direction that gravity is acting on the object); the opposite of zenith. In observational astronomy it can also refer to the lowest point in the sky in a celestial object's apparent orbit.
NASA	The National Aeronautics and Space Administration, the USA's national space agency. NASA's remit includes: human and robotic space exploration; Earth observation satellites; studying the Sun and the space environment; astrophysics; and operating space observatories.
nebula (plural nebulae)	An interstellar cloud made from dust or gas. Historically the term was also used to describe any faint or fuzzy astronomical object and so included galaxies and star clusters, as well as actual nebulae.
Neptune	The eighth planet from the Sun. It is an ice giant, about 17 times the mass of Earth, orbiting the Sun at 30 AU. Its atmosphere is made up of hydrogen and helium, along with some other gases. Photographs reveal Neptune's atmosphere to take on a blue color due to traces of methane.
neutron star	The tiny dense remnant formed when a massive star collapses in a supernova explosion, toward the end of its life cycle. They are believed to be composed entirely of neutrons, a type of subatomic particle.
Newton, Sir Isaac (1642–1727)	English physicist, mathematician and philosopher, lauded as the greatest scientist of his time. He made important discoveries in optics, classical mechanics and gravity, among other topics. He showed that the same physical principles governing motions on Earth also apply to the motions of celestial bodies.
Newtonian telescope	A reflecting telescope of a type originally designed by Isaac Newton featuring a concave primary mirror and a secondary mirror mounted diagonally near the top of the telescope tube.
nova (plural novae)	A white-dwarf star in a binary system that is observed to flare up, becoming thousands of times brighter, before fading back to its usual luminosity over a number of weeks. The cause is material (mainly hydrogen and helium) from the companion star causing runaway nuclear fusion reactions on the white dwarf's surface.
observatory	A location used for observing. Astronomical observatories are often sited at high altitudes and dry locations, such as Mauna Kea in Hawaii, to get the clearest views through Earth's atmosphere. Radio telescopes can see through cloud, but have to be sited well away from urban areas so as to minimize interference from man-made radio sources.
Olbers' Paradox	The idea that the dark sky we see at night suggests that the Universe is not infinite: if it were, we would see light from stars in every direction. The idea that dark clouds could be blocking this light does not help, as over time the clouds would themselves heat up until they start radiating light themselves. This problem is attributed to German amateur astronomer Heinrich Wilhem Olbers (1758–1840).

Oort cloud	A hypothetical cloud of comets stretching out thousands of astronomical units (AU) from the Sun, far beyond the Kuiper Belt. It is thought to be the remnants of the protoplanetary disk from which the Solar System formed and is the source of long-period comets. It is named for Dutch astronomer Jaan Oort (1900–1992).
orbit	The path through space of an object subject to gravitational forces, for example, the planets orbiting the Sun. Most orbits in the Solar System are ellipses.
parallax	The apparent shift in the location of a distant object due to the changing line of sight; the closer the object, the larger the shift. Astronomers use this principle to measure the distance to the closest stars. For all but the closest thousands of stars, the parallax angle is too small to measure.
parsec	A measure of distance, corresponding to a parallax angle of one arcsecond. It is equal to 3.26 light years or approximately 31 trillion kilometers (19 trillion miles). Distances between stars are usually measured in parsecs.
penumbra	A ring of partial shadow around the full shadow during a solar eclipse. Observers looking at the Sun from within the penumbral zone will see a partial solar eclipse, where only part of the Sun is obscured by the Moon. In solar physics, the penumbra is the lighter part around the edges of a sunspot.
perihelion	The point on Earth's (or another planet's) elliptical orbit where it is at its closest point to the Sun. The opposite is aphelion.
period	The time taken to complete a single orbit cycle. Earth's orbital period is one year.
photosphere	The outer layer of the Sun (or any star) from where the light that we see originates. The Sun's photosphere has a granular appearance on scales of about 1000 miles, due to convection cells in the Sun's plasma.
planet	A large non-stellar object orbiting a star. In antiquity, five planets were known, visible as "stars" that moved slowly and regularly across the night sky. Within the Solar System, the defined planets are: Mercury, Venus, Earth, Mars, Jupiter, Saturn, Uranus and Neptune. Pluto was formerly classed as a planet between its discovery in 1930 through to a ruling on planet definitions in 2006. Since the mid 1990s, hundreds of extrasolar planets have been discovered.
planetary nebula	The ring-shaped opaque clouds of gas ejected by some types of stars late in their life. The name is a misnomer, as they are not actually related to planets, but in early telescopic observations they looked similar to the giant planets.
planetesimal	The building blocks from which planets were made in the early Solar System. Planetesimals are the smallest bodies that are able to attract one another through their mutual gravity. Larger, Moon-sized bodies involved in planet formation are called "protoplanets."
Pluto	A dwarf planet orbiting the Sun with an inclined orbit lying, on average, beyond Neptune's. Pluto was discovered in 1930 by Clyde Tombaugh and until 2006 it was considered the ninth planet in the Solar System. In 2006 a resolution was passed by the International Astronomical Union (IAU) that reclassified Pluto as a dwarf planet and, later, the first in a new group of objects classed "plutoids." The debate about what makes a planet continues.
pole star	The star Polaris, or Alphae Ursae Minoris, which lies about one degree away from the North Celestial Pole. Over thousands of years Earth's rotational pole precesses, similar to how a spinning top rotates. This means that while the Earth rotates around its axis, the axis also rotates in a cycle that take around 25 770 years to complete. As this precession changes how the pole is aligned, by AD 3000 Polaris will no longer be the pole star.

population I	In cosmology, population I stars are metal-rich young stars containing a higher percentage of metal elements (those heavier than hydrogen and helium). This is a result of the enrichment of the interstellar medium by nuclear fusion in earlier generations of stars.
population II	Older metal-poor stars, which formed at an older time in the Universe than population I stars. These might have been preceded by the first population III stars, which had no metal elements at all.
prominence	In solar physics, a large loop or column of plasma stretching away from the Sun's surface above the photosphere.
Ptolemy, Claudius (c. AD 90–168)	A Roman–Egyptian mathematician, astronomer and astrologer who compiled the *Almagest*. He is also known to have worked on mapping and optics.
pulsar	A rotating neutron star, which emits a narrow beam of electromagnetic radiation. We can only detect the radiation as pulses when the beam is pointing toward Earth, analogous to a ship seeing a lighthouse's beam.
quasar	A distant point-like energy source originating from a massive black hole at the center of a distant galaxy. They are among the most energetic objects in the Universe. The name comes from a shortening of "quasi-stellar radio source."
radar astronomy	Astronomy using reflected microwave signals to learn about nearby objects, mainly the planets and asteroids in the Solar System. It complements visual observations by providing very accurate distances and relative speed measurements using the Doppler effect.
radio astronomy	The branch of astronomy using radio and microwave parts of the electromagnetic spectrum. There are a number of astronomical sources of radio waves, ranging from the Sun, to stars and galaxies, and the Cosmic Background Radiation.
radio telescope	A dish or antenna used to detect astronomical radio sources. Many small radio telescopes may be combined using interferometry to achieve higher resolution.
Reber, Grote (1911–2002)	American amateur astronomer and pioneer of radio astronomy, who built a 9-meter radio dish in his back yard. He conducted the first sky survey in radio frequencies and played an important role in establishing radio astronomy in the 1950s.
red giant	A low-mass giant star in the late stages of stellar evolution. Having used up most of the hydrogen fuel in its core, it starts fusing hydrogen in its outer layers.
redshift	According to the Doppler effect, something is described as redshifted if its relative motion is away from the observer. Light and other forms of radiation appear to have a lower frequency (or longer wavelength) than they do in reality. The opposite of this is blueshift. The concept of redshift is important in cosmology as due to the expansion of the Universe objects at larger distances have higher redshifts.
resolving power	The ability of a telescope (or other instrument) to see small details clearly. For instance, to the naked eye most binary stars will look like a single star but when seen through a small telescope or binoculars you might be able to resolve them into a pair of stars.
resonance	In physics, it is a property of vibrating systems to act more strongly at certain frequencies. In orbital mechanics, resonance in the period of planetary bodies acts to enhance the gravitational effects between them. Sometimes the result will be unstable; for example, the resonance between Jupiter and the asteroid belt leads to the Kirkwood gaps. Other times they will form stable orbits, such as the orbital periods of Pluto and Neptune, which are locked in the ratio 2:3.

rings	A flat disk of dust, ice or other small particles orbiting around the gas giant planets. The most obvious ring system is around Saturn, which extends for thousands of kilometers but is estimated to be only about ten meters thick.
Roche limit	A satellite orbiting a planet is elongated by the planet's tidal gravitational force: within this limiting distance the tidal force is sufficient to rip the satellite apart.
satellite	A celestial object that orbits a planet or other body larger than itself. The natural satellites of planets and dwarf planets are called moons. Man-made or artificial satellites are objects put into orbit by mankind for any of a number or reasons, such as Earth and weather observation, communications, navigation or research.
Saturn	The sixth planet from the Sun. It is a gas giant, the second largest planet after Jupiter, with a prominent ring system visible through a small telescope or binoculars. Like Jupiter, it has an atmosphere made up mainly of hydrogen. Much of our knowledge of Saturn and its moons comes from the Cassini–Huygens mission.
Schwarzschild radius	In a black hole, the radius at which light cannot escape.
seasons	The changes in weather and daylight resulting from the tilt of the Earth's axis relative to its path around the Sun. The northern hemisphere is tilted toward the Sun in the summer months of May, June and July, and away from the Sun in winter, November, December and January. The southern hemisphere has the opposite seasons.
SETI	The Search for Extraterrestrial Intelligence, the scientific search for evidence of intelligent life elsewhere in the cosmos. The most common searches are for narrow-frequency interstellar communications. Most SETI projects are privately funded and are not connected to those who believe in UFOs.
Solar and Heliospheric Observatory (SOHO)	A solar observation satellite launched in 1995 as a collaboration between NASA and ESA. Its main aims are to investigate the Sun's outer layers, to measure the solar wind and to probe the Sun's interior through helioseismology.
solar cycle	The approximately 11-year cycle of variation in solar radiation and magnetic activity (including sunspots, solar flares, etc.). It was first observed in 1843 by German astronomer Samuel Heinrich Schwabe (1789–1875), who had been counting the number of sunspots.
Solar System	The Sun and all of the objects orbiting under its gravitational influence, extending out to a distance of about two light years. Much of the outer regions have not been well explored.
solar wind	A stream of charged particles flowing outward from the Sun. The solar wind interacting with the Earth's magnetic field gives rise to the auroras.
solstice	The two times per year when the Sun reaches its northernmost or southernmost points in the sky. The solstices usually occur around 21 June and 21 December.
Space Shuttle	NASA's reusable manned orbital spacecraft, which operated between 1981 and 2011. The shuttles launched numerous satellites, including the Hubble Space Telescope, and helped construct and service the manned space stations.
spectral line	In spectroscopy, emission lines or absorption lines corresponding to changes in the energy levels of atoms or molecules. Observing spectra can help us learn about the chemical compositions of the emitter in question and its environment as the energy levels are characteristic of specific atoms or molecules.
spectrum (plural spectra)	In optics, an unbroken continuum of colors producing the rainbow. By analogy, the principle may be applied to other wavelengths and frequencies. Spectroscopy, the study of spectra and identifying, is an important tool in astronomy.

spiral galaxy	A galaxy type defined by its shape, being a flattened disk with two or more spiral arms of stars connected at the galaxy's bulging center. Barred spiral galaxies have an additional feature, a bar of stars connecting the arms running through the galaxy's center.
star	A massive ball of gas held together by its own gravity, which radiates energy due to nuclear fusion reactions in its core. The Sun is our closest star. From the Earth, distant stars shine as pinpoints of light in the night sky.
star cluster	A collection of stars bound by their mutual gravitational attraction. High-density star clusters with a compact core of stars are called globular clusters.
stellar evolution	The process describing the life cycle of a star. Stars are formed when a dense molecular cloud collapses under its own gravity, releasing heat energy, and nuclear fusion reactions begin. Most normal stars will continue to burn their hydrogen fuel for several billion years. Stars end their lives in one of a number of ways, depending on their mass and composition as they start to run out of fuel. These include white dwarfs, neutron stars and black holes.
Stonehenge	A prehistoric monument in southern England, dating from around 2500 BC, made up of concentric rings of large standing stones, some topped by lintels. Its original function is unknown but the alignments of some of the stones suggest its use as an astronomical observatory to predict equinox and solstice events.
Sun	The star at the center of the Solar System. Its diameter is 1392 000 km, 109 times that of Earth, and it contains more than 99% of the total mass of the Solar System.
sunspot	A temporary "dark" spot on the Sun's surface caused by magnetic activity within the Sun. Although they appear darker than the Sun's surface, they are still actually very bright! Historic records show they were observed by Chinese astronomers as early as 28 BC but the first telescopic observations were not until the seventeenth century. Modern photographs show that they have a dark center, the umbra, but are lighter around the edges, called the penumbra.
supernova	A stellar explosion caused when an aging star collapses to form a neutron star or black hole, releasing prodigious amounts of energy that blows the outer layers of the star out into space. Observationally, a supernova increases in luminosity to the point where a single star can temporarily outshine its host galaxy, before fading away over weeks or months.
supernova remnant	The outer layers of gas and dust that are blown into space during a supernova explosion. Multiwavelength pictures show them as approximately spherical with intricate filaments and bubble shapes. X-ray images show that there is superheated gas at the shock front where the expanding gas meets the interstellar medium.
T Tauri star	A type of young variable star found in star-forming regions. Many have protoplanetary disks, which might be solar systems being formed. They are named after the prototype star, T Tauri.
telescope	An optical instrument used to collect and focus light through lenses and/or mirrors. The power of a telescope is determined by the area of its primary lens or mirror. A typical amateur telescope might have an aperture of 10–15 centimeters, but the largest optical telescopes have primary mirrors around 10 meters. By analogy, instruments that do this for other parts of the electromagnetic spectrum are also called telescopes, although the means of collecting and imaging the radiation might be very different.
terrestrial planet	The four inner planets of the Solar System: Mercury, Venus, Earth and Mars. These are sometimes also called rocky planets. The other planets are termed gas giants.

tidal force	A result of differential gravitational forces acting on two bodies. The pull of the Moon gives rise to the ocean tides observed in the seas on Earth. Tidal forces also act elsewhere in the Solar System, for example, causing heating in Jupiter's moon Io.
time zone	A set of internationally agreed boundaries that maintain an agreed time around the world. Most follow geopolitical boundaries, for example, North America is divided into Eastern Time (5 hours behind the agreed "zero" point in Greenwich, England), Central Time (-6 hours), Mountain Time (-7 hours) and Pacific Time (-8 hours). Most of continental Europe lies within the same time zone.
Titan	Saturn's largest moon and the only moon in the Solar System to have an appreciable atmosphere. Titan was discovered in 1655 by Dutch astronomer Christiaan Huygens (1629–1695). Much of what we know about Titan comes from the Huygens probe, part of the Cassini–Huygens mission, which descended to Titan's surface in 2005.
Tombaugh, Clyde (1906–1997)	American astronomer best known for his discovery of Pluto in 1930 while a young researcher at the Lowell Observatory in Arizona. He also discovered hundreds of asteroids.
transit	In astronomy, an event where one celestial body passes in front of another as viewed from Earth. Approximately once per decade, Mercury transits the Sun. Much rarer is a Transit of Venus – the transit of June 2012 will not be repeated until the year 2117. The discovery of extrasolar planets raises the possibility of detecting and characterizing them when they transit their parent stars.
ultraviolet astronomy	The branch of observational astronomy using the ultraviolet (UV) part of the electromagnetic spectrum. Ultraviolet radiation is characteristic of objects hotter than typical stars, including massive stars and old galaxies.
umbra	The area under full shadow during a solar eclipse. In solar physics, the umbra is the dark central region of a sunspot.
Uranus	The seventh planet from the Sun. It is a gas giant or an "ice giant" like its neighbor Neptune. Although Uranus is visible with the naked eye, it was not recognized as a planet until its discovery by William Herschel in 1781.
Van Allen radiation belts	Two donut-shaped regions of space around Earth where charged particles from the solar wind are trapped by Earth's magnetic field. Man-made satellites must be shielded from these ions, which can cause damage to electronics and circuitry.
variable star	A star whose apparent magnitude from Earth varies with time. This can be due to changes in the star's actual luminosity or due to an eclipsing companion star that blocks our view temporarily. Variable stars are studied to determine their brightnesses, periodicities, regularities and spectra.
Venus	The second planet from the Sun. Venus is a terrestrial planet, the closest in size and bulk composition to Earth. However, unlike Earth, it has a dense carbon dioxide atmosphere, a higher atmospheric pressure and a searing surface temperature of 735 K (460 °C). From Earth, Venus is visible as the brightest object after the Moon in the morning or evening sky.
Very Large Array (VLA)	A radio astronomy observatory near Socorro, New Mexico. It consists of 27 antennas, each with a 25-meter dish, spread out along special tracks in a Y-shape. The antennas may be used together as a radio interferometer.
Very Long Baseline Interferometry (VLBI)	A technique in radio astronomy where radio telescopes separated by large distances can be used together as a radio interferometer. The resolution achievable through this method is proportional to the distance between the farthest telescopes.

Viking mission

A pair of space probes sent to Mars in 1975, each comprising an orbiter and a lander, to map and study the Martian surface. The landers carried a suite of instruments, including experiments to detect potential signs of life in Martian soil, all of which returned negative results.

Voyager mission

A pair of US space probes launched in 1977 to study Jupiter and Saturn. The probes continued into the outer Solar System and have now passed beyond the limit of the solar wind into interstellar space. Voyager 1 has traveled the farthest of any man-made object.

white dwarf

A small dense star that has exhausted its nuclear fuel. It is the end result of stellar evolution for stars not massive enough to become neutron stars. White dwarfs approaching 1.4 solar masses, the Chandrasekhar limit, may explode as supernovae and collapse into neutron stars.

X-ray astronomy

The branch of observational astronomy that uses the X-ray part of the electromagnetic spectrum. X-rays are emitted from hot gases at around a million kelvin, so are indicative of high-energy processes around pulsars and black holes or in supernova explosions.

year

The Earth's orbital period, measured compared to the fixed reference point of the stars, is 365.256 days.

zenith

The point vertically above an object (usually defined by the direction away from which gravity is acting on the object); the opposite of nadir. In observational astronomy it can also refer to the highest point in the sky in a celestial object's apparent orbit.

zodiac

The twelve divisions of the ecliptic, which marks the fundamental plane when using an ecliptical coordinate system. The 30° divisions of the zodiac have been replaced in modern astronomy by 360° measurement of longitude.

FIGURE CREDITS

Figure 1.32	Courtesy of Whipple Museum of the History of Science, University of Cambridge.
Figure 1.33	skvoor/Shutterstock.com.
Figure 1.45	Courtesy of Stefan Seip.
Figure 1.49	Courtesy of Stefan Seip.
Figure 1.50	Olinchuk/Shutterstock.com.
Figure 2.2	rnl/Shutterstock.com.
Figure 2.3	Redrawn from R. C. Bless, *Discovering the Cosmos*.
Figure 2.4	Redrawn from R. C. Bless, *Discovering the Cosmos*.
Figure 2.19	Nicolaus Copernicus Museum, Frombork, Poland / Giraudon/ The Bridgeman Art Library.
Figure 2.20	Redrawn from R. C. Bless, *Discovering the Cosmos*.
Figure 2.24	AIP Emilio Segre Visual Archives, Brittle Books Collection.
Figure 2.25	© DeAgostini / SuperStock.
Figure 2.26	© Newberry Library / SuperStock.
Figure 2.27	Steven Wynn / photos.com.
Figure 2.35	Courtesy AIP Emilio Segre Visual Archives, W. F. Meggers Collection.
Figure 2.36	Courtesy of Jim & Rhoda Morris.
Figure 3.1	Georgios Kollidas/Shutterstock.com.
Figure 3.3	Andrew Lambert Photography/Science Photo Library.
Figure 3.21	Courtesy of NASA.
Figure 5.1	Courtesy of Robert Gendler.
Figure 5.25	Courtesy of Paul Hirst.
Figure 5.26	Andre Nantel/Shutterstock.com.
Figure 5.28a	Science Photo Library.
Figure 5.28b	Bell Laboratories/ Alcatel-Lucent USA Inc., courtesy AIP Emilio Segre Visual Archives, Physics Today Collection.
Figure 5.28c	Emilio Segre Visual Archives/American Institute of Physics/Science Photo Library.
Figure 5.29	Courtesy of the NAIC – Arecibo Observatory, a facility of the NSF.
Figure 5.31	Courtesy of NASA/NRAO/AUI/NSF.
Figure 5.33a	Courtesy of NASA/CXC/SAO.
Figure 5.33b	Courtesy of NASA/CXC/SAO.
Figure 5.34	Courtesy of NASA.
Figure 5.35	Courtesy of NASA.
Figure 5.36	Courtesy of NASA: KSC-03PD-2349.
Figure 5.40	Courtesy of NASA: MSFC-75-SA-4105-2C.
Figure 5.41	Courtesy of NASA/JPL-Caltech.
Figure 5.42	Courtesy of Photolab, JPL (Jet Propulsion Laboratory), 1996.
Figure 5.43	Courtesy of NASA/JPL-Caltech.
Figure 7.1	Courtesy of NASA.
Figure 7.2	Courtesy of Apollo 16, NASA.
Figure 7.4	Courtesy of Apollo 12/NASA.
Figure 7.17	Courtesy of NASA.
Figure 7.18	Courtesy of NASA National Space Science Data Center.

Figure 7.23	Courtesy of Calvin J. Hamilton.
Figure 7.24	Courtesy of NASA.
Figure 7.31	Courtesy of USGS.
Figure 7.32	Courtesy of USGS.
Figure 7.38	MaxPhoto/Shutterstock.com.
Figure 7.40	Courtesy of NASA.
Figure 7.41	Courtesy of NASA/JPL/MSSS.
Figure 7.42	Courtesy of NASA.
Figure 7.43	Courtesy of NASA.
Figure 8.1	Courtesy of NASA/JPL/University of Arizona.
Figure 8.2	Courtesy of NASA/JPL.
Figure 8.6	Courtesy of NASA/JPL/University of Arizona.
Figure 8.7	Courtesy of NASA.
Figure 8.8a	Courtesy of NASA/JPL/DLR.
Figure 8.8b	Courtesy of NASA/JPL/University of Arizona.
Figure 8.9a	Courtesy of NASA/JPL/Ted Stryk.
Figure 8.9b	Courtesy of NASA/JPL/DLR.
Figure 8.10a	Courtesy of NASA/JPL/Ted Stryk.
Figure 8.10b	Courtesy of NASA.
Figure 8.12	Courtesy of NASA/JPL/Space Science Institute.
Figure 8.16	Courtesy of NASA/JPL/Space Science Institute.
Figure 8.17	Courtesy of NASA/JPL/USGS.
Figure 8.18	Georgios Kollidas/Shutterstock.com.
Figure 8.19a	Photos.com.
Figure 8.19b	Photos.com.
Figure 8.20	Photograph by R. Sterling Trantham, New Mexico State University, courtesy AIP Emilio Segre Visual Archives.
Figure 8.21	JCEIv/Shutterstock.com.
Figure 8.24	Courtesy of NASA/JPL.
Figure 8.25	Courtesy of NASA/JPL.
Figure 8.26	Courtesy of NASA/JPL.
Figure 8.27	Courtesy of NASA.
Figure 8.29a	Courtesy of US Naval Observatory.
Figure 8.29b	Courtesy of NASA, ESA, H. Weaver (JHU/APL), A. Stern (SwRI) and the HST Pluto Companion Search Team.
Figure 9.3	Courtesy of D. Roddy (US Geological Survey), Lunar and Planetary Institute.
Figure 9.4	Courtesy of NASA.
Figure 9.9a	Courtesy of NASA.
Figure 9.9b	Courtesy of JHUAPL.
Figure 9.9c	Courtesy of ISAS, JAXA.
Figure 9.9d	Courtesy of NASA/JPL-Caltech.
Figure 9.24a	Courtesy of NASA.
Figure 9.24b	Courtesy of NASA.
Figure 9.32a	(a) Courtesy of Andrew Cooper.
Figure 9.32b	Photo by Dr. Paolo Candy (www.hesnet.net/candy).
Figure 9.32c	Courtesy of Jure Skvarc, Bojan Dintinjana, Herman Mikuz (Crni Vrh Observatory, Slovenia).
Figure 9.33	Georgios Kollidas/Shutterstock.com.
Figure 9.37	Courtesy of Halley Multicolor Camera Team, Giotto Project, ESA. Copyright: MPAE.
Figure 9.38	Courtesy of NASA/JPL/Caltech.

Figure 15.21c	X-ray: courtesy of NASA/CXC/SAO; infrared: courtesy of NASA/JPL-Caltech; Optical: courtesy of MPIA, Calar Alto, O. Krause *et al.*
Figure IV.1a	Courtesy of NASA, The Hubble Heritage Team, STScI, AURA.
Figure IV.1b	Courtesy of NASA, ESA and The Hubble Heritage Team (STScI/AURA).
Figure IV.1c	Courtesy of NASA/JPL-Caltech/STScI.
Figure IV.1d	Courtesy of Hubble Heritage Team, ESA, NASA.
Figure IV.2	Courtesy of NASA, ESA, G. Illingworth (University of California, Santa Cruz), R. Bouwens (University of California, Santa Cruz and Leiden University) and the HUDF09 Team.
Figure 16.1	Courtesy of Jimmy Westlake (Colorado Mountain College).
Figure 16.2a	http://www.spacetelescope.org/images/heic0506a. Courtesy of NASA, ESA, S. Beckwith (STScI), and The Hubble Heritage Team (STScI/AURA).
Figure 16.2b	Courtesy of Ken Crawford (Rancho Del Sol Observatory).
Figure 16.2c	Courtesy of NASA and The Hubble Heritage Team (STScI/AURA)
Figure 16.2d	Courtesy of NASA, ESA, and the Hubble Heritage (STScI/AURA)-ESA/Hubble Collaboration.
Figure 16.5	Courtesy of Royal Astronomical Society – Herschel's galactic model – C001/2064.
Figure 17.1	Courtesy of NASA, ESA and T. M. Brown (STScI).
Figure 17.3	Courtesy of NASA.
Figure 17.4a	Courtesy of NASA, ESA, Hubble Heritage (STScI/AURA); Acknowledgement: A. Aloisi (STScI/ESA) *et al.*
Figure 17.4b	Courtesy of ESO/S. Brunier
Figure 17.4c	Courtesy of Local Group Galaxies Survey Team, NOAO, AURA, NSF.
Figure 17.8	Courtesy of NASA, ESA and the Hubble SM4 ERO Team.
Figure 17.9	Courtesy of Brad Whitmore (STScI) and NASA.
Figure 17.10	Courtesy of NASA/JPL-Caltech.
Figure 17.16	Courtesy of NASA Jet Propulsion Laboratory (NASA-JPL).
Figure 17.19	Reproduced by permission of the AAS.
Figure 17.20	Image courtesy of NRAO/AUI.
Figure 17.21	Courtesy of Jack Schmidling.
Figure 17.27	Courtesy of Dr. John Hutchings, Dominion Astrophysical Observatory, NASA/ESA.
Figure 17.28	Courtesy of NASA, Robert Gendler http://www.robgendlerastropics.com/.
Figure 17.36a	Robert Gendler / Nasa.
Figure 17.36b	NASA and The Hubble Heritage Team (STScI/AURA).
Figure 17.36c	Holland Ford, Space Telescope Science Institute/Johns Hopkins University; Richard Harms, Applied Research Corp.; Zlatan Tsvetanov, Arthur Davidsen, and Gerard Kriss at Johns Hopkins; Ralph Bohlin and George Hartig at Space Telescope Science Institute; Linda Dressel and Ajay K. Kochhar at Applied Research Corp. in Landover, Md.; and Bruce Margon from the University of Washington in Seattle.; NASA.
Figure 17.36d	X-ray: NASA/CXC/KIPAC/N. Werner et al Radio: NSF/NRAO/AUI/W. Cotton.
Figure 18.1	Courtesy of Freshwater and Marine Image Bank.
Figure 18.2	Courtesy of Biblioteka Jagiellonska, manuscript BJ 10000, f. 9v.
Figure 18.18	Courtesy of WMAP science team/NASA.
Figure 18.21	Courtesy of WMAP Science Team/ NASA.
Figure 18.24	Courtesy of W. Schaap (Kapteyn Institute, U. Groningen) *et al.*, 2dF Galaxy Redshift Survey.
Figure 18.25	Courtesy of Springel *et al.* (2005).
Figure 19.1	Courtesy of NASA.
Figure 19.2	Courtesy of Jarrod Boord/Shutterstock.com.
Figure 19.3	Courtesy of J. William Schopf, University of California, Los Angeles.
Figure 19.5	Courtesy of NASA JPL.
Figure 19.6a	Courtesy of NASA Johnson Space Centre.
Figure 19.6b	Courtesy of NASA.

Figure 19.13	Courtesy of Seth Shostak, SETI Institute.
Figure 19.14	Courtesy of NASA.
Figure 19.16	Courtesy of the NAIC – Arecibo Observatory, a facility of the NSF.

Table credits

Table 12.1	*Source*: Impey, C. & Hartmann, W. K. (2000). *The Universe Revealed*, Appendix B-2, Instructor's edition. Place: Brooks-Cole.
Table 12.2	*Source*: Adapted from Impey, C. & Hartmann, W. K. (2000). *The Universe Revealed*, Appendix B-2. Brooks-Cole.
Unnumbered Table 14.1	*Source*: Romer, R. H. (1985). *Energy: Facts and Figures*. Place: Spring Street Press.
Table 15.1	*Source*: Adapted from Bless, R. C. (1996). *Discovering the Cosmos*. University Science Books, p. 398.
Table 19.1	*Source*: Jakosky, B. (1998). *The Search for Life on Other Planets*. Cambridge: Cambridge University Press, p. 111.

INDEX

Page numbers in *italic* denote figures. Page numbers in **bold** denote tables. Letters A and B indicate references to appendices and textboxes, respectively.